Magnetism and Magnetic Materials – 1972

Part 2

AIP Conference Proceedings

Series Editor: Hugh C. Wolfe

Number 10, Part 2

Magnetism and Magnetic Materials – 1972
(18th Annual Conference-Denver)

Editors

C. D. Graham, Jr.
University of Pennsylvania

and

J. J. Rhyne
U. S. Naval Ordnance Laboratory

American Institute of Physics
New York   1973

Copyright © 1973 American Institute of Physics, Inc.
This book, or parts thereof, may not be
reproduced in any form without permission

L.C. Catalog Card No. 72-623469

ISBN 0-88318-109-6

AEC CONF-721114

American Institute of Physics

335 East 45th Street

New York, N.Y. 10017

Printed in the United States of America

Section 26.   Tricritical Points - Theory

THEORY OF TRICRITICAL PHASE TRANSITIONS

Eberhard K. Riedel
Duke University, Durham, N. C. 27706

ABSTRACT

A theory for tricritical phase transitions in two-component systems (such as He^3-He^4 mixtures and antiferromagnets in a uniform magnetic field) is reviewed. The theory is based on a three-dimensional model for a class of tricritical points, which is solved by using the renormalization-group approach. This yields all tricritical exponents (including logarithmic correction factors) and the tricritical scaling fields. The relations of the theory to a tricritical scaling approach and to tricritical experiments are also discussed.

INTRODUCTION

The renormalization-group approach (RGA) and the parameter scaling theory have been extremely successful for discussing tricritical phenomena in two-component systems (such as He^3-He^4 mixtures and antiferromagnets in a uniform magnetic field). Here some features of the theory of tricritical phase transitions by the author[1] and Wegner and the author[2,3] are reviewed and their consequences for tricritical experiments discussed.

Terminology and Notation. The tricritical point separates, in the space of thermodynamic fields,[4] a first-order transition line from a single line of critical points.[5] The phase transition lines constitute singled-out directions in the field space.[6] It is crucial to incorporate this fact into a theory of tricritical phase transitions.[1] The theory presented here is formulated in terms of pairs of "conjugate" scaling fields μ and scaling densities Q (each characterized by a scaling index $y = d-x$, where d is the dimension of the system).[1-3] The scaling fields are microscopically defined by the renormalization-group procedure[2,3,7] and are related to the special features of the phase diagram.[1,8]

MICROSCOPIC THEORY

Model. Consider a three-dimensional model with (n-component) classical spins interacting with ferromagnetic nearest neighbor interactions. The reduced Hamiltonian H_0, in the sense of Wilson's RGA,[9] has the general form

$$H_\ell = - \int d^3x \, [1/2 \, |\nabla z(x)|^2 + Q_\ell\{z(x)\}]. \qquad (1a)$$

$$Q_\ell\{z\} = r_\ell z^2 + u_\ell z^4 + v_\ell z^6 + \ldots, \tag{1b}$$

with, for $\ell = 0$, initial coefficients r_0, u_0, and v_0 that depend on the thermodynamic fields.[4] Here z denotes the reduced spin variable. The Landau-Ginsberg Hamiltonian for a tricritical transition,[10] the Blume-Emery-Griffiths spin-1 model for He^3-He^4 mixtures,[11] the Ising spin-1/2 model for an antiferromagnet in a uniform magnetic field,[12,13] and the isotropic Ising model ($|J_F| = J_{AF}$) for a metamagnet[14] can be "reduced" to the form (1). Thus Eqs. (1) represent a class of models exhibiting tricritical points.

Tricritical Exponents. In a RGA the tricritical instability corresponds to a "twice" unstable fixed point.[2] We have used the approximate RGA equations [9,15] to calculate the tricritical exponents of model (1) for the Gaussian tricritical fixed point in three dimensions. The results for the exponents related to the three relevant scaling densities Q_i (with $y_i > 0$) are summarized in Table 1.[2] As expected, the tricritical exponents, which are molecular-field-like, differ from the critical-line exponents. This establishes that critical exponents depend on the degree of instability of the corresponding fixed point. For the tricritical transition three scaling densities, $Q_{1/2}$, Q_1 and Q_2, are relevant, i.e., exhibit critical fluctuations. (This is the basic difference to ordinary second-order transitions where Q_2 is irrelevant. Hypercritical points of order higher than three are characterized by more than three relevant scaling densities.) The tricritical scaling density Q_3 is "marginal", i.e., is characterized by a vanishing scaling exponent $y_3 = 0$. Without changing the tricritical exponents, it changes the asymptotic tricritical form of several thermodynamic functions from a pure power law to a power law <u>times</u> a fractional power of a logarithmic correction factor.

Table I. Tricritical Exponents

i	$\nu_t x_i$	$\nu_t[d-2x_i]$	$\nu_t[d-x_i]$	$d/x_i - 1$	$2[1+x_i-1/2\,d]$
1/2	$\beta_t = 1/4$	$\gamma_t = 1$	$\Delta_t = 5/4$	$\delta_t = 5$	$\eta_t = 0$ [a]
1	$\omega_t = 1/2$ [b]	$\lambda_t = a_t = 1/2$ [b]	1 [c]	$\delta_{n,t} = 2$	$\eta_{n,t} = 1$
2	1	$-1/2$	$\phi_t = 1/2$ [b]	$1/2$	3

[a] The value $\eta_t = 0$ is enforced by the approximations made in the RGA.

[b] The exponents ω_t, λ_t, and ϕ_t were defined in Ref. 1.

[c] This result follows from the condition $\nu_t[d-x_1] = 1$; hence $\nu_t = 1/2$, and $\Delta_{n,t} = 1$.

Scaling Fields. For each ℓ the Hamiltonian (1) can be expanded in terms of the scaling densities Q_i with coefficients $\mu_i(\ell)$, i.e., $H_\ell = H^* + \Sigma_i \mu_i(\ell) Q_i$. The μ_i are called scaling fields since the free energy as a function of the μ_i satisfies the exact, generalized scaling relation[3,7]

$$F\{\mu_i(0)\} = e^{-d\ell} F\{\mu_i(\ell)\} . \qquad (2)$$

The factor $e^{-\ell}$ is an arbitrary scale factor of the cutoff momentum for the renormalized Hamiltonian (1). The tricritical point (in the physical plane[16]) is determined by the condition that the two relevant scaling fields $\mu_1(0)$ and $\mu_2(0)$ vanish.

Logarithmic Corrections. Wegner and the author proposed a general scheme to calculate logarithmic corrections to the critical behavior from a set of coupled differential equations for the μ_i.[3] In the tricritical case the exponent relations $y_3 = 0$ and $2y_2 = y_1 = 2$ both imply logarithmic factors in the asymptotic tricritical form of thermodynamic functions. For the tricritical transition of model (1) the following scaling fields were obtained:[3]

$$\mu_1 = g_1(0) e^{y_1 \ell} , \quad \mu_2 = g_2(0) e^{y_2 \ell} (\ell + \ell_0)^{p_2 + 1/2} . \qquad (3)$$

where the $g(0)$'s are functions of the thermodynamic fields[4] and $p_2 = -2(n+4)/(3n+22)$.[16] These results yield logarithmic corrections in the ordering density, $m \propto |g_1(0)|^{1/4} \ln |g_1(0)|^{1/4}$, and the equations for the critical line and the first-order line.

Critical Line. The equation for the critical line in the tricritical region is

$$g_{1,c} \propto g_{2,c}^2 \left| \ln |g_{2,c}| \right|^{2p_2 + 1} \qquad (4)$$

Hence the g_2 scaling field is tangentially parallel to the critical line at the tricritical point, whereas the g_1-field and the g_h-field[16] denote weak and strong directions in the sense of Ref. 6. In terms of the physical fields $\delta T = (T - T_t)/T_t$ and $\delta g = (g - g_t)/g_t$, Eq. (4) can be written

$$\delta g = A\delta T + B(\delta T)^2 (\ln(\delta T/\theta_m))^{(6-n)/(3n+22)} \qquad (5)$$

with a change-over temperature $\theta_m = (T_m - T_t)/T_t$.

SCALING THEORY

Principle of Competition. In the tricritical region <u>two</u> scaling hypotheses can be made, one with respect to the tricritical point and one with respect to the critical line.[1] The actual scaling variables were chosen to be related to certain geometric features of the tricritical phase diagram. They are microscopically defined by the RGA as in Eq. (2). The measurable exponents depend on the direction of approach toward the tricritical point and the critical line.[1,13] If, in particular, the experiment is performed at the tricritical value of the nonordering density[4] (instead of, for example, at the tricritical value of the nonordering field), then the tricritical exponents are "renormalized" by the tricritical crossover exponent ϕ_t, i.e., $\omega_t^* = \omega_t/\phi_t, \ldots$. In the double-scaling region close to the critical line thermodynamic functions can be characterized by a critical-line exponent and an amplitude exponent, i.e., data in this region can be reduced to a single point.[1,17,16] Corrections of logarithmic order to the phenomenological scaling approach are obtained by substituting the tricritical scaling fields (3) into Eq. (2) and determining ℓ from $\mu_1(\ell) = 1$.[3]

Exponent Relations. Relations among the tricritical exponents can be easily derived[1] and are evident from the first row of Table I. Of new interest are several relations for exponents related to the nonordering density/field variables, like

$$\omega_t + \lambda_t = 1, \quad \omega_t = \lambda_t/(\delta_{n,t}-1), \quad \lambda_t = (2-\eta_{n,t})\nu_t . \tag{6}$$

Scaling predicts moreover $\omega_t = 1 - \alpha_t$ and $\lambda_t = \alpha_t$. Hence the only additional exponent needed to characterize tricritical phenomena is the crossover exponent ϕ_t. It characterizes the additional "critical degree of freedom" in tricritical systems. Experimentally, the crossover exponent can be determined via Eq. (5), from the amplitude exponents in the double-scaling region, or by using exponent relations for "renormalized" exponents, such as $\omega_t^* + \lambda_t^* = 1/\phi_t$. The crossover exponent also determines the regions of different critical behavior in the field space, i.e., first-order, second-order, and tricritical regions.[1]

Tricritical Experiments. The theory gives a consistent description of tricritical experiments in He^3-He^4 mixtures. The situation for magnetic systems has not yet been discussed in detail.

REFERENCES AND FOOTNOTES

1. E. K. Riedel, Phys. Rev. Lett. **28**, 675 (1972).
2. E. K. Riedel and F. J. Wegner, Phys. Rev. Lett. **29**, 349 (1972). (In the present paper fluxion dots on tricritical exponents are omitted for simplicity.)
3. F. J. Wegner and E. K. Riedel, Phys. Rev. B (to be published).
4. The behavior of systems near tricritical points (as in He^3-He^4 mixtures, antiferromagnets like $FeCl_2$, systems undergoing structural transitions like NH_4Cl) can be described by three thermodynamic densities and their conjugate fields: the entropy

s and temperature T, the ordering density m and ordering field h (the superfluid order parameter etc.), and the nonordering density n and nonordering field g (the He^3-molar concentration etc.).

5. R. B. Griffiths, Phys. Rev. Lett. <u>24</u>, 715 (1970).
6. R. B. Griffiths and J. C. Wheeler, Phys. Rev. A <u>2</u>, 1047 (1970).
7. F. J. Wegner, Phys. Rev. B <u>5</u>, 4529 (1972).
8. This puts also several of the geometrical considerations of Ref. 6 on a microscopic basis.
9. K. G. Wilson, Phys. Rev. B <u>4</u>, 3184 (1971).
10. L. D. Landau, Phys. Z. Sowjetunion <u>11</u>, 26 (1937).
11. M. Blume, V. J. Emery, and R. B. Griffiths, Phys. Rev. A <u>4</u>, 1071 (1971).
12. D. P. Landau, Phys. Rev. Lett. <u>28</u>, 449 (1972).
13. E. K. Riedel, to be published.
14. F. Harbus and H. E. Stanley, Phys. Rev. Lett. <u>29</u>, 58 (1972).
15. F. J. Wegner, Phys. Rev. B <u>6</u>, 1891 (1972).
16. The scaling field related to the ordering field h is given by $\mu_h = g_h(0) \exp(y_h \ell)$ with $y_h = 5/2$ and can be easily included in the discussion.
17. This is a general feature of the parameter scaling theory of Riedel and Wegner [Z. Phys. <u>225</u>, 195 (1969), and Phys. Rev. Lett. <u>24</u>, 730, 930(E) (1970)] and applies, for example, also to anisotropic systems (anisotropic magnets, magnets with a layered structure) and systems with geometric restrictions (finite size effects) like films.

MONTE CARLO STUDIES OF TRICRITICAL PHENOMENA[†]

B. L. Arora[*] and D. P. Landau
University of Georgia, Athens, Georgia 30601

ABSTRACT

We have used a Monte Carlo technique to study two Ising magnetic systems which exhibit tricritical points. The critical behavior of a spin-1, Blume-Emery-Griffiths model on a square lattice was studied over a wide range of zero field splitting parameters Δ. As expected, the critical exponents β, γ, δ, all retained their $\Delta = 0$ values until quite near the tricritical point where all three exponents changed dramatically to take on new "Tricritical values". In addition, the critical field curve for a spin-½ metamagnet on a simple cubic lattice was calculated. A tricritical point was observed at $T_t = 0.58 T_N$ with the critical field being smooth and continuous at the tricritical point. The behavior of the magnetization and susceptibility was contrasted with that found previously for a simple cubic antiferromagnet, but good agreement was found with recent series expansion results on the identical metamagnet.

INTRODUCTION

It has been known for some time that the nature of the antiferromagnetic-paramagnetic phase transition sometimes changed order along the critical field curve. Only recently, however, has Griffiths[1] pointed out that there was an entire class of heretofore unexplored critical phenomena of which the simple antiferromagnet was but one example. Griffiths considered both an "ordering field", η, (staggered magnetic field for an antiferromagnet) and a "disordering field" Δ (applied magnetic field for an antiferromagnet) and showed that the phase diagram in Δ-η-T space contained three thermodynamic surfaces which intersected along a single "critical curve" in the Δ-T plane. Moreover, the boundaries to the three surfaces came together at a single point, the "tricritical point", T_t. In the experimentally accessible Δ-T plane the order of the observed phase transition changes at T_t. As a result of this work, there have been numerous recent investigations[2-18] of the thermodynamic behavior of various systems near T_t. In this paper we shall concentrate on two particular systems, the Blume-Emery-Griffiths[2] (BEG) model of $S = 1$ Ising spins subject to a zero field splitting[19] and a simple $S = ½$ metamagnet of the type investigated by Harbus and Stanley[4]. In both cases our studies have been carried out using an importance sampling Monte Carlo technique. This method has been previously described in some detail[10,21] and we shall not elaborate on the technique here.

[†]Research supported in part by the National Science Foundation
[*]On leave from A.R.S.D. College, Dhaula Kuan, New Delhi, India

RESULTS - BEG MODEL

Recent interest in the BEG model has been high since it can be shown to be a magnetic analogue of the phase separation in He^3-He^4 mixtures. The Hamiltonian for this model can be written:

$$\mathcal{H} = D\sum_i (S_{iz}^2 - 2/3) - J \sum_{(i,j)} S_{iz} S_{jz} + g\mu_B H \sum_i S_{iz} \qquad (1)$$

where the ferromagnetic exchange J acts only between the z nearest neighbors and we shall now refer to the "reduced zero field splitting" $\Delta = D/zJ$. It should be noted that if we define $X = 1 - \langle S_z^2 \rangle$, which is a measure of the quadrupolar moment, then X is equivalent to the He^3 concentration in an He^3-He^4 mixture. The ordering field in this model is the applied field H and Δ plays the role of the disordering field. We have studied the behavior of the magnetization, susceptibility and quadrupole moment as a function of both Δ and H for a 40×40 square lattice with periodic boundary conditions. As earlier results[8,9] had indicated, the transition curve shows a kink near T_t in the Δ-T plane. We have also determined the X-T phase diagram and show our values along with the mean field result in Fig. 1. Not only are X_t and T_t significantly different from mean field predictions[2], but the λ-line does not seem to join smoothly with the large X phase separation boundary. If one assumes that the discontinuity in X obeys a simple law of the form:[22]

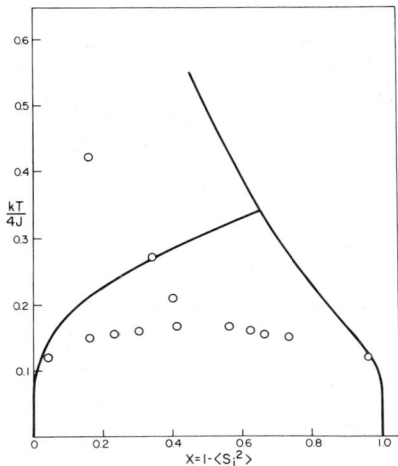

Fig. 1. Phase diagram for the S=1 BEG model on the square lattice. Solid line is the mean field curve, open circles are the Monte Carlo data.

$$\Delta X \propto \left| \frac{T-T_t}{T_t} \right|^{\beta_u} \qquad (2)$$

one finds from our data that $\beta_u = 0.65\pm 0.10$ in disagreement with the mean field result $\beta_u = 1$.

We have also studied the behavior of the spontaneous magnetization and susceptibility near the critical curve. In particular one expects that as $T \to T_c$ then

$$\begin{aligned} M &\propto \varepsilon^\beta \\ \chi &\propto \varepsilon^{-\gamma} \\ M &\propto H^{1/\delta} \end{aligned} \qquad (3)$$

where $\varepsilon = |(T-T_c)/T_c|$. Near T_c we would expect the exponents to take on their normal "critical values". As $T_c \to T_t$ however, the exponents should assume new "tricritical" values (which are probably different from the "critical" exponents). From our data we have determined the critical exponents β, γ^-, γ^+ and δ and find that for $\Delta=0$ they have the same values (see the Table) as for an $S=\frac{1}{2}$ system,

Table I Critical and Tricritical Exponents for BEG Model

	β	γ^-	γ^+	δ
$\Delta=0$, M.C.	0.125±0.005	1.65±0.10	1.60±0.15	14.8±0.4
Spin ½, critical	1/8	7/4	7/4	15
M.C., tricritical	.09 ±.02	1.0±0.3	1.1±0.4	10.8±0.7

in conformity with the universality hypothesis. As Δ increases the exponents remain essentially unchanged until $\Delta \approx 0.4$. (For a 40×40 square lattice $\Delta_t = 0.485$ and for $\Delta > 0.5$ no phase transition occurs). Beyond this point the exponents change quite rapidly and assume the tricritical values shown in the Table. Due to the nature of the "crossover curve" which separates the critical and tricritical regions it appears that for $0.4 \lesssim \Delta \lesssim \Delta_t$ we are in the tricritical region for large ϵ but then enter the critical region very near to the critical field curve. Thus, the "exponent" derived from this data would not be typical of either the critical or tricritical behavior. For this reason only the values for $\Delta = \Delta_t$, $T \to T_t$ are significant and we show only these in the table. We can now test the exponent equality

$$\beta_t(\delta_t - 1) = \gamma_t^- \qquad (4)$$

From our data we find $\beta_t(\delta_t - 1) = 0.88 \pm 0.24$ which within experimental error equals γ_t^-.

RESULTS - SIMPLE CUBIC METAMAGNET

Harbus and Stanley[4] have recently considered a simple metamagnet on a simple cubic lattice with

$$\mathcal{H} = J_{xy} \sum_{(i,j)} S_{iz} S_{jz} + J_z \sum_{(i,k)} S_{iz} S_{kz} + g\mu_B H \sum_i S_{iz} \qquad (5)$$

where the first sum is over the nearest neighbors in the x-y plane and the second is over nearest neighbors coupled in the z-direction. To simulate a metamagnet we have chosen J_{xy} ferromagnetic and $J_z = -J_{xy}$. This is then exactly the same model considered by Harbus and Stanley in their series expansion study. As the series expansion analysis produced a non-physical upwards hook in the critical field curve near T_t, the location of the tricritical point as well as the study of the tricritical behavior became quite complicated. We have, therefore, considered this Hamiltonian on a 12×12×12 simple cubic lattice with periodic boundary conditions

With the intent of independently determining both the critical field curve and the tricritical temperature. As a result of our studies we have located the critical field curve from T = 0 to T = T_N and have found a tricritical point at T_t/T_N = 0.58±0.01 which is the same as that obtained by Harbus and Stanley. In addition, our critical field curve agreed quite well with theirs as long as we were above T_t and showed no sign of any kink at T_t (as in the BEG model). In order to compare our results with those obtained earlier[7] on the simple cubic antiferromagnet with nearest- and next nearest neighbor interactions, we looked carefully at the behavior of the magnetization. The observed discontinuity in the magnetization, m, across the phase boundary was fitted to a law of the form

$$\Delta m \propto \left|\frac{T_t - T}{T_t}\right|^{\beta_u} \qquad (6)$$

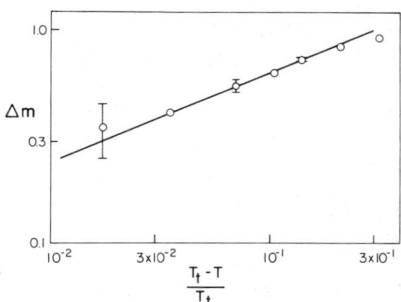

Fig. 2. Discontinuity in the magnetization at the critical field vs distance in temperature from T_t.

and we show the results on a log-log scale in Fig. 2. The best fit to the data yields a value β_u = 0.5±0.1, which is substantially lower than that obtained for the simple cubic antiferromagnet[7] (β_u = 0.78±0.20) and surprisingly close to the result for the square lattice antiferromagnet (β_u = 0.58±0.11). In addition, we have examined the behavior of the susceptibility near T_t along two paths, one at constant m = m_t and the other along constant (H/T = H_t/T_t). If we assume a divergence of the form

$$\chi \propto \left|\frac{T-T_t}{T_t}\right|^{-\gamma} \qquad (7)$$

we find γ_t^+ = 0.55±0.15 for both paths. Again, lack of knowledge about the "crossover curve" make the significance of this result uncertain. If both paths lie wholly within the tricritical region, we would, in fact, expect the two exponents to be identical. If, however, the constant m path approached T_t asymptotically coincident with the critical field curve the exponents determined along the two paths should disagree. (The scatter in the present data makes it impossible to determine if the critical field curve and constant m path are coincident.) Nonetheless this value for γ along a path of constant (H/T) is in excellent agreement with that found by Harbus and Stanley. Moreover the values obtained along the two paths disagree with those obtained for the simple cubic antiferromagnet but are surprisingly near to the result for the antiferromagnet on a square lattice. An analysis for γ_t^- seems to indicate a much larger value than for γ_t^+ but the errors are quite large.

SUMMARY AND CONCLUSION

The results of our study of the BEG lattice have confirmed that the critical exponents β, γ, and δ retain their normal "critical" values along the phase boundaries, but do assume new and, in all cases, different tricritical values at $T = T_t$. Within the regrettably large experimental errors these tricritical exponents have been shown to obey one of the usual exponent equalities. The results on the simple cubic metamagnet agree with the earlier series expansion results and seem to imply that the tricritical behavior could be two dimensional in nature.

Further work is in progress to reduce the errors on the tricritical exponents already obtained and to extract values for others not considered here. We wish to thank Mr. F. Harbus and Drs. H. E. Stanley, E. K. Riedel, and D. Stauffer for helpful discussions.

REFERENCES

1. R. B. Griffiths, Phys. Rev. Lett. **24**, 715 (1970).
2. M. Blume, V. J. Emery, and R. B. Griffiths, Phys. Rev. **A4**, 1071, (1971).
3. J. F. Nagle and J. C. Bonner, J. Chem. Phys. **54**, 729 (1971).
4. F. Harbus and H. E. Stanley, Phys. Rev. Lett. **29**, 58 (1972).
5. L. Reatto, Phys. Rev. **B5**, 204 (1972).
6. J. C. Bonner and J. F. Nagle, J. Appl. Phys. **42**, 1280 (1972).
7. D. P. Landau, Phys. Rev. Lett. **28**, 449 (1972).
8. D. M. Saul and M. Wortis, AIP Conf. Proc. **5**, 349 (1972).
9. E. K. Riedel, Phys. Rev. Lett. **28**, 675 (1972).
10. B. L. Arora and D. P. Landau, Bull Amer. Phys. Soc. **17**, 300 (1972).
11. E. K. Riedel and F. J. Wegner, Phys. Rev. Lett. **29**, 349 (1972).
12. D. Stauffer, Phys. Rev. **B6**, 1839 (1972).
13. O. K. Rice and D. R. Chang, Phys. Rev. **A5**, 1419 (1972).
14. R. Bausch, Z. Physik **254**, 81 (1972).
15. P. J. Kortman (to be published).
16. L. S. Schulman (to be published).
17. F. J. Wegner and E. K. Riedel (to be published).
18. E. K. Riedel (to be published).
19. Capel (Ref. 20) first investigated some properties of this model but did not really explore the tricritical region.
20. H. W. Capel, Physica **32**, 966 (1966).
21. D. P. Landau, J. Appl. Phys. **42**, 1284 (1971).
22. We have adopted the notation due to R. B. Griffiths (to be published).

METAMAGNETISM OF $FeCl_2$ NEAR THE TRICRITICAL POINT

Richard P. Kenan *
Battelle Memorial Institute, 505 King Ave., Columbus, Ohio 43201

Roger E. Mills
Physics Department, University of Louisville, Louisville, KY 40208

ABSTRACT

Properties of the metamagnet $FeCl_2$ near the tricritical point, T_{3C}, have been calculated as functions of both temperature and uniform magnetic field. Sublattice magnetizations and anisotropy energies are obtained from self-consistent calculations based on a complete basis operator theory and are used to calculate the free energies of the magnetic phases. The width of the ferromagnetic-antiferromagnetic coexistence curve is found to be linear in $T_{3C}-T$. The susceptibility and the specific heat are obtained along both sides of the coexistence curve for $T<T_{3C}$, and in the vicinity of the line of second-order transitions for $T>T_{3C}$. The single-ion anisotropies have been incorporated consistently in the calculations over the T-H plane.

INTRODUCTION

A variety of systems exhibit phase transitions which are of the first order over a range of temperatures, but which become of second order above a certain tricritical temperature, T_{3C} or T_3^*. This type of behavior was studied by Griffiths[1] in the case of ^{3}He-^{4}He liquid mixtures, and a modified Ising model was used by Blume, Emery, and Griffiths[2] in a further study of the same materials. It was noted by Griffiths[1] that metamagnetic material, in particular $FeCl_2$, exhibit similar behavior. An approach to a scaling theory of the thermodynamic properties of such materials was developed by Riedel[3].

A series of experimental results[4-7] have shown that the states which give rise to the low-temperature behavior of $FeCl_2$ can be classified as a triplet. One of the three levels is raised somewhat above the others by a combination of L·S coupling and crystal field effects. It is convenient to describe this set of levels as those of a spin s=1 together with a single-ion anisotropy term. A brief account of an application of mean-field theory to such a model was given earlier[8] and it was shown that the single-ion anisotropy term could shift appreciably the position of the isotherms for the model. It was also shown that within the limitations of the theory, a simple scaling of the isotherm and other properties with respect to the single-ion anisotropy energy appeared to hold.

In view of the possibilities that spin waves might be

*This research was supported in part by a grant, AF-AFOSR-68-1535.

observable, an attempt was made to improve the calculation by casting it in the RPA form, but difficulties in the consistency of the decoupling left the results in doubt. More recently, work by Murao and Matsubara[9] appeared in which was developed a theory based upon a complete basis set of operators. In M&M it was shown that a consistent decoupling scheme could be obtained if all of the spherical tensor operators developable from a given spin operator are treated on a common basis. Although M&M found that extraneous roots might appear, we have found that with slightly different basis operators, the extraneous roots do not appear in the case where the easy axis of magnetization and the axis of the single-ion anisotropy coincide. It is the purpose of this paper to indicate the nature of the calculation and to describe some of the results which have been obtained from the calculations. A fuller, more quantitative description will be presented elsewhere.

THE SELF-CONSISTENT CALCULATION

The Fe^{++} ions in the free state have a ^5D ground state, but in the crystalline $FeCl_2$ environment this level is split by crystal field effects and spin-orbit coupling, so that the levels which dominate the low-temperature magnetic phenomena can be described by the Hamiltonian

$$H = -\Sigma_\alpha \{ \Sigma_i [\omega_0 s^z_{\alpha i} + D(s^z_{\alpha i})^2 + J\Sigma_{nn}(s^z_{\alpha i} s^z_{\alpha i+\delta_{nn}} + \xi s^+_{\alpha i} s^-_{\alpha i+\delta_{nn}}) - K\Sigma_{nn}(s^z_{\alpha i} s^z_{\beta i+\delta_{nn}} + \zeta s^+_{\alpha i} s^-_{\beta i+\delta_{nn}})]\} \quad (1)$$

where $\omega_0 = g\mu H$, and the sums are over sublattices α, sites i, and nearest neighbors, nn. Where β is used as a sublattice label in an expression with an α, it implies the opposite sublattice. J represents the longitudinal portion of an anisotropic, intralayer exchange tensor, and ξJ is the corresponding transverse part. K is the longitudinal part of the interlayer exchange, and ζK the transverse part. The factor D is the single-ion anisotropy energy, about $15 cm^{-1}$. The exchange is taken to be anisotropic because of the $R\bar{3}m$ symmetry of $FeCl_2$. This seems to be well-substantiated by experiments. We assume for simplicity that the only contribution to dispersion of the spin wave spectrum comes from the intralayer coupling.

The spin s in (1) is taken to be $s=1$. Hence one needs nine basis operators, including the identity. However if H is restricted only to the z-axis, the only Green's functions which need to be considered are

$$G^{\alpha\alpha'}(ij;t) = -\Theta(t)<[s^+_{\alpha i}(t), s^-_{\alpha' j}(0)]_-> = <<s^+_{\alpha i}; s^-_{\alpha' j}>> \quad (2)$$

and

$$\Gamma^{\alpha\alpha'}(ij,t) = <<A^+_{\alpha i}; s^-_{\alpha' j}>>, \quad (3)$$

where $A_{\alpha i}^{+}$ is the operator $s_{\alpha i}^{z} s_{\alpha i}^{+} + s_{\alpha i}^{+} s_{\alpha i}^{z}$. It is also useful to employ a quadrupole operator

$$Q_{\alpha i} = 3(s_{\alpha i}^{z})^2 - 2 \quad . \tag{4}$$

The equations of motion calculated for these Green's functions involve higher order Green's functions. If the higher order functions involve several operators from the same site, the complete basis set permits reduction to combinations of the original basis operators. If the multiple operators involve different sites, the M&M decoupling is used. As a consequence of the use of the complete basis set, all the operators from a single site which have non-zero expectation values have those values determined self-consistently. Here $\sigma_\alpha = <s_{\alpha i}^z>$ and $q_\alpha = <Q_{\alpha i}>$ are the only non-trivial values. After the Green's functions' equations of motion are found, decoupled, and Fourier-transformed, one arrives at length at a set of implicit equations for σ_α, σ_β, q_α, and q_β:

$$\frac{4}{3} - \frac{1}{3} q_\alpha - \sigma_\alpha = \sigma_\alpha (S_1^\alpha - S_2^\alpha) + q_\alpha S_3^\alpha \quad , \tag{5}$$

where

$$\sigma_\alpha - q_\alpha = q_\alpha (S_1^\alpha - S_2^\alpha) + \sigma_\alpha S_3^\alpha \quad , \tag{6}$$

$$S_1^\alpha \equiv \frac{1}{N} \Sigma_k [f(E_+^\alpha) + f(E_-^\alpha)] \quad , \tag{7}$$

$$S_2^\alpha \equiv \frac{1}{N} \Sigma_k \left(\frac{E_1^\alpha}{E_+^\alpha - E_-^\alpha}\right) [f(E_+^\alpha) - f(E_-^\alpha)] \quad , \tag{8}$$

$$S_3^\alpha \equiv \frac{1}{N} \Sigma_k \left(\frac{2D}{E_+^\alpha - E_-^\alpha}\right) [f(E_+^\alpha) - f(E_-^\alpha)] \quad , \tag{9}$$

where the k-sums are over the Brillouin zone, $f(X)$ is a Bose distribution function, and the dispersion curves are

with

$$E_{\mp} = \omega_0 + 2Jz\sigma_\alpha - 2Kz'\sigma_\beta - \frac{1}{2}\{E_1^\pm [(E_1)^2 + 4D(D - \Omega_1^\alpha)]^{\frac{1}{2}}\} \tag{10}$$

$$E_1^\alpha = 2Jz\xi\sigma_\alpha \gamma(k) \quad , \quad \Omega_1^\alpha = 2Jz\xi q_\alpha \gamma(k) \quad . \tag{11}$$

Here z is the number of nearest neighbors, and $\gamma(k)$ is the structure factor. Similar equations hold for σ_β and q_β upon interchange of α and β. The equations (5)-(11) involve implicitly the opposite sublattice only through a mean field term as a consequence of the neglect of ζK, but as a set provide a basis for determination of the mean values.

SOLUTION OF THE SELF-CONSISTENT EQUATIONS

The actual solution of the self-consistent equations involves rather extended computer calculations. The performance of this solution is greatly aided, particularly in the vicinity of T_{3C}

to the Néel temperature, T_N, by extracting derivative equations which give the susceptibility and the temperatures characteristic of the stable solutions of the equations. It is found that two such solutions exist for zero field, a ferromagnetic (F) and an antiferromagnetic (AF) solution. Both transistions are second order in the temperature, and for parameters like those used, $T_N>T_C$. The existence of two stable solutions makes necessary the calculation of the free energy per spin, F, so that the thermodynamically preferred solution can be determined. The calculation of F is carried out using the Callen and Shtrikman moment-generating theorem[10], generalized to include the effects of the single-ion anisotropy energies. The generalization requires the determination of effective anisotropy factors in addition to the determination of the usual mean field. A bonus feature of the use of this procedure is that it facilitates calculation of the specific heat. The entropy can be directly expressed using the single-site density function, and the specific heat then follows by numerical differentiation.

Self-consistent solutions for both the F and AF states have been generated by computer for temperatures below T_N, fields up to the T=0 transition field, exchange anisotropy, ξ, up to about 0.7, and single-ion anisotropies up to 2Jz in magnitude, all for a ratio K/J equal to 0.1, chosen for comparison with $FeCl_2$. Magnetization curves thus obtained suggest that in the M-H plane, there should be a scaling law for the single-ion anisotropy for <u>fixed</u> ξ, for the boundary curves are independent of D even though they do depend appreciably on ξ. The discontinuity in M appears to be linear in $T_{3C}-T$. Above T_{3C}, the difference $\sigma_\alpha - \sigma_\beta$ goes like $|H-H_C|^{1/2}$, where H_C is the transition field.

In the region $T_N>T>T_{3C}$, iterative calculations are impractical, so the derivative expansions mentioned above are used to determine the locus $H_C(T)$. It was found that for fixed D the values of H_C for different values of ξ lie on a single curve, suggesting again a scaling law. Susceptibility calculations have been carried out extensively for zero ξ and for a few non-zero values. It is found that $\partial M/\partial H$ diverges at (T_{3C}, H_{3C}) only, while the sublattice magnetization difference vanishes with vertical tangent at H_C for all $T \geq T_{3C}$.

The calculations reported here were intended for comparison with experiments on $FeCl_2$. Qualitatively, the results are satisfactory, but quantitatively, comparisons with experiment are disappointing. Even in MFT, $\sigma_\alpha - \sigma_\beta$ in zero field falls off too rapidly, and the introduction of dispersion via the spin-wave modes does nothing to help this situation. Further, the ratio T_{3C}/T_N is much higher (.95 to .97) than is observed (about .87). This value would be reduced for larger ratios K/J than 0.1, but the value of this ratio for $FeCl_2$ seems fairly well established.

The use of the present theory as a model of tricritical phenomena is attractive, however, for more of the elements of real materials are included. The uniaxial anisotropy is especially important, since without it the spin-flop state will have a lower free energy than the AF state in any nonzero field. This is not the case for Ising systems which have effectively an infinite uniaxial anisotropy. Models of this latter type have been studied in MFT by Blume et al.[2], by series techniques by Stanley[11], and with Monte Carlo techniques by Landau[12]. The neutron scattering experiments of Birgeneau et al.[6] show clearly the existence of spin waves with dispersion in the spectra, and so the need for consideration of the transverse terms in the exchange interactions is obvious. The present theory shows that such calculations are possible, but that a more thorough consideration of spin correlation energies is needed.

REFERENCES

1. R. B. Griffiths, Phys. Rev. Letters $\underline{24}$, 715 (1970).
2. M. Blume, V. J. Emery, and R. B. Griffiths, Phys. Rev. $\underline{A4}$, 1071 (1971).
3. E. K. Riedel, Phys. Rev. Letters $\underline{28}$, 675 (1972).
4. I. S. Jacobs and P. E. Lawrence, Phys. Rev. $\underline{164}$, 866 (1967).
5. K. Ono, A. Ito, and T. Fujita, J. Phys. Soc. Japan $\underline{19}$, 2119 (1964).
6. R. J. Birgeneau, W. B. Yelon, E. Cohen, and J. Makovsky, Phys. Rev. $\underline{B5}$, 2607 (1972).
7. P. Carrara, Thesis, Centre d'Etudes Nucleairs, Saclay, Report No. CEA-R-3535 (1968) (Unpub.).
8. R. P. Kenan, R. E. Mills, and C. E. Campbell, J. Appl. Phys. $\underline{40}$, 1027 (1969).
9. T. Murao and T. Matsubara, J. Phys. Soc. Japan $\underline{25}$, 352 (1968).
10. H. B. Callen and S. Shtrikman, Sol. State Comm. $\underline{3}$, 5 (1965).
11. F. Harbus and H. E. Stanley, Phys. Rev. Letters $\underline{29}$, 58 (1972).
12. D. P. Landau, Phys. Rev. Letters $\underline{28}$, 449 (1972).

SCALING HYPOTHESIS AND DATA COLLAPSING FOR CRITICAL POINTS OF THIRD ORDER

T. S. Chang[*], Alex Hankey[+], and H. E. Stanley
Physics Department, Massachusetts Institute of Technology,
Cambridge, Massachusetts 02139

ABSTRACT

At tricritical points thermodynamic functions scale with respect to three independent field variables. Such points have been called critical points of the third order. We present, in this paper, a general scaling hypothesis for such critical points and demonstrate how the concept of invariant spaces leads naturally to the prediction of data collapsing from volumes to lines in terms of "double-power scaling functions".

After the discovery of tricritical points,[1] Riedel[2] presented a scaling hypothesis for them. As has been discussed in a companion paper[3] at this conference, the termination point of a line of ordinary critical points in a 3-dimensional field space [e.g., a tricritical point] may be called a critical point of third order ($^3CRS_0$). Along a critical line, three distinct types of directions (labelled x_1, x_2, x_3 in Fig. 1a) may be defined. These are, respectively, the strong and weak directions of Griffiths and Wheeler[4], and a third direction locally parallel to the critical line. At any point P away from the $^3CRS_0$, the scaling hypothesis for the critical line is normally stated in terms of a generalized homogeneous function

$$G(\mu^{a_1} x_1, \mu^{a_2} x_2; x_3) = \mu G(x_1, x_2; x_3) , \qquad (1)$$

where G is the singular part of the Gibbs potential, $\mu(>0)$ is an arbitrary parameter, (a_1, a_2) are the scaling powers for the critical line, and x_3 is an inactive variable which does not scale. Because only two variables scale, the critical line is a line of critical points of second order and will be denoted by $^2CRS_1$. (There are three such lines in Fig. 1).

At a $^3CRS_0$, we hypothesize that the Gibbs potential scales with all three directions $\bar{x}_i$ (the limiting directions of x_i at the $^3CRS_0$ as shown in Fig. 1a) such that[2,5]

$$G(\lambda^{\bar{a}_1} \bar{x}_1, \lambda^{\bar{a}_2} \bar{x}_2, \lambda^{\bar{a}_3} \bar{x}_3) = \lambda G(\bar{x}_1, \bar{x}_2, \bar{x}_3) , \qquad (2)$$

---

[*] Also at North Carolina State University.
[+] Supported by a Lindemann Fellowship. Present address: Stanford Linear Accelerator Center.

where $\lambda(>0)$ is an arbitrary parameter, and $\bar{a}_i$ are the scaling powers associated with the $^3CRS_o$. Eq. (2) is equivalent to the statement that any thermodynamic function $f = F(\bar{x}_1, \bar{x}_2, \bar{x}_3)$, related to G through appropriate differentiations, is an invariant equation under the one-parameter continuous group of transformations:

$$f' = \lambda^{\bar{a}_f} f, \quad \bar{x}_i' = \lambda^{\bar{a}_i} \bar{x}_i, \quad (i=1,2,3), \tag{3}$$

where $\bar{a}_f$ is the scaling power of f associated with the $^3CRS_o$ and is expressible in terms of $\bar{a}_i$.

Under the transformations $\bar{x}_i \to \bar{x}_i'$, certain functions $y(\bar{x}_1, \bar{x}_2, \bar{x}_3)$ are absolute invariants, i.e., $y(\bar{x}_1, \bar{x}_2, \bar{x}_3) = y(\bar{x}_1', \bar{x}_2', \bar{x}_3')$. It is known that all such invariants are expressible in terms of a basic set of two absolute invariants, e.g.,

$$y_1 \equiv \bar{x}_1/\bar{x}_3^{\bar{a}_1/\bar{a}_3}, \quad y_2 \equiv \bar{x}_2/\bar{x}_3^{\bar{a}_2/\bar{a}_3}. \tag{4}$$

It is easily demonstrated from Eq. (2) that appropriately scaled thermodynamic functions near a $^3CRS_o$ can be expressed as functions[6] of these invariants, e.g.,

$$y_o \equiv f/\bar{x}_3^{\bar{a}_f/\bar{a}_3} = F_2(y_1, y_2). \tag{5}$$

Near a critical line ($^2CRS_1$), thermodynamic functions must also satisfy the scaling requirement of Eq. (1). But, near the $^3CRS_o$, scaled thermodynamic functions already satisfy expressions such as Eq. (5). It is therefore more convenient to make an equivalent statement of Eq. (1) in the two-dimensional invariant plane of (y_1, y_2) such that the resulting expression will be manifestly invariant with respect to the symmetry requirements of both the $^3CRS_o$ and $^2CRS_1$. If we adopt the strong requirement that a point in one phase remains in that phase under the scale transformation of $\bar{x}_i \to \bar{x}_i'$, the coexistence surface (CXS) bounded by a $^2CRS_1$ near a $^3CRS_o$ becomes a coexistence curve in the invariant plane (cf. Fig. 1b). It is then possible to choose the relevant strong and weak directions for scaling for a $^2CRS_1$ (now a point in the invariant plane) by forming appropriate linear combinations of (y_1, y_2), e.g.,

$$y_i = \sum_{j=1}^{2} R_{ij}(y_i - k_j), \quad (i=1,2) \tag{6}$$

where R_{ij} are constants, and (k_1, k_2) are the values of (y_1, y_2) of the $^2CRS_1$. We note that the new scaling variables (which are absolute invariants under $\bar{x}_i \to \bar{x}_i{}'$) appropriately vanish at the critical line, (Fig. 1b).

In place of Eq. (1), we now hypothesize that at the ${}^2CRS_1$ and and near the ${}^3CRS_o$,

$$y_o(\mu^{a_1}\tilde{y}_1, \mu^{a_2}\tilde{y}_2) = \mu^{a_f} y_o(\tilde{y}_1, \tilde{y}_2), \qquad (7)$$

where a_f is the scaling power of the thermodynamic function f associated with the ${}^2CRS_1$ and is expressible in terms of a_1 and a_2. This functional relation reduces immediately to the following invariant form:[6]

$$z_o = F_1(z_1), \qquad (8)$$

where $z_o \equiv y_o/\tilde{y}_2^{a_f/a_2}$ and $z_1 \equiv \tilde{y}_1/\tilde{y}_2^{a_1/a_2}$.

Double-power scaling functions such as that given in Eq. (8) predict that data near both a ${}^3CRS_o$ and a ${}^2CRS_1$ will collapse from volumes to lines. The region within which these functions are valid is probably bounded by some "crossover" curve in the invariant plane (cf., Fig. 1b),

$$f_c(y_1, y_2) = 0, \qquad (9)$$

which describes a conical surface surrounding the ${}^2CRS_1$ with generators emanating from the ${}^3CRS_o$ in the $\bar{x}_i'$ space (cf., Fig. 1a).

For the critical line lying in the magnetic field-temperature (H-T) plane of the metamagnet whose phase diagram is depicted in Fig. 1a, we have $R_{ij} = \delta_{ij}$, $(k_1, k_2) = (0, -k)$, and Eqs. (8, 9) become

$$f\bar{x}_3^{(a_f/a_2 - \bar{a}_f/\bar{a}_2)/\varphi} [T - T_c(H)]^{-a_f/a_2}$$

$$= F_1\{H_{st} \bar{x}_3^{(a_1/a_2 - \bar{a}_1/\bar{a}_2)/\varphi} [T - T_c(H)]^{-a_1/a_2}\}, \qquad (10a)$$

$$f_c[H_{st} \bar{x}_3^{-\bar{a}_1/\bar{a}_3}, (T - T_t)\bar{x}_3^{-\varphi}] = 0, \qquad (10b)$$

where H_{st} is the staggered field and $\varphi \equiv \bar{a}_3/\bar{a}_2$. It will be gratifying if Eqs. (10) are borne out by future model calculations and experimental measurements.

We are grateful to R. B. Griffiths, F. Harbus, E. K. Riedel, and J. C. Wheeler for interesting discussions and for sending us their preprints. This research is supported by NSF, ONR, AFOSR, and NASA.

REFERENCES

1. R. B. Griffiths, Phys. Rev. Lett. 24, 715 (1970).
2. E. K. Riedel, Phys. Rev. Lett. 28, 675 (1972).
3. A. Hankey, T. S. Chang, and H. E. Stanley, to be published in the proceedings of this conference.
4. R. B. Griffiths and J. C. Wheeler, Phys. Rev. A2, 1047(1970).
5. A. Hankey, H. E. Stanley, and T. S. Chang, Phys. Rev. Lett. 29, 278 (1972).
6. Scaling functions may change their values or functional forms as the scaled variables change signs.

Figure 1†

(a) Schematic diagram of a metamagnet showing a $^3CRS_0$ at T_t where three $^2CRS_1$ terminate (a tricritical point). Cross-hatched areas are coexistence surfaces. $T_c(H)$ and L_F are the second and first order lines in the H-T plane, respectively. Only one crossover cone is shown.

(b) The invariant (y_1, y_2)-plane of (a). Dashed curves are coexistence surfaces. The three $^2CRS_1$ map into points in this invariant plane. A typical set of scaling directions $(\tilde{y}_1, \tilde{y}_2)$ for one of the $^2CRS_1$ is shown.

†In this figure, the subscripts of $^3CRS_0$ and $^2CRS_1$ have been suppressed.

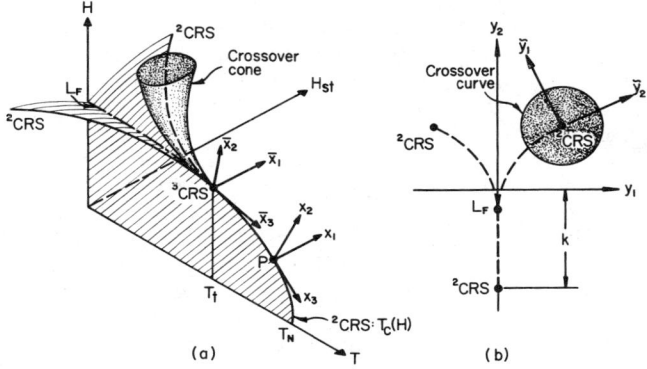

TRICRITICAL ISING MODELS AND THE SMOOTHNESS POSTULATE*

F. Harbus and H. E. Stanley
Massachusetts Institute of Technology, Cambridge, Mass. 02139

ABSTRACT

As applied to magnetic systems with tricritical points, the smoothness postulate predicts constant critical-point exponents along the second order portion of the phase boundary. We consider here the behavior of the critical-point exponent γ_{st} characterizing the staggered susceptibility $\chi_{st} \sim [T-T_c(H)]^{-\gamma_{st}}$ in two Ising model antiferromagnets with tricritical points. The evidence indicates that γ_{st} remains 5/4 for a wide range of fields, and appears to rule out a continuously changing index along the critical line.

In contrast to ferromagnets, antiferromagnets display a phase transition which persists even in the presence of an external field H. The field serves to oppose the intrinsic antiferromagnetic ordering of the system, resulting in a line of critical points in the H-T plane $T=T_c(H)$, with the critical temperature a decreasing function of H. One of the consequences of the "smoothness" postulate posed by Griffiths[1] is that the critical indices characterizing the phase transition will not change as one moves down along the critical line, at least for small critical fields near the Néel point.

In metamagnetic systems, however, it is observed that for low enough temperatures, the nature of the phase transition does in fact change drastically--the external field induces a first order rather than second order transition to the paramagnetic phase. Griffiths[2] called the point of changeover from second to first order behavior the "tricritical point" (TCP), and went on to suggest that such a point is a likely possibility for the breakdown of smoothness.[1] The TCP is the terminus of three critical lines and represents a special symmetry point in the metamagnetic phase diagram spanned by H, T, and a staggered magnetic field H_{st}. Some reports of TCP exponents have been made for both real and model magnetic systems.[3-5]

Rapaport and Domb[6] applied series extrapolation techniques to test the smoothness predictions for a 2-sublattice Ising model antiferromagnet with nearest neighbor (negative) exchange only, uniform with respect to lattice direction. The phase transition in such a "simple" antiferromagnet is expected to remain second order all the way to absolute zero. To study models with tricritical behavior, we have applied high-temperature series expansions to two different Ising model antiferromagnets which incorporate ferro-

*Work supported by NSF, ONR, and AFOSR.

magnetic interactions within each sublattice. In the context of these models, the smoothness postulate predicts that exponents will stay the same along the second order portion of the phase boundary, and change discontinuously at the TCP. Of course, close to the TCP crossover effects between the critical line exponents and the TCP exponents will occur. Once close enough to the critical line, however, the critical line exponents are predicted to dominate.

Our calculated phase boundaries for the two Ising tricritical models and estimates of their respective TCP susceptibility exponents are given elsewhere.[5] In this paper we focus upon the behavior of the exponent γ_{st} characterizing the divergence of the staggered susceptibility χ_{st} along the critical line,

$$\chi_{st} \sim [T-T_c(H)]^{-\gamma_{st}} . \qquad (1)$$

Our evidence over a wide range of field values H supports the smoothness prediction that γ_{st} is independent of H. Although it was impossible to verify this arbitrarily close to the estimated TCP due to increasing irregularity of the series, we believe that our results cover a sufficiently wide range of fields to argue convincingly against a continuous variation of γ_{st} along the critical line.

It is not surprising that closer to the TCP a finite number of series terms will begin to fail to reveal true asymptotic critical behavior. A closely analogous, and conceptually simpler, situation occurs in the application of series to study critical behavior when a three-dimensional lattice crosses over to a two-dimensional lattice as the exchange parameter linking adjacent layers goes to zero.[7]

The first tricritical Ising model treated has a Hamiltonian

$$\mathcal{H} = -J_{xy} \sum_{\langle ij \rangle}^{xy} s_i s_j - J_z \sum_{\langle ij \rangle}^{z} s_i s_j - \mu H \sum_i s_i \qquad (2)$$

where the first sum is over nearest neighbor (nn) spins coupled within an x-y plane on the sc lattice, while the second sum is over nn spins coupled along the z direction. To simulate a metamagnet, we take $J_{xy} > 0$ (ferromagnetic) and $J_z < 0$ (antiferromagnetic). We generate high-temperature series expansions to eighth order in inverse temperature for the two-spin correlation function; these series are exact in the external field. Fig. 1 shows estimates of γ_{st} based upon ratio analysis methods applied to the series for χ_{st} for the parameter choice $J_{xy} = 1$, $J_z = -1$. Represented are "1/n" sequences for several values of the variable $h \equiv \mu H/k_B T$, the natural variable in which the field enters the expansions. The plots of Fig. 1 are actually determined from the series after a bilinear transformation on the original expansion variable is carried out in order to mitigate the effect of oscillations in the ratio plots.

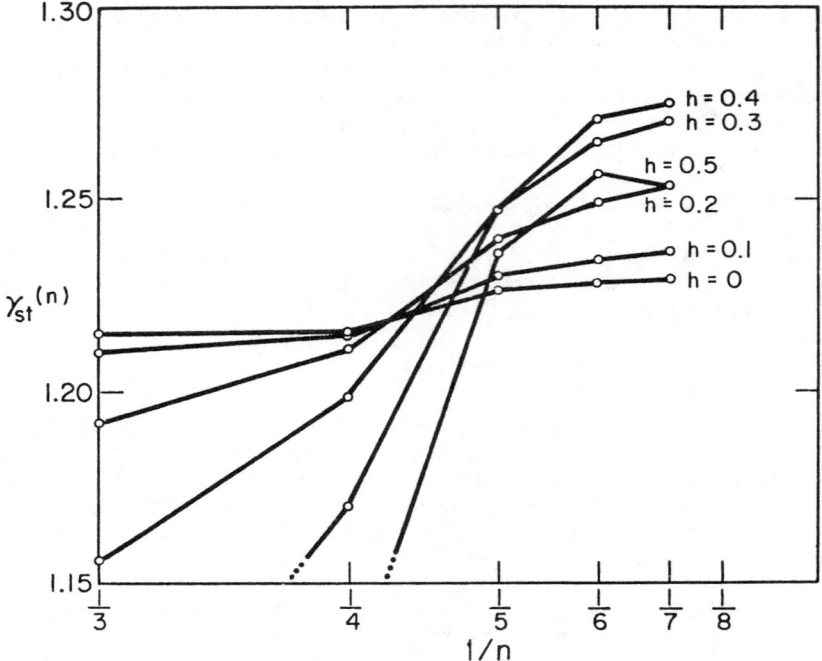

Fig. 1. Sequences of estimates $\gamma_{st}^{(n)}$ vs. $1/n$ for Hamiltonian (2) with $J_{xy}=1$, $J_z = -1$ along various $h \equiv \mu H/k_B T$ paths. The $\gamma_{st}^{(n)}$ are calculated from ratio method on χ_{st} series after transformation from expansion variable $\beta \equiv 1/k_B T$ to $\beta* = \beta/(1+1.5\beta)$.

The estimates in Fig. 1 for γ_{st} are closely scattered on both sides of the smoothness postulate prediction of 5/4. The final two estimates for any h path fall in the range of $1.250 \pm .025$. To gauge the extent of the fields covered in this plot, we note that the $h = .5$ path corresponds to a critical field of 1.7 (in dimensionless units), to be compared with the maximum T=0 critical field of 2. Adding to the evidence that the exponent 5/4 indeed correctly characterizes the behavior of χ_{st} in this range is the Padé approximant analysis carried out on the same set of series. Each χ_{st} series is raised to the 4/5 power and the dominant physical singularity located in the resulting Padé table. There is very good agreement between the critical temperatures given by the ratio method and those located by the Padé technique. Such consistency would not be expected if the χ_{st} series did not in fact diverge with a 5/4 power law.

The second tricritical model has nearest neighbor antiferromagnetic bonds and next nearest neighbor ferromagnetic bonds on the sc lattice. In an obvious notation,

$$\mathcal{H} = -J_1 \sum_{\langle ij \rangle}^{nn} s_i s_j - J_2 \sum_{\langle ij \rangle}^{nnn} s_i s_j - \mu H \sum_i s_i \qquad (3)$$

with $J_1 < 0$ and $J_2 > 0$. Fig. 2 gives estimates for γ_{st} for this model with $J_1 = -1$ and $J_2 = +1/2$. The sequences here are based upon a slightly modified version of the ratio method in which one forms sequences of linear extrapolants of an original set of estimates. As can be seen from the rather fine scale in Fig. 2, the final estimates for γ_{st} are extremely close to 1.250. There is some upward curvature, but this is present even in the h=0 sequence. In zero field the staggered susceptibility will have the same exponent as the uniform susceptibility in the corresponding ferromagnet, well established to be 5/4 for the three dimensional Ising model.

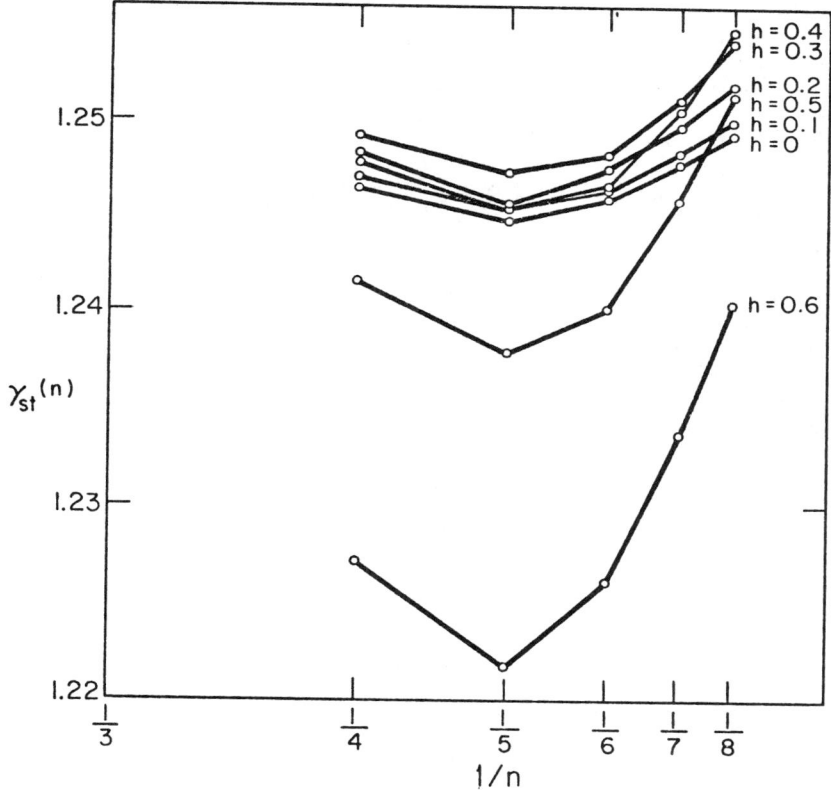

Fig. 2. Sequences of estimates $\gamma_{st}^{(n)}$ vs. $1/n$ for Hamiltonian (3) with $J_1 = -1$, $J_2 = +1/2$ along various h paths. The $\gamma_{st}^{(n)}$ are a set of linearly extrapolated exponents. No bilinear transformation on the expansion variable β is necessary.

The plots certainly give little indication of γ_{st} moving away from 5/4 to any significant degree for the fields shown. In this model, the T=0 critical field is 6, while the h= .6 path corresponds to critical field of 4.7. Again, consistency between ratio and Padé estimates for the critical temperatures in this range (and somewhat beyond) is excellent. Closer to the estimated TCP the ratios on the χ_{st} series give apparently decreasing values for γ_{st}. However, we believe this to be a spurious effect which would not be observed if sufficiently longer series were available, and conclude that our overall evidence supports the smoothness postulate prediction for both tricritical models.

REFERENCES

1. R. B. Griffiths, in Critical Phenomena in Alloys, Magnets, and Superconductors, edited by R. E. Mills, E. Ascher, and R. I. Jaffee (McGraw-Hill, New York, 1971), pp. 377-391.

2. R. B. Griffiths, Phys. Rev. Lett. 24, 715 (1970).

3. (a) D. P. Landau, B. E. Keen, B. Schneider, and W. P. Wolf, Phys. Rev. B3, 2310 (1971); (b) W. B. Wolf, B. Schneider, D. P. Landau, and B. E. Keen, Phys. Rev. B5, 4472 (1972).

4. D. P. Landau, Phys. Rev. Lett. 28, 449 (1972).

5. F. Harbus and H. E. Stanley, Phys. Rev. Lett. 29, 58 (1972); see also Proc. International Conference on Padé Approximants, Canterbury, England, July 1972, P. R. Graves-Morris, Ed. (Academic Press, London, 1973).

6. D. C. Rapaport and C. Domb, J. Phys. C4, 2684 (1971).

7. F. Harbus and H. E. Stanley, submitted for publication.

TWO SOLUBLE MODELS OF MAGNETIC SYSTEMS SHOWING CRITICAL POINTS OF HIGHER ORDER

Alex Hankey[*], T. S. Chang[†], and H. E. Stanley
Physics Department, Massachusetts Institute of Technology,
Cambridge, Massachusetts 02139

ABSTRACT

We give two soluble models which contain critical points of order four — one order more complex than tricritical points or the tetracritical point presented to this conference two years ago by Nagle and Bonner. The models are one dimensional Ising models with a long-range interaction; one contains one staggered magnetic field, the other two such fields. The formal solutions are given.

The geometric theory of phase transitions and critical phenomena[1] has led to the discovery of systems where several lines of critical points intersect.[2] Points where three lines of critical points meet have been called tricritical points[2], and points where four lines of critical points meet, tetracritical points[3] and so on[4].

It has been shown[5], however, that in the particular case considered by Nagle and Bonner, which was reported on at this conference[6] two years ago, the tetracritical point is a point on a single smooth curve of tricritical points from which it is topologically indistinguishable. This is because tricritical points are the endpoints of lines of points where three phases are in equilibrium and the same is true of the tetracritical point. A more complex point will be the endpoint of a line of points where four (or more) phases are in equilibrium, and which is also the endpoint of lines of tricritical points. A necessary condition for the existence of a more complex point is, therefore, a line of points where four phases coexist and this is not satisfied by the Nagle-Bonner model[5].

A new classification of critical points using strictly topological quantities has therefore been proposed[5,7]. Critical points are classified by an index called the order $\mathscr{O}$, which turns out to be equal to the dimensionality n of the total space[8] of field variables, minus the dimensionality d of the critical space considered: $\mathscr{O} = n-d$. Thus for ordinary critical points $\mathscr{O} = 2$, and for tricritical points (and the Nagle-Bonner tetracritical point) $\mathscr{O} = 3$.[9] To find a critical point of order 4, it is necessary to find a system where four phases coexist on a line of points for varying temperature; furthermore, the end of this line of points must also be the endpoint of four lines of tricritical points. This critical point will be topologically different from and one dimension less than the lines of tricritical points.

---

[*] Supported by a Lindemann Fellowship. Present address: Stanford Linear Accelerator Center.

[†] Also at North Carolina State University.

Motivated by these considerations we have found several systems containing lines of tricritical points. Among them are two exactly soluble one dimensional Ising models[10] which contain intersecting lines of tricritical points and thus[7] a critical point of order four. The solutions of these models are presented here.

The first model is a variation on the model of Nagle and Bonner. To make four phases stable in the region T > 0 (two ferromagnetic and now two antiferromagnetic instead of a single disordered phase), we split the long-range interaction so that it acts separately on the two sublattices of odd numbered and even numbered spins. The Hamiltonian is:

$$\mathcal{H} = -J_{SR} \sum_i s_i s_{i+1} - \sum_{i,r} J(r) s_{2i} s_{2i \pm 2r} - \sum_{i,r} J(r) s_{2i+1} s_{2i \pm 2r+1}$$

$$- H \sum_i s_i - H_2 \sum_i (-)^{i+1} s_i \quad (1)$$

Here H_2 is the staggered magnetic field of wavelength 2 lattice sites and $J(r)$ is the usual long-range interaction $J(r) = \text{Lim} (\gamma \to 0) a \gamma e^{-\gamma r}$. This Hamiltonian has an important discrete symmetry which will necessarily be reflected by the phase diagram of the solution of the model. It is defined by the operation $s_i \to (-)^i s_i$, $H \to H_2$, $H_2 \to H$, $J_{SR} \to -J_{SR}$. That means that the solution for positive J_{SR} is related to that for negative J_{SR} by

$$G(H, H_2, T; +J_{SR}) = G(H_2, H, T; -J_{SR}) \quad (2)$$

so that $M(H, H_2, T; +J_{SR}) = M_2 (H_2, H, T; -J_{SR})$. In particular the well-known[3] lines of tricritical points in the J_{SR}, H, T hyperplane for $J_{SR} < 0$ will be complimented by lines of tricritical points in the J_{SR}, H_2, T hyperplane for $J_{SR} > 0$: these will intersect on the T axis which therefore passes through a critical point of order four. The partition function for the Hamiltonian (1) is easily calculated using the same methods as Nagle and Bonner[3]. The additional complication of a second long-range interaction is easily dealt with by defining magnetisations. $M^O = \sqrt{2} \sum s_{2i+1}$, $M^E = \sqrt{2} \sum s_{2i}$ and the corresponding fields $H^O = (H + H_2)/\sqrt{2}$, $H^E = (H - H_2)/\sqrt{2}$. The Gibbs function can now be thought of either as a function of H, H_2 or of H^O, H^E. The Gibbs function is first calculated with zero long-range interaction. This gives $G_{NN}(H_0, H_{20}, T)$ (the same as in reference 3); the corresponding Helmholtz potential is then given by $A_{NN}(M, M_2, T) = G_{NN}(H_0, H_{20}, T) + MH_0 + M_2 H_{20}$. Note that by our definitions $MH + M_2 H_2 = M^O H^O + M^E H^E$. The effects of the long-range interaction can now be subtracted off as $E_{LR} = -a(M^{O2} + M^{E2}) = -a(M^2 + M_2^2)$ Hence the solution of the problem is given by

$$A(M, M_2, T) = CE \left\{ G_{NN}(H_0, H_{20}, T) + MH_0 + M_2 H_{20} - a(M^2 + M_2^2) \right\} \quad (3)$$

where CE denotes convex envelope. The physical fields H, H_2 minimise A−MH−$M_2 H_2$ giving

$$H = H_0 - 2aM \tag{4a}$$

$$H_2 = H_{20} - 2aM_2 \tag{4b}$$

At the special point 4 lines of tricritical points, bounding 6 surfaces of critical points intersect. For $J_{SR} = 0$ the interaction is the same as in the Weiss model — and the same exponents result. This is in contradiction with the analysis of multicritical point exponents provided by Theumann and Hoye[11].

The second model follows a suggestion of reference 4 and introduces a second staggered magnetic field — one of wavelength four lattice sites H_4. This simplifies the calculation because we only have to treat a block of four spins.[12] The Hamiltonian is now

$$\mathcal{H} = -J_{SR} \sum_i s_i s_{i+1} - \sum_{i,r} J(r) s_i s_{i \pm r}$$
$$- H \sum_i s_i - H_2 \sum_i (-)^{i+1} s_i - H_4 \sum_i c_i s_i \tag{5}$$

where $c_i = +1$ if $i = (4n+1), (4n+2)$; $c_i = -1$ if $i = (4n+3), 4n$. This is solved by the usual methods applied to one dimensional Ising models.[3,4] First it is solved without the long-range interaction by introducing four 2×2 transfer matrices representing interactions between spins $(4n+1, 4n+2) \ldots (4n, 4n+1)$.

$$T^1 = \begin{bmatrix} \text{Exp}[\beta(J+H+H_4)] & \text{Exp}[\beta(-J+H_2)] \\ \text{Exp}[\beta(-J-H_2)] & \text{Exp}[\beta(J-H-H_4)] \end{bmatrix} \quad T^2 = \begin{bmatrix} \text{Exp}[\beta(J+H)] & \text{Exp}[\beta(-J-H_2+H_4)] \\ \text{Exp}[\beta(-J+H_2-H_4)] & \text{Exp}[\beta(J-H)] \end{bmatrix}$$

$$T^3 = \begin{bmatrix} \text{Exp}[\beta(J+H-H_4)] & \text{Exp}[\beta(-J+H_2)] \\ \text{Exp}[\beta(-J-H_2)] & \text{Exp}[\beta(J-H+H_4)] \end{bmatrix} \quad T^4 = \begin{bmatrix} \text{Exp}[\beta(J+H)] & \text{Exp}[\beta(-J-H_2-H_4)] \\ \text{Exp}[\beta(-J+H_2+H_4)] & \text{Exp}[\beta(J-H)] \end{bmatrix} \tag{6}$$

These are multiplied together to give the 2×2 matrix M acting between neighboring blocks of four spins. The larger eigenvalue λ^+ of this matrix will give the partition function when the number of blocks of spins becomes infinite. The Gibbs function per spin is then given by

$$G(H, H_2, H_4, T) = \frac{kT}{4} \ln \lambda^+ = \tfrac{1}{4} kT \ln(X + \sqrt{X + \text{Det}}) \tag{7}$$

where $\text{Det} = \text{Det M} = (e^{2\beta J} - e^{-2\beta J})^4$.

$$X = e^{4\beta J} \text{Ch}(4\beta H) + e^{-4\beta J} \text{Ch}(4\beta H_2) + 1 + \text{Ch}(4\beta H_4) +$$
$$+ 4\text{Ch}(2\beta H)\text{Ch}(2\beta H_2)\text{Ch}(2\beta H_4) \tag{8}$$

The same method as for the first model is now applied, but as for Nagle and Bonner, one must take the convex envelope not of the full Helmholtz potential $A(M, M_2, M_4, T)$ but of the mixed function $G_1(M, H_2, H_4, T)$ so that

$$G_1(M, H_2, H_4, T) = \left\{ CE \ G(H_0, H_{20}, H_{40}, T) + H_0 M - aM^2 \right\} \quad (9)$$

where the three magnetic fields are related to the bare fields by

$$H = H_0 - 2aM, \quad H_2 = H_{20}, \quad H_4 = H_{40} \quad (10)$$

An analysis of the $T=0$ plane of this model[5] reveals points where 5 and 6 phases coexist. The former is the end of 3 lines where 3 phases coexist and one line where 4 coexist. As T is increased these lines sweep out surfaces of points which must be bounded for increasing T. The boundaries should be lines of critical points of order 3 and they will intersect at a critical point of order 4.

REFERENCES

1. R. B. Griffiths and J. C. Wheeler, Phys. Rev. A , 1047 (1970).
2. R. B. Griffiths, Phys. Rev. Letters 24, 715 (1970).
3. J. F. Nagle, Phys. Rev. A2, 2124 (1970).
   J. F. Nagle and J. C. Bonner, J. Chem. Phys. 54, 729 (1971).
4. W. K. Theumann and J. S. Hoye, J. Chem. Phys. 55, 4159 (1971).
5. A. M. A. Hankey, Ph.D. Thesis, MIT (1972).
6. J. C. Bonner and J. F. Nagle, J. App. Phys. 42, 1280 (1971).
7. T. S. Chang, A. Hankey and H. E. Stanley (a preprint).
8. The word space means here, a smooth connected subspace of the total space of thermodynamic field variables.
9. Generally $\mathscr{O}$ is equal to the number of independent vector directions out of a critical space and may in principle be equaled to the number of directions of scaling.
10. The models were proposed by considering the $T=0$ phase diagrams, a method that is reported in detail in reference 5.
11. Specifically, Theumann and Hoye suggested on the last page of their paper (reference 4) that where s critical points converge $\beta = 1/2s$. But in the case given here $s = 6$ and $\beta = 1/2$.
12. The number of lattice sites that must be considered is the LCM of all the wavelengths of the staggered fields introduced. The smallest LCM of two positive integers greater than one is $4 = LCM(2,4)$. (LCM means lowest common multiple.)

RESISTIVE ANOMALIES AT FERROMAGNETIC CRITICAL POINTS

T.G. Richard and D.J.W. Geldart
Department of Physics, Dalhousie University
Halifax, Nova Scotia, Canada.

The effect of spin fluctuations on the electrical resistivity, $\rho(T)$, is studied in a single band model. It is shown that $\rho'(T)=d\rho(T)/dT$ varies as the magnetic specific heat for $T<T_c$; a previously suggested term of the form $\varepsilon^{2\beta-1}$ is absent[1]. Just above the Curie point, $\rho'(T) \propto \varepsilon^{-\alpha}$ with a coefficient which is positive *if* the Fermi surface of the dominant current carriers is contained within the first Brillouin zone. In their temperature range of validity, typical "long range" correlation functions[2] also yield $\rho'(T)>0$. These conclusions yield a consistent picture of isotropic Nickel-like ferromagnets.

1. M.E. Fisher and J.S. Langer, Phys. Rev. Letters, 20, 665 (1968).

2. M. Ferer, M.A. Moore, and M. Wortis, Phys. Rev. Letters, 22, 1382 (1969).

MAGNETO-OPTIC STUDY OF THE CRITICAL PROPERTIES OF YIG[*]

D. D. Berkner and J. D. Litster[†]

Center for Materials Science and Engineering and Physics Dept.
Massachusetts Institute of Technology
Cambridge, Massachusetts 02139

ABSTRACT

We have used the Faraday rotation of 3.39μ laser light to measure the reduced total magnetization, σ, of yttrium iron garnet as a function of internal magnetic field H. Our measurements are for 53 isotherms over the temperature range $T_c - 109°C < T < T_c + 13°C$, where we find the critical temperature to be $T_c = 550.41 \pm 0.02°K$. We find the critical exponents to be given by $\beta = 0.370 \pm 0.005$, $\delta = 4.65 \pm 0.1$, and $\gamma = 1.35 \pm 0.02$. We also find that all data for $\sigma < 0.50$ are consistent with the scaling laws and an equation of state such as that proposed by Widom. We have experimentally determined the scaled magnetization $m = \sigma |1 - T/T_c|^{-\beta}$ as a function of scaled magnetic field $h = (g\mu_B SH/kT_c)/|1 - T/T_c|^{\beta\delta}$ over the range $0.018 < m < 8.12$ and $3 \times 10^{-3} < h < 5 \times 10^5$.

TEXT

We report here the result of our measurements of the equation of state of the insulating ferrimagnet yttrium iron garnet (YIG) in the region near the critical point. A number of studies of the critical properties of YIG have been reported during the past four years[1-3]. The results of these experiments, carried out by rather classical methods are not entirely consistent; therefore we have carried out a detailed study using Faraday rotation to measure the magnetization.

In a simple ferromagnet we expect the Faraday rotation to be proportional to the total sample magnetization[4]. The rotation in ferrimagnetic YIG is somewhat more complicated. It is theoretically predicted[5] that each sublattice contributes separately to the

---

[*]Supported by the Advanced Projects Agency under contract DAH1567C0222 and National Science Foundation under Grant GH-33635.

[†]John Simon Guggenheim Memorial Fellow 1971-72.

Faraday rotation. In the critical region, where the correlation length for the order parameter becomes long compared to the range of inter- and intra-sublattice interactions, both sublattice magnetizations are proportional to the total magnetization and we expect the Faraday rotation also to be proportional to the total magnetization.

In a ferro- or ferrimagnet, the internal magnetic field H_i is related to the external field H_e and the total magnetization M by $H_i = H_e - DM$, where D is a demagnetizing factor dependent upon sample geometry. For $T < T_c$ and M less than its saturation value for that isotherm H_i is zero (in the absence of coercive field and hysteresis effects) and so $M = H_e/D$. Writing the Faraday rotation $\Theta = VM$, we have $\Theta = VH_e/D$ and can therefore determine the value of V/D for our sample by making measurements at less than saturation magnetization. We found V/D to be independent of temperature from T_c to $T_c - 109°C$ (within about a 2% error caused by some hysteresis at the lower temperatures) when H_e was applied along the easy direction of magnetization, [111]; this experimentally confirms the expectation that the rotation is proportional to the total magnetization. Absolute calibration was by comparing saturation rotation for the lower temperature isotherms with reduced magnetization values obtained by NMR measurements[6].

Our experiments were carried out at 3.39μ (where YIG is quite transparent) on a single crystal disk 1 cm in diameter and 2 mm thick along the [111] direction. We found the rotation to be $\Theta = (VM_0/D)\sigma$ where σ is the reduced total magnetization and $(VM_0/D) = 0.395$ radian for our sample. We could measure Θ with an accuracy of 10^{-4} radian, corresponding to a sensitivity of 3 parts in 10^4 of the magnetization at 0°K. The sample temperature was stabilized to better than 10 mdeg in an oven and temperatures were measured with a platinum resistance thermometer. External fields from zero to 8 kG were used, and the maximum rotation observed was 0.20 radian (11.1°) corresponding to $\sigma = 0.50$. Measurements were made for 53 isotherms ranging from $T_c - 109°C$ to $T_c + 13°C$, and we found $T_c = 550.41 \pm 0.02°K$.

By analyzing data along the critical isochore, coexistence curve (obtained by extrapolation to zero field), and along the critical isotherm we obtained the exponents $\gamma = 1.35 \pm 0.02$, $\beta = 0.370 \pm 0.005$, and $\delta = 4.65 \pm 0.1$. We then examined the consistency of all of our data with the scaling hypothesis[7] by plotting the scaled magnetization $m = \sigma|1 - T/T_c|^{-\beta}$ against the scaled magnetic field $h = (g\mu_B S/kT_c)H_i|1 - T/T_c|^{-\beta\delta}$. (For YIG $4\pi M_0 = 2459$ G and $g\mu_B S/kT_c = 6.11 \times 10^{-7}$ G^{-1}). We found that all of our 53 isotherms fell on one universal curve with two branches — one for $T > T_c$ and the other for $T < T_c$. Our data confirm the scaling hypothesis, within our experimental errors, over the range $0.018 < m < 8.12$ and $3 \times 10^{-3} < h < 5 \times 10^5$. This range includes temperatures more than 100°C below T_c, external fields up to 8 kG and values of

σ up to 0.50. We show in
Fig. 1 a scaling plot of
h/m vs. m for 22 of our 53
isotherms. This plot only
includes the range $0.05 < m < 5$ and $1.0 < h/m < 10^4$;
however, the data points
not shown fall equally
well on the universal
curve. The exponents of
course also fit the scaling
law $\delta = 1 + \gamma/\beta$.

We may compare our
results for the exponents
with those obtained by conventional methods. The
agreement is best with the
data of Ohbayashi and Iida[1]
who found $\gamma = 1.32 \pm 0.04$,
$\beta = 0.35 \pm 0.02$, and $\delta = 4.6 \pm 0.2$. We also find
reasonably good agreement
with the values $\gamma = 1.31 \pm 0.01$, $\beta = 0.380 \pm 0.005$,
and $\delta = 4.42 \pm 0.05$ obtained
by Miyatani and Yoshikawa;
the greatest discrepancy is
just outside the error
limits assigned, perhaps
optimistically, by these
authors. Anderson has
reported[3] a value of $\beta = 0.36$ for $T < T_c - 1°C$ and
$\beta = 0.47$ for $T_c - T < 1°C$.

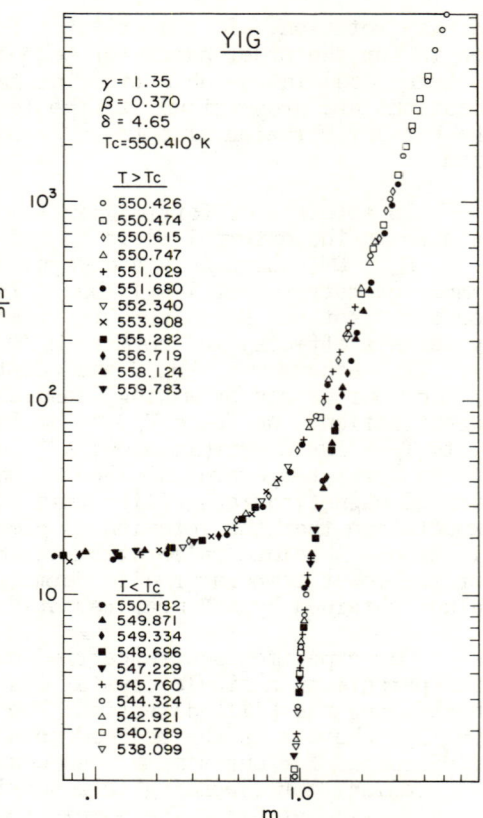

Fig. 1. Scaling plot for YIG (see text for discussion of variables plotted).

Such behavior was not observed by the two previously-mentioned sets
of authors[1,2], nor do we find a different value of β from 0.37 when
our five isotherms within the first degree below T_c are extrapolated
to zero field. We also find that all of the data points for these
five isotherms scale correctly using the same exponents as our
other isotherms. Anderson and his colleagues analyzed their data by
the so-called "kink point" method[8]; this is not a reliable way to
analyze data and the apparent change in the value of β very close to
T_c is probably an artifact of the data analysis. The same opinion
has also been expressed by Arrott[9], and more recently by Anderson
et al.[10].

We have also analyzed our data using the parametric approach
suggested by Josephson[11] and Schofield[12]. In particular we chose
the transformation suggested by Schofield:

$$\hat{H} = (g\mu_B S/kT_c)H_i = a\theta(1 - \theta^2)r^{\beta\delta} \qquad (1)$$

$$t = (T - T_c)/T_c = (1 - b^2\theta^2)r \qquad (2)$$

$$\sigma = (M/M_o) = g(\theta)r^\beta \qquad (3)$$

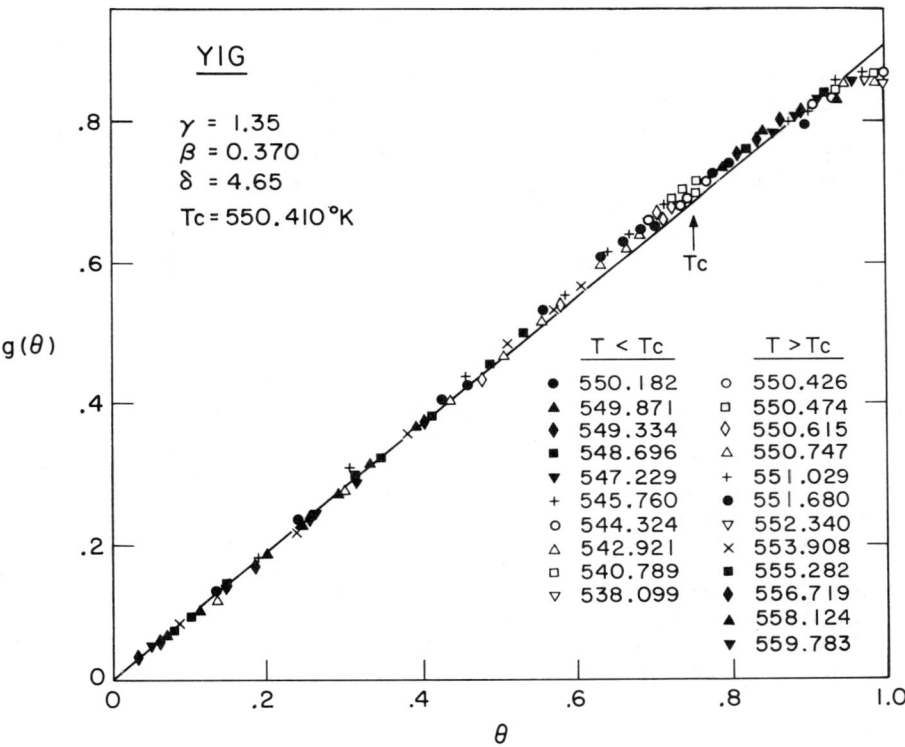

Fig. 2. Comparison of YIG data with the linear model parametric equation of state.

The parameter r represents the "distance" from the critical point in the H,T plane. We chose the value of $b^2 = (\delta-3)(\delta-1)^{-1}(1-2\beta)^{-1}$ because if $g(\theta) = k\theta$ with this value of b^2, paths of constant r in the H,T plane also correspond to paths of constant specific heat at constant magnetization[13] — thus r plays a physically meaningful role if $g(\theta)$ is indeed linear. We solved equations (1) and (2) for the values of r and θ corresponding to the H and t of each data point; from σ/r^β we then obtained $g(\theta)$ as a function of θ. The result for 22 of our isotherms is shown in Fig. 2. (In the figure, θ = 0 is the critical isochore, θ = 1 the coexistence curve, and the critical isotherm also maps into one point as indicated.) For θ ≤ 0.95, $g(\theta)$ appears to be very closely a linear function of θ, however closer to the coexistence curve a systematic departure (well outside of experimental error and rising to ~7% near $\theta = 1$) is observed. Thus the simple linear model[12] equation of state is not adequate for the critical behavior of YIG. Similar, but much smaller, departures can be seen for $CrBr_3$,[4] although the linear model appears to provide a very good approximation to the equation of state for pure fluids[4,14].

Our concluding remarks are as follows. We find the results of our Faraday rotation measurements of the critical properties of YIG to be in good agreement with data obtained by more classical methods. Our results are consistent with the scaling law hypothesis and show no sign of a change in exponent β very close to T_c. We find the linear model form of the parametric equation of state does not adequately represent our data. We plan, in a future publication, to present our data in tabular form, to discuss the experimental details, and to present the results of our search for an equation of state for YIG.

REFERENCES

1. K. Ohbayashi and S. Iida, J. Phys. Soc. Japan 25, 1187 (1968) and J. Appl. Phys. 41, 1265 (1970).
2. K. Miyatani and K. Yoshikawa, J. Appl. Phys. 41, 1272 (1970).
3. E. E. Anderson, H. J. Munson, S. Arajs, A. A. Stelmach, and B. L. Tehan, J. Appl. Phys. 41, 1274 (1970).
4. J. T. Ho and J. D. Litster, Phys. Rev. B2, 4523 (1970).
5. W. A. Crossley, R. W. Cooper, J. L. Page, and R. P. van Stapele, J. Appl. Phys. 40, 1497 (1969) and Phys. Rev. 181, 896 (1969).
6. J. D. Litster and G. B. Benedek, J. Appl. Phys. 37, 1320 (1966).
7. B. Widom, J. Chem. Phys. 43, 3898 (1965); L. P. Kadanoff, Physics 2, 263 (1966); R. B. Griffiths, Phys. Rev. 158, 176 (1967).
8. M. Rayl and P. J. Wojtowicz, Phys. Lett. 24A, 142 (1968).
9. A. Arrott, J. Appl. Phys. 42, 1282 (1971).
10. E. E. Anderson, S. Arajs, A. A. Stelmach, B. L. Tehan, and Y. D. Yao, Phys. Lett. 36A, 173 (1971).
11. B. D. Josephson, J. Phys. C2, 1113 (1969).
12. P. Schofield, Phys. Rev. Letters 22, 606 (1969).
13. P. Schofield, J. D. Litster, and J. T. Ho, Phys. Rev. Letters 23, 1098 (1969).
14. C. C. Huang and J. T. Ho, private communication.

MAGNETOCALORIC EFFECT IN NICKEL AT ITS CURIE TEMPERATURE

John E. Noakes
Scientific Research Staff, Ford Motor Company, Dearborn, Michigan

Anthony S. Arrott
Department of Physics, Simon Fraser University, Burnaby, B.C., Canada

ABSTRACT

Measurements of the magnetocaloric effect in a single crystal nickel sphere have been carried out for applied fields from 90 to 900 oersteds for temperatures close to T_c. The field dependence is consistent with a parametric equation of state fitted to magnetization data. For this equation the adiabatic exponent $\pi=(\gamma+2\beta-1)/\beta=2.78$. The data are normalized to and compared with the specific heat data of Connelly, Loomis and Mapother [Phys. Rev. B $\underline{3}$, 924 (1971)].

We have attempted to extend the work of Weiss and Forrer[1] in a study of single crystal nickel to lower fields and to smaller temperature intervals about T_c. Our magnetization measurements have been reported previously[2]. Our initial work on the magnetocaloric effect is presented here. We use field changes in a solenoid to produce the adiabatic changes in magnetization. The field is zero except during a 6 sec. (100 msec. rise time) current pulse. Ten different current pulses producing from 90 to 900 oersted are used, one pulse every 5 minutes during a 50 minute cycle. In addition the susceptibility is monitored every 30 seconds in the low field of 9 oersteds. This enables us to find T_c within ± 0.01 deg. K.

In the measurements reported here the furnace drifted slowly through T_c at ~ 01. deg./hr. The crystal, 1 cm dia, was in vacuum surrounded by a graphite coated quartz tube heater and 10 concentric tantalum radiation shields. The furnace temperature is monitored in the intervals between pulses by a means of a Pt - Pt(.13 Rh) thermocouple. The sample in an Al_2O_3 retainer is balanced on the tip of a differential thermocouple, 0.1 mm Pt - Pt(.10 Rh) which is used to measure the temperature rise during the pulses.

The reproducibility of the readings from a Keithly model 149 nanovoltmeter is ± 1 mdeg., which is slightly better than that obtained by Weiss and Forrer. As Weiss and Forrer found, there is a problem of the heat capacity of the thermocouple and the flow of heat down the leads. In our case it is quite serious. We have used a correction factor (f) in order to bring results into thermodynamic consistency with specific heat results as we explain below. In the preliminary results shown in Fig. 1, the measured temperature rises have been multiplied by the factor f = 1.32.

The thermodynamic relations we require are for paths shown in Fig. 2. The lines AC and BD are adiabatics. The line BC is isothermal. We consider a system with fixed volume and take as dependent intensive variables S the entropy per unit volume and M the magnetic moment per unit volume.

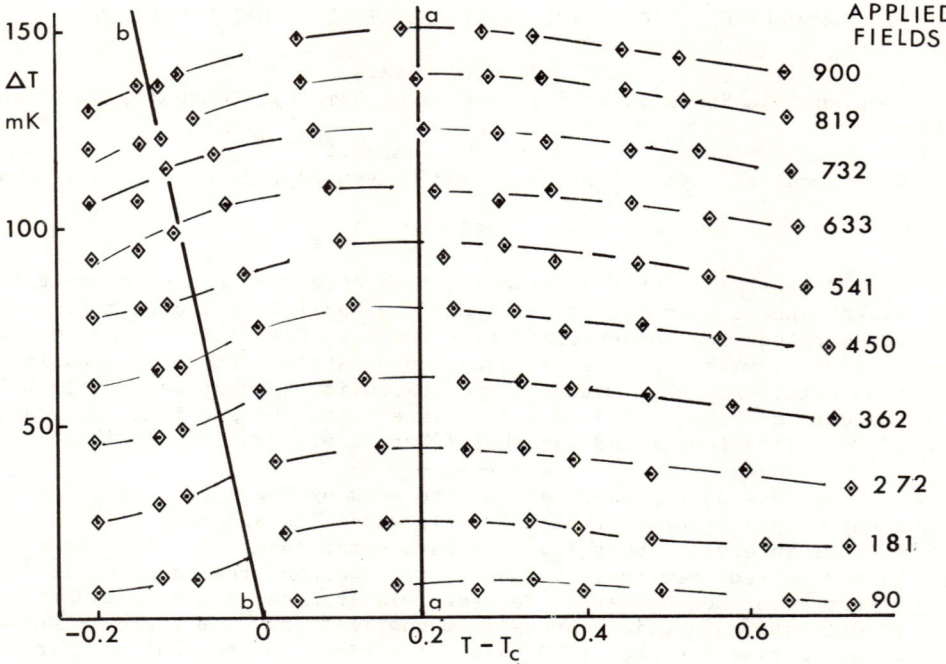

Fig. 1. The magnetocaloric temperature rise $\Delta T(T,H_1)$ for various initial temperatures T when the external field is increased from zero to H_1 = 90, 181, 272, 362, 450, 541, 634, 732, 819 and 900 oersteds. T_c is determined from concurrent susceptibility measurements. Data along a (constant initial temperature) and along b (constant final temperature) are used in Figs. 3a and 3b, respectively.

Fig. 2. See text.

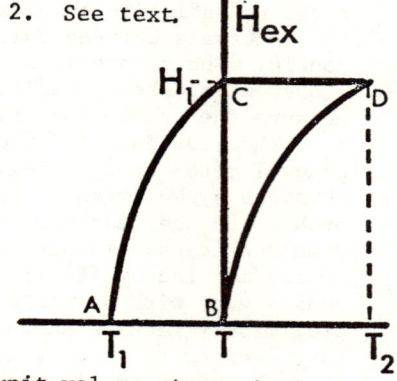

$$\Delta S_{AB} = - \Delta S_{BC} = \Delta S_{CD} \qquad (1)$$

$$S_{AB} = \int_{T_1}^{T} \frac{C_o}{T} dT \qquad (2)$$

and

$$\Delta S_{CD} = \int_{T}^{T_2} \frac{C_{H_1}}{T} dT \qquad (3)$$

Note that C_{H_1} is the specific heat per unit volume at constant volume and constant external field, H_1. The connection with magnetism is

$$\Delta S_{BC} = \int_0^{H_1} \left(\frac{\partial M}{\partial T}\right)_{H_{ex}} dH_{ex} = -\int_0^{M(H_1)} \left(\frac{\partial H}{\partial T}\right)_M dM \qquad (4)$$

To evaluate these expressions we fit our magnetization data to a parametric equation of state in a modification of the approach of Schofield[3]. This modification is Eq. 8 of ref. 4 with parameters $\gamma = 1.295$, $\beta = .376$, $M_1 = 42.5$ gauss, $T_1 = .340$ deg. K, and $b = .208$ and with the choice $n = 1$.

There are two convenient regions in which to check the consistency between the magnetization data and the magnetocaloric data without the necessity of exact knowledge of the specific heat. In one of these the specific heat is rather independent of field. This is particularly useful in that then the correction factor f should not be field dependent either. Near $T = T_c + 0.2$ deg. it is approximately true for fields from 90 to 960 oersteds that the magnetocaloric effect is a maximum, $(\partial^2 M/\partial T^2)_{H_{ex}}$ vanishes, and C_{H_1} is independent of H_1. We denote the specific heat at this temperature by $\bar{C}_H$ and using Eqs. (1), (3) and (4) write

$$\Delta T(T_c + 0.2, H_1) = - \frac{T_c}{\bar{C}_H} \int_0^{H_1} \left(\frac{\partial M}{\partial T}\right)_{H_{ex}} dH_{ex} \bigg]_{T=T_c+0.2} \qquad (5)$$

The comparison between the measurements of ΔT and the prediction from the magnetic equations of state are shown in Fig. 3a. The fit

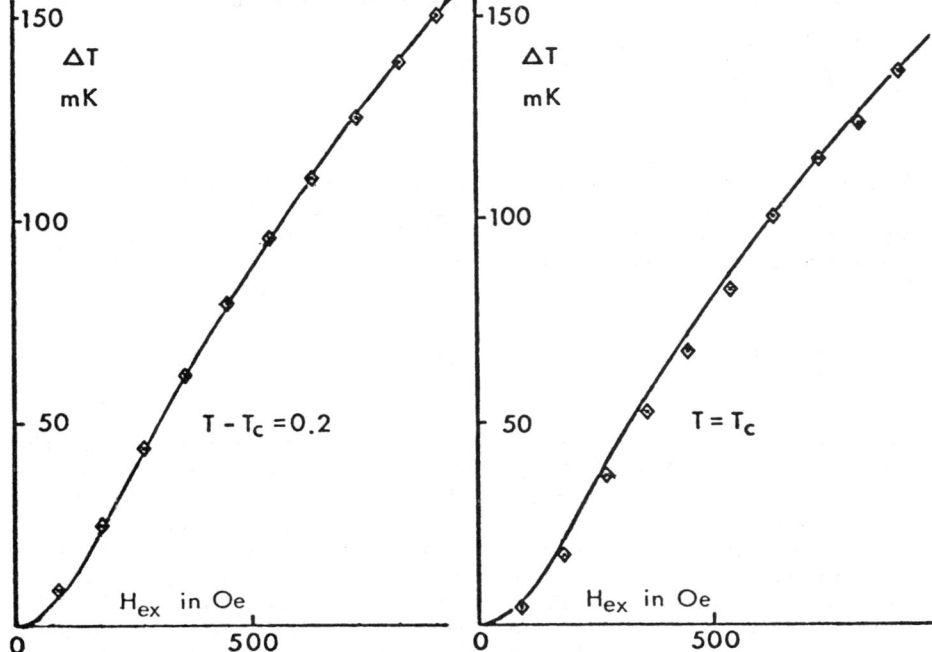

Fig. 3. Calculated and measured (normalized) field dependence of the magnetocaloric effect. a. $\Delta T(T_c+0.2,H_1)$. b. $\Delta T(H_1,T_c)$.

requires a normalization constant given by $f \cdot \bar{C}_H$. We use the definitive work of Connelly, Loomis and Mapother[5], hereafter referred to as CLM, to obtain values for the specific heat. After converting their value $\bar{C}_H' = 37.6$ joules/mole/deg. to volume units we find $f = 1.32$

The other convenient region is just below T_c where the specific heat in zero field goes through a maximum. We denote this value by $\bar{C}_o$. We use the notation $\Delta T(H_1, T)$ when the temperature ends at T with increasing field. From Eqs. (1), (2) and (4)

$$\Delta T(H_1, T_c) = -\frac{T_c}{\bar{C}_o} \int_0^{H_1} \left(\frac{\partial M}{\partial T}\right)_{H_{ex}} dH_{ex} \bigg]_{T=T_c} \qquad (6)$$

To obtain the final temperature from the data it is necessary to have the value of f. We use the value 1.32 as obtained above but with the caution that the change in the heat capacity of the sample may well influence f. CLM find $\bar{C}_o' = 39.9$ joules/mole deg. To obtain the fit shown in Fig. 3b we used the value $\bar{C}_o' = 40.6$ joules/mole/deg. That the fit is not as good in Fig. 3b as in Fig. 3a may well reflect the dependence of f upon the specific heat of the sample.

A further check of the data is found using the relation

$$\int_{T_1}^{T_c} \frac{C_o}{T} dT = \int_{T_c}^{T_2} \frac{C_{H_1}}{T} dT \qquad (7)$$

For H = 540 oersteds the magnetization in our sphere is comparable to that in the center of the flat plate (demagnetizing factor $4\pi/40$) used by CLM in an applied field of 240 oersteds. As C_o and C_{H_1} are both temperature independent in the above integrals, we write

$$\bar{C}_o = C_{H_1} \frac{T_2 - T_c}{T_c - T} = C_{H_1} \frac{\Delta T(T_c, H_1)}{\Delta T(H_1, T_c)} \qquad (8)$$

In evaluating $\Delta T(H_1, T_c)$ there is only a slight dependence upon the choice of f. We find $\Delta T(T_c, H_1)/\Delta T(H_1, T_c) = 1.075$ for H = 540 oersteds. From CLM for their applied field of 240 oersteds $C_H' = 37.9$ joules/mole/deg. Using this value, we find $\bar{C}_o' = 40.8$.

As the equation of state used here also fits the data of Weiss and Forrer with a precision of better than 0.3 percent, one can carry out a similar analysis on their data. Mathon and Wohlfarth have also analysed the Weiss and Forrer data using the last expression in Eq. (4) to calculate the dependence of the temperature rise from T_c for various values of the magnetization. This expression conveniently does not involve the demagnetizing effect. They compare their analysis with the scaling law prediction that $\Delta T(T_c, M) = M^\pi$ where $\pi = (\gamma + 2\beta - 1)/\beta$. They deduce that $\pi = 2.75 \pm 0.15$. The equation of state we use gives $\pi = 2.78$. The values of the magnetocaloric effect found by Weiss and Forrer at their lowest field, 430 oersted applied, are too low in comparison to their values in higher fields. This we attribute to an effect on the demagnetizing factor because they imbedded their thermocouple in a small hole drilled perpendicular to the field axis near one of the poles of their sphere. Weiss and Forrer had their thermocouple in contact with the one region which was less magnetized than the rest of their sphere. This effect is more serious at low fields. In avoiding this difficulty we have accentuated the problem of the correction factor f.

At present these experiments are not sufficiently accurate to determine more than one can deduce from the magnetization data and thermodynamics. It is hoped that our more recent measurements using phase sensitive detection at 2/3 Hz with an increase in sensitivity by a factor of 30 will prove to be more determining once we have found a solution to the correction factor problem. The aim of these experiments is to exploit the fact that a precise determination of π would help reduce the uncertainties in γ and β which arise when one fits data with a six parameter equation of state.

REFERENCES

1. P. Weiss and R. Forrer, Ann. Phys. (Paris) 5, 153 (1926).
2. A. Arrott and J.E. Noakes, Phys. Rev. Letters 19, 786 (1967); J.E. Noakes and A. Arrott, J. Appl. Phys. 39, 1235 (1968); A. Arrott, J. Appl. Phys. 42, 1282, (1971).
3. P. Schofield, Phys. Rev. Letters 22, 606 (1969).
4. A. Arrott and J.E. Noakes, J. Appl. Phys. 42, 1288 (1971).
5. D.L. Connelly, J.S. Loomis and D.E. Mapother, Phys. Rev. B. 3, 924 (1971).
6. J. Mathon and P. Wohlfarth, J. Phys. C 2, 1647 (1969).

CRITICAL MAGNETIC BEHAVIOR OF GIANT MOMENTS IN Ni-Rh

W. C. Muellner and J. S. Kouvel
Department of Physics
University of Illinois, Chicago, Ill. 60680

ABSTRACT

Detailed magnetization-field-temperature data were taken on a weakly ferromagnetic Ni-Rh alloy of ~35 at.% Rh near its Curie point ($T_c \simeq 44°K$). The critical exponents determined for the temperature dependences of the spontaneous magnetization (σ_{sp}) and the initial paramagnetic susceptibility (χ_o) and for the critical magnetic isotherm are respectively as follows: $\beta = 0.476$ (±0.015), $\gamma = 1.50$ (±0.05), and $\delta = 4.15$ (±0.08). These exponent values, though anomalous in comparison to various model predictions (unlike those for pure nickel, which are remarkably Heisenberg-like), nevertheless obey the scaling relation, $\delta = 1 + \gamma/\beta$. Moreover, all the data in the critical region are found to conform to a homogeneous equation of state.

From the values of the coefficients A and B in the expressions, $\sigma_{sp} = A(T_c - T)^\beta$ and $\chi_o = B(T - T_c)^{-\gamma}$, it is deduced that the elementary moments involved in the ferromagnetic-paramagnetic transition of this alloy have an average magnitude of ~20 μ_B. This is consistent with the size of the giant moments found in some paramagnetic Ni-Rh alloys (38-42 at.% Rh), as deduced from their Curie-Weiss susceptibility constants and low-temperature saturation magnetizations.[1] In all respects, the critical magnetic behavior of this Ni-Rh alloy resembles very closely that of dilute PdFe, the classic giant-moment system.[2] The existence of giant moments, presumably in the form of magnetic polarization clouds centered around statistical Ni-rich local regions[1], would appear also to be a basic intrinsic phenomenon in the Ni-Rh system.

1. W. C. Muellner and J. S. Kouvel, Magnetism and Magnetic Materials, AIP Conf. Proc., No. 5, 1972, p. 487.
2. J. S. Kouvel and J. B. Comly, "Critical Phenomena in Alloys, Magnets, and Superconductors", R. E. Mills, E. Ascher, and R. I. Jaffee, eds. (McGraw-Hill, N.Y., 1971) p. 437.

MAGNETIC SPECIFIC HEAT ANOMALY IN $GdNi_2$

J.A. Cannon
Fordham University*, Bronx, N.Y. 10458

J.I. Budnick+
National Science Foundation, Washington, D.C.

R.S. Craig, S.G. Sankar, D.A. Keller
University of Pittsburgh, Pittsburgh, Pa. 15312

ABSTRACT

We have accurately determined the critical behavior of the magnetic contribution to the heat capacity in $GdNi_2$, using both a phase sensitive A.C. technique in the critical region and a calorimetric technique from helium temperature up to room temperature. The deduced values of critical exponents are $\alpha = 0.36$ and $\alpha' = 0.026$.

INTRODUCTION

$GdNi_2$, a cubic Laves phase intermetallic ferromagnet has been the subject of considerable interest since Kawatra[1] and co-workers reported that the temperature dependence of the resistivity, relatively near the critical temperature, T_c, appears to behave according to the spin-disorder scattering theory of DeGennes and Friedel[2]. The theory predicts a cusp in the resistivity at the critical point on the basis of an Ornstein-Zerneke spin-spin correlation function. The large well localized Gd moment and the filled Ni 3d band which characterize $GdNi_2$ should result in temperature dependence predicted by a local moment theory. For reduced temperature $\varepsilon = (T-T_c)/T_c < 0.1$ however, the dominance of short range fluctuations cited by Fisher and Langer[3] appears to determine the temperature dependence. Zumsteg and Parks[4] have found that correcting the resistivity for the measured lattice expansion does not significantly change the critical behavior of the resistivity. Later resistivity and susceptibility measurements[5] indicate that molecular field predictions appear to apply for $\varepsilon > 0.1$.

EXPERIMENTAL

We have measured the specific heat of $GdNi_2$ using a steady state calorimetric technique described elsewhere[6]

*Supported in part by N.S.F.
+On leave from Fordham University

from 5 to 300°K and using an a.c. chopped light technique. We have made the measurement in the critical region on 2 different samples.

The calorimetric data (figure 1) shows a critical temperature, Tc, of 73.7°K. The calorimeter was operated using typical ΔT values of 1.5°K. Also shown in the figure (crosses) is the non-magnetic contribution to the specific heat which is discussed in the next section.

The sample measured by the steady state technique was prepared from stoichiometric amounts of high purity gadolinium (99.9% with respect to metallic impurities) and Johnson-Matthey spectroscopic standard nickel. The sample was annealed at approximately 800°C for four days. The sample on which a.c. measurements were made was arc melted from stoichiometric amounts of 99.999% Ni and 99.9% gadolinium sponge in an argon atmosphere. X-ray spectra show similar absorption patterns for both samples.

The A.C. sample (0.024" thick) was mounted on a copper base within an evacuated ($\sim 10^{-6}$mm Hg) space. The sample was insulated electrically from the base by a thin mica strip and thermally fastened to the base by thermal conducting compound. A copper-constantan (0.003") thermocouple was spot welded to the sample at the interface with the base. The other side of the sample was exposed to the chopped light through a quartz window. The copper base was in thermal contact with a nitrogen bath, the temperature of which could be varied by pumping. Absolute temperature measurement was by a platinum resistance thermometer imbedded in the copper base. The sample was irradiated by light from a tungsten filament lamp powered by a Sorenson Nobatron current supply regulated to 0.1% of output current. The A.C. temperature response was detected on a PAR HR-8 lockin-amplifier after being amplified by a PAR-190 input transformer. The lockin zero offset was used to increase resolution in the critical region. The HR-8 output was recorded on an X-Y recorder whose x-axis was driven through a D.C. amplifier by an independent thermocouple in thermal contact with the copper base.

The solution of the heat conduction equation for the boundary conditions of this experiment shows that for proper frequency and sample thickness values the temperature response is linear in $1/ftC_p$ where f is the frequency; t, the sample thickness; and C_p, the specific heat at constant pressure. Our sample temperature response at 77.4°K is linear in 1/f up to 19 Hz. The A.C. temperature response then is inversely proportional to the specific heat for frequencies below 19 Hz, since the 1/f dependence is clearly established. Measurements were made at 7Hz.

The result of a typical A.C. experiment performed at 7 Hz is seen in Figure 2, which shows Tc = 72.5°K, comparable to Tc determined by resistivity and susceptibility

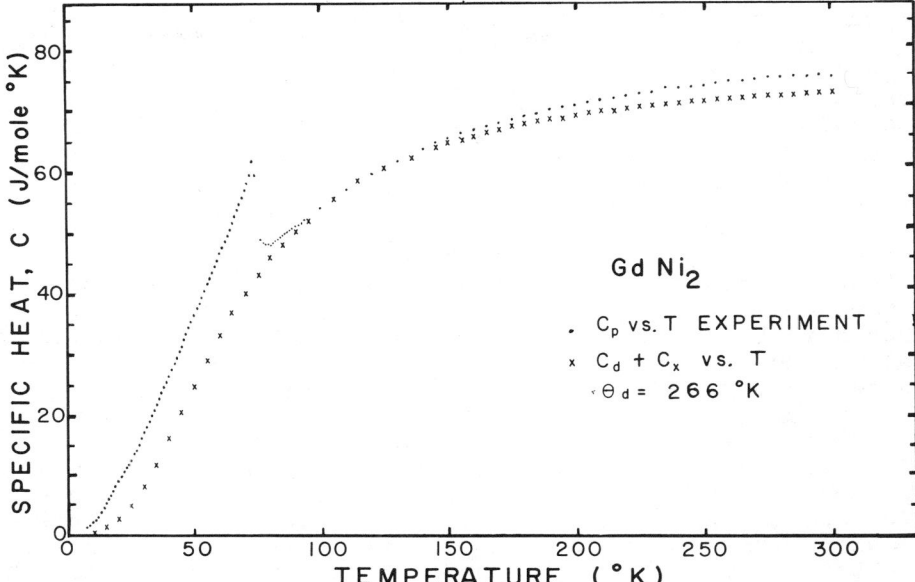

Fig.1. Temperature dependence of the heat capacity of GdNi$_2$ (dots). Lattice Debye (Cd) and linear temperature (Cx) correction vs. T (crosses).

measurements on the same sample. The trace is generated by allowing the system to drift up from 65°K, after pumping the bath down to less than 100mm Hg. The average drift rate is 1°K/8min. The absolute temperature is marked at 1°K intervals by reading the platinum resistance thermometer.

ANALYSIS

In order to extract the magnetic contribution from the total specific heat we have considered lattice and electronic contributions to the specific heat. Observing that the calorimetric C_p data (figure 1) is almost flat at room temperature, we assume, following a Debye model that the lattice specific heat, Cd, is 9R at 300°K. The measured specific heat at this temperature is 1.42 J/mole °K higher than 9R. We describe this contribution, Cx, as being due to other effects (electronic, for example) which are linear in T. Cx is small (less than 1% of Cd) throughout the critical region. Using these assumptions one can derive the Debye temperature, Θ_D, for the system as a function of temperature. It is found that Θ_D is 266°K to within 1% in the region 100°K to 150°K below which the effects of magnetic ordering become significant.

Figure 1 shows the match between the measured specific heat and the calculated non-magnetic contributions. In the region 100 to 135°K the deviations are less than 0.2%. A similar estimate of the heat capacity of non-magnetic LaNi$_2$ matches measured values of C_p with comparable accuracy from 245°K to below 70°K. It is estimated that the GdNi$_2$

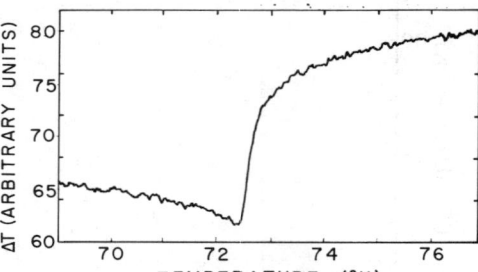

Fig.2. GdNi$_2$ a.c. temperature response, ΔT, vs T in critical region.

non-magnetic specific heat is approximated to within 1% throughout the critical region.

We have compared the A.C. data to the calorimetric data by shifting the magnetic peak by the difference in Tc and by assuming that the non-magnetic contributions (Cd and Cx) are known functions of temperature. We then took the product ($\Delta T \cdot C_p$) of the two sets of data in the critical region. We find that the product is constant to within 0.6% except for those 2 C_p data points nearest Tc which bracket the transition. We do not expect that C_p points determined using ΔT ranges which span the transition would be reliable. The 0.6% error in ($\Delta T \cdot C_p$) is comparable to the accuracy of the calorimetric measurement in this temperature range[6]. The fact that the product is constant supports the fact that $\Delta T \sim 1/C_p$ as indicated by the frequency dependence and can be used to convert the ΔT data to absolute C_p values in the critical region.

The magnetic contribution, C_M, derived by subtracting Cd+Cx from the critical region C_p is assumed to obey an exponent law $C_M = A \, \varepsilon^{-\alpha}$ for T > Tc and $C_M = A \, \varepsilon^{-\alpha'}$ for T < Tc. The data were "least squares" fitted to this form as before with α (α') and Tc as variable parameters. Figure 3 gives the fit of the data to a log-log plot of C_M vs ε. The critical exponent α is 0.36 with Tc = 72.5°K. Similar analysis for C_M vs ε below Tc give $\alpha' = 0.026$.

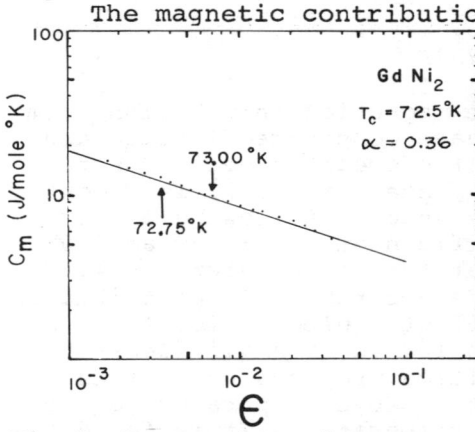

Fig.3. Temperature dependence of magnetic contribution to heat capacity, GdNi$_2$
$\varepsilon = (T-Tc)/Tc$.

DISCUSSION

The asymmetry in α and α' should be considered in view of certain aspects of the lattice expansion coefficient data. The strict lattice anharmonic contribution to the specific heat is quite small (<1%) in the critical region. However, when one compares the critical behavior of the thermal expansion coefficient by separating the anomaly from ordinary thermal variations the data shows similar critical exponent asymmetry. Such asymmetry has been found by Golding[7] in precise measurements of α and α' for $RbMnF_3$. It is notable that the anomaly in the expansion coefficient indicates a stiffening at the critical point which persists just below the ordering temperature. Also the large temperature range over which magnetic energy is significant is consistent with such asymmetry.

The data reported here cannot be used to derive the critical behavior of $GdNi_2$ at $\varepsilon > 0.1$ since the lattice correction is large compared to the magnetic contribution at this temperature. However, the critical behavior for $\varepsilon < 0.03$ is characterized by $\alpha = 0.36$ and below Tc by $\alpha' = 0.026$. The results are not according to mean field theory for $\alpha \neq \alpha' \neq 0$. There does appear to be some asymmetry between ordered and disordered phases while the material is well characterized by Curie-Weiss behavior over a wide range of temperatures in the paramagnetic region.

Although we do not interpret these results in terms of a microscopic theory of critical exponents there are some spin-orbit theoretical treatments mentioned in the current literature as having possible application in $GdNi_2$. Burzo and LaForest[8] mention the "interband mixing" work of Watson[9] and co-workers as a possible explanation for the negative g shift they observe in E.P.R. studies of $GdNi_2$. Taylor[10] in a recent review article suggests that $(Re)Ni_2$ compounds could have some pair interaction coupling as proposed by Levy[11]. The authors wish to thank Mr.W.Rodger and Mr.M.King for technical assistance.

REFERENCES

1. M.P.Kawatra,S.Skalski,J.A.Mydosh,J.I.Budnick,Phys.Rev. Letters 23,83 (1969).
2. P.G.DeGennes,J.Friedel,J.Phys.Chem.Solids 4,71 (1958).
3. M.E.Fisher,J.S.Langer,Phys.Rev.Letters 20,665 (1968).
4. F.C.Zumsteg,R.D.Parks,J.Physique 32,C1-534(1971).
5. J.A.Cannon,J.I.Budnick,M.P.Kawatra,J.A.Mydosh,S.Skalski Phys.Letters 35A,247 (1971).
6. W.E.Wallace,C.Deenadas,A.S.Thompson,R.S.Craig,J.Phys. Chem.Solids 32,805 (1971).
7. B.Golding,Phys.Rev.Letters,27,1142 (1971).
8. E.Burzo,J.LaForest,Intern.J.Magnetism,3,171 (1972).
9. R.E.Watson,S.Koide,M.Peter,A.J.Freeman,Phys.Rev.139, A167 (1965).
10. K.N.R.Taylor,Adv.Phys.20,551 (1971).
11. P.M.Levy,J.Appl.Phys.41,902 (1970).

EFFECT OF THE ORDER-DISORDER TRANSITION ON THE ELECTRICAL AND MAGNETIC PROPERTIES OF Fe_3Al*

G. A. Thomas,[†] J. M. Lawrence, P. M. Horn and R. D. Parks
Department of Physics and Astronomy
University of Rochester, Rochester, New York 14627

ABSTRACT

Over a range of compositions in the vicinity of the composition, Fe_3Al, the Fe-Al system undergoes an order-disorder transition (at $T = T_o$) which is of second order. This is followed by a ferromagnetic transition at a lower temperature. We have measured both the electrical resistivity and magnetic susceptibility in a large temperature interval which embraces both transitions. An anomaly occurs in the electrical resistivity in the vicinity of the order-disorder transition which has the general character of the anomaly predicted for an ideal localized antiferromagnet. The anomaly is characterized by a monotonically decreasing resistivity with increasing temperature in the vicinity of T_o and a divergence in the temperature derivative of the resistivity at T_o, and can be explained as arising from a combination of superzone-induced gap effects and concentration fluctuations. An anomaly is observed in the magnetic susceptibility in the vicinity of T_o which reflects a changing spin-spin exchange interaction in the vicinity of T_o.

INTRODUCTION

Over the years considerable effort has been focused on the study of lattice disorder and its effects on the properties of alloy systems. Near 500°C the alfenol alloys of composition near that of Fe_3Al undergo a second order, order-disorder transition from the DO_3 to $L2_o$ structure.[1] At a temperature of roughly 150-250°C lower, depending upon the exact composition, these alloys undergo a second phase transition, viz., from the paramagnetic to ferromagnetic state.[2] Thus, the Fe_3Al system is a unique laboratory in which to examine the effects of gradually changing disorder on magnetic properties (as manifested in the precursive effects in the susceptibility above the Curie point) as well as electrical properties.

EXPERIMENTAL PROCEDURE

Studies of the phase diagram indicate that a transition to a homogeneous ordered state does not occur at the stoichiometric

---

*Work supported by the U. S. Office of Naval Research.

[†]Present address: Bell Labs, Murray Hill, New Jersey 07974

composition Fe_3Al but does occur in the range from 27 to 34 atomic percent Al, where the Fe_3Al structure persists, probably with vacancies on Fe sites.[2] [3]A sample with 28.63 atomic percent Al was chosen for the present study.[3] Several pieces were cut from the same boule and were machined into either spheres (with diameter about 5 mm) for the magnetic susceptibility measurements or thin strips (approximately 5 cm x 5 mm x 1 mm) for the resistivity measurements. The resistivity R was measured by the conventional four probe method, wherein leads were spot-welded to the sample. The magnetic susceptibility χ was measured by the Faraday method, utilizing a Cahn microbalance. For both measurements the samples were first "blued" by briefly flaming them in air and then placed in a high purity argon at a pressure of one atmosphere, where they resided throughout the course of the measurements. These two procedures were found to adequately inhibit the emanation of Al from the samples which was observed to occur at temperatures of approximately 600°C and above. This decomposition was monitored directly in a separate experiment by observing the variation in resistivity (due to the deposition of Al) of an insulating strip placed near coated and uncoated test samples. In the resistivity measurements discussed below an additional check was made to verify that there was no measurable variation in ordering temperature with time.

Both the χ and R measurements were made using standard techniques but with care to allow for the slow annealing times. Initially, all samples were annealed in the experimental apparatus at T ~ 600°C for about one week. The measurements were then taken either at constant T or at drift rates of 1-2K/hr and satisfactory equilibrium conditions were verified periodically by monitoring χ or R at constant T as a function of time. Near the Curie point T_M an 8-day anneal preceded the R measurements and particular care was taken in following the above procedure.

RESULTS

The results of the R and χ measurements are shown in Figs. 1 and 2. The resistivity measurements representing equilibrium are indicated by solid circles; the open circles represent readings taken at drift rates above 10K/hr which are sufficiently slow to maintain equilibrium for T ~ 600°C but are not at lower temperatures. The variation in measured values of $1/\chi$ due to the long equilibrium time was comparable to the measuring accuracy in $1/\chi$ in the temperature region far below the order-disorder transition T_o, so that satisfactory results were obtained over the entire T-range studied.

The resistivity data are normalized to the peak value R_p which occurs at 490°C. The resistivity of our sample at this point was (250 ± 30) $\mu\Omega$-cm, in reasonable agreement with earlier studies.[4] There is no R anomaly within the accuracy of our measurements at T_M (with T_M determined from χ), indicating that scattering from spin fluctuations is swamped by scattering from order-disorder fluctuations, phonons and vacancies. The anomaly associated with

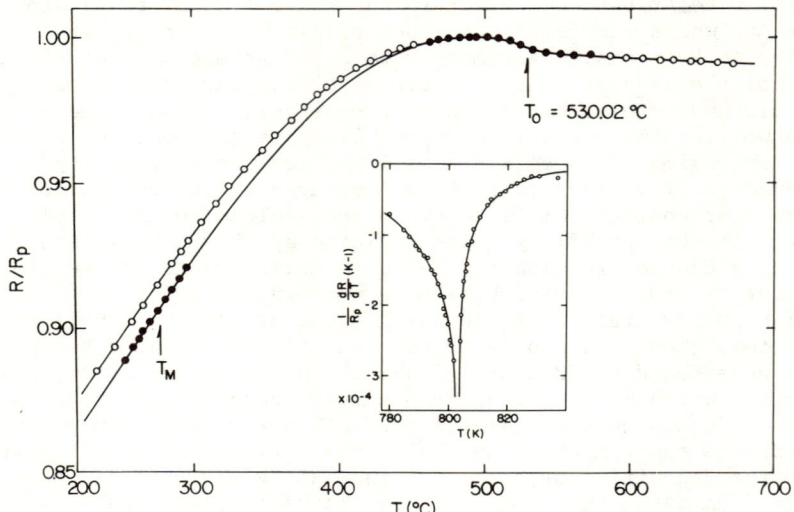

Fig. 1. Measured resistance of Fe_3Al. Insert shows temperature derivative near T_o.

the atomic ordering in the vicinity of $T_o = 530°K$ is qualitatively similar to that predicted by Suezaki and Mori[5,6] for an antiferromagnet. This behavior is consistent with the statistical equivalence of these two ordering processes within the context of the Ising model. In the vicinity of T_o, Suezaki and Mori predict contributions arising from superzone gap formation and critical scattering giving

$$dR/dT = -B_g \varepsilon^{-(\alpha'+\gamma')/2} + B_- \varepsilon^{-(\alpha'+\gamma'-1)} \quad T < T_o$$

$$-B_+ \varepsilon^{-(\alpha+\gamma-1)} \quad T > T_o \quad (1)$$

where $\varepsilon = (T_o - T)/T_o$, and α' and γ' are the critical exponents of the specific heat and the susceptibility. Equation (1) qualitatively accounts for the broad maximum below T_o and predicts a sharp negative peak in dR/dT at T_o. This latter peak is observed in dR/dT, as shown in the insert to Fig. 1, and accurately identifies T_o.

The magnetic susceptibility measurements, Fig. 2, clearly mirror the magnetic ordering at $T = T_M = 310°C$, but also reveal an anomaly near T_o. The measured values of T_o and T_M for the alloy composition studied are in good agreement with earlier measurements by various experimenters.[2] The behavior of $1/\chi$ as T is increased above T_M is characteristic of a simple ferromagnet, but as T approaches T_o an anomaly ensues. The variation of $1/\chi$ over the wide temperature interval shown suggests that both above and below T_o, the data tends to become linear with the same slope, as

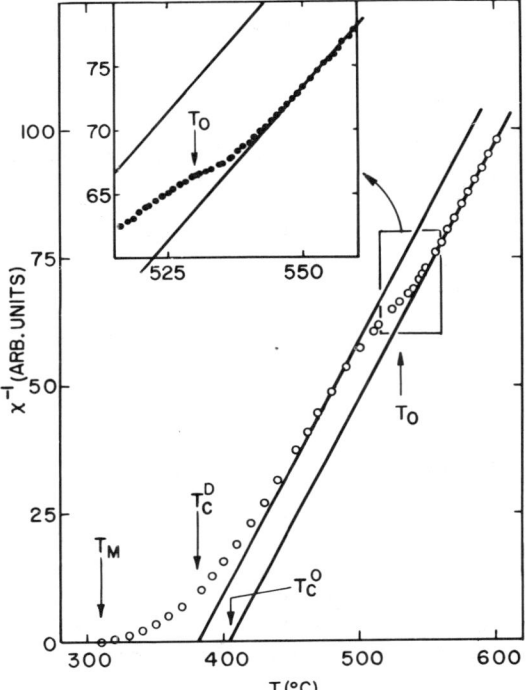

Fig. 2. Measured magnetic susceptibility of Fe_3Al. Insert shows expanded view of anomaly near T_o.

suggested by the two straight lines. This indicates the same Curie constant, and hence the same moment per Fe atom above and below T_o, but a change in T_M, and also the mean field transition temperature T_c, as a result of the ordering. Thus, we may write

$$\chi^{-1} = \begin{cases} \chi_0^{-1}(1 - T/T_c^D), & \text{for } T \gg T_o \\ \chi_0^{-1}(1 - T/T_c^O), & \text{for } T \ll T_o \end{cases} \quad (2)$$

where T_c^D and T_c^O are the mean field Curie points in the disordered and ordered phases, respectively. In the region near T_o, more complex behavior is clearly indicated in the insert to Fig. 2 and can be thought of as mirroring a changing spin-spin exchange interaction which is manifested in the shift from T_c^O to T_c^D.

If we assume that the magnetic behavior results from a Ruderman-Kittel interaction between approximately localized spins via the conduction electrons, then, the effect of the atomic ordering will influence χ as it does R to the extent that atomic ordering changes the effective number of conduction electrons (as a consequence of the appearance of superzone-induced gaps). The changing coordination numbers in the vicinity of T_o of the Fe atoms on the disordered sites might be expected to contribute also to the magnetic anomaly near T_o.

CONCLUSION

While the general nature of both the resistivity and magnetic anomalies observed in the vicinity of the order-disorder transition can be understood on the basis of the above discussion of superzone-induced gap effects and coordination number fluctuations, a more quantitative understanding must await the unraveling or dismissal of other possible contributing mechanisms. In particular, the effect of anomalous lattice effects, as determined from measurements of the lattice parameter near T_o and the dependence of the resistivity and the magnetic exchange interaction on lattice parameter, must be considered. This question will receive high priority in our continuing work on the Fe_3Al system.

REFERENCES

1. E.g., L. Guttman and H. C. Schnyders, Phys. Rev. Letters **22**, 520 (1969).
2. M. Hansen, Constitution of Binary Alloys (McGraw-Hill, 1958); R. P. Elliott, First Supplement (to above) (McGraw-Hill, 1965); F. A. Shunk, Second Supplement (McGraw-Hill, 1969).
3. L. Guttman, generously supplied this sample, which was made by Texas Instruments, Inc.
4. A. J. Bradley and A. H. Jay, Proc. Roy. Soc. (London) **A136**, 210 (1932); C. Sykes and H. Evans, Proc. Roy. Soc. (London) **A145**, 529 (1934) and J. Iron and Steel Inst. **131**, 229 (1935); C. Sykes, Proc. Roy. Soc. (London) **A148**, 422 (1935); H. Saito, Nippon Kinzoku Gakkai-Si **B14**, 1 (1950).
5. Y. Suezaki and H. Mori, Progr. Theor. Phys. **41**, 1177 (1969).
6. R. D. Parks, A.I.P. Conf. Proc. **5**, 630 (1971).

SUSCEPTIBILITY OF THE FERROMAGNET $GdNi_2$*

P. M. Horn and R. D. Parks
Department of Physics and Astronomy
University of Rochester, Rochester, New York 14627

ABSTRACT

Measurements have been made of the low field critical magnetic susceptibility of the intermetallic $GdNi_2$ in the reduced temperature interval $10^{-3} \lesssim \varepsilon \lesssim 4$, where $\varepsilon = (T - T_c)/T_c$ and T_c is the ferromagnetic transition temperature. The observation of mean field behavior, viz., $\chi = \chi_0(T - T_c^{MF})^{-1}$ over an unusually large temperature interval allowed a precise determination of the mean field parameters T_c^{MF} and χ_0. This in turn allowed a unique comparison with high temperature Ising and Heisenberg expansions, including second nearest neighbor interactions. The experimental results imply an anomalously long interspin force range for $GdNi_2$, which is manifested not only in the behavior of χ for large ε, but in the small apparent value of the critical index γ (viz., $\gamma = 1.19$) in the interval $10^{-3} \lesssim \varepsilon \lesssim 10^{-1}$. This analysis implies that universality, if it holds for this material, does so in a temperature interval sufficiently narrow about the critical point to render it experimentally non-demonstrable.

*Work supported by U.S. Office of Naval Research.
A complete report on this work, co-authored by
D. N. Lambeth and H. E. Stanley, will be published elsewhere.

THERMOELECTRIC POWER AND ELECTRICAL RESISTIVITY OF SOME Ni-BASED ALLOYS NEAR THE CURIE POINT

I. Nagy and L. Pál

Central Research Institute for Physics, Budapest, Hungary

ABSTRACT

The anomalous behaviour of the absolute thermoelectric power /ATP/ near the critical point was investigated in Ni and its dilute alloys with some 3d transition elements. Both the ATP and the temperature coefficient of the electrical resistivity /TCER/ in the vicinity of the Curie point were measured in Ni/Fe/, Ni/Mn/ and Ni/Cr/ alloys with different impurity concentrations. It was found that the critical behaviour of the transport properties depends strongly on the impurity.

In Ni/Mn/ alloys an oscillatory behaviour of the ATP was observed around the Curie point, while there was also a very significant difference in the critical behaviour of the TCER from that in pure Ni. Fe and Cr impurities, on the other hand, did not alter the monotonous temperature dependence of the ATP at the Curie point, and the singularity of the TCER did not differ from that in pure Ni. It is suggested that the peculiar behaviour of Ni/Mn/ alloys originates from virtual bound states near the Fermi level, since a shift of the Fermi level at the Curie temperature could drastically change the density of states at the Fermi energy and this might account for the observed anomalies of the transport properties.

INTRODUCTION

In two earlier papers [1,2] we have reported experimental results on the behaviour of the absolute thermoelectric power /ATP/ and the temperature coefficient of the electrical resistivity /TCER/ of very high purity Ni and Fe near the Curie point. It was found that in a narrow temperature interval around this point the TCER displays a stronger than logarithmic divergence which accords well with the observations of Craig et al.[3] but conflicts with the experiments of Kraftmakher[4] and Zumsteg et al.[5]. This anomalous behaviour of the TCER was accompanied by an oscillation of the ATP in the same temperature interval. Since it could be assumed the observed anomalies are attributable to the presence of different impurities, it was therefore decided to measure the ATP and TCER in Ni-based alloys of various Mn, Fe and Cr content.

EXPERIMENTAL PROCEDURE AND RESULTS

The alloys were prepared from high purity starting metals by vacuum melting. The samples were cold-rolled and annealed in vacuum at $600^\circ C$ for two hours. The composition of the alloys was determined by atomic absorption analysis. Table I presents some of the data characterising the samples.

Table I

Alloys	N^o	c at. %	$T_c C^o$	λ
Ni/Fe/	1	0.24	355	0.4
	2	0.63	359	0.4
	3	1,25	364	0.4
Ni/Cr/	1	0.12	349	0.4
	2	0.51	333	0.4
	3	1.13	302	0.5
Ni/Mn/	1	0.50	344	0.7
	2	1.17	338	1.0
	3	1.82	332	1.3

The TCER and ATP measurements were carried out by the previously used direct current method. The Curie points shown in Table I were defined by the maximum of the TCER curves.

On Fig. 1 we have drawn the temperature dependence of the TCER and the ATP of Ni/Fe/ alloys in the vicinity of the Curie point. The TCER curves have the same shape as for pure Ni, except for an increasing broadening with rising impurity concentration. The critical exponent /given in the last column of Table I/ seems to be independent of the Fe concentration, but it must be noted that the calculation of λ involved a definite uncertainty owing to the broadening of the TCER curves. The ATP curves do not show any anomaly at the Curie point.

The Ni/Cr/ alloy gave the same TCER behaviour as the Ni/Fe/ alloys, as can be seen from Fig. 2. Here too no anomalous behaviour of the ATP was observed at the Curie point, except that its value exhibited a strong dependence on the Cr content.

In the case of Mn impurity both the TCER and the

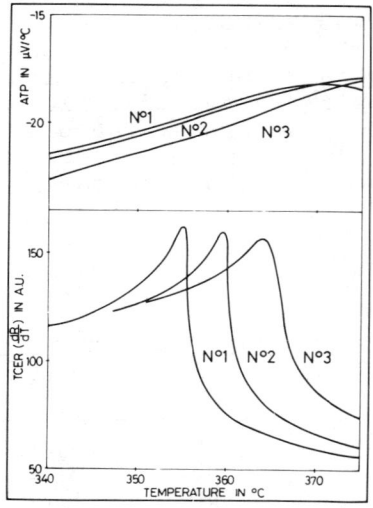

Fig. 1. The TCER and the ATP of Ni/Fe/ alloys in the vicinity of the Curie point.

ATP manifested very peculiar behaviour in the vicinity of the Curie point /Fig. 3/. The critical exponent gets larger with rising Mn content, and as the Mn impurity brings about an increase in the TCER values below the Curie point, a large flat maximum is observed. Nevertheless the most striking feature is seen on the ATP curves; these show a minimum followed by a maximum around the Curie point. The difference between the maximum and minimum values and the temperature intervals between them widen with mounting Mn concentration.

DISCUSSION

An explanation of the above experimental results can be supplied by assuming the existence of a virtual bound state associated with the impurity. According to the Friedel theory [6] and much other experimental evidence [7,8], 3d transition element impurities bring about the formation of s↑d↑ virtual bound states in the vicinity of the Fermi level. The presence of a virtual level results in an additional contribution to both the electric resistivity and the ATP. These contributions are described satisfactorily by Friedel's formula

$$\rho_{o\uparrow} = \frac{20\pi c}{nK_F} \cdot \frac{1}{1 + \left(\frac{E_F - E_1}{\Delta}\right)^2} \quad (1)$$

and making use of Mott's equation [9] for the diffusion ATP,

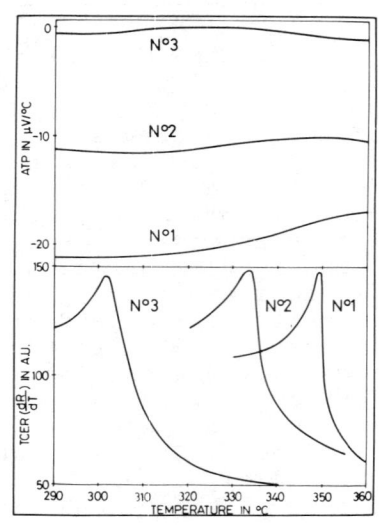

Fig. 2. The TCER and the ATP of Ni/Cr/ alloys in the vicinity of the Curie point.

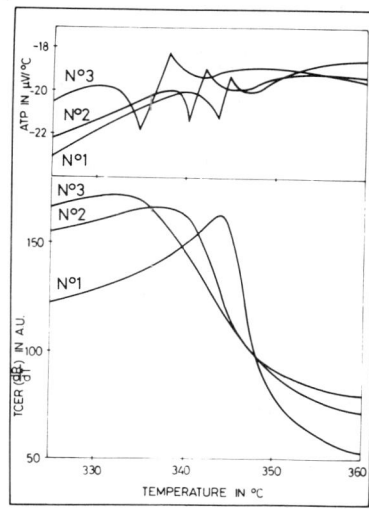

Fig. 3. The TCER and the ATP of Ni/Mn/ alloys in the vicinity of the Curie point.

$$S_{o\uparrow} = \frac{-\pi^2 k^2 T}{3e} \left(\frac{d \ln \rho_{o\uparrow}}{dE} \right)_{E_F} \quad (2)$$

where E_F is the Fermi energy, E_1 the average energy of the virtual bound state and Δ its width /assuming a Lorentzian shape/.

According to the Mott band model [10], Ni has a $\bar{d}$ band split into two sub-bands by the two spin directions. Since the splitting disappears above the Curie point, it must be accompanied by a shift of the Fermi level to a lower value. This implies that $\rho_{o\uparrow}$ and $S_{o\uparrow}$ may be temperature dependent in the vicinity of the Curie point, where an intensive variation of E_F is expected. According to measurements of ρ_o and S_o in some dilute Ni-based alloys at low temperatures [7,8], it seems fairly likely that the virtual level lies far below the Fermi level in the case of Ni/Fe/, and above the Fermi energy for Ni/Cr/ alloy. With Mn impurity the virtual level may be located below, but very near to, the Fermi level.

In the latter case the movement of the Fermi energy may be sufficient for it to cross the virtual level. This would entail a change in sign of the impurity contribution to the ATP at some critical temperature, T_K, where $E_F = E_1$. Below this temperature S_o is negative because $E_F > E_1$, but it becomes positive if $T > T_K$, where E_F shifts below E_1. An analogous behaviour of the ATP can be observed on Fig. 3.

With Fe or Cr impurities the oscillation of the ATP does not appear, according to our model, because in Ni/Fe/ alloy E_1 is too far from E_F for the Fermi and virtual levels to reach a crossing point, whereas in Ni/Cr/ alloy the two energies are actually diverging during the shift of E_F.

The TCER in the critical region has been calculated by Fisher and Langer [11] by considering the critical fluctuation of the short-range spin order, and they obtained a logarithmic divergence at the critical point. However, in the presence of transition metal impurities ρ_o also may give a contribution to the TCER - a factor which is missing from the calculation.

From eq. 1 it is immediately apparent that the sign of $d\rho_o/dT$ depends on E_F-E_1. Hence, in the case of Ni/Mn/ alloy $d\rho_o/dT$ is positive for $T<T_K$ but with decreasing value as the temperature approches T_K /i.e. where $E_F=E_1$/, while for $T>T_K$ the contribution is negative. The change in sign of $d\rho_o/dT$ therefore causes both the flattening of the TCER curve below T_K and the apparent rise in the critical exponent above T_c.

With Fe and Cr impurities, on the other hand, $d\rho_o/dT$ does not alter sign and furthermore its absolute value may be negligible compared to the magnetic part of the TCER, because of the distance of the virtual bound state from the Fermi level.

REFERENCES

1. I. Nagy, L. Pál, Phys. Rev. Letters, 24, 894 /1970/.
2. I. Nagy, L. Pál, Journ. de Phys. C, 32, 531 /1971/.
3. P. P. Craig, W. I. Goldburg, T. A. Kitchens, J. I. Budnick, Phys. Rev. Letters, 19, 1334 /1967/.
4. Ya. A. Kraftmakher, Fiz. Tverd. Tela, 9, 1528 /1967/
5. F. C. Zumsteg, R. D. Parks, Phys. Rev. Letters, 24, 520 /1970/.
6. J. Friedel, Nuovo Cimento /Suppl.2./, 7, 287 /1958/.
7. T. Farrel, D. Greig, J. Phys. C, 3, 138 /1970/.
8. M. C. Cadeville, J. Roussel, J. Phys. F, 1, 686 /1971/.
9. N. F. Mott, H. Jones, The theory of the properties of metals and alloys. /Diver Publications, Inc., New York, 1958/.
10. E. P. Wohlfarth, Proc. of the Int. Conf. on Magnetism, Nottingham, England /1964/.
11. M. E. Fisher, J. S. Langer, Phys. Rev. Letters, 20, 665 /1968/.

Section 28.   Co-Fe Alloys; Walls in Whiskers

MAGNETIC AND CRYSTALLINE PROPERTIES OF HEXAGONAL Co-Fe ALLOYS

T. Wakiyama

Department of Electrical Engineering, Tohoku University, Sendai, Japan. 980

ABSTRACT

The X-ray study revealed that the Co-Fe alloy system in the range 0 to 8 at.% Fe is classified into three regions with different phases ; (1) the ABAB-h.c.p. below 1 at.% Fe, (2) the ABAC-double h.c.p. between 1 and ~4 at.% Fe and (3) the ABC-f.c.c. above ~4 at.% Fe. It was found that discontinuous increases in saturation moment take place at these phase boundaries. The value of the saturation moment extrapolated to the zero Fe content for the double h.c.p. phase was intermediate between values for the h.c.p. and f.c.c. phases. The observed change in saturation moment per Fe atom can be explained by the rigid band model. The direction of easy magnetization of the magnetocrystalline anisotropy was found to change from the c-axis for the h.c.p. phase into the c-plane for the double h.c.p. phase. Domain patterns with double walls were observed on the c-plane of the double h.c.p. crystals. The temperature dependence of Co^{59} NMR for domain and domain wall in 0.43 at.% Fe alloy revealed the spin rotations due to a change in anisotropy. For 1.04 at.% Fe alloy, the observed two different resonances are thought to be due to Co^{59} nuclei on two different lattice sites in the double h.c.p. structure.

INTRODUCTION

According to the old literature[1], the Co-Fe alloy system has been reported to consist of three phases, i.e., b.c.c., f.c.c. and h.c.p. phases depending upon Fe content. Weiss and Forrer[2] first measured the composition dependence of the saturation magnetization of these alloys. No definitive data were presented, however, on magnetic properties in the hexagonal region.

The magnetocrystalline anisotropy of the hexagonal phase of this alloy system was first investigated by Chikazumi, Yosida and the present author[3]. They found that 1.04 at.% Fe-Co single crystal exhibits an anomalously large negative uniaxial anisotropy ($K_{u1} = -10.2 \times 10^6$ erg/cm^3 for 32 kOe at 4.2 K), which holds the spins perpendicular to the hexagonal c-axis, in contrast with hexagonal pure Co and Co-base alloys below 1 at.% Fe. It was also found by X-ray analysis that this anomalous alloy has a hexagonal crystal structure with a stacking sequence of the ABAC type, instead of the ABAB type prevailing in hexagonal pure Co and Co-base alloys below 1 at.% Fe[3]. Recently, the present author[4] found that the magnetic moment depends on composition anomalously at low Fe concentration, due to a change in the h.c.p. stacking sequence.

The present investigation has been started for the purpose of clarifying these anomalous magnetic and crystalline properties of hexagonal Co-Fe alloys. The crystal structures, saturation magnetization, magnetocrystalline anisotropy, magnetic domain patterns and nuclear magnetic resonance were investigated for available single crystals and a number of polycrystals of the Co-Fe alloys in the range 0 to 8 at.% Fe. The magnetic and crystalline properties were found to have a close relation with each other in these alloys.

EXPERIMENTAL PROCEDURE

Polycrystal specimens were prepared by melting raw materials of 5 to 7 g in a plasma jet flame furnace using high purity argon gas. The raw materials used were Co of 99.99 % purity and Fe of 99.99 % purity. The specimens were melted two or three times after being turned over up-side-down on a water cooled copper hearth to ensure the homogeneity. For the specimens with dilute content of

Fe less than 1 at.%, the Fe raw material was buried in a capsule made of pure Co ingot to avoid the evaporation of a small amount of Fe in the process of melting. The change of the weight after alloying was as small as of 0.2 to 2 mg. The error of the Fe content was estimated to be at most ±0.02 at.% Fe for the specimens with Fe less than 1 at.% and ±0.04 for more than 1 at.%.

Some of the single crystals used in this study were the same single crystals used in the earlier investigation[3]. They were prepared by using 99.99 % pure Co and 99.99 % pure Fe. Several additional single crystals were also grown from the melt in vacuum by the Bridgeman method, and then cooled slowly to room temperature through a point at which the crystal transforms from f.c.c. to hexagonal structure. The content of Fe was determined by chemical analysis. The single crystals with 1.53, 1.58 and 2.11 at.% Fe contain about 0.15 at.% Ni, because in these cases the raw materials used were electrolytic Co of 99.8 % purity and electrolytic Fe of 99.9 % purity.

The lattice parameters were measured by X-ray diffraction as a function of concentration using needle-like polycrystalline specimens. Crystal structures were investigated for the single crystals utilizing a precession camera with Mo Kα radiation. In this technique an undistorted section of reciprocal lattice is recorded on a film, then we can examine the stacking variants of close-packed structure. Great care was taken to avoid introducing stacking disorder in the specimens examined.

For the saturation magnetization measurements, a number of the polycrystalline specimens were shaped by the lapidary method into spheres of 5 to 6 mm in diameter with an accuracy of ±2 μm. The magnetization was measured with the induction method at 4.2 K in the field up to 45 kOe produced by a superconducting magnet with a persistent current. The time integral of the induced voltage was measured by an integrating digital voltmeter. Calibration was carried out by performing the same experiment on a standard pure

Co sample, the saturation value of which was adopted as 162.55emu/g[5] at 4.2 K.

The magnetocrystalline anisotropy constants were determined from magnetization curves measured on single-crystal spheres in the principal crystallographic directions of the hexagonal structure. Dimensions of the specimens used for this purpose were 3 to 5 mm in diameter with an accuracy of a few micrometers. The orientation of the crystal was determined first by the light-figure method and then checked by the back-reflection Laue technique.

For the observation of magnetic domains, a disk single crystal of 4.6 mm in diameter and 0.76 mm in thickness with the (0001) plane and a rectangular specimen of $4.3 \times 3.6 \times 3.6$ mm^3 with the (0001), the ($10\bar{1}0$) and the ($\bar{1}2\bar{1}0$) planes were prepared. Domain patterns were observed by means of the well-known powder pattern technique.

The NMR measurements were made by means of a marginal oscillator with a frequency-modulation device. The resonance signal was displayed on an oscilloscope or a recorder chart. For the specimens several plates of approximately $0.5 \times 10 \times 10$ mm^3 were cut out from the ingot with large grains which has the same purity as single crystals used in the anisotropy measurements[3]. They were ground with fine emery paper and then the ground surfaces were removed by the electropolishing. This process was found to be unavoidable to get the well-defined resonance signals. The specimens thus prepared were sealed in evacuated glass tubes before use.

EXPERIMENTAL RESULTS

1. Lattice Constant and Crystal Structures[6]

The lattice constants obtained at room temperature for the Co-Fe alloys in the range 0 to 8 at.% Fe are plotted in Fig.1 as a function of the Fe concentration. Both lattice parameters a and c increase linearly with increasing Fe content. It was found that discontinuous changes in lattice parameters, the values of c/a and

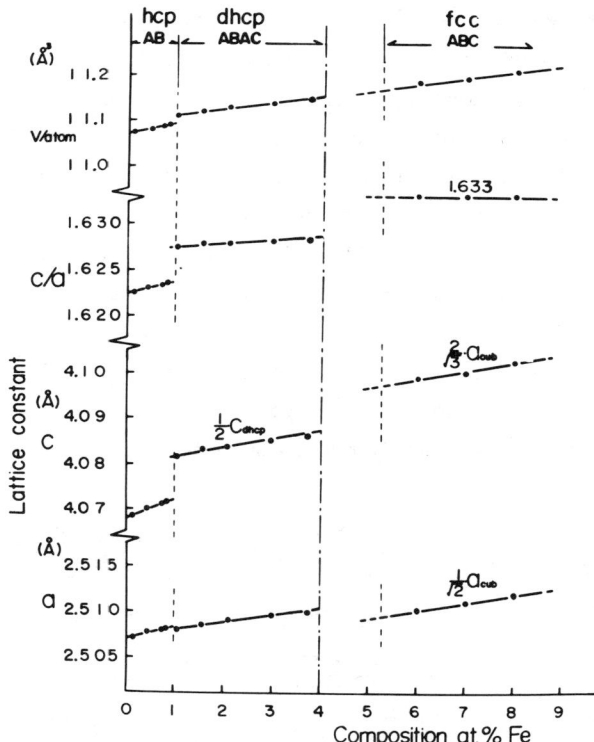

Fig.1 Lattice constants, c/a and volume per atom for Co-Fe alloys at room temperature. (Onozuka, Yamaguchi, Hirabayashi and the present author)

Fig.2 Zero-level precession photographs of the single crystals of (a) 0.83 at.% Fe-Co and (b) 2.11 at.% Fe-Co. The indices refer to the fundamental h.c.p. cell. (Onozuka, Yamaguchi, Hirabayashi and the present author)

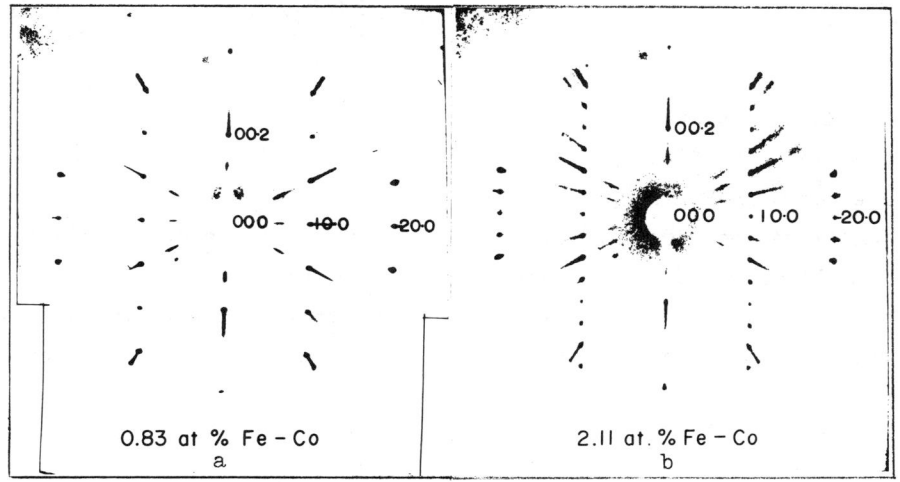

volume per atom take place at about 1 and ~4 at.% Fe corresponding to the phase boundaries between the h.c.p. and double h.c.p. and the double h.c.p. and f.c.c. structures (Fig.1). In order to investigate the stacking sequence of the close packed planes, the series of diffraction spots with the indices 10.ℓ and 20.ℓ were observed. Figures 2 (a) and (b) show zero-level precession photographs of the single crystals with 0.83 and 2.11 at.% Fe. The diffraction spots in Fig.2 (a) correspond to those of h.c.p. structure with the stacking sequence of the ABAB type. In Fig.2 (b)[7], however, the spots along the 10.ℓ and 20.ℓ rows were found to be indexed with ℓ= 0, 1/2, 1, 3/2 and so on. Thus the spacings of the adjacent spots were equal to a quarter of the distance between the origin and 00.2 reflection of the h.c.p. cell. In addition to this fact, the observed relative intensity was also found to be explained as the double h.c.p. structure with the stacking sequence of the ABAC type. Figure 3 shows both ABAB-h.c.p. and ABAC-double h.c.p. crystal structures.

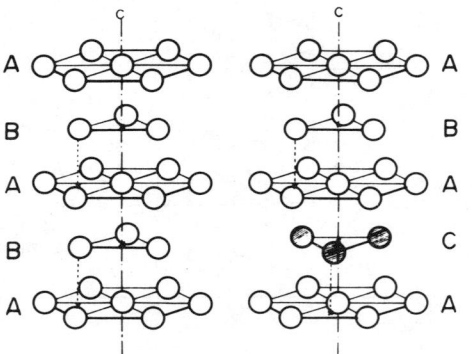

Fig.3 ABAB-h.c.p. and ABAC-double h.c.p. crystal structures.

2. Saturation Magnetization

The high-field portions of the magnetization curves at 4.2 K are shown in Figs.4 (a) and (b). As the field increases, the magnetization increases even up to 45 kOe in the case of the alloys with lower Fe concentration, while it remains almost constant or saturates for the alloys with higher Fe concentration. The saturation magnetization was obtained by the law of approach to saturation expressed by

$$M = M_s (1 - a/H - b/H^2 - \ldots) + \chi_0 H. \qquad (1)$$

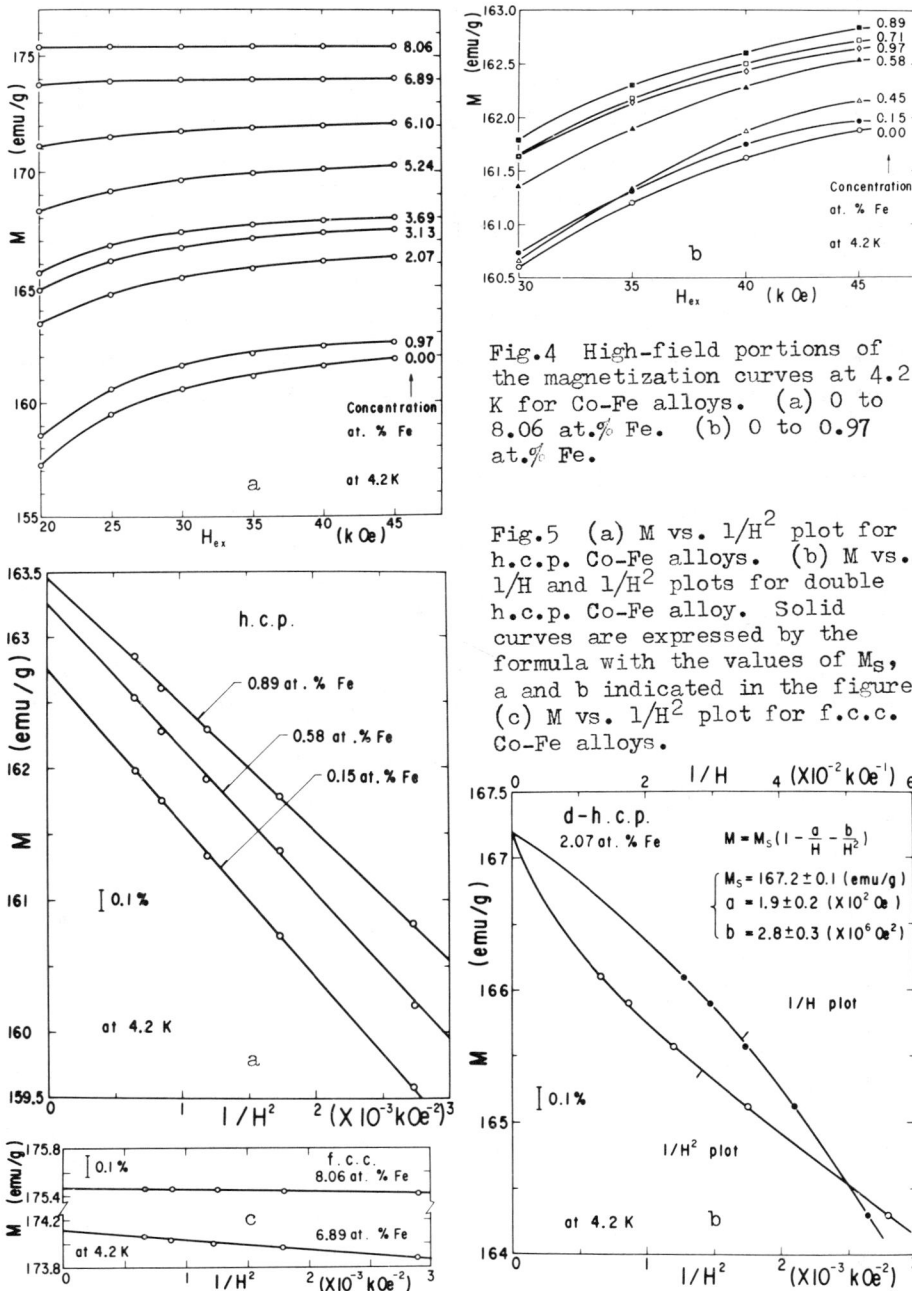

Fig.4 High-field portions of the magnetization curves at 4.2 K for Co-Fe alloys. (a) 0 to 8.06 at.% Fe. (b) 0 to 0.97 at.% Fe.

Fig.5 (a) M vs. $1/H^2$ plot for h.c.p. Co-Fe alloys. (b) M vs. $1/H$ and $1/H^2$ plots for double h.c.p. Co-Fe alloy. Solid curves are expressed by the formula with the values of M_s, a and b indicated in the figure. (c) M vs. $1/H^2$ plot for f.c.c. Co-Fe alloys.

In the present cases, the last term $\chi_4 H$ was found to be negligibly small. It was found that the data for the alloys below 1 at.% Fe with the h.c.p. structure were fitted with straight lines when M vs. $1/H^2$ plot was used, as shown in Fig.5 (a). This means that the second term in (1) can be neglected, thus a = 0. In these alloys the saturation values were obtained by b/H^2 extrapolations and the values of the coefficient b were determined from the slopes of straight lines. For the alloys between 1 and 4 at.% Fe which have the double h.c.p. structure, it was found that both $1/H$ and $1/H^2$ plots deviate from linear relationship as shown in Fig.5 (b). This fact indicates that neither of these terms can be neglected. In these cases, we analyzed the data graphically by trial and error method assuming appropriate values of a and b. An example of results of analysis is given in Fig.5 (b), where both curves of the $1/H$ and $1/H^2$ plots are very well expressed by the formula (1) with the values of M_s, a and b indicated in the figure. In the case of the alloys with higher Fe concentration such as 6.89 and 8.06 at.% Fe which have the f.c.c. structure, it was found that the experimental data obey either the a/H law or the b/H^2 law. Figure 5 (c) shows examples of the $1/H^2$ plot.

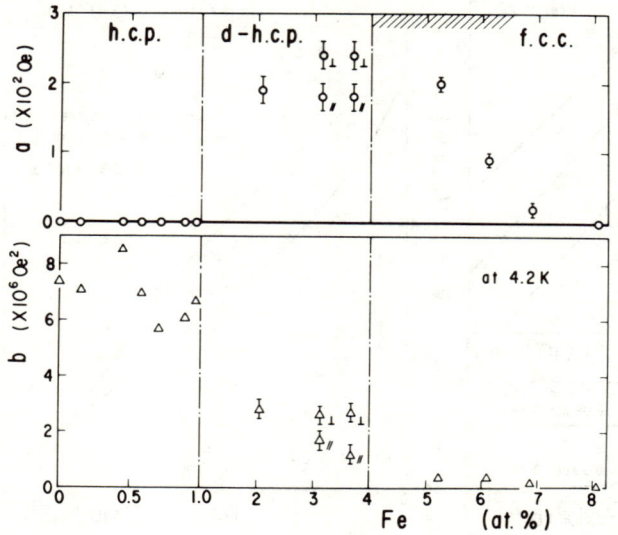

Fig.6 Values of a and b in the law of approach to saturation as a function of concentration for Co-Fe alloys.

The values of a and b thus obtained are plotted against composition in Fig.6, where the data with marks ∥ and ⊥ were obtained from the magnetization measurements parallel and perpendicular to elongated grains of the specimens, respectively. The saturation magnetic moments thus determined are plotted as a function of the Fe concentration in Fig.7 (a). In more detail for the alloys below 1 at.% Fe, the concentration dependence of the saturation moment is shown in Fig.7 (b).

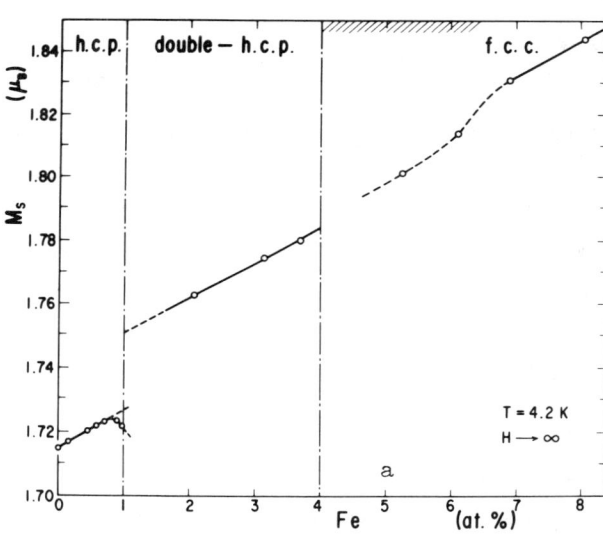

Fig.7 Saturation magnetic moment per atom at 4.2 K as a function of concentration for Co-Fe alloys. (a) h.c.p., double h.c.p. and f.c.c. Co-Fe alloys. (b) Enlarged graph in h.c.p. region.

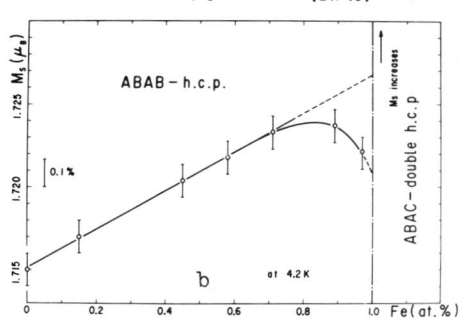

3. Magnetocrystalline Anisotropy

The magnetization curves measured at 300 K in the principal crystallographic directions of the hexagonal structure are shown in Fig.8 (a) and (b), for the alloys with 1.58 and 2.11 at.% Fe respectively. It is well known that h.c.p. Co exhibits a uniaxial anisotropy which makes the direction of easy magnetization parallel to the c-axis of the crystal at room temperature. In the

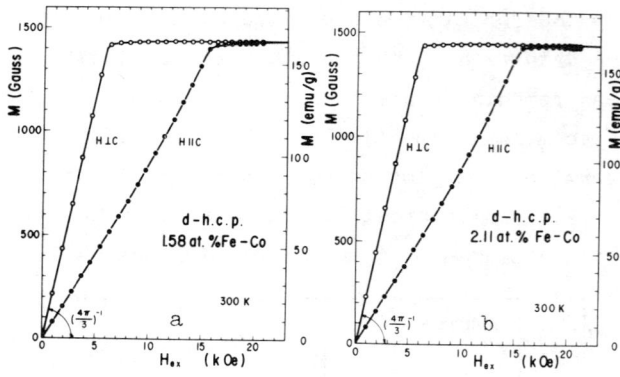

Fig.8 Magnetization curves at 300 K for spherical single crystals of double h.c.p. (a) 1.58 at.% Fe-Co and (b) 2.11 at.% Fe-Co.

case of Co-Fe crystals with the double h.c.p. structure, however, it was found that the direction of easy magnetization was changed to be in the hexagonal c-plane, as seen in Figs.8 (a) and (b). The anisotropy constants K_{u1} and K_{u2}, in the expression of the anisotropy energy

$$E_a = K_{u1} \sin^2\theta + K_{u2} \sin^4\theta + \ldots \qquad (2)$$

where θ is the angle between the internal magnetization and the c-axis, can be determined by the

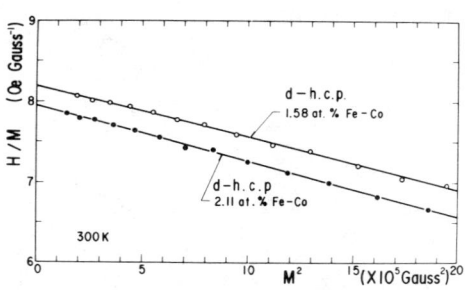

Fig.9 H/M vs. M^2 plot for double h.c.p. 1.58 and 2.11 at.% Fe-Co.

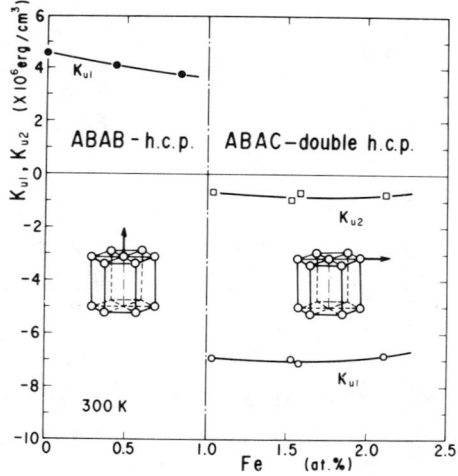

Fig.10 Magnetocrystalline anisotropy constants at 300 K as a function of concentration for hexagonal Co-Fe alloys. Closed circles were obtained by the torque method[3].

method of Sucksmith and Thompson[8]. The magnetization curve in the direction of hard magnetization parallel to the c-axis is expressed by the formula

$$H/M = (4K_{u2}/M_s^4)M^2 - 2(K_{u1} + 2K_{u2})/M_s^2, \qquad (3)$$

where H is the effective field. Figure 9 shows H/M vs. M^2 plots, which give good straight lines. From the gradient of these straight lines and the extrapolated point on the lines to the ordinate, the values of K_{u1} and K_{u2} can be separately determined. The anisotropy constants thus obtained are plotted as a function of the Fe concentration in Fig.10, where the values of K_{u1} measured by the torque method[3] are also plotted.

4. Magnetic Domain Patterns

As well known, for h.c.p. Co with positive K_{u1} at room temperature, the domain pattern observed on the c-plane of the hexagonal crystal is rather similar to the checkerboad or circular pattern, which reflects internal wedge-type reverse domains. In contrast with h.c.p. Co, for the double h.c.p. Co-Fe alloy, its uniaxial anisotropy is negative and large in magnitude as shown in Fig.10, then the observation of domain patterns may be interesting. Some examples of domain patterns observed on 1.53 at.% Fe-Co single crystals are shown in Figs.11 (a) to (d). Figure 11 (a) and (b) show typical domain patterns observed on the c-plane of the double h.c.p. crystal. From these patterns, it is evident that the c-plane is the plane of easy magnetization. It was found that double walls and rhomboid domains similar to the domain often observed for a thin film[9] are formed even in the case of a bulk crystal. The fact that the domains observed on the c-plane are elongated along the ($\bar{1}2\bar{1}0$) direction may indicate the basal anisotropy. The domain pattern observed on the (10$\bar{1}$0) plane is shown in Fig.11 (c), where a stripe pattern along the ($\bar{1}2\bar{1}0$) direction and a small fern-like pattern can be seen. The pattern observed on the ($\bar{1}2\bar{1}0$) plane is given in Fig.11

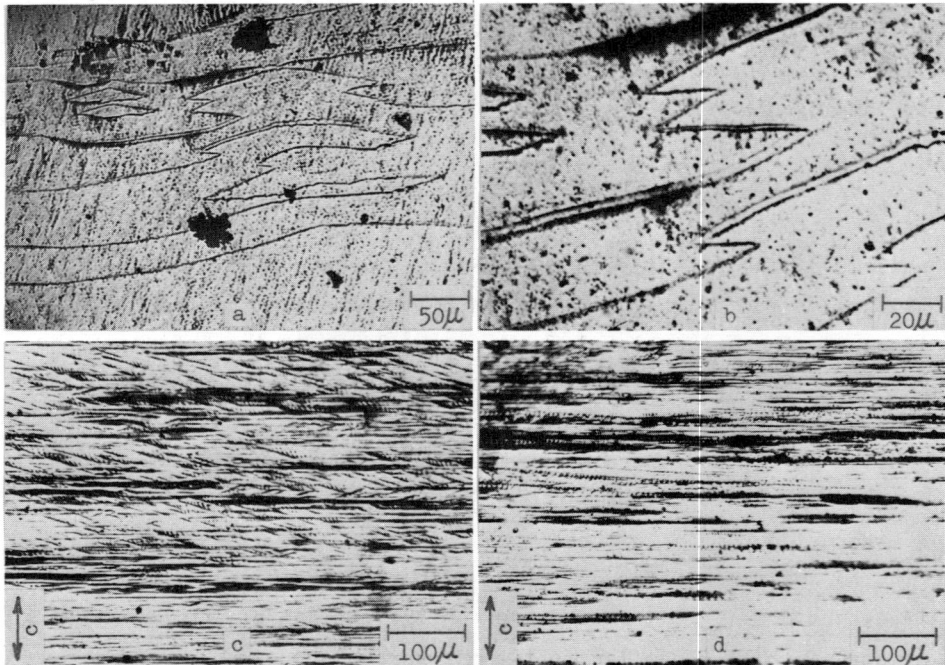

Fig.11 Domain patterns observed on double h.c.p. 1.53 at.% Fe-Co single crystals. (a) On the c-plane. (b) Double walls on the c-plane. (c) On the $(10\bar{1}0)$ plane. (d) On the $(\bar{1}2\bar{1}0)$ plane.

(d), which shows also a stripe pattern lying along the $(10\bar{1}0)$ direction.

5. Nuclear Magnetic Resonance[10]

In order to investigate magnetic and crystalline properties of Co-Fe alloys with the ABAB-h.c.p. and ABAC-double h.c.p. structures from a microscopic point of view, the nuclear magnetic resonances of Co^{59} in these alloys were observed as a function of temperature. In Fig12 (a), an example of recorder traces of the zero-field nuclear resonance is shown for 0.43 at.% Fe alloy with the ABAB-h.c.p. structure at room temperature. Additional signals associated with faults were not observed in the present electropolished specimens. The observed line shapes are similar to those for pure h.c.p. Co^{11}, which spread over a wide frequency range with two peaks at

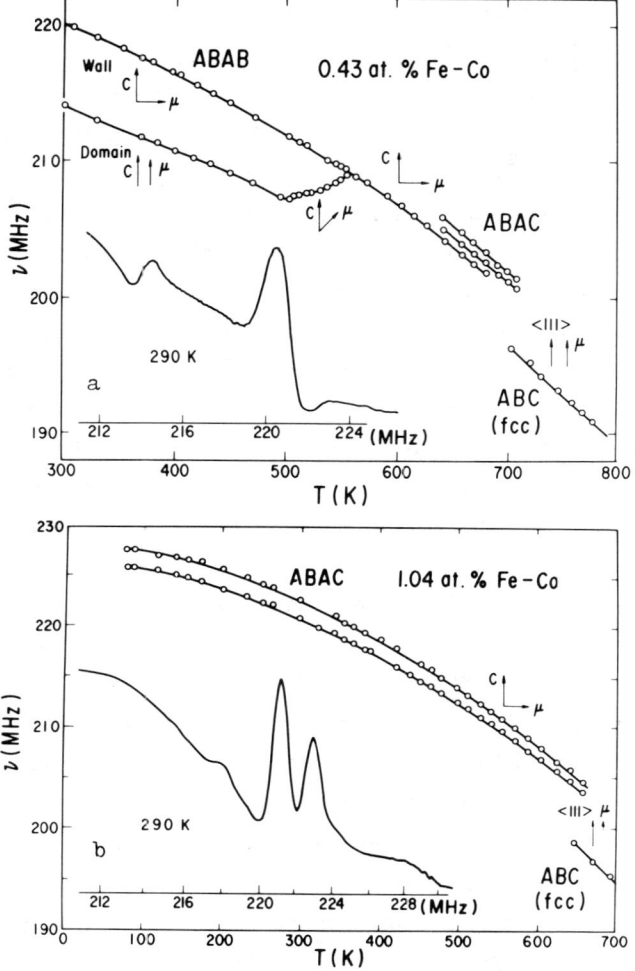

Fig.12 Recorder trace of the Co^{59} NMR and temperature dependence of the resonance frequency. (a) 0.43 at.% Fe-Co. (b) 1.04 at.% Fe-Co. (Kawakami, Kōi and the present author)

214 and 221 MHz. As in the case of pure h.c.p. Co, the low-and high-frequency end of the spectrum observed are attributable to the domain and wall resonance, respectively. Figure 12 (a) shows also these resonance frequencies as a function of temperature.

In the case of 1.04 at.% Fe alloy with the ABAC-double h.c.p. structure at room temperature, the line shapes of the resonance are entirely different from those in 0.43 at.% Fe alloy with the ABAB-h.c.p. structure, as shown in Fig.12 (b). The resonance frequencies of 1.04 at.% Fe alloy are found to be higher than those of 0.43 at.% Fe alloy. In Fig.12 (b), the frequencies of the two peaks are plotted as a function of temperature for 1.04 at.% Fe alloy.

DISCUSSION

The X-ray study revealed that the Co-Fe system in the range 0 to 8 at.% Fe is classified into three regions with different phases; (1) the ABAB-h.c.p. structure below 1 at.% Fe, (2) the ABAC-double h.c.p. structure between 1 and ~4 at.% Fe and (3) the ABC-f.c.c. structure above ~4 at.% Fe, although an ambiguity still remains in the range ~4 to ~6 at.% Fe. A remarkable discontinuous increase in lattice constant c was found to occur at the phase boundaries, where a change in the lattice constant a is small. The obtained c/a ratio increases linearly with the increase of the Fe concentration and shows discontinuous changes at the phase boundaries. In the hexagonal region, the c/a values were found to be smaller than the ideal value 1.633. It may be interesting to note that the c/a value extrapolated to the zero Fe content for the double h.c.p. phase is equal to approximately a mean value between the value for the h.c.p. pure Co and that for the f.c.c. pure Co. It should be noted that in Fig.2 (b) no streaks of the diffraction spots due to stacking faults are observed, accordingly the almost perfect periodicity is thought to be formed in the double h.c.p. structure. The crystal, however, may not be completely free from stacking faults, because in the present alloys stacking faults are readily formed during their crystal transformation.

In connection with these crystalline properties, the constants in the a/H and b/H^2 terms in the expression (1) for the law of approach to saturation change with the Fe concentration as shown in Fig.6. The origin of the various terms in the expression (1) continues to be an interesting problem[12]. Recently, the $\chi_0 H$ term has been investigated experimentally[13] and explained by a theory of the high-field susceptibility or in terms of paraprocesses. As well known, the b/H^2 term usually arises from magnetocrystalline anisotropy[14], and also from the influence of internal strain on the approach to saturation as derived by Becker and Polley[15]. The most controversial problem is the origin of the a/H term[12]. Brown[16] showed that this term is attributable to the stress field about dislocations, while Néel[17] considered a contribution to this term from nonmagnetic imperfections or voids. In any case, the a/H term is thought to be a measure of the inhomogeneity of magnetic substances. According to the present investigation, in the h.c.p. phase, the values of a are zero, while the values of b are large. This fact indicates that the alloys with the h.c.p. phase are homogeneous. The observed values of b agree fairly well with those estimated from the uniaxial anisotropy constants. The most interesting case is the double h.c.p. phase, where the values of a are fairly large as compared with those in the other phases. This fact would imply the presence of any inhomogeneity, perhaps stacking faults associated with the double h.c.p. structure. The obtained values of b are smaller than those in the h.c.p. phase, in spite of the larger magnetocrystalline anisotropy in the double h.c.p. phase. In this case, however, the uniaxial anisotropy constant is negative as shown in Fig.10, then the magnetic hardness is reduced by the existance of the plane of easy magnetization. For the alloys containing the higher Fe concentration, both the values of a and b are reduced to zero because of the homogeneous f.c.c. phase and also because of the cubic symmetry of the magnetocrystalline anisotropy, although in the range ~ 4 to ~ 6 at.% Fe some finite values of a, probably due to some retained hexagonal phase, were observed.

The change of saturation magnetic moment with concentration for the Co-Fe alloy is of interest with regard to understanding the magnetic behaviour of this typical ferromagnetic alloy. No hexagonal Co-base dilute alloys containing Fe impurities have so far been investigated and the points shown on the well-known Slater-Pauling curve correspond to the data for the f.c.c. and b.c.c. phases. Figure 7 (a) shows the saturation moment per atom as a function of Fe concentration obtained from the present investigation. It was found that even in the hexagonal region the saturation moment increases with Fe concentration almost in parallel to the Slater-Pauling curve. Furthermore, it was confirmed that the abrupt increase in saturation moment takes place when the Fe content increases beyond 1 at.% Fe, where the change in crystal structure from h.c.p. to double h.c.p. type occurs. This abrupt change in saturation moment has been also found in the previous study[4], and was thought to be due to the presence of the double h.c.p. structure. However, an anomalous decrease in magnetization in the h.c.p. phase reported in the previous study[4] (at 300 and 77 K in the field up to 21 kOe) is thought to be due to the effects of a decrease in Curie temperature by the addition of Fe and of a relatively weak field.

In the present results shown in Fig.7 (a), the value of the saturation moment extrapolated to the zero Fe content for the double h.c.p. structure should be different from those for the h.c.p. and f.c.c. structures. This means that the double h.c.p. pure Co, if it exists, should have a different value of the saturation moment from those for the h.c.p. and f.c.c. pure Co. It is interesting to note that the extrapolated value of the saturation moment, that is the saturation moment of the double h.c.p. pure Co, is $1.74 \mu_B$, in contrast with $1.715 \mu_B$ for the h.c.p. pure Co[5] and $1.751 \mu_B$ for the f.c.c. pure Co[18]. This fact may indicate that the mechanism of the appearance of the magnetic moment in Co depends on its crystal structure. In the case of the ABAC-double h.c.p. structure, there are two kinds of lattice sites in equal weight ; one is surrounded by its nearest neighbours in the h.c.p. manner, and the other is surrounded in the f.c.c. manner, as shown in Fig.3. If the electronic states of Co atoms on these two lattice sites would be different from each other, it may be expected that two different resonances with an equal intensity from Co^{59} nuclei can be observed in the NMR experiment. The two different resonance frequencies observed for 1.04 at.% Fe with the double h.c.p. structure (Fig.12 (b)) are thought to correspond to the above-mentioned expectation. This would imply that the microscopic mechanism of appearance of the magnetic moment in Co is closely related to its local environment.

Dilute alloys are suitable for the investigation about the magnetic behaviour of metals and alloys, because in this case the solute atoms play a role of atomic-scale probes perturbing the solvent metal. In this sense a more detailed investigation was carried out at dilute Fe content in the Co-Fe alloys. Figure 7 (b) shows the saturation moment as a function of Fe concentration below 1 at.% Fe. It was found that the saturation moment increases linearly with the slope of $1.1 \pm 0.3 \mu_B$/Fe atom in the h.c.p. Co-base dilute alloys. The present results can be well explained by the rigid band model[19] which predicts the linear increase of saturation moment of $1 \mu_B$/Fe. It may be interesting to note that the saturation moment shows a tendency to decrease in the region close to the phase boundary as seen in Fig.7 (b). This anomalous behaviour is similar to that of the Invar alloys[20].

Among 3d-transition metals and alloys, as far as we know, the double h.c.p. Co-Fe alloy is a unique magnetic material which shows the easy basal-plane anisotropy with the large negative K_{u1} at room temperature and below. Even for this material, the c/a value was found to be smaller than the ideal value 1.633. Furthermore, both the ABAB-h.c.p. and ABC-f.c.c. Co are known to have the easy axis parallel to the c-axis (or ⟨111⟩ axis). Therefore, it is not so easy to explain the large negative K_{u1} of this material with the ABAC stacking sequence in terms of the nearest neighbour interaction. The negative K_{u1} might be due to the influence of a hexagonal crystalline field produced by further neighbours of the f.c.c. sites. The origin of magnetic anisotropy of metals and alloys continues to be a controversial problem, because the details of 3d electron states and the exchange interaction between them have not yet been clarified. Yosida, Okiji and Chikazumi[21] proposed a model based on the Anderson theory[22] to explain the uniaxial magnetic anisotropy for iron-group impurities in the h.c.p. Co. However, a further development in the theory of ferromagnetism is necessary to discuss the magnetic anisotropy of 3d-transition metals and alloys in more detail.

The large negative uniaxial anisotropy observed in the double h.c.p. Co-Fe crystals makes the investigation of magnetic domains in this material particularly interesting. One of the most interesting features in observed domains is the double walls on the c-plane, as shown in Figs.11 (a) and (b). These double walls are thought to be due to the Néel-type wall[23] as has been observed on thin films[9]. However, the mechanism of appearance of the Néel-type wall in the present bulk crystal may be different from that in the case of thin films. In the present case, it is likely that the large energy of negative uniaxial anisotropy plays an important role to change the plane of spin rotation inside the wall from parallel to the wall surface to parallel to the c-plane. The patterns observed on the $(10\bar{1}0)$ and $(\bar{1}2\bar{1}0)$ planes are rather complicated. Hubert[24] reported that, in a conical anisotropy range in pure Co at elevated temperature, a layered structure in the remanent state with a periodicity of $10 \sim 20 \mu$ along the c-axis is induced by internal stresses, possibly due to stacking faults resulting from the crystal transformation. In the present case, although the occurrence of conical anisotropy is not expected from the measured values of K_{u1} and K_{u2}, the observed stripe patterns may be due to stacking faults possibly present in the double h.c.p. crystals. This corresponds to the fact that the constant a in the a/H term is large in the double h.c.p. region as seen in Fig.6.

The temperature dependence of the Co^{59} NMR was found to reflect that of the magnetocrystalline anisotropy and a crystal transition in Co-Fe alloys. As already mentioned, the observed spectrum shown in Fig.12 (a) are attributable to the resonances of nuclei in the domains and domain walls, in the presence of the anisotropy of hyperfine field[11]. In the case of 0.43 at.% Fe alloy with the easy c-axis, the magnetic moments μ are aligned parallel and perpendicular to the c-axis in the domains and at the center of the domain walls, respectively, as schematically shown in Fig.12 (a). The domain NMR frequency was observed to increase with temperature gradually between 500 and 557 K and coincide with the wall NMR frequency above this temperature. This fact indicates the spin rotation from the c-axis to the direction in the c-plane. It was also found that two new resonances having a nearly equal intensity appear at 640 K (Fig.12-a) and their line shapes are similar to those for 1.04 at.% Fe alloy with the ABAC-double h.c.p. structure (Fig.12-b). This change in the NMR with a thermal hysteresis is thought to be due to a crystal transition from the ABAB-h.c.p. to the ABAC-double h.c.p.. In addition to this, the well-known transition from hexagonal to f.c.c. occurs at higher temperature.

On the other hand, in the case of 1.04 at.% Fe alloy with the easy c-plane, it may be expected that the domain and wall resonances take place at a same frequency even at room temperature. The observed two different resonances were believed to be due to Co^{59} nuclei on two different lattice sites and this situation was already discussed. The transition from hexagonal to f.c.c. also occurs in this alloy as shown in Fig.12 (b).

ACKNOWLEDGEMENTS

The author is indebted to Dr. T. Onozuka, Dr. S. Yamaguchi and Professor M. Hirabayashi for the X-ray investigation, and to Dr. M. Kawakami, Professor T. Hihara and Professor Y. Kōi for the NMR investigation.

The author would like to express his thanks to Professor T. Anayama and Professor M. Takahashi for their encouragement. Thanks are also due to Professor S. Chikazumi for his valuable discussion throughout the course of this work and also for his kind inspection of the manuscript.

REFERENCES

1. R. M. Bozorth, Ferromagnetism (D. Van Nostrand Co., Prinston, N. J., 1951), p.190.
2. P. Weiss and R. Forrer, Ann. Phys. 12, 279 (1929).
3. S. Chikazumi, T. Wakiyama and K. Yosida, Pro. of the Inter. Conf. on Magnetism, Nottingham, 756 (1964).
4. T. Wakiyama, J. Phys. Cl 32, 340 (1971).
5. H. P. Myers and W. Sucksmith, Proc, Roy. Soc. A207, 427 (1951).
6. T. Onozuka, S. Yamaguchi, M. Hirabayashi and T. Wakiyama, to be published.
7. T. Onozuka, S. Yamaguchi, M. Hirabayashi and T. Wakiyama, J. Phys. Soc. Japan 33, 857 (1972).
8. W. Sucksmith and J. E. Thompson, Proc, Roy. Soc. 225, 362 (1954).
9. H. J. Williams and R. C. Sherwood, J. Appl. Phys. 28, 548 (1957).
10. M. Kawakami, Y. Kōi and T. Wakiyama, to be published.
11. M. Kawakami, T. Hihara, Y. Kōi and T. Wakiyama, J. Phys. Soc. Japan 33, No.6 (1972).
12. D. E. Grady, Phys. Rev. B4, 3982 (1971).
13. for instance cf. T. Wakiyama, J. Phys. Soc. Japan 32, 1222 (1972).
14. N. S. Akulov, Z. Physik 69, 882 (1931).
15. R. Becker and H. Polley, Ann. Physik 37, 534 (1940).
16. W. F. Brown, Phys. Rev. 60, 139 (1941).
17. L. Néel, J. Phys. Radium 9, 184 (1948).
18. J. Crangle, Phil. Mag. 46, 499 (1955).
19. J. Friedel, Nuovo Cimento Suppl. 7, 287 (1958).
20. S. Chikazumi, T. Mizoguchi, N. Yamaguchi and P. Beckwith, J. Appl. Phys. 39, 939 (1968).
21. K. Yosida, A. Okiji and S. Chikazumi, Progr. Theoret. Phys. 33, 559 (1965).
22. P. W. Anderson, Phys. Rev. 124, 41 (1961).
23. L. Néel, Compt. Rend. 241, 533 (1955).
24. A. Hubert, J. Appl. Phys. 39, 444 (1968).

941

DOMAIN CONFIGURATIONS, BLOCH WALLS AND
MAGNETIZATION PROCESSES IN IRON WHISKERS
FROM D.C. TO 200 kHz. THEORY AND EXPERIMENT II

A.S. Arrott, B. Heinrich, and D.S. Bloomberg
Simon Fraser University, Burnaby 2, B.C., Canada

ABSTRACT

The a.c. susceptibility of iron whiskers with the Landau domain structure has been studied over a wide range of frequencies, for various d.c. bias fields, with several coil configurations, and as a function of temperature and sample dimensions. The behavior is approximately accounted for by a simple model which treats the long domain wall as an elastic membrane and the closure domains as simple springs attached to the ends of that membrane. The reasons for the success of such a crude approximation have been sought by carrying out calculations of the magnetostatics of the problem. Almost all the free magnetic poles are on the surfaces of the whiskers. The magnetization within each domain is almost parallel to the bowed long domain wall at the wall, has a divergenceless pattern within the domain, but is not parallel to the surface at the surfaces. These calculations explain the low frequency behavior and show the degree to which the membrane model is an adequate approximation. The effects of eddy current damping are treated by the use of an analogy with diffusion theory which predicts well the measured propagation of pulses along domain walls. The temperature dependence from 650 to 1050 deg. K has been examined in detail. The behavior is well accounted for by the models except very close to T_c where the magnetic losses exhibit several new anomalies.

INTRODUCTION

Detailed studies of domain structures in iron whiskers have been carried out by a number of investigators within the last 20 years[1-9]. These studies have provided clear pictures of the configurations of domain walls and how these depend on magnetic field. They have shown the existence of the simple Landau structure (see fig. 1) for whiskers with the surfaces parallel to (100) planes. We have studied such whiskers as cores of transformers.

We use two coils coupled to iron whiskers and measure the in phase and out phase signal from the secondary for various amplitudes and frequencies applied to the primary. The output signal is studied as a function of whisker dimensions, applied d.c. biasing field, temperature and gaseous impurities. The coils have been used in three configurations:
    a. two uniformly wound solenoids longer than the whiskers;
    b. one uniform long solenoid and one short coil (small compared

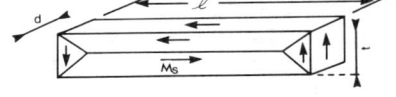

Fig. 1. The Landau Structure

to the whisker length) with the short coil in various positions along the length of the whisker; and

c. two short coils, one in the center and the other at various positions along the length of the whisker.

Typical whiskers are approximately rectangular bars (they often show a slight taper along the length) with dimensions in the range $10^4 \times 10^2 \times 10^2$ in μm. Such whiskers can be successively cut to shorter and shorter length. In a typical length dependence experiment the range studied is about a factor of four.

The a.c. fields from the primary coils have been varied from 0.01 oersted for the long solenoids to as large as 20 oersted in some pulse experiments with the short coils. Typically a field of 0.1 oersted is used. Frequencies from 1 Hz to 7 MHz have been used, but most of the data have been in the range 5 to 200 kHz. For most of the measurements a phase sensitive detection system was used[10]. The d.c. biasing fields were usually produced in the gap of a large electromagnet driven by a bipolar power supply to facilitate continuous field reversals.

Temperatures from the helium boiling point to above the Curie temperature of iron have been used. The regions most studied are room temperature and from 700 deg. K to 1050 deg. K. An intensive investigation in the region within one degree of T_c has been carried out. The results of the latter investigation are only touched on in this report and will appear elsewhere[11].

We discuss first the low frequency in phase response (d.c. behavior). The experimental results at d.c. are interpreted in terms of the almost complete dominance of magnetostatic energy in determining the behavior. The magnetostatics is discussed in some detail. We also give a simplified model of the magnetostatics which has proved useful for discussing the behavior at higher frequencies.

We next discuss the role of eddy currents. The striking difference between the in phase and out phase behavior at low frequencies is that the former is almost completely independent of the domain configuration (there are many possibilities in addition to the Landau structure) while the latter is a sensitive indicator of the domain configuration. In particular the Landau structure has its own distinctive signature in the dependence of the out phase component of the a.c. susceptibility upon d.c. bias field (see fig. 2). It was indeed just this observation which made this work possible[12 - 15]. At high frequencies the eddy current effects are dominant and result in changes of the magnetic configuration of the domain walls in the whisker. These things are determined by direct measurements and are explained reasonably well by our approximate treatment of the problem. We formulate a model in which the magnetic response obeys a diffusion equation. Studies of transmission of pulses from one short coil to another along the whisker length confirm the appropriateness of the analogy.

Measurements of the temperature dependence show that the behavior is in accordance with expectations from the models used to

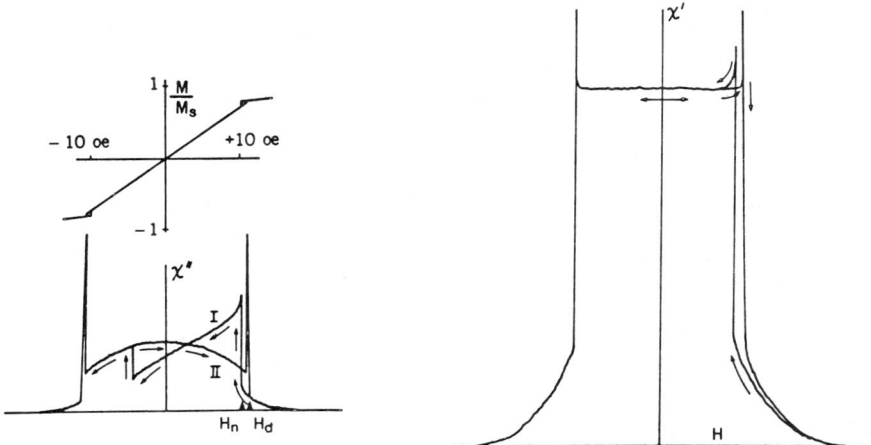

Fig. 2. D.C. magnetization loop and the in phase χ' and out phase χ" response at 1 kHz. H_n is the nucleation field. H_d is the departure field. The Landau phase, I, is formed from the Coleman phase II. The results are for long coils.

account for the behavior at room temperature. These expectations are that:
- a. the d.c. susceptibility is almost completely independent of the temperature as well as the magnetic field (up to a critical field called the departure field H_d, fig. 2);
- b. the departure field is proportional to the spontaneous magnetization and tracks its dependence upon temperature;
- c. the out phase component of the low frequency susceptibility is proportional to the electrical conductivity.

Two additional topics are mentioned in a final section. The behavior of iron whiskers is drastically modified by the presence of gaseous impurities. This can be used to study diffusion of gases and to deduce properties of the domain walls. The behavior of iron whiskers near the Curie temperature provides a unique look at the nature of the critical phenomena on the scale of micromagnetics. As discussed for the room temperature behavior, the loss component of the susceptibility is quite sensitive to the details of the magnetization process. We report briefly some of the discoveries we have made.

It continues to be our hope that the special properties of iron whiskers will lead to some useful devices. The data and the models presented here should be of help to designers.

D.C. BEHAVIOR

Some principle features of the d.c. behavior are:
- a. the magnetization is linear with applied field for fields up to the departure field (There is a small deviation from linearity very close to the departure field.);

b. this susceptibility does not depend on domain configuration (again except near the departure or nucleation fields);
c. for a uniform external magnetic field the magnetic moment per unit length varies rather quadratically along the whisker and shows a finite drop at the end of the whisker;
d. for an approximately delta function field applied at the center of the whisker the magnetic moment per unit length varies rather linearly again with a finite drop at the end of the whisker; and
e. for the short coil the drop at the ends is half what it is for the long coil if both produce the same magnetization at the center.

The first conclusion that one might draw from these observations is that the behavior is similar to that of a rectangular bar with infinite permeability. Unfortunately this was the last conclusion that we came up with. When we were finally led to it and after we deduced the behavior of such a bar, we found that much of this was well known experimentally and theoretically by German workers in the years 1925-1939[16-21]. We should have started with the recognition of the close relation between the homogeneous rectangular bar with infinite permeability and the iron whisker with highly mobile domain walls. We will take that approach here and work back to the simple phenomenological model we have previously presented for understanding the observations on whiskers[12-15]. That model treats the domain walls as elastic membranes.

A schematic comparison of the magnetization pattern for the infinite permeability rectangular bar, hereafter usually referred to as the bar, and for the iron whisker is shown in fig. 3.

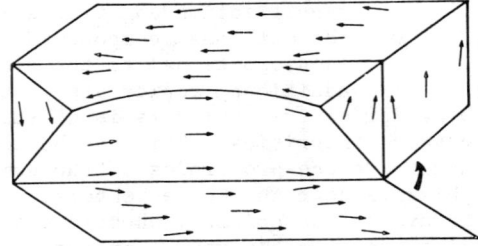

Fig. 3. The top, side, and bottom magnetization pattern for a whisker. All sides of the bar of infinite permeability resemble the bottom of the whisker.

The patterns for the bar and that part of the whisker which is magnetized along the field are quite similar. That part of the whisker magnetized against the field has the same x components as the bar but the z and y components are reversed. Contrary to the usual representations of whisker magnetization, the magnetization within the domains is not constant or parallel to the whisker edges. For iron the pattern is almost parallel to the curved wall near the wall, and near the surfaces it is somewhat less curved. The essential feature is that the two patterns produce almost the same distribution of magnetic charge density. For the bar the charge is all on the surface while for the iron whisker all but about 2 percent of the charge is on the surface.

We make no apology for adopting the language of magnetic charge[22]. We first treat the bar. The problem is to find a distribution of charge on the surface that will everywhere inside the bar give a demagnetizing field H_D which is equal and opposite to the externally applied field, H_{ex}. The details of this and other magnetostatic calculations are to be found elsewhere[23]. Here we sketch the approach and show some of the results. We first assume that a rectangular bar and a cylindrical rod of the same length and the same cross section have the same charge per unit length along the axis. We then return to the problem of the actual distribution of this charge over the rectangular cross section. For the rod we evaluate the field H_D on axis and equate it to $-H_{ex}$. Thus we write

$$H_D(z) = 2\pi r \int_{-\ell/2}^{\ell/2} \frac{\sigma(z') \cdot (z-z') dz'}{\{r^2 + (z-z')^2\}^{3/2}}$$

$$- \pi r^2 \sigma_e \{(z-\ell/2)^{-2} + (z+\ell/2)^{-2}\}, \quad (1)$$

where $\sigma(z)$ is the charge density on the cylindrical surface and σ_e is the average charge density over the positive end of the rod of radius r and length ℓ. In writing Eq. (1) we are taking $\nabla \cdot M = 0$ inside. To obtain the magnetization from the charges we assume that $\partial M_z/\partial z$ is essentially constant on a given cross section, then

$$\sigma(z) = M_r(z) = -\frac{r}{z}\frac{\partial M_z}{\partial z}, \quad (2)$$

for the cylindrical surface and at the end surface

$$\sigma_e = M_z(\ell/2) \quad (3)$$

The magnetization is found from

$$M_z(z) = \sigma_e + \frac{2}{r}\int_z^{\ell/2} \sigma(z) dz \quad (4)$$

Eq. (1) is solved by breaking up the charge distribution into 2n blocks of charge on the sides plus one on each end. By symmetry this reduces to n+1 variables which can be found by evaluating $H_D(z) = -H_{ex}$ at n+1 points and solving the n+1 linear equations using standard double precision library routines. Fig. 4 shows the calculated curves for M_z along rods of infinite permeability and a "best fit" quadratic curve. The data points are for iron whiskers of the same cross section to length ratios.

For the whiskers the magnetization is interpreted as the magnetic moment per unit length divided by the cross sectional area. In terms of the displacement y of the domain wall for the Landau structure

$$M_z = 2yM_s/t \quad (5)$$

Fig. 4. The calculated variation of M_z along rods of infinite permeability for several ratios of diameter to length (.0368, .0194, .0099 top to bottom) in a uniform field. The points are experimental results on one whisker successively cut to shorter and shorter length (see also fig. 7). The smooth curves are "best fit" quadratics as given in Eq. (12).

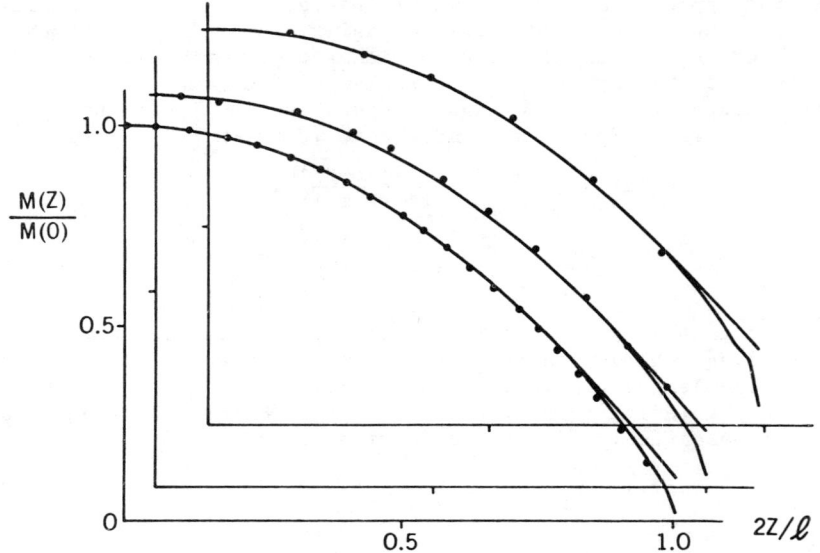

where t is the whisker thickness in the direction of y and M_s is the spontaneous magnetization. We call the width of the wall d. Here we treat mostly the case d=t. The charge per unit length of the whisker is given by

$$q_\ell = -2M_s d \cdot dy/dz = -t d \partial M_z/\partial z \qquad (6)$$

while for the bar it is given by

$$q_\ell = 2\pi r\sigma = -\pi r^2 \partial M_z/\partial z \qquad (7)$$

The presence of a wall with a slope dy/dz implies a charge per unit length as given above, but it does not answer the question of how that charge is distributed in the cross section. The answer is that this depends on the ratio of the anisotropy field H_K to $4\pi M_s$. If $H_K/4\pi M_s$ is very large, the charges are all on the wall and the magnetization is everywhere parallel to the surfaces. In that case the magnetostatic energy is somewhat greater ($\sim 4/3$) than that calculated for the bar. The charges in general divide between the wall and the surfaces in the ratio H_K to $4\pi M_s$. For iron this is in the ratio 1 to 50. With only 2 percent of the charge on the wall the magnetic response of the bar and the whisker should be very close. In all cases the volume charge is negligible. If $\theta = dy/dz$ is the angle made by the wall and φ is the angle made by the magnetization, both with respect to the z axis in the y direction then

$$\varphi = H_D/H_K = (4\pi M_s/H_K)(\theta-\varphi) \tag{8}$$

and to a good approximation the magnetization components in a square cross section are given by

$$M_y = \pm \varphi M_s [1-|y|/d]$$

$$M_x = \varphi M_s x/d \tag{9}$$

A better approximation is given in fig. 5 which shows the results of a calculation on a grid of the indicated density. A more detailed calculation is possible with the methods used if one assumes that $\nabla \cdot M = 0$ in the domain volume, instead of

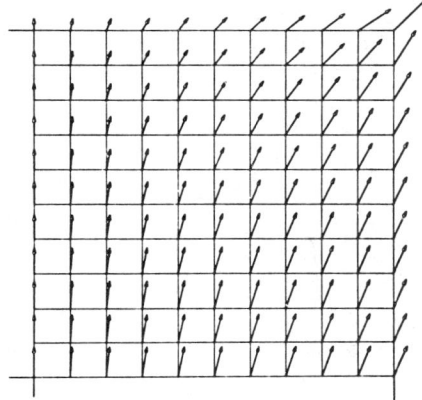

Fig. 5. Magnetization components in one quadrant of the plane perpendicular to the long axis of a whisker. $H_K/4\pi M_s = .02$.

showing that it is very close to zero as was done for fig. 5. The more detailed result for the charge density variation along the sides of the whisker in the cross section is given in fig. 6. The calculations are carried out by replacing integral equations by simultaneous linear equations in the charge per unit region. The

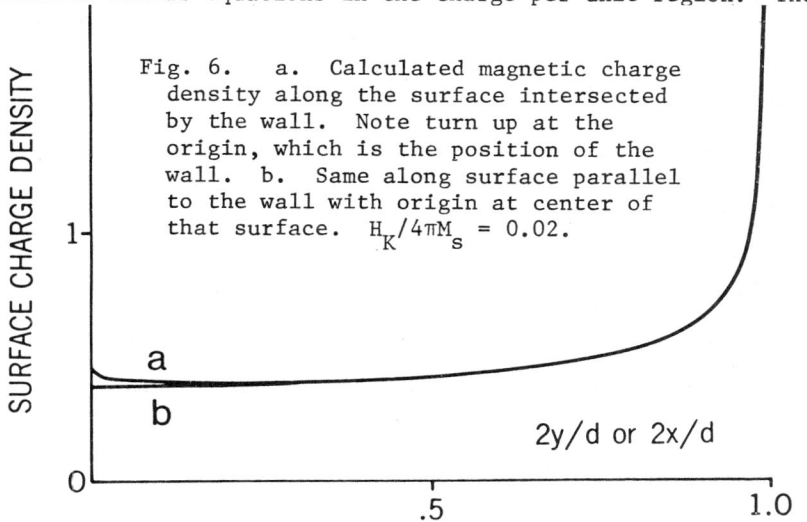

Fig. 6. a. Calculated magnetic charge density along the surface intersected by the wall. Note turn up at the origin, which is the position of the wall. b. Same along surface parallel to the wall with origin at center of that surface. $H_K/4\pi M_s = 0.02$.

approximation made is that the particular cross section is part of a long whisker with constant charge per unit length. In the limit of small H_K these calculations also give the charge distribution for the rectangular bar of infinite permeability. For large anisotropy the charge on the surface decreases and the intersection of the wall with the side surface becomes relatively more charged. All this is to show the similarity of the iron whisker to the rectangular bar.

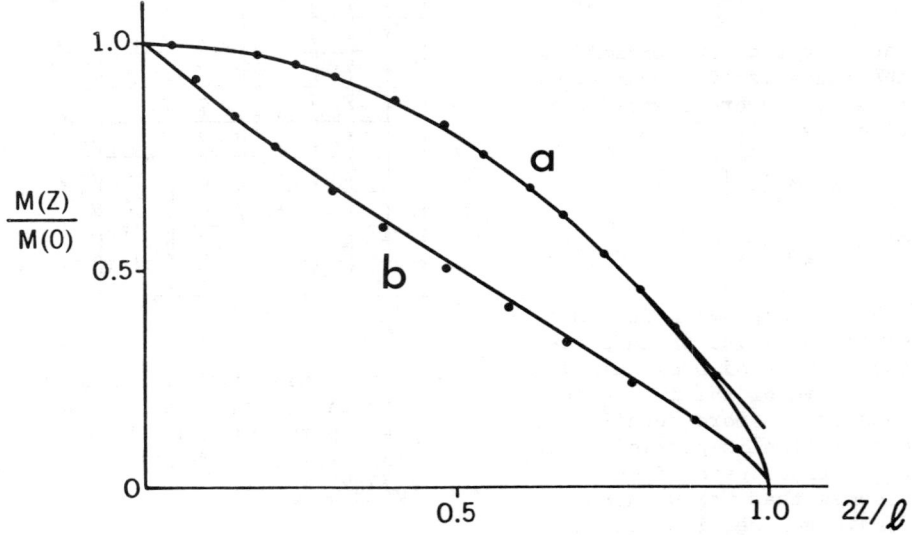

Fig. 7. a. Same as fig. 4 for another ratio of diameter to length (0.0139) of the same whisker. b. Same as "a" except that the field is from a coil much shorter than the whisker.

It is also possible to show the response of these systems to a field applied from a small coil at the center of the whisker. Eq. (1) is solved for the condition $H_D(z) = - H_{ex}(z)$ where $H_{ex}(z)$ was calculated for the coil geometry used. The comparison of the experiment with the computer calculations are shown in fig. 7. Again the evidence is quite strong that magnetostatics can account for the behavior of the d.c. susceptibility of iron whiskers. Note that these results are not quite so well approximated by linear behavior as the results for the homogeneous driving coil are approximated by a quadratic behavior. The computer solutions for other coil dimensions show that by chance we had chosen the radius so as to produce the most linear curve. A larger coil, of course, makes the wall more quadratic with negative curvature, but a shorter coil would have produced positive curvature on each side of a cusp.

We have previously given the argument by which we considered only the interaction of the charges resulting from the slope of the wall with the magnetization in the near region of those charges[12-15]. While this model gave a good account of the initial experiments and led us to confirm many of its predictions, we could not understand why it was possible to replace a non-local interaction with a local one[14]. This situation remains. We have not been able to see analytically why the magnetostatics mocks the behavior of an elastic membrane attached to springs on its ends. The local model gives the variation of the magnetization along the whisker as the solution to the differential equation

$$4\pi M_s^2 \xi t d(d^2y/dz^2) = -2dM_s H(z) \qquad (10)$$

where ξ is a dimensionless parameter to be adjusted either to fit experiment or to agree with a proper magnetostatic calculation. This gives a quadratic variation for a uniform field and a linear variation for a highly localized field. This equation is solved with the boundary condition that the ends of the whisker act as springs attached to the elastic membrane. The spring constant is written as

$$\gamma = 2\pi M_s^2 d\eta \qquad (11)$$

where η is a dimensionless parameter to be treated as discussed for ξ. The result is

$$M(z) = 2M_s \frac{y}{t} = \frac{H}{2\pi} \left\{ \left[\left(\frac{\ell}{2}\right)^2 - z^2 \right] / \xi t^2 + 2\ell/\eta t \right\} \qquad (12)$$

The form of Eq. (12) is seen in fig. 4 to be a good approximation to the experimental results and to the solution of the magnetostatic problem of the rod with infinite permeability. As far as we can tell this and a number of other agreements are the sole justifications for the use of the analogy with the elastic membrane. The analogy is particularly useful when the complications of eddy currents are considered.

A.C. SUSCEPTIBILITY

We have found it convenient to discuss the a.c. response of the whisker as a whole according to the expression for the response of a damped spring (no mass), that is

$$\chi(\omega) = (\alpha_{eff} - i\omega\beta_{eff})/(\alpha_{eff}^2 + \omega^2\beta_{eff}^2) \qquad (13)$$

As the response of the whisker is not quite this simple, we treat α_{eff} and β_{eff} as functions of frequency and d.c. bias fields. For the most part we are concerned with the simpler case of no bias field. We find experimentally that α_{eff} increases with frequency by a factor of 2 and that β_{eff} decreases with frequency by about 10 percent. The damping arises from eddy currents. These are easily incorporated into the simplified local model by adding a term to

Eq. (10) which is proportional to wall velocity, thus

$$\alpha \frac{\partial^2 y}{\partial z^2} + \beta_1 \frac{\partial y}{\partial t} = -2dM_s H(z,t) \tag{14}$$

where

$$\alpha = 4\pi \xi M_s^2 dt \tag{15}$$

as discussed above, and the damping coefficient is approximated by

$$\beta_1 = q\left[\left(\frac{16M_s}{c}\right)^2 \frac{d^2}{\pi\rho}\right] \cdot \tfrac{1}{2} \sum_{\text{odd } n} n^{-3} \left[\tanh\frac{n\pi}{d}\left(\frac{t}{2}-y\right) + \tanh\frac{n\pi}{d}\left(\frac{t}{2}+y\right)\right] \tag{16}$$

where ρ is the resistivity and q is a dimensionless factor to cover up the approximations made in the model. Eq. (16) is the result of treating each segment of the wall as if it were part of an infinitely long wall. β_1 is a damping coefficient per unit length. The infinitely long wall in the center of the whisker was treated by Williams, Shockley and Kittel[24]. Agarwal and Rabius[25] extended the problem to treat the displaced wall. Their computer solutions suggest that to a fair approximation the eddy currents on one side of the wall are independent of what happens on the other side of the wall. This observation is the basis for writing Eq. (16). The boundary condition for the end springs can be modified to include damping on the ends, thusly

$$\pm 2\alpha \frac{\partial y}{\partial z}\left(\pm \frac{\ell}{2}\right) - \beta_2 \frac{\partial y}{\partial t}\left(\pm \frac{\ell}{2}\right) = \gamma\, y\left(\pm \frac{\ell}{2}\right) \tag{17}$$

where γ and β_2 represent the combined spring constant and damping coefficient for the two ends, respectively. Note that the dimensions of γ and β_2 differ from those of α and β_1 by a factor of length. The solution of Eq. (14) with the boundary conditions of Eq. (17) with β_1 taken as independent of wall position (valid for the wall in the center of the whisker) is expressed in complex notation by

$$\frac{y}{2dM_s H} = \left\{\frac{\cosh\phi - \cosh\frac{2x}{\ell}\phi}{\phi^2 \cosh\phi}\right\} \left\{\left(\frac{\ell}{2}\right)^2 \frac{1}{\alpha} - \frac{\phi^2}{i\omega\beta_1 + \frac{(\gamma+i\omega\beta_2)\phi}{\ell\tanh\phi}}\right\} + \left\{\frac{1}{i\omega\beta_1 + \frac{(\gamma+i\omega\beta_2)\phi}{\ell\tanh\phi}}\right\} \tag{18}$$

where

$$\delta = \frac{(1+i)\ell}{2\phi} = \left(\frac{2\alpha}{\omega\beta_1}\right)^{\!\frac{1}{2}} \tag{19}$$

plays the role of a "magnetic skin depth". It measures the length of "penetration" of the disturbance in from the ends of the wall, if we define disturbance to be a slope of the wall, hence the appearance

of magnetic charge. The use of complex notation does not make the
result more obscure than other forms. We have found that the best
way to penetrate this obscurity is with a computer. For that purpose
the complex notation is ideal. We have previously shown that the
frequency dependence can be accounted for within the range of
behaviors allowed by the four parameters of the model[14]. Two of
these parameters are fixed by the d.c. behavior as discussed above.
The low frequency out phase component gives one relationship between
the two damping parameters. Thus there is just one parameter free
to fit the full frequency range for both the in phase and out phase
components. Qualitatively what happens is that the damping of the
end springs is chosen so as to remove the contributions of the end
springs to both the in phase and out phase components at high
frequencies. Thus α_{eff} increases and β_{eff} decreases. It is hard to
know if there is any real physical significance to the picture
generated by the success of this model.

Even if the damping of the end spring were neglected, there
would be another reason why the solution to the diffusion Eq. (14)
gives an increase in α_{eff}. The shape of the long domain wall
changes with frequency because the eddy current damping tends to
flatten out the response. For the frequency range studied, however,
the calculated effect of this flattening on α_{eff} is much less than is
experimentally observed. But the wall flattening itself can be
directly measured using the small detector coil moved along the wall.
In fig. 8 we show the wall shape as measured and as computed for a
range of frequencies.

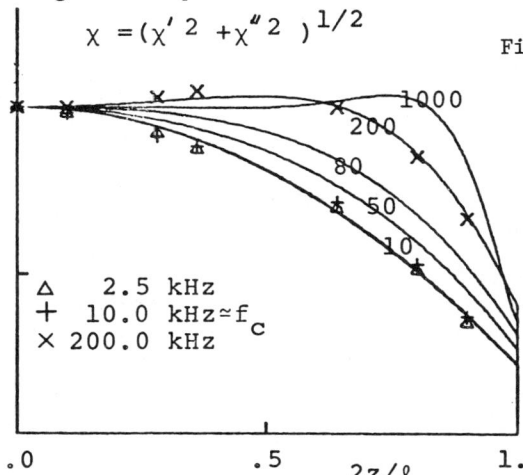

Fig. 8. The total response of a whisker to a uniform driving field as determined by a short coil moved along its length for frequencies of 2.5, 10, and 200 kHz. The lines are calculated from the local model for 2.5, 10, 50, 80, 200, and 1000 kHz from bottom to top. The lines for 2.5 and 10 kHz are practically indistinguishable. The rise in χ away from the center of the whisker is predicted by the model and is seen experimentally.

A further test of the diffusion analogy was to use two small
coils and to send pulses along the domain wall from one to the other.
A field was established in one coil at the center of the whisker and
then turned off suddenly in approximately 10^{-8} seconds. A signal
proportional to dy/dt was measured in the other coil for various
positions of that coil. For a 6.6 mm whisker it takes less than 10^{-5}
sec. for the disturbance to propagate from the center to the end.

The time is significantly shorter when a d.c. bias field is used to place the wall nearer the surface and thus to reduce the eddy current damping. This latter case is too difficult to analyze, so we have calculated the behavior for the case where the d.c. bias field is zero. The solution to the diffusion equation is given by Carslaw and Jaegar[26]. The solution can be carried out including the end springs[15]. Fig. 9 shows the behavior of a 12 mm whisker for which the effect of the end springs is negligible.

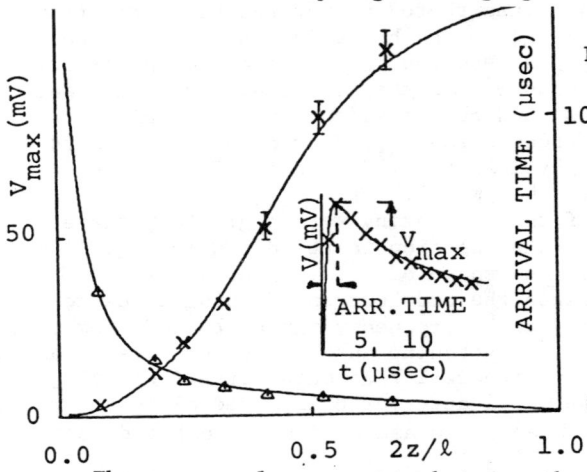

Fig. 9. The insert shows the form of a pulse received at a small detector coil after the field of a small magnetizing coil is turned off. The main curves show dependence of the peak signal (left scale) and its arrival time (right scale) upon position of the small detector coil. The solid curves are calculated using the diffusion analogy.

These several agreements between the model and experimental results are somewhat baffling. The more we think about the nature of the approximations made, the less reason we have for expecting them to work. Neither the magnetostatics nor the eddy currents are local phenomena. On the other hand, we can see no satisfactory method of handling the two non-local problems simultaneously.

As discussed in some detail in ref. 14 there are other configurations of domains which are easily distinguished from the Landau structure by observing the dependence of the out phase component on d.c. bias field. That the in phase components are essentially identical is explained by the dominance of the magnetostatic energy. No matter what the domain wall configuration, the charge per unit length remains the same and is distributed only slightly differently over a given cross section.

Though two different configurations would be moving walls so as to sweep out the same volume per unit length, the losses will depend on the area of wall in motion. The area of wall in motion would be greater if there were a pair of similar walls in motion. In this case the losses would be halved. This can be seen by writing for the magnetization of the whisker

$$\chi' = \frac{2M_s d}{h} \int_0^\ell y\,dz \qquad (20)$$

where y is the position of one wall or the sum of the displacements of two walls and h is the amplitude of the a.c. field.

The loss component can be written as

$$\chi'' = \frac{\beta_1 \omega}{h^2} \int_0^\ell y^2 dz. \qquad (21)$$

Then

$$\frac{\chi''}{\chi'^2} = \frac{\beta_1 \omega}{(2M_s d)^2} \frac{\int_0^\ell y^2 dz}{\left[\int_0^\ell y dz\right]^2} \qquad (22)$$

This is very crude because β_1 would not be quite the same for one wall as for two. One could increase the loss by having less than "one whole wall" in motion. An example has been provided by Coleman and Scott[1]. They show a wall that goes from one side of the whisker to an adjacent (rather than opposite) side. In this case d becomes much less as the wall approaches a corner and the losses accordingly would increase. This is in qualitative agreement with the behavior we have previously discussed[27]. A detailed calculation of the losses for this situation has not been carried out.

This argument can be used to explain the very high losses right at departure (see fig. 12, ref. 14). Usually the Landau structure jumps over to what we refer to as the Coleman structure just before departure. The Coleman structure then exhibits very high losses just as the wall is driven into the corner. Presumably the wall can stay together when flattened into the corner over a considerable distance, much as DeBlois has shown for walls pushed to the edge of nickel-cobalt platelets[7]. The losses observed just at departure can increase by as much as a factor of five for a d.c. field change of a few hundredths of an oersted.

TEMPERATURE DEPENDENCES

The measurements have been extended to high temperatures by constructing the transformer coils from ceramic insulated nickel wire, which previous experience[28] had shown to be very stable at 1000°K. A single layer inner coil provides the a.c. driving field, while a multilayer outer coil is used for detection. The transformer is placed in a cavity, diameter 1 cm, length 2 cm, in a silver cylinder, diameter 3 cm, length 8 cm. The cylinder axis is vertical while the axis of the coils and whisker is horizontal. The cylinder is surrounded by a heater and four radiation shields concentric with the cylinder but of greater length, 10 cm and 20 cm respectively. The spaces above and below the cylinder are filled with stacks of disc shaped shields. The whole furnace is enclosed in a water cooled vacuum jacket. The power input, 100 watts at 1000 deg. K, is controlled to better than 2 parts in 10^6 and the water jacket is maintained to 0.01 deg. C. The whisker, encapsulated in a quartz tube, is constant in temperature to within millidegrees for long periods of time. Gradients are, hopefully, minimized by the high symmetry of the furnace. Designed for the purpose of studying critical phenomena in whiskers, this system is more than adequate for the studies of the temperature dependence over a wide range, but cannot be used below 650 deg. K because, unfortunately, nickel becomes ferromagnetic.

The three simple predictions for the temperature dependence which follow from the discussion of the room temperature behavior are:
 a. The departure field should track the spontaneous magnetization;
 b. The effective stiffness parameter α_{eff} should be independent of temperature; and
 c. The effective damping parameter β_{eff} should follow the conductivity directly.

While these three things follow directly from the simplified model, they have a more general validity. If magnetostatics is the completely dominant factor in determining the magnetic response, then it follows that it is only the whisker geometry that counts. There is, however, a small correction from the presence of the anisotropy. A magnetization pattern which is parallel to the bowed wall and puts charges on the surface necessarily makes angles with respect to the preferred direction. This increase in energy throughout the domain volume is a local energy and is therefore easily incorporated in the local model. The stiffness of the wall given by Eq. (15) becomes

$$\alpha = 4\pi M_s^2 \, dt \left(\xi + \frac{2}{3} \frac{H_K}{4\pi M_s} \right) \qquad (23)$$

From the magnetostatic calculations ξ for a typical whisker is about 2 so that for iron at room temperature one might anticipate a 2/3 percent effect. There is in addition the effect that the anisotropy gives a torque at the wall so that some of the charge appears there. The change in magnetostatic energy between the case where all the charge is on the wall and the case where it is all on the surface is a factor of 5/4. It is this increase in energy that provides the "spring" which keeps most of the charge on the surfaces. This "spring" is only slightly compressed. Only two percent of the charge is on the wall for iron whiskers at room temperature. Thus the term in Eq. (23) is almost the whole effect from anisotropy.

Our most accurate data only cover the range from 650 deg. K to above 1050 deg. K. If a 2/3 percent effect on α_{eff} were to be expected from anisotropy between T_c and 300 deg. K, then only a 1/6 percent effect should be seen from T_c to 650 deg. K. The change observed in α_{eff} is larger than this and is attributed to the temperature dependence of the ratio of frequency to characteristic frequency through the conductivity[14]. α_{eff} is slightly dependent upon this ratio for the frequency used.

The effective damping parameter β_{eff}, like α_{eff}, does not depend on the magnetization. The only parameter on the right side of Eq. (18) is the magnetic skin depth, δ, given in Eq. (19). As α and β_1 are both proportional to M_s^2, only the conductivity and the dimensions of the whisker come into the calculation of β_{eff}, which then is directly proportional to the conductivity. A comparison of the temperature dependence of β_{eff} and the conductivity is shown in fig. 10. The agreement is satisfactory until about 1 deg. K below T_c where the rapid change is attributed to a change to a magnetization process quite different than the simple movement of a sharp domain wall in the Landau structure.

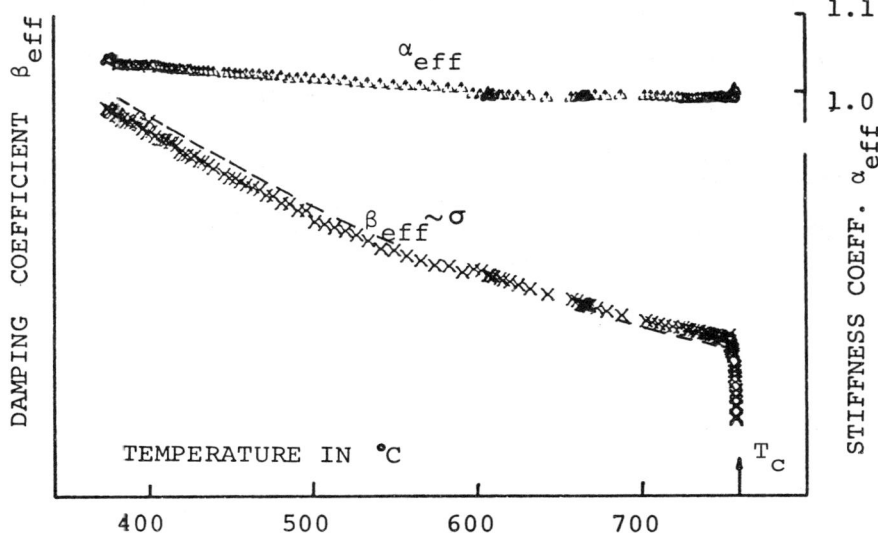

Fig. 10. Temperature variation of α_{eff} and β_{eff} at 10 kHz. Note the suppression of the zero to show the slight variation of α_{eff}, Δ's. The solid curve compares the conductivity of iron with the measurements of β_{eff}, x's.

If we were to accept the equivalence of the whisker with the rectangular bar of infinite permeability, we would equate the departure field with the field at which the center of the bar just reached saturation. Then to the extent that α_{eff} is independent of temperature the departure field would track M_s. We have compared the temperature dependence of the departure field to the equivalent expression used in analysing magnetization data just below the critical temperature, that is

$$H_d(T) = H_d(T_c - T_2) \cdot \left(\frac{T-T_c}{T_2}\right)^\beta \tag{24}$$

As can be seen in fig. 11-12 the fit is excellent for the critical exponent $\beta = .368$. A fit carried out only using the data within

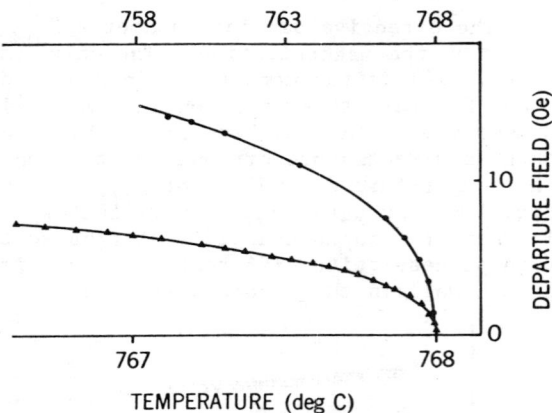

Fig. 11. The temperature dependence of the departure field below T_c. Points for the range to 1.4 deg. below T_c, lower curve, were fit to Eq. (24) to give $\beta = .368$. Eq. (24) with the same parameters is compared with the data for the range to 10 deg. below T_c in the upper curve.

Fig. 12. Logarithmic plot to compare data for the range to 100 deg. below T_c with Eq. (24). The solid line is drawn through the data. the dashed line at the right is a continuation of a straight line with slope $\beta = .368$. The dashed line at the left is a continuation of a straight line with slope $\beta = .345$.

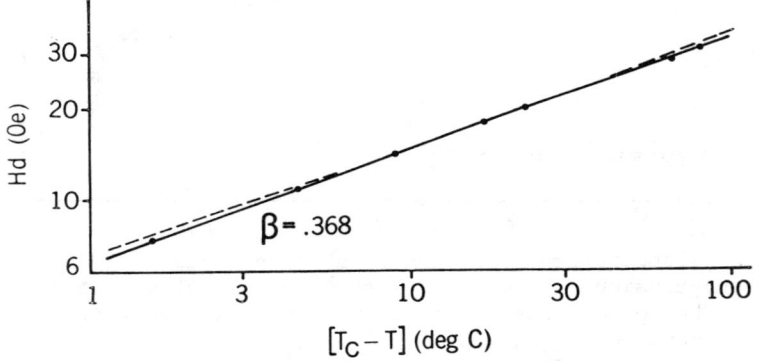

1 deg. K of T_c is in good agreement with fits over much wider temperature ranges, despite the considerable change in magnetization process indicated by the change in the losses within that last degree. This is a further indication that magnetostatic energy determines the whisker response right up to T_c.

GASEOUS IMPURITIES

We have previously shown that certain whiskers show a time dependent decrease in susceptibility when the wall is moved to a new position. Time constants of several seconds are typical. The effect can be as large as an order of magnitude decrease in the in phase component of the susceptibility. This is the well known magnetic after effect[29]. Some of the whiskers exhibit the after effect in the "as prepared" state. All of the whiskers that have been at T for

several days in the studies of critical phenomena show the effect
when cooled to room temperature, but for these the effect goes away
after several days at room temperature. While a theory exists[30]
which can be applied to the motion of single domain walls, we have
not yet carried out the necessary analysis. We have made a preliminary study that shows how the susceptibility decrease depends on the
amplitude of the a.c. driving field (see fig. 13).

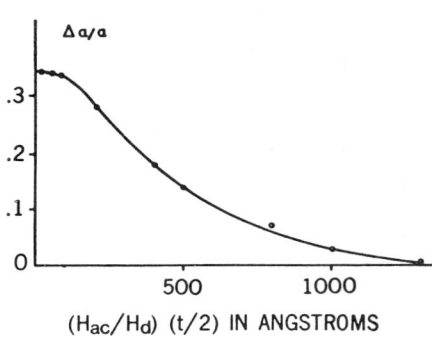

Fig. 13. Effect of amplitude of a.c. field, $H_{a.c.}$, on trapping of gaseous impurities. α is the stiffness of the wall for large amplitude or for any amplitude when the d.c. field is swept and gas does not follow the wall. $\Delta\alpha$ is the increase in stiffness for small amplitude at constant d.c. field ($H_{d.c.} = 0$). $H_{a.c.}$ is converted to maximum displacement of the wall using the relation $y = (t/2) \cdot (H_{ac}/H_d)$.

BEHAVIOR NEAR T_c

The d.c. susceptibility above T_c is in accordance with previous studies of polycrystalline iron spheres[28]. The effective stiffness is given by

$$\frac{\alpha_{eff} - \alpha^0_{eff}}{\alpha^0_{eff}} = \frac{1}{4\pi D}\left(\frac{T-T_c}{T_1}\right)^\gamma \qquad (25)$$

for T greater than T_c. α^0_{eff} is the stiffness below T_c. The factor
$4\pi D$ is the demagnetizing factor for a bar of the same dimensions of
the whisker, but with a susceptibility (internal) given by the other
factor on the right side of the equation. One can show from the
magnetostatic calculations that this dependence of the demagnetizing
factor is not a serious consideration. Obtaining a reliable value
for $4\pi D$, however, requires putting the measurements on an absolute
basis for the coil geometry used.

The out phase component behavior is rather more exciting.
There are a series of anomalies. Some of these we think we can
explain, but some are quite baffling. One of these anomalies is a
rapid increase in the loss component giving a peak in the temperature dependence with a width of the order of several hundredths of a
degree about T_c, fig. 14. The anomaly among anomalies appeared in
one sample which showed an additional loss superimposed upon the
"usual anomalous loss". This loss peaked 0.01 deg. K above T_c
and was present only in d.c. bias fields of less than 0.02 oersted.

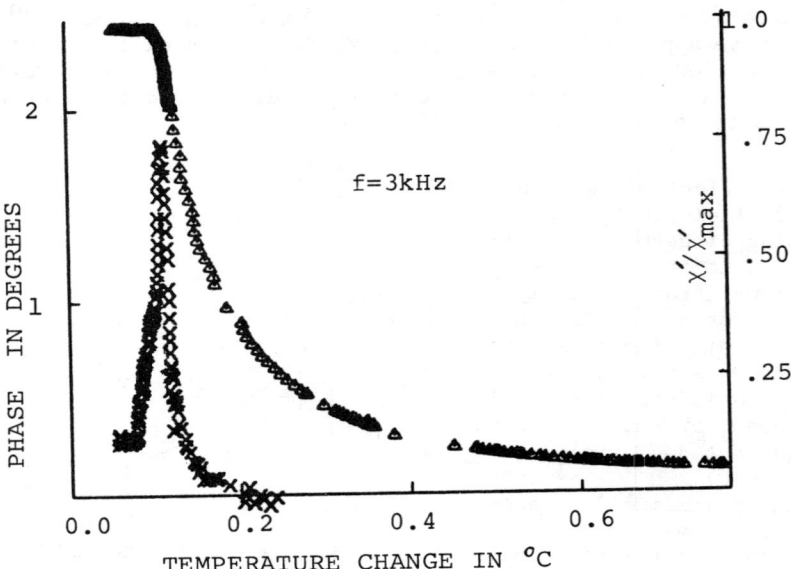

Fig. 14. Temperature dependence of χ' and phase near T_c for 3 kHz.

This very strange loss which was 12 times the "usual anomalous loss" (see fig. 14) is shown in fig. 15. The field dependence of the "usual anomalous loss" is seen with the very strange loss superimposed. The in phase response also shows an anomaly in the region below 0.02 oersted. The behavior is not unlike that seen in fig. 2 at the departure field.

Fig. 15. Anomaly in the very low field behavior of one whisker just above the critical temperature.

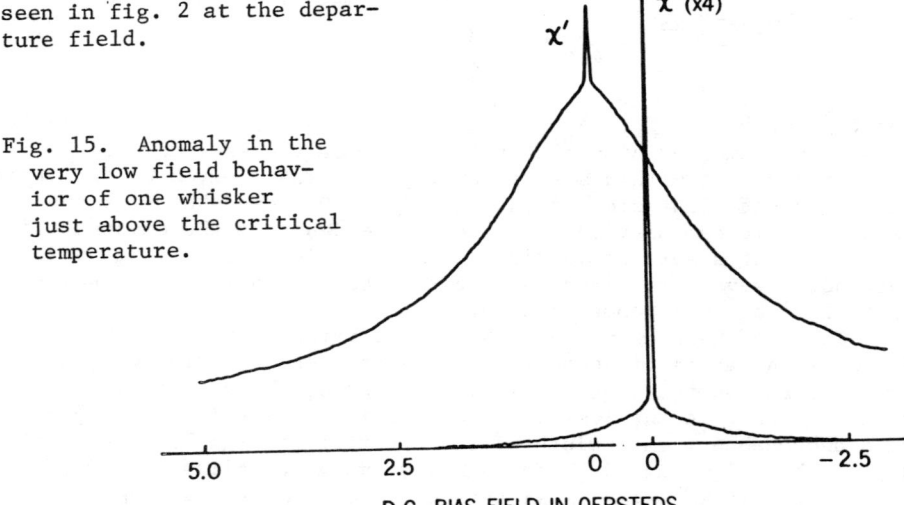

Further results and details are to be presented elsewhere[11].

ACKNOWLEDGMENTS

This work has been supported by the National Research Council of Canada. The high temperature work was carried out in collaboration with Dr. John E. Noakes of the Ford Motor Company. Mr. Vaclav Macura was responsible for much of the instrumentation. To the above we are much indebted.

REFERENCES

1. R.V. Coleman and G.G. Scott, Phys. Rev. $\underline{107}$, 1276 (1957).
2. G.G. Scott and R.V. Coleman, J. Appl. Phys. $\underline{28}$, 1512 (1957).
3. G.G. Scott and R.V. Coleman, J. Appl. Phys. $\underline{29}$, 526 (1958).
4. R.W. DeBlois and C.D. Graham, Jr., J. Appl. Phys. $\underline{29}$, 528 (1958).
5. R.W. DeBlois, J. Appl. Phys. $\underline{36}$, 1647 (1965), Tech. Rep. No. AFCRL-68-0414 (General Electric Co., Schenectady, New York).
6. R. Gemperle, Phys. Stat. Sol. $\underline{14}$, 21 (1966); Czech. J. Phys. B21, $\underline{89}$ (1971).
7. R. Gemperle and J. Kaczer, Phys. Stat. Sol. $\underline{34}$, 255 (1969).
8. W. Hagedorn and H.H. Mende, Z. Angew. Phys. $\underline{30}$, 68 (1970).
9. P.W. Shumate, R.V. Coleman, and R.C. Fivaz, Phys. Rev. B, $\underline{1}$, 394 (1970).
10. Princeton Applied Research Model 124 (two of them) outputs read on Hewlett-Packard digital voltmeter 2402A in data acquisition system.
11. A.S. Arrott, B. Heinrich, and J.E. Noakes; this conference and to be published.
12. A.S. Arrott and B. Heinrich, Physics in Canada $\underline{27}$, No. 4, 34 (1971).
13. A.S. Arrott, B. Heinrich, and D.S. Bloomberg, "Magnetism and Magnetic Materials in 1971", 896.
14. B. Heinrich and A.S. Arrott, Can. J. Phys. $\underline{50}$, 710 (1972).
15. A.S. Arrott and B. Heinrich, IEEE-Mag. 8, Sept. 1972.
16. C.G. Lamb, Phil. Mag. $\underline{48}$, 262 (1899).
17. K. Warmuth, Arch. f. Electotech. $\underline{30}$, 761 (1935) see in particular fig. 2; $\underline{31}$, 124 (1937); $\underline{33}$, 747 (1939) see in particular formulas on p. 748.
18. F. Stablein and H. Schlechtweg, Zeit. f. Phys. $\underline{95}$, 630 (1935).
19. R.M. Bozorth and D.M. Chapin, J. Appl. Phys. $\underline{13}$, 320 (1942).
20. T. Okoshi, J. Appl. Phys. $\underline{36}$, 2382 (1965).
21. G. Deitz and R.G. Meingast, Zeit. Angew. Physik $\underline{31}$, 77 (1971).
22. W.F. Brown, Jr., "Magnetostatic Principles in Ferromagnetism", North-Holland, Amsterdam (1962) (in particular p. 4); "Micromagnetics", Interscience, New York (1963).
23. D.S. Bloomberg Thesis, Simon Fraser University (1973).
24. H.J. Williams, W. Shockley, and C. Kittel, Phys. Rev. $\underline{80}$, 1090 (1950).
25. P.D. Agarwal and L. Rabius, J. Appl. Phys. S$\underline{31}$, 2465.
26. H.S. Carslaw and J.C. Jaegar, "Conduction of Heat in Solids", Oxford (1947).
27. Those structures which we referred to as phase II and phase III in ref. 14 can be interpreted in terms of this structure we refer to here as the Coleman structure. We are indebted to Prof. Coleman for this identification.
28. J.E. Noakes, N. Tornberg, and A. Arrott, J. Appl. Phys. $\underline{37}$, 1264 (1966).
29. L. Néel, J. Physique Rad. $\underline{13}$, 249 (1952).
30. F. Schreiber, Zeit, F. Angew. Physik $\underline{9}$, 203 (1957).

IRON-NICKEL-SILICA FERROMAGNETIC CERMETS

J. J. HANAK and J. I. GITTLEMAN
RCA Laboratories
Princeton, N. J. 08540

ABSTRACT

We have developed a method for quickly determining the compositional dependence of permeability and resistivity in three-component sputtered cermets and report results obtained for $[Fe_y Ni_{1-y}]_{1-x}(SiO_2)_x$. Films of variable composition were sputtered onto large square substrates from disc-shaped targets formed from unequal sectors of Fe, Ni, and SiO_2. Resistivity and permeability were determined using a commercial four-point probe and magnetic recording head. From these data, contours of constant permeability and resistivity were obtained. A particularly prominent permeability peak was observed at $[Fe_{.7}Ni_{.3}]_{.55}[SiO_2]_{.45}$ (by vol.) where $\mu \sim 170$ and $\rho \sim 10^{-1} \Omega$-cm. This composition has proved useful in designing sputtering targets which yield films with predetermined properties.

INTRODUCTION

Ferromagnetic cermets formed by co-sputtering a ferromagnetic metal and an insulator are an array of very small (30 Å - 300 Å) metal grains embedded in an insulating matrix. Fig. 1 shows the compositional dependence of the resistivity of the $Ni-SiO_2$ cermet system. The rapid rise of resistivity with increasing insulator concentration and the transition from metal-like to semiconductor-like behavior (at about 35 vol. % SiO_2 in $Ni-SiO_2$) seem to be general properties of sputtered cermets.[1] Magneto resistance and magnetization[2] measurements show that the Curie temperature of $Ni-SiO_2$ cermets fall rapidly in the range 45-55 vol. % SiO_2; while for $Fe-SiO_2$, high values of T_c seem to persist to much higher concentrations of SiO_2.[3] Fig. 2 is a plot of the initial permeability vs. composition for $Fe-SiO_2$ and 80-permalloy-SiO_2 cermets. These data are given to show the wide range of permeabilities available in a multicomponent cermet system such as $(Ni_y Fe_{1-y})_{1-x}(SiO_2)_x$. Fig. 1 shows that a wide range of resistivities is also available. In as much as there may be many applications for ferromagnetic films each requiring a different permeability and resistivity, an interesting question arises. What is the quickest way to determine the relative concentrations of the components of a complex cermet system which exhibits the desired properties? In the following a method is described for quickly surveying the resistivity and initial permeability of a three component system $([Ni_y Fe_{1-y}]_{1-x}[SiO_2]_x)$.

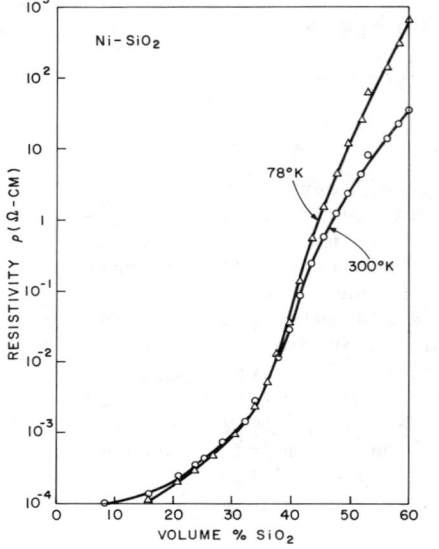
Fig. 1. Resistivity vs. Composition in Ni-SiO$_2$ Cermets.

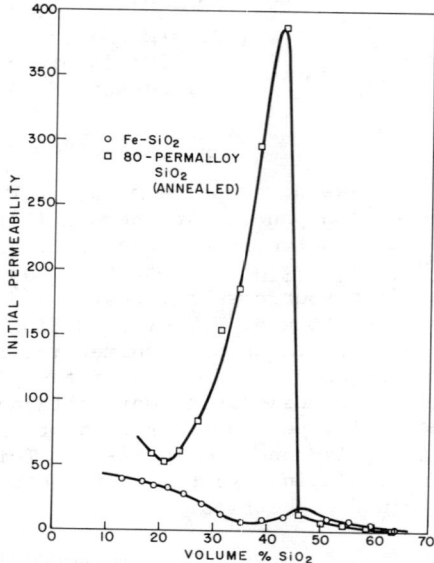
Fig. 2. Permeability vs. Composition in Fe and 80 Permalloy Cermets

SPECIMENS

The sputtering target was in the form of a circular disk nominally 6 inches in diameter. The circle was divided into three sectors, one sector for each of the components of the final specimen. The central angle of each sector was determined by the range of compositions to be represented in the specimen.

The substrate was a square piece of glass nominally 4.5 inches on a side. The film sputtered on the substrate from the sectored target was a ferromagnetic cermet, the composition of which was a function of the position coordinates of the substrate. Because of the masking arrangement, the specimen consisted of a square 3.5 inches on a side with a series of regularly spaced, rectangular film patches on its periphery. By measuring the thickness of each of the patches and utilizing the known target geometry and composition, the thickness and composition of any point in the specimen could be computed.[4]

PERMEABILITY AND RESISTIVITY MEASUREMENTS

The sample was mounted on a small vertical milling machine by means of a vacuum chuck. For measuring the permeability, a commercial recording head was mounted on the milling head. The recording head contained a pair of windings, one for the record/

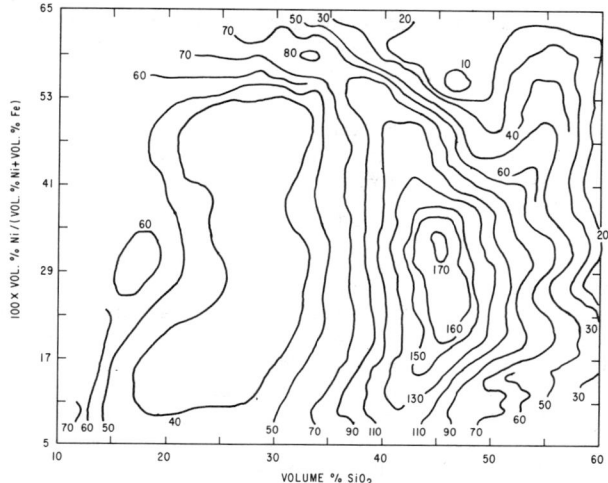

Fig. 3. Contours of Constant Permeability for Fe-rich Fe-Ni-SiO$_2$ Cermets. (As sputtered)

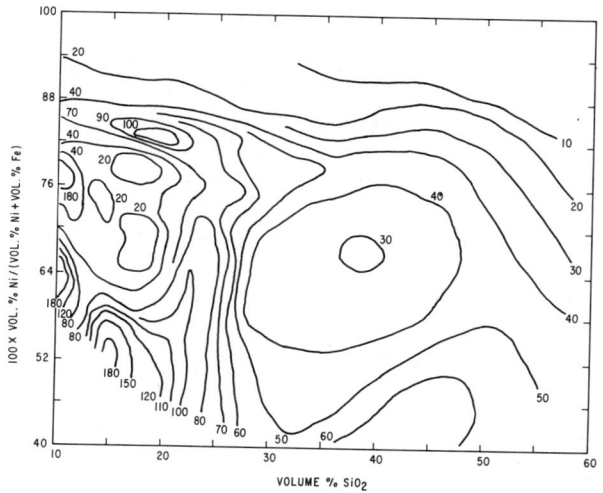

Fig. 4. Contours of Constant Permeability for Ni-rich Fe-Ni-SiO$_2$ Cermets (As sputtered)

playback function and one for the erase function. One of the windings was driven by the reference source of a PAR HR-8 phase-sensitive detector. The HR-8 then measured the change in signal from the other winding when the head was brought in contact with the specimen. The signal is linearly related to the magnetic reluctance of that portion of the sample beneath the gap of the recording head. The reluctance, in turn, is inversely proportional to µd, the product of permeability and thickness. By using two standards the permeability at any point (x,y) of the film can be determined.

The resistivity as a function of position is obtained by replacing the recording head with a commercial four-point probe. A simple dc supply in conjunction with a digital voltmeter is used to determine resistance.

With this apparatus, the permeability and resistivity as a function of the specimen coordinates can be quickly obtained. Since the composition as a function of specimen coordinates is known, a computer program was prepared to yield contours of constant permeability and constant resistivity vs. composition. Fig. 3 gives the contours of constant permeability for Fe-rich Fe-Ni-SiO$_2$ cermets. The abscissa is the volume percentage of SiO$_2$ in the specimen, and the ordinate is the volume percentage of Ni in the Ni-Fe alloy grain. Fig. 4 is a similar plot for the Ni-rich cermets. Fig. 5 gives the contours of constant resistivity using the same abscissae and ordinates. These specimens were annealed in vacuum for 2 hours at 250°C. In general

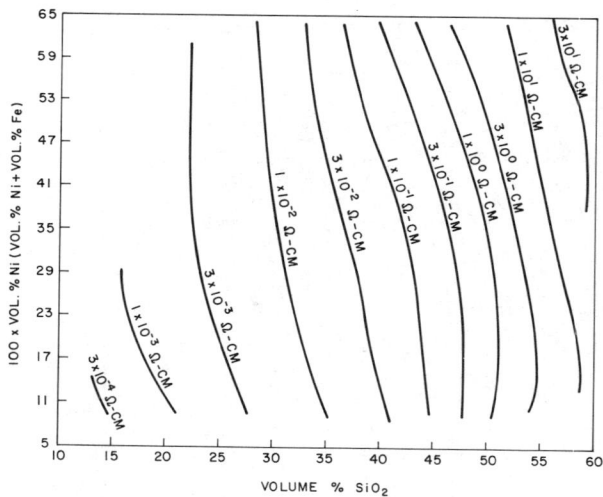

Fig. 5. Contours of Constant Resistivity for Fe-rich Fe-Ni-SiO$_2$ Cermets. (As sputtered)

the effect of annealing on the resistivity contours was small. The positions of the peaks and valleys of the permeability contours remained unchanged after the anneal. However, the permeability in the neighborhood of the peaks was enhanced while the permeability in the neighborhood of the valleys and depressions remained substantially unchanged.

These results indicate that the resistivity increases with increasing SiO_2 concentration in a manner similar to other cermets (roughly exponentially) and is only slightly dependent on the Fe/Ni ratio of the metal grains. Furthermore, the permeability peak in the vicinity of the cermet $(Ni_{.3}Fe_{.7})_{.55}(SiO_2)_{.45}$, by virtue of its prominence and isolation from other peaks, provides a convenient means for testing our ability to make uniform large area films of a given cermet. Two sputtering targets were made; one in which powders of Fe, Ni and SiO_2 were mixed in the appropriate proportions, pressed and sintered; and one in which an iron target was overlayed with SiO_2 strips and small Ni squares the areas of which were chosen on the basis of relative sputtering rates. Films made from these targets exhibited permeabilities which were within about 25% of that of the peak and resistivities which agreed within a factor of two.

The relative errors in the measured permeability and resistivity are about 10% and 1% respectively. The composition is computed from thickness measurements as described in ref. 4. The error in the concentration of any component is less than 5 vol. % and is least at the center of the specimen (approximately the center of Figs. 3-5). Furthermore the concentration of minority components tend to be underestimated and that of the majority component overestimated[4] so that the cermet $(Fe_{0.90}Ni_{0.10})_{.80}(SiO_2)_{.20}$ as determined from the figures might actually be $(Fe_{0.85}Ni_{.15})_{0.75}(SiO_2)_{.25}$.

The authors thank J. P. Pellicane and S. Bozowski for valuable assistance in sample preparation and measurements.

REFERENCES

1. C. A. Neugebauer, Thin Film Solids 6, 443 (1970); B. Abeles and Ping Sheng, Proc. of 13th Int. Conf. on Low Temperature Physics, Boulder, Colo. (1972).
2. J. I. Gittleman, Y. Goldstein, and S. Bozowski, Phys. Rev. B1 5, 3609 (1972); M. Rayl, P. J. Wojtowicz, M. Abrahams, R. Harvey and C. Buiocchi, Phys. Letters 36A, 477 (1971).
3. M. Rayl, unpublished data.
4. J. J. Hanak, H. W. Lehmann, and R. K. Wehner, J. Appl. Phys. 43, 1666 (1972).

SUPERPARAMAGNETIC FERRITE PARTICLES PRODUCED BY MILLING

A. E. Berkowitz and J. A. Lahut
General Electric Corporate Research and Development
Schenectady, New York 12301

ABSTRACT

Superparamagnetic particles of Ni-, Co-, Fe-, and NiZn ferrites were prepared by the milling techniques described by Kaiser and Miskolczy. The particles were extracted from the carrier liquid and structural and magnetic properties were measured before and after removing the organic surfactant layer. The lattice spacings and Curie temperatures of the particles were not significantly changed by the milling. However, the presence of the surfactant layer reduced σ_s by >50%, indicating that the chemical bonding of the organic molecules to the ferrite decreased the magnetization of the surface layers of the particles. Removal of the surfactant restored σ_s to nearly bulk values. In addition, the presence of an anomalously high anisotropy was indicated by very large coercive forces at 10°K. The possibility that a surface anisotropy is responsible for this behavior is discussed.

INTRODUCTION

Kaiser and Miskolczy[1] have recently described a method for producing stable suspensions of magnetite particles that are superparamagnetic at room temperature. The procedure involved milling the magnetite for long times (>1000 hours) with a carrier fluid and relatively large amounts of an organic surfactant. This produced particles in the 100Å diameter range which were coated with the organic surfactant and remained in indefinite suspension in the carrier fluid. They interpreted some of their magnetic measurements on the liquid suspensions as indicating a decrease in the magnetic moment of the particles resulting from the interaction between the surfactant molecules and ions on the surface of the magnetite particles. We have applied their technique to a variety of simple and mixed ferrites and report on the magnetic and structural properties of these particles as measured after extracting them from the carrier fluid.

EXPERIMENTAL DETAILS

Starting materials were coarse, crushed powders of Ni-, Fe-, Co-, $Ni_{.4}Zn_{.6}$-, and $Ni_{.6}Zn_{.4}Fe_2O_4$ prepared by standard ceramic techniques. MgO was also milled under the same conditions in order to establish the quantity and magnetic behavior of the mill contaminant. Kerosene was the carrier and Enjay[2] was the surfactant. After milling for ~1000 hours, the resulting liquid was centrifuged at 17,000 gees for 20 minutes. After this treatment the particles were very well dispersed and in stable suspension in the kerosene. Eleccron micrographs (at 300,000X) of the particles in suspension indicated that they were almost equiaxed with irregular surfaces and average diameter <350Å. The particles were collected by flocculating the suspension with acetone. Excess surfactant was removed by repeated washings in methylene dichloride and collected by centrifuging at 144,000 gees. We will refer to particles in this condition as 'as-washed'. The 'as-washed' particles could be readily resuspended in methylene dichloride. Heating the 'as-washed' particles in air to ≥ 500 C resulted in a weight loss of 15-30%; the weight loss was proportional to the specific area of the samples. Thus, the 'as-washed' particles still retained an effective organic coating. The organic coating was removed by heating the 'as-washed' particles in air. At 200 °C, ~10% of the net weight loss took place; at 250 °C, ~80%, at 400 °C, ~90%; at 500 °C, ~100%. We shall report primarily on samples heated at 400 °C and 500 °C as well as the 'as-washed' samples.

Chemical analysis of the milled MgO samples indicated the presence of a substantial (~14 w/o) amount of Fe from the wear of the C-steel balls. This Fe was almost entirely oxidized to α-Fe_2O_3 (from x-ray and magnetic measurements). Chemical analysis of the milled ferrite samples indicated Fe contaminant of the same order as in the MgO by comparing the metal ion ratios in the milled and starting materials. Thus it was possible to correct the measured magnetic properties of the milled ferrites for the mill contaminant by using the chemical analyses and the magnetic behavior of the milled MgO in the 'as-washed' and annealed conditions. The Ni/Zn ratio was increased slightly in the milled mixed ferrites indicating that some Zn was preferentially dissolved from the surface of the particles during the milling. When the Fe contaminant in the milled $NiFe_2O_4$ was determined from the Fe/Ni ratio, a value larger than for any of the other samples was obtained. Furthermore, when this correction was applied to the magnetic data, a moment higher than that for the bulk ferrite was obtained. It was surmised that Ni was preferentially dissolved from the surface of these particles, contributing to the high Fe/Ni ratio. Therefore in this case we applied a correction for Fe contaminant that was the average for the other samples. This average correction was also used for the Fe_3O_4 sample since excess Fe could not be determined from chemical analysis.

From x-ray and electron diffraction it was established that the milled particles retained the same cubic structure and cell edge lengths as the corresponding bulk materials. Average particle diameters measured from the line broadening of x-ray diffraction patterns of 'as-washed' samples were somewhat smaller than those determined from the electron micrographs. This could be due to missing very small particles and/or counting clumps as single particles; we have relied on the average sizes from line broadening.

Magnetic measurements were made in a VSM from 4.2 °K to T_c in fields up to 24 kOe. Hysteresis measurements were made on particles immobilized in a suitable binder. Since all samples were superparamagnetic in their respective critical regions, T_c was determined[3] from the peak in $d[(\chi_o T)^{1/2}]/dT \propto d[\sigma_s(T)]/dT$ where χ_o is the susceptibility measured in low (<50 Oe) fields, T is the temperature and $\sigma_s(T)$ is the spontaneous magnetization. Saturation magnetization of the samples was obtained by 1/H extrapolation of the data from 8 to 24 kOe. The data were corrected for a small (10^{-4} emu/gm-Oe) high field susceptibility when required to get a fit of <0.1% to the 1/H extrapolation from 8 to 24 kOe. The extrapolations were <5% from the 24 kOe value.

DISCUSSION OF DATA

The data on particle size and magnetic properties of the samples are collected in Table I. Sample treatment is indicated by the

Table I - Particle Properties

Formula	D(Å)	σ_s/σ_s (Bulk)			H_c(Oe)		σ_R/σ_s	
		293	77	4.2°K	77	10°K	77	10°K
Fe_3O_4 (A)	65	.78	.84	.84	24	310	.04	.26
$CoFe_2O_4$ (A)	61	.38	.39	---	4340	>12000	---	---
$CoFe_2O_4$ (B)	90	.60	---	---	>11000	>16000	---	---
$CoFe_2O_4$ (C)	151	.83	---	---	>12500	>16600	---	---
$NiFe_2O_4$ (A)	70	.30	.36	.39	523	2980	.22	.42
$NiFe_2O_4$ (B)	90	.74	.80	.80	610	1285	.33	.44
$NiFe_2O_4$ (C)	142	.97	.97	.93	460	640	.43	.45
$Ni_{.6}Zn_{.4}Fe_2O_4$ (A)	84	.31	.31	.29	268	1700	.23	.52
$Ni_{.6}Zn_{.4}Fe_2O_4$ (B)		.80	.79	.75	294	670	.33	.40
$Ni_{.6}Zn_{.4}Fe_2O_4$ (C)		.99	.95	.94	346	675	.32	.42
$Ni_{.4}Zn_{.6}Fe_2O_4$ (A)	84	.27	.25	.23	180	1575	.17	.47
$Ni_{.4}Zn_{.6}Fe_2O_4$ (B)	105	.76	.66	.63	209	857	.20	.43
$Ni_{.4}Zn_{.6}Fe_2O_4$ (C)	140	1.0	.94	.88	246	550	.28	.39

letters A, B, and C: A, 'as-washed'; B, 400°C anneal; C, 500°C anneal. Average particle diameters, as determined from x-ray line broadening, are shown in the second column. The samples were loosely packed during the anneals, and it appears that appreciable particle growth required temperatures >400°C. In the third column are shown the ratio of the saturation magnetization of the samples to that of the respective bulk materials at 293, 77 and 4.2°K. σ_s values of the samples were corrected for the magnetization and weight of mill contaminant and for the weight of surfactant as explained above. Low temperature values for σ_s of $CoFe_2O_4$ could not be obtained because of the large hysteresis. The most interesting aspect of these data is that all samples in the 'as-washed' condition, except Fe_3O_4 have σ_s values <40% of their respective bulk values. Even after the

400 °C anneal (B), σ_s is still reduced by ~20%. The decrease in σ_s is definitely associated with the presence of the organic surfactant (as determined by weight loss on annealing). As soon as some surfactant is removed, at temperatures as low as 200°C, σ_s begins to increase. σ_s values remain stable when no further weight loss on annealing occurs (~500°C). Thus, our data confirm the suggestion[1] that a decrease in σ_s results from the complexing of the surfactant molecules with ions on the surface of the ferrite particles, producing a non-ferrimagnetic surface layer. If the thickness of this 'dead' layer is taken to be a cell edge, very good agreement with the measured σ_s values for the 'as-washed' samples is obtained by using the average particle diameters and recognizing that the specific area of the samples is increased by the non-spherical shape of the particles. It is interesting to note that the decrease in σ_s for Fe_3O_4 is less than half of that for the other ferrites, although the surfactant coverage is essentially the same. This suggests that the state of Fe^{2+} is less significantly altered by the interaction with the organic molecules than are the Ni^{2+} or Co^{2+}. It is, of course, possible that atomic surface disorder or uncoupled surface spins can contribute to the decrease in σ_s. The data, however, indicate that the major cause is the surfactant interaction. The temperature dependence of σ_s for the NiZn ferrites indicates a higher T_c than in the bulk materials. This is consistent with the observation of a slightly higher Ni/Zn ratio in the milled samples. However, the T_c for these samples were within 10 K of the values for the bulk materials. T_c for the $NiFe_2O_4$ and $CoFe_2O_4$ were the same as bulk values.

In the fourth and fifth columns are the coercive forces and ratios of remanence to saturation at 77 and 10° K. All samples except $CoFe_2O_4$ were superparamagnetic at room temperature. The large coercive forces of the $CoFe_2O_4$ samples are consistent with the high anisotropy field.[4] More unexpected are the large coercive forces of the Ni and NiZn ferrites at 10°K. We may inquire whether these values are consistent with the anisotropies usually assumed to govern single domain particle behavior. We consider randomly oriented, non-interacting single domain particles whose magnetization rotates coherently. In this case, $H_c \simeq H_K/2$, where H_K is an anisotropy field arising from shape, magnetocrystalline or stress anisotropy.[5] We calculate H_c for $NiFe_2O_4$ for which the magnetocrystalline anisotropy and magnetostriction have been determined at 4.2 K.[6] From the electron micrographs of the 'as-washed' samples we find that the axial ratio is <2. Thus, $H_c(\text{shape}) = 1/2(N_b - N_a)M_s \simeq$ 450 Oe, where N_b and N_a are the demagnetizing coefficients normal and parallel to the shape axis and M_s is the spontaneous magnetization at 4.2°K. For magnetocrystalline anisotropy, $H_c = K/M_s = 300$ Oe. For uniaxial stress, $H_c = 3\lambda\sigma/2M_s = 2 \times 10^{-7}\sigma$, where λ is the magnetostriction constant and σ the stress. Therefore, a stress of $>10^{10}$ dynes/cm^2 is required to account for the measured H_c of the 'as-washed' $NiFe_2O_4$ particles at 10°K. This enormous stress would imply an unacceptably high dislocation or point defect density in these small particles. Another possibility to account for this high H_c is surface anisotropy. We may estimate the required surface anisotropy as follows. The surface anisotropy K_s (ergs/cm^2) can be normalized to an effective volume anisotropy.

$$K_V = K_s A/V = 6K_s/d \qquad (1)$$

where A and V are the surface area and volume of a particle of diameter d. We now put

$$K_V/M_s = 6K_s/M_s d > H_c \qquad (2)$$

since H_c will be reduced by particles which are not single domains and by the likelihood that only a fraction of the surface of each particle is involved. This gives

$$K_s > H_c M_s d/6 \qquad (3)$$

We apply this relation to the data in Table I after reducing the measured d-values of the 'as-washed' samples by two cell edges (2 x 8.3A) to account for the 'dead' layer discussed above. This gives $K \geq .08$, .12, and .12 ergs/cm^2 for the 'as-washed' 10°K data for Ni-, Ni$_{.6}$Zn$_{.4}$-, and Ni$_{.4}$Zn$_{.6}$Fe$_2$O$_4$, respectively. These values are in good agreement with estimates for surface anisotropy[7] which suggests that this mechanism is more reasonable than a stress of >10^{10} dynes/cm^2.

ACKNOWLEDGEMENTS

We are grateful to C. E. Van Buren for preparing the samples, to E. F. Koch for the electron micrographs, to D. H. Mitchell for x-ray analysis and to B. R. Cooper, I. S. Jacobs and W. H. Meiklejohn for useful discussions.

REFERENCES

1. R. Kaiser and G. Miskolczy, J. Appl. Phys. <u>41</u>, 1064 (1970).
2. Enjay Chemical Co., New York, New York.
3. E. Kneller, in Magnetism and Metallurgy, Vol. I, edited by A. E. Berkowitz and E. Kneller (Academic Press, N. Y., 1969), p. 405.
4. H. Shenker, Phys. Rev. <u>107</u>, 1246 (1957).
5. E. C. Stoner and E. P. Wohlfarth, Phil. Trans. Roy. Soc. (London, <u>A 240</u>, 599, 1948).
6. A. B. Smith and R. V. Jones, J. Appl. Phys. <u>37</u>, 1001 (1966).
7. I. S. Jacobs and C. P. Bean, in Magnetism, Vol. III, edited by G. T. Rado and H. Suhl (Academic Press, New York, 1963) p. 314.

EFFECT OF COATING INDUCED STRESSES ON PROPERTIES OF CUBE TEXTURED 3% Si-Fe SHEETS

K. Foster, J. Seidel, and J. W. Shilling
Westinghouse Research Laboratories, Pittsburgh, Pa. 15235

ABSTRACT

Glass coatings having varying coefficients of thermal expansion were applied to Epstein strips cut in the rolling direction from cube textured 3% Si-Fe having two levels of directional orientation. A low expansion glass resulted in substantial reductions in loss and magnetostriction. Allowing coated samples to cool under applied tensile stress resulted in further loss and magnetostriction reductions.

INTRODUCTION

Cube textured 3% Si-Fe sheets have been shown to have higher losses[1,2] and magnetostriction values[3] measured in the rolling direction than cube-on-edge oriented sheets having similar degrees of directional orientation. Furthermore, Littmann[1] has indicated a much smaller dependence of losses on directional orientation, as indicated by the induction reached for a field of 10 Oe (B_{10}), for cube textured material produced by a surface energy controlled secondary recrystallization process than for cube-on-edge oriented material. The higher losses of the cube textured material have been attributed to its larger grain size[1] and 90° domain wall mechanisms of magnetization[2], while high magnetostriction values have been related to the presence of domains in the cross direction.[3] Thus, although some application of cube textured Si-Fe to rotating machinery has been made,[4] little effort has been made to use cube textured material in power transformers, where low losses and magnetostriction in given sheet directions are required. In the present work, glass coatings which impart residual tensile stresses to the metal have been used to substantially reduce both losses and magnetostriction in the rolling direction of cube textured 3% Si-Fe.

EXPERIMENTAL

Two commercially melted 0.3 mm thick cube textured 3% Si-Fe samples were investigated, one having a B_{10} of 17.3 kG and the other a B_{10} of 18.6 kG. All tests were carried out on 30 cm Epstein samples, cut in the rolling direction. The samples had been annealed at 1200°C with a loose Al_2O_3 separator coating for secondary grain growth, after which the Al_2O_3 was removed. Prior to application of glass coatings, samples were reannealed at 1200°C with a MgO coating to form a simulated "mill coating", which provided adherence for the subsequent glass coatings. Two glasses were used, one, designated H, having a coefficient of thermal expansion of 13×10^{-6} in/in/°C, essentially the same as that of 3% Si-Fe. The other glass, designated L, had an expansion coefficient of 7.2×10^{-6} in/in/°C, considerably lower than that of 3% Si-Fe. The coatings were applied as alcohol slurries and fired in air at 800°C, as described previously.[5] Total coating thicknesses of about 0.014 mm (0.007 mm per side) were obtained after firing. Some coated strips were retreated by heating to 550°C and cooling to room temperature under an applied tensile stress of 350 psi.

Core loss measurements were made at 60 Hz on samples of 12 strips each using a 30 cm Epstein frame. DC magnetostriction measurements were made on individual strips using a strain gauge method.[5] Magnetostriction measurements were also made as function of applied tensile stress on uncoated strips.

RESULTS AND DISCUSSION

Core loss and dc magnetostriction data for glass coated samples are given in Table I.

Table I Effect of Glass Coatings and Applied Tension During Cooling on Properties of Cube Textured 3% Si-Fe

B_{10}[a) (kG)	Glass	Applied Stress (psi)	60 Hz Core Loss (W/lb) 15 kG	17 kG	λ at 15 kG[b) $(in/in \times 10^6)$
17.3	None	None	0.81	1.21	14.0
17.3	H	None	0.78	1.17	12.6
17.3	H	350	0.75	1.09	7.7
17.3	L	None	0.82	1.28	3.3
17.3	L	350	0.75	1.07	-2.4
18.6	None	None	0.79	0.96	22.1
18.6	L	None	0.67	0.88	9.1
18.6	L	350	0.64	0.85	2.2

a) Average of batch prior to glass coating.
b) Average of two strips.

The uncoated samples both have rather high losses, and there is very little difference in loss at 15 kG as a function of directional orientation. Both samples also have high positive dc magnetostriction (λ) values, with the higher permeability sample having a considerably higher λ value. Glass H, which should not impart significant stress to the steel, was only applied to the lower permeability sample. This coating alone reduced losses slightly, but had only a negligible effect on magnetostriction. Cooling coated strips under applied tension further reduced losses and resulted in a fairly large decrease in magnetostriction.

Glass L, which had a lower thermal expansion than the metal, was applied to both steel samples. Application of glass L to the lower B_{10} sample did not reduce losses, but did result in a large decrease in magnetostriction. Cooling strips with this coating under applied tension resulted in a decrease in loss and a further decrease in λ at 15 kG to a negative value. Application of the low expansion glass to the higher permeability sample alone significantly decreased both losses and magnetostriction. Cooling these strips under tension further reduced losses and magnetostriction.

The results of Littmann[1] on the effect of permeability on core loss for oriented materials are plotted in Fig. 1. The uncoated

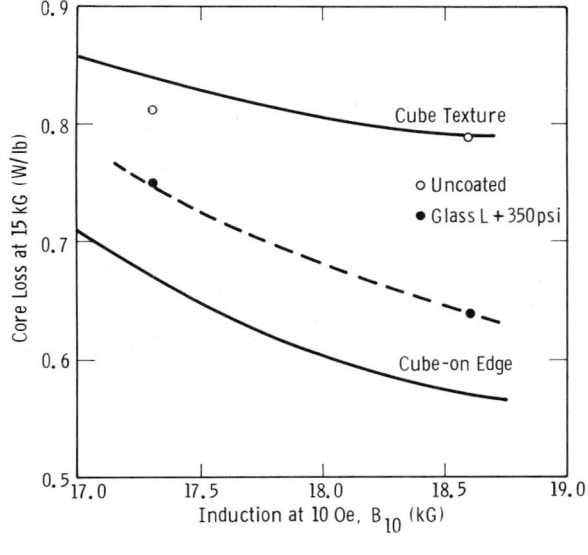

Figure 1 - Effect of B_{10} on core loss of oriented Si-Fe (solid curves from Littmann)

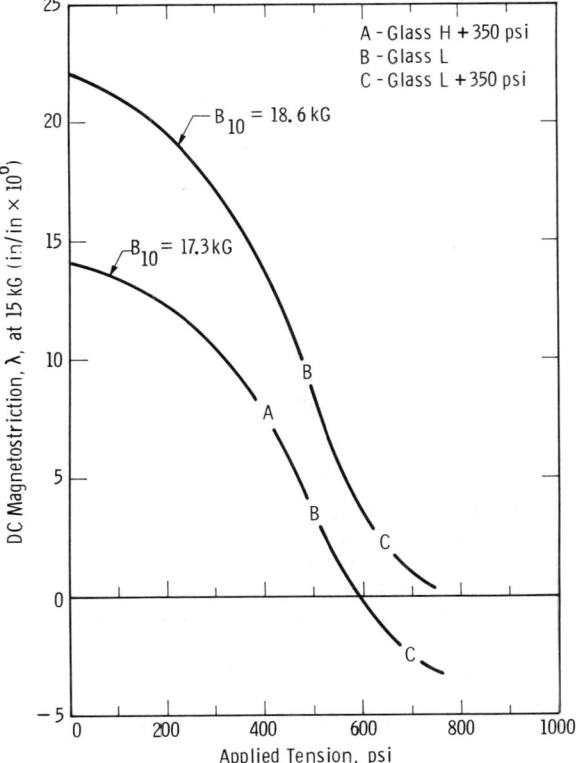

Figure 2 - Effect of applied tension on DC magnetostriction at 15 kG of uncoated cube textured 3% Si-Fe strips

samples from the present investigation are in good agreement with the curve for cube texture. The best loss values obtained with glass coating are considerably lower than the cube texture curve, and the slope of the curve for the coated samples is very similar to that of the cube-on-edge material. The higher loss for the coated samples is now likely due almost entirely to the grain size difference (25 mm for cube texture versus 2 mm for cube-on-edge).

The effect of applied tensile stress on magnetostriction (at 15 kG) of uncoated cube textured samples is shown in Fig. 2. For both steels, a sharp reduction in magnetostriction occurred between 200 and 700 psi, with the lower B_{10} sample becoming negative above 600 psi and the higher B_{10} sample approaching zero at 750 psi. The magnetostriction values of coated samples have been superimposed on the respective curves. Application of glass H to the lower permeability steel and cooling under applied tension resulted in a λ value equivalent to 400 psi applied tension. Application of glass L to both steels resulted in λ values equivalent to about 500 psi applied tension, while cooling this coating under 350 psi tensile stress resulted in λ values equivalent to 600 to 700 psi applied tension.

Low thermal expansion glass coatings are known to impart isotropic tensile stresses in 3% Si-Fe sheet.[6] In (110)[001] grain oriented 3% Si-Fe, an isotropic tensile stress will favor domains oriented along the rolling direction, and the effects observed are similar to those produced by a uniaxial tensile stress applied along the rolling direction.[6] In the case of (100)[001] grain oriented 3% Si-Fe, however, domains along the rolling and cross directions will be equally favored by an isotropic tensile stress and thus the effects of glass coating should not be the same as those resulting from a uniaxial tensile stress applied along the rolling direction. Nevertheless, the results of this experiment clearly indicate that when low thermal expansion glass coatings are applied to cube oriented 3% Si-Fe, effects very similar to those produced by a uniaxial tensile stress result, i.e., a substantial decrease in magnetostriction occurs, regardless of permeability; and core loss is reduced when B_{10} is large (18.6 kG).

The fact that the magnetostriction was large and positive during magnetization of bare samples indicates that a substantial amount of domain structure in the demagnetized state occupied transverse easy directions, i.e., directions other than the rolling direction;[2,7] during magnetization, a significant amount of this transverse structure was transferred by 90° wall motion to the direction of applied field. Thus, the decrease in magnetostriction produced by glass coating indicates that the amount of transverse domain structure in the demagnetized state was greatly reduced and that more of the magnetization process in glass coated samples occurred by 180° wall motion. Then, in order to explain the observed results, one must determine the mechanism(s) whereby low thermal expansion glass coatings reduce the amount of transverse domain structure. Residual compressive stresses are believed to produce transverse domain structure. The observed behavior could be explained, therefore, if glass coating reduced the amount of compressive residual stress present. Since the glass coatings were fired at 800°C, annealing effects necessary for the removal of unfavorable residual stresses may have occurred.

An additional improvement in properties was observed in both high and low permeability samples when the glass coated samples were cooled under a tensile load. It is interesting to note that this improvement also occurred in samples coated with a high thermal expansion glass (H), which by itself produced negligible improvement. Several investigators have shown that annealing 3% Si-Fe sheets under an external load can improve properties quite apart from glass coating effects.[8,9] The results of this experiment are consistent with this finding, although the manner in which this process introduces a uniaxial anisotropy along the tensile axis is not clear.

It is not surprising that the core loss of glass coated samples cooled under stress correlated well with their orientation. This was not true in the uncoated state. The fact that glass coating alone resulted in magnetostriction reduction in all samples, but only loss reduction in high permeability samples, merely reflects the fact that the losses of well oriented samples will be lower than poorly oriented samples in the absence of complicating residual stresses.

CONCLUSIONS

1. Low expansion glass coatings, which produce residual tensile stresses in the metal, substantially reduced the magnetostriction in the rolling direction of cube textured 3% Si-Fe by the elimination of 90° domain processes.
2. Low expansion glass coatings significantly reduced the rolling direction losses of cube textured samples with a high degress of directional orientation, but had a negligible effect on losses of samples with poorer directional orientation.
3. Cooling glass coated strips under an applied tensile load resulted in further decreases in loss and magnetostriction.

REFERENCES

1. M. F. Littmann, J. Appl. Phys. 38, 1104 (1967).
2. G. W. Wiener, K. Foster and D. S. Shull, J. Appl. Phys. 38, 1102 (1967).
3. J. L. Walter and H. C. Fiedler, U.S. Patent 3,034,935 (1962).
4. R. M. Frost, R. A. Larson, L. S. Myers, D. M. Pavlovic and G. L. Purdy, J. Appl. Phys. 42, 1798 (1971).
5. K. Foster and J. Seidel, AIP Conf. Proc. No. 5, 1514 (1971).
6. P. Banks and E. Rawlinson, Proc. IEE 114, 1537 (1967).
7. J. W. Shilling, IEEE Trans. Mag. M7, 557 (1971).
8. F. N. Dunayev, V. Druzhzhinin, N. Malev and T. Prasonva, Fiz. Met. Met. 20, 458 (1965).
9. A. Moses, S. Pegler and J. Thompson, Proc. IEE 119, 1222 (1972).

RELATIONSHIP BETWEEN STATISTICAL DISTRIBUTION OF
GRAIN ORIENTATIONS AND B_{10} IN POLYCRYSTALLINE (110)[001]
3% Si-Fe SHEET

W. M. Swift, W. T. Reynolds and J. W. Shilling
Westinghouse Research Laboratories, Pittsburgh, Pa. 15235

ABSTRACT

The distribution of angular deviations from perfect (110)[001] orientation was determined by the Schulz X-ray reflection method for eight polycrystalline 3% Si-Fe grain-oriented sheets as a function of B_{10} (dc induction at H = 10 Oe) between 16.4 and 19.3 kG. The angular deviations were equally and normally distributed with the root-mean-square angular deviation, ω_{rms}, increasing with decreasing B_{10}.

INTRODUCTION

The relationship between B_{10} and crystallographic orientation of polycrystalline (110)[001] grain-oriented 3% Si-Fe sheet has been studied using grain-by-grain orientation measurements made by etch pit and optical goniometer techniques.[1,2] One study[1] has shown that B_{10} is inversely proportional to the variance of the angular deviations ω_c and ω_n, i.e., rotation of the (110) plane about the sheet cross and normal directions, respectively, but not significantly dependent on the angular deviation ω_r (rotation of the (110) plane about the sheet rolling direction). The other study[2] showed that B_{10} was approximately inversely proportional to the average value of $(\omega_c + \omega_n)$ which is in accordance with the findings of McCarty et al.[1] However, no previous investigation of the relationship of orientation to induction in 3% Si-Fe has determined the distribution of grain orientations, correlated the statistics of the distribution with B_{10} and compared the results with theoretical models of magnetization. This investigation describes a method for measuring orientations of coarse-grained sheet, gives the results of the method for samples of 3% Si-Fe having a range of B_{10} values and discusses the observed correlation.

EXPERIMENTAL PROCEDURE

Epstein test pieces sheared from eight commercial 0.011 in. thick cold rolled, decarburized 3% Si-Fe strip were degreased, given various annealing treatments in dry hydrogen to obtain various degrees of (110)[001] textures and tested for B_{10} according to ASTM Method A139. Specimens of these test pieces were prepared for X-ray analysis by the Schulz reflection method in the following way. About 120 rectangular pieces having 1-1/2 inch length along the rolling direction by 1-1/8 inch width along the cross direction were sheared, deburred and stacked one upon another to a height of 1-1/4 inch. The pieces in each stack were firmly clamped and electron beam welded together using minimal penetration and heat affected zone. The stacks were sectioned by electro-discharge machining to provide laminated

X-ray specimens with normal directions either along the rolling direction or along the cross direction. Cutting marks were removed as necessary by grinding; disturbed metal was removed from all specimens by chemical polishing. Specimens prepared in this manner presented many grains to the X-ray beam. Since the area of the sample within the X-ray beam approaches a minimum (about 150 mm^2) as the position of the sample approaches that corresponding to the middle of its pole figure the minimum number of grains exposed to the beam can be calculated. In this study the minimum number of grains ranged from 270 to 50 since grain size ranged from 2 to 11 mm. Each specimen was scanned for (200) and (110) reflections by conventional apparatus used in the Schulz method.[3] Intensities of reflection were continuously recorded as angle between reflecting plane normal and the sample surface normal and the azimuth angle were changed at constant rates.

RESULTS AND DISCUSSION

Computer plotted pole figures and associated intensity histograms for (110)[001] secondary recrystallized 3% Si-Fe sheet having a B_{10} value of 18.5 kG are shown in Figs. 1 and 2. These pole figures and intensity histograms are representative of the results of the texture analyses obtained in this study. Figure 1 shows a (110) pole figure in which the specimen surface normal parallels the rolling direction. Contours represent deviation of {110} poles from the four positions corresponding to ideal {110}<001> texture. Deviation is expressed as rotation, ω_c, of {110} about the cross direction, CD, and rotation, ω_n, of {110} about the sheet normal, SN. For the same sample of steel, Fig. 2 shows a (110) pole figure in which the specimen surface normal parallels the cross direction. Here contours in the central region of the pole figure show deviation of {110} poles from ideal {110}<001> texture; this deviation is expressed as rotation, ω_n, and rotation, ω_r, of {110} about the rolling direction. Contours in both figures are symmetric which implies that ω_c, ω_n and ω_r are equally distributed.

Histograms in Figs. 1 and 2 show the fraction of intensity versus the polar angle, alpha, between (110) plane normals and the specimen normal for each 2.5° alpha x 360° azimuth (annular area) of their respective pole figures. These histograms represent the distribution of grain orientations in the specimen. A goodness of fit test revealed that the orientations are normally distributed at the 99% confidence level for both histograms. Furthermore, since variances of the distributions in Figs. 2 and 3 were $(49.2°)^2$ and $(52.2°)^2$ respectively, they did not differ at the 99% confidence level.

Grain orientation distributions of eight 3% Si-Fe samples having B_{10} values from 16.4 to 19.3 kG were analyzed in the foregoing manner. It was found in all the samples that the grain orientations were normally distributed about {110}<001> as a mean and had a standard deviation equal to the root-mean-square deviation, ω_{rms}. B_{10} values decrease as ω_{rms} values increase as shown in Table I and Fig. 3. These observations generally agree with the findings of McCarty et al.[1]

Domain models have previously been used to predict B_{10} values in 3% Si-Fe.[4,5,6] In these studies, B_{10} appeared to be in good agreement with an equation derived by Becker and Döring[7] for the magnetization occurring in single crystals at the knee of the magnetization curve, or

$$B_{10} = \frac{B_s}{\alpha_1+\alpha_2+\alpha_3} \quad , \tag{1}$$

where B_s is the saturation magnetization and $\alpha_1, \alpha_2, \alpha_3$ are the direction cosines of the three easy directions closest to the applied field. These studies were in poor agreement with the assumption that domains at 10 Oe were along the easy direction closest to the applied field, in which case B_{10} in a single crystal would be given by,

$$B_{10} = \alpha_1 B_s \quad . \tag{2}$$

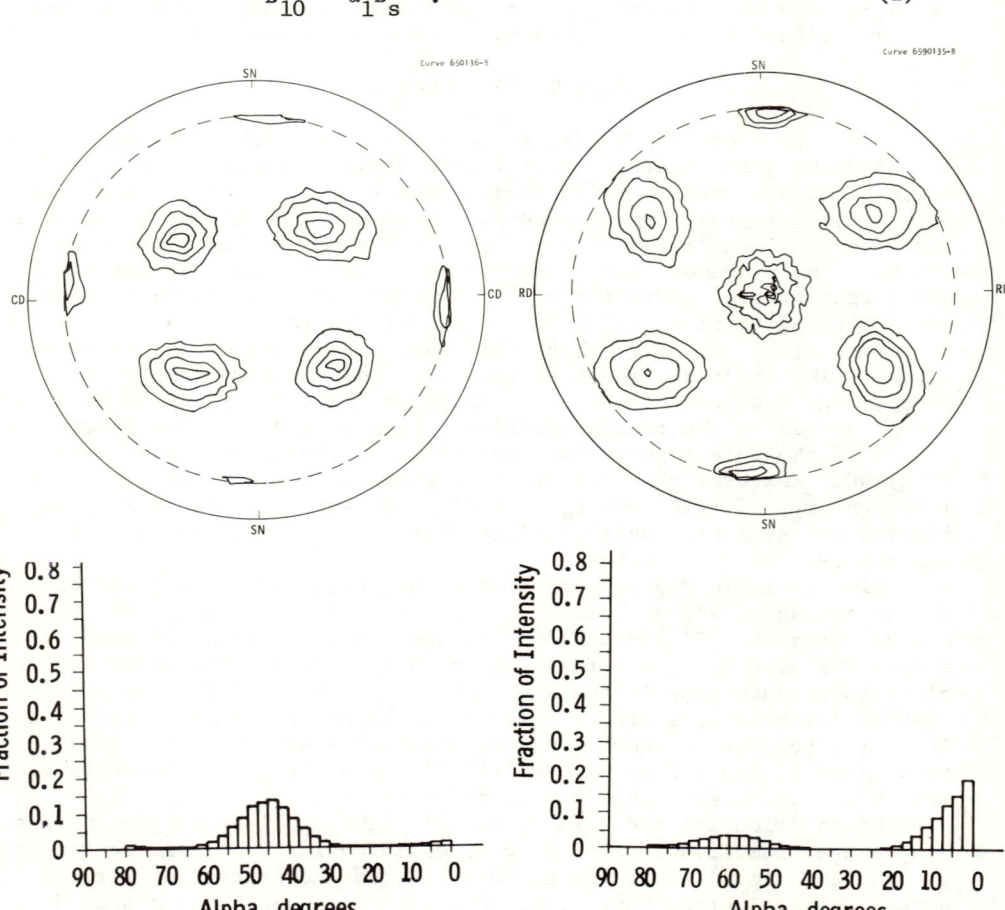

Fig. 1 — (110) pole figure and intensity histogram for 3% Si-Fe with B_{10}=18.5 kG. Specimen normal parallels the sheet rolling direction.

Fig. 2 — (110) pole figure and intensity histogram for 3% Si-Fe with B_{10}=18.5 kG. Specimen normal parallels the sheet cross direction.

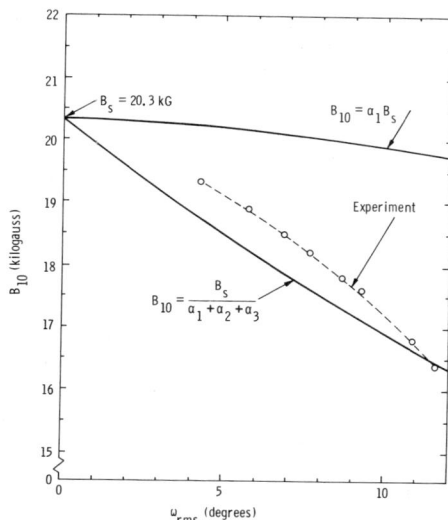

Fig. 3 - Relationships between B_{10} and root-mean-square angular deviation, ω_{rms}, from perfect (110)[001] orientation in polycrystalline grain-oriented 3% Si-Fe.

These equations are difficult to apply to polycrystals since grains cannot magnetize independently (c.f. ref. (5)). Nevertheless, Eqs. (1) and (2) have been applied to polycrystals using average values of ω_r, ω_n, and ω_c in the expressions for the direction cosines. Such a calculation assumes that all grains magnetize to the same induction, and that the induction is determined by the average ω_r, ω_n and ω_c of the sample.[4] In the present study, an attempt was made to calculate a statistical average of B_{10} for a given sample using the ω_r, ω_n and ω_c distributions and relations for B_{10} in terms of the direction cosines given by Eqs. (1) and (2). In this calculation, it was assumed that ω_r, ω_n, and ω_c were not correlated which is in accord with the pole figure data. This calculation was in poorer agreement with the data than when average values of ω_n, ω_r, and ω_c were used in Eqs. (1) and (2) and thus was evidently a less reasonable extension of Eqs. (1) and (2) to the behavior of polycrystals. Implicit in this calculation was the assumption that grains magnetized to different inductions depending on their particular values of ω_r, ω_n and ω_c which presumably resulted in poor agreement with experiment.

In Fig. 3, the results of the present experiment are compared to B_{10} values calculated using Eqs. (1) and (2). Average values of ω_{rms} from Table I were used in constructing the experimental curve. The direction cosines used in Eqs. (1) and (2) for a particular value of ω_{rms} were calculated using average values of ω_n, ω_r and ω_c assuming a normal distribution.

Values of B_{10} calculated using average values of ω_r, ω_n and ω_c in Eqs. (1) and (2) were not in good agreement with experiment, especially for small values of ω_{rms}. Eq. (1) was a much better approximation to the data than Eq. (2), although the two equations seemed more to represent upper and lower bounds than accurate predictions for B_{10}. Certainly all domains cannot be parallel to the closest easy direction of magnetization since large magnetostatic energies would exist and thus agreement with Eq. (2) was not expected. B_{10} values larger than that predicted by Eq. (1) would occur if domains rotated out of easy directions toward the direction of applied field. An applied field of 10 Oe was significantly beyond the knee of the magnetization curve in all samples, and thus the internal field was not zero and some rotation was expected.[8] Since any experimental errors in fabricating or aligning the pole figure sample lead to an increase in ω_{rms}, measured values of ω_{rms} were slightly larger than actual values. This effect would tend to bring the data into better agreement with Eq. (1). We have estimated this error to be at most 1° and more likely ~1/2° in which case measured values of B_{10} were still significantly in excess of values calculated using Eq. (1). Thus it appears that B_{10} is not in good agreement with either Eq. (1) or Eq. (2). Rotation, experimental error, or the inability to apply Eq. (1) to polycrystals may all be responsible in part.

CONCLUSIONS

1. Angular deviations from perfect (110)[001] orientation in polycrystalline (110)[001] 3% Si-Fe sheet are equally and normally distributed.
2. The root-mean-square angular deviation from perfect (110)[001] orientation in polycrystalline 3% Si-Fe increases with decreasing B_{10} over the range of B_{10} values from 16.4 to 19.3 kG.
3. No simple domain theory accurately predicts the observed dependence of B_{10} on angular deviation from perfect (110)[001] orientation in polycrystalline 3% Si-Fe.

REFERENCES

1. M. McCarty, G. L. Houze, and F. A. Malarari, J. Appl. Phys. 38, 1096 (1967).
2. M. F. Littman, J. Appl. Phys. 38, 1104 (1967).
3. L. G. Schulz, J. Appl. Phys. 20, 1030 (1949).
4. K. Foster and J. J. Kramer, J. Appl. Phys. 31S, 233 (1960).
5. J. W. Shilling, IEEE Trans. Mag. M-7, 557 (1971).
6. D. J. Craik and D. A. McIntyre, IEEE Trans. Mag. M-5, 378 (1969).
7. R. Becker and W. Döring, Ferromagnetism, Berlin: Springer, 119 (1939).
8. H. Lawton and K. Stewart, Proc. Roy. Soc. A193, 72 (1948).

ACKNOWLEDGMENT

The authors acknowledge Dr. D. R. Thornburg for his assistance with computer programming.

Table I   Experimental Root-Mean-Square Angular Deviation in (110)[001] 3% Si-Fe as a Function of B_{10}

B_{10} (kG)	ω_{rms} (200)Scan||RD	ω_{rms} (110)Scan||RD	ω_{rms} (110)Scan||CD	ω_{rms} (110)Scan||SN
16.4	11.4°	11.6°	11.8°	11.8°
16.8	10.7°	10.9°	11.2°	--
17.6	8.76°	9.15°	9.90°	8.81°
17.8	8.65°	8.67°	8.83°	--
18.2	7.77°	7.69°	7.70°	--
18.5	6.73°	6.64°	7.25°	--
18.9	5.61°	5.73°	6.01°	--
19.3	4.05°	4.41°	4.50°	--

MAGNETOSTRICTION AND MICROMAGNETISM: MAGNETIZATION REVERSAL OF CYLINDRICAL PARTICLES AND DOMAIN NUCLEATION IN UNIAXIAL PLATES*

M. W. Muller and R. W. Patterson**
Department of Electrical Engineering
Washington University, Saint Louis, Missouri 63130

ABSTRACT

We extend micromagnetic nucleation theory by removing the customary constraint of rigidity. The magnetoelastic interaction is added to the magnetic free energy as a perturbation and it is evaluated for various competing nucleation patterns or their linear combinations to determine the pattern yielding the lowest energy.

We apply this technique to the initiation of magnetization reversal in long cylindrical particles and we compute the reduction of the nucleation field for curling and buckling as a function of particle radius. We find that the radial ranges over which curling and buckling are favored are nearly unaffected by magnetostriction if the particles are free, but that buckling is dominant over a slightly extended range of radii for particles imbedded in a matrix, as in a permanent magnet.

We determine the nucleating domain pattern of a platelet with easy axis perpendicular to the faces, initially saturated by a large in-plane field. The domain nucleation threshold depends on the thickness, anisotropy, and magnetoelastic coupling of the plate. The nucleation mode has incipient stripe domains parallel to the field (the mode always predicted in the absence of magnetostriction) in thin plates, perpendicular to the field in thick plates, and inclined with respect to the field in the intermediate thickness range.

INTRODUCTION

Micromagnetostatics attempts to find the configurations of a magnetized body that reduce its free energy to a minimum. When nonuniform distributions of the magnetization must be considered, this is generally too difficult a task to carry out even for the standard classical (continuum rather than atomistic) model, and two approximations are commonly used:
1. It is assumed that the magnitude of the magnetization is independent of position;
2. The body is taken to be rigid.

These approximations simplify the energy minimization, since every constrained parameter is removed from the variational procedure.

---

* Work supported by the National Science Foundation under grants number GK-15701 and GH-32000
**Present Address: Westinghouse Research Laboratories, Pittsburgh, Pennsylvania 15235

In the present work we retain the first constraint (which limits the validity of the micromagnetic description to temperatures well below the Curie temperature[1]) but we permit the body to deform. In order to avoid excessive mathematical complications, we treat the magnetoelastic interaction as a perturbation; that is to say, we first minimize the energy with respect to the distribution of the magnetization subject to the constraint of rigidity, and we then correct the rigid-body solutions for magnetostrictive effects. In applying this technique to the description of magnetization reversal and domain nucleation phenomena we find that the relatively small magnetoelastic energy may in some instances discriminate among qualitatively dissimilar but energetically nearly degenerate states. Under these conditions magnetostriction may give rise to gross qualitative effects.

THEORY

The theory of domain nucleation or switching in a rigid ellipsoidal magnet seeks to determine the geometry of the possible infinitesimal deviations from uniform magnetization (modes) and the values of the external field at which the establishment of each mode becomes energetically favorable. The mode encountered first in traversing a magnetization cycle (normally that associated with the largest - or least negative - applied field) is assumed to initiate domain formation or switching.

If we now remove the constraint of rigidity and assume that this does not affect the magnetization, then the elastic deformation produced by the magnetostrictive forces (which we will call the magnetoelastic relaxation) can only decrease the energy associated with the nucleation of any mode. Such a change in the energy of a mode will manifest itself experimentally as a corresponding change in the threshold field for nucleation of the mode. In some instances this change of threshold will not differ much from mode to mode; this need not, however, be true in general. Differences may arise because the deformation that can be produced by a given distribution of forces depends not only upon the magnitude of these forces but also upon their geometry. Thus for example it was pointed out by Hubert[2] in discussing the magnetoelastic relaxation of specimens with domain structure that strains can develop only if their tangential components are continuous across domain walls. Because of this requirement of continuity of displacements (or compatibility conditions upon the strains) the degree of magnetoelastic relaxation and hence the change in nucleation field may differ significantly among the various modes.

Thus the magnetoelastic interaction may, in first order, give rise to any of the following effects:
1. A small change in the threshold field of the nucleation mode, without any change in the energetic ordering of the modes;
2. "Mode crossing": a sufficiently large differential shift in threshold field to change the nucleation mode;

3. Mode mixing: creation of a new nucleation mode from a superposition of (nearly) degenerate unperturbed modes.

We have published elsewhere[3] details of the calculation sketched here. In the present note we give only the results of two applications of the theory.

MAGNETIZATION REVERSAL OF CYLINDRICAL PARTICLES

The micromagnetic modes initiating magnetization reversal of a long cylindrical particle in an applied field parallel to the axis are magnetization curling and buckling[4]; magnetization buckling (which for very slender particles becomes indistinguishable from coherent rotation) is favored when the particle radius is smaller than a critical value R_c, curling when it is larger.

If the crystal structure of the particle has uniaxial symmetry, with the principal axis along the cylinder axis, then these modes will generate magnetostrictive forces that produce deformations as follows: buckling produces a shear strain that imparts a periodic lateral displacement to the cylinder axis, leaving the cross section unchanged; curling produces a helical twist that also leaves the cross section unchanged but progressively rotates successive cross sections. The magnetoelastic relaxation associated with the (approximate but very accurate) buckling mode is complete; that is, each volume element can deform as if it were free. The magnetoelastic relaxation associated with curling is not quite complete; the radial dependence of the strain cannot match that of the magnetization optimally, but it does achieve 98% of the maximum possible energy reduction. Thus for free particles magnetostriction affects the curling and buckling modes almost equally, and produces practically no change in R_c. The change in the nucleation field is independent of radius and, in terms of the quantity D defined below, is given by $\Delta H_c \cong 2\pi MD$ for both curling and buckling. Its direction is such as to make H_c less negative; thus, in the absence of magnetocrystalline anisotropy, it leads to the prediction of positive H_c for large radii.

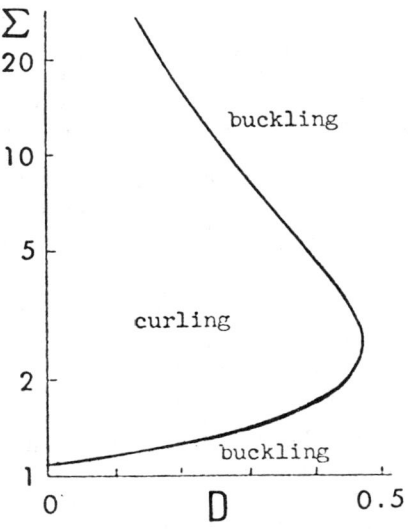

Figure 1. Critical radius for curling and buckling vs magnitude of the magnetoelastic coupling

If the particles are imbedded in a relatively rigid matrix as in a permanent magnet, we would expect the strain associated with buckling, which does not produce a cumulative displacement, to be affected very little. However the strain associated with curling, which for long particles would produce a large relative rotation of the ends, will be largely prevented from developing. Hence we expect the buckling mode to be favored under these conditions. For the results shown in Figure 1 it is assumed that the strain associated with buckling can develop fully, that associated with curling not at all. The quantities plotted are

$$\Sigma = R_c MC^{-1/2} \quad \text{versus} \quad D = 2b_{44}^2 (M^2 c_{44})^{-1}$$

where M is the saturation magnetization, C is the exchange constant, and c_{44} and b_{44} are elastic and magnetoelastic tensor components respectively. Typical values of D are of order 0.01 to 0.1 (D must be small compared with unity if the perturbation theory is to be valid). Thus the effect of magnetostriction on the critical radius is expected to be finite but quite small. There is another critical radius at large values of Σ, but the magnitude of the nucleation field is very small at large radii and the effect of imperfections (dislocations, etc.) is dominant.

DOMAIN NUCLEATION IN UNIAXIAL PLATES

If a rigid uniaxial platelet with easy axis perpendicular to the faces is saturated by a large in-plane field and the field is then reduced, an incipient domain pattern will nucleate. It was found long ago[5] that the micromagnetic equations are satisfied by incipient stripe domain patterns, and that the lowest energy mode has its stripes oriented parallel to the field. Incipient stripe domain patterns that are oriented differently have a higher energy. The energy difference arises from the dipolar term (flux closure is possible only for stripes parallel to the field); the difference in energy _density_ among modes differently oriented is a decreasing function of thickness.

When one examines the geometry of the magnetostrictive forces generated by the nucleation mode one finds[3] that the requirement of continuity of the displacements completely prevents any magnetoelastic relaxation. Some reduction of energy is possible, however, if the stripes are inclined with respect to the field, and complete relaxation (each volume element deforming as though free) is allowed for the mode with stripes normal to the field, which in the absence of magnetostriction has the highest energy. Thus the magnetoelastic energy density is a function of the orientation of the mode. It is not, however, a function of the plate thickness.

From these considerations we can predict qualitatively the effect of magnetostriction on domain nucleation: in thin plates the dipolar energy will be dominant and the stripes will nucleate parallel to the field; in thicker plates the minimum energy will be realized for stripes at increasing angles to the field, and beyond a certain thickness the magnetoelastic energy will be dominant and the stripes will nucleate perpendicular to the field. In the language of the theoretical discussion above, we expect to find mode crossing in thick specimens, and mode mixing (superposition of parallel and perpendicular stripes to form an inclined-stripe pattern) for intermediate thicknesses.

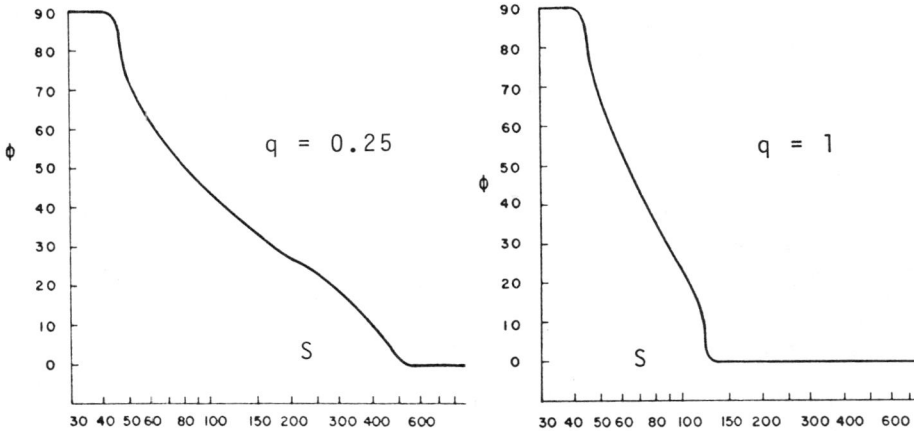

Figures 2 and 3. Orientation of nucleating stripe domains as a function of thickness.

These predictions are borne out by detailed calculations some of whose results are shown in Figures 2 and 3. In these figures we plot the angle ϕ between the normal to the stripes and the applied field against the dimensionless measure of thickness T

$$S = TM(2C)^{-1/2}$$

The figures are drawn for materials with a magnetoelastic coupling characterized by $D = .006$ and with two values of the magnetocrystalline anisotropy $q = K/2\pi M^2$, where K is the first order anisotropy constant. The curve for $q = 0.25$ is appropriate for hexagonal cobalt; that for $q = 1$ represents the lower limit (typically $q = 5$ to 10) for garnet materials used in bubble devices.

It will be observed in these figures that the transitional thickness range in which the nucleating pattern passes from parallel to perpendicular is narrower in the more anisotropic material. This effect can be readily interpreted in terms of the previously known result that the energy difference between the "parallel" and "perpendicular" modes is smaller in the more anisotropic materials. The small magnetoelastic energy is more effective in discriminating among these more nearly degenerate modes.

REFERENCES

1. N. Minnaja, Phys. Rev. B1, 1151 (1970)
2. A. Hubert, Phys. Stat. Sol. 22, 709 (1967)
3. R. W. Patterson and M. W. Muller, Int. Jl. Magnetism, in press
4. E. H. Frei, S. Shtrikman and D. Treves, Phys. Rev. 106, 446 (1957); A. Aharoni and S. Shtrikman, Phys. Rev. 109, 1522 (1958)
5. M. W. Muller, Phys. Rev. 122, 1485 (1961); W. F. Brown, Jr., Phys. Rev. 124, 1348 (1961)

SINGLE CRYSTAL MnBi PLATELET BY VACUUM DEPOSITION

Shigeo HONDA, Susumu KONISHI, and Tetsuzo KUSUDA
Department of Electronics, Hiroshima University
Hiroshima, Japan

ABSTRACT

The magnetic properties of single crystal MnBi platelet were studied with the aid of the Kerr effect. Other physical properties were analysed by the X-ray microanalyser and the differential interferometrical microscopy. The hysteresis loop has the reentrant characteristics and the coercive force is very small. The crystal appears to grow by a layer growth mechanism and the growth is strongly affected by the doped impurity.

INTRODUCTION

Ferromagnetic MnBi films were first prepared by Williams et al.[1] by vacuum evaporation. Many basic properties of this medium have been studied during the last decade. More recently, the utilization of the MnBi films for optical memory application has been investigated[2,3]. The magnetic easy axis of the films prepared by the evaporation technique is parallel to the crystalline c-axis and perpendicular to the plane of the film. Almost all the films reported hitherto consisted of fine crystallites.[3]

This paper reports the magnetic and some physical properties of single crystal MnBi platelets which have surprisingly different magnetic properties from those of the polycrystalline films.

PREPARATION

MnBi single crystal platelets were prepared by the conventional evaporation technique. The Bi layer was evaporated prior to the Mn layer on a glass substrate with room temperature.[3] Both Bi and Mn were evaporated at a rate of about 2 Å/sec under a pressure of 10^{-7} Torr. The double layers were annealed at 270°C for 6 hours under the pressure of 0.5 to 1 x 10^{-8} Torr.

The platelets prepared were too thick to allow adequate light transmission for observation by Faraday effect; therefore, the polar Kerr effect was used for the study of the magnetic domains. Figure 1 shows the single crystal platelet with the typical hexagonal shape and the beautiful stripe domains with about 1.0 μm-width[4,5] viewed by the polar Kerr effect at zero applied field.

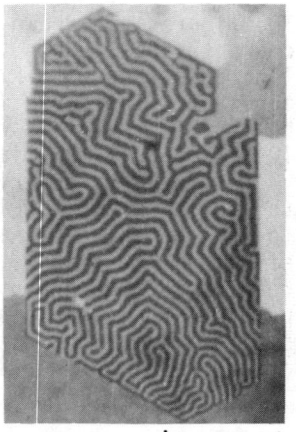

Fig. 1. Single crystal MnBi platelet with the hexagonal shape and stripe magnetic domains with about 1.0 μm-width.

Many hexagonal single crystal platelets with their c-axis oriented perpendicular to the film plane were found to have grown on the glass substrate. The thickness of the platelets is almost identical and is approximately 0.6 μm. The dimensions in the film plane are different from platelet to platelet, ranging from a few microns up to 100 μm in diameter.

PHYSICAL AND MAGNETIC PROPERTIES

Figure 2 shows the crystal grains and the magnetic domains viewed by the polar Kerr effect. The fine grains at the circumference are the fine MnBi crystallites, 0.2 μm thick and about 2 μm in diameter. The hexagonal platelet with the beautiful stripe domains

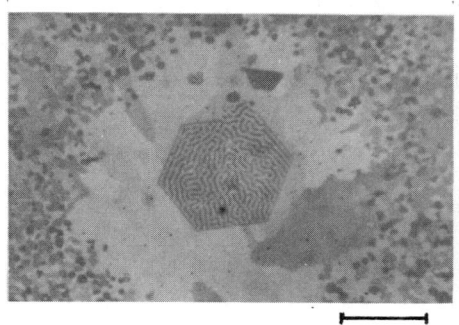

Fig. 2. Hexagonal single crystal platelet and fine crystallites of MnBi.

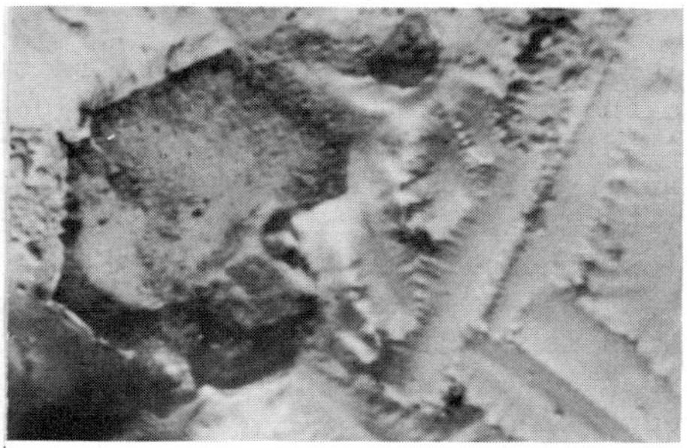

Fig. 3. Differential interferometrical image of the MnBi platelet and Bi crystals.

in the center is the MnBi single crystal, 0.6 μm thick and about 20 μm in diameter. The large "light and shade" pattern of the background is caused by the non-magnetic birefringent Bi crystal. Indeed, the pattern is not affected by the magnetic field and the X-ray microanalyses show that its region contains only Bi.[6] A common phenomenon is that the hexagonal platelet is found growing inside of the fine MnBi crystallites but it is magnetically isolated from the fine crystallites by the bismuth matrices as shown in Fig. 2.

Figure 3 shows the single crystal platelet and its vicinity observed by the differential interferometrical microscopy. In Fig. 3, the growth steps of the platelet and the flows of the bismuth in the vicinity are observed. It should be suggested from Fig. 3 and from the X-ray microanalyses[6] that the melted bismuth and the manganese runned to the crystal seed from the surrounding and then the thicker platelet was formed parallel to the substrate surface and grown along the c-axis by the layer growth mechanism during the annealing process, and also that the excess bismuth was left unreacted in the vicinity.

When the evaporated atomic ratio of Bi to Mn is unity or less, the vicinity of the hexagonal platelet is empty as shown in Fig. 4.

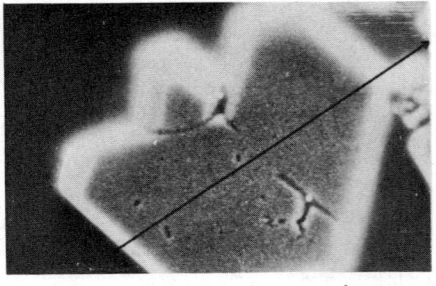
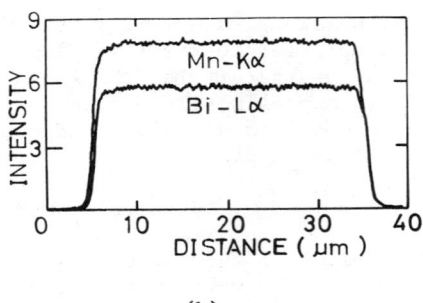

(a)   (b)

Fig. 4. X-ray microanalyses; (a) secondary electron image, (b) intensity distributions of Mn-Kα and Bi-Lα X-rays.

Fig. 4(a) shows an image pattern of secondary electrons of the hexagonal platelet and Fig. 4(b) shows the intensity distributions of Mn-Kα and Bi-Lα characteristic X-rays scanned along the line in Fig. 4(a). Note that the variation of Bi and Mn contents is parallel and that both Bi and Mn decrease abruptly at the grain boundaries.

The flux reversals were observed using high magnification Kerr apparatus. The hysteresis loop has the reentrant characteristics and the coercive force is small (Fig. 5). This sample reaches saturation at about 7.0 kOe (point A). Increasing the field further to 12 kOe (point B), the nucleation of the reverse domain

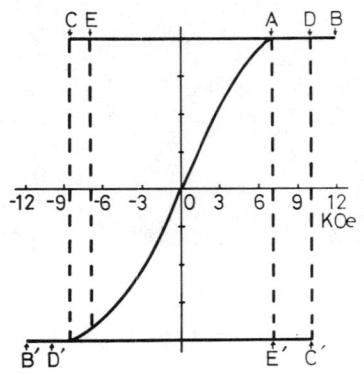

Fig. 5. Hysteresis loop in the MnBi platelet.

occurs at -8.5 kOe (point C). The flux reversal under this field strength is almost perfect, and the single domain state with opposite polarity was maintained when decreasing the field to zero. Again setting the field to 10 kOe (point D), the nucleation occurs at -7 kOe (point E). The flux reversal under this field strength is not perfect, and the stripe domains were formed when decreasing the field to zero.[6] Figure 5 shows that the hysteresis loop is unsymmetrical. Similar unsymmetric hysteresis loops were observed in $SmCo_5$[7]. The reason for the lack of symmetry is not known at present.

It is found from Fig. 5 that the nucleation field of the reverse domain varies according to the setting field strength.[7] Even though the same setting field is applied, the nucleation field is different from platelet to platelet. Figure 6 shows an example; when applying a field of -12 kOe to the virgin sample, two platelets (α and β) reached saturation as indicated by dark color as shown in (a). Subsequent application of a positive filed of 12 kOe causes only one platelet (α) to saturate in the reversed direction as indicated by

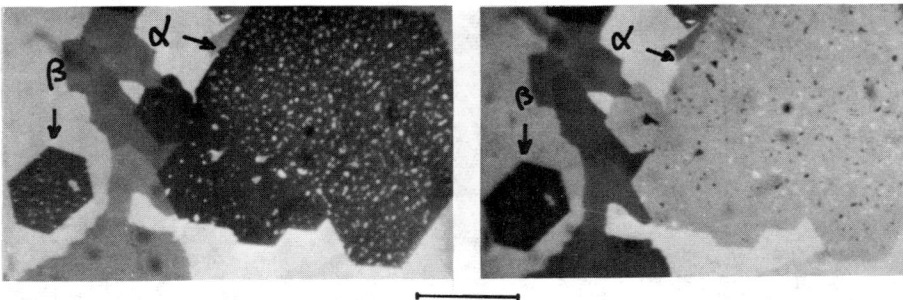

Fig. 6. Crystal dependence of the nucleation field; (a) domains at zero field after applying -12 kOe, (b) after applying +12 kOe.

the white color, whereas the other platelet (β) remains unchanged as shown in (b). These suggest that the nucleation field is sensitive to the imperfection of the crystal.

The reentrant characteristics of hysteresis loop and low value for coercive force show the peculiar writing characteristics on the platelet by laser pulse.[3] With no bias field, a spot nucleated by thermal writing continues to expand by wall motion and finally the platelet becomes demagnetized. This was demonstrated beautifully by laser-beam-writing as shown in Fig. 7; the black stripe shows a Curie-point switched area with the continuous domain.

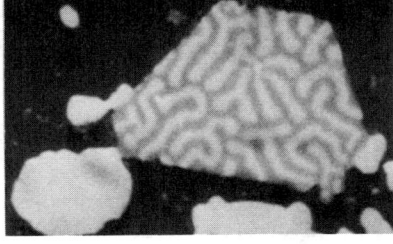

Fig. 7. Pattern of Curie-point-switching by the laser beam.

IMPURITY EFFECTS

The doped impurity effect on the crystal growth and on the magnetic properties of the single crystal platelets were studied. The samples were prepared by evaporating the bismuth, the impurity metal and the manganese, successively. Ten to twenty percents of the manganeses were substituted by the impurity metals. Figure 8(a) shows the typical "snow's pattern" observed by the differential interferometrical microscopy in the MnBi:Cr film. The X-ray microanalyses suggest that the "snow's pattern" consists of some hexagonal single crystal platelets of MnBi:Cr which are much thicker than the background with only a little Bi and Cr. Two different magnetic domain structures are observed in the "snow's pattern"; one is a labyrinth structure with uniform domain width as shown in Fig. 1 and the other consists of the essentially parallel domains of which the width is irregular as shown in Fig. 8(b). The latter may be caused by the decrease of the magnetic anisotropy energy due to the impurity metal Cr.

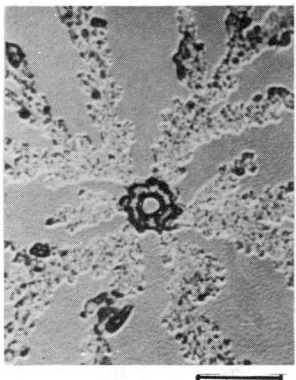 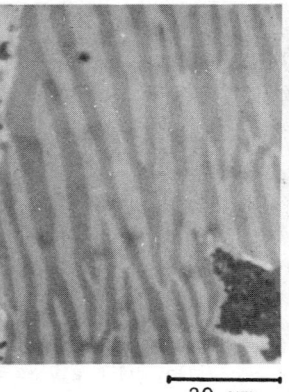 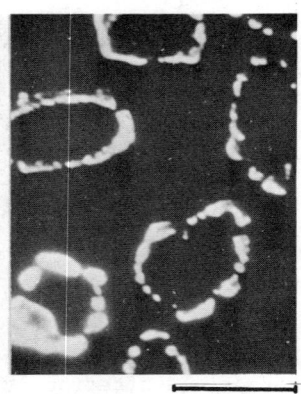

(a) 200 μm   (b) 20 μm   20 μm

Fig. 8. MnBi:Cr crystals; (a) differential interferometrical image, (b) magnetic domain pattern.

Fig. 9. Secondary electron image in the MnBi:Ti film.

The secondary electron image of the MnBi:Ti film is shown in Fig. 9.[8] The bright "doughnut structure", which is much thicker than the background with only a little Bi and Ti, consists of some hexagonal single crystal of MnBi:Ti. The "doughnut structure" crystal also has two domain patterns as shown in Fig. 1 and Fig. 8(b). These show that the doped impurities strongly affect on the growth mechanism and on the magnetic properties of the MnBi crystal.

Single crystal MnBi platelet shows beautiful stripe domain pattern. Its flux reversal takes place in essentially the same way as in platelets of other uniaxial materials, eg., cobalt, rare-earth compounds,[5] ferrites and orthoferrites. More detailed investigations for the physical properties of the MnBi single crystal will furnish further information for the magnetization processes in this hard magnetic materials. Efforts are being directed toward the investigation of wall structures of bubbles and also of Curie-point switching process in MnBi single crystals.

The authors would like to thank Mr. Yoshiteru Hosokawa for preparing the samples and Miss Noriko Chokai for editorial assistance.

REFERENCES

1. H. J. Williams, R. C. Sherwood, F. G. Foster and E. M. Kelley, J. Appl. Phys. 28, 1181 (1957).
2. D. Chen, J. Appl. Phys. 37, 1486 (1966).
3. W. K. Unger and R. Räth, IEEE Trans. Mag., MAG-7, 885 (1971).
4. C. Kittel, Phys. Rev. 70, 965 (1946).
5. Z. Málek and V. Kamberský, Czech. J. Phys. 8, 416 (1958).
6. S. Honda, S. Konishi and T. Kusuda, Appl. Phys. Lett. 21, 441 (1972).
7. J. J. Becker, IEEE Trans. Mag., MAG-7, 644 (1971).
8. W. K. Unger, E. Wolfgang, H. Harms and H. Haudek, J. Appl. Phys. 43, 2875 (1972).

ANISOTROPY OF DEMAGNETIZATION

R. Vergne, Z. Blazek,* and J.L. Porteseil
Laboratoire de Magnétisme du C.N.R.S.
B.P. 166 - 38042 GRENOBLE-Cedex - France

ABSTRACT

The authors have measured the magnetization of a polycrystal of an iron-like material in weak fields. They show that the Rayleigh coefficients a and b depend on the type of demagnetization used, and they give the angular dependence and field behaviour of a and b. A model is suggested to account for these results, which are discussed in terms of field evolution of several populations of Bloch walls.

INTRODUCTION

One generally demagnetizes a ferromagnetic sample by an alternating magnetic field, the amplitude of which slowly decreases to zero. Several authors[1-3] have shown that this leads to an anisotropic demagnetized state when the substance has several easy axes of magnetization. This effect disappears in a uniaxial substance.

When the magnetizing field is very weak compared to the coercive force H_c, the magnetization follows a law of the form given by Rayleigh[4]:

$$J = AH + BH^2 \qquad (1)$$

L. Néel[5] accounted for this law by showing that a single Bloch wall moving in a random potential follows such a law, and calculated the macroscopic constants A and B for a polycrystalline sample of a uniaxial crystal. We try to extend these results to the case of a cubic iron-like material.

EXPERIMENTAL RESULTS

Measurements were performed on a thin disc (ϕ = 60 mm, t = 1.2 mm) which was a transverse slice cut from a cylindrical rod. Its weight composition was: Fe 94.3%; Ni 4.3%; Cr 1.5%; C 0.3%. The sample was annealed by heating to 650°C and slowly cooled; its coercive force was 14.7 Oe.

The specimen was located inside two groups of coils which could be rotated, one of them providing the demagnetizing field $\mathcal{H}$ and the other the measuring field H. It was possible to make an "isotropic" demagnetization (denoted in what follows by the symbol ($\mathcal{V}$)) by rotating the sample at a constant angular speed in the field $\mathcal{H}$ as $\mathcal{H}$ decreases to zero.

The magnetization J was measured by an electronic fluxmeter[6]. For this disc, the Rayleigh law was obeyed to an accuracy of 0.1% between 7×10^{-2} Oe and 2.2 Oe, and it was found to be magnetically isotropic in the disc plane following a rotational demagnetization. In very weak

*Permanent address: SVUM
PRAGUE 1 - Opletalova 25, Czechoslovakia

magnetic fields, the magnetic thermofluctuation after-effect appears. We determined the macroscopic constants A and B by fitting the values of the susceptibility $\chi = \vec{J}/H = A + BH$ on a linear law by a least-square method. The results were reproducible to $\pm 0.1\%$ for A and $\pm 2.5\%$ for B. We found: $A = 4.405$ emu/Oe and $B = 0.112$ emu/Oe^2 for the "isotropic" demagnetized state.

1) Influence of the peak amplitude $\mathcal{H}_m$ of the demagnetizing field.

After an isotropic demagnetization we perform a linear demagnetization of peak amplitude $\mathcal{H}_m$ and measure in the direction of $\mathcal{H}$ the quantities $A_\parallel$ and $B_\parallel$. These quantities increase with $\mathcal{H}_m$ up to limiting values reached when $\mathcal{H}_m = \mathcal{H}_{kr} \sim 3H_c$ (Fig.1).

2) Influence of the direction of $\mathcal{H}$.

When $\mathcal{H}_m > 3H_c$ (large amplitude demagnetization) we found that A and B decrease when γ increases from 0 to $\pi/2$ where γ is the angle in the disc plane between H and $\mathcal{H}$. Figs.2 and 3 show the relative angular variations of A_γ and B_γ, where $A_\parallel$, $A_\perp$ refer to $\gamma = 0$, $\pi/2$ respectively.

3) Memory effects.

A variety of sequential demagnetizations were investigated, followed by measurements of the coefficients A,B. When $\mathcal{H}_m > 3H_c$, we found that the last demagnetization erases the effects of earlier ones. However, when $\mathcal{H}_m \lesssim H_c$, the demagnetizations are incomplete. Here the early demagnetization is remembered in part but the last demagnetization is most important.

THEORETICAL MODEL AND DISCUSSION

To account for these results, we made the following assumptions:

- the crystal axes are at random in the polycrystalline sample;
- in the crystal, there are two types of regions: in the regions of type I the features of the magnetic domains depend on the demagnetization, while they are independent of it in the regions of type II;
- the regions II give only isotropic macroscopic effects;
- in the regions I, there 180° and 90° Bloch walls;
- in the Rayleigh region, the magnetization processes are due to wall displacements only;
- these displacements are independent and the total area of the walls is constant[5].

Let $\vec{J}_1$ and $\vec{J}_2$ be the spontaneous magnetizations on opposite sides of a wall, and ΔV the volume swept by the moving wall. The magnetization $\Delta \vec{J}$ is:

$$\Delta \vec{J} = (\vec{J}_2 - \vec{J}_1) \Delta V \tag{2}$$

$\vec{J}_2 - \vec{J}_1$ is the polarization $\vec{P}$, considered for either type of wall.

Space limitations prevent a detailed elaboration[7]. In brief, we consider a cubic crystal whose easy axes are $\langle 100 \rangle$, making angles $\beta_i (i=1,2,3)$ with the direction of $\mathcal{H}$, where $\beta_1 < \beta_2 < \beta_3$. The probability distributions for a particular $\langle 100 \rangle$ to be the first, second or third angular neighbor of $\mathcal{H}$ are calculable. Six wall types are possible in each crystallite: 180° walls designated (1,-1), (2,-2) and (3,-3); and 90° walls designated (1,2), (1,3) and (2,3), where the indices refer to the angular neighbor relationship.

The contribution of the motion of these different walls to the total magnetization is anisotropic but calculable in the Rayleigh region. To these six anisotropic Rayleigh law expressions valid for regions of type I, we must add one isotropic expression for type II regions.

1) <u>Large amplitude demagnetizations</u>.

One often assumes in this case that the vectors J settle down along the easy axes which are the nearest neighbours of $\mathcal{H}$. So, we shall assume that the 180° and 90° walls in the regions I are of types (1,-1) and (1,2). The magnetization law is then the sum of 2 anisotropic laws and one isotropic. We identify the experimental coefficients A_γ and B_γ with weighted sums of the theoretical variations.

Fig.2 shows the <u>relative</u> variations of A_γ and the calculated curves for walls of types (1,-1) and (1,2), which variations are the same in this case. There is a slight disagreement ($\sim 3 \times 10^{-3}$ of the total magnetization) that we are presently trying to explain. On the other hand, the calculated variations of B_γ are different for the walls (1,-1) and (1,2). Figure 3 shows a comparison of these coefficients with experiment. We found that the measured curve is best described by the following "active" populations:

        10% 180° walls of the (1,-1) type
        90% 90° walls of the (1,2) type.

2) <u>Demagnetization with $\mathcal{H}_m \sim H_c$ or less</u>.

In this case the six anisotropic populations can exist. We can show that the evolution of A_γ and B_γ with $\mathcal{H}_m$ is consistent with our assumptions. The unfavourably oriented walls (2,-2) and (1,3) (type p_{II}) and (3,-3) and (2,3) (type p_{III}) progressively disappear and are replaced by (1,-1) and (1,2) walls (type p_I) (Fig.4).

CONCLUSIONS

The results quoted shows that one must take into account the anisotropy of demagnetization when studying a cubic material;

The suggested model provides background for the empirical recipe of reproducible demagnetization, i.e., a peak amplitude $\sim 3H_c$ or more, and for the assumption according to which J_s settles down on the axes close to the direction $\mathcal{H}$;

It seems that the 90° walls are most important in the reorientation processes.

REFERENCES

1. J.C. Barbier, B. Ferlin Guion, J. Appl. Phys. Supp. <u>33</u>, 1226 (1962).
2. J.C. Barbier, H. Ruby, R. Vergne, C.R. Acad. Sc. <u>260</u>, 3014 (1965).
3. J. Covo, Thèse de 3ème Cycle, Grenoble (1960).
4. Lord Rayleigh, Phil. Mag. <u>23</u>; 225 (1887).
5. L. Néel, Cahiers Phys. <u>12</u>, 1 (1942).
6. R. Vergne, J.L. Porteseil, Rev. Phys. Appl. <u>6</u>, 95 (1971).
7. Z. Blažek, Thèse, Grenoble (1972).

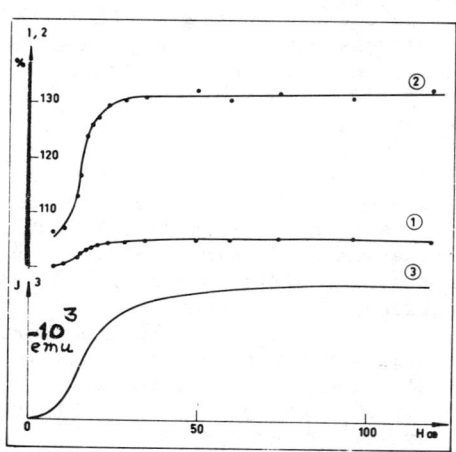
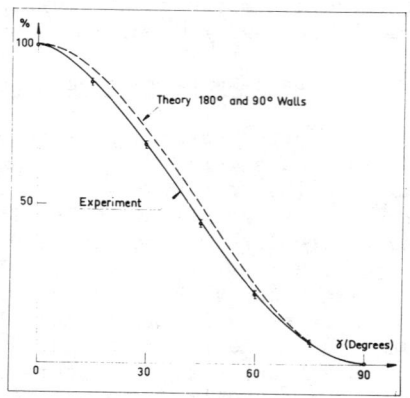

Fig.1 (1) Field dependence of normalized $A_\|$: $10^2 [A_\|(\mathcal{H}_m) - A(\mathcal{H}_{kr})] / A(\mathcal{H}_{kr})$; (2) field dependence of normalized $B_\|$ in same form; (3) total sample magnetization $\mathcal{J}(H)$.

Fig.2 Relative angular variation of A_γ for $\mathcal{H}_m \geq \mathcal{H}_{kr}$: $10^2 (A_\gamma - A_\perp)/(A_\| - A_\perp)$.

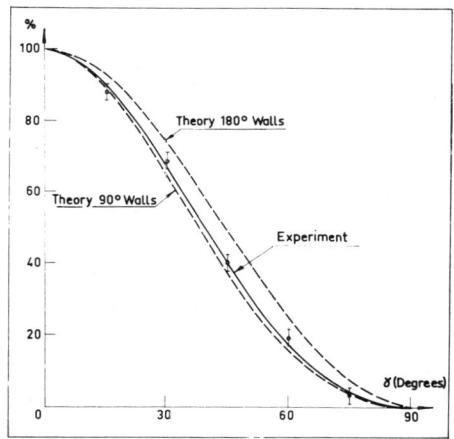

Fig.3 Relative angular variation of B_γ for $\mathcal{H}_m > 3H_c$ as in Fig.2.

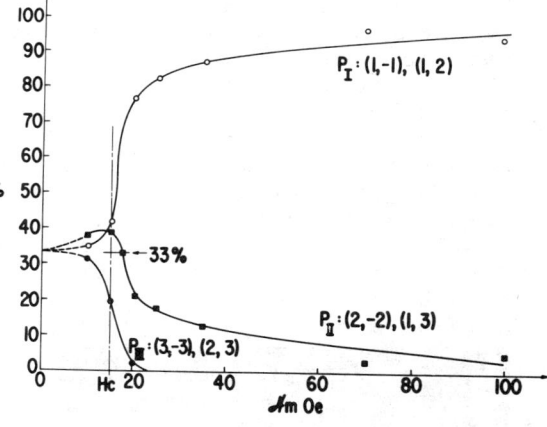

Fig.4 Evolution of the relative populations of various types of walls.

STABILITY CONDITIONS FOR NÉEL WALLS AND CROSS-TIE WALLS IN THIN MAGNETIC FILMS

L. J. Schwee
Naval Ordnance Laboratory, Silver Spring, Md. 20910

ABSTRACT

The stability of Néel walls and cross-tie walls has been studied in 80-20 Ni-Fe thin films between 200Å and 640Å. It was found that well defined, overlapping ranges of stability exist for Néel and cross-tie walls and that walls can be transformed from one type to another by the application of a magnetic field along the hard direction.

INTRODUCTION

The type of stability conditions required for a memory are found to exist in the walls of thin magnetic films. Knowledge of the stability conditions is also very helpful in understanding the peculiarities of creep.

EXPERIMENTAL PROCEDURE

Walls were placed on the film using a flat coil which had a bifilar winding. This flat coil was placed on the film and about an ampere was sent through the coil which was wound with wire 87 μm in diameter. The coil was skewed about 5° to the easy axis of the film. This caused the magnetization to rotate to 5° from the hard axis of the film but directed oppositely under each neighboring winding. When the current was shut off, the magnetization relaxed to the easy axis with walls placed as shown in Fig. 1. Using this technique walls could be placed at angles from 5° to about 45° from the easy axis. It was found that the walls at 5° to the easy axis behaved very much like those placed along the easy axis by another technique.[1] Since the bifilar coil method was much easier to use and the results were the same, the results that are given can be considered true for walls directed along the easy axis. Walls directed 45° to the easy axis have somewhat different stability ranges but behave in a similar fashion.

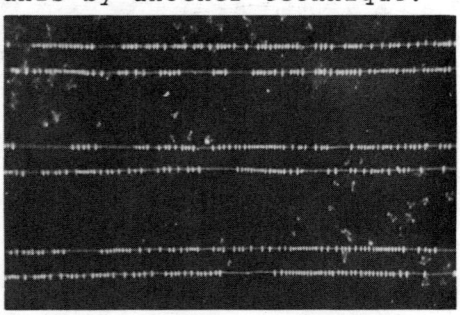

Fig. 1. Walls placed on film using a bifilar winding.

The walls were observed using a wet Bitter solution under a cover glass illuminated by two lights directed at oblique angles. One light was directed along the direction of the walls, the other perpendicular to the walls. A reversal of the polarity of a Néel wall caused a sudden change in the intensity of the light reflected from the domain wall. Using the other light cross-tie nucleation and annihilation could be observed. Whenever a section of wall inverted, a cross tie could be found at one end of the inverted section. Bloch-line motion could be inferred by observing wall reversal, and cross-tie movement by the change in location from one end of the newly inverted section to the other. A magnification of 200X was found to be convenient. The cross-tie model used by Middelhoek[2] was found sufficient to explain the results. However, the following analogy was helpful.

A HELPFUL ANALOGY

It is common for single domain reversal to occur by the following mechanism. First small reversal domains are nucleated, then they grow by domain wall motion until the remaining small domains of the original orientation pop and reversal is accomplished. In a similar manner, a Néel wall of one polarity reverses by the following mechanism. First inverted Néel wall sections which are bounded by a cross tie and Bloch line are nucleated. This can be called cross-tie wall nucleation. Then the inverted sections grow by Bloch line motion until the small sections of Néel wall of original polarity pop and reversal is complete.

The process is shown graphically in Fig. 2 for a 400Å thick film. The boxes represent field ranges over which the walls are stable as a function of field applied along the hard direction of the film. Suppose first that we have a field of $-.6\ H_k$ applied to the film. Then the wall must be a negative Néel wall as long as fields between $-.7\ H_k$ and $.12\ H_k$ are applied. Once a field larger than $.12\ H_k$ is applied the cross-tie wall will be nucleated and it is stable as shown. To obtain a positive Néel wall a field larger than $.4\ H_k$ must be applied. At this thickness with no applied field the wall can be

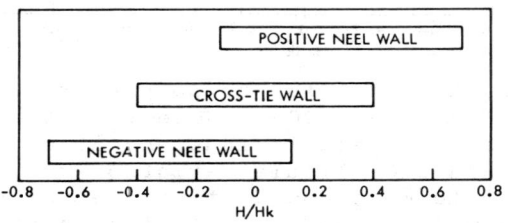

Fig. 2. Stability conditions for a 400Å thick film.

cross tie, positive Néel, or negative Néel depending on the magnitude of the last field applied and the state previous to the application of that field.

WALL STABILITY AND FILM THICKNESS

To study the wall stability as a function of thickness a film was chosen which varied in thickness from 200Å to 640Å along its 3-inch length. Thickness was measured using a light transmission technique described in the appendix.

The film anisotropy varied from 4 Oe at 200Å to 4.8 Oe at 640Å. This was taken into account in the normalization used in Fig. 3 where the field applied along the hard direction is normalized to H_k. Stability conditions for Néel and cross-tie walls as a function of thickness are plotted.

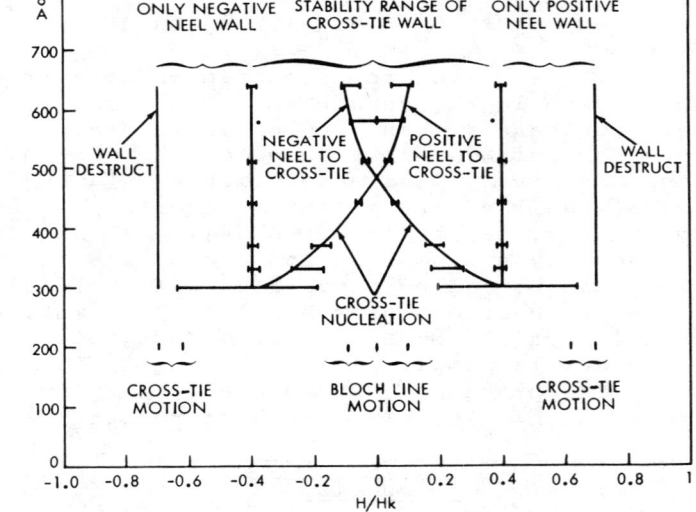

Fig. 3. Stability conditions as a function of thickness.

To understand Fig. 3 it is helpful to compare Fig. 2 with the values shown in Fig. 3 at 400Å. Note that the stability range for a cross-tie wall is independent of thickness. Between 300Å and 640Å little variation was seen in the field required to move and destroy the walls. Generally two neighboring walls would join to form a tip which would then propagate along the easy axis annihilating the two walls. The only parameter that varies with thickness is the cross-tie nucleation field. Above 480Å cross ties are always present at zero applied field. The repeated use of the Bitter solution may have increased the scatter in the data shown in Fig. 3. An attempt was made to draw the cross-tie nucleation field lines through points where nucleation seemed most probable.

Below 440Å it is possible for a negative Néel wall to invert by the motion of a Bloch line which can be generated at the film edge or at a defect. Below 400Å it became difficult to find the nucleation of cross-tie walls because of Bloch line motion. For example, if we had a negative Néel wall and the field was slowly changed from negative values to positive, at about 0.05 H_k several walls would reverse polarity by Bloch line motion. Only on those walls where such reversal did not occur could cross ties eventually be nucleated. And then it appeared that only cross ties that were simultaneously nucleated could be found. This results from the fact that cross-tie nucleation implies Bloch line generation and at these fields reversal by Bloch line motion is favored. In other words, as soon as a cross tie and Bloch line were generated the Bloch line would move close to the neighboring cross tie. Here Bloch line motion can be observed between cross ties. Occasionally they hang up on imperfections but very small additional fields send them along again.

At 300Å some walls would not reverse until fields in excess of the cross-tie stability range were applied. In this case a part of a wall would reverse and one cross tie could be found. It was stable only because it had no Bloch line available with which it could annihilate itself. By increasing the field the cross tie would sometimes move thus causing Néel wall reversal. Sometimes it would not move but the lower energy Néel wall would. The cross tie then ends up as the tip of a domain. This is the beginning of the double wall phenomenon found in thinner films.

Below 300Å the probability of nucleating a cross tie by the application of quasi-static, hard-axis fields is very low except at a defect. However, on walls placed with a field along the easy axis cross ties appear in films as thin as 120Å.

Results of measurements on other Ni-Fe films compared favorably with the data shown in Figure 3. When a small easy axis field was present creep was noticed to occur as a result of Bloch line and cross-tie motion and when a cross-tie wall is nucleated.

DYNAMIC BEHAVIOR

We found that cross ties could be created and annihilated using 2 nsec wide pulses along the hard axis. They also appeared to move but this was probably through annihilation and nucleation. Also in thicker films the cross ties became beautifully periodic when such pulses were applied. This also happens when a 60 Hz field with a peak amplitude of about 0.4 H_k is applied. The cross ties apparently creep to equidistant positions in films thicker than 500Å. No cross ties could be generated in films thinner than 300Å using the ac field except at defects where half-cross ties appeared which had a tie only on one side of the wall.

WALL ENERGIES

The thickness at which a cross-tie wall becomes energetically more favorable than a Néel wall with no field applied has long been disputed. It has been pointed out[3] that the Néel wall and cross-tie wall energies are relative minima separated by an energy barrier. This is confirmed by the data shown in Fig. 3. However, knowledge of the stability conditions does not give information on the relative depth of energy minima at zero field.

Because of the nucleation energy of the cross-tie wall and possibly an annihilation energy, it is difficult experimentally to pinpoint the thickness where the cross-tie wall becomes energetically favored at zero applied field. However, we can guess that this must occur at about 300Å.

APPENDIX

It has been found by H. R. Irons of this laboratory that film thickness could be determined very accurately and simply by measuring light transmitted through the film with a photodiode. The apparatus was calibrated using the Tolansky interferometer technique. The current through the photodiode is given by the equation

$$I = Ce^{-bt} \qquad (1)$$

where I is current, t is film thickness, and C and b are constants. An area of film 0.1 inch in diameter is used in the measurement.

ACKNOWLEDGMENT

The author wishes to thank H. R. Irons for supplying the films and for many helpful discussions.

REFERENCES

1. L. J. Schwee, IEEE Trans. Magn. MAG-8, (1972).
2. S. Middelhoek, Thesis, Univ. of Amsterdam, Holland (1961).
3. D. S. Lo and M. M. Hanson, IEEE Trans. Magn. MAG-5, 115 (1969).

STABILITY CONDITIONS FOR SAW-TOOTH WALLS BETWEEN HEAD-ON DOMAINS

N. Minnaja and M. Nobile

Honeywell Information Systems Italia, 20010 Pregnana Milanese, Italy

ABSTRACT

The boundary between head-on magnetized domains in a continuous ferromagnetic film takes the form of a saw-tooth wall, along which a distribution of magnetic poles is localized. The purpose of this paper is to derive theoretically the significant parameters of the sawtooth (angle and wavelength) from the properties of the material and the film thickness. This has been accomplished by minimization of the total energy per unit length, given by exchange, anisotropy, dipoles and poles; the last term is divergent, but it has been properly renormalized. With no applied field, for each wavelength the energy exhibits a minimum at one tip angle, but it is always a decreasing function of the wavelength. If, however, we take into account the fact that a wall can be displaced only if the effective field acting on it exceeds coercivity, we can find out a condition of frozen equilibrium which determines the stable wavelength also in zero field.

INTRODUCTION

Head-on magnetization patterns with saw-tooth domain boundaries have been reported by Curland and Speliotis[1] in recording experiments on thin metal films. Hsieh and Soohoo[2] have given an extended evidence of such a configuration and attempted to correlate the tip angle to the parameters of the material and its thickness. Their approach was essentially based on the minimization of the total energy of the wall vs. the tip angle, this energy resulting to be independent of the wavelength of the pattern; this last conclusion is an immediate consequence of the basic assumption, that the magnetostatic contribution to the energy can be evaluated by schematizing the wall as a dipole line. On the other hand the separation region between two head-on domains contains a distribution of poles, and taking into account their interactions changes significantly the total energy; it is easy to accept this conclusion, because the magnetostatic energy per unit length of a single pole line is divergent also in the case of finite width. First purpose of this paper is to show that this divergent contribution to the energy can be renormalized in a well defined way, and evaluate it. This evaluation will allow the minimization of the total energy per unit length in a proper way.

EVALUATION OF THE MAGNETOSTATIC ENERGY

Let us consider the simplest possible model of the transition between two head-on magnetized regions in an infinitely thin film (in the sense that its thickness, t, is much smaller than all other lengths to be met in the model); it consists of a wall with a uniform distribution of poles in an infinitely long strip of width 2a, the pole density value being M/a if M is the saturation magnetization. When we evaluate the magnetostatic energy per unit length of such a configuration as

$$W_M = -\frac{t}{2} \int \vec{M} \cdot \vec{H} \, dx \qquad (1)$$

x being a coordinate normal to the wall, we find out that W_M diverges; this divergence is removed only if we assume that a complementary distribution of poles exists somewhere in the film. A consistent way to insert this complementary distribution is to place it at distance L from the considered transition, L being considered a constant length much larger than all lengths to be met in the model besides the wall length, 2h. Accordingly we find out

$$W_M = 4 M^2 t^2 \left(\frac{3}{2} + \ln \frac{L}{2a} \right) \qquad (2)$$

Let us now consider the energy coming from a piece of wall of length b, which is bW_M. This energy can be split into three terms,

$$E_1 = b \cdot 4M^2 t^2 \left(\frac{1}{2} + \ln \frac{b}{a} \right) \qquad (t \ll a \ll b \ll L \ll h) \qquad (3)$$

$$E_2 = b \cdot 4M^2 t^2 \left(1 + \ln \frac{h}{b} \right) \qquad (4)$$

$$E_3 = b \cdot 4M^2 t^2 \cdot \ln \frac{L}{2h} \qquad (5)$$

E_1 is the self-energy of the poles in the considered piece of wall, E_2 is the interaction energy of the piece of wall with the residual wall, E_3 is the interaction with the "far" poles.

This splitting is particularly convenient for an extension to the case in which the transition region consists of a saw-tooth wall of width 2a, with wavelength 2p and with tip angle $\pi - 2\vartheta$ (see fig.1). Let us now consider the energy coming from a segment of wall of length $b = p/\cos\vartheta$, starting from the previous splitting. Retaining the same meaning for the subscripts, we obtain easily

$$E_1 = p \cdot 4M^2 t^2 \cos\vartheta \left(\frac{1}{2} + \ln \frac{p}{a \cos\vartheta} \right) \qquad (6)$$

$$E_3 = p \cdot 4M^2 t^2 \cdot \ln \frac{L}{2h} \qquad (7)$$

while E_2 must be evaluated by summing the interactions of the chosen segment with all others, with the result

$$E_2 = p \cdot 4M^2 t^2 \left[f(\vartheta) + \ln \frac{h}{p} \right] \qquad (8)$$

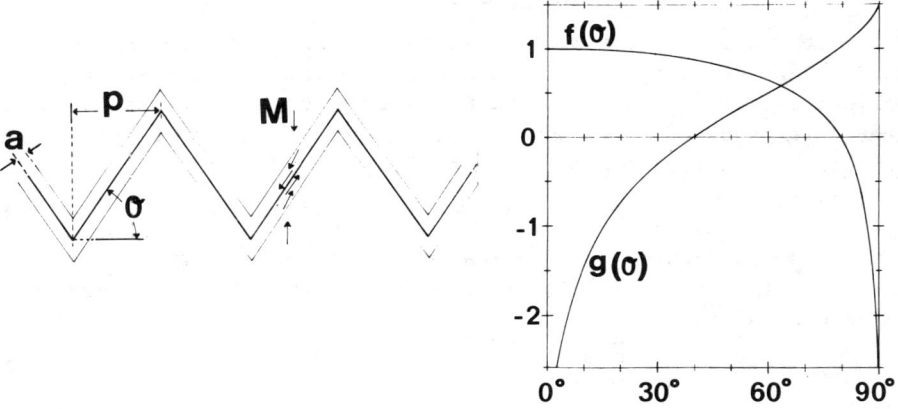

Fig.1: Model of the saw-tooth wall

Fig.2: Plot of $f(\vartheta)$ and $g(\vartheta)$ vs. ϑ; $g(\vartheta) = f(\vartheta) + \ln(1/2 \tan \vartheta)$.

where $f(\vartheta)$ has been evaluated numerically and is sketched in fig. 2. The magnetostatic energy per unit length (normal to the easy axis) takes the form

$$W_M = 4M^2 t^2 \left[f(\vartheta) + \frac{1}{2} \cos\vartheta + \ln\frac{L}{2p} + \cos\vartheta \ln\frac{p}{a \cos\vartheta} \right] \quad (9)$$

Note that the dependence on h has disappeared, whereas W_M depends on p if $\vartheta \neq 0$.

TOTAL ENERGY

The total energy of a segment of the saw-tooth wall also includes the contributions due to exchange, anisotropy and dipole interaction. A model which can be taken consists in considering a 180° wall, with its associated energy containing the mentioned contributions, in the middle of a transition region wide 2a (significantly larger than the wall itself) which contributes to the energy through its poles and its anisotropy[3] (see fig.1). If K is the anisotropy constant and σ is the 180° wall energy per unit length (parallel to the wall itself) we obtain for the total energy per unit length (normal to the easy axis)

$$W_T = W_M + Kat \cos\vartheta + \sigma t/\cos\vartheta \quad (10)$$

From Eqs.(9) and (10), after minimization with respect to a, it is obvious that W_T does not exhibit any minimum in plane (p,ϑ) because W_M is a decreasing function of p for $\vartheta \neq 0$.

On the other hand, if p is fixed due to some physical mechanism, it is possible to determine the value of ϑ yielding the lowest W_T.

FREEZING OF THE WAVELENGTH

Let us pay some attention to the mechanism giving origin to the head-on arrangement, assuming a physical situation similar to the experiments of Hsieh and Soohoo[2]. As a matter of fact we assume a linear dependence of an external field Ha vs. x, in a range of x larger than the transition region, such as

$$Ha = \gamma x \qquad (11)$$

Apart from an unessential constant coming from the region in which the saturation magnetization is uniformly attained, the expression of W_T is changed to

$$W_H = W_T + \frac{1}{3} \frac{Mt\gamma}{\cos^2 \vartheta} \left(\frac{p^2}{4} \sin^2 \vartheta + a^2 \right) \qquad (12)$$

For each value of $\gamma (\neq 0)$ W_H takes one minimum value for one definite set (a, p, ϑ). In particular the half-wavelength corresponding to this minimum is related to ϑ through

$$p = 2 \cos \vartheta \left(\frac{6 Mt}{1 + \cos \vartheta} \right)^{1/2} \qquad (13)$$

Let us assume now that we start with a very high γ and we reduce it towards zero. From the previous treatment we could expect that p and ϑ readjust themselves accordingly, so that W_H takes always its minimum value. A change of p, however, results in a displacement of the whole wall, and is therefore hindered, if the effective field acting on it does not exceed Hc, the coercivity of the material. We come therefore to the assumption, that p is frozen when the highest value of the component of the applied field parallel to the actual wall (which acts on the tip) coincides with the coercivity, namely that the value of p for any lower γ is determined by equation

$$\frac{Hc}{\sin \vartheta} = \frac{\gamma \cdot p \tan \vartheta}{2} \qquad (14)$$

which together with (13) yields the stable value

$$p = \frac{12 Mt}{Hc} \cos \varphi (1 - \cos \varphi) \qquad (15)$$

φ being the value of ϑ which minimizes W_H with respect to p and for the value of γ determined by (14).

For values of γ lower than this one, p has to be considered as a constant, given by (15), in the minimization of W_H. In particular, for zero applied field, we must minimize W_T with this value of p.

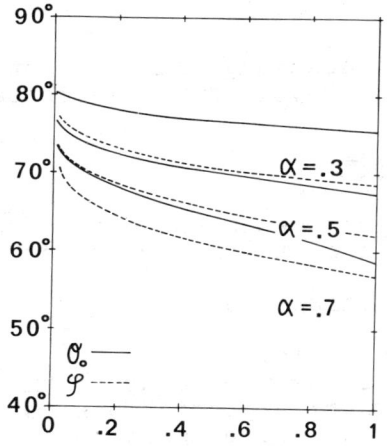
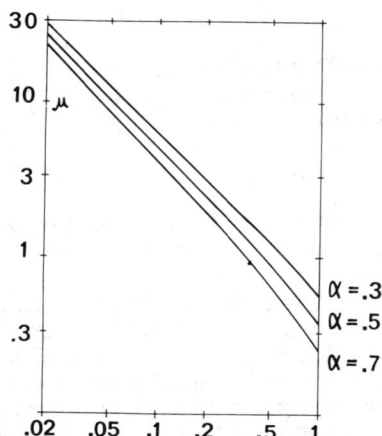

Fig.3: φ and ϑ_0 vs. H_C/H_K for some values of α

Fig.4: Width of the transition region= $p \tan \vartheta_0$ vs. H_C/H_K for some values of α

$$M = 800 \text{ emu}$$
$$t = 10 \text{ nm}$$
$$H_K = 200 \text{ oe}$$

An inspection of the form of all written equations shows that φ and the final value of ϑ for zero applied field, ϑ_0, depend only on two parameters describing the film, namely $H_C/H_K = MH_C/2K$ and $\alpha = \sigma/4M^2 t$. In fig.3 these dependences are plotted. On the other hand p is proportional to M and t and inversely proportional to H_K, and depends upon H_C/H_K and α as shown in fig.4.

REFERENCES

1. N. Curland and D.E. Speliotis, J. Appl. Phys. <u>41</u>, 1099 (1970)
2. E.J. Hsieh and R.F. Soohoo, AIP Conf. Proc. <u>5</u>, 727 (1971)
3. G.A. Jones and B.K. Middleton, Phys.Stat.Sol. (a) <u>3</u>, K259 (1970)

Section 30.   Domain Wall Behavior

MICROSTRUCTURE AND MICROMAGNETISM

H. Kronmüller

Max-Planck-Institut für Metallforschung and Institut für Theoretische und Angewandte Physik der Universität Stuttgart, 7 Stuttgart (BRD)

ABSTRACT

The interaction between lattice imperfections and ordered spin structures gives rise to the following effects:
1. Modification of the domain structure.
2. Pinning of the domain walls.
3. Reduction of the nucleation field.

The theoretical backgrounds for the treatment of the long range and short range magnetic interactions is developed within the framework of micromagnetism and of the Heisenberg model. Special topics to be considered are:
  a) The pinning of very narrow domain walls by point defects dislocations and antiphase boundaries.
  b) The pinning of bubble domains.
  c) The effect of dislocations on the nucleation field of high coercivity materials.

I. INTRODUCTION

The effect of lattice imperfections on the magnetic properties of ferromagnets has been studied extensively in the last twenty years [1,2,3]. It is a well known fact that lattice imperfections may influence significantly the characteristic properties of the hysteresis loop. This statement is applicable to single domain particles as well as to macroscopic specimens which show a domain structure. Which characteristic properties of the hysteresis loop are influenced markedly, depends on the type of lattice defects present as well as on the material properties such as the crystalline anisotropy, the exchange constant and the magnetostriction. Usually, a number of different types of lattice defects must be considered which are classified as follows:

1. Planar defects: Grainboundaries, stacking faults, antiphase boundaries in ordered alloys.
2. Line defects: Dislocations.
3. Point defects: Interstitials, vacancies, vacancy clusters, substitutional or interstitial impurity atoms.

4. Atomic disorder: Incompletely ordered intermetallic compounds.

A theoretical treatment of the interaction between the spontaneous magnetization and lattice defects is based on the continuum theory of micromagnetism [4,5,6] and the Heisenberg description. Continuum theory may be applied if the defect in question produces a long range perturbation, as, e.g., the dislocation, the stress field of which decreases according to $1/r$. Defects which are related to the atomic order, e.g., stacking faults and antiphase boundaries, must be treated by the Heisenberg model. Furthermore, the Heisenberg model must be used in the case of extremely hard magnetic materials where the change in the spin orientation takes place within a distance of a few lattice constants. The parameter governing the problem of extremely hard magnetic materials is given by the exchange length

$$\delta_o = (A/K_1)^{1/2}, \qquad (1)$$

where A is the exchange length and K_1 is the crystalline anisotropy constant. The parameter δ_o in soft magnetic materials usually has a value in the range of 10^2 to 10^4 Å whereas in hard magnetic materials such as Dy, Co_5Sm or Co, δ_o generally falls in the range of 5 to 50 Å. In the second type of materials, the crystalline anisotropy constant K_1 is of the order of magnitude $10^7 - 10^8$ erg/cm^3.

It is intended within this paper to present a brief review on the different types of interactions between lattice defects and the ferromagnetic state. In particular, the basic theoretical ideas which are necessary in studying these types of interactions will be outlined and applied to some problems of current interest. Special topics to be treated are the role of lattice defects in affecting the magnetic hardening particularly in the case of narrow domain walls. Also in the case of bubble domains we wish to evaluate that portion of the coercive field which is due to dislocations.

II. MICROMAGNETISM OF THE IMPERFECT CRYSTAL

II.1. General aspects

The effect of lattice defects on magnetic properties usually can be described by the following properties:
1. The difference

$$\Delta \phi_t(\vec{H}_{ext}) = \phi_t - \phi_o \qquad (2)$$

between the magnetic Gibbs free energy, ϕ_t of the crystal with defects and the Gibbs free energy, ϕ_0, of the ideal crystal.[+] $\Delta\phi(\vec{H}_{ext})$ depends on the applied field, $\vec{H}_{ext}$. The anisotropic part of $\Delta\phi_t(\vec{H}_{ext})$ defines an induced anisotropy which can be determined from torsion experiments.

2. The change of the magnetic moment parallel to the applied field

$$\Delta M_z(\vec{H}_{ext}) = \int (J_s(\vec{r},\vec{H}_{ext}) - J_z(\vec{r},\vec{H}_{ext}))d^3r \qquad (3)$$

describes the inhomogeneity of the spontaneous magnetization, $\vec{J}_s(\vec{r},\vec{H}_{ext})$, due to the lattice defects. In the approach to ferromagnetic saturation for example the suitable quantity measured is the susceptibility

$$\chi = \frac{d(\Delta M_z(\vec{H}_{ext}))}{d\,H_{ext}}$$

which is directly related to the effect of perturbations.

3. At small magnetic fields the magnetization curve $J(\vec{H})$, of a demagnetized specimen obeys the Rayleigh law

$$J = \chi_0 H + \alpha H^2 , \qquad (4)$$

in the field range $0 < H < H_0$, where H_0 is of the order of magnitude $0.5\,H_c$. In general the values of the initial susceptibility, χ_0, the Rayleigh constant, α, the field, H_0, and the coercive field, H_c, all depend sensitively on the lattice defects present.

II.2. The magnetic free enthalpy

Any theoretical treatment of the interaction between lattice defects and the spontaneous magnetization is based on the magnetic part of the free enthalpy. Magnetic energy terms contributing to the Gibbs free energy are the exchange energy ϕ_{Ex}, the magnetocrystalline energy ϕ_K, and the magnetostatic dipole energies ϕ_S and ϕ_{Hext}, of the stray field, H_S, and the external field respectively. When lattice defects are present in the ferromagnet all of the above cited energy terms are affected. Furthermore, if the lattice defects are centers of stress, in addition the elastic energy ϕ_{El} must be taken into account. The total

[+] For definition of ϕ_t see eq. (5) to eq. (7).

magnetic Gibbs free energy then may be written as the sum of five terms:

$$\phi_t = \phi_{Ex} + \phi_K + \phi_{El} + \phi_S + \phi_{H_{ext}} . \qquad (5)$$

Since a lattice defect influences in general all of these energy terms, it is not useful to discuss the coupling mechanisms in terms of these individual energies. A more general concept is based on a division of the different types of interactions into long range interactions and short range interactions. In principle all defects with long range stress fields lead to a long range interaction, whereas those defects which produce a perturbation within atomic dimensions lead to short range interactions. Also there may exist defects which produce both short range as well as long range interactions. All of these interaction types will be now discussed in detail

a) Long range interactions

1. Straight dislocation lines have a long range stress field decreasing according to $1/r$. Due to the magnetoelastic coupling energy the spontaneous magnetization becomes inhomogeneous, i.e., ϕ_{Ex}, ϕ_K and ϕ_S are increased.
2. Straight dislocation dipoles and point defects have stress fields decreasing according to $1/r^2$ and $1/r^3$; thus leading also to long range interactions.
3. Nonmagnetic inclusions or small precipitates with magnetic moment other than the matrix are surrounded by a magnetic stray field which in general is inhomogeneous, and gives rise to an increase of the stray field energy as well as the exchange energy.

b) Short range interactions

1. Point defects change locally the exchange integrals and the spin orbit coupling. Thus the origin of the perturbation is of atomic dimensions, whereas the perturbation in the spin structure may extend over several hundred Ångströms due to the coupling between the spins by the exchange energy.
2. In the center of dislocations it is suggested that the exchange integrals as well as the spin orbit coupling are changed significantly; thus leading to a short range interaction.
3. Planar defects such as stacking faults and antiphase boundaries may be described by local changes of the exchange integrals and the spin orbit coupling constant.

II.3. Theoretical treatment of magnetoelastic long range interactions

The effect of elastic stresses on the spontaneous magnetization can be treated by the continuum approach of micromagnetism. In this approximation the total Gibbs free energy may be written as [5]

$$\phi_t = \int F' d^3r - \frac{1}{2} \int \vec{H}_s \cdot \vec{J}_s \, d^3r - \int \vec{H}_{ext} \cdot \vec{J}_s \, d^3r , \qquad (6)$$

where the last two terms correspond to the magnetostatic energies of the stray field and the external field. The free energy density F' as a function of the direction consines γ_i of $\vec{J}_s$ is given by [3], +

$$F' = A(\nabla \gamma_i)^2 + \phi_K(\gamma_i) - \overset{\Rightarrow}{e^M} \cdot \overset{\Rightarrow}{\sigma}^{(i)} + \frac{1}{2} \overset{\Rightarrow}{e^E} \cdot \overset{\Rightarrow}{c} \cdot \overset{\Rightarrow}{e^E} . \qquad (7)$$

In eq. (7) the first two terms correspond to the exchange energy and the magnetocrystalline energy of the undeformed crystal. The third term is known as the magnetoelastic coupling energy [7] which describes the interaction between the spontaneous magnetostrictive strains e^M and the internal stresses $\sigma^{(i)}$ of the defects. The fourth term corresponds to the elastic energy associated with the magnetostrictive strains e^M [3]. The elastic energy of each individual defect within the framework of this approximation is assumed to be independent of the spontaneous magnetization and therefore is not included in eq. (7). In order to write eq. (7) in a more explicit form, further information concerning the magnetostrictive strains e^M and the elastic strains e^E are required. In general e^M, according to the relation

$$\overset{\Rightarrow}{e^M} = \overset{\Rightarrow}{e^K} + \overset{\Rightarrow}{e^{Ex}} , \qquad (8)$$

is composed of contributions from the strain dependence of $\phi_K(e^K)$ and of $\phi_{Ex}(e^{Ex})$. The spontaneous magnetostrictions e^{Ex} and e^K are derived from tensors of rank four according to

$$e^{Ex}_{ij} = \Lambda_{ijkl} \nabla \gamma_k \nabla \gamma_l \; ; \quad e^K_{ij} = \lambda_{ijkl} \gamma_k \gamma_l , \qquad (9)$$

where Λ_{ijkl} and λ_{ijkl} are the so-called magnetostriction tensors of the exchange and the magnetocrystalline energy. The strains e^{Ex} and e^K are quantitatively different from each other because the exchange magnetostriction becomes of importance only in extremely hard magnetic materials, where large gradients

+ In the following the double arrows refer to tensors.

in the spin directions can occur, e.g., within domain walls.

The elastic strains e^E which are due to inhomogeneous distributions of the spontaneous magnetostriction e^M can be determined from the compatibility condition [8,+]

$$\text{Ink}(\overline{\overline{e^E}} + \overline{\overline{e^M}}) \equiv \nabla \times (\overline{\overline{e^E}} + \overline{\overline{e^M}}) \times \nabla = 0, \quad (10a)$$

and the equilibrium condition

$$\text{Div}\,\overline{\overline{\sigma^E}} = 0. \quad (10b)$$

$\overline{\overline{\sigma^E}}$ corresponds to the elastic stress tensor

$$\overline{\overline{\sigma^E}} = \overline{\overline{c}} \cdot \overline{\overline{e^E}}. \quad (11)$$

Micromagnetic equilibrium conditions are obtained from the variational problem $\delta \phi_t = 0$. In a compact form they may be written as

$$[\vec{J}_s \times \vec{H}_{eff}] = 0 \quad (12)$$

for static problems. The effective field is defined as

$$\vec{H}_{eff} = -(2A/J_s^2)\Delta\vec{J}_s + \vec{H}_K + \vec{H}_S + \vec{H}_{ext} + \vec{H}_{El} + \vec{H}_{\sigma(i)}, \quad (13)$$

where

$$\vec{H}_K = -\nabla_{\vec{J}}\phi_K \,; \quad \vec{H}_{El} = -\nabla_{\vec{J}}(\tfrac{1}{2}\overline{\overline{e^E}} \cdot \overline{\overline{c}} \cdot \overline{\overline{e^E}})\,;$$

$$\vec{H}_{\sigma(i)} = -\nabla_{\vec{J}}\left\{\overline{\overline{e^M}} \cdot \overline{\overline{\sigma^{(i)}}}\right\}. \quad (14)$$

From eq. (12) the spin distribution can be determined for any given arrangement of internal stresses. It should be noted that eq. (12) has been used widely to study the effect of dislocations on both the saturation magnetization and the domain structure [9,10,11].

II.4. Theoretical treatment of the exchange and spin orbit short range interactions

Magnetic short range interactions can result from point defects or planar defects such as stacking faults or antiphase boundaries. The origin of the magnetic perturbation in these cases is the local change of both the exchange integrals and the spin orbit coupling. Fig. 1 represents the interaction of an impurity atom of spin S_f and with exchange integrals $J_{fi}(|\vec{R}_i - \vec{R}_f|)$ with the matrix atoms of spin S_i. The spin orbit coupling also will be changed so that the total magnetic energy of the substitutional atom will be given by

+ For definition of the operator Ink see Kröner [8].

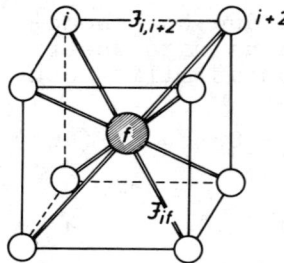

Fig. 1. Magnetic impurity atom in a bcc lattice.

$$\emptyset_f = -2\sum_i^{n.n} J_{fi}\vec{S}_f\cdot\vec{S}_i + \sum_i^{n.n}\lambda_i\vec{L}_i\cdot\vec{S}_i + \lambda_f\vec{L}_f\cdot\vec{S}_f \quad , \quad (15)$$

where λ_i and λ_f are the spin orbit coupling constants of the angular moments L_i and L_f of the neighbouring matrix atoms and the impurity atoms respectively. Eq. (15) also can be applied to defects of the host lattice itself. For example, in the case of a self-interstitial the index f would refer to the interstitial atom. For the vacancy defect case the index pair (i,f) must be replaced by the index pair (i,i+2), since the exchange couplings between next nearest neighbours are believed to be altered.

When the spin directions change slowly within the local neighbourhood of the point defect, the matrices (15) can be replaced by the direction cosines, γ_i, of the spontaneous magnetization. Within the framework of this continuum approach the energy of the defect can be written as

$$\emptyset_f = \emptyset^{(h)} + \sum_{m,n=1}^{3} C_{mn} \sum_{i=1}^{3} (\nabla_m\gamma_i)(\nabla_n\gamma_i) + W(\gamma_1,\gamma_2,\gamma_3), (16)$$

where the exchange matrix C_{mn} and $W(\gamma_i)$ must reveal the symmetry of the lattice defect. The energy term $\emptyset^{(h)}$ corresponds to the exchange energy of the defect in a homogeneous magnetization field. As an example let us consider the energy, $\emptyset_f$, of a point defect with tetragonal symmetry. The tetragonal axis of the defect is taken to be parallel to the z-axis of a cartesian coordinate system and the x- and the y-axis are aligned parallel to the two-fold axes of the defect. If the direction cosines of J_s are taken with respect to the coordinate system defined above, we obtain

$$\emptyset_f = \emptyset^{(h)} + C_{11}\sum_{i=1}^{3}\left\{(\partial\gamma_i/\partial x_1)^2 + (\partial\gamma_i/\partial x_2)^2\right\} +$$
$$+ C_{33}\sum_{i=1}^{3}(\partial\gamma_i/\partial x_3)^2 + k(1-\gamma_3^2) \quad , \quad (17)$$

Eq. (16) can be used to determine the interaction energy between point defects and domain walls. This requires only a replacement of $(\nabla \gamma_j)^2$ by $(d\varphi/dz)^2$ where φ is the angle of spin rotation within the domain wall. In general the influence of point defects upon the spin structure of the domain wall is neglected. In the case of planar defects such as stacking faults or antiphase boundaries (APB), however this influence should be taken into account. The atomic arrangement of a stacking fault in an hcp lattice is shown in Fig. 2. Fig. 2 shows that stacking faults change the next nearest neighbour exchange interactions which are denoted by the shaded atoms. Fig. 3 and Fig. 4 show examples where the nearest neighbour exchange integrals at APB are different from those in the ordered matrix. For example

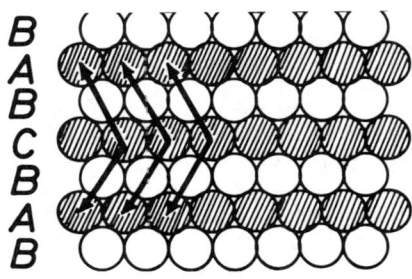

Fig. 2. Stacking fault in an hcp lattice.

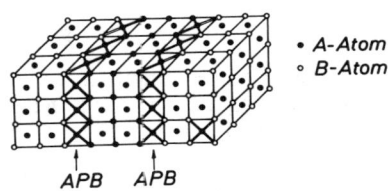

Fig. 3. Antiphase boundaries in an ordered alloy of type AB.

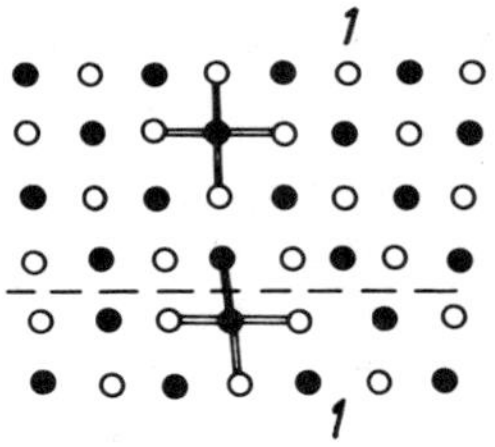

Fig. 4. Generation of an APB by dislocation glide.

in a diatomic alloy AB, the ordered matrix has only A-B nearest neighbour couplings present, whereas within the APB nearest neighbour A-A- and B-B-coupling appears. Fig. 4 shows an APB generated by a moving edge dislocation which has moved into the ordered region from the left and stopped at position 1.

Micromagnetic equilibrium conditions are of particular interest when the interaction between short range perturbations and domain walls is considered. The theoretical method which can be applied to the case of a planar defect parallel to the domain wall of a uniaxial crystal will now be discussed. Within the planar defect, it is assumed the exchange integral is smaller than in the matrix, i.e., the spin rotation in or at the planar defect is more abrupt than in the matrix as shown in

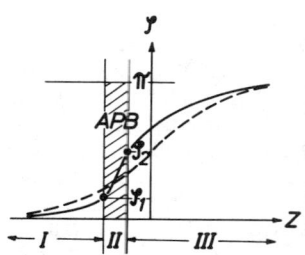

Fig. 5. Change of spin distribution due to an APB. The broken line denotes the spin configuration of the unperturbed domain wall.

Fig. 5. In order to evaluate the interaction energy between the domain wall and the planar defect, we divide the domain wall into three regions (Fig. 5). The external regions I and III can be treated in the continuum theory approximation whereas the internal region II is more accurately described by the Heisenberg model. In regions I and III the domain wall equation

$$A(d^2\varphi/dz^2) - k_1 \sin 2\varphi = 0 \qquad (18)$$

with the solution

$$\sin^2\varphi_o = 1/\text{ch}^2(z/\delta_o) \,. \qquad (19)$$

The domain wall energy per unit area in regions I and III is given by

$$\gamma_{I,III} = 2\sqrt{A\,K_1}\,(1-\cos\varphi_{1,2}) \,. \qquad (20)$$

The exchange energy of region II if two atomic layers are involved may be written as

$$\gamma_{II} = (2A/a)p\cos(\varphi_1 - \varphi_2) \,, \qquad (21)$$

where p corresponds to the ratio J'/J (J' = exchange in the planar defect, J = exchange integral in the matrix, a = distance between the two atomic layers). The total energy

$$\gamma_f = \gamma_I(\varphi_1) + \gamma_{III}(\varphi_2) + \gamma_{II}(\varphi_1 - \varphi_2) \qquad (22)$$

must be minimized with respect to the angles φ_1 and φ_2. The detailed calculation shows that for p<1 the exchange energy, γ_{II}, of the planar defect increases whereas the energies γ_I and γ_{III} are found to decrease. The total energy difference between the perturbed and unperturbed wall is found to be

$$\Delta\gamma = 4\sqrt{A\,K_1} - \gamma_f = \sqrt{A\,K_1}\,(1 - 1/p)(a/\delta_o)\sin^2\varphi_o \,, \qquad (23)$$

where $\sin^2\varphi_o$ is given by eq. (19).

III. APPLICATIONS OF MICROMAGNETISM IN THE FIELD OF LATTICE IMPERFECTIONS

III.1. Modification of the domain structure

The effect of dislocations on the ferromagnetic domain structure has been investigated most extensively for plastically deformed Ni-, Co-, and Fe-single

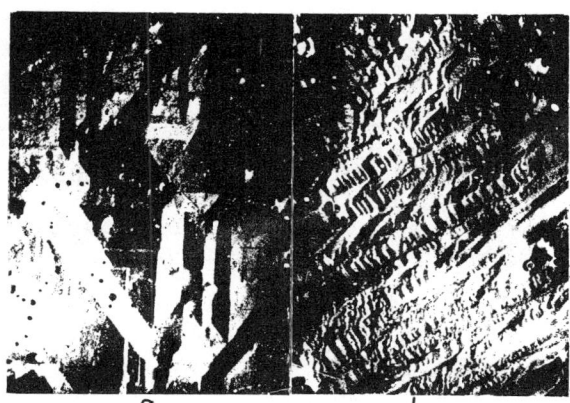

Fig. 6. Domain structure on the ($\bar{1}$01)-plane of Ni-single crystal [11].
6a. Undeformed crystal, magnification x 50.
6b. Applied flow stress = 2 kg/mm^2, magnification x 50.

crystals. Fig. 6 shows an example of the change in the domain structure on the ($\bar{1}$01)-plane of a deformed Ni-crystal. To analyse these effects, it is useful to separate the internal stress $\vec{\sigma}^{(i)}$ according to

$$\vec{\sigma}^{(i)} = \vec{\sigma}^{(l.r.)} + \vec{\sigma}^{(s.r.)} \qquad (24)$$

$\vec{\sigma}^{(l.r.)}$ represents the long range contribution of $\vec{\sigma}^{(i)}$ with a wavelength λ larger than the linear dimensions of the domains (D), and $\sigma^{(s.r.)}$ represents the contributions with associate wavelength $\lambda < D$. The ratio of the amplitude of these two terms depends upon the arrangement of the dislocations present. The long range term exists only if the dislocation structure shows some long range order of dislocations of the same sign.

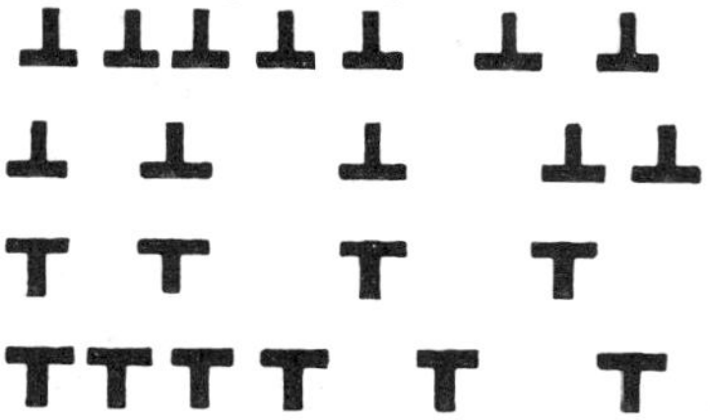

Fig. 7. Model for a layered structure of dislocation groups.

As an example Fig. 7 shows a representation of a dislocation structure, which is composed of dislocation layers. The effect of $\sigma^{(s.r.)}$ on the orientation of the spontaneous magnetization is qualitatively different from the effect of $\sigma^{(l.r.)}$. In the case of $\sigma^{(s.r.)}$, the volume average of the magnetoelastic coupling energy vanishes for a homogeneous magnetization, whereas the volume average remains finite for $\sigma^{(l.r.)}$. The average easy directions of magnetization therefore must be determined from the equilibrium conditions

$$\frac{\partial}{\partial \gamma_i} \left\{ \phi_K(\gamma_i) + \phi_S(\gamma_i) + \phi_{H_{ext}}(\gamma_i) - \vec{e}^M \cdot \vec{\sigma}^{(l.r.)} \right\} = 0, \qquad (25)$$

where $(-)\vec{e}^M \cdot \vec{\sigma}^{(l.r.)}$ corresponds to the magnetoelastic coupling energy of the long range stresses. Eq. (25) for $H_{ext} = 0$ in general leads to a number of easy directions with energy minima. The absolute value of the energy of these minima is in general different. It turns out that the easy directions with the lowest absolute energy is determined by the energy change associated with the magnetoelastic energy, $\Delta\phi_M^{(s.r.)}$ of the short range stresses. It has been shown previously [10,12] that within the framework of micromagnetism

$$\Delta\phi^{(s.r.)} = \sum_i A_{iiii} \beta_i^2 (1-\beta_i^2) - \sum_{i \neq j} A_{iijj} \beta_i^2 \beta_j^2 +$$

$$+ \sum_{k \neq i,j} A_{ijij} (1 - \beta_k^2 - 4\beta_i^2 \beta_j^2) \quad . \qquad (26)$$

The coefficients A_{ijkl} in the case of magnetostrictive isotropy ($\lambda_{100} = \lambda_{111} = \lambda_s$) are defined as

$$A_{ijkl} = \frac{9 J_s^2 \lambda_s^2}{16 A\, V_D} \int \frac{|\vec{\tilde{\sigma}}_{ij}(\vec{k}) \cdot \vec{\tilde{\sigma}}_{kl}(\vec{k})|\, d^3\vec{k}}{k^2 (k^2 + \varkappa_{H_{eff}}^2)^2} \quad . \qquad (27)$$

In eq. (26) the direction cosines β_i of the easy directions refer to the $(\vec{b}, \vec{l}, \vec{n})$ coordinate system of an edge dislocation ($\vec{b}$ = Burgers vector, $\vec{l}$ = line direction, $\vec{n}$ = glide plane normal). In particular we have: $\beta_1 = \cos \sphericalangle(\vec{J}_s^o, \vec{b})$, $\beta_2 = \cos \sphericalangle(\vec{J}_s^o, \vec{l})$, $\beta_3 = \cos \sphericalangle(\vec{J}_s^o, \vec{n})$. Furthermore, in eq. (27) V_D is taken as the volume of the domain, $\varkappa_{H_{eff}} = |H_{eff}| J_s^2 / 2A$, where $|\vec{H}_{eff}|$ is the average effective field and $\vec{\tilde{\sigma}}(\vec{k})$ corresponds to the Fourier transform of the internal stress tensor

$$\vec{\tilde{\sigma}}(\vec{k}) = (2\pi)^{-3/2} \int \vec{\sigma}(\vec{r})\, e^{i \vec{k}\cdot\vec{r}}\, d^3\vec{r} \quad . \qquad (28)$$

Eq. (26) shows that the energy change due to dislocations is highly anisotropic as a consequence of the complicated inter-play between the anisotropies of the stress tensor and the magnetoelastic coupling energy. $\Delta\phi^{(s.r.)}$ can be considered as an additional anisotropy of orthorhombic symmetry with three two-fold axes parallel to $\vec{b}$, $\vec{l}$, and $\vec{n}$. $\Delta\phi^{(s.r.)}$ has associated with it the following properties:
A. Edge dislocations.
1. $\Delta\phi^{(s.r.)}$ exhibits a minimum in the direction of the dislocation line.
2. Maxima of $\Delta\phi^{(s.r.)}$ are observed in the $\vec{b}$- and $\vec{n}$-direction (Fig. 8).

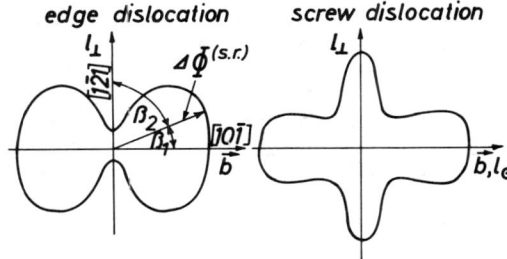

Fig. 8. The induced anisotropies of edge and screw dislocations. $l_\perp$ denotes the line direction of edge dislocations.

3. If the dislocations are arranged in a layer-like structure such as depicted in Fig. 7, it is found all three directions ($\vec{b}, \vec{l}, \vec{n}$) become easy directions.

B. Screw dislocations.
1. Maxima of $\Delta \phi^{(s.r.)}$ are observed in the $\vec{b}$- and the $\vec{l}$-direction, whereas a minimum is observed in the $\vec{n}$-direction. The change of the domain structure depicted on the $(\bar{1}01)$-plane of a plastically deformed Ni-crystal as seen in Fig. 6 is due to the induced anisotropy $\Delta \phi^{(s.r.)}$ of a layered dislocation structure (see Fig. 7). As previously discussed such a dislocation structure produces an induced anisotropy with minima in the $\vec{b}$, $\vec{l}$, and $\vec{n}$-direction. This effect is due to the reduction of the stress components σ_{13} and σ_{33}. The existence of $\Delta \phi^{(s.r.)}$ now renders the original easy directions of the perfect crystal to become energetically non-equivalent. Only the $[111]$-direction, which is perpendicular to the glide plane and parallel to the $\vec{n}$-direction, remains in its low energy state. Consequently, the domain structure reveals this preference for the (111)-direction by the formation of ladder rungs. The ladders are clearly visible in Fig. 6.

III.2. Pinning of plane domain walls

In general, a domain wall (d.w.) can interact with a large number of dislocations and point defects.[+] The long range interactions between the internal stresses associated with these defects and the elastic stresses σ^E of the d.w. can be most easily described in terms of the elastic stresses σ^E of the d.w.. The elastic stresses σ^E originate from the inhomogeneous magnetostrictive strains, e^M, within the wall. In order to avoid the discontinuities introduced by e^M

[+] The interaction between dislocations and d.w. lately was made visible [13].

in the strain field, elastic strains, e^E, are created which may be determined according to the method outlined in section II.2. The stress tensor σ^E has been determined previously by Rieder [14] for various types of d.w.. In the following dislocations will be described by their line element, $d\vec{l}_i$, and their Burgers vector, $\vec{b}_i$. The elastic properties of the point defects will be characterized by their dipole tensor Q which describes both the volume dilatation and the shear strains introduced by the point defect locally. For vacancies, the components of Q extend over the range of some tenth of an atomic volume, whereas in the case of interstitial atoms the Q_{ij}'s can extend over a range comparable with the atomic volume. The magnetic short range interactions of the point defects are characterized by the perturbation energy $\emptyset_f$ as derived in section II.4. Antiphase boundaries are described by their interaction energy given by eq. (23).

The forces $\vec{p}_i$ exerted by different kinds of lattice defects on a d.w. are the following:

1. The long range interaction between a d.w. and dislocations is determined from Peach and Koehler's [15] formula:

$$\vec{p}_1 = \sum_i \int d\vec{l}_i(\vec{r}) \times (\vec{\sigma}^E(\vec{r}) \cdot \vec{b}_i) \quad , \qquad (29)$$

where the integration extends over all dislocation lines.

2. The total interaction force between point defects and a d.w. is given by

$$\vec{p}_2 = \sum_j \{ Q_j \cdot \nabla \vec{\sigma}^E(\vec{r}_j) + \nabla \emptyset_{f,j}(\vec{r}_j) \} \quad , \qquad (30)$$

where $\vec{r}_j$ denotes the position of the j'th point defect. The first term in eq. (30) corresponds to the long range elastic interaction whereas the second term takes into account the short range magnetic interactions.

3. For the interaction force between an APB and a d.w. we obtain

$$\vec{p}_3 = -F_{APB}\sqrt{A K_1} \, (1-1/p)(a/\delta_o)\sin 2\varphi_o (d\varphi_o/dz). \qquad (31)$$

The various interaction forces give rise to a statistical field of force, the properties of which[1,2] must be determined by means of statistical methods. According to our previous results the magnetic field required in order to displace a 180°-wall configuration (plane d.w. or cylindrical bubble domain) a distance L_3 parallel to the z-direction, is given by

$$H_c = \frac{1}{2J_s} \left[\frac{2 \ln(L_3/2L_0)}{F_B L_3} \right]^{1/2} \left[\sum_i N_i \int p_i^2(z) dz \right]^{1/2} . \quad (32)$$

In eq. (32) N_i corresponds to the average volume density of the ith type of defect, F_B denotes the d.w. area to be displaced and $2L_0$ corresponds to the average wavelength of the field of force. Equations for the initial susceptibility χ_0, and the Rayleigh constant, α, also were derived [16] previously. These properties will not be discussed in detail here.

To carry out a qualitative discussion of the different contributions to H_c, we shall restrict ourselves to the dependence of H_c on the domain wall width δ_B. For simplicity, we shall consider the contributions $H_c^{(i)}$ of the different lattice defects separately. The discussion is complicated since the way in which H_c depends upon δ_B is influenced somewhat by the type of interaction as well as by the type of defect under consideration. In general, the spontaneous magnetostriction includes terms e^K of the magnetocrystalline energy, and terms e^{Ex} due to the exchange energy. The latter terms are proportional to $(\nabla \varphi)^2$ and become important in narrow domain walls. In the case of 180°-walls, $\vec{\sigma}^E$ may be written as

$$\vec{\sigma}^E = \vec{B}_1^K \sin^2\varphi + \vec{B}_2^K \sin 2\varphi + \vec{B}_3^{Ex}(\nabla\varphi)^2 , \quad (33)$$

where φ corresponds to the angle of rotation of J_s in the d.w.. In previous discussions usually only the first two terms in (33) were taken into account which led to the following results for the Rayleigh region [16]

$$\chi_0 \propto \sqrt{\delta_B/N_1} \quad ; \quad H_c \propto \sqrt{N_1 \delta_B} \quad ; \quad \alpha \propto \delta_B/N_1 \quad . \quad (34)$$

The results of (34) hold for magnetic materials such as Ni and Fe, with d.w. widths larger than 100 Å. This occurs because it is justified to neglect the exchange terms in wide domain wall materials. In materials with narrow d.w., however, this approximation no longer holds, and the coercive fields $H_c^{(i)}$ are given by:

1. Dislocations:

$$H_c^{(1)} = \sqrt{N_1} \left\{ (C_K^{(1)} \delta_B + C_{Ex}^{(1)}/\delta_B^3) \right\}^{1/2} . \quad (35)$$

2. Point defects:

$$H_c^{(2)} = \sqrt{N_2} \left\{ (C_K^{(2)} + C_{K;f}^{(2)})/\delta_B + (C_{Ex}^{(2)} + C_{Ex;f}^{(2)})/\delta_B^5 \right\}^{1/2} . \quad (36)$$

3. Antiphase boundaries:
$$H_c^{(3)} = \sqrt{N_3} \, (C_{Ex;f}^{(3)}/\delta_B^5)^{1/2} \quad . \tag{37}$$

The parameters $C_K^{(i)}$ and $C_{Ex}^{(i)}$ refer to the long range interactions of e_K and e_{Ex}, and are quadratic functions of the magnetostriction tensors λ_{ijkl} and Λ_{ijkl} respectively. The parameters $C_{K;f}^{(i)}$ and $C_{Ex;f}^{(i)}$ take care of the short range interactions. In Fig. 9, the way in which each $H_c^{(i)}$ depends upon δ_B is shown qualitatively where the logarithmic values are taken. In soft magnetic materials where the d.w. are rather extended, the magnetostrictive interactions of the e^K-terms give rise to the strongest pinning mechanism. In hard magnetic materials, i.e., narrow d.w., the exchange magnetostriction, e^{Ex}, predominates. The dominant interaction forces are due to point defects and APB which lead to a $\delta_B^{-5/2}$-term. In an intermediate region the exchange terms in eq.(35) occuring from the presence of dislocations and which vary as $\delta_B^{-3/2}$ may become important. In conclusion we may summarize as follows: In soft magnetic materials, dislocations are responsible for the pinning of domain walls, whereas in hard magnetic materials point defects, atomic disorder, and antiphase boundaries dominantly pin the d.w..

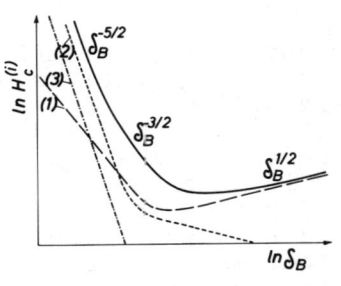

Fig. 9. The dependence of various H_c-contributions on the domain wall width.

III.3. Pinning of bubble domains

In bubble domain materials usually the domain wall width is larger than 100 Å [17]. Magnetostrictive stresses therefore are supposed to be due exclusively to e^K. As a consequence of this only the long range interaction with dislocation should be important. In the following we consider a platelet of thickness, h, of a hexagonal crystal with its easy axis of magnetization (c-axis) perpendicular to the plane of the platelet. In Fig. 10, a cylindrical bubble domain (b.d.) of radius r_o and d.w. width δ_B, is represented and is assumed to be interacting with a straight dislocation of Burgers vector $\vec{b}$ at a distance, d, from the centre of the b.d.. Within the d.w. $\vec{J}_s$ rotates from the positive direction of the easy axis into the opposite direction.

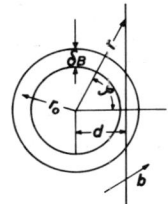

Fig. 10. The geometry of a b.d. interacting with a dislocation.

The angle Θ of rotation is given by
$$\sin^2\Theta(r) = 1/\text{ch}^2[(r-r_0)/\delta_B] \quad . \quad (38)$$

In a polar coordinate system (r, φ, z) with the z-axis parallel to the easy direction, the tensor e^K of the spontaneous magnetostriction is given by

$$e^K = \begin{pmatrix} \lambda_{12}\sin^2\Theta(r) & 0 & 0 \\ 0 & \lambda_{11}\sin^2\Theta(r) & \frac{1}{4}\lambda_{44}\sin 2\Theta(r) \\ 0 & \frac{1}{4}\lambda_{44}\sin 2\Theta(r) & -\lambda_{33}\sin^2\Theta(r) \end{pmatrix}, (39)$$

where the λ_{ik} are the components of the magnetostriction tensor for a hexagonal crystal. The elastic stresses, σ^E, and the strains, e^E, may be determined according to the method outlined in section II.2. It turns out that only a few types of dislocations interact with the b.d.. One of these dislocation types is the edge dislocation with its line direction parallel to the easy axis and its Burgers vector lying in the basal plane. The force exerted on the b.d. is given by

$$\vec{p} = h\sigma_{rr}^E(\vec{r})\,\vec{b}\sin^2\psi \cdot \vec{r}/|\vec{r}| \quad , \quad (40)$$

where $\vec{r}$ represents the radius vector from the center of the b.d. to the dislocation line, and ψ corresponds to the angle between $\vec{r}$ and $\vec{b}$.

Another type of dislocation interacting with the b.d. is the edge dislocation lying in the basal plane with a Burgers vector $\vec{b}$ parallel to the easy direction. The interaction force in this case is given by

$$\vec{p} = \int \sigma_{zz}^E(\vec{r}) \cdot [\vec{dl}(\vec{r}) \times \vec{b}] \quad . \quad (41)$$

The two types of dislocations discussed so far lead to quantitatively different results for the total interaction force. This is due to the fact that the range of interaction of the dislocation with the b.d. is quite different between the two cases considered. The dislocations which are perpendicular to the basal plane have an interaction range of the order of δ_B, whereas the dislocations lying in the basal plane have an interaction range of the order of $2r_o$. Since in general the relations $\delta_B \ll 2r_o$ and $h < 2r_o$ hold, the interaction force due to the second type of dislocation is believed to be much stronger. We therefore restrict our attention to this type of dislocation. With the stress component (G = shear modulus, ν = Poisson's ratio)

$$\sigma_{zz}(r) = \tfrac{2G}{1+\nu}(\lambda_{33}-\lambda_{12})/\operatorname{ch}^2\{(r-r_o)/\delta_o\}, \quad (42)$$

the total force exerted on the b.d. by a straight dislocation line which lies in the basal plane is given by

$$p(d) = \tfrac{2Gb}{1+\nu}(\lambda_{33}-\lambda_{12})\int \frac{dy}{\operatorname{ch}^2\{\delta_o^{-1}(\sqrt{d^2+y^2}-r_o)\}}. \quad (43)$$

The integration of eq. (43) cannot be performed explicitly. In Fig. 11, the results of a computer calculation of the force p(d) for different ratios r_o/δ_o are represented. The main results may be summarized as follows:
1. The interaction force shows a maximum at distances $d \sim r_o$.
2. The maximum interaction force, p_{max}, increases linearly with r_o (see Fig. 12).
3. The force becomes half of its maximum value in a range of the order of $2\delta_o$.
4. The pinning force p(0) for d = 0 is given by $[4Gb\delta_o/(1+\nu)][\lambda_{33}-\lambda_{12}]$.

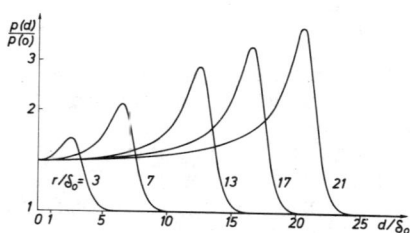

Fig. 11. Force-distance curves of bubble domains for various parameters r_o/δ_o.

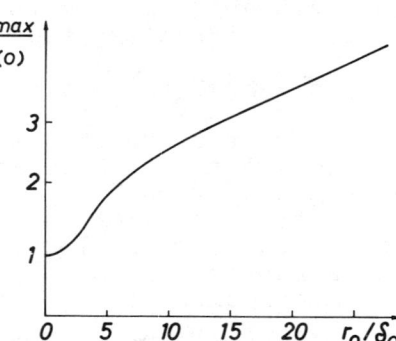

Fig. 12. The maximum force of interaction as a function of the bubble domain radius.

The force-distance curves represented in Fig. 11 can be used to discuss the coercive fields for two types of b.d. motion: 1. When the radius r_o of the b.d. is increased or decreased by a homogeneous field. 2. When the b.d. is linearly displaced by a field gradient ∇H. Using the results developed above, the coercive field for mode one or the critical field gradient ∇H_c for mode two displacements according to eq. (32), with $F_B = 2hr_o$, can be written as

$$\left. \begin{matrix} H_c^{(1)} \\ \\ \frac{\pi}{2} r_o \nabla H_c^{(2)} \end{matrix} \right\} \sim \frac{2Gb(\lambda_{33} - \lambda_{12})}{J_s(1+\nu)} \sqrt{12.5 N_1 r_o} \left\{ \frac{\ln(L_3/2L_i)}{2hr_o L_3} \right\}^{1/2}, \quad (44)$$

where $2L_i$ corresponds to the average wavelength of the statistical field of force corresponding to the appropriate displacement modes $i=1$ and $i=2$. It is of interest to note that the coercive field of each displacement mode is independent of r_o. This occurs because both the interaction force and the b.d. surface area increase linearly with increasing r_o. The difference in the two coercive fields occurs as a result of the different wavelength, $2L_i$, of the statistical field of forces for each of the two displacement modes. According to our previous results [2,10], the wavelength is given by four times the linear displacement of a d.w. configuration which is necessary to shift the d.w. into a statistically independent position. This length for the first displacement mode corresponds to $2L_1 \sim 4\delta_B$ and for the second one to $2L_2 \sim 8r_o$. Since in general the relation $\delta_B \ll 2r_o$ holds, this means that the coercive force which must be overcome is larger for either increasing or decreasing the b.d. diameter when compared to the coercive force required for a linear displacement of the whole b.d.. This difference in the coercive forces may be rather appreciable. For example in the case where L_3 is taken to be $4L_i$, the ratio of the coercive forces is given by

$$\frac{\frac{\pi}{2} r_o \nabla H_c^{(2)}}{H_c^{(1)}} = \left(\frac{\delta_B}{2r_o} \right)^{1/2}. \quad (45)$$

Since this ratio is at least of the order of 1/10 our result means that the linear displacement of a b.d. should be possible by much smaller magnetic forces than the force required to blow-up or constrict the bubble domain.

III.4. The nucleation field in imperfect crystals

As is well known the coercive field of a ferromagnet depends sensitively on its spatial dimensions. In macroscopic specimens H_c is determined by the pinning of d.w. by lattice defects or the Peierls force exerted by the lattice on the d.w.. Microscopic particles, with dimensions smaller than a critical diameter, d_c, may consist on one individual domain. In this case H_c increases with decreasing diameter, d, and the magnetization reversal takes place by an inhomogeneous rotation (curling). At still smaller diameters the coercive field becomes constant and is determined by the nucleation field for homogeneous rotation. Finally at very small diameters the thermal fluctuations in the particle assist the magnetization reversal, and the coercive field again decreases. The general feature of the variation of H_c with d is outlined in Fig. 13. Here the coercive field of the ideal crystal is compared with that of the imperfect crystal. It is of interest to note that the role of the defects concerning H_c is a twofold. In particles with many domains lattice defects increase H_c, whereas in single domain particles lattice defects decrease H_c. This latter effect is observed in all high coercivity materials such as Ba-ferrites or the Co_5Sm-compound. It is this effect which must be considered as the main obstacle for the realization of the ideal coercive field, which for cylindrical particles is given by

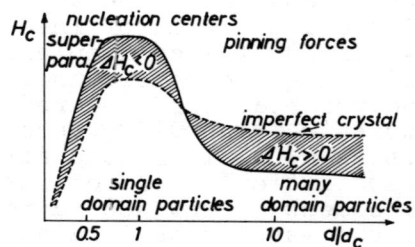

Fig. 13. Qualitative dependence of the coercive field on the particle diameter d. (———, perfect crystal; - - - -, imperfect crystal.)

$$H_c = 2K_1/J_s + 2\pi J_s \qquad (46)$$

It must be supposed that point defects are not the source for the observed smaller nucleation fields because this type of defects only give perturbations within an atomic scale, whereas the nucleation processes take place within dimensions of the order of some δ_B. Therefore it is suggesting itself that the stress fields of dislocations are responsible for a local reduction of the effective crystalline anisotropy constant K_1. As was shown in section III.1 the induced magnetoelastic anisotropy $\Delta \phi^{(s.r.)}$ may increase or decrease the total energy of the easy directions, and so give rise to a locally reduced effective crystal anisotropy constant. Particularly in materials with large magnetostriction constants, such as high coercivity materials usually are, this mechanism may be of importance.

ACKNOWLEDGEMENTS

The author wishes to thank Professor Dr. J. Roberts for several helpful discussions and reading the manuscript; Dr. A. Hubert for his assistance with numerical calculations; and Dipl.-Phys. H.R. Hilzinger for many interesting discussions.

REFERENCES

1. A. Seeger, H. Kronmüller, H. Rieger and H. Träuble, J. Appl. Phys. 35, 740 (1964).
2. H. Träuble, Moderne Probleme der Metallphysik, vol. 2, edited by A.Seeger (Springer-Verlag, Berlin-Heidelberg-New York, 1965).
3. H. Kronmüller, Moderne Probleme der Metallphysik, vol. 2, edited by A.Seeger (Springer-Verlag, Berlin-Heidelberg-New York, 1965).
4. W.F. Brown,Jr., Micromagnetics (J. Wiley, New York, 1963).
5. W.F. Brown,Jr., Magnetoelastic Interactions, Springer Tracts in Natural Philosophy, vol. 9 (Springer-Verlag, Berlin-Heidelberg-New York, 1966).
6. H. Kronmüller, Z.angew.Physik 23, 130 (1967).
7. R. Becker and W. Döring, Ferromagnetismus (Springer-Verlag, Berlin, 1938).
8. E. Kröner, Kontinuumstheorie der Versetzungen und Eigenspannungen (Springer-Verlag, Berlin, 1958).
9. H. Kronmüller and A. Seeger, J.Phys.Chem.Solids 18, 93 (1961).
10. H. Kronmüller, Int.J. of Nondestructive Testing 3, 315 (1972).
11. H. Willke, Phil. Mag. 25, 397 (1972).
12. H. Kronmüller, J. Appl. Phys. 38, 1314 (1967).
13. A.J. Kurtzig and J.R. Patel, Phys.Lett. 33A, 123 (1970).
14. G. Rieder, Abhandl.Braunschweig.Wiss.Ges. 11, 20 (1959).
15. M. Peach and J.S. Koehler, Phys.Rev. 80,436 (1950).
16. H. Kronmüller, Z.angew. Physik 30, 9 (1970).
17. A.A. Thiele, J.Appl.Phys. 41, 1139 (1970).

MOTION OF 180° DOMAIN WALLS IN UNIFORM MAGNETIC FIELDS

N. L. Schryer and L. R. Walker
Bell Laboratories, Murray Hill, N. J. 07974

ABSTRACT

The equations of motion of a 180° domain wall in an infinite, uniaxially anisotropic medium exposed to an instantaneously applied uniform dc magnetic field, H_o, have been integrated numerically. Media resembling YIG and an orthoferrite have been studied with Gilbert α's from .001 to .03. Fields above and below the critical value, $H_c = 2\pi QM_o$ (M_o, the saturation magnetization), below which a steady state solution is known to exist have been considered.

For all these cases the turn angle of the wall is substantially a function of time alone and the polar angle, θ, contrives to satisfy the relation: $\ln \tan \theta/2$ linear in x, the displacement normal to the wall. For $H_o < H_c$, the motion tends asymptotically to the steady state solution. For $H_o > H_c$, the velocity passes through a maximum, decreases, becomes briefly negative and reverses again. It is zero when the wall has turned through 180°. The motion is periodic within the accuracy of the solution.

The equations of motion may have solutions consistent with the <u>observed</u> shape characteristics of the wall provided one term may be dropped. The velocity, angle and wall thickness then follow automatically. The criterion for neglecting the single term may then be examined self-consistently and appears to be: $\frac{2K}{M_o}$, the anisotropy field, should be much greater than H_c. Some YIG runs with $\alpha = 0.1$ where this condition is marginal do indicate new features. The solution for $H_o < H_c$ agrees with one given by Bourne and Bartram (to be published). The characteristic time involved in these solutions is

$$\frac{1+\alpha^2}{\gamma \sqrt{H_c^2 - H_o^2}}$$ and the half-period in the high field solutions is

similarly $$\frac{\pi(1+\alpha^2)}{\gamma \sqrt{H_o^2 - H_c^2}}.$$

A full account of this work will be published elsewhere.

IMPORTANCE OF OVERLAP CONTRIBUTIONS TO THE AXIAL ANISOTROPY OF Mn^{2+}

V. J. Folen
Naval Research Laboratory, Washington, D.C. 20390

ABSTRACT

Electron-spin-resonance (ESR) measurements at 24 GHz have been made on single crystals of K_2ZnF_4 doped with Mn. The spectra obtained in these measurements correspond to Mn^{2+} in the Zn^{2+} sites (point group 4/mmm) in this material. The measured values of the spin-Hamiltonian parameters corresponding to the fine and hyperfine structure are $g_\parallel = 2.0030$, $g_\perp = 2.0028$, $A = -88.7 \times 10^{-4} cm^{-1}$, $B = -89.3 \times 10^{-4} cm^{-1}$, $D = 36.0 \times 10^{-4} cm^{-1}$, $a = 5.6 \times 10^{-4} cm^{-1}$, and $F = 1.9 \times 10^{-4} cm^{-1}$. In addition, the transferred hyperfine interaction (THI) parameters were obtained from the ESR measurements. The measured values of these parameters [which result from interactions between the Mn^{2+} 3d electrons and the F^- and which refer to the four equivalent nearest-neighbor F^- ions in the (001) plane (type I) and the two equivalent nearest-neighbor F^- ions along the fourfold axis (type II)] are $A_s^I = 18.1$, $A_s^{II} = 14.9$, $A_\sigma^I = 0.4$, $A_\sigma^{II} - \frac{1}{2} A_\pi^{II} = 0.3$, and $A_\pi^I \approx 0$, where the subscripts s, σ, and π refer to s, p_σ and p_π electronic bonding, respectively, and all quantities are in units of $10^{-4} cm^{-1}$. From these THI parameters, the fractions of unpaired spin densities were calculated. The results of the THI measurements were then used in conjunction with LCAO theory (involving spin-orbit and intraionic spin-spin interactions) to determine the local electronic overlap contributions to the spin-Hamiltonian D-parameter of Mn^{2+}. For these contributions, the overlap enters the calculation via the normalizing factors of the molecular orbitals. Using overlap integrals determined from the experimental values of the THI tensor, this theory yielded the value of $D = +35.9 \times 10^{-4} cm^{-1}$ which can be compared with the experimental value of $+36.0 \times 10^{-4} cm^{-1}$. Thus it is seen that overlap makes an important contribution to the physical origin of the axial anisotropy of Mn^{2+}, and that local overlap contributions can not only yield the correct sign of this anisotropy but also its correct magnitude. Since the overlaps used for this result are measured quantities, they represent the overlaps that actually exist when the dopant (Mn^{2+}) is embedded in a host lattice (K_2ZnF_4). In the present work, therefore, it was not necessary to assume that the lattice geometry in the vicinity of the Mn^{2+} ion is in exact correspondence with that in the vicinity of the Zn^{2+} ion. Part of the detailed results of this work has been published [See V. J. Folen, Phys. Rev. B6, 1670 (1972)] and additional results have been submitted for publication.

SPIN-WAVE SPECIFIC HEAT OF RbNiF$_3$ BY STEADY-STATE AC-TEMPERATURE CALORIMETRY *

W. Stutius**, J. R. Dillinger, and D. L. Huber

Dept. of Physics, University of Wisconsin, Madison, Wis. 53706

ABSTRACT

The specific heat of a small single crystal of ferrimagnetic RbNiF$_3$ has been measured at temperatures between 1.5 and 4 K in magnetic fields up to 32 kOe with the c-axis of the crystal both parallel and perpendicular to H by steady-state, ac-temperature calorimetry. We also present calculations of the frequency of the acoustic spin-wave modes as a function of magnitude and direction of the external magnetic field with respect to the crystal axes. The experimental results are compared with theoretical predictions based on recent measurements of the anisotropy field H_A and the exchange constants J_{AA} and J_{AB}.

INTRODUCTION

RbNiF$_3$ is a transparent ferrimagnet with a Curie temperature T_c=133 K. The Ni^{2+}ions occupy two nonequivalent lattice sites A and B. The magnetic moments are aligned collinearly in planes perpendicular to the hexagonal c-axis with a stacking sequence BBABBA where the A spins are antiparallel to the B spins. A detailed investigation of the magnon energies in RbNiF$_3$ was carried out by Chinn, Zeiger and O'Connor[1] and more recently by Als-Nielsen, Birgeneau and Guggenheim[2] using neutron diffraction techniques. Both groups, however, neglected the influence of the anisotropy field H_A and the external field H in their calculations, and we will take both into account in the derivation of the dispersion relation for long-wavelength acoustical magnons.

THEORY

We introduce an effective Hamiltonian whose anisotropy field and spin-wave stiffness are adjusted to equal the values of those parameters in the real system

$$H = \frac{A}{N}\sum_q S_z(q)S_z(-q) - g\mu_B H \cdot (S_z(0)\cos\alpha + S_x(0)\sin\alpha) - \frac{1}{N}\sum_q J(q)\vec{S}(q)\cdot\vec{S}(-q) \quad (1)$$

Here A is a positive anisotropy constant, N is the number of spins, H the magnitude of the external field, α the **angle** between the direction of H and the crystallographic c-axis. $\vec{S}(q)$ and $J(q)$ are the Fourier transforms of the spin and exchange interaction, respectively.

The spin-wave energies are obtained by linearizing the equations of motion for the spin components perpendicular to the equilibrium orientation. With the axes as defined in (1) the average spin lies in the x-z plane making an angle θ with respect to the c-axis. θ is obtained by minimizing the sum of the anisotropy and Zeeman energies.

With relations $g\mu_B H_A = 2A(S-\tfrac{1}{2})$, $Da^2q^2 = 2S(J(0)-J(q))$

we obtain the following spin-wave dispersion relations

a. for $\alpha = \pi/2$ ($H \perp c$)
$$(\hbar\omega_q)^2 = (Da^2q^2 + g\mu_B(H+H_A)) \cdot (Da^2q^2 + g\mu_B H) \tag{2}$$

b. for $\alpha = 0$ ($H // c$)
$$(\hbar\omega_q)^2 = (Da^2q^2) \cdot (Da^2q^2 + g\mu_B H_A \cdot (1-(H/H_A)^2)) \quad \text{for } H \leq H_A \tag{3}$$

and
$$\hbar\omega_q = Da^2q^2 + g\mu_B(H-H_A) \quad \text{for } H \geq H_A \tag{4}$$

There is no gap for $H//c$ as long as H is less than H_A. The magnon specific heat is given by

$$C/k_B = V_c/(2\pi)^3 \iiint_{\text{Br.Zone}} (\hbar\omega_q/k_B T)^2 \frac{\exp(\hbar\omega_q/k_B T)}{(\exp(\hbar\omega_q/k_B T)-1)^2} d^3q \tag{5}$$

EXPERIMENT

The electronics used to measure the absolute magnitude of the ac heat capacity is set up in essentially the same way as described by Sullivan and Seidel[3]. The sample was about 0.5 cm long having a square cross section with sides of approximately 0.4 cm and weighed 0.36 grams. A small heater was wound from 1.2 mil diam. Evanohm wire. The sample thermometer consisted of a chip of a nominal 47 Ω Allen-Bradley carbon resistor which was ground down giving a resistance of about 200 Ω at room temperature and ca 2000 Ω at 4.2 K. It weighed less than 1 milligram. Both the heater and thermometer were attached to the sample with GE 7031 varnish. The thermometer was calibrated as a function of temperature and magnetic field. Provided that the sample, heater and thermometer reach thermal equilibrium in a time short compared to $1/\omega$, where $\omega/2$ is the frequency of the oscillator driving the heater on the sample, and provided that the sample-to-bath relaxation time is long compared to $1/\omega$, then the specific heat C of the sample is given by[3]

$$C = \dot{Q}_o/(2\omega T_{ac}) \tag{6}$$

Here $\dot{Q}_o$ is the amplitude of the heater power and T_{ac} the amplitude of the ac temperature variation.

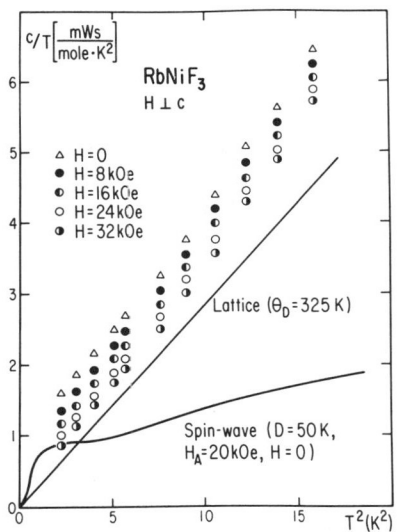

Fig. 1. C/T vs. T^2 for a RbNiF$_3$ single crystal with H$\perp$c. The spin-wave specific heat (solid line) is calculated with our model using the values H$_A$=20 kOe and D=50 K. The lattice heat capacity is taken as the difference between the zero field experimental points and the calculated spin-wave specific heat.

Fig. 1 shows the measured specific heat for the crystal oriented with its c-axis perpendicular to the magnetic field (easy direction) for various magnetic field strengths and temperatures. Instead of analyzing the total heat capacity in terms of the magnetic and lattice contributions we plot the difference $\Delta C=C(H,T)-C(0,T)$ as shown in Fig. 2. It is the inherent advantage of the ac calorimetry that changes in the specific heat as a function of an external parameter can be measured as accurately as the absolute magnitude itself, limited only by the signal-to-noise ratio, the thermometer calibration, and the stability of the electronics and temperature. The best fit between the calculated curves and the experimental points in Fig. 2 is obtained using D=50($\pm$2) K for the spin-wave stiffness and H$_A$=20 kOe for the anisotropy field. This compares favorably with D=65.2K obtained from a formula given by Keffer[4] using the exchange constants reported by Als-Nielsen et al.[2]

With these values we calculate the spin-wave specific heat for H$\perp$c from eqs. (2) and (5) and subtract this from the total heat capacity. We are then left with the lattice specific heat and find Θ_D=325($\pm$5) K for the Debye temperature (cf. Fig. 1).

To illustrate the directional dependence of the magnetic specific heat we show in Fig. 3 curves typical for the change ΔU in the ac voltage across the sample thermometer as a function of H for two orientations of the sample. These curves were taken keeping the resistance of the sample thermometer constant, and not at a constant temperature, which can change several hundredths of a degree going from H=0 to H=32 kOe depending on the temperature. Magnetoresistive corrections have to be applied for proper evaluation of the data.

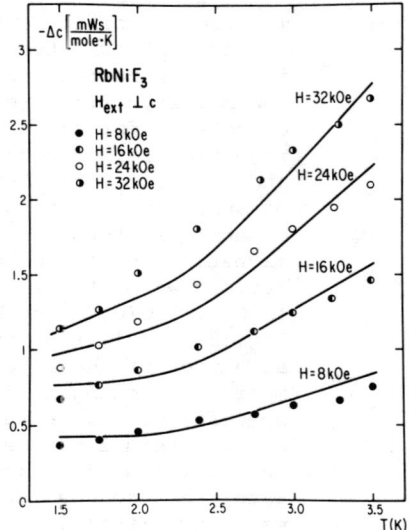

Fig. 2. $\Delta C(H,T) = C(H,T) - C(0,T)$ for various values of the magnetic field and temperature. The solid curves are calculated using eq.(2) and represent the best fit.

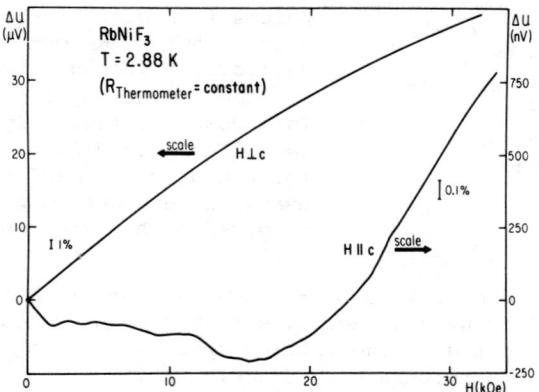

Fig. 3. Change of the ac voltage across the sample thermometer as a function of magnetic field for two different orientations of the $RbNiF_3$ crystal.

Nevertheless, we see that the effect of an external field is very anisotropic as expected form comparing eq. (2) with eqs. (3) and (4). The curve for H//c agrees qualitatively with the predictions of our model, but a quantitative fit could not be obtained, since a slight misalignment has profound effects for H//c contrary to $H \perp c$.

CONCLUSIONS

We have shown that the steady-state, ac-temperature calorimetry can readily be used to measure changes in heat capacity occuring in magnetically ordered materials in high magnetic fields. By employing a mechanism to rotate the sample in an external magnetic field the directional dependence of the magnetic specific heat could be observed directly. The limiting factor in this type of experiment is mainly the magnetoresistance of the sample thermometer which can be mastered by careful calibrations.

* Work supported by the U.S. Atomic Energy Commission
** Present address: Xerox Palo Alto Research Center, Palo Alto, California 94304

REFERENCES

1. S.R. Chinn, H.J. Zeiger, and J.R. O'Connor, Phys. Rev. $\underline{B3}$, 1709 (1971).

2. J. Als-Nielsen, R.J. Birgeneau, and H.J. Guggenheim, Phys. Rev. $\underline{B6}$, 2030 (1972).

3. P.F. Sullivan and G. Seidel, Phys. Rev. $\underline{173}$, 679 (1968).

4. F. Keffer, in: Handbuch der Physik XVIII/2, ed. S. Flügge, Springer Verlag (1966), p. 159.

LIGHT SCATTERING BY MAGNONS IN THE WEAK FERROMAGNET NaNiF$_3$

P. Moch, R.V. Pisarev* and C. Dugautier
Laboratoire de Physique des Solides, associé au C.N.R.S.
Université de Paris VI
E.N.S., 24 rue Lhomond, Paris V, France

ABSTRACT

One and two magnons Raman scattering was observed in NaNiF$_3$: the results, which include experiments with an applied magnetic field, allow the determination of most of the coefficients describing the magnetic properties : isotropic exchange, antisymmetric exchange, symmetric anisotropic exchange, single ion anisotropy. Temperature dependence is also discussed.

INTRODUCTION

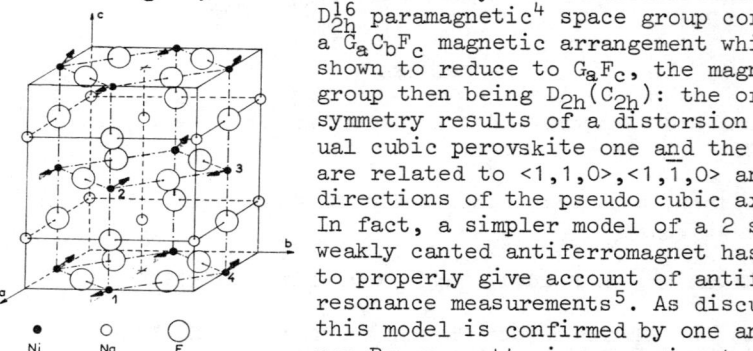

Fig.1: Unit cell of NaNiF$_3$

Below $T_N \simeq 150°K$[1,2,3], NaNiF$_3$ is a weakly canted 4 sublattices antiferromagnet; its orthorhombic crystal structure described by the D_{2h}^{16} paramagnetic[4] space group corresponds to a $G_a C_b F_c$ magnetic arrangement which has been shown to reduce to $G_a F_c$, the magnetic point group then being $D_{2h}(C_{2h})$: the orthorhombic symmetry results of a distorsion of the usual cubic perovskite one and the $\underline{a},\underline{b},\underline{c}$ axes are related to $<1,1,0>, <1,\bar{1},0>$ and $<0,0,1>$ directions of the pseudo cubic axes (Fig.1). In fact, a simpler model of a 2 sublattices weakly canted antiferromagnet has been shown to properly give account of antiferromagnetic resonance measurements[5]. As discussed below, this model is confirmed by one and two magnon Raman scattering experiments; moreover, we show that, as far as one is concerned with two magnon scattering, which involves high energy magnons near the Brillouin zone boundaries, NaNiF$_3$ is completely equivalent to a simple cubic spin 1 antiferromagnet, with first neighbours isotropic exchange interaction only, which allows a precise determination of J. On the other hand, the low frequency $\vec{k}=0$, one magnon spectrum provides a determination of most of the other exchange and single ion anisotropy coefficients which enter into the free energy expression.

EXPERIMENTAL

Relatively well-polarized spectra were studied using samples cut with each face perpendicular to $\underline{a},\underline{b}$ or $\underline{c}$, the principal axes of the dielectric tensor. Their temperature dependence was studied and, at 2°K, we made measurements with an applied magnetic field up to 50 kG. Note that the low frequency one magnon scattering in the

* Present address: Laboratory of Magnetism, A.F. Ioffe Physica-Technical Institute, Academy of Sciences USSR, Leningrad K-21, USSR.

2-20 cm⁻¹ range was observed, using the 5145Å line of a single mode Ar laser coupled with an Iodine cell, in order to reabsorb the elastically-scattered light. In the following, we shall not discuss the phonon scattering, although it was also studied and, at least partially interpreted.

TWO MAGNON SCATTERING

At $2°K$, there is a broad Raman line in the 400-500 cm⁻¹ range (Fig.2). Its shape fits exactly with the calculated Γ_3^+ spectrum for an isotropic Heisenberg antiferromagnet with next neighbours exchange interaction between spin 1 [6], if one takes $J<S> = 41.8\pm0.3$ cm⁻¹ (here the cut-off frequency for the two magnon line is written $2zJ<S>=12J$, since there are 6 neighbours and since $<S>=1$). The relative intensities of the polarized spectra agree with the calculation using the Γ_3^+ symmetric tensor related to O_h and remembering that a,b and c respectively correspond to <1,1,0>, <1,$\bar{1}$,0> and <0,0,1> in the cube. The observed shape not only implies that any term other than isotropic exchange has a negligible effect on the two magnon spectrum, which could be expected from other experimental data such as antiferromagnetic resonance, or that second neighbours interaction is small compared to J, which could be expected too; it proves that the exchange coupling between two next-neighbour Ni^{2+} ions is practically the same for any pair, which allows us to use a 2 sublattices model whereas a 4 sublattices one would lead to two different exchange integrals (J_{12} and J_{14}), which would significantly modify the density of magnon states[7], and consequently, the shape of the line. The relative frequency (normalized to 1 at $T = 0°K$), varies with the normalized temperature T/T_N exactly in the same way in $NaNiF_3$, and in the cubic antiferromagnet $KNiF_3$ [6] if one takes $T_N(NaNiF_3)=156°K$ [3], and $T_N(KNiF_3)=246°K$ [8]. Also note that using the exchange integrals derived from two magnon scattering, one finds $[J_{(NaNiF_3)}/J_{(KNiF_3)}]$. $T_{N(KNiF_3)}= 146°K$, very near the measured $T_{N(NaNiF_3)}$. (149^2 or $156°K^3$) Finally, we have tentatively attributed a broad temperature-dependent line, centered at 710 cm⁻¹ at $T = 2°K$, to four magnon scattering.

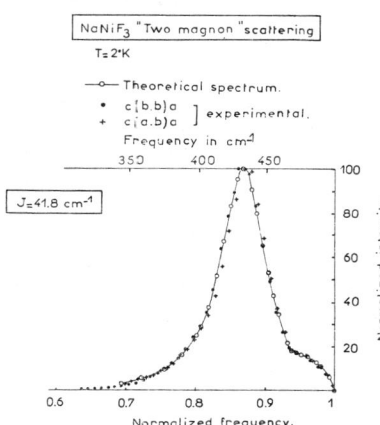

Fig.2: Comparison of the experimental 2 magnon spectrum shape with the theoretical one[6] and determination of J.

ONE MAGNON SCATTERING

The low magnon frequencies at $\vec{k} = 0$ depend strongly on terms other than J. Since a two sublattices model provides a convenient approximation, from symmetry arguments, the free energy, following

E.I. Golovenchitz et al[5], is expressed by :

$$F = \alpha^2 \left[E \vec{S}_1 \cdot \vec{S}_2 + \vec{D} \cdot (\vec{S}_1 \wedge \vec{S}_2) + J_{xx} S_{1x} S_{2x} + J_{zz} S_{1z} S_{2z} - A_{xz}(S_{1x}S_{1z} - S_{2x}S_{2z}) \right.$$
$$\left. - A_{xx}(S_{1x}^2 + S_{2x}^2) - A_{zz}(S_{1z}^2 + S_{2z}^2) \right] - \alpha \mu_B (\vec{g}_1 \vec{S}_1 + \vec{g}_2 \vec{S}_2) \cdot \vec{H} \quad (1)$$

$$\text{with} \quad \vec{\vec{g}}_i = \begin{pmatrix} g_{xx} & 0 & \pm g_{xx}\tau_1 \\ 0 & g_{yy} & 0 \\ \pm g_{zz}\tau_2 & 0 & g_{zz} \end{pmatrix} \begin{array}{l} + \text{ for } i = 1 \\ - \text{ for } i = 2 \end{array} \quad (2)$$

$\vec{S}_1$ and $\vec{S}_2$ are unit vectors (spin 1), α is the relative sublattice magnetization (at T=0°K we take α=1). E,$\vec{D}$ (here $\vec{D}//\underline{c}$), J_{xx} (and J_{zz}) respectively stand for the isotropic, antisymmetric and symmetric anisotropic exchange. A_{xx} and A_{zz} correspond to the single ion anisotropy and the last term gives the effect of a magnetic field $\vec{H}$. We take E = ZJ = 6J. In (1), F is normalized to a pair of neighbouring spins. α is strongly temperature-dependent, the other coefficients, at least the exchange terms, are expected to show a smaller temperature variation. The equilibrium positions are found by minimizing (1) : in zero magnetic field, each sublattice magnetization lies along $\underline{a}$ with a canting along $\underline{c}$; a field along $\underline{a}$ induces a spin flip transition for a critical value H_c. Solving the equation of motion[5]:

$$\alpha \hbar \dot{\vec{S}}_i = - \vec{S}_i \wedge \frac{\partial F}{\partial \vec{S}_i} \quad (3)$$

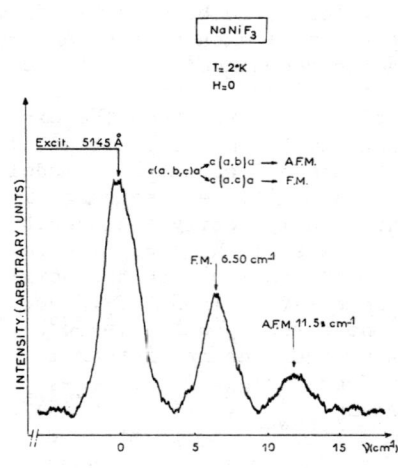

Fig.3: One magnon Raman spectrum of NaNiF$_3$ at 2°K. Here the scattered beam is unpolarized to show simultaneously FM and AFM lines.

provides two resonance frequencies [ferromagnetic (FM) and antiferromagnetic (AFM)]. The solutions are expressed for $\vec{H}//\underline{a}$ (if H>H$_c$), $\vec{H}//\underline{b}$ and $\vec{H}//\underline{c}$ in ref.5 and 9; for $\vec{H}$=0, we found, using Loudon's method[10], that the two magnon representations are Γ_1^+ and Γ_2^+ of $D_{2h}(C_{2h})$.

At 2°K, there are two Raman lines centered at 6.5±0.3cm^{-1} (FM), and 11.5 ±0.3cm^{-1} (AFM) for $\vec{H}$=0 (Fig.3). The FM line scatters strongly only in α_{xz} and α_{yz} polarizations, and the AFM in α_{xy} in agreement with the expected selection rules for antisymmetric Raman tensors Γ_2^+ and Γ_1^+ in C$_{2h}$. Both lines show an integrated intensity smaller than the two magnon line, by a factor around 30. They are observed until T≈140°K with a width at half-maximum smaller than 0.5 cm^{-1}. Their intensity increases with temperature until 100°K and then decreases: the initial increase is well understood, considering that the thermal population variation is the domin-

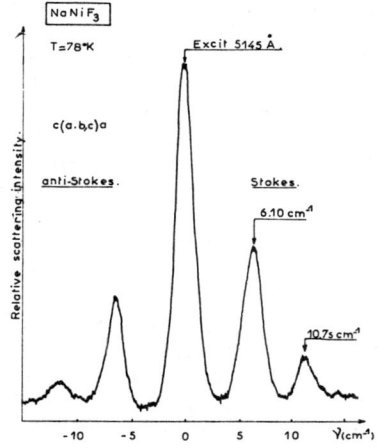

Fig.4: Stokes and anti-Stokes one magnon spectrum at 78°K.

Fig.5: $\vec{H}(//a)$ dependence of the FM line.

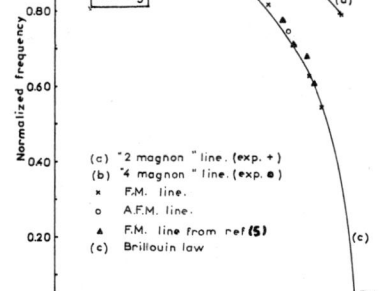

Fig.6: One, two and four magnon line temperature dependence.

ant effect; similarly the anti-Stokes spectrum, completely absent at 2°K, is easily observed at 20°K and above.(Fig.4). A magnetic field was applied along the three main crystallographic directions. Fig.5 shows the spin flip transition for $H_c \simeq 20 kG$: the imperfect crystal orientation prevents the FM frequency shifting to 0 at $H=H_c$. A_{xx} and J_{xx} (or A_{zz} and J_{zz}) cannot be independently evaluated from the magnetic field dependence of the spectra, but only $A'_{xx}=A_{xx}+J_{xx}/2$; with these restrictions the derivation of the coefficients is possible, but practically too imprecise for some of them. The g tensor, for a Ni^{2+} ion in a nearly cubic fluorine environment, is nearly isotropic, however, even small off-diagonal terms significantly modify the frequencies[5]: we then took $g_{xx}=g_{yy}=g_{zz}=2.3$, which is expected to be exact within 1% [11], and following ref.5, we put $\tau_1=\tau_2=\tau$. The coefficients resulting from the analysis of our experimental data are reported on Table I (first line). For comparison with other experiments, we have also expressed them in magnetic field units putting :

$$H_E = \frac{\alpha E}{g\mu_B} \; ; \; H_D = \frac{\alpha D}{g\mu_B} \; ; \; H_\tau = \frac{\alpha E \tau}{g\mu_B}$$

$$H_{A_{xz}} = \frac{\alpha A_{xz}}{g\mu_B}; \; H_{A'_{xx}} = \frac{\alpha A'_{xx}}{g\mu_B};$$

$$H_{A'_{zz}} = \frac{\alpha A'_{zz}}{g\mu_B}; \; H_{A_{x-z}} = \frac{2\alpha (A'_{xx}-A'_{zz})}{g\mu_B}$$

As expected, the D value is much higher than the other terms, except the isotropic exchange. D is responsible for the canting.
Golovenchitz et al have observed the FM line at $T>77°K$ [5]. Additionally, using the results of static magnetic measurements, they also derived some of the coefficients of (1),(Table I,3rd line). The main advantage of the present study is the possibility of a calculation

completely independent of other measurements (with an exception for g)
Unfortunately, except for E, where our determination is very accurate
the precision is low, due to the $0.3 cm^{-1}$ indetermination in absolute
Raman frequencies. We think that D, A'_{xx} and A'_{zz} are exact within $\sim 10\%$,
but that the uncertainty on A_{xz} and E_T is not smaller than $0.5 cm^{-1}$.

T (°K)	α	g	E (cm⁻¹)	H_E (kG)	Z	EZ (cm⁻¹)	H_Z (kG)	D (cm⁻¹)	H_D (kG)	A_{xz} (cm⁻¹)	H_{xz} (kG)	A'_{xx} (cm⁻¹)	H'_{xx} (kG)	A'_{zz} (cm⁻¹)	H'_{zz} (kG)	$H_{n=d}$
2	1	2.30	251	2,350	0.0025	-0.6	-6	16.9	157	+0.1s	+1.5	-0.15s	-1.4s	-0.19	-1.7s	+0.6
77*	0.93	2.30	251	2,180	0.0025	-0.6	-6	16.9	146	+0.1s	+1.5	-0.15s	-1.3s	-0.19	-1.6	+0.5
77**	0.93	2.14	226	2,100	0.0120	-2.7	-25	17.4	162	+1.3	+12	n.m.▲	n.m.▲	n.m.▲	n.m.▲	-1.1

* Calculated from our measurements at 2°K supposing that only α is temperature dependent.
** From ref. [5].
▲ Not measured.

TABLE I : Final results (see text)

The relative FM and AFM frequencies (normalized to 1 at T=0°K) show exactly the same temperature dependence (Fig.6). This temperature dependence fits perfectly with the theoretical α(T) variation in the molecular field approximation, taking $T_N = 156°K$. It suggests that all the coefficients of (1), except α, are temperature independent. This assumption allows comparison with ref.5, at 77°K (Table I, lines 2 and 3). In our opinion, the 2.14 g value used in ref.5 is underestimated and may partly explain some residual discrepancies between the two sets of coefficients. Finally, using our values, we calculated at 77°K, the spontaneous magnetic moment $\sigma_0 = 0.068 \mu_B$ per Ni ion, a value near the $0.059 \mu_B$ experimental one [3], and a susceptibility $\chi_c = 79 \cdot 10^{-6}$ in excellent agreement with the $77 \cdot 10^{-6}$ experimental results [3].

Acknowledgements. It is a pleasure to thank P.P. Syrnikov who grew the $NaNiF_3$ crystal.

REFERENCES

1. S. Ogawa, J. Phys. Soc. Japan, 15, 2361 (1969).
2. A. Epstein, J. Makovski, M. Melamud and H. Shaked, Phys. Rev., 174, 560 (1968).
3. V.M. Judin and A.B. Sherman, Phys. Stat. Sol., 20, 759 (1967).
4. W. Rudorff, J. Kandler and D. Babel, Z. Anorg. All. Chem., 317, 261 (1962).
5. E.I. Golovenchitz, V.A. Sanina and A.G. Gurevitch, Sov. Phys. Solid State, 11, 516 (1969).
6. S.R. Chinn, H.J. Zeiger and J.R. O'Connor, Phys. Rev. B3, 1709 (1971).
7. J.P. Van Der Ziel and L.G. Van Uitert, Phys. Rev., 179, 343 (1969).
8. R.V. Pisarev et al, (to be published).
9. G.F. Herrman, J. Phys. Chem. Solids, 24, 597 (1963).
10. R. Loudon, Adv. in Phys., 17, 243 (1968).
11. P.A. Fleury, p.157 in Proc. Inter. Conf. of Light Scattering in Solids, edited by Flammarion (1972).

ANISOTROPIC EXCHANGE EFFECTS IN THE OPTICAL SPECTRUM
OF FERROMAGNETIC Er^{3+}:$Tb(OH)_3$*

R. L. Cone[†] and W. P. Wolf
Becton Center, Yale University, New Haven, Ct. 06520

ABSTRACT

The effects of anisotropic exchange have been investigated using the optical spectrum of isolated Er^{3+} impurities in ferromagnetic $Tb(OH)_3$. Due to simplifications resulting from the nature of the system, the exchange effects can be isolated from those of other interactions and analyzed using an effective operator applicable to all levels of the Er^{3+} ion. A least squares fit using four parameters has provided a good description of both the signs and magnitudes of eleven observed splittings. The resulting contributions of the anisotropic terms are generally an order of magnitude larger than that of the isotropic exchange in this case, indicating that any realistic analysis of ion-ion interactions involving ions with large orbital admixtures must consider the effects of anisotropy in the exchange interaction.

INTRODUCTION

A number of past experiments[1,2] have clearly shown that the exchange interaction can be anisotropic; yet, detailed analysis of the results based on the theoretical work of Levy[3] and Elliott and Thorpe[4] has been hampered by two main problems: the competing effects of other interactions and the lack of sufficient experimental information to determine all of the parameters in the theory. In certain carefully chosen systems,[2] however, it is possible to separate the contributions of the exchange interaction from those of other interactions and to analyze the effects of anisotropy in a general way. This is illustrated by analysis of the splitting of isolated doublet energy levels of Er^{3+} impurity ions in the Ising-like ferromagnet $Tb(OH)_3$.

ENERGY LEVELS OF Er^{3+} IN $Tb(OH)_3$

When an Er^{3+} ion is introduced into a $Tb(OH)_3$ crystal, a number of new energy levels result which may be referred to as localized magnons or excitons[5] and can be further classified as highly-localized modes involving excitation of the Er^{3+} ion or so-called shell modes involving excitations of the neighboring Tb^{3+} ions. Analysis of these energy levels must include the effects of four types of ion-ion interaction in addition to the free ion and crystal field effects.

* Supported in part by U.S. Atomic Energy Commission and National Science Foundation.
[†] Present address: Department of Physics and Astronomy, University of Georgia, Athens, Georgia 30601.

These interactions are the electronic exchange, the magnetic dipole-dipole interaction, the electric multipole interaction, and virtual-phonon coupling, each of which can be written in terms of products of two spherical tensor operators, one on each of the respective ions.[1-4] Since the highest rank for these operators is 7 for f electrons,[1] it follows that the above interactions cannot couple the low-lying states of a Tb^{3+} ion in $Tb(OH)_3$: $|^7F_6, J_z=6>$ and $|^7F_6, J_z=-6>$. All other crystal field states of the Tb^{3+} ions are above 118 cm^{-1}; hence, the admixture of these states into the low-lying states due to the above interactions should be negligible. It is thus reasonable to think of the crystal states as products of a state of the impurity ion and states of the individual Tb^{3+} ions where only the low-lying states of Tb^{3+} need to be considered.[6]

The energy levels can then be determined using perturbation theory, and for the highly-localized modes the result is striking. The corrections to the energy levels of the impurity ion are then a first order effect with all higher-order terms negligible to a very good approximation. The magnetic dipole-dipole interaction and the electronic exchange interaction will split the levels in first order, while the electric multipole interaction and virtual-phonon coupling will only contribute equal shifts to both components which can be incorporated into the crystal field. Since the contribution of the magnetic dipole-dipole interaction can be calculated using the experimental Zeeman splitting factors[7] and the dipole field for $Tb(OH)_3$,[8] we can isolate the contributions of the exchange interaction.

The energies of the shell modes can be estimated to be higher than that of the ground state by an amount equal to the ground state splitting of Tb^{3+}, 6.4 cm^{-1}. At 1.3°K these modes will not be appreciably populated, thus we need to consider them no further.

CONTRIBUTION OF THE EXCHANGE INTERACTION TO THE SPLITTINGS

Levy[3] has derived an expression which represents the exchange interaction between two ions in states Ψ_1 and Ψ_2, which are antisymmetrized only with respect to the n_1 electrons on ion 1 and the n_2 electrons on ion 2, by an effective operator:

$$<\Psi_1' \Psi_2' | \sum_{i=1}^{n_1} \sum_{j=1}^{n_2} - \frac{e^2}{r_{ij}} P_{ij} | \Psi_1 \Psi_2> \quad (1)$$

$$= <\Phi_1' \Phi_2' | \sum_{i,j}^{n_1 n_2} \sum_{\substack{k_1 k_2 \\ q_1 q_2}} - \Gamma_{q_1 q_2}^{k_1 k_2} u_{q_1}^{(k_1)}(i) u_{q_2}^{(k_2)}(j) \{\tfrac{1}{2} + 2\vec{s}_i \cdot \vec{s}_j\} | \Phi_1 \Phi_2>.$$

To determine the effect of this operator on the highly-localized modes, we use the following constraints:[2]
(1) We require time reversal symmetry for the crystal and a Hermitian energy matrix.
(2) The nature of the Tb^{3+} states requires that $q_2=0$.
(3) The final operator which acts on the Er^{3+} ion must transform like the identity representation of the magnetic point group C_{3h}.

The resulting effective operator involves eight parameters α_{kq} and β_{k0}, which are linear combinations of the Γ's and Tb^{3+} matrix elements.

$$\mathcal{H}_{exch} = \sum_{i=1}^{11} \left(2\left[\alpha_{00} u_0^{(0)}(i) + \alpha_{20} u_0^{(2)}(i) + \alpha_{40} u_0^{(4)}(i) \right. \right.$$
$$\left. + \alpha_{60} u_0^{(6)}(i) + \alpha_{66} \{ u_6^{(6)}(i) + u_{-6}^{(6)}(i) \} \right] s_0^{(1)}(i) \quad (2)$$
$$\left. + \frac{1}{2} \left[\beta_{10} u_0^{(1)}(i) + \beta_{30} u_0^{(3)}(i) + \beta_{50} u_0^{(5)}(i) \right] \right).$$

$\mathcal{H}_{exch}$ is expressed as a sum of single electron operators[9] and is valid for all Er^{3+} states in the f^{11} configuration. Matrix elements of the tensor operators were calculated using Racah's methods.[2]

EXPERIMENTAL RESULTS AND ANALYSIS

As we mentioned above, the shell modes are not appreciably populated at 1.3°K due to the large Tb^{3+} ground state splitting and thus they do not contribute to the absorption spectrum. On the other hand, the Er^{3+} splitting turns out to be much smaller (0.87 cm^{-1}) and transitions from both components are observable in optical absorption experiments.[7] Using the selection rules for linearly polarized light determined by the C_{3h} site symmetry and comparing the relative intensities of the transitions, it was possible to determine not only the magnitudes of the splittings but also the <u>relative signs</u>. Since the theory predicts these signs, a much more rigorous test of the theory results.

In order to choose states for which the Er^{3+} wavefunctions should be most reliable, only states for which good agreement was found between theory and experiment for the parallel Zeeman effect were chosen.[7] The eleven experimental levels which met this requirement are given in Table I.[7]

A straightforward least squares procedure was employed since the splittings depend linearly on the parameters. When all terms in Eq 2 were included, the results indicated that four of the parameters were essentially zero. The results of the final fit using the remaining four parameters are given in Table II, while the final parameters are given in Table III. From Table II it can be seen that the theory provides a good description of both the signs and magnitudes of the eleven splittings. By contrast, a fit which included only isotropic exchange (the α_{00} term in Eq 2) could not reproduce even the signs.

Table I Spontaneous magnetic splittings of Er^{3+} in ferromagnetic $Tb(OH)_3$ at 1.3°K, with dipolar and exchange contributions. All energies are in cm^{-1}.

Level (a)	μ	E_{av} (b)	Splitting ΔE (c)	ΔE_d (d)	ΔE_{exch} (e)
$^4I_{15/2}$	5/2	0.00	-0.87±0.01	-0.77±0.01	-0.10±0.02
$^4I_{11/2}$	3/2	10275.14	0.92±0.01	1.06±0.02	-0.14±0.03
	1/2	10279.22	0.05±0.01	0.22±0.02	-0.17±0.03
	1/2'	10322.80	-4.71±0.01	-4.39±0.02	-0.32±0.03
	3/2'	10341.11	-3.47±0.01	-3.57±0.03	0.10±0.04
$^4I_{9/2}$	3/2	12434.06	-0.61±0.01	-0.69±0.06	0.08±0.07
	1/2	12581.25	0.49±0.01	0.45±0.03	0.04±0.04
$^4F_{9/2}$	3/2	15294.42	-4.01±0.01	-4.31±0.04	0.30±0.05
	1/2	15362.01	0.66±0.01	0.51±0.02	0.15±0.03
$^4S_{3/2}$	3/2	18445.89	2.33±0.01	2.11±0.03	0.22±0.04
	1/2	18497.78	0.50±0.01	0.83±0.03	-0.33±0.04

(a) J manifold and μ quantum number (irreducible representation of site group C_{3h}).
(b) $E_{av} = [E(\mu) + E(-\mu)]/2$, where $E(\mu)$ and $E(-\mu)$ indicate the energies of the two components of a doublet level.
(c) $\Delta E = E(\mu) - E(-\mu)$
(d) Calculated magnetic dipole-dipole contribution of the splitting. $\Delta E_d = -2\mu_z H_d$ where $H_d = 9.33 \pm 0.13$ kilogauss.
(e) $\Delta E_{exch} = \Delta E - \Delta E_d$

The inadequacy of the isotropic exchange alone to describe the splittings is further emphasized by the fact that the contributions of the anisotropic terms involving α_{20}, α_{40}, and α_{60} to the various splittings are generally an order of magnitude larger than those of the isotropic term.

CONCLUSION

By careful choice of the system, it has been possible to take into account the anisotropy present in the exchange interaction without introducing unjustified simplifications to reduce the number of parameters. Moreover, the agreement between theory and experiment has provided support for the theoretical work of Levy[3] and Elliott and Thorpe[4] and added to the experimental evidence indicating that any realistic analysis of ion-ion interactions involving ions with large orbital admixtures must take into account the anisotropy of the exchange interaction.

Further details are being published elsewhere.[2]

Table II   Comparison of theory and experiment for 4 parameter fit to exchange splittings of Er^{3+} doublet levels in $Tb(OH)_3$ at 1.3°K.

Level		Exp.	Calc.	Error
$^4I_{15/2}$	5/2	-0.10±0.02	-0.13	-0.03
$^4I_{11/2}$	3/2	-0.14±0.03	-0.13	0.01
	1/2	-0.17±0.03	-0.12	0.05
	1/2'	-0.32±0.03	-0.35	-0.03
	3/2'	0.10±0.04	0.05	-0.05
$^4I_{9/2}$	3/2	0.08±0.07	0.21	0.13
	1/2	0.04±0.04	0.18	0.14
$^4F_{9/2}$	3/2	0.30±0.05	0.21	-0.09
	1/2	0.15±0.03	0.14	-0.01
$^4S_{3/2}$	3/2	0.22±0.04	0.15	-0.07
	1/2	-0.33±0.04	-0.22	0.11

Table III Exchange parameters for fit to exchange splitting of Er^{3+} doublet levels in $Tb(OH)_3$ at 1.3°K.

$\alpha_{00} = -0.069 \pm 0.022$ cm^{-1} $\alpha_{40} = 1.97 \pm 0.44$ cm^{-1}

$\alpha_{20} = -1.45 \pm 0.10$ cm^{-1} $\alpha_{60} = -1.14 \pm 0.11$ cm^{-1}

ACKNOWLEDGEMENTS

The authors wish to thank Professor P. M. Levy and Professor M. F. Thorpe for many helpful discussions.

REFERENCES

1. Reviews have been given by W. P. Wolf, J. Phys. (Paris) 32, C1-26 (1971) and J. M. Baker, Rep. Prog. Phys. 34, 109 (1971).
2. R. L. Cone and W. P. Wolf (to be published).
3. P. M. Levy, Phys. Rev. 177, 509 (1969).
4. R. J. Elliott and M. F. Thorpe, J. Appl. Phys. 39, 802 (1968).
5. R. A. Cowley and W. J. L. Buyers, Rev. Mod. Phys. 44, 406 (1972).
6. An additional requirement is that none of the Er^{3+} excited states are near a Tb^{3+} energy level.
7. R. L. Cone, J. Chem. Phys. 57 (to be published Dec. 1, 1972).
8. A. T. Skjeltorp, Ph.D. Thesis, Yale University, 1971, Univ. Microfilm 71-31013.
9. The spherical tensor operators $u^{(k)}$, which act on the orbital angular momentum of a single electron, are defined by the reduced matrix elements $(\ell||u^{(k)}||\ell) = 1$.

EXCHANGE INTERACTIONS OF SOME DIVALENT IRON GROUP IONS IN HYDRATED COMPLEXES[†]

C. L. Francis and J. W. Culvahouse
Department of Physics and Astronomy, University of Kansas
Lawrence, Kansas 66044

ABSTRACT

Measurements of the non-dipolar part of the spin-spin interactions of divalent iron group ions in double nitrate crystals by the pair technique have been made. The exchange interaction between pairs of ions in the X and Y sites can be described by an antiferromagnetic isotropic interaction $K_i \vec{S}_1(i) \vec{S}_2(i)$ between the ionic spins. The values of K_i have been determined for the following pairs: Mn(X)-Ni(Y), K_i = 0.0428 cm^{-1}; Co(X)-Ni(Y), K_i = 0.0618 cm^{-1}; Co(Y)-Ni(X), K_i = 0.0679 cm^{-1}. These values plus measurements reported by other investigators on Co(X)-Co(Y) and Ni(X)-Ni(Y) pairs agree with an accuracy of the order of 10% with the values predicted by a single parameter model in which only e_g orbitals are involved in the exchange process. Exchange interactions with both ions in the X site have been measured for pairs of Co(X)-Ni(X) ions. The non-dipolar component $J_\parallel S_1^z S_2^z$ of the anisotropic interaction between the effective spins has been measured. The value $J_\parallel$ = -0.010 cm^{-1} (ferromagnetic) along with the previously reported measurements of Co(X)-Co(X) and Ni(X)-Ni(X) leads us to conclude that for the X-X interaction exchange between e_g-e_g orbitals is ferromagnetic while exchange between t_{2g}-e_g orbitals is antiferromagnetic.

INTRODUCTION

The general chemical formula for the double nitrates is $[M(NO_3)_6]_2 [M' \cdot 6H_2O]_3 \cdot 6H_2O$ where M is a trivalent rare earth ion and M' is a divalent iron group ion. We use the abbreviated notation MM'N with the chemical symbol for the elements M and M'. The double nitrates provide an appropriate material for the investigation of spin-spin interactions since it is possible to substitute a wide variety of divalent paramagnetic ions and since there exist two different divalent sites (the X and the Y sites). Each divalent ion is surrounded by an octahedron of water molecules. The bonding is primarily through hydrogen bonds so the resulting superexchange is weak but still stronger than the dipole-dipole interaction. Culvahouse and Schinke[1] and Dixon and Culvahouse[2] have used EPR pair spectra to measure the non-dipolar spin-spin interactions of Co-Co and Ni-Ni pairs in LaZnN and LaMgN. They found that the only non-dipolar interactions were between pairs of ions in the nearest neighbor (NN) X-X and the NN X-Y sites. In this paper we report measurements of the spin-spin interactions between the following pairs of ions: Co(X)-Ni(Y) and Co(Y)-Ni(X) in LaZnN, LaMgN, and BiMgN; Mn(X)-Ni(Y) in LaZnN; and Co(X)-Ni(X) in BiMgN.

X-Y PAIRS

References 1 and 2 have shown that for Co-Co and Ni-Ni X-Y pairs the non-dipolar interaction can be explained quite precisely by an isotropic antiferromagnetic interaction between the ionic spins of the form

$$H_{12} = K_i \vec{S}_1^{(i)} \cdot \vec{S}_2^{(i)} \tag{1}$$

In Ref. 1 the validity of isotropic exchange between ionic spins of Co^{2+} was shown to hinge upon involvement of only the e_g orbitals in the exchange process. This interaction can be expressed in terms of effective spin operators as

$$H_{12} = J_\| S_1^z S_2^z + J_\perp (S_1^+ S_2^- + S_1^- S_2^+)/2, \tag{2}$$

where

$$J_\| = (1/4) g_{s_\|}(X) g_{s_\|}(Y) K_i, \tag{3}$$

and

$$J_\perp = (1/4) g_{s_\perp}(X) g_{s_\perp}(Y) K_i, \tag{4}$$

and g_s is the spin contribution to the g tensor. Further, for any pair of ions A and B containing configurations with two e_g holes

$$K_i = \frac{k f_A f_B}{4 \left|\vec{S}_A^{(i)}\right| \left|\vec{S}_B^{(i)}\right|}, \tag{5}$$

where k is an exchange constant for the two e_g holes and f is a factor which depends on admixture of other configurations. If k does not vary from ion to ion, the exchange between ionic spins is given by a single parameter. Measurements for cobalt pairs and nickel pairs reported in Ref. 1 and 2 yield $K_i(Co) = 0.0438 \pm 0.0010$ cm^{-1} and $K_i(Ni) = 0.096 \pm 0.002$ cm^{-1}. From the known ionic configurations one calculates $f_{Co} = 0.985$ and $f_{Ni} = 1.00$ which yields k values of 0.406 cm^{-1} and 0.384 cm^{-1} respectively. If one assumes that the differences are independent of the ion in the other site and are due entirely to the changes in the transfer integrals and overlap integrals between the e_g orbitals on the ions in the X and Y sites and the ligands, then the results for mixed pairs should be related to those for identical pairs by

$$K_i(A(X)-B(Y)) K_i(B(X)-A(Y)) = K_i(A(X)-A(Y)) K_i(B(X)-B(Y)). \tag{6}$$

In order to test the validity of (6), measurements of K_i(Co-Ni) were undertaken. The experiments were performed using a superheterodyne spectrometer operating at 16-17 GHz. All measurements were made at 4.2K and below. Crystals were grown from solutions containing 3.0% Co^{2+} and 0.2% Ni^{2+} in LaZnN, LaMgN, and BiMgN. The low concentration of Ni relative to Co facilitated the observation of Ni-Co spectra near isolated Ni spectra without interference from Ni-Ni pairs. Spectra of Co(X)-Ni(Y) pairs were identified about the Ni(Y) $|1\rangle \rightarrow |0\rangle$ low field isolated ion transition and Co(Y)-Ni(X) pair spectra were identified about the Ni(X) $|-1\rangle \rightarrow |0\rangle$ isolated ion transition in LaZnN, LaMgN, and BiMgN. In addition Co(Y)-Ni(X) pair spectra were identified about the Ni(X) $|0\rangle \rightarrow |1\rangle$ transition in LaZnN. The data were analyzed in the same manner used in Ref. 1 and 2. The spin Hamiltonian parameters for Co^{2+} are listed in Ref. 1 for all three salts and those for Ni^{2+} in LaZnN and LaMgN are listed in Ref. 2. The spin Hamiltonian parameters for Ni^{2+} in BiMgN were measured in this experiment with the results: g = 2.245 ± 0.008, D(X) = 0.300 ± 0.003 cm^{-1}, and D(Y) = 2.46 ± 0.04 cm^{-1}. The values of $J_{\parallel}$ and K_i derived from the pair spectra are listed in Table I. Comparison of the intensity of the pair spectra at 4.2K and 1.2K verified that the interaction is antiferromagnetic. Since K_i(Co(X)-Ni(Y)) and K_i(Co(Y)-Ni(X)) differ by much more than the experimental error, these results show that the exchange interaction between dissimilar ions in the X and Y sites depends on which ion is in which site. It is notable that the geometric mean of these values, 0.0648 ± 0.0008 cm^{-1}, is in excellent agreement with the prediciton of 0.0648 ± 0.0010 cm^{-1} obtained from the RHS of (6) and the data in Ref. 1 and 2. Thus, the validity of (6) is verified. If one assumes that the major contribution to the exchange constant k in (5) is proportional to the product of the transfer integrals between the e_g orbitals and ligands and is independent of the ion in the other site, then the values for the Co-Co, the Ni-Ni, and the two types of Co-Ni pairs are consistant to within 0.2% with the result that the transfer integral at the Y site increases by 7.6% from Ni to Co whereas the transfer integral at the X site decreases by 1.8% from Ni to Co.

TABLE I. Experimental Values of $J_{\parallel}$ and K_i for X-Y Pairs

Type of pairs	Host	$J_{\parallel}$ (cm^{-1})	K_i (cm^{-1})
Co(X)-Ni(Y)	LaZnN	0.104 ± 0.002	0.0618 ± 0.0012
	LaMgN	0.098 ± 0.002	0.0619 ± 0.0012
	BiMgN	0.113 ± 0.002	0.0708 ± 0.0014
Co(Y)-Ni(X)	LaZnN	0.176 ± 0.002	0.0680 ± 0.0010
	LaZnN	0.175 ± 0.002	0.0677 ± 0.0010
	LaMgN	0.176 ± 0.002	0.0679 ± 0.0010
	BiMgN	0.197 ± 0.003	0.0761 ± 0.0011

The similarity of the results for LaZnN and LaMgN suggests that the exchange is not very sensitive to changes in the divalent ions in the host crystal. The exchange interactions listed in Table I for Ni-Co pairs indicate a large sensitivity to the trivalent ions used in the host. From the geometric mean of the two types of Ni-Co pairs in BiMgN and using (5) and (6), one estimates that the values of K_i for Ni pairs and Co pairs in this material would be 0.112 ± 0.002 cm^{-1} and 0.048 ± 0.001 cm^{-1} respectively, a 10% increase over the La salts.

Note added in proof: A measurement of Co(X)-Co(Y) pairs in BiMgN yields K_i(Co) = 0.0472 ± 0.0012 cm^{-1} which is in good agreement with the prediction of 0.048 ± 0.001 cm^{-1}, providing further evidence of the validity of (6).

The complexity of the Mn^{2+} spectra makes it impractical to analyze the Mn-Mn pair spectra. Therefore we have used the Mn(X)-Ni(Y) pair interactions to investigate the exchange constant for Mn pairs. Crystals were grown from solutions containing 3.0% Mn^{2+} and 0.2% Ni^{2+} in LaZnN. The search for pair spectra was made near the transitions of the Ni(Y) ion with the field along the crystal symmetry axis. This region is well removed from that occupied by the Mn^{2+} spectra. Since the spin of Mn^{2+} is 5/2, six pair lines are expected about the isolated Ni^{2+} line. Five of these were observed, the sixth being obscured by the Ni^{2+} line. The value of K_i(Mn(X)-Ni(Y)) is 0.0428 ± 0.0014 cm^{-1}. If one uses (6), one obtains K_i(Mn(X)-Mn(Y)) = 0.0191 ± 0.0014 cm^{-1}. Analysis by Culvahouse[3] of susceptibility measurements made by Sapp[4] leads to an experimental value of 0.0174 ± 0.003 cm^{-1} which is in adequate agreement. Using the value of k = 0.406 cm^{-1} appropriate to Co pairs (5) yields K_i(Mn(X)-Mn(Y)) = 0.0162 ± 0.0004 cm^{-1}. Assuming that the changes in transfer integrals observed for Ni and Co can be linearly extrapolated to Mn^{2+} one predicts K_i(Mn(X)-Ni(Y)) = 0.0363 cm^{-1}. This value is 16% below the experimental value and discredits the idea that the transfer integrals show a linear variation with atomic number.

X-X PAIRS

Reference 1 reported that the interaction between NN X-X Co^{2+} pairs was antiferromagnetic and anisotropic whereas Ref. 2 reported that the interaction between NN X-X Ni^{2+} pairs was ferromagnetic and isotropic. In order to explain the anisotropy of the Co^{2+} pairs Ref. 1 assumed that t_{2g} orbitals as well as e_g orbitals were involved in the exchange process. However, no model which explained the magnitude or sign of the interaction was found. Therefore we undertook the measurement of Co(X)-Ni(X) pairs with the hope of gaining additional insight into the X-X interaction.

Since the two ions are dissimilar the analysis of the pair spectra proceeds in the same manner as for X-Y pairs. The pair spectra about the Ni(X) ion should consist of two lines corresponding to the two orientations of the NN Co(X) spin. In LaZnN and LaMgN it was possible to observe only one of these lines due to a shift of the pair spectra relative to the isolated ion which causes one of the pair lines to fall under the isolated ion spectra. In BiMgN it was possible to observe both pair lines, but only when oriented along the symmetry axis. Thus we were able to measure only the constant $J_{\|}$ of (2). The value we obtained is $J_{\|} = -0.010 \pm 0.002$ cm^{-1}. The sign of the interaction was determined to be ferromagnetic by comparison of the intensity of the pair lines at 4.2K and 1.2K.

Table II lists the values of $J_{\|}$ and the ground state orbitals for the three X-X interactions which have been measured. From an examination of the values listed in Table II, we conclude that exchange between two e_g orbitals is ferromagnetic while exchange between e_g-t_{2g} orbitals is antiferromagnetic. However, we have not been able to find a model which adequately explains the X-X interaction.

TABLE II. Values of $J_{\|}$ for X-X Pairs

Type of pairs	Ground state orbitals	$J_{\|}$ (cm^{-1})
Co-Co	$t_{2g}^1 e_g^2 - t_{2g}^1 e_g^2$	0.064 ± 0.003
Ni-Ni	$e_g^2 - e_g^2$	-0.095 ± 0.003
Co-Ni	$t_{2g}^1 e_g^2 - e_g^2$	-0.010 ± 0.002

REFERENCES

† Research supported in part by National Science Foundation Grants Nos. GP-15256 and GH-34582. Helium gas supplied by Office of Naval Research Contract No. NOnr-2775(00).
1. J. W. Culvahouse and David P. Schinke, Phys. Rev. 187, 671 (1969).
2. R. T. Dixon and J. W. Culvahouse, Phys. Rev. B3, 2279 (1971).
3. J. W. Culvahouse, Private Communication.
4. R. C. Sapp, Private Communication.

WEAK FERROMAGNETISM IN BiFeO$_3$ - NdFeO$_3$ SOLID SOLUTIONS*

Van E. Wood and A. E. Austin
Battelle-Columbus Laboratories, Columbus, Ohio 43201

ABSTRACT

There are three phases in the pseudobinary solid solution Bi$_{1-x}$Nd$_x$FeO$_3$ - the rhombohedral ferroelectric BiFeO$_3$ phase for $x \leq 0.1$, a tetragonal phase near $x = 0.15$, and the orthorhombic NdFeO$_3$ structure for $x \geq 0.2$. Intrinsic weak ferromagnetism is present in all three phases. In rhombohedral Bi$_{.9}$Nd$_{.1}$FeO$_3$, the room temperature moment is about 3 mμB/formula unit after thermomagnetic annealing. The moment increases fairly smoothly with increasing Nd content through all three phases, at least up to $x = 0.5$. The presence of weak ferromagnetism in the BiFeO$_3$ phase is explained on the basis of the effect of Nd in regularizing the Fe-O distances, thus affecting superexchange at least locally. These results are compared with those found by others for Bi$_{1-x}$La$_x$FeO$_3$ and Bi$_{1-x}$Pr$_x$FeO$_3$.

INTRODUCTION.

We have been investigating materials which may be both ferroelectric and ferromagnetic at or near room temperature. Such materials may possess unique memory or signal-processing capabilities. It has been reported that a weak magnetic moment may be induced in ferroelectric-antiferromagnetic[1] BiFeO$_3$ by partial replacement of Bi by a rare earth,[2,3] Pr or La. Neodymium appeared to us to be a more attractive ion from the standpoint of ionic size and unique valency. Accordingly, we investigated properties of the pseudobinary solid solutions Bi$_{1-x}$Nd$_x$FeO$_3$.

MATERIAL PREPARATION AND PHASE ANALYSIS

The solid solutions were made by reacting the desired ratio of Nd$_2$O$_3$ and Fe$_2$O$_3$ with excess Bi$_2$O$_3$ in platinum under one atmosphere of oxygen. This was done at 750 C, below the decomposition temperature of BiFeO$_3$, to avoid formation of Bi$_2$Fe$_4$O$_9$. Starting materials were reagent grade Bi$_2$O$_3$, 99.9% pure Nd$_2$O$_3$ and 99.999% pure FeCl$_2$ converted to Fe$_2$O$_3$. To promote reaction and homogenization, the materials were given multiple firings after intermediate grinding and cold pressing at 80,000 psi. Excess Bi$_2$O$_3$ was removed by multiple leaching of ground material in concentrated nitric acid. This method was found necessary to produce single-phase materials as determined by X-ray diffraction, optical microscopy and electron-probe microanalysis.

---

* Work supported by Advanced Research Projects Agency, U.S. Department of Defense, and monitored by U.S. Army Missile Command under Contract No. DAAH01-70-C-1076.

Reaction of nominal stoichiometric compositions did not yield single-phase materials. Under these conditions, another Bi-deficient ternary phase, of composition $Bi_{.66}Nb_{.25}FeO_{3-\delta}$ dominated. This previously unreported phase is tetragonal with $a = 3.90$ Å and $c = 9.02$ Å. Minor α-Fe_2O_3 and $Bi_2Fe_4O_9$ phases also occurred in some runs.

Along the pseudo-binary $Bi_{1-x}Nd_xFeO_3$, three solid solution phases exist at room temperature. These are rhombohedral $BiFeO_3$ for $0 \leq x \leq 0.1$, a tetragonal phase for $Bi_{.85}Nd_{.15}FeO_3$, and the orthorhombic orthoferrite for $0.2 \leq x \leq 1.0$. Table 1 lists unit cell parameters and molar volumes. In the rhombohedral $BiFeO_3$ phase, there is only a slight decrease in a_R and volumes Ω with addition of Nd up to $x = 0.1$.

Table 1 Crystal structure data on $Bi_{1-x}Nd_xFeO_3$

| | Unit Cell Parameters (Å) and Molar Volume (Ω, Å^3) | | | | | |
| | Rhombohedral | | Orthorhombic | | | Ω |
Composition	a_R	α	a_o	b_o	c_o	
$BiFeO_3$	5.638	59°18'	-	-	-	62.3
$Bi_{.95}Nd_{.05}FeO_3$	5.631	59°18'	-	-	-	62.1
$Bi_{.9}Nd_{.1}FeO_3$	5.624	59°22'	-	-	-	62.0
$Bi_{.85}Nd_{.15}FeO_3$	-	-	5.60	5.60	7.80	61.2
$Bi_{.80}Nd_{.20}FeO_3$	-	-	5.43	5.59	7.90	59.9
$Bi_{.75}Nd_{.25}FeO_3$	-	-	5.43	5.57	7.90	59.7
$Bi_{.65}Nd_{.35}FeO_3$	-	-	5.44	5.59	7.80	59.3
$Bi_{.5}Nd_{.5}FeO_3$	-	-	5.42	5.58	7.80	59.0
$NdFeO_3$*	-	-	5.44	5.57	7.75	58.8

* S. Geller and E. A. Wood, Acta Cryst. 9, 563 (1956).

This pseudobinary system is similar to that reported by Viskov[2] for $BiFeO_3$-$PrFeO_3$, except that the limits of phase fields in $Bi_{1-x}Nd_xFeO_3$ are at smaller x. In the $Bi_{1-x}La_xFeO_3$ system,[3] the rhombohedral solid solution extends to $x \approx 0.18$ and the tetragonal phase from $x \approx 0.18$ to 0.54. This shift in the limits of solid solutions can be accounted for by the decrease in ionic radii from 1.18 Å for La^{3+} and 1.14 Å for Pr^{3+} to 1.12 Å for Nd^{3+}, and the corresponding shift from 12-fold coordination toward eight-fold coordination with oxygen in the orthoferrites.[4] Thus the solid solution of Bi^{3+} with radius 1.11 Å is greatest in $NdFeO_3$ because of the closer ionic size. However, the volume decreases from $BiFeO_3$ to $NdFeO_3$ primarily because of the change in the coordination of the rare-earth (or Bi) ion with oxygen, 12-fold coordination in $BiFeO_3$ going to 8 closer oxygens at 2.4 to 2.6 Å and 4 further-removed oxygens at 3.2 to 3.3 Å in $NdFeO_3$ by rotation and tilt of the oxygen octahedra.

The tetragonal phase in $Bi_{1-x}Pr_xFeO_3$ is asserted to be ferroelectric.[2] Piezoelectricity tests at 77 and 300 K on $Bi_{1-x}Nd_xFeO_3$ solid solutions, which were positive for all the rhombohedral-phase materials, showed no evidence of acentricity in the tetragonal phase sample $Bi_{.85}Nd_{.15}FeO_3$, and the possibility of ferroelectricity in this phase is therefore remote.

MAGNETIC PROPERTIES

Since the magnetic transition temperatures for these materials were expected to be around 400 C, magnetic susceptibility measurements were made from room temperature to around 500 C. Much above 500 C, there is some danger of dissociation of the compounds. All the samples showed strong thermomagnetic effects. It was thus necessary to specify a measuring procedure to compare samples of different compositions. The procedure adopted was that of gradually increasing the temperature and at each temperature of measurement cycling the magnetic field until a linear variation of total magnetization with field was obtained. Once the ordering temperature was surpassed and the maximum measurement temperature reached, the sample was cooled either in the residual field of the magnet (a few oersteds) or a strong field (10 kOe) and the susceptibility measured in the same way during cooling. The only significant change from the heating part of the cycle was the expected increase in the moment at all temperatures below the Curie point. Cooling in the residual field led to an increase in the room-temperature magnetization by a factor of around 1.2 to 2, while cooling in a high field could increase it by a factor of 3 or so.

The high-field susceptibility data showed that in all compositions the magnetic transition took place in the temperature range between about 360 and 410 C. The transition, substantially antiferromagnetic but with weak associated ferromagnetism in most cases, appears to be of first order for $x \leq 0.2$. Differential thermal analysis on samples with $x = 0$ and $x = 0.15$ did not, however, reveal any evidence of a latent heat. The largest jump in susceptibility at the ordering temperature occurs for the orthorhombic sample $x = 0.2$.

In Fig. 1 apparent spontaneous magnetic moments are shown for the initial heating cycle. The strong diminution in the magnitude of the moment at the magnetic ordering temperature in all cases except $x = 0$ indicates that it is weak ferromagnetism of the (Bi, Nd) ferrite that is observed, and not some impurity effect. The large moment of the x = 0.25 sample above the transition temperature results from a small amount of garnet impurity. The apparent small moment in $BiFeO_3$ is not intrinsic, but is also the result of an impurity, probably $\alpha\text{-}Fe_2O_3$; we believe there is no moment as large as 10^{-4} μ_B per Fe^{3+} in pure $BiFeO_3$. The moment increases monotonically with x. This is strongly at variance with results of measurements[4] on $Bi_{1-x}Pr_xFeO_3$, where the moment reached a maximum for x = 0.25 and then decreased rapidly at least to x = 0.6. In $Bi_{1-x}La_xFeO_3$, there is again a small moment[3] in the rhombohedral phase, not found in $Bi_{1-x}Pr_xFeO_3$, and a moment increasing monotonically with x in the orthoferrite phase.

DISCUSSION

The most interesting result of these measurements is the occurrence of small moments in samples having the rhombohedral, pyroelectric, and presumably ferroelectric, $BiFeO_3$ structure. This moment, about 1 $m\mu_B$ per magnetic ion in $Bi_{.9}Nd_{.1}FeO_3$, can be increased to about 3 $m\mu_B$ by thermomagnetic annealing. A definite drop in the magnetization near 400 C leads us to think there is also a moment of around 0.1 μ_B in $Bi_{.95}Nd_{.05}FeO_3$. Since the space group of $BiFeO_3$ has finally been established[1] as R3c, the magnetic space group of the solid solutions is apparently R3c', and the magnetic and electric moments both lie along the trigonal axis.

The limit of solid solution in rhombohedral $BiFeO_3$ for the three systems occurs at the same volume, that corresponding to a pseudo-cubic cell edge of 3.95 Å, a decrease from 3.965 Å for $BiFeO_3$. Shifts in cation displacement and oxygen positions with rare-earth solid solution could account for the volume decrease. Such shifts should be in the direction of reducing the disparity between Bi-O and Fe-O distances in $BiFeO_3$, thus reducing ferroelectric polarization.

In the orthoferrite phase, the near-neighbor Fe-O-Fe superexchange angles (calculated on the basis of constant average Fe-O distance) are near constant for $0.35 \leq x \leq 1.0$, a range in which the magnetization doubles, showing that the magnitude of the weak moment is not entirely determined by near-neighbor superexchange. In the range $0.20 \leq x \leq 0.35$, the c(b') axis superexchange decreases sharply. The corresponding rapid magnetization increase reflects the tilting of the oxygen octahedra, which is not allowed in the rhombohedral structure. In the rhombohedral phase, the cation and oxygen shifts with x very likely lead to some initial increase in the near-neighbor superexchange angles. The resulting increase in the magnitude of the antisymmetric exchange might account for the observed small moment. As an alternative explanation, localized anisotropy might occur as a result of distortions in the oxygen octahedra around the rare earth.

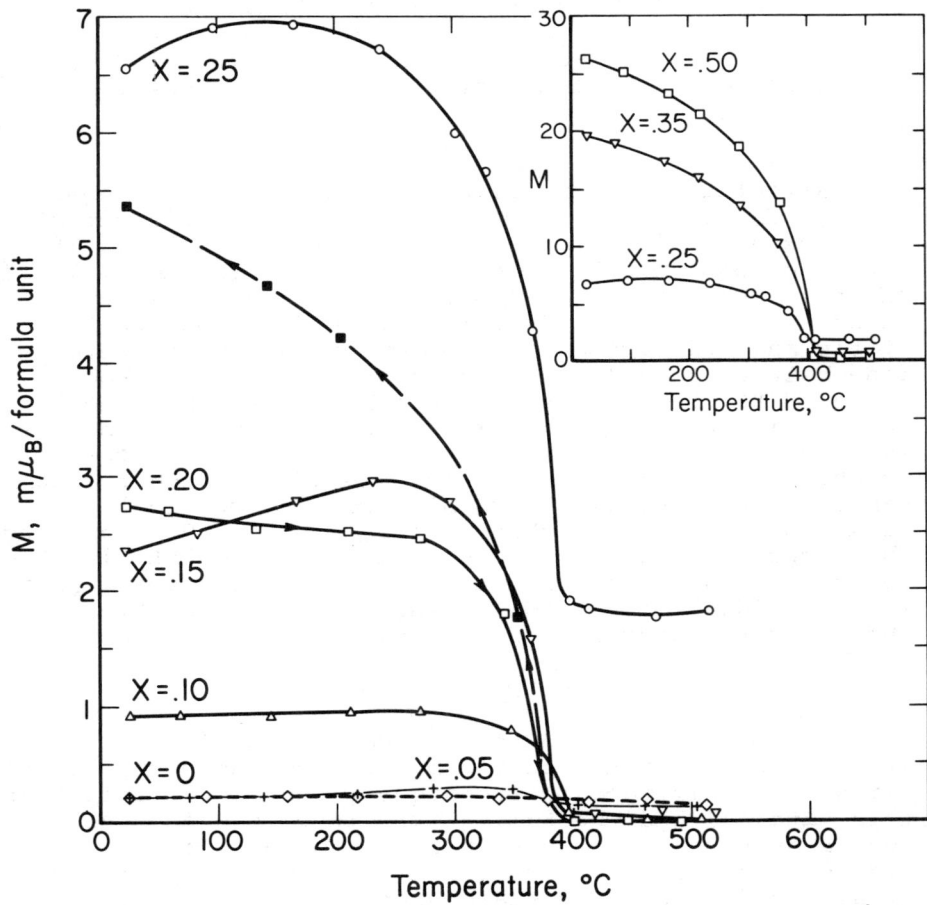

Figure 1 Remanent magnetization of $Bi_{1-x}Nd_xFeO_3$, temperature increasing. For x = 0.2, thermoremanence on cooling also shown.

We wish to thank J. F. Miller, R. Smith, N. F. Hartman, K. C. Brog and E. W. Collings for their contributions to this work.

REFERENCES

1. J. M. Moreau, C. Michel, R. Gerson and W. J. James, J. Phys. Chem. Solids 32, 1315 (1971) and references therein.
2. A. S. Viskov, Yu. N. Venevtsev, V. M. Petrov, and A. F. Volkov, Inorganic Materials 4, 71 (1968).
3. Yu. E. Roginskaya, Yu. N. Venevtsev, S. A. Fedulov and G. S. Zhdanov, Soviet Phys. Crystallog. 8, 490 (1964).
4. M. Marezio, J. P. Remeika and P. D. Dernier, Acta. Cryst. B26, 2008 (1970).

MAGNETIC PROPERTIES OF Sr_2FeO_3F, A HIGHLY ANISOTROPIC OXI-FLUORIDE

J. H. Schelleng
Naval Research Laboratory, Washington, D.C. 20390

ABSTRACT

Strontium iron oxi-fluoride has the "two-dimensional" K_2NiF_4 structure in which the Fe^{3+} ions are situated in an axially distorted octahedron of ligand ions. The fluorine ions, which prefer the axial sites and fill half of them, subject the Fe^{3+} ions to a large axial crystal field. To study the effect that this field has on the static magnetic properties, moment measurements were made between 5 and 700°K and in fields up to 19 kOe. Above 50°K, the susceptibility exhibits an apparent Curie-Weiss paramagnetism with a Curie constant that is less than one-half that expected for free Fe^{3+} ions. It is shown that this behavior is consistent with the dominance of the crystal field anisotropy energy in the paramagnetic state of this material. Below 50°K, although the temperature dependence of the susceptibility in this region is again unusual, the appearance of a small parasitic residual moment due to slight Fe^{2+} impurities indicates the presence of long range magnetic order.

INTRODUCTION

The preparation and electrical properties of Sr_2FeO_3F were first reported by Galasso and Darby[1]. Using X-ray diffraction techniques, they determined that this material had the K_2NiF_4 structure in which the Fe ion is surrounded by an axially distorted octahedron of ligand ions. The fluorine ions were found to prefer the axial sites which they share with an equal number of oxygen ions. The distribution of the fluorines on these sites is said to be random. This would lead to the presence of three different types of Fe sites with zero, one and two fluorine neighbors. In the latter two types, the magnetic ion would be in a site having considerable axial asymmetry. If ordered antiferromagnetically, the exchange between iron ions in adjacent basal planes of this structure cancel giving this material a two-dimensional magnetic character. The presence, in any order, of different ligand ions in the axial sites will not destroy this property.

MEASUREMENTS

A sintered sample of the material was synthesized

following the Galasso and Darby procedure[1]. The structure of the sample was checked by X-ray powder diffraction and was found to agree with their results. No noticeable second phase was observed. The chemical composition of the material was checked by wet chemical analysis and by fluorine activation analysis and was found to be correct within the experimental errors of these techniques, although 0.5% of the iron present was in the form of Fe^{2+}. The magnetic measurements were made on solid pieces as well as on powders ground from sintered disks. The magnetic properties of the powdered samples were found to change with time and X-ray studies of these powders showed that slight shifts in the positions of the diffraction peaks and considerable changes in their relative intensities occurred as a result of this aging. The changes are believed to be due to atmospheric moisture that causes the fluorine ions in the crystal structure to be replaced by hydroxyl radicals. On the other hand, the solid samples were found to be quite stable.

Using a PAR Model 155 vibrating sample magnetometer, the magnetization was measured as a function of applied field (from 0 to 19 kOe) and temperature (between 5 and 300°K). The moment was found to vary linearly with applied field at each temperature. The resulting inverse susceptibility is shown as a function of temperature in Fig. 1.

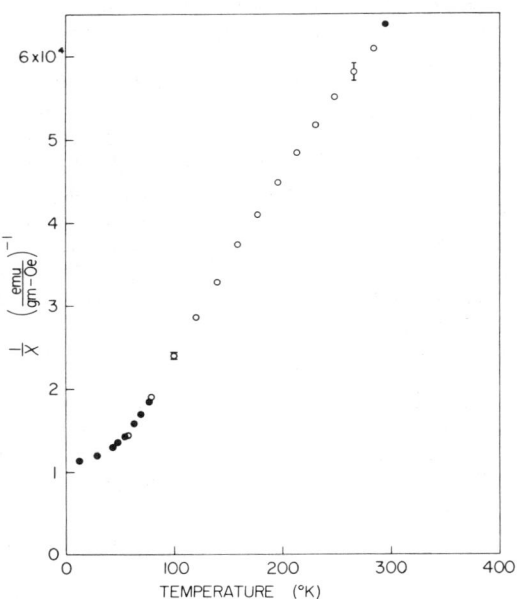

Fig. 1
$1/\chi$ vs T for Sr_2FeO_3F. Open and closed circles show different runs.

The data can be divided into two segments. Above approximately 50°K the curve is nearly linear, with only a slight curvature. If this is taken to indicate paramagnetic behavior, the Curie constant per Fe^{3+} ion, as calculated from the slope just above 50°K was found to be 0.210×10^{-23} emu - °K/gauss while that taken from the slope just below 300° was found to be 0.294×10^{-23} emu - °K/gauss. These values are to be compared to the value of 0.727×10^{-23} emu - °K/gauss that is theoretically expected for $S = 5/2$ ions. Moment measurements were also made between 300 and 700°K. Although quantitative results in this temperature range were made difficult by the small signal and the high noise level introduced by the oven, no sharp transitions were noted, and the Curie constant remained at about one-half that expected for $S = 5/2$ ions.

Below 50°K the inverse susceptibility curve deviates strongly from a straight line, with decreasing slope as the temperature approaches zero. A small residual moment also appears below 50°K. This moment was found to depend both on temperature and upon past conditions of magnetic annealing. The largest value noted for this residual moment was 0.48% of the saturation moment calculated for Fe^{3+} ions in the material.

DISCUSSION

Sr_2FeO_3F appears to have a magnetic transition at about 50°K between an antiferromagnetic and a paramagnetic state. This conclusion, and the $S = 5/2$ character of the ions, are supported by Mössbauer measurements made at this laboratory.[2] The small residual moment seen in the antiferromagnetic state is probably due to the small Fe^{2+} impurity. In contrast with the behavior of a typical simple antiferromagnet, Sr_2FeO_3F exhibits a gradual magnetic transition and its powder susceptibility continues to increase with decreasing temperature. A final analysis of the low temperature region has not yet been made but possible models based on a two-dimensional magnetic structure and on a high uniaxial anisotropy are being considered in the present work.

A high uniaxial anisotropy might provide an explanation for the anomalous value of the Curie constant found in the paramagnetic region. The appropriate single ion hamiltonian, in the molecular field approximation, is given by

$$\mathcal{H} = -AS_z^2 - \mu_B g \vec{H} \cdot \vec{S} + J \vec{S} \cdot (\vec{S}_{av})$$

where the z axis is taken to be the anisotropy axis and

where A, J and H, respectively, are the anisotropy and exchange constants and the applied field. In contrast to the normal situation, the anisotropy term will be taken as the dominant term, and the Zeeman and exchange terms taken as the perturbations. Note that this does not necessarily mean that the constant A is large with respect to the constant J. In the paramagnetic state at suitably low applied fields, S_z^2 can be large compared to $\vec{S} \cdot (\vec{S}_{av})$, which involves the average spin of the neighbors. This average spin will be given by

$$\vec{S}_{av} = \frac{\sum_i \langle \vec{S} \rangle_i\, e^{-\mathcal{H}/kT}}{\sum e^{-\mathcal{H}/kT}}$$

where $\langle \vec{S} \rangle_i$ is the expectation value of the spin in the state i. The anisotropy term splits the ground state sextet into three $|\pm m\rangle$ doublets and it will be assumed that perturbation interactions connecting these doublets can be neglected. When one has the applied magnetic field and, consequently, the exchange field directed along the z-axis, the $|\pm m\rangle$ doublets will be split. The magnetic susceptibility for this case can then be described by a Curie constant $C_\parallel$ and a paramagnetic Curie temperature $\theta_\parallel$, the form given by

$$C_\parallel = \eta_\parallel \frac{N(g\mu_B)^2}{kT} \quad ; \quad \theta_\parallel = \eta_\parallel \frac{J}{k} \qquad (1)$$

where

$$\eta_\parallel = \frac{\sum_m m^2 e^{-A m^2/kT}}{\sum_m e^{-A m^2/kT}}$$

If $kT \gg A$, then $\eta_\parallel$ becomes $\frac{35}{12}$ and $C_\parallel$ and $\theta_\parallel$ reduce to the values normally found for the low anisotropy case. When the external magnetic field is applied along the x-axis, however, the perturbations will mix the $|\pm\tfrac{1}{2}\rangle$ doublet components. The magnetic susceptibility for this case can then be described by formulae analogous to Eq. (1), except that

$$\eta_\perp = \frac{3\, e^{-A/4kT}}{\sum_m e^{-A m^2/kT}}$$

which in the limiting case of $kT \gg A$ becomes $9/12$. The parallel and perpendicular susceptibilities are combined in the usual way to find the powder susceptibility. Again for $kT \gg A$ and $kT \gg J$, this leads to

$$\chi = \tfrac{1}{3}\chi_\parallel + \tfrac{2}{3}\chi_\perp \approx \frac{C}{1.98}\,\frac{1}{T - 2.18\, J/k} \qquad (2)$$

where C is the normal low anisotropy Curie constant.

Thus for S = 5/2, the high temperature Curie constant observed in a powder sample of a material having high uniaxial anisotropy would be very nearly one-half that normally obtained in the low anisotropy case. It should be noted that Eq. (2) need not exactly apply to the case of Sr_2FeO_3F as the conditions of $kT \gg A$ and J may not be strictly valid. The situation is further complicated by the probable presence of several types of ligand configurations. For this reason no attempt has been made to fit this powder data to get values for A and J. It would seem more important at this point to test the basic theory by measuremnts on single crystals and at high fields. Both these approaches are underway in the present work. A word might be said about the order of magnitude of the exchange J. An extrapolated slope of the $1/\chi$ vs. T curve taken just below 300°k gives a paramagnetic Curie temperature of -68°K. Application of Eq. (2) indicates an antiferromagnetic total exchange with $J/k = -31°K$. This is reasonably close to the exchange that one might deduce from the ordering temperature as well. It must be remembered however, that while the exchange applicable to the paramagnetic state includes the interactions between ions in adjacent planes, the exchange applicable to the ordered state will not, because of the two dimensionality of the lattice. The difference, however, should be small.

The author would like to thank Dr. V. J. Folen for suggesting the problem and Dr. D. W. Forester for helpful discussions and for permission to mention his Mössbauer results.

[1] F. Galasso and W. Darby, J. Phys. Chem. **67**, 1451 (1963).
[2] D. W. Forester; Private communication.

CRYSTALLINE ELECTRIC FIELD LEVELS IN DILUTED RARE EARTH ALUMINUM CUBIC LAVES PHASES

A. Furrer, W. Buehrer, W. Haelg, and H. Heer
Delegation AF, Swiss Federal Institute for Reactor Research
CH-5303 Wuerenlingen, Switzerland

J. Kjems
A.E.C. Research Establishment Risø, DK-4000 Roskilde, Denmark

H. G. Purwins and E. Walker
Departement physique de la matiere condensee, Universite
CH-1200 Geneva, Switzerland

ABSTRACT

The crystalline electric field levels of intermetallic Laves phase compounds of type $R_xY_{1-x}Al_2$ (R = Pr, Er, Tm) have been determined by neutron spectroscopy. The polycrystalline samples have been diluted with yttrium in order to reduce magnetic interactions between the rare earth ions and thereby reducing the widths of the crystal field transition lines. However, other relaxation mechanisms and the small crystal field splittings made it impossible to separate the crystal field transition peaks completely. The intensity arising from crystal field transitions has been identified by its temperature dependence as well as by comparison with corresponding "non-magnetic" YAl_2 spectra. A least squares fitting procedure has been applied to the experimental data in order to obtain the best possible crystal field parameters. At temperatures $T > 50$ K and momentum transfers $Q > 1$ $Å^{-1}$ two-phonon processes have to be taken into account in analyzing the measured energy spectra. The resulting crystal field level schemes agree qualitatively with estimates based on a nearest neighbour point charge model and quantitatively with measured susceptibility data.

MAGNETIC PROPERTIES OF SOME RARE-EARTH GOLD COMPOUNDS*

L. R. Sill and S. R. Snow
Northern Illinois University, DeKalb, Illinois 60115

A. J. Fedro
Northern Illinois University, DeKalb, Illinois 60115
and
Argonne National Laboratory, Argonne, Illinois 60439

ABSTRACT

A magnetic susceptibility study of the b.c. tetragonal $MoSi_2$-type intermetallic compounds RAu_2, where R is one of the heavy rare earths, has been made at temperatures from 2.5 to 300°K in applied fields up to 26 kOe. The temperature dependence of the inverse susceptibility at high temperatures for these compounds followed the Curie-Weiss law giving effective moments close to those of the trivalent rare-earth ions. At low temperatures antiferromagnetic behaviour was observed in all of these compounds. In addition, a first order magnetic transition was observed below their Neel temperatures for $TbAu_2$ and $DyAu_2$. These observations are in agreement with the neutron diffraction results of Atoji. The results are interpreted by using the RKKY interaction along with an anisotropic field term.

INTRODUCTION

The heavy rare earth metals (R = Gd,Tb,Dy,Ho,Er & Tm) form a series of compounds RAu_2 which crystallize with the body-centered tetragonal $MoSi_2$-type structure.[1] The AB_2 compounds where A is one of the heavy rare earths and B one of the magnetic transition elements ordinarily exhibit Laves-phase type crystal structures and have been studied extensively.[2] Much less is known about the AB_2 compounds where A is one of the heavy rare earth elements and B is a non-magnetic element. For the latter compounds there is a competition between crystal structures of the type AlB_2 and its varient $CeCu_2$ and the type $MoSi_2$. The compounds AB_2 where A is one of the heavy rare earths and B is Cu are isostructural with $CeCu_2$ and were found to be antiferromagnetic and to exhibit metamagnetic behaviour.[3]

We have been pursuing a systematic study of the magnetic properties of the system RAu_2. We recently reported the results of magnetic susceptibility and Mossbauer measurements on $TbAu_2$.[4] We now

---

*Based on work performed under the partial support of a NSF grant and partially under the auspices of the U.S. Atomic Energy Commission.

report the results of magnetic susceptibility measurements on the entire series of RAu$_2$ compounds. Neutron-powder-diffraction measurements by Atoji on the compounds RAu$_2$ (R = Tb,Dy,Ho,Er & Tm) have shown that all of these compounds order antiferromagnetically with an initial incommensurate moment alignment at low temperatures.[5-9] Since Gd has an extremely large neutron absorption cross section, a neutron diffraction measurement has not been possible for GdAu$_2$. Our measurements give new information on the magnetic properties of GdAu$_2$.

EXPERIMENTAL

The compounds were made by arc melting a stoichiometric mixture of the heavy rare earth (99.9+% pure) and gold (99.99% pure). To insure homogeneity the buttons were melted several times, the buttons being turned between each melting. Weight losses were less than 1%. These buttons were homogenized at about 900°C in an evacuated capsule for about 3 days. Spheres approximately 1/8 inch in diameter were prepared from the buttons. The spheres were annealed in evacuated capsules at a temperature of about 600°C for about 3 hours and furnace cooled. X-ray photographs showed that the samples had the proper structure and were free of a second phase.

A commercial version of the vibrating sample magnetometer was used to make the measurements. Image effects were corrected for in the calibration process. The magnetization data were least square fitted to straight lines of $1/\chi$ vs T giving the Curie constant, paramagnetic Curie temperature and a temperature independent susceptibility term. The errors in the effective moment and temperature measurements are estimated to be ±1% and ±0.5% respectively.

RESULTS

The temperature dependence of the magnetization of the RAu$_2$ compounds is shown in Figures 1-5. A summary of the results of the susceptibility measurements is shown in Table 1. The inverse susceptibility vs temperature curves for all of the compounds except GdAu$_2$ deviated considerably from linearity in the region immediately above their respective transition temperatures. The table lists the effective moments calculated from the linear part of the $1/\chi$ vs temperature curves where the Curie-Weiss law is valid.

The susceptibility for each of the compounds at the lowest temperature was field independent, therefore, the magnetization peaks are interpreted as indicating antiferromagnetic transitions and the discontinuity in magnetization found in TbAu$_2$ and DyAu$_2$ as indicating a first-order antiferromagnetic transition.

Metamagnetic behaviour was observed by Miura et al. in $DyAu_2$ at 4.2°K in applied fields up to 100 kOe.[10] The first critical field was reported as being about 35 kOe. Apparently the critical fields, if they exist, in the compounds RAu_2 where R is given by Ho, Er & Tm are large enough so as to not manifest themselves through a reduction in the magnitude of the peaks in the magnetization curves in applied fields up to 26 kOe.

The strong deviation from Curie-Weiss behaviour over a limited temperature region above their respective transition temperatures for R = Tb, Dy, Ho, Er & Tm is apparently due to the crystal field acting on the magnetic ions since it is not present in $GdAu_2$ where one would expect such an effect to be negligible since the Gd ion is in an all-spin state. An effective magnetic moment which is larger than the free ion value is almost always observed when Gd metal is combined chemically with a non-magnetic metal and is usually attributed to the polarization of the conduction electron spins by the exchange interaction between spins of the tripositive Gd ions. For example, GdCu & GdDy exhibit effective moments of 8.46^{11} and 8.57^{12} Bohr magnetons respectively. We observed such a phenomenon in the case of $GdAu_2$ too. The large effective moment found on the tripositive Ho ion cannot be explained. We point out, however, that in the cases of Ho_3Al_2, HoAl and $HoAl_3$ the effective moments are $10.9,^{13}$ 11.2^{14} and 10.9^{15} Bohr magnetons respectively.

DISCUSSION

The general features of the magnetic behaviour of the RAu_2 compounds can be understood using a slightly modified version of the RKKY theory assuming the s-f exchange integral is element dependent due to changes in the 4f wave function as one goes through the rare earth series. Considerations of stability are related to $I(\vec{q})$, the Fourier transform of the effective f-f exchange interaction with $\vec{q}$ the pitch parameter. Previous results by Fedro and Shaffer have suggested the feasibility of a metastable incommensurate phase at high temperatures, where exchange forces are dominant, which is driven to the commensurate phase at lower temperatures by anisotropic crystalline field effects similar to that in rare earth metals.[16] Such behaviour is found in $TbAu_2$ and $DyAu_2$. The absence of a second transition in $GdAu_2$ suggests that the system at the Neel temperature is already in the commensurate phase and remains locked in that state as the anistropic crystalline field terms grow when the temperature is lowered. The magnetic structure of the RAu_2 compounds where R is Ho, Er and Tm is complicated just as in the case of the rare earth metals themselves since the exchange energy is getting smaller through its dependence on the deGennes factor and competition with crystal field effects becomes important.

Table I Summary of magnetic susceptibility results

	T_{N_α} (°K)	T_{N_β} (°K)	$\mu_{eff}(\mu_\beta)$
GdAu$_2$	48	--	8.38
TbAu$_2$	55	42.5	9.83
DyAu$_2$	32	24	10.52
HoAu$_2$	9	--	10.97
ErAu$_2$	6	--	9.45
TmAu$_2$	3.5	--	7.62

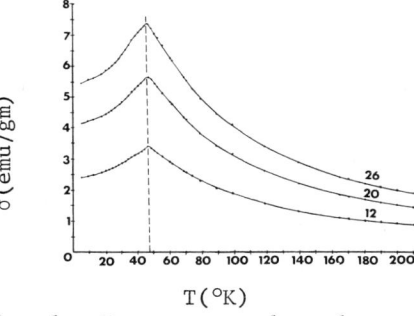

Fig. 1. Temperature dependence of the magnetization of GdAu$_2$

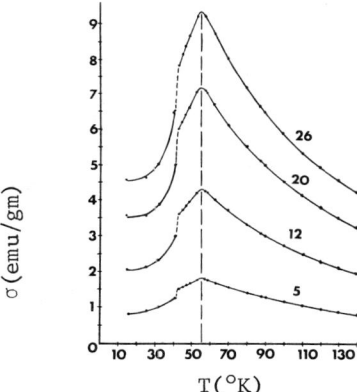

Fig. 2. Temperature dependence of the magnetization of TbAu$_2$

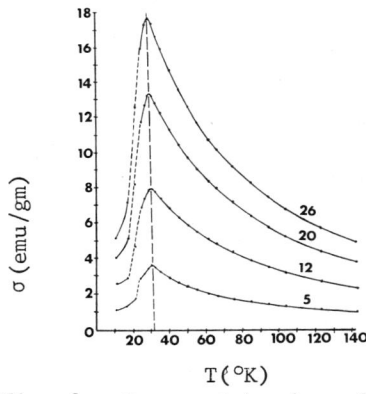

Fig. 3. Temperature dependence of the magnetization of DyAu$_2$

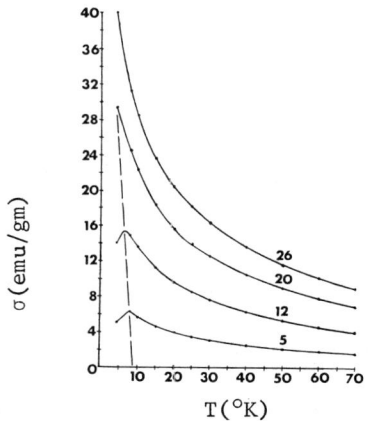

Fig. 4. Temperature dependence of the magnetization of HoAu$_2$

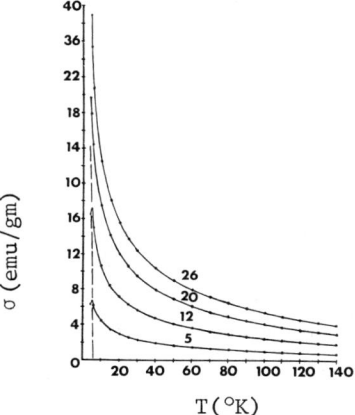

Fig. 5. Temperature dependence of the magnetization of ErAu$_2$

REFERENCES

1. A. E. Dwight and R. A. Conner, Jr., Act. Cryt. $\underline{22}$, 745 (1967).
2. W. E. Wallace, "Progress in the Science and Technology of the Rare Earths," Pergamon, New York, Vol. III, p. 1 (1968).
3. R. C. Sherwood, H. J. Williams and J. H. Wernick, J. App. Phy., $\underline{35}$, 1049 (1964).
4. L. R. Sill, A. J. Fedro and C. W. Kimball, Intern. J. Magn., $\underline{1}$, 319 (1971).
5. M. Atoji, J. Chem. Phy., $\underline{48}$, 560 (1968).
6. M. Atoji, J. Chem. Phy., $\underline{51}$, 3877 (1967).
7. M. Atoji, J. Chem. Phy., $\underline{52}$, 6433 (1970).
8. M. Atoji, J. Chem. Phy., $\underline{57}$, 2402 (1972).
9. M. Atoji, J. Chem. Phy., $\underline{57}$, 2407 (1972).
10. S. Miura, T. Kaneko, M. Ohoski and K. Kamigaki, J. Phy. Paris, $\underline{32}$, C1-1124 (1971).
11. R. E. Walline and W. E. Wallace, J. Appl. Phy., $\underline{42}$, 604 (1965).
12. R. E. Walline and W. E. Wallace, J. Appl. Phy., $\underline{41}$, 3285 (1964).
13. B. Barbara, C. Bechle, R. Lemaire and R. Panthenet, J. Appl. Phy., $\underline{39}$, 1084 (1966).
14. A. M. van Diepen, H. W. de Wijn and K. H. J. Buschow, Phy. Stat. Soc., $\underline{29}$, 189 (1968).
15. K. H. J. Buschow and J. F. Fast, Z. Phy. Chem., $\underline{50}$, 1 (1966).
16. A. J. Fedro and J. C. Shaffer, AIP Conf. Proc. $\underline{5}$, 1390 (1971).

MAGNETIC PROPERTIES OF $Y_{1-x}Th_xFe_3$ AND $Lu_{1-x}Th_xFe_3$

C. J. Kunesh, K. S. V. L. Narasimhan, and R. A. Butera
Department of Chemistry
University of Pittsburgh
Pittsburgh, Pennsylvania 15213

ABSTRACT

The pseudo-binary systems $Y_{1-x}Th_xFe_3$ and $Lu_{1-x}Th_xFe_3$ have been investigated with respect to magnetic properties. All compounds studied exhibit the $PuNi_3$-type structure. For $x \leq .3$, the magnetization vs. temperature behavior for both systems is consistent with a simple ferromagnetic iron sublattice. For $x > .3$ (excluding $x = 1$), these systems show anomalies in M vs. T and M vs. H data. These anomalies arise from a non-collinear to collinear spin structure transition within the iron sublattice. The present results are explained on the basis of a hybrid local moment-band model originally proposed by Friedel.

INTRODUCTION

The $Gd_{1-x}Th_xFe_3$ system has been found to exhibit anomalous magnetic behavior[1]. Compounds in this system with $x > .5$ show abnormally low magnetization values at temperatures below approximately 230°K and applied fields of less than 10 kOe. Subsequently, a study of the $Dy_{1-x}Th_xFe_3$ system and the compounds $Y_{.1}Th_{.9}Fe_3$, $Lu_{.1}Th_{.9}Fe_3$, and $Sc_{.1}Th_{.9}Fe_3$ was made[2] in which similar anomalous behavior was observed[3]. In the present study, the previous work on $Y_{.1}Th_{.9}Fe_3$ and $Lu_{.1}Th_{.9}Fe_3$ has been extended across the entire range of composition. These systems allow the magnetic behavior of the iron sublattice to be investigated directly as a function of composition without the interference of other moment-carrying atoms.

EXPERIMENTAL

All compounds were prepared from stoichiometric amounts of the metals (Y, Lu 99.9%, Fe Johnson-Matthey spectroscopic grade and Th 99.99%) by induction melting in a water-cooled copper boat under an atmosphere of flowing purified argon. The ingots were wrapped in tantalum foil and sealed in evacuated quartz tubing and annealed for 3-4 weeks at 950°C. After annealing, the compounds were examined by X-ray diffraction and thermomagnetic analysis for the presence of secondary phases. All but one of the $Y_{1-x}Th_xFe_3$ compounds reported in this study were determined to be single phase. $Y_{.3}Th_{.7}Fe_3$ contained a very small amount of 2-7 phase. For the $Lu_{1-x}Th_xFe_3$ system, single phase samples were obtained for $x = .1$ to $.45$, while small amounts of 2-7 phase were present for $x = .5$ to $.9$. Magnetic measurements were performed on powder samples by means of the Faraday force balance technique.

RESULTS AND DISCUSSION

The variation of magnetization with temperature for all compounds studied is shown in Figures 1-2. These figures show the onset of the anomalous magnetic behavior as a function of increasing Th content in the $Y_{1-x}Th_xFe_3$ and $Lu_{1-x}Th_xFe_3$ systems. The variation of magnetization with applied magnetic field below the transition temperature (77°K) is shown for the various compositions of the $Y_{1-x}Th_xFe_3$ system in Figure 3. This data shows that in the anomalous composition range (x > .3) the magnetic transition can be induced by a critical applied field ranging from ~1 to 16 kOe., depending on composition. The critical field increases with increasing Th content (x) to about .7 and then decreased with further increase in x. The M vs. H curves for the anomalous compounds all show initial ferromagnetic behavior for low field values followed by the onset of the field-induced transition. The complete magnetization curve appears to be a composite of ferromagnetic and antiferromagnetic magnetizations.

In a previous paper [2] a model was proposed for these anomalous compounds based upon a Friedel oscillation mechanism.[3] The model postulates that the alignment of magnetic moment between two neighboring transition metal atoms (say A and B) will be ferromagnetic or antiferromagnetic, depending upon the sign of the Friedel oscillation produced by atom A at the site of atom B. The Friedel oscillation is a spherical fluctuation in the d-band charge density arising from the perturbation produced by the transition metal ion core on the d-band free electron wave functions. The wavelength (λ) of the Friedel oscillation is inversely proportional to the d-band Fermi wave vector (k_F). In turn k_F is large for half filled bands and small for almost empty or almost filled bands. Upon formation of $ThFe_3$, the holes in the iron d-band are partially filled by thorium electrons. This corresponds to a decrease in k_F and an increase in λ since the Fe d-band is more than half filled. If we now substitute trivalent metals for Th we increase k_F (decrease d-band concentration) and decrease λ. Therefore, there could be a transition from antiferromagnetic to ferromagnetic alignment with increasing temperature for compounds with the proper combination of iron-iron distances and λ.

According to the above model the M vs. H data for the anomalous $Y_{1-x}Th_xFe_3$ compounds should show antiferromagnetic behavior below the transition temperature. However, the 77°K M vs. H data for the anomalous $Y_{1-x}Th_xFe_3$ compounds appear to show ferromagnetic contributions. This behavior can be accounted for within the general framework of the above mentioned model by considering the $PuNi_3$ structure in greater detail. It is known that the $PuNi_3$ structure contains three cyrstallographically non-equivalent iron sites.[4] We assume that the 3(b), 6(c) and 18(h) sites each constitute a magnetic sublattice, and that the 3(b) and 6(c) sublattices are ferromagnetic and aligned parallel to each other in all compounds studied and at all temperatures below T_c. Using the Friedel model this is equivalent to assuming that for all compounds and temperatures the 3(b) and 6(c) sites are always within the main (positive) peak of the Friedel oscillation produced by the nearest neighbors in their magnetic sublattice. Or, more precisely, that the vector sum of the Friedel oscillations of all surrounding Fe atoms is always positive at the 3(b) and 6(c) sites. We assume, however, that the 18(h) sublattice behaves differently. Figure 4 shows hypothetical Friedel oscillations about an 18(h) iron site in $ThFe_3$, $Y_{.1}Th_{.9}Fe_3$, and $Y_{.9}Th_{.1}Fe_3$ as a function of 18(h) - 18(h) distance. In $ThFe_3$ and $Y_{.9}Th_{.1}Fe_3$, atom B is in a positive portion of the Friedel oscillation and the 18(h) sublattice will be ferromagnetically aligned at all temperatures and apparently parallel to the 3(b)-6(c) sublattice moments. For $Y_{.1}Th_{.9}Fe_3$, there will be a transition from antiferromagnetic to ferromagnetic alignment within the 18(h) sublattice with increasing temperature.

Good support for this model comes from the shape of the M vs. H curves. Figure 5 shows the M vs. H curve for $Y_{.1}Th_{.9}Fe_3$ at 4.2°K and the hypothetical M vs. H curves for the 3(b)-6(c) sublattice and 18(h) sublattice resolved from the experimental curve as follows.

We assume that magnetization observed up to a field of 8 kOe arises mainly from the 3(b)-6(c) ferromagnetic sublattice since the 18(h) sublattice is antiferromagnetically coupled and contributes a near zero moment. Upon increasing the field beyond 8 kOe the 18(h) sites become ferromagnetically aligned and we observe an increase in the total moment. Since there are two 18(h) iron atoms for every 3(b)-6(c) atom, it might be expected that the 3(b)-6(c) M vs. H curve should saturate at half the magnetization value for the 18(h) site curve. However, Blow[4] has found that in $ThFe_3$ the 3(b)-6(c) sites have a hyperfine field 20% less than that of the 18(h) sites. Mössbauer studies performed in this laboratory on $Y_{.1}Th_{.9}Fe_3$ and $Gd_{.2}Th_{.8}Fe_3$[5] give similar results. Since hyperfine fields are in general proportional to localized magnetic moments, it appears that the 3(b)-6(c) iron atoms carry a lower moment than do the 18(h) sites. In Figure 5 the experimental curve appears to saturate at a value of 3.0 Bohr magnetons. Using this value and the Mössbauer findings a saturation moment of 0.86 Bohr magnetons was obtained for the 3(b)-6(c) sublattice contribution. The magnetization curve for the 18(h) sublattice was obtained by subtracting the 3(b)-6(c) curve from the experimental curve.

While the proposed model explains the main features of the bulk magnetic measurements, the final determination of the spin structure of these materials must await the results of neutron diffraction experiments which are currently in progress.

REFERENCES

1. J. E. Greedan, AIP Conf. Proc., No. 5 Part 2, 1425 (1971).
2. C. J. Kunesh, K. S. V. L. Narasimhan and R. A. Butera, J. Phys. Chem. Solids (in press).
3. J. Friedel, G. Leman, and S. Olszewski, J. Appl. Phys. $\underline{32}$, 3255 (1961).
4. S. Blow, J. Phys. C., $\underline{3}$, 159 (1970).
5. C. J. Kunesh, unpublished work.

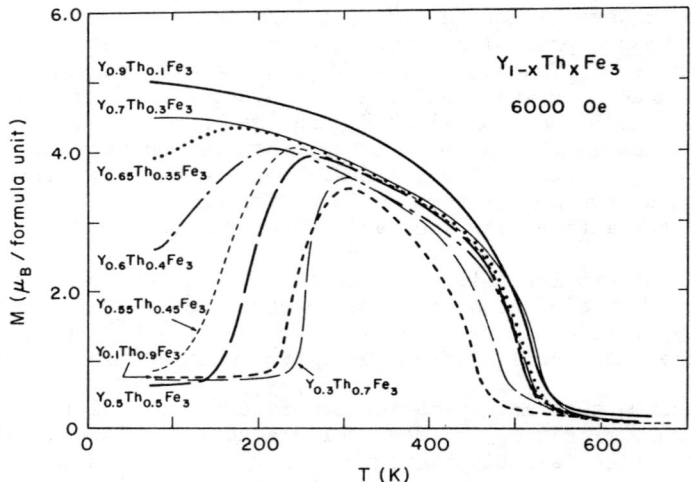

Fig. 1. Magnetization versus temperature for $Y_{1-x}Th_xFe_3$.

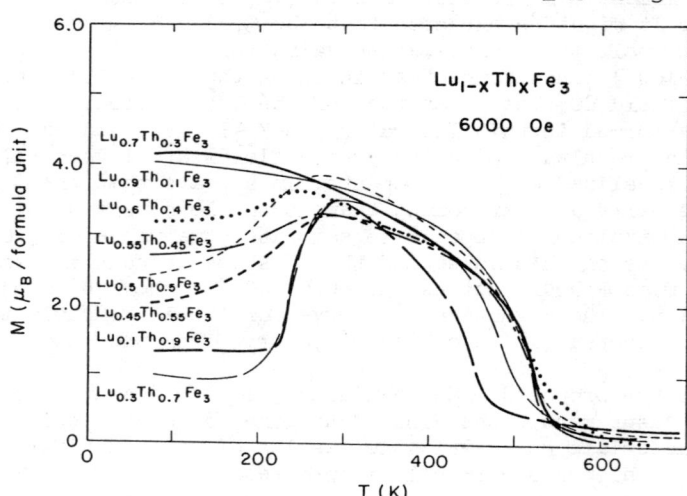

Fig. 2. Magnetization versus temperature for $Lu_{1-x}Th_xFe_3$.

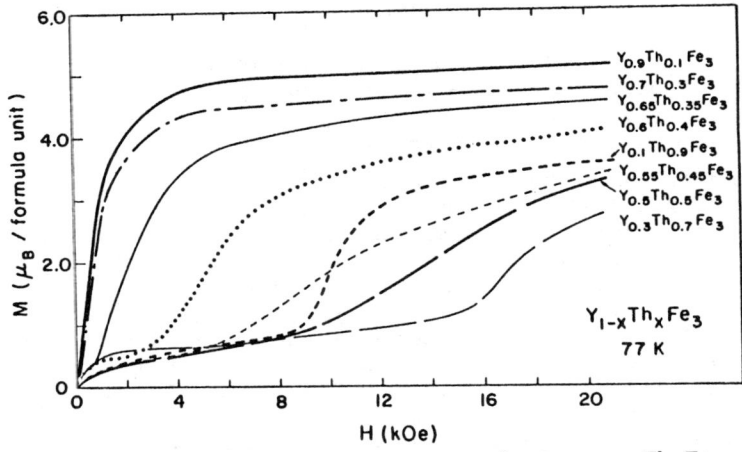

Fig. 3. Magnetization versus field at 77°K for $Y_{1-x}Th_xFe_3$.

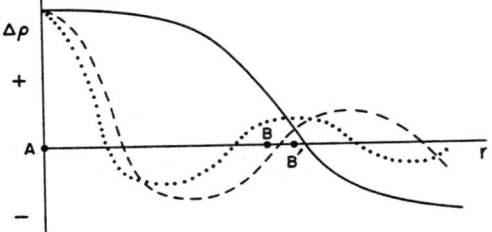

Fig. 4. Hypothetical Friedel waves for $ThFe_3$ (solid), $Y_{.1}Th_{.9}Fe_3$ (dashed), and $Y_{.9}Th_{.1}Fe_3$ (dotted).

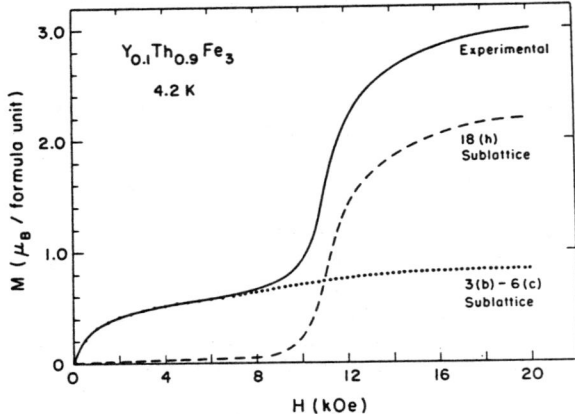

Fig. 5. Resolution of M vs. H curve for $Y_{.1}Th_{.9}Fe_3$ at 4.2°K into 3(b)-6(c) and 18(h) sublattice contributions.

THEORY OF MAGNETIC PHASE TRANSITIONS IN METALLIC ACTINIDE COMPOUNDS*

Paul Erdös and John M. Robinson
The Florida State University, Tallahassee, Fla. 32306

ABSTRACT

The intriguing sharp multiple magnetic phase transitions in UP and NpC are explained on the basis of a model which involves an exchange-aided electron delocalization transition. Quantitative agreement with sublattice magnetization, magnetic susceptibility, electron effective mass, and latent heat data is obtained.

Several of the actinide compounds exhibit a variety of unexplained first-order magnetic and electronic phase transitions.[1] The experimental data of[2,3,4] Figs. 1 and 2 are typical examples. No spin canting or change in lattice structure is observed at these transitions. Previous theories[5,6] have assumed a well-defined valence for the actinide ions and the validity of crystal field concepts for a localized magnetic 5f electron, but were unable to account for magnetization and susceptibility data simultaneously. The mechanism of the first-order phase transition remained unknown, except perhaps in UO_2.[6]

We propose a model owing much to the work of Falicov, et al. on the metal-insulator transition.[7,8] It takes into account recent experimental evidence[9] that the energy levels of the ion with 5f electrons overlap those of the itinerant 6d-7s band electrons, so that the 5f electron may be delocalized at low temperatures[10] or under pressure.[11] A valence-change model for rare-earth compounds was proposed previously.[12]

The model involves the following 5 assumptions: 1) There are highly correlated, localized, electronic states derived from the 5f actinide orbitals and described for a given valence as crystal field states of the ion; 2) There are itinerant electronic states described by band theory. For simplicity, but not from necessity, the dispersion relation is taken to be $\varepsilon_k = \hbar^2 k^2/2m^*$, where m^* is the effective mass; 3) Each of the N actinide (A) ions may be in one of two possible valence states. If all the ions are in the state of lower valence, there are z electrons per A-ion in the band. In general, there are z+p band electrons per A-ion, where p ($0 \leq p \leq 1$) is the probability that an A-ion is in the higher valence state. When p=1 and there is no magnetic ordering, the energy needed to raise one electron from the bottom of the conduction band to a localized state is Δ, an adjustable parameter (See Fig. 3); 4) There is a Coulomb repulsion between an ion in the lower valence state and a band electron, if the two are in the same Wigner-Seitz cell. This interaction energy is G, an adjustable parameter; 5) The magnetic exchange interaction, $\lambda(p)$, between ions is treated in the molecular field approximation.

Turning now to the case of UP and NpC, we assume the two valence states allowed for a U or Np ion to correspond to the $5f^2$ (U^{4+}, Np^{5+}) and $5f^3$ (U^{3+}, Np^{4+}) configurations. The lowest level of the $5f^3$ ions is assumed to be the E_1' (or Γ_6) doublet, (as explained

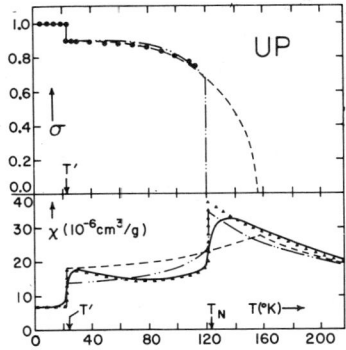

Fig. 1. Relative sublattice magnetization (top diagram) and magnetic powder susceptibility χ of UP as functions of temperature T. $T'=22.5°$ K and $T_N=121°$ K.

Top: Dots--experiment (See Ref. 2); Dashed line--theory, col. A, Table I; Dash-dotted line--theory, cols. B and C, Table I.

Bottom: Heavy line--experiment (See Ref. 3); Dashed line--theory, col. A, Table I; Dash-dotted line--theory, col. B, Table I; Triangles--theory, col. C, Table I.

later, it could also be a quartet) with a magnetic moment μ_3. The $5f^2$ ion is treated as a Γ_{2g} (or Γ_5) magnetic triplet, with the value $\mu_4 = 2.00\mu_B$, deduced from neutron diffraction work on UP[13] and NpC.[4] The parameter $r=\mu_3/\mu_4$ is allowed to vary slightly, as crystal field theory is not reliable for its prediction.

Note that the "valence" states are here used as basis states, the occupation numbers of which determine thermodynamic properties in the mean field approximation. Because of covalency and hybridization,[9] these ionic states are not the exact electronic eigenstates of the system. The charge density of itinerant electrons is concentrated at the actinide sites, because the band states are derived mainly from the 6d and 7s actinide atomic orbitals. This is evidenced by the fact that no significant volume change at $T=T'$ is reported for UP and NpC. In several other compounds, volume discontinuities accompanying first-order electron delocalization transitions are also very small. Examples are Ti_2O_3,[14] several rare earth cobaltates,[15] and Fe_3O_4.[16]

Let us denote the number of U^{3+} ions in the two magnetic sublevels $m'=\pm\frac{1}{2}$ by β^+ and β^-, and the number of U^{4+} ions in the three magnetic sublevels $m'=\pm1,0$ by α^+, α^- and α^0, respectively. The free energy, F, of the system is

$$F = F_b(p) + N(1-p)\Delta - N(1-p)^2 G - \tfrac{1}{2}N\lambda\sigma^2 - Tk_B \ln W$$
$$+ \rho(\alpha^+ + \alpha^0 + \alpha^- - Np) + \tau(\beta^+ + \beta^- - N + Np) + \zeta(zN + pN - \int_0^\infty D(\varepsilon)f(\varepsilon)d\varepsilon) . \quad (1)$$

Here F_b is the free energy of the band electrons. The ordered magnetic moment per ion is

$$\mu_4 \rho = N^{-1}\mu_4(\alpha^+ - \alpha^- + r\beta^+ - r\beta^-) . \quad (2)$$

The thermodynamic probability, W, is the total number of ionic microstates:

$$W = N!(\alpha^+!\alpha^-!\alpha^0!\beta^+!\beta^-!)^{-1} . \quad (3)$$

$D(\varepsilon)$ is the density of states in the band and $f(\varepsilon)$ is the occupation probability of the band states. ρ, τ and ζ are Lagrange multipliers which allow for the conservation of the number of ions and electrons.

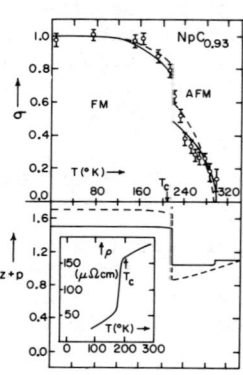

Fig. 2. Relative sublattice magnetization σ (top diagram) and conduction electron number z+p per Np-ion as functions of temperature T for $NpC_{0.93}$ $T_c \simeq 220°K$.

Top: Circles--experiment (See ref. 4); Dashed line--theory, cols. D and E, Table I; Solid line--theory, col. F. Table I.

Bottom: Dashed line--theory, cols. D and E, Table I; Solid line--theory, col. F, Table I.

Insert: Specific resistivity vs. temperature (experiment on $NpC_{0.96}$, Ref. 9) which shows the localization transition at $T_c \simeq 200°K$.

Minimizing F with respect to $f(\epsilon)$, $\alpha^+, \alpha^-, \alpha^0, \beta^+, \beta^-$ and p leads to two self-consistent equations. They are

$$p^{-1} = 1 + 2Z^{-1}ch(rx)\exp\{-\beta[\Delta-\zeta(p)-2G(1-p) + \tfrac{1}{2}\sigma^2 d\lambda/dp]\} \quad , \quad (4)$$

and

$$\sigma = pB_1(x) + r(1-p)B_{\tfrac{1}{2}}(rx) \quad , \quad (5)$$

where $x = \lambda(p)\sigma\beta$, B_J is the Brillouin-function, $\beta = (k_BT)^{-1}$, $Z=1+2ch(x)$, and $\zeta(p)$ is the Fermi level when the band is occupied by $N(z+p)$ electrons. For a given T, there are in general several solutions of Eqs. (4) and (5). The physical solution is the one of lowest free energy.

The "moment jump" transitions at T' and T_c in Figs. 1 and 2 are due to the following factors: a) At T=0°K there are two competing physical solutions to Eqs. (4) and (5). One represents the "delocalized" phase (p=1, σ=1), and has a lower free energy than the second "localized" phase (p=0, σ=r). b) In the delocalized phase, the energy $E_1 \simeq \Delta-\zeta(1)+\lambda(1)(1-r) \gg k_BT'$ is required for localization. In the localized phase it requires only $E_0 \simeq 2G-\Delta+\zeta(0)-\lambda(0)(1-r)r \ll E_1$ to excite an electron to the band. c) As T is raised from T=0°K, the entropy of the localized phase rises much more rapidly than that of the delocalized phase, due mostly to the increasing number of higher valent ions, left behind by thermal excitation of electrons in the former phase. Therefore, the free energy of the localized phase falls below that of the delocalized phase at T=T', whereupon the magnetization drops sharply from σ≃1 to σ≃r.

Neither the magnetic entropy nor the entropy of the band electrons is very important for the "moment-jump" transition at T=T', and in these respects this transition has a rather different cause than the transitions described in Refs. 7-8. The exchange λ is, however, important in removing the degeneracy of the localized states. In fact if k_BT' is less than $\lambda(0)$ and $\lambda(1)$, the transition is nearly independent of the multiplicities assumed for the magnetic manifolds of the two valence states.

The parallel and perpendicular magnetic susceptibilities from which the powder susceptibility in the AFM state is calculated, are (in an approximation valid if $\mu_3 \simeq \mu_4$) per ion:

$$\chi_{\|} = \mu_4^2 \phi [\lambda_{FM}(1-\phi)]^{-1} \quad , \quad (6)$$

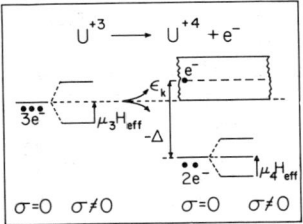

Fig. 3. Schematic diagram of the delocalization transition in certain actinide compounds such as UP and NpC. The symbols Δ, ε_k, σ, μ_3, μ_4 are explained in the text, and H_{eff} is the molecular field.

and
$$\chi_\perp = \mu_4^2 (\lambda - \lambda_{FM})^{-1} \text{ for } T<T_N, \quad \chi_\perp = \chi_{||} (\sigma = 0) \text{ for } T>T_N, \quad (7)$$

with
$$\phi = \lambda_{FM} \beta [r^2 (1-p)(1-\tanh^2 (rx)) + 3pz^{-2} + pz^{-1}] \quad (8)$$

Note that the exchange in the FM state, λ_{FM}, enters here as a new adjustable parameter.

The p-dependence of λ and λ_{FM} comes from RKKY[1]-type indirect coupling via the conduction band, calculated for the NaCl lattice for the AFM-I and FM states in Ref. 1, and also from the (unknown) variation with valence of the direct and/or superexchange interactions of A-ions. However, for UP, in the ordered state ($T<T_N$) we need only specify the exchange for p=0 and for p=1. In the simplest model, (Table I, Col. A and Fig. 1, dashed curves, $\lambda(p)$ is assumed to be a constant, i.e., $\lambda(0)=\lambda(1)$; and the simplification $\lambda_{FM}(0)=0$ is made. In the second model (Table I, Col. B and Fig. 1 dot-dashed curves), the steep decrease of the experimental magnetization near $T_N=121°K$ (T_N is obtained from Ref. 17) is theoretically reproduced by introducing one additional parameter, $\lambda'(0)=-0.029$ eV, through

Table I. Parameters (in eV, except r, z and m*/m, which are dimensionless) of the electron delocalization theory of the magnetic actinide compounds UP and NpC, used in Figs. 1 and 2. The models A-F are explained in the text.

	UP			NpC		
	A	B	C	D	E	F
Δ	0.515	0.421		3.12	1.00	0.717
G	0.211	0.145		0.707	0.263	0.214
r	0.89	0.89	same as B	0.7	0.7	0.28
z	0.1	0.2		0.7	0.7	0.5
m*/m	6.85	8.40		1.76	5.76	7.4
$\lambda(0)$	0.018	0.025		0.043	0.047	0.017
$\lambda_{FM}(0)$	0.0	0.0	0.011	–	–	–
$\lambda_{FM}(1)$	-0.029	-0.020	same as B	–	–	–

$\lambda(p)=\lambda(0)+\lambda'(0)p$, for small p. Finally, in the third model (Table I, Col. C and Fig. 2, triangles) two additional parameters $\lambda'_{FM}(0)=-0.126$ eV and $\lambda''_{FM}(0)=0.255$ eV are introduced to refine the theoretical susceptibility through $\lambda_{FM}(p)=\lambda_{FM}(0)+\lambda'_{FM}(0)p+\lambda''_{FM}(0)p^2$, for small p. Low temperature (T<T') electronic specific heat data[17] on UP yield in the free-electron approximation, $m^*/m=8.04$ and 7.83 for the parameters of models A and B (or C) respectively, in good agreement with the entries of Table I. The experimental values for the heat of transition, $T\Delta S$, at T=T' and T=T_N are 10.2 cal/mole and 87.3 cal/mole, respectively,[17] in good agreement with the theoretical values from model B of 8.1 cal/mole and 87.7 cal/mole, respectively. For NpC, the RKKY calculation[1] correctly predicts the change from FM to AFM-I magnetic ordering at T_c (Fig. 2) for the choice of parameters listed in Table I, Col. F. Two other "fits" with λ assumed independent of p (Table I, Cols. D and E, and Fig. 2) are shown to illustrate that the magnetization curve is not much affected by the detailed p-dependence of λ or by the effective mass, m^*. The sharp decrease in the occupation of the band predicted by the theory at T_c corresponds very well to the observed increase in the resistivity near T=T_c (See Fig. 2, insert). We have not taken into account any effects of the carbon vacancies present in the NpC_x samples. The temperature T_c is practically independent of x for $x \gtrsim 0.85$, but T_N depends strongly on x.[9]

Further work is in progress on the actinide compounds UAs and US and on the phase diagrams of $UP_{1-x}S_x$, $UAs_{1-x}P_x$, and $UAs_{1-x}S_x$. This theory may apply to certain rare-earth semiconductors, particularly to EuO and $Li_xMn_{1-x}Se$.

REFERENCES

*This research was sponsored by the Air Force Office of Scientific Research, U.S. Air Force, under grant number AFOSR-70-1940.

1. J. Grunzweig-Genossar, et al., Phys. Rev. 173, 562 (1968).
2. S. L. Carr, et al., Phys. Rev. Letters 23, 786 (1969).
3. J. M. Gulick and W. G. Moulton, Phys. Letters 35A, 429 (1971).
4. G. H. Lander, et al., J. Phys. Chem. Solids 30, 733 (1969).
5. Chris Long and Yung-Li Wang, Phys. Rev. B3, 1656 (1971).
6. S. J. Allen, Phys. Rev. 166, 530 (1968); 167, 492 (1968).
7. L. M. Falicov, et al., Phys. Rev. Letters 22, 297 (1969); Phys. Rev. B2, 3383 (1970); Solid State Commun. 10, 455 (1972).
8. B. Alascio, et al., Phys. Rev. B5, 3708 (1972).
9. M. B. Brodsky, in A.I.P. Conf. Proc. 5, Magnetism and Magnetic Materials, C. D. Graham and J. J. Rhyne, Eds., N.Y. (1972).
10. A. J. Arko, et al., Phys. Rev. B5, 4564 (1972).
11. B. T. Matthias, discourse, Int. Conf. on Magnetism, Grenoble, France, 1970.
12. G. Busch and O. Vogt, Phys. Letters 20, 152 (1966).
13. N. A. Curry, Proc. Phys. Soc. 89, 427 (1966).
14. D. Adler, Rev. Mod. Phys. 40, 714 (1968).
15. V. A. Bhide, et al., Phys. Rev. Letters 28, 1133 (1972).
16. L. R. Bickford, Jr., Rev. Mod. Phys. 25, 75 (1953).
17. J. F. Counsell, et al., Trans. Faraday Soc. 63, 72 (1967).

PRESSURE INDUCED LOSS OF FERROMAGNETISM IN UPt[*]

J. G. Huber, M. B. Maple and D. Wohlleben
University of California, San Diego
La Jolla, California 92037

ABSTRACT

The magnetization of the weak ferromagnet U_xPt_{1-x} was studied from $x = 0.48$ to $x = 0.54$, at temperatures between 4.5°K and 300°K, fields up to 8 kGauss and pressures up to 20 kbar. The magnetization curves were analyzed in terms of the differential paramagnetic susceptibility χ_d at 8 kGauss and the saturation magnetization M_s. The saturation moment μ_s per U atom at $p = 0$ and $T = 4.5°K$ is a function of x, typically of order 0.5 to $0.2\mu_B$, but only $0.07\mu_B$ at $x = 0.52$. At $T = 4.5°K$, μ_s drops at first steeply with increasing pressure and then approaches zero asymptotically; at 20 kbar it is reduced by 94% at $x = 0.50$ and by 98% at $x = 0.52$. The Curie temperature is about 30°K and does not depend on pressure within experimental accuracy. The differential paramagnetic susceptibility has a relatively broad maximum near 17°K at all pressures and concentrations. Plots of χ^{-1} vs. T above 30°K curve towards the T-axis with decreasing T for all x and p, and exhibit two nearly linear portions between 30°K and 50°K and between 70°K and 300°K. The high temperature linear portion gives an effective moment of about $\mu_{eff} \sim 2.7 \pm 0.2\mu_B$ per uranium atom, and the low temperature portion $\mu_{eff} \sim 1.3\mu_B$, depending on concentration and pressure.

---

[*]Research supported by the U. S. Air Force Office of Scientific Research, Grant no. AF-AFOSR-71-2073.

STABILIZATION OF THE 5f ENERGY BAND IN ACTINIDE-Rh$_3$ INTERMETALLIC COMPOUNDS*

W. J. Nellis
Argonne National Laboratory, Argonne, Ill. 60439
and
Monmouth College, Monmouth, Ill. 61462

A. R. Harvey and M. B. Brodsky**
Argonne National Laboratory, Argonne, Ill. 60439

ABSTRACT

The magnetic susceptibilities and electrical resistivities of ThRh$_3$, URh$_3$, NpRh$_3$, and PuRh$_3$ have been measured from 2–300°K. These compounds have the cubic AuCu$_3$-type structure with a total lattice parameter variation of only 0.5% for the latter three compounds. These materials are metallic and have room-temperature resistivities of about 50 μΩ-cm. The susceptibility of ThRh$_3$ is temperature-independent. The URh$_3$ susceptibility decreases weakly with temperature. The susceptibility of NpRh$_3$ has a large 0°K-value of 5.5×10^{-6} emu/g, decreases as $-T^2$ up to 125°K, and then fits a Curie-Weiss law with $p_{eff} = 3.63$ μ$_B$/NpRh$_3$ and $\Theta = -483$°K. PuRh$_3$ is antiferromagnetic below $T_N = 6.6$°K. The resistivities of ThRh$_3$, URh$_3$, and NpRh$_3$ increase from their residual values as $\rho = \rho_0 + AT^n$, where n = 5, 3, and 2, respectively. The PuRh$_3$ resistivity follows $\rho = \rho_0 + AT$ above T_N. The data are interpreted by a model in which the 5f level is above the Fermi energy in ThRh$_3$, a broad, hybridized 6d-5f band overlaps the Fermi level in URh$_3$, NpRh$_3$ is nearly magnetic with a spin fluctuation temperature $T_S \simeq 15$°K, and PuRh$_3$ is weakly magnetic with a moment of about 1 μ$_B$ subject to spin fluctuations above T_N.

INTRODUCTION

Actinide metals have been characterized by the stabilization of the 5f energy level across the series. That is to say, the 5f band is above the Fermi level in actinium and thorium. Hybridized 6d-5f bands occur at the Fermi level in protactinium, uranium, neptunium, and plutonium. Americium has a nearly-filled, nonmagnetic j = 5/2 5f level, and rare earth-like localized magnetic moments form in curium and berkelium.[1] While this model is consistent with available data, it is difficult to understand theoretically because of the complex crystal structures and experimentally because of numerous anomalies in physical properties, especially for the elements uranium, neptunium, and plutonium. Thus, it would be advantageous to find an atomically ordered, cubic environment in which various actinide atoms

---

*This work was performed under the auspices of the U. S. Atomic Energy Commission.

**Temporary address: Imperial College, London (until Sept. 1973).

could be placed without appreciably changing the lattice parameter. Such a system would perhaps be sufficiently simple to stimulate theoretical calculations and have physical properties varying in a systematic fashion. Actinide-Rh$_3$ intermetallic compounds comprise such a system.

EXPERIMENTAL PROCEDURE

The compounds were made by arc-melting stoichiometric amounts of high-purity metals in an argon atmosphere. These compounds melt congruently near 1500°C, and measurements were performed on as-cast specimens whose lattice parameters, a_0, are listed in Table I. These

Table I Lattice parameters, room-temperature susceptibilities and resistivities, and results of low-temperature resistivity fits to $\rho = \rho_0 + AT^n$ for actinide-Rh$_3$ compounds

Compound	ThRh$_3$	URh$_3$	NpRh$_3$	PuRh$_3$
a_0 (Å)	4.110±0.004	3.988±0.001	4.007±0.005	4.010±0.003
χ_{300} (10^{-6} emu/g)	0.40	1.78	3.85	1.77
n	5	3	2	1
A ($\mu\Omega$cm/°K^n)	2.6×10^{-7}	2.53×10^{-5}	5.09×10^{-3}	1.82×10^{-1}
ρ_0 ($\mu\Omega$cm)	18.5	1.26	3.78	16.8
$\rho_{300} - \rho_0$ ($\mu\Omega$cm)	33.1	54.9	56.5	38.4

values are in good agreement with the literature values which exist for ThRh$_3$,[2] URh$_3$,[2] and PuRh$_3$.[3] No second phases were detected by X-ray diffraction. The high-angle reflections in the Debye-Scherrer powder patterns were somewhat diffuse, causing the relatively large uncertainties in a_0. However, the low residual resistivities, $\rho(0°K)$, of 1.26, 3.78, and 3.0 $\mu\Omega$-cm for URh$_3$, NpRh$_3$, and PuRh$_3$, respectively, imply that effects of inhomogeneity and strain are small in the resistivity and susceptibility specimens. A small amount of rhodium was lost in melting the ThRh$_3$ ingot, so that this sample is about two percent off stoichiometry. This condition is reflected in a high residual resistivity of 18.5 $\mu\Omega$-cm. Electrical resistivity and magnetic susceptibility measurements were made on specimens cut by spark erosion from the same ingot.

RESULTS

The magnetic susceptibilities in the range 2.5-300°K are shown in Fig. 1. The susceptibilities of ThRh$_3$, URh$_3$, and NpRh$_3$ are field-independent in the range 10-14.5 kOe and these high-field values are shown in the figure. The susceptibilities of PuRh$_3$ were obtained by fitting χ to a two-term expression linear in H^{-1} and extrapolating to infinite field. Two PuRh$_3$ samples were studied. Both showed an antiferromagnetic transition at 6-7°K. The ThRh$_3$, URh$_3$, and NpRh$_3$

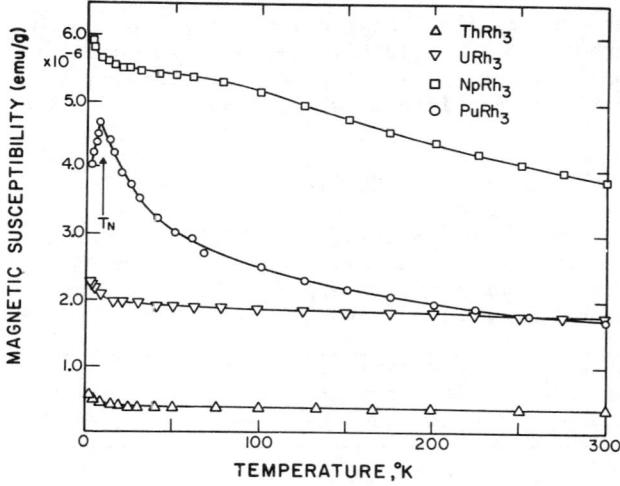

Fig. 1. Magnetic susceptibilities of ThRh$_3$, URh$_3$, NpRh$_3$, and PuRh$_3$.

specimens have small impurity contributions to the susceptibility below 15°K, which increase with the susceptibility of the matrix. Above 20°K the ThRh$_3$ susceptibility is temperature-independent and the susceptibility of URh$_3$ decreases from 2.0×10^{-6} emu/g at 20°K to 1.78×10^{-6} emu/g at room temperature. The susceptibility of NpRh$_3$ fits $\chi = 5.53 [1 - 0.635 \times 10^{-5} T^2] \times 10^{-6}$ emu/g to within 0.5% in the range $20 < T < 125°K$. For $125 < T < 300°K$, χ fits a Curie-Weiss dependence to within 0.3% with $p_{eff} = 3.63\ \mu_B$/NpRh$_3$ and $\Theta = -483°K$. The data for PuRh$_3$ were reported earlier.[4] The paramagnetic data fit $\chi = \chi_0 + C/(T-\Theta)$ with $p_{eff} = 1.0\ \mu_B$/PuRh$_3$, $\Theta = -63°K$, and $\chi_0 = 1.1 \times 10^{-6}$ emu/g. The values of χ at 300°K, χ_{300}, are listed in Table I.

The electrical resistivities were measured in the range 2-300°K. The resistivities of ThRh$_3$, URh$_3$, and NpRh$_3$ increase from their residual values as T^n, i.e., $\rho = \rho_0 + AT^n$, where n = 5, 3, 2, respectively. These power dependences are followed up to 16, 45, and 15°K, respectively, and results for URh$_3$ and NpRh$_3$ are shown in Fig. 2. The resistivity of the better PuRh$_3$ sample, as judged by a lower residual value, indicates a Néel temperature of $T_N = 6.6°K$. In the range $7 < T < 13°K$ the PuRh$_3$ resistivity fits $\rho = \rho_0 + AT$. Since T_N is in the residual resistance temperature range of ThRh$_3$ and URh$_3$, phonon scattering is assumed negligible near T_N, and $\rho(T_N) = \rho(0°K) + \rho_m$, where ρ_m is close to the spin-disorder value. Thus, $\rho_m \simeq 15$ μΩcm for PuRh$_3$ (see below). The T dependence and slope change at T_N for PuRh$_3$ are illustrated in Fig. 2. The values of ρ_0, n, A, and $\rho_{300}-\rho_0$ for all the compounds are listed in Table I.

DISCUSSION

The contraction in the volume per formula unit between ThRh$_3$ and URh$_3$ is analogous to that which occurs in the pure actinides, and suggests that a hybridized 6d-5f band overlaps the Fermi level in URh$_3$, but not in ThRh$_3$. Such a band would produce a higher density of states at the Fermi level for URh$_3$ than for ThRh$_3$ and explain the increase in both magnetic susceptibility and ($\rho_{300}-\rho_0$) in URh$_3$ over ThRh$_3$. The susceptibility of NpRh$_3$ indicates an even higher, energy-

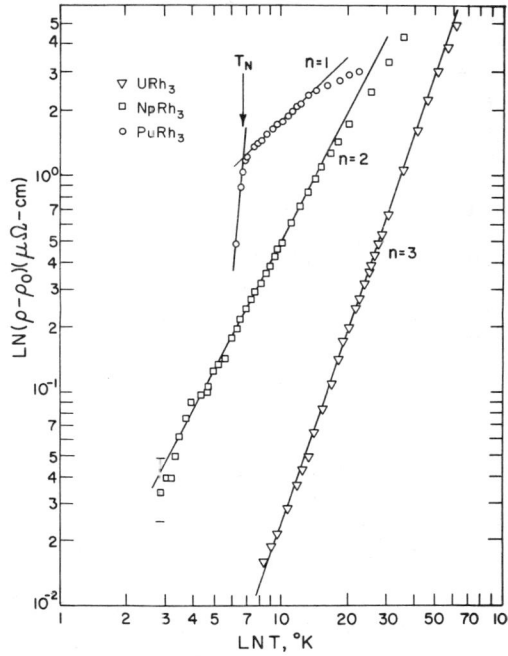

Fig. 2. Bilogarithmic plots of the low-temperature electrical resistivities of URh$_3$, NpRh$_3$, and PuRh$_3$.

dependent density of states at the Fermi level as manifested by the large value of $\chi(0°K)$ and by the $-T^2$ temperature dependence up to 125°K. Comparison with isostructural NpPd$_3$ with T_N = 55°K and an ordered moment of 2 μ_B/Np[5] indicates that the 6d and 5f states must be hybridized to a large extent in NpRh$_3$. Thus, it is improbable that the effective moment can be ascribed to an integral number of 5f electrons. However, it can be noted that p_{eff} = 3.63 μ_B/NpRh$_3$ is close to the value 3.87 μ_B for the 5f^3 state with quenched orbital angular momentum. In PuRh$_3$ the 5f level is sufficiently stabilized to produce weak antiferromagnetism at T_N = 6.6°K and p_{eff} = 1.0 μ_B/Pu atom. The isostructural compound PuPd$_3$ is also a weak antiferromagnet with T_N = 25°K, p_{eff} = 1.0 μ_B,[4] and an ordered moment of 0.8 μ_B/Pu atom.[6] Assuming that PuRh$_3$ has approximately the same ordered moment, then plutonium in PuRh$_3$ appears to be trivalent since Hund's rules yield paramagnetic and ordered moments of 0.85 μ_B and 0.72 μ_B, respectively, for the 5f^5 electronic configuration.

The low-temperature resistivities of these compounds vary in a systematic fashion. The T^5 dependence of ThRh$_3$ is due to intraband (s-s) electron-phonon scattering. The T^3 dependence of URh$_3$ is explained by interband (s-d or s-f) electron-phonon scattering which can be expected with the occupation of a broad 6d-5f band with a high density of states. The T^2 dependence for NpRh$_3$ and the T dependence above T_N for PuRh$_3$ can be explained by the localized spin fluctuation (l.s.f.) model.[7,8] In this model 6d-5f electrons occupy a nonmagnetic ground state at T = 0°K but electrons at the Fermi level are thermally excited locally from one spin band to the other as the temperature increases. The l.s.f. resistivity in atomically ordered materials increases initially as $\rho \propto T^2$, for T ≃ T_S there is a region in which $\rho \propto T$, and then the l.s.f. resistivity tends to level off well above T_S, the characteristic temperature which corresponds to the peak in the spin fluctuation spectrum. Such a model has been proposed for the T^2 resistivity of pure plutonium[9] and of uranium intermetallic compounds.[10] Thus, the l.s.f. model explains the T^2 resistivity of NpRh$_3$ and the T resistivity of PuRh$_3$, where it is assumed that magnetic ordering quenches the spin fluctuations below T_N.[10] Hence, for PuRh$_3$, $T_S < T_N = 6.6°K$ and it is difficult to determine the usual spin-disorder resistivity because the plutonium magnetic moment is temperature-dependent just abive T_N. For NpRh$_3$ $T_S \simeq 15°K$, assuming that $T_S \simeq T_1$, the upper limit of the T^2 region.[11] The region in which $\rho \propto T$ is not observed in NpRh$_3$, presumably because of the phonon resistivity which increases rapidly above 20°K, as indicated by the ThRh$_3$ and URh$_3$ data. Implicit in the theoretical result that the l.s.f. resistivity levels off for $T >> T_S$ is the idea that this nearly constant resistivity is essentially a spin-disorder effect and that above T_S spin fluctuations may have a local moment behavior.[12] The Curie-Weiss dependence of the susceptibility above $T = 125°K$ is, therefore, indicative of the magnetic behavior expected for temperatures well above T_S.

REFERENCES

1. For reviews of these ideas see, for example, H. H. Hill, Physica 55, 186 (1971); R. Jullien, E. Galleani d'Agliano, and B. Coqblin, Phys. Rev. B 6, 2139 (1972).
2. A. E. Dwight, J. W. Downey, and R. A. Conner, Acta Cryst. 14, 75 (1961).
3. V. I. Kutaitsev, N. T. Chebotarev, M. A. Andrianov, V. N. Konev, I. G. Lebedev, V. I. Bagrova, A. V. Beznosikova, A. A. Kruglov, P. N. Petrov, and E. S. Smotritskaya, Atomnaya Energiya 23, 511 (1967) [translation: Sov. Atomic Energy 23, 1279 (1967)].
4. W. J. Nellis and M. B. Brodsky, in Magnetism and Magnetic Materials 1971, edited by C. D. Graham, Jr. and J. J. Rhyne (American Institute of Physics, New York, 1972) p. 1483.
5. G. H. Lander, M. B. Brodsky, B. D. Dunlap, W. J. Nellis, and M. H. Mueller, Bull. Am. Phys. Soc. 17, 338 (1972).
6. G. H. Lander and M. H. Mueller (unpublished).
7. A. B. Kaiser and S. Doniach, Intern. J. Magnetism 1, 11 (1970).
8. S. Doniach, in Magnetism and Magnetic Materials 1971, edited by C. D. Graham, Jr. and J. J. Rhyne (American Institute of Physics, New York, 1972), p. 549.
9. W. J. Nellis and M. B. Brodsky, in Plutonium 1970 and the Other Actinides, edited by W. N. Miner (The Metallurgical Society of the American Institute of Mining, Metallurgical, and Petroleum Engineers, New York, 1970), p. 346; A. J. Arko, M. B. Brodsky, and W. J. Nellis, Phys. Rev. B 5, 4564 (1972).
10. K. H. J. Buschow and H. J. van Daal, in Magnetism and Magnetic Materials 1971, edited by C. D. Graham, Jr. and J. J. Rhyne (American Institute of Physics, New York, 1972) p. 1464.
11. It is unlikely that the T^2 resistivity of NpRh$_3$ is due to magnetic impurities in a strongly exchange-enhanced host because of the size of the coefficient A, which is 1/4 that of Pu (Ref. 9) but 50 times greater than that of Pd(0.5% Ni) (A. I. Schindler and B. R. Coles, J. Appl. Phys. 39, 956 (1968).
12. For a discussion of this idea see, for example, J. W. Loram, R. J. White, and A. D. C. Grassie, Phys. Rev. B 5, 3659 (1972).

LOW TEMPERATURE SPECIFIC HEAT AND MAGNETIC PROPERTIES OF RPt_5 (R = La, Ce, Pr, Nd) INTERMETALLIC COMPOUNDS

K. S. V. L. Narasimhan, V. U. S. Rao and R. A. Butera
Department of Chemistry, University of Pittsburgh
Pittsburgh, Pennsylvania 15213

ABSTRACT

The heat capacity (C_p) of the RPt_5 (R = La, Ce, Pr, Nd) intermetallic compounds having the hexagonal $CaCu_5$ structure has been measured from 1.4 to 20K. The magnetic susceptibility (χ) has been measured from 2 to 300K. C_p of $LaPt_5$ followed $C_p = \gamma T + \beta T^3$ behavior. C_p of $CePt_5$ showed an upturn below 2.5K while that of $NdPt_5$ exhibited a pronounced λ-type anomaly indicating the onset of magnetic order at 1.8K. The entropy associated with the excess heat capacity was nearly Rln2, showing that only the lowest crystal field doublet of Nd^{3+} was popuulated at low T. C_p of $PrPt_5$ revealed a large Schottky anomaly. It was possible to associate this anomaly with an overall splitting of about 380K of the J = 4 ground multiplet of Pr^{3+}, with a predominant B_2^o term. The χ of $PrPt_5$ calculated on the basis of this CF splitting was in reasonable agreement with the observed susceptibility behavior. The electrical resistivity of the compounds did not reveal any striking anomalies.

INTRODUCTION

The present work was aimed at understanding the crystal field and exchange interactions in the RPt_5 (R = La, Ce, Pr, Nd) intermetallic compounds with the aid of low temperature (1.4 to 20K) heat capacity measurements and susceptibility measurements in the range 2 to 300K. The RPt_5 compounds crystallize in the hexagonal $CaCu_5$ type structure[1] which is also exhibited by the technically important RCo_5 compounds. The influence of crystalline electric field is very important[2] in the understanding of the magnetic anisotropy of the RCo_5 compounds, and hopefully, because of the sameness of crystal structure, the study of crystal field interactions in RPt_5 might throw some light on the magnetic behavior of the former. As revealed from our magnetic investigations, the exchange interactions in the RPt_5 compounds (R = Ce, Pr, Nd) are quite weak; magnetic ordering was actually observed only in $NdPt_5$ at 1.8K. These compounds therefore provide the opportunity to study the crystal field effects without interference from exchange interactions.

EXPERIMENTAL

The compounds were prepared from the best commercially available elements by induction melting in a water cooled copper boat under purified argon. The ingots were cast into hemispherical buttons for the heat capacity measurements and annealed at 950°C for a week.

X-ray and metallographic examination revealed the compounds to be single phase of the $CaCu_5$ structure. Magnetic measurements were carried out using a Faraday balance[3] from 2 to 300K.

Heat capacity measurements were carried out in the temperature range 1.4-20K. A computer controlled calorimeter[4] was used in which energy was supplied in pulses of duration varying from 1 to 10 seconds and measured by an integrating digital voltmeter. Points were taken every .1K up to 4.2K and .2K above 4.2K.

RESULTS

(a) <u>Heat Capacity</u>:

The results of heat capacity measurements are shown in Figs. 1-3. LaPt$_5$ exhibits a C_p vs. T curve typical of metallic systems without any evidence for anomalous behavior. For this compound at temperatures below ~8°K one can express $C_p = \gamma T + \beta T^3$ with $\gamma = .014$ Joules/Mole-K^2 and $\beta = 6.347 \times 10^{-4}$ Joules/M-K^4. In the compound CePt$_5$, C_p/T vs. T^2 curve shows a sharp upturn at low temperatures, perhaps indicative of magnetic order below 1.4K. In the case of PrPt$_5$ no evidence for a λ-type anomaly was found. However a very prominent Schottky anomaly was observed. In the case of NdPt$_5$ a very clear λ-type anomaly was found indicating the onset of magnetic order at 1.8K.

(b) <u>Magnetic Susceptibility</u>

Although at elevated temperatures the susceptibilities showed nearly free-ion behavior, at low temperatures, marked deviations were found, especially in the case of PrPt$_5$ which seemed to indicate behavior typical of a van Vleck paramagnet (Fig. 4). LaPt$_5$ was found to be diamagnetic.

Resistivity measurements were made in the temperature range 2 to 300K, but no significant anomalies were found.

DISCUSSION

(a) <u>PrPt$_5$</u>

Since the heat capacity and susceptibility of this compound showed very interesting features, a crystal field model was constructed to explain the results.

Several years ago Bleaney[5] studied the crystal field interaction in the isostructural RNi$_5$ compounds and gave arguments to show that in the total crystal field interaction represented by

$$\mathcal{H} = B_2^0 O_2^0 + B_4^0 O_4^0 + B_6^0 O_6^0 + B_6^6 O_6^6, \tag{1}$$

$B_2^0 O_2^0$ was the dominant term. This circumstance is an outcome of the fact that in calculating the crystal field one needs to consider only the other rare-earth ions since the Ni or Pt atoms possess d^{10} configurations and can be nominally ascribed zero net charge. A point charge calculation of the crystal field at the rare earth site reveals that in PrPt$_5$, $B_2^0 \sim 5K$ and that the fourth and sixth order contributions can be ignored in comparison with it. The dominance of the B_2^0 term over the others arises because of the two nearest rare-earth neighbors at distance c_0 (~4.39A) along the hexagonal axis. The six next nearest rare-earth neighbors at a distance a_0 (~5.35A) in the basal plane give a smaller negative contribution to B_2^0.

The Schottky heat capacity ΔC_p of PrPt$_5$ was obtained by subtracting the measured C_p of LaPt$_5$ from that of PrPt$_5$. ΔC_p (meas) vs. T for

PrPt$_5$ is plotted in Fig. 2. With only a second order contribution to the crystal field, agreement with experiment is obtained in the temperature range 1.4 to 12K if the crystal field overall splitting (CFOAS) is ~380K. This corresponds to $B_2^o \sim$ 8K, in reasonable agreement with the point-charge estimate. However ΔCp (meas) is seen to deviate from ΔCp (calc) above 12K. This could be due to several reasons, one of which may be that the C_p of LaPt$_5$ does not completely represent the electronic and lattice contributions to the C_p of PrPt$_5$.

With a CFOAS of 380K and considering only the B_2^o term, the expected susceptibility of a polycrystalline sample was calculated using the method described by Holmes and Schieber[6], where one treats the applied field as a perturbation to the crystal field interaction. $\chi_{\parallel}$ and $\chi_{\perp}$, the susceptibilities parallel and perpendicular to the hexagonal axis were calculated. For a polycrystalline sample, $\chi = 1/3\chi_{\parallel} + 2/3\chi_{\perp}$. $1/\chi_{calc}$ obtained by this procedure is compared with $1/\chi$meas for PrPt$_5$ in Fig. 4. Inclusion of a Weiss molecular field alters the calculated susceptibility: $1/\chi'_{calc} = 1/\chi_{calc} - \lambda$. Agreement between the calculated and measured susceptibilities is obtained for $\lambda = 3.2$ emu.mole^{-1} which corresponds to a Weiss temperature $\theta = 4.8$K. We therefore conclude that a CFOAS of ~380K dominated by a positive B_2^o term gives a fair accounting of heat capacity and susceptibility characteristics of PrPt$_5$.

(b) <u>NdPt$_5$ and CePt$_5$</u>

By making an extrapolation of ΔCp vs. T. curve for NdPt$_5$, the third law entropy associated with the λ-type anomaly was found to be about 5.9 Joules deg^{-1} mole^{-1}. This value is quite close to Rln2 (=5.76) and indicates the crystal field ground state of Nd^{3+} in NdPt$_5$ is a doublet. At temperatures below T_c = 1.8K, the ground state degeneracy is removed by the exchange field.

In the case of CePt$_5$, it is not possible to deduce the entropy since the peak in the λ-anomaly appears to be below our temperature range of measurement. The magnetic susceptibilities of NdPt$_5$ and CePt$_5$ also reveal the influence of crystal field interaction. Efforts are being made to interpret these on a basis similar to that employed for PrPt$_5$.

Acknowledgement: This work was assisted by the National Science Foundation through the University Science Development Program. The authors would like to thank Dr. W. E. Wallace and Dr. R. S. Craig for helpful discussions.

REFERENCES

1. A. E. Dwight, Trans. ASM, 53, 479 (1961).
2. J. E. Greedan and V. U. S. Rao, J. Solid State Chem. (in press).
3. R. A. Butera, R. S. Craid and L. V. Cherry, Rev. Sci. Instr. 32, 708 (1961).
4. R. S. Craig, S. Nasu and H. H. Neumann (to be published).
5. B. Bleaney, Proc. Phys. Soc. 82, 469 (1963).
6. L. Holmes and M. Schieber, J. Phys. Chem. Solids 29, 1663 (1968).

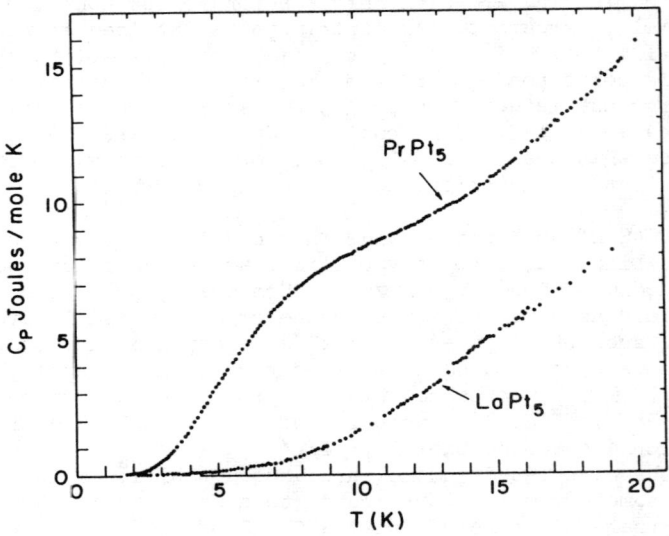

Fig. 1. C_p vs. T measurements for LaPt$_5$ and PrPt$_5$

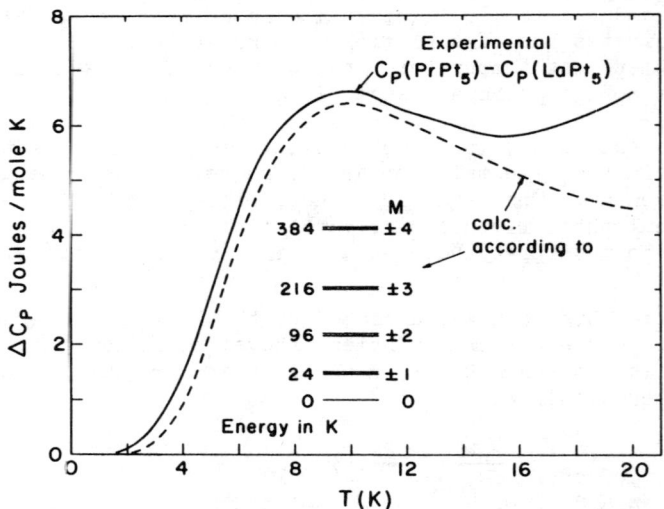

Fig. 2. Crystal field heat capacity (ΔC_p) of PrPt$_5$ (solid line) compared with the calculated ΔC_p (dashed line)

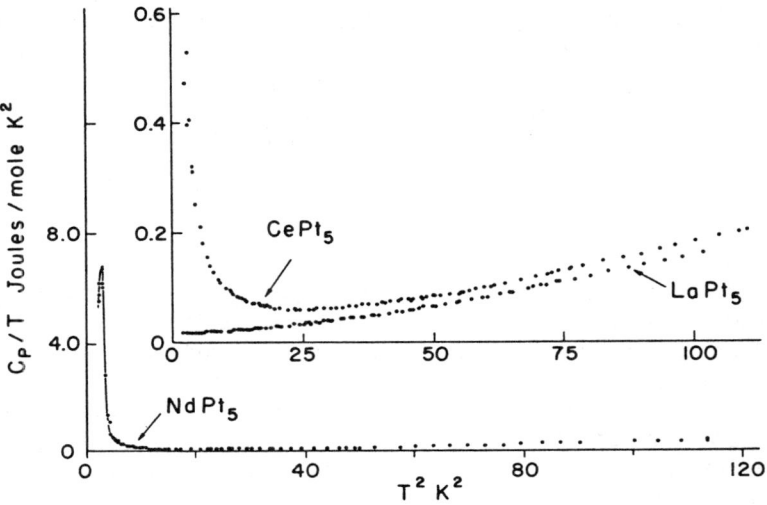

Fig. 3. Measured C_p/T vs. T^2 for $LaPt_5$, $CePt_5$ and $NdPt_5$.

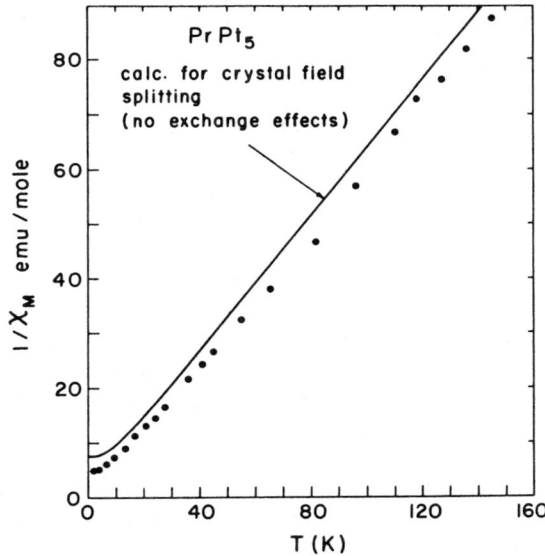

Fig. 4. Measured $1/\chi$ of $PrPt_5$ (dots) compared with the calculated values (solid line).

Section 33.
Tape Duplication and Magnetic Printing

MAGNETIC TAPES FOR CONTACT DUPLICATION BY ANHYSTERETIC AND THERMAL TRANSFER METHODS

H. Sugaya
Matsushita Electric Industrial Company, LTD.
Products Development Laboratory
1006, Kadoma, Osaka, Japan

ABSTRACT

According to the increasing demands for magnetic tape printing especially for video tape, the magnetic tape duplication using a contact process is one of the most attractive solutions. There are two major methods, i.e. anhysteretic transfer and thermal transfer. The coercivity of the master tape should be 2.5 times higher than the slave by computer simulation for anhysteretic transfer, but is not important for thermal transfer. The highest coercivity of the slave tape will be decided by the master tape which, in turn, is limited by the recording head saturation for anhysteretic transfer, however, the thermal characteristics of coercivity and magnetic induction under T_c are more dominant for thermal transfer. The tape surface roughness as well as the coercivity is very important for short wavelength printing.

In this paper, the properties of the master and slave tapes for anhysteretic and thermal transfer are discussed with theoretical and experimental results.

1. INTRODUCTION

Since magnetic recording techniques have been developed, record and simultaneous playback functions were one of the biggest features of the magnetic tape recorder. However, recording and storage functions for mass communication, such as printed characters and figures on paper, are becoming the next biggest feature of magnetic recording. Thus, video tape recorders(VTR), for instance, have also been changing from record-and-playback to playback-only units as their main application, according to the development of mass communication systems through television receivers. The demands for mass duplication of pre-recorded video tapes are increasing very rapidly. The contact duplication of magnetic tape by the anhysteretic method was conceived by R. Müller-Ernesti[1] in 1941. R. Herr and M. Camras published their experiments with audio tape recording by the contact printing method in 1949[2][3]. The mass duplication of audio tape, which has since increased, was solved by the techniques of high frequency instrumentation recording and by the connection to many slave recorders. When it comes to video tape mass duplication, the contact duplication method is considered to be the best way at this moment, because it is not dependent on a complicated rotating head mechanism or a very high (megahertz region) recording frequency. The development activities in video tape contact duplication with this background were started with our report presented to the Intermag Conference in 1969[4]. The research and development activities in this field have been quite vivid thereafter: R. Van Den Berg[5] in 1969. C. G. Ginsburg[6], W. B. Hendershot III[7], D. Esterly[8] and J. E. Dickens[9] reported their experimental results in 1970, and F. Kobayashi[10)11], J. C. Mallinson[12], D. L. A. Tijaden[13] and J. Hokkyo[14] reported their theoretical analysis of contact duplication in 1971. Contact duplication can be divided into two methods; the anhysteretic transfer method which is carried out with a magnetic bias field, and the thermal transfer method which is carried out at the Curie temperature of the slave tape. The general aspects of anhysteretic and thermal transfer methods were discussed at the Conference on Advances in Magnetic Recording in 1971, and reported in the Annals of the New York Academy of Sciences in 1972[15]. In this paper, the physical properties of the master and slave tapes for anhysteretic and thermal transfer processes are mainly discussed.

2. MASTER AND SLAVE TAPES FOR ANHYSTERETIC TRANSFER PROCESS

2.1 PRINCIPLE OF THE ANHYSTERETIC TRANSFER PROCESS

Since we have reported a more precise analysis of the anhysteretic transfer process[10)11] previously, only an outline of this process is presented here. In contact duplication by the anhysteretic transfer process, a pre-recorded high coercivity master tape and a conventional blank slave tape are placed in close contact in a magnetic transfer field. As a magnetic transfer field, an AC magnetic field is generally applied, but a DC magnetic field also can be used[21]. In the contact printing process, a signal field is located on the master tape and only the bias field is subsequently decreased. The magnetization process utilizing anhysteretic magnetic transfer is, therefore, called the "Ideal anhysteretic magnetization process." The principle of the contact printing machine was used as a dynamic transfer method, and is shown in Fig. 1a, but in order to obtain stable and uniform

duplication at short wavelengths, the bifilar method has more advantages, as follows:
1) No slippage between the master and the slave tapes when a bias field is applied.
2) The space between the master and the slave tape can be minimized with the pressure roller.
3) Very highly reliable duplication utilizing a simplified mechanism.

On the other hand, a major disadvantage of the bifilar method is accidental printing (print-through) from the master tape to the adjacent magnetic layers at long wavelengths; but, fortunately, the frequency-modulated video signal for the VTR is located in the 2-20 μm wavelength range which is short enough to avoid accidental printing problems.

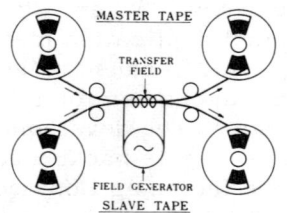

(a) Conventional tape transport for contact printing (Dynamic transfer)

(b) Bifilar tape winding method developed for video tape duplication

Fig. 1 Principle of contact printing machine

There are, of course, many method with which to solve accidental printing that occurs at the longer wavelengths of audio signals [4], and, therefore, this problem is no longer considered serious.

2.2 COERCIVITIES OF THE MASTER AND THE SLAVE TAPES

The maximum field strength on a master tape, which differs according to wavelength, is about 250 oe, but the practical linear range of the field strength in the video range is more or less 150 oe. In order to magnetize the slave tape sufficiently, the apparent magnetic susceptibility of the slave tape has to be increased with the bias field. Fig. 2 is an ideal magnetization curve of normalized slave tape coercivity and remanence. The curve was calculated with the aid of a hysteresis curve simulator, which was made possible by the combination of an analog computer and closed magnetic circuitry having a Hall element and magnetic tape material in series [16]. The influence of the demagnetization field is not considered in Fig. 2. The magnetization mechanism at short wavelengths, is, however, influenced by the demagnetization field. A bias field strength sufficient for contact duplication should be more than one and a half times the slave tape coercivity from Fig. 2, but the bias field as a transfer field will also erase the recorded signal on the master tape as was calculated by the hysteresis curve simulator (Fig. 3).

The magnetization of the duplicated slave tape is very much dependent on the coercivity ratio of the master tape to the slave tape. This relationship was obtained using the hysteresis curve simulator (Fig. 4). The coercivity of the master tape should, therefore, be more than two and half times that of the slave tape, according to Fig. 4. The coercivity of the master tape should be at least 750 oe when a conventional video tape (H_c= 250-300 oe) is used as a slave tape. If the so-called high energy tapes, CrO_2 (H_c= 450 oe) or Co·γ-Fe_2O_3 (H_c= 400-500 oe), are used for the slave tape, the coercivity of the master tape has to be at least 1300 oe. Such high coercivity tape can be made from Co·γ-Fe_2O_3 or metal powder, but it is difficult to find a good binder system and a process which will cause good dispersion especially for metal powder. Thus, the upper limit for master tape coercivity will mainly be decided by the difficulty in producing the tape itself and the saturation problem of the master recording head. Generally speaking for video signal recording, a ferrite head will start to become saturated with over 600 oe coercivity tape. Even a Sendust head will start to become saturated with 1500 oe tape, and a new magnetic head material has to be developed in order

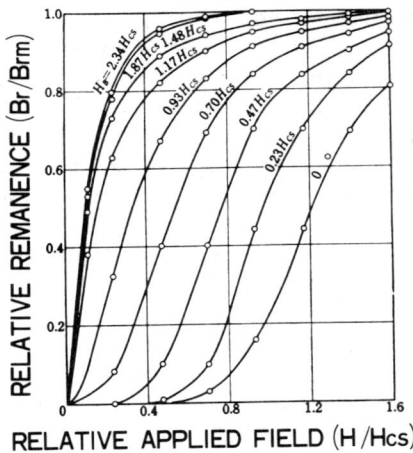

Fig. 2 Ideal magnetization curve with different bias field (H_B)

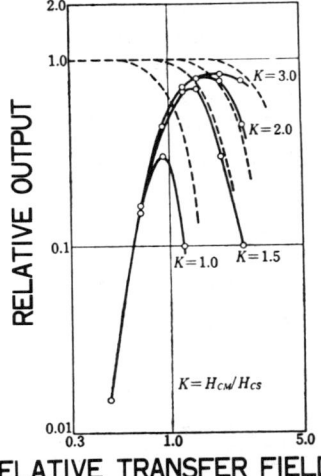

Fig. 3 Recorded master tape output duplicated slave tape output under changing transfer field. (using the simulator).

to use tapes with higher coercivity than 1500 oe. There exist high saturation materials, such as Co-Fe-alloy, but the mechanical as well as the soft magnetic properties are not satisfactory for a magnetic head. At any rate, we must compromise many conditions before we decide which master and slave tapes to use.

Using two kinds of master tapes and four kinds of slave tapes (Table 1), some experiments in video signal duplication were carried out using the bifilar method. A higher output even at short wavelengths can be obtained from the combination of a higher coercivity of master and slave tapes (Fig. 5). The maximum transfer output by various combinations of master and slave tapes is shown in Fig. 6, which references the transfer output from master tape M_2 to a conventional video tape (H_c= 300 oe) as 0 db. High energy slave tapes (437-520 oe) in combination with the super high energy master tape (1300 oe) have a few db higher output than the conventional slave tape (300 oe) in combination with a 770 oe master tape. In other words, these experiments indicate the possibility of using

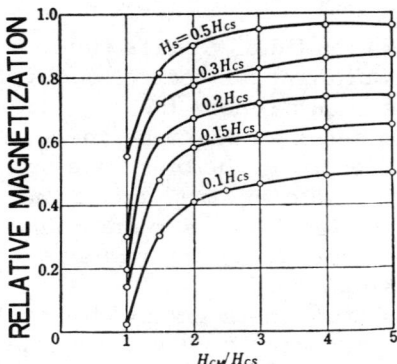

Fig. 4 Relative magnetization of duplicated slave tape vs. the coercivity ratio of master tape (H_{CM}) to slave tape (H_{CS}) at different signals (H_S).

	MASTER TAPE		SLAVE TAPE			
	M 1	M 2	Co·γ—Fe_2O_3	CrO_2		γ-Fe_2O_3
Hc (Oe)	1300	770	520	437	455	300
4π Ir (Gauss)	1380	1140	770	1430	1530	940
Ir / Im	0.79	0.79	0.75	0.80	0.85	0.75
Ir' / Ir	0.97	0.90	0.88	0.78	0.39	0.58
COAT. THICK.	4.7	7.5	8.3	4.5	3.8	4.8
SURF. ROUGH.	0.055	0.166	0.127	0.098	0.112	0.197

4π Ir (=Br) : LONGITUDINAL REMANENT MAGNETIC INDUCTION
4π Ir' (=Br') : TRANSVERSE REMANENT MAGNETIC INDUCTION
Im : LONGITUDINAL MAXIMUM MAGNETIC INDUCTION

Table I. Physical properties of master and slave tapes used in Fig. 5 and Fig. 6. Coating thickness and surface roughness are in μm.

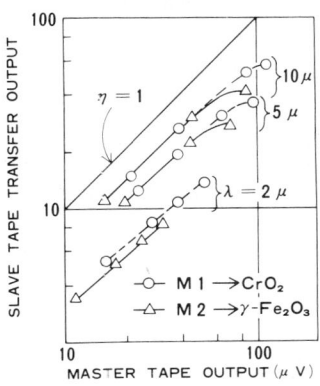

Fig. 5 Slave tape transfer output vs. master tape output with two different combinations of tapes.

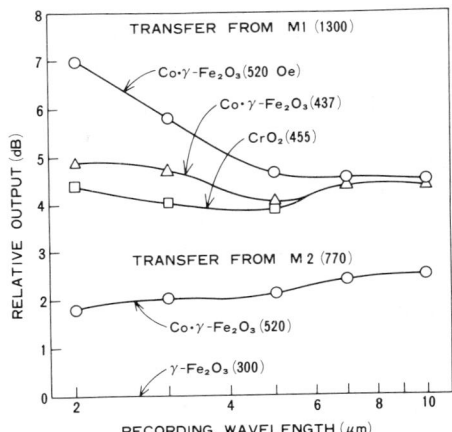

Fig. 6 Relative transfer output of various slave tapes from two kinds of master tapes (M_1, M_2). ( ) indicates coercivity of tape in oe.

high energy tape as a slave tape for anhysteretic contact printing, and also the limitations of anhysteretic contact printing at the present stage.

2.3 SURFACE ROUGHNESS OF THE MASTER AND SLAVE TAPES

Another very important result of the previous experiments is that the transferred slave tape output is not always related to the coercivity of the slave tape (Fig. 6), but is also dependent on the surface roughness of the master and the slave tape at short wavelengths. If the spacing between the master and the slave tape is simply considered to be the sum of the master and the slave tape surface roughnesses, the transfer efficiency of a 3 μm wavelength signal under the linear region corresponds to the equation for spacing

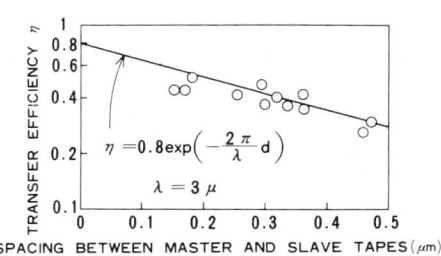

Fig. 7 Transfer efficiency as a function of spacing between the master and slave tapes due to the tape surface roughness.

loss and is shown in Fig. 7. Thus, the surface roughness of the master and slave tapes for anhysteretic transfer affects the transfer efficiency as well as the coercivity of the tapes at short wavelengths, and surface roughness influences the tape runability. If only the tape surface roughness becomes smooth without improving the binder system and tape transport (such as, tape guiding materials, surface finishing and construction), the tape will cause a "stick-slip" vibration which will distort the playback picture due to the time base jittering on VTR, and, at times, the tape will not run at all under bad environmintal conditions, such as high humidity.

2.4 Br AND COATING THICKNESS OF THE SLAVE AND MASTER TAPES

The remanent magnetic induction (B_r) and the coating thickness of the slave and master tapes (δ_s and δ_m) are other important factors which determine the transfer output at long wavelengths. Considering that the transfer efficiency η at long wavelengths is not influenced by demagnetization,

$$\eta = 4\pi^2 S \frac{\delta_s}{\lambda} \qquad (1)$$

where S is the apparent initial susceptibility of an ideal magnetization process. The transfer efficiency at long wavelengths will relate to δ_s and S of the slave tape. S will be approximately related to the maximum remanent magnetic induction, provided the bias field strength is one and a half times higher than the coercivity of the slave tape. The experimental results by M. Sato[17] and our computed results[10)11)] are compared and show good coincidence in Fig. 8 and Fig. 9. The larger B_r and δ_s, the higher the transfer efficiency at long wavelengths (>1 mm), but B_r is more dominant than δ_s at medium wavelengths ($\approx 100\,\mu m$). B_r and δ_s are not important to the transfer efficiency at wavelengths shorter than $100\,\mu m$, according to a model which considers the influence of the demagnetization field.

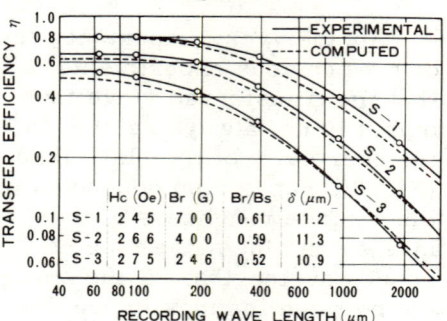

Fig. 8 Experimental and computed transfer efficiencies η using various slave tapes (γ-Fe_2O_3) having different B_r. (Experimental data by M. Sato)

The remanent magnetic induction and coating thickness of the slave tape are mainly discussed herein. B_r and the coating thickness of the master tape are also important factors, but are generally a lesser influence to the transfer output than the coercivity. However, in order to obtain a higher output at long wavelengths, of course, both B_r and δ_m should be as large as possible, but there is a point of compromise.

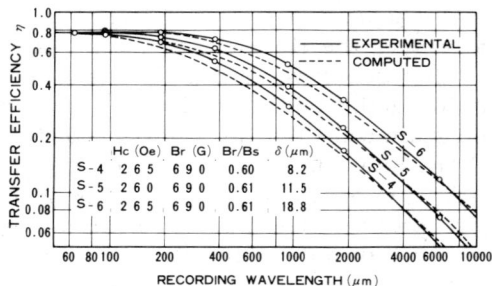

Fig. 9 Experimental and computed transfer efficiencies using various slave tapes (γ-Fe$_2$O$_3$) having defferent coating thicknesses δ.
(Experimental data by M. sato)

2.5 ANHYSTERETIC TRANSFER CONCLUSIONS

The conclusions drawn from the experimental and thoeretical results of the properties of the master and slave tapes are as follows:
1) H_c of the master tape should be 2.5 times higher than that of the slave tape.
2) The bias field strength should be at least 1.5 times the slave tape coercivity and should not exceed that of the master tape.
3) In order to obtain a higher output at short wavelengths, higher coercivities and as smooth a surface as possible are desirable along with the conditions listed above.
4) The coercivity of the master tape is mainly limited by the recording head material and tape coating techniques.
5) In order to obtain a higher transfer efficiency at longer wavelengths, δ_s and B_r of the slave tape should be large.
6) In order to obtain the highest duplicated output at long wavelengths, both B_r and δ_m of the master tape should be as large as possible.

In actual video tape duplication, other mechanical factors, such as the uniformity of the coating thickness, the accuracy of tape slitting, the slitted edge condition, and the homogeneity of the backing material, are very important in order to obtain exact positioning of the master and slave tape when contact duplication is performed, and the toughness of the magnetic coating, tape transportation and the environment of the location are important factors to consider in minimizing the dropout on a duplicated slave tape. These conditions are also very important factors affecting the life of the master tape. The master tapes discussed in this paper were mainly made of cobalt doped gamma ferric oxide cubic powder. These types of master tapes are demagnetized somewhat by the influence of temperature and pressure, and, therefore, the tapes have to be used under the proper condition[18) 19)]. According to the increase in master tape coercivity, magnetic coating techniques are improving, because it is becoming more and more difficult to obtain good dispersion. One of the possibilities to solve this problem is that a mirror image signal be first recorded on a conventional tape and then a high coercivity magnetic material be plated on the recorded tape[20)]. Thus, a signal can be recorded on any high coercivity tape without using a special high saturation head. An AC transfer field was used as a bias to obtain the results described in this paper, but there is also a possibility of using a DC transfer field[21)]. DC bias is not suitable for video and audio signal duplication in which a low noise figure is important, but may be for digital recording.

3. MASTER AND SLAVE TAPES FOR THE THERMAL TRANSFER METHOD

3.1 PRINCIPLE OF THERMAL TRANSFER

The apparent initial susceptibility of hard magnetic materials can be increased at their Curie temperatures (T_c) instead of the bias field as in anhysteretic transfer. The T_c of CrO_2 is exceptionally lower (125°C) than that of other magnetic materials, and so tape backing materials such as a polyester film, can be maintained without serious deformation at T_c. Contact duplication utilizing T_c was conceived by G. Akashi[22)] in 1960, and improved by J. Greiner[23)] and A. Kumada[24)]. The duplication of video signals by thermal transfer was reported by J. E. Dickens[9)] and W. B. Hendershot III[7)] in 1970. In principle, tape transportation of the master and slave tape is the same as shown in Fig. 1a and Fig. 1b, although Fig. 1a is preferable for thermal transfer by replacing the field generator with a temperature increasing device on the slave tape. CrO_2 tape was placed in an oven where the temperature was increased to the T_c of CrO_2 and then cooled to room temperature in different DC magnetic fields (10, 15, 20 and 50 oe). The magnetization of the tape was observed using a magnetometer while changing the ambient temperature (Fig. 10). The magnetization axis during the heating process shown in Fig. 10 has been expanded four times to show greater detail. The saturation of magnetization starts at an applied field of 50 oe.

However, when shorter wavelength signals than 10 μm are actually duplicated by thermal means, a 150 oe signal field will be neccesary due to the large influence of the demagnetization field. In order to make the blocking temperature clear after contacting with the master tape, the applied field(10 oe) is removed from the tape at temperatures different from those of the previous experiment(Fig. 11). The master tape can be separated from the slave tape at 110°C without a decrease in the duplicated output according to the results shown in Fig. 11, provided the applied field is DC or of a very long wavelength. In the case of a shorter wavelength than 10 μm, for instance, the frozen temperature will be in the 80-90°C range due to the demagnetization effect. The cooling time of the slave tape from T_c to 80°C is on the order of a few m sec just after contact with the master tape at room temperature. The temperature ascent of the master tape in contact with the heated slave tape is, therefore, nothing serious even though CrO_2 tape is used for the master tape. This is one of the very important features of thermal transfer. A complicated mirror image machine is no longer neccesary due to the simple thermal transfer process. The relationship between the thermoremanent magnetization of CrO_2 and the applied field at different temperatures is shown in Fig. 12. The thermoremanent magnetization is higher than anhysteretic magnetization so that the thermal transfer efficiency is generally higher than for the anhysteretic one, especially at longer wavelengths than 200 μm

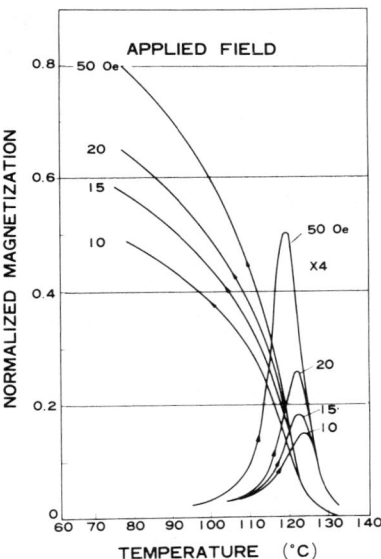

Fig. 10 Normalized magnetization of CrO_2 tape with different DC magnetic fields at high temperatures (under T_c)

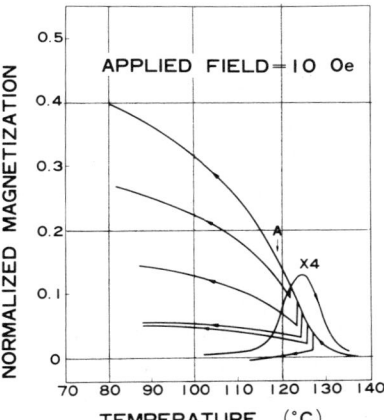

Fir. 11 Normalized magnetization of CrO_2 tape by a 10 oe DC magnetic field when the tape is cooled at different temperatures.

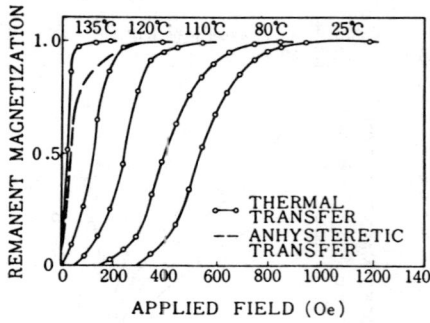
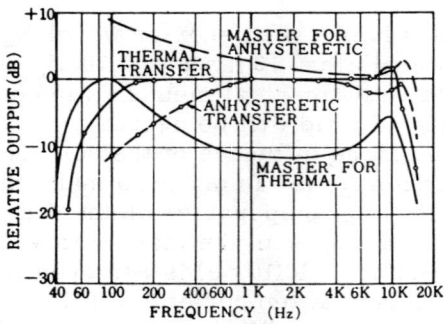

Fig. 12 Thermoremanent magnetization of CrO_2

Fig. 13 Comparison between thermal and anhysteretic transfer at long wavelengths (Tape speed: 7.5 i.p.s.)

where the signal field becomes very small (Fig. 13).
The magnetization mechanism of thermoremanent duplication can be explained very well by the theory of thermal fluctuation after-effect[25)26)27)28)]. The time change of magnetic particles, whose magnetization is parallel or anti-parallel to the signal field direction, at a temperature under T_C is:

$$\frac{dN_+}{dt} = -C\left[N_+\exp\left(-\frac{V(2K_u+I_sH)^2}{4kT\,K_u}\right) - N_-\exp\left(-\frac{V(2K_u-I_sH)^2}{4kT\,K_u}\right)\right] \quad (2)$$

where,
N_+ and N_- : number of particles parallel and anti-parallel to the signal field. V: particle volume. K_u: anisotropy constact.
I_s : saturation magnetization. H: applied field. k: Boltzman's constant. C: constant. T: absolute temperature.
The time constant τ of equation (2) is

$$\tau = \frac{1}{C}\exp\left(\frac{V(2K_u-I_sH)^2}{4kT\,K_u}\right) \quad (3)$$

Equation (3) indicates that the time constant related to the relaxation time is exponentially related to the applied temperature and field, and inversely related to the particle volume. When CrO_2 particles are actually magnetized by an applied field H, a demagnetization field will arise from the magnetized particles. The H in equations (2) and (3) has to be considered as the sum of externally applied field H_O and demagnetization field H_D. This is of special importance at short wavelengths. In view of these facts, a γ-Fe_2O_3 master tape recording a 95 μm wavelength signal at different levels and a CrO_2 slave tape were wound together using the bifilar method shown in Fig. 1b, exposed for 20 minutes in an oven at varying temperatures, cooled to room temperature, and the transfer signals on the slave tape were observed (Fig. 14). The transferred signal will relate to the signal field on the master tape and the applied temperature even at higher temperatures than the T_C of CrO_2. The difference between Fig. 10 and Fig. 14 may chiefly be due to the influence of the demagnetization field.

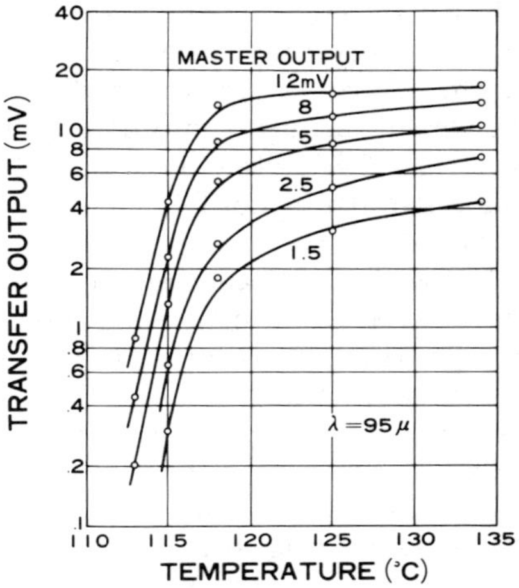

3.2 SLAVE TAPE FOR THERMAL TRANSFER METHOD

The slave tape for thermal transfer is more important than the master tape. The T_C of the slave tape for thermal transfer should be low, but not as low as room temperature in order to keep good magnetization stability. The thermal characteristics of CrO_2 tape, which is the only available slave tape for thermal transfer at the present stage, are shown in Fig. 15. In order to minimize the demagnetization effects during the cooling process, H_C should be recovered faster than B_r. On such a point of view, CrO_2 has preferable thermal characteristics. The backing material for video tape, at present, is a polyester film, and substitute products are not available as yet from the point of view of strength, flexibility, surface smoothness, stability, and cost performance. The polyester film is, however, deformed at 80°C, and shrinks 2-3% at 150°C. The track width of a video head used in a helical-scanning type video tape recorder (VTR) is, for instance, 100 μm. The tracking tolerance will be plus or minus 10 μm at best. Considering these situations, the ideal thermal characteristics of a slave tape will be something like the dotted line

Fig. 14 Thermal transfer output from $\gamma \cdot Fe_2O_3$ master tape recorded at different levels to CrO_2 tape at different temperatures.

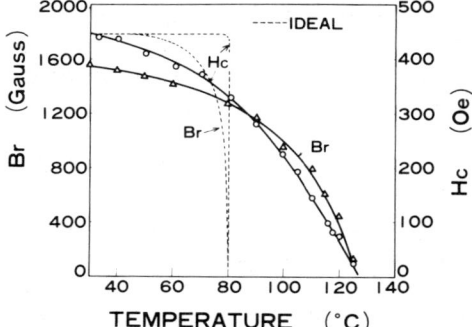

Fig. 15 Temperature dependency of remanent induction B_r and coercivity H_C of CrO_2, and ideal thermal characteristics.

shown in Fig. 15, and so the T_C of conventional CrO_2 is still high. In order to lower the T_C of CrO_2, a sulfur modified CrO_2 was tried 29) 30). The 5% by weight sulfur modified CrO_2, for instance, can decrease the T_C to 85°C without changing the acicularity and other magnetic properties considerably (Fig. 16). However, it has a fairly large temperature dependency on H_C even at room temperature. Considering the lowering of saturation magnetization and other factors, the sulfur modified CrO_2 tape may be used at as low a temperature as 110°C utilizing present techniques.

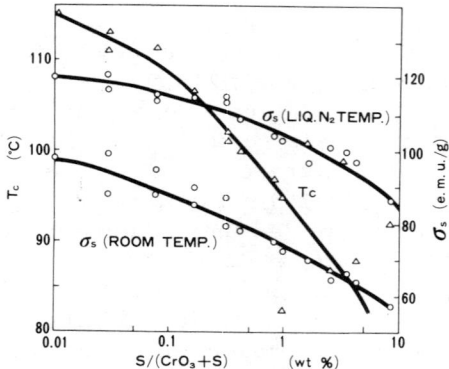

Fig. 16 Curie temperature T_C and saturation magnetization σ_s vs. additive amount of Sulfur in CrO_2
(E. Hirota et. al. 1972)

3.3 MASTER TAPE FOR THERMAL TRANSFER METHOD

The master tape for thermal transfer is not as critical, provided the T_C is not lower than that of the slave tape, and, therefore, the slave tape can also be used for the master tape as was discussed previously. The B_r should be preferably high as was also mentioned before and the coated surface should be as smooth as possible. The specially designed master tape for anhysteretic transfer, therefore, yields better results than conventional video tape. In order to minimize the shrinkage of the master tape backing, non-magnetic, endless material can be used 9). Other important features are no special mirror image machine, simultaneous duplication from an original tape to a copy tape, and flexibility in tape width.

3.4 THERMAL TRANSFER CONCLUSIONS

There are no choices available for slave tapes except CrO_2, at present, and tape deformation due to the applied temperature is a major problem in thermal transfer rather than the magnetic properties of the master and slave tapes. The sulfur modified low T_C CrO_2 is, therefore, one of the most attractive approaches to minimize this fatal problem. Thermal transfer, itself, has many advantageous points though its development is somewhat behind that of anhysteretic transfer. For instance, the transfer efficiency is higher than that of anhysteretic transfer, and the same tape can be used for both the slave and master tapes. These factors imply the possibility of the realization of a system which can duplicate simultaneously from an original recorded tape to a copy tape utilizing a special endless recording medium that does not require a special mirror image VTR. At any rate, the thermal transfer process is still in a premature stage and a better recording media for thermal transfer is expected to be developed in the future.

4. CONCLUDING REMARKS

The contact duplication techniques in practical use are quite new to either anhysteretic or thermal transfer processes. The technical demands on magnetic tapes for contact duplication are somewhat different from those previously developed for magnetic materials. The temperature characteristics of H_c and B_r, the particle shape, the particle size distribution and so on, have to be improved without changing the magnetic properties as well as the non-magnetic mechanical properties of the tape in many cases. I hope, however, that even better magnetic tapes for contact duplication will be rapidly developed in answer to the increasing demands for mass-duplication utilizing magnetic tape.

5. ACKNOWLEDGMENT

The author wishes to thank Prof. S. Iwasaki for his guidance, and to express his appreciation to T. Nakao, S. Nishimura, and Dr. T. Nasu for giving him the opportunity to write this paper. He would also like to thank F. Kobayashi and M. Ono for their suggestions and comments.

REFERENCES

1) R. Müller-Ernesti, German Patent, No. 910602 (Oct. 31, 1941)
2) R. Herr, "Duplication of magnetic tape recording by contact printing", Tele-Tech, vol. 8, 11, 28-30 (1949)
3) M. Camras and R. Herr, "Duplicating magnetic tape by contact printing", Electronics, vol. 22, 78-83 (1949)
4) H. Sugaya et al., "Magnetic tape duplication by contact printing at short wavelength", IEEE Trans. on Mag., vol. MAG-5, 3, 437-441 (1969)
5) R. Van Den Berg, "The design of a machine for high speed duplication of video records", SMPTE 105th Tech. Conf. (April, 1969)
6) C. P. Ginsburg, "Contact duplication of quadraplex video tapes", SMPTE Winter TV Conf. (1970)
7) W. B. Hendershot III, "Thermal contact duplication of video tape", Proc. Inter. Broad. Conf. (London), 204 (1970)
8) D. Esterly, "Contact duplication of transverse video tape recording", J. SMPTE, vol. 79, 903-907 (1970)
9) J. E. Dickens and L. K. Jordan, "Thermoremanent duplication of magnetic tape", 108th SMPTE Conf., No. 18 (1970)
10) F. Kobayashi and H. Sugaya, "Theoretical analysis of contact printing on magnetic tape", IEEE Trans. on Mag., vol. MAG-7, 2, 244-248 (1971)
11) F. Kobayashi and H. Sugaya, "Computer simulation of contact printing process", IEEE Trans. on Mag., vol. MAG-7, 3, 528-531 (1971)
12) J. C. Mallinson et al., "A thoery of contact printing", IEEE Trans. on Mag., vol. MAG-7, 3, 524-527 (1971)
13) D. L. A. Tijaden and A. M. A. Rigckert, "Theory of anhysteretic contact duplication", IEEE Trans. on Mag., vol. MAG-7, 3, 532-537 (1971)
14) J. Hokkyo and N. Ito, "Theoretical analysis of the process of contact printing of magnetic recording", Proc. of 3rd Conf. of Magnetic Recording, Hungary (1970)
15) H. Sugaya and F. Kobayashi, "Magnetic tape duplication by contact printing", Ann. N. Y. Academy of Sciences, vol. 189, 214-238 (1972)
16) H. Sugaya and F. Kobayashi, "Simulator for magnetic recording process", (in Japanese) Telev. Eng. of Japan, vol. 22, 4, 289-295 (1968)
17) M. Sato, Doctor Thesis, Tokyo Insti. of Tech. (1961)
18) H. Sugaya, "Magnetic recording head/tape interface, practical problems B", Intermag. Conf. (Kyoto), 47.3 (1972)
19) A. Kuroe, F. Kobayashi and H. Sugaya, "A study of de-

magnetization of video signal by repeated playback", (in Japanese) Proc. of Tech. Group of Magnetic Recording of IECE of Japan, vol. MR-72-19 (1972)
20) G. Akashi et al., "New duplication method by plating for magnetic recording", Intermag Conf. in 1972 to be published in IEEE Trans. on Mag., vol. MAG-8, 3 (1972)
21) J. R. Morrison and D. E. Speliotis, "The magnetic transfer process", IEEE Trans. on Mag, vol. MAG-4, 3, 290-295 (1968)
22) G. Akashi, Japanese Patent, S 39-4259 (April, 1964)
23) J. Geiner et al., U. S. Patent, 3,364,496 (Jan., 1968)
24) A. Kumada and F. Hayama, U. S. Patent, 3,465,105 (Sept., 1968)
25) L. Neél, Phycica, vol. 15, 225-234 (1949)
26) J. R. Morrison and D. E. Speliotis, "Thermoremanent magnetization properties of CrO_2", IEEE Trans. on Mag., vol. MAG-7, 3, 536-537 (1971)
27) M. Ono, F. Kobayashi and H. Sugaya, "Theoretical investigation of thermal printing process", (in Japanese) Proc. Tech. Group of Magnetic Recording of IECE of Japan, vol. MR-71-10 (1971)
28) M. Ono, F. Kobayashi and H. Sugaya, "A study of thermal transfer process", Intermag. Conf. in 1972, 19. 4 to be published in IEEE Trans. on Mag., vol. MAG-8, 3 (1972)
29) E. Hirota et al., "Sulfur modified chromium dioxide", Intermag. Conf. (Kyoto), 23. 3 (1972)
(30) T. Kawamata et al., "Sulfur modified chromium dioxide", to be published in Japan. J. of Appl. Phys. (1973)

A REVIEW OF MAGNETIC PRINTING

W. H. Meiklejohn
General Electric Corporate Research and Development
Schenectady, New York 12301

ABSTRACT

Magnetic printing is quite similar to electrostatic printing. The basic steps consist of recording a magnetic latent image, development of the recording with a magnetic ink and finally transferring the ink to a sheet of paper. Since papers on magnetic printing were first published in 1951, each step of this process has become highly developed. Impactless magnetic printers have been displayed at several conferences during the past few years. This review of the work on magnetic printers shows that printing speeds and print quality equal to electrostatic and ink jet printers can be achieved.

INTRODUCTION

Magnetic printers function in a manner quite similar to the more familiar electrostatic printers. A magnetic latent image is recorded instead of an electrostatic image and a dry magnetic ink is used to develop the magnetic image. The magnetic ink on the developed image is then transferred to a sheet of paper to produce the hard copy.

The main features of magnetic printers are that they are impactless and therefore quiet, and almost any kind of paper can be used in the printer. The latent image is not degraded by high humidity and, with the proper font, the printed document can be read with a magnetic head. Although duplicate copies cannot be made in the initial printout, the latent magnetic image can be reused to consecutively print out additional copies without any additional input. Printing speeds of 240 characters per second are commercially available and printing speeds of 30,000 characters per second have been achieved in the laboratory.

Development work on magnetic printers has been performed in several laboratories [1-9] during the past decade. This work showed that magnetic printers are technically feasible and have some advantages over other types of printers.

PRINTING BY MAGNETISM

"Printing by Magnetism" was the title of an article

published in 1839 by W. Jones.[10] He considered all the essential elements of a magnetic printer and said that it was "capable of producing impressions fully equal to lithography." He apparently never assembled the printer but considered using a blackened iron plate and a recording head consisting of an iron needle aligned in the direction of the earth's magnetic field. The magnetic ink of iron filings "washed in highly rectified spirits of wine". He pointed out that many impressions could be made from a single recording and that to prepare the plate for a new recording "it will require to be heated when the magnetic virtue will be expelled". And finally he pointed out that if this printer were constructed "we should have the magnetic fluid not only guiding our course o'er the deep blue ocean but through the still thicker mazes of science".

Nearly 150 years later we have a magnetic printer available which is manufactured by Data Interface.[11] This machine will print a 10 x 12 dot matrix at 240 characters per second. A sample of the printing is shown in Fig. 1a. Standard Telecommunication Laboratories Limited[8] has done printing at 30,000 characters per second. A sample of this high speed printing is shown in Fig. 1b. Fig. 1 is a photograph of the actual printed material.

HOW IT'S DONE

A schematic diagram of a magnetic printer is shown in Fig. 2. The rotating drum has a high coercivity magnetic material on its surface similar to computer recording discs. A row of recording heads (approx. 1200) extend completely across the recording surface. These heads may be on 6 or 7 mil centers at a density of about 150 heads per inch. The incoming binary information passes through a small memory and a character translator. The recording heads are serially addressed to form the magnetic latent image on the rotating drum in the form of a dot matrix. If a 10 x 12 matrix is used to produce the characters, then the row of heads must be addressed twelve times to form a line of characters.

The recorded surface is then passed over a fluidized bed and picks up a magnetic ink powder on the latent image. The powder consists of resin coated magnetic particles of approximately 10 micron diameter. The ink on the magnetic drum is then transferred to paper by pressure contact between the drum and the paper. The magnetic ink is then sealed to the paper by application of heat which melts the resin coating on the magnetic particles.

The drum must then be cleaned by a brush or magnetic pick-off to remove the particles which were not transferred to the paper. Finally the magnetic recording on the drum is erased and the new information is recorded.

RECORDING HEADS

Magnetic recording has become so highly developed in the audio, video and computer fields that there should not be any problems in applying it to magnetic printing. The speeds at which recordings are made on magnetic drums and magnetic discs in the computer field would produce latent images for printing at the rate of one million characters per second.

What is needed are low cost multiple heads. A magnetic printer using a 9 x 15 dot matrix and printing 80 columns requires 720 heads. In the magnetic printer described by Allen and Watson[9] the heads consist of a 4 mil diameter iron wire wound with 50 SWG enameled copper wire. The magnetic field at the recording surface was produced by energizing two adjacent coils such as to produce opposite magnetic polarities at the tips. This produces an in-plane magnetization as do conventional recording heads. Drive currents of one ampere produced good magnetic recordings with a head-to-drum separation of 1 mil. These wire recording heads were arranged to produce 150 tracks to the inch.

Another method of obtaining low cost multiple heads is to use batch fabricated heads that are being developed for head-per-track disc systems in the computer field. These heads are designed to both write and read the recorded bits. However, for use in the magnetic printer the heads are used only to write. Hence, the requirements placed on the batch fabricated heads for use in the printer are much less stringent and this should lower costs.

Batch fabricated heads have been described by a number of workers.[12-22] The magnetic recording head described by Barton and Stockel did not use any magnetic material. The recording field was created by passing a current of 1.5 amperes through a wire. The addition of a thin film magnetic circuit was proposed to reduce the magnitude of the drive current but was not tried.

Valstyn and Kosy[13] calculated the magnetic field produced by a single turn magnetic-film recording head shown in Fig. 3. The calculations predict that adequate recording fields can be obtained with a few hundred milliamperes. In a specific case a field greater than 790 Oe throughout a 1μm thick recording media was calculated for 200 ma through a 25μm wide strip conductor. The magnetic circuit consisted of a 2μm thick film and a 2μm gap. The head to recording surface spacing was 1μm.

Watanabe and Matsumoto[14] fabricated thin film magnetic recording heads that required between 60 ma and 150 ma recording current with a 30μm gap. They estimated that the cost per head would be about one thousandth that of a conventional recording head.

Tchon and Rodbell[15] fabricated small recording heads by rolling a composite wire containing a copper core and a magnetic shell. The rolled wire was lapped on one edge to expose the copper core and produce a 10μm recording gap. By attaching current leads to the side of the rolled wire at short intervals, a line of heads is obtained.

More complicated multiturn thin film heads have been constructed by Lazzari and Melnick[16] These heads should require less drive current than the one turn thin film heads, but Lazzari and Melnick found that relatively large currents of about 500 ma are required.

Romankiw[17] reported on thin film batch fabricated heads which had a width of 200μm. Although the write current was not discussed, a current of 1 amp was required to saturate the magnetic film near the gap. They indicated that the width of the head could be decreased by an order of magnitude which would certainly be sufficient for use in magnetic printing.

Sims[3] suggested that in-plane magnetization normal to the motion of the head would be desirable. The advantage proposed is that no modulated carrier is needed and the electrical circuits are simpler. However, I believe that a major disadvantage is that batch fabricated heads of the required design would be difficult to make.

All of these heads produce in-plane magnetization of the recording material. To produce magnetization perpendicular to the plane of the film would require magnetic recording media with a large magnetic anisotropy such as MnBi. Correspondingly the head would be required to produce large magnetic fields and hence require large drive currents.

It appears that there is sufficient interest in making low cost integrated recording heads for use in disc storage that these heads will be available for use in magnetic printers.

RECORDING MEDIA

Magnetic recording materials have been extensively investigated for discs and tapes in the computer field. These materials are ideally suited for magnetic printing. For computer applications the recording material must have a large coercive force in order to generate a sufficient external field that can be detected during the read process. Such a large external field is required in magnetic printing to attract and hold the magnetic ink particles. The density of magnetic recordings on discs and tapes are an order of magntiude greater than is needed for magnetic printing. Recording densities of 4000 flux reversals per inch have been written in thin magnetic films. Magnetic printing requires only about 400 flux reversals per inch.

Standard magnetic recording media consists of particulate material such as γFe_2O_3 in a suitable binder and electrodeposited thin metallic films. The particulate material is used on both tape and discs, while the metallic films are confined to drum memories.

There has been considerable work done on other particulate materials such as Fe_3O_4, cobalt substituted iron oxides, CrO_2 and some metal particle work. An excellent review of the work on these materials is given by Bate and Alstad.[22] Such materials can be used as a magnetic printing media and are used in the Data Interface printer.

A metallic film was used in the high speed printer described by Brewster.[8] Since metallic recording films of excellent magnetic properties are easily produced by electrochemical deposition, autocatalytic deposition and vacuum deposition, it appears that metallic films will be more widely used in magnetic printers. Since the required density of recording in magnetic printing is an order of magnitude less than in computer disc applications the metallic films can be thicker. This added thickness will create much larger magnetic fields and field gradients to pick up the magnetic ink particles. The magnetization of these metallic films are an order of magnitude greater than for particulate materials and have coercive forces five times that of particulate materials. An excellent review of these metallic recording materials is given by Bate and Alstad.[22]

DEVELOPMENT

Development of the magnetic latent image, i.e., the application of the magnetic ink to the recording, has involved both wet and dry processes. Most investigators have found that it is much easier to obtain a clean background by using the wet process.

The wet process development consists of passing the recorded surface through an agitated suspension of fine ferromagnetic particles. In general, this process has been used when the recording is on a flexible tape material. A typical liquid inking assembly has been described by Begun.[6] The liquid ink consists of a low viscosity, highly volatile carrier such as alcohol, carbon tetrachloride, freon 113, or petroleum ether containing mechanically suspended magnetic particles.

The forces acting on the magnetic particles suspended in the liquid are gravitational, centrifugal, electrostatic, mechanical, and viscous, as well as the magnetic forces. Some of these forces can be used to an advantage in getting the particles to adhere to the magnetic latent image and the others in keeping the particles from adhering to the background regions where there is no magnetic latent image.

The magnetic force is given by

$$\vec{F}_p = (\vec{M}_p \cdot \nabla) \vec{H}$$

where F_p = force on the particle/unit volume
H = magnetic field
M_p = magnetic moment/unit volume of the particle.

The total magnetic force per unit mass of the particle (F_T) is given by

$$\vec{F}_T = \frac{1}{\rho} (\vec{M}_p \cdot \nabla) \vec{H} \qquad (1)$$

where ρ = density of the particle.

Although the total magnetic force acting on the particle does depend on the size of the particle, equation (1) shows that the acceleration of the particle toward the magnetized latent image due to the magnetic force does not depend upon the size of the particle. It is this acceleration of the particle which is important in accumulating the particles on the latent image.

We can make a calculation of the acceleration of the particles toward the latent image. The model for making these calculations is shown in Fig. 4. We will calculate the acceleration toward the surface of a particle located along a perpendicular line passing through a magnetic pole. This is the line of the maximum gradient and therefore maximum acceleration.

In order to determine the acceleration of the magnetic particle we must calculate the magnetic field and the magnetic field gradient. We will assume that the magnetic field produces a magnetization (M) in the particle such that the internal demagnetizing field ($4\pi M/3$ for a sphere) is equal to the magnetic field at the center of the particle due to all sources of M except the polarization of the particle.

For an infinite number of poles of infinite extent in the z direction the analytic function representing the field is given by:

$$\phi + i\psi = 2Mh \; [\ln \tan \frac{\pi(x+iy)}{4a}] \qquad (2)$$

where ϕ is the scalar magnetic potential and ψ is the conjugate harmonic of ϕ. The spacing between the poles is 2a and the thickness of the magnetized recording media is h.

If we assume that the recording media is magnetic tape of 12μm thickness and the magnetization of 100 emu, and the particle located at y_o = 15μm has a density of 5 gm/cm^3 then the solution of eq. (1) and (2) yield an acceleration of 3×10^5 cm/sec^2.

The magnitude of the ratio of the magnetic force to the mass of the particle can only be appreciated when compared to the electrostatic case. The force to mass ratio of a charged toner particle in an electrostatic printer is calculated from the data given by Thourson[24] to be 10^6 cm/sec^2.

The acceleration of the magnetic particles is about 1/3 that of the electrostatic particles. However, as the magnetic particles accumulate on the surface, the magnetic fields and field gradient above the surface increase due to the increased magnetization of the particles. In the electrostatic case there is also some polarization of the toner particles but there is also a decrease in the original E field due to the charge neutralization as the toner particles of opposite polarity accumulate on the latent image.

Of course all of the forces acting on the magnetic particles must be calculated before a good comparison can be made with the electrostatic printer. The objective of these simple magnetic calculations is to show that contrary to general opinions given in the literature the magnetic forces are not much smaller than the electrostatic forces for the electrostatic printer.

DRY INK DEVELOPMENT

The most serious problem in the development of a magnetic latent image is to keep the ink particles from adhering to the regions of the surface where there is no recording. Electrostatic forces are primarily responsible for the adherence of ink particles in the background regions. If various means can be employed to keep the recording surface and the particles from having an electrostatic charge, then the background will be relatively free of particles.

There is very little published information on dry ink development for magnetic printers. However, dry inking systems which use a fluid bed or a magnetic brush have been developed which produce a satisfactory background.

The methods that have been used to apply the dry ink to the recorded surface include the use of a fluid bed, a magnetic brush, and a cascade process. A schematic diagram of a fluid bed being used to apply the dry magnetic ink to the recorded surface is shown in Fig. 2. The magnetic ink used in the fluid bed must be free flowing for satisfactory operation. In order to obtain a free flowing magnetic ink the magnetic particle must be coated with a resin in order to reduce the magnetic forces between particles.

The use of a magnetic brush may appear to be self-defeating for magnetic printing, but it has been used in a Creed Ferrodot magnetic printer.[25] A schematic diagram of a printer using a magnetic brush is shown in Fig. 5. The magnetic and viscous forces holding the particles in the magnetic brush must be less than the forces attracting the particles to the magnetic latent image on the drum. It would seem that it would be difficult to achieve this imbalance at all times and therefore some areas might not be properly inked. On the other hand, this system might insure a clean background, which is probably the most difficult problem to solve.

Cascade development which has been so successfully used in electrostatic printers has also been used in magnetic printers.[6] However, more sophisticated cascade development schemes are probably required. Cascade development consists of pouring the ink on the surface of the drum and letting the excess fall off as the drum surface rotates below the axis of the drum. In the case of electrostatic printers, the ink particles (toner) are carrier by a larger particle, which may be a glass bead, which has assumed a charge of the opposite sign from the toner particle by virtue of triboelectric effects.

INK

The ink used in magnetic printing consists of fine ferromagnetic particles ($\sim 10\mu m$) which are coated with a low melting point resin so that the particles can be heat sealed to the paper after transfer from the recorded surface.

One might use soft magnetic materials of high permeability for the ink particles or hard magnetic materials with a high remanence. Since the force attracting the particles to the magnetic latent image is proportional to the dot product of the magnetic moment of the particle and the magnetic field gradient of the latent image, the magnetized hard magnetic materials might seem to be a better choice. However, the difficulty of presenting the ink particles to the magnetic image in the proper orientation and the fact that the particles would adhere to the recording surface where there is no latent image would rule out the hard magnetic materials. In addition the whole inking process would be aggravated by the clumping of the ink particles due to their magnetized state.

Soft ferrite magnetic particles are readily available in the industry. These have the desirable characteristics of being black and in a high oxidation state. It might appear to be difficult to coat these particles with a low melting point resin. However, encapsulation techniques have been highly developed during the past decade. The Creed Ferrodot printer[25] exhibited at the 1969 Intermag Conference in Amsterdam and the Data Interface printer[11] exhibited at the 1972 SJCC in Atlantic City made use of encapsulated magnetic particles.

Not much information has been published on the optimum particle size or the most desirable magnetic properties of the ink particles. Begun[6] has reported that for wet development, particles having a diameter $1\mu m$ produce about ten times the background as particles having a size of $10\mu m$. The optimum properties for the particles depend upon many factors such as how well they adhere to the recorded image, how much background they produce, and how well the particles transfer to paper.

Another approach to obtaining a clean background is to remove the particles from the background region after the application of the ink. Sims[3] has suggested that this might be done by spinning the drum recording surface at a high speed. The centrifugal force would be enhanced by using large particles. Of course the magnetic force must be much larger than the centrifugal force so that the particles will remain on the magnetized regions.

Most investigators have found that it is much more difficult to achieve a clean background with dry ink development, but some have been successful.

TRANSFER AND FIXING

Transfer of the magnetic ink to a sheet of paper has been enhanced by many investigators by using a specially prepared paper. Such schemes as dampening the paper, applying a "sticky" coating to the paper or using a coarse paper have been successfully tried.

It is most desirable however to have a magnetic printer which can use any kind of paper and particularly ordinary teletype paper. It would appear that simple pressure contact between the inked drum and a roll of paper would not be a reliable method of transfer. However the Creed Ferrodot printer[25] and the Data Interface printer[11] are purported to transfer the ink by simple pressure contact between the drum and the paper.

It would appear that reliable transfer of the ink to the paper might be as difficult to solve as the problem of maintaining a clean background.

Once the resin coated ink particles are transferred to the paper, they are fixed by the simple application of heat to soften or melt the resin coating.

CONCLUSIONS

The work that has been reported on magnetic printers indicates that a standard 132 characters/line printer can be developed which will print matrix type characters of excellent legibility at a speed of 30,000 characters per second. The major factors in the development of a viable magnetic printer appear to be the development of low cost integrated recording heads, including the drive electronics, and the achievement of a clean background.

ACKNOWLEDGEMENTS

I am indebted to Mr. M. A. Lowry of Data Interface for Fig. 1a and to Mr. A. E. Brewster of Standard Telecommunication Laboratories Limited for Fig. 1b.

REFERENCES

1. R. B. Atkinson and S. G. Ellis, Jour. of the Franklin Institute, Nov. 1951, pp. 373-381.
2. T. M. Berry and J. P. Hanna, General Electric Review, July 1952, p. 20.
3. J. C. Sims, Proc. West. Computer Conf., 1953, pp. 1 pp. 160-166.
4. J. P. Hanna, 5th Annual Tech. Meeting. Tech. Assoc. of Graphic Arts, Wash., D. C., April 1953, pp. 22-27.
5. J. B. Geham, IRE National Convention Record, 6, Pt. 5 1958, pp. 198-203.
6. S. J. Begun, IRE National Convention Record, 6, Pt. 5 1958, pp. 190-197.
7. J. Seehof, et al, Proc. Eastern Joint Computer Conf., 1958, pp. 243-250.
8. A. E. Brewster, Electronics and Power, Feb. 1968, pp. 62-64.
9. N. A. Allen and C. A. Watson, IEEE Trans. on Magnetics, Sept. 1969, p. 450.
10. W. Jones, Mechanics Magazine, Vol. 31, 1839, p. 342.
11. Datamation, July 1972, p. 59.
12. J. C. Barton and C. T. Stockel, The Radio and Electronic Eng., Jan. 1964, pp. 11-15.
13. E. P. Valstyn and D. W. Kosy, IEEE Trans. Mag. Vol. Mag 5, No. 3, Sept. 1969, pp. 442-445.
14. Y. Watanabe, S. Matsumoto, and N. Yajima, IEEE Trans. Mag. Mag 5, 1964, pp. 918-920.
15. W. E. Tchon and D. S. Rodbell, IEEE Trans Mag. Mag 6, No. 3, Sept. 1970, pp. 593-597.
16. J. P. Lazzari and I. Melnick, IEEE Trans. Mag. Mag 6, No. 3, Sept. 1970, pp. 601-602.
17. L. T. Romankiw, I. M. Croll, and M. Hatzakis, IEEE Trans. Mag. Mag 6, No. 3, 1970, pp. 597-601.
18. J. P. Lazzari and I. Melnick, IEEE Trans. Mag. Mag 7, No. 1, 1971, pp. 146-150.
19. D. Augier and J. P. Lazzari, IEEE Trans. Mag. Sept. 1971, pp. 679-683.
21. G. F. Sauter, et al, IEEE Trans. Mag. Mag 8, No. 2, 1972, pp. 194-200.
22. E. P. Valstyn, Annals New York Academy of Science, 1972, pp. 191-205.
23. G. Bate and J. K. Alstad, IEEE Trans. Mag., Mag 5, No. 4, 1969, pp. 821-839.
24. T. L. Thourson, IEEE Trans. on Electron Devices, Vol. Ed-19, No. 4, 1972, pp. 495-511.
25. Electronics, Nov. 23, 1970, Electronics Int. Section.

THIS IS THE NEW DI-240 MAGNETIC PRINTER. THE CHARACTERS
ARE FORMED WITHIN A 10X12 MATRIX ALLOWING THE PRINTER TO USE
UPPER AND LOWER CASE, FULL PUNCTUATION AND.......

(a)

N N N N N N N N N N N N N N N N N N N N N N N N
e e e e e e e e e e e e e e e e e e e e e e e e
8 8 8 8 8 8 8 8 8 8 8 8 8 8 8 8 8 8 8 8 8 8 8 8
2 2 2 2 2 2 2 2 2 2 2 2 2 2 2 2 2 2 2 2 2 2 2 2

(b)

FIGURE 1

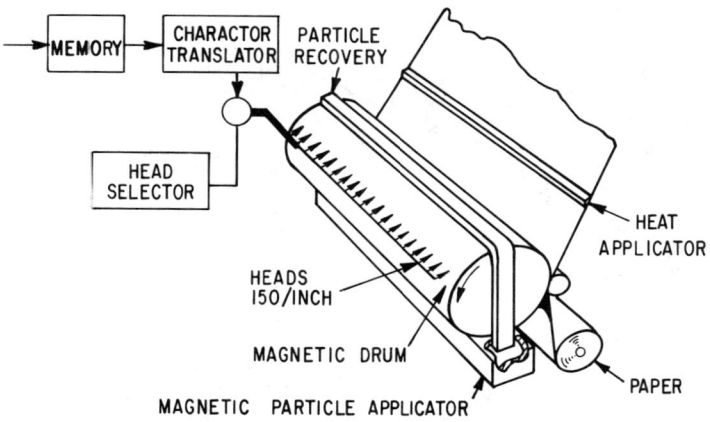

FIGURE 2

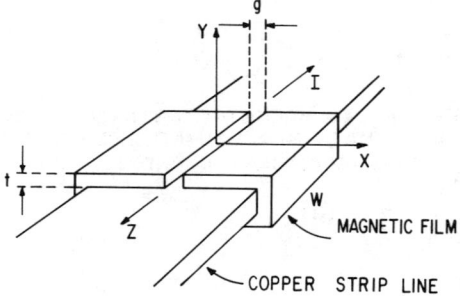

FIG. 3

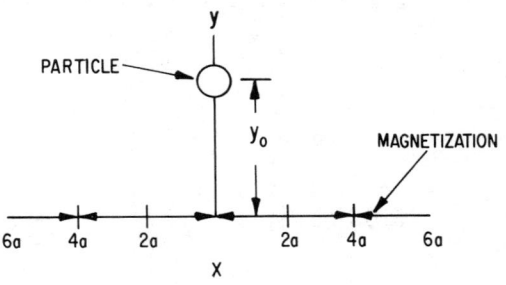

FIG. 4

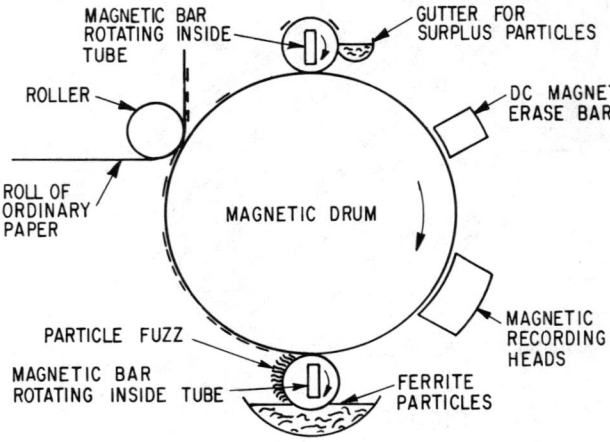

FIG. 5

MAGNETIC PROPERTIES OF RARE EARTH ATOMS IN PALLADIUM -- CRYSTAL FIELD EFFECTS EPR AND MAGNETIC MOMENT

H.C. Praddaude
Francis Bitter National Magnet Laboratory,[†] Massachusetts Institute of Technology, Cambridge, Massachusetts 02139

R.P. Guertin[*]
Tufts University, Medford, Massachusetts 02155

S. Foner and E.J. McNiff, Jr.
Francis Bitter National Magnet Laboratory,[†] Massachusetts Institute of Technology, Cambridge, Massachusetts 02139

ABSTRACT

Recent studies of the magnetic properties of rare earth (RE) atoms dissolved in metals and particularly in Pd are reviewed. The effects of the crystalline electric field on the RE and the corresponding magnetic behavior and electron paramagnetic resonance (EPR) are discussed. Selected examples are chosen from an extensive series of magnetic measurements on dilute alloys of $Pd_{1-x}R_x$ where R = Ce, Pr, Nd, Sm, Eu, Gd, Tb, Dy, Ho, Er, Tm, Yb, or Lu in order to illustrate the behavior of the RE in Pd. The magnetic properties of the RE in Pd are well described by a semiempirical formula which yields the RE moment and hence the valence state, and host susceptibility. The temperature and field dependence of the magnetic moment also are compared with calculations based on crystal field parameters derived from a reanalysis of recent EPR measurements of RE ions in single crystal Pd by the University of Geneva group. This analysis shows that for Dy an effective spin Hamiltonian is not appropriate for describing the EPR data, but it is adequate for Er in Pd.

INTRODUCTION

In this paper we review the magnetic properties of rare earth (RE) ions in a Pd host. Both the electron paramagnetic resonance (EPR) measurements of RE ions and the temperature and field dependence of the RE ion magnetic moment will be discussed. For the discussion of EPR we will draw upon experiments and interpretation of the results of other workers. For the discussion of the RE magnetization we will refer mainly to an extensive series of measurements carried out at the Francis Bitter National Magnet Laboratory.

A large bibliography exists for the magnetic properties of 3d and 4f ion dopants in insulators. In this case the free ion magnetic properties of the dopant, perturbed by the crystalline electric field, are a good first order approximation. This is not the case for atoms dissolved in metals, where there is a strong interaction between the host-metal conduction electrons and the outer electrons of the impurity ion. As a result the free

---
[†] Supported by the National Science Foundation.
[*] Visiting Scientist at the Francis Bitter National Magnet Lab, MIT.

ion properties may be extensively modified and new experimental phenomena observed. In the case of RE atoms dissolved in metals the interaction between the conduction electrons and the magnetic 4f electrons is shielded by the $5s^2 5p^6$ closed outer electron shell of the RE ions. Here free ion magnetic properties are a good first order approximation. Investigations of the magnetic properties of RE ions in metals are hampered because many of the standard experimental techniques used with insulators are difficult or impossible to use. For example, ordinary optical absorption is not possible in metals and EPR is difficult to observe. In spite of these difficulties, EPR data have become available for several RE atoms dissolved in metals, e.g. in Au, Ag, Cu, Pt and Pd. The simplest example of RE ion EPR is that of Gd^{3+}, which is an S state ion. More recently non-S state EPR lines have been observed in the noble metals, and, within the last few months, EPR of RE ions has also been detected in single crystal Pd.

The EPR and magnetization data for RE ions in metals have been analyzed assuming that the $(2J+1)$ degenerate ground state levels of the $4f^n$ configuration are split by the crystalline electric field of the host lattice. The total level splitting in a metal is reduced compared to the splitting in an insulator because the electrostatic shielding afforded by the conduction electrons reduces the crystalline electric field seen by the RE ion. However in the case of a RE ion in a Pd host, we find that fitting the magnetization data to crystal field theory (CFT) may not be adequate. In general the magnetic data does not have enough structure to provide a sharp measure of the CFT parameters. Furthermore, even if additional information is obtained from EPR, cases are found when a CFT fit is not adequate. We will discuss a method of fitting magnetization data accurately to a three parameter semi-empirical formula which yields values for the saturation moment of the RE ion and the host susceptibility.

A summary of the status of the CFT calculations, EPR and magnetic moment results and their limitations is presented.

ELECTRON PARAMAGNETIC RESONANCE OF RARE EARTH IONS IN METALS

Paramagnetic resonance of non-S state RE ions in dilute alloys of Cu, Ag, Au, Pt and Pd have been observed by Griffiths and Coles,[1] Hirst et al,[2] Davidov et al,[3,4,5] and Devine et al.[6,7] The EPR of S-state Gd^{3+} has been reported by Peter et al[8] and recently the fine structure of Gd^{3+} in single crystal Pd was resolved by Devine et al.[9] Discussions of these results employ a simplified model that assumes that the $(2J+1)$ degenerate ground state levels of the RE ions are split by the crystalline electric field into a number of multiplets. These levels arise from the $4f^n$ configuration and satisfy Hund's rules. For cubic symmetry the CFT Hamiltonian according to Lea, Leask and Wolf,[10] is

$$\mathcal{H} = g_J \mu_B \vec{H}_o \cdot \vec{J} + W \left\{ x \frac{O_4}{F_4} + (1 - |x|) \frac{O_6}{F_6} \right\} \qquad (1)$$

where H_o is the external magnetic field, g_J is the Lande g-factor, W is an overall scaling parameter for the crystal field strength, x is proportional to the ratio of the strengths of the fourth and sixth order tensor operators O_4 and O_6, and F_4 and F_6 are numerical constants. The z-axis is taken along a fourfold symmetry axis. If the magnetic field is absent, Lea, Leask and Wolf have calculated the splitting of the (2J + 1) levels normalized to W = +1, and catalogued them according to the irreducible representations Γ_i of the cubic group. Figure 1 reproduces their results for the case of J = 15/2, which is appropriate for Dy^{3+} and Er^{3+}. If a very small magnetic field H_o is applied, to first order in H_o the ground state Γ_i degenerate levels will couple and the splitting will be independent of the remaining multiplets. That is, the g-factors of the Γ_i multiplet will be independent of the parameter W which determines the overall splitting of the (2J + 1) levels. This is the effective spin Hamiltonian approximation. At sufficiently low temperatures EPR will occur only between the ground state levels of the Γ_i representation, and the anisotropy of the resonances will involve these levels only. As the temperature increases, the higher multiplets will become populated and EPR within these excited states may be observed.

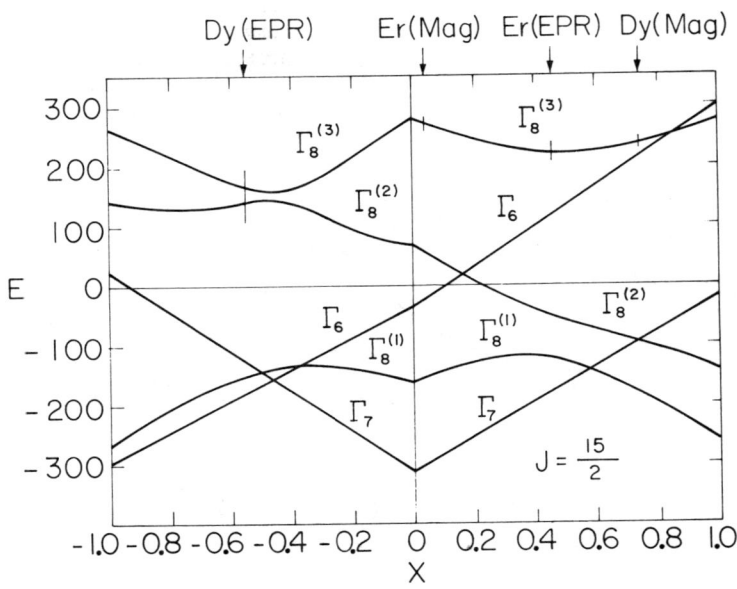

Fig. 1 — Energy in units of W versus parameter x (Eq. (1)) for J = 15/2, for W = +1 and for H_o = 0 (see Ref. 10). The arrows show the values of x obtained from fits of crystal field theory to EPR or to magnetic moment measurements. The latter values are discussed in Section IV where it is pointed out that the error bars associated with x may be large.

We have applied the CFT analysis to the resonance results for Dy^{3+}, Er^{3+} and Gd^{3+} ions in single crystal Pd reported by Devine et al.[6,7,9] For Dy^{3+} we find that the above spin Hamiltonian approximation fits the anisotropy of the most anisotropic resonance line but fails to predict the position of the remaining resonances. However, the available information is sufficient to assign the resonance to one of the quadruplets $\Gamma_8^{(i)}$ and not to the doublets Γ_6 or Γ_7. From Fig. 1 we see that the value of x must be between -1 and +0.8 if W is negative, and that x must be between +0.6 and +1 if W is positive. On the basis of a point charge calculation, Devine et al decided that W and x should be negative. Instead of using the spin Hamiltonian approximation we diagonalized the complete 16 x 16 matrix and calculated[11] the energy levels, matrix elements, temperature dependent transition probabilities and magnetization versus applied field. Surprisingly, for Dy^{3+} we find that the g-factors are strongly magnetic field dependent, even for magnetic fields as small as a few hundred gauss (the EPR data was taken with fields up to 3500 gauss). This allowed us to obtain both W and x by fitting theory to the most anisotropic resonance which occurs between levels 2 and 3 of the $\Gamma_8^{(3)}$ multiplet, see Fig. 2. In this way we obtained the values W = -89.3 mK and x = -0.5418. Also in Fig. 2 we show the calculated position of two other resonances between the levels 1-2 and 3-4.

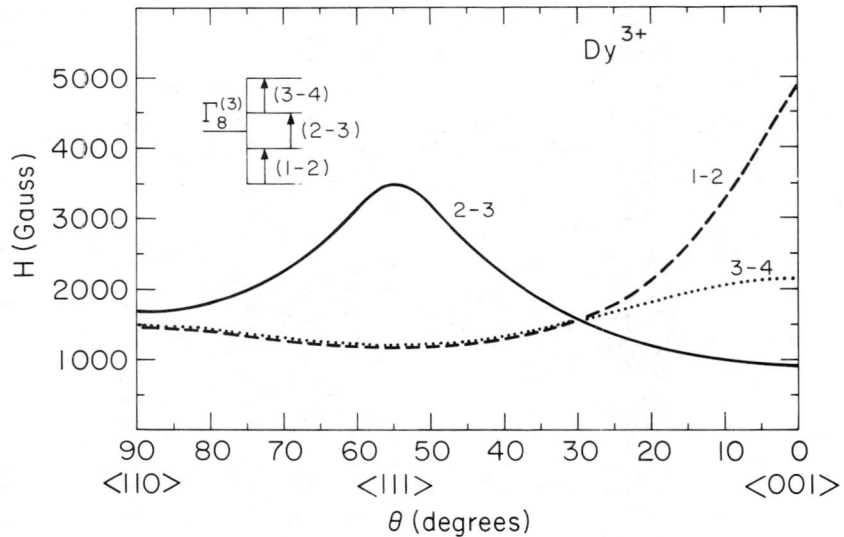

Fig. 2 — Calculated EPR resonance field, H, versus angle, θ, for Dy^{3+} and for a frequency of 9.500 GHz. The ground state level manifold $\Gamma_8^{(3)}$ is shown in the insert. The most intense line occurs between levels 2 and 3.

Next we consider the Er^{3+} resonances which were reported recently.[7] In this case we agree that the spin Hamiltonian provides a good description of the resonances. From the Dy^{3+} results it was deduced that the crystal field splitting is small so that the g-factors are strongly magnetic field dependent. If the value of W is similar for Dy and Er, and this is a reasonable assumption, also we may expect difficulties with the spin Hamiltonian approximation in Er. However, the calculations show that the g-factors for Er^{3+} are remarkably insensitive to the magnetic field even for values of W a factor of three smaller than that of Dy. This result can be understood by referring to Fig. 1. The value of x = -0.5418 for Dy occurs in a region where the two quartets $\Gamma_8^{(3)}$ and $\Gamma_8^{(2)}$ repel each other and consequently the magnetic field perturbation couples them strongly. In this case W must be very large for these two levels to be well separated. On the other hand x = +0.465 for Er^{3+}. The excited doublet Γ_6 is further separated from the $\Gamma_8^{(3)}$ quartet and, because it has different symmetry from the $\Gamma_8^{(3)}$ quartet, the matrix elements connecting both multiplets are smaller than in the previous case. Consequently for magnetic fields smaller than 5 kG the g-factors are not strongly perturbed. There is one other important difference between the Er and Dy analysis. For Er all the observed resonances are very well accounted for and the transition probabilities correlate quite well with the observed EPR intensity. For Dy, the values of W and x calculated to fit the strongest resonance do not fit the remaining weaker resonances. The EPR data for Dy^{3+} fall between the predictions of the full CFT and the effective spin Hamiltonian approximation. This disagreement is also reflected in the fit to the magnetization which will be discussed in the next section.

The magnetic properties of Gd in Pd are puzzling. The Gd^{3+} ion has the very stable configuration $^8S_{7/2}$ and consequently the crystal field is inoperative to first order. Higher order perturbations produce a fine structure splitting which was observed recently by Devine et al.[9] The g-factor is concentration dependent,[12,13] but for concentrations lower than 1500 ppm of Gd in Pd the g-factor stabilizes at approximately 1.8. This large negative g-shift is difficult to explain. Furthermore, as we will discuss later, the magnetic moment of the Gd ion appears to increase from $7\mu_B$ for Gd concentrations larger than 0.5% to $\sim 10\mu_B$ for concentrations of the order of 0.1%.

LOW FIELD MAGNETIC SUSCEPTIBILITY OF Pd(RE) ALLOYS

EPR measurements at low temperature provide very detailed information about the low lying levels of the ground state configuration. Increasing the temperature would, in principle, provide similar information about the higher levels of the (2J + 1) manifold if the broadened resonances were still observable. Bulk magnetic moment measurements provide information about the positions of the higher levels. Either the temperature dependence of the low field magnetic susceptibility of the RE ion, $\chi_R(T)$, or the magnetization of the RE ion as a function of the applied magnetic field may be employed. The susceptibility and magnetization are smooth functions of the temperature and the magnetic field respectively. It should be emphasized that the CFT fit to these data is insensitive to small variations of the CFT parameters. The susceptibility measurements have the added drawback that the susceptibility of the metallic host, $\chi_{mat}(T)$, must be subtracted from the measured susceptibility, $\chi(T)$, in order to obtain the RE ion contribution, $\chi_R(T)$. This subtraction procedure is reliable when $\chi_{mat}(T) \ll \chi(T)$.

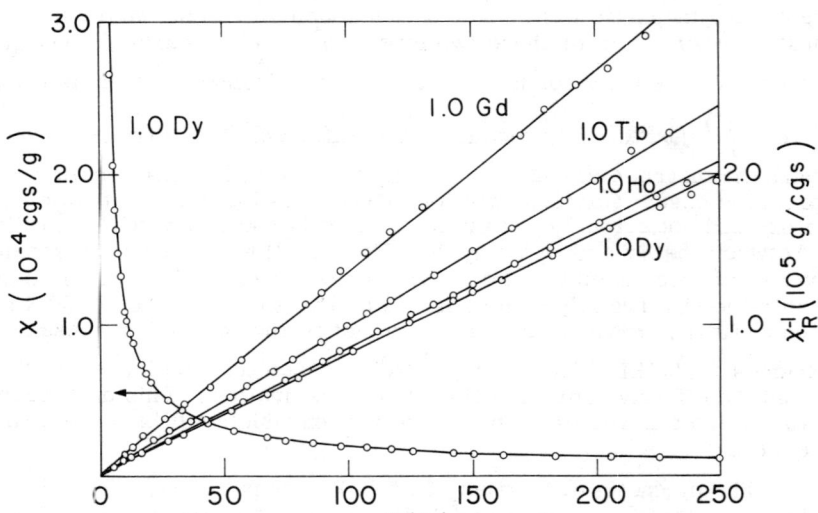

Fig. 3 — The rare earth inverse susceptibility, $\chi_R^{-1}(T)$, vs temperature, T, for $Pd_{1-x}R_x$ alloys where R = Gd, Tb, Ho or Dy, and the measured susceptibility, $\chi(T)$, for $Pd_{0.99}Dy_{0.01}$. Here $\chi_R(T) = \chi(T) - \chi_{mat}(T)$ where $\chi_{mat}(T)$ is the susceptibility of $Pd_{0.99}Lu_{0.01}$. The slope of $\chi_R^{-1}(T)$ vs T determines the effective moment per R atom.

We have measured $\chi(T)$ for many RE atoms in Pd over the temperature range $4.2 \leq T \leq 250$ K. As shown by the examples in Fig. 3, plots of $\chi_R^{-1}(T)$ vs T show a Curie law behavior except for small deviations for $T \lesssim 20$ K. These deviations are too small to determine accurately the values of the crystal field parameters. For Dy the values of W and x obtained from our analysis of the EPR results were used to calculate $\chi_R^{-1}(T)$ vs T. The calculated deviations from a Curie law for T below 20 K have the same sign as observed experimentally but are ~3 times smaller.

The moment per RE atom obtained from the slope of $\chi_R^{-1}(T)$ vs T data agree with the moments calculated from the high field magnetization results for a 1% RE concentration. In all these cases the calculated RE^{3+} free ion moment is obtained.

IV. HIGH FIELD MAGNETIZATION OF Pd(RE) ALLOYS

An extensive series of magnetic moment measurements[14] vs field, $\sigma(H_o)$, was made in the $Pd_{1-x}R_x$ alloys where R = Ce, Pr, Nd, Sm, Eu, Gd, Tb, Dy, Ho, Er, Tm, Yb or Lu. These measurements were done mostly on samples with x ≃ 0.01, but in order to examine concentration effects in some cases x ranged from ~0.001 to 0.05. For magnetic fields smaller than 55 kG the measurements were done in a superconducting magnet using a PAR vibrating sample magnetometer. These results were used to normalize the high magnetic field measurements made up to 220 kG in water-cooled Bitter solenoids with a very low frequency vibrating sample magnetometer. The measurements indicate that Lu, Ce and Eu do not show a magnetic moment when dissolved in Pd. The result for Lu (which has a full 4f shell) is expected, but Ce and Eu may or may not have a moment depending on the valence state of the RE ion in the host matrix. The observed zero moments of both Eu and Ce indicate that Eu goes into solution in Pd as Eu^{3+}, but that Ce goes into solution as Ce^{4+}. The latter result is in disagreement with previous results[15] which suggested $1.1\mu_B$/Ce atom. All the other RE atoms in solution in Pd show a magnetic moment (see, for example, Fig. 4). It should be noted that even at fields higher than 150 kG and T ~ 1.5 K there is still an appreciable curvature in the magnetization vs field data. Also, the approach to saturation is only weakly dependent on temperature. Linear extrapolation of the high field data to zero field is not adequate: the resulting band susceptibility is systematically too large and the RE magnetic moment is too small. Fitting the data to a Brillouin function with adjustable parameters also yields large systematic deviations.[16,14] We find that the semiempirical formula

$$\sigma(H_o) = \frac{AH_o}{1+B|H_o|} + CH_o \quad (2)$$

fits all the high field data to better than 1%. For a fixed temperature, the first term on the right hand side of Eq. (2) describes the field dependence of the rare earth moment; it is the simplest rational Padé approximant. In the limit of large applied fields, H_o, Eq. (2) becomes $(A/B) + CH_o$ so that C is just the matrix susceptibility and A/B is proportional to p_{SAT}, the saturation moment per rare earth atom. In the limit of small H_o, Eq. (2) approaches $(AH_o + CH_o)$, thus A is a measure of the low

Table I. Examples of the results of the analysis of magnetic moment data to 200 kG at 4.2 K on $Pd_{1-x}R_x$ alloys. The parameters C, B, and A are determined from an rms fit to Eq. (2). C is the band susceptibility, B is a measure of the ease of magnetic saturation (low B indicates an impurity that is very difficult to saturate) and $p_{SAT}(\alpha A/B)$ is the moment of R. The gJ is the moment expected for an R^{3+} ion (except Ce and Pr which are R^{4+}).

Rare Earth	Conc. at.%	C (10^{-6}/gm)	B (10^6G)$^{-1}$	$p_{SAT}(\alpha A/B)$ (μ_B/R atom)	gJ
Ce	1.00	5.66[a]		<10^{-3}	0
Pr	1.00	6.3	20	2.1	2.1
Nd	1.00	5.4	25	3.3	3.3
Sm	1.00	5.9	6	0.8	0.7
Eu	1.30	6.1[b]		0	0
Gd	0.10	5.9	35	10	7
Gd	0.29	5.6[b]	60	8.0	7
Gd	0.68	5.0[b]	90	7.5	7
Gd	1.00	4.6	100	7.2	7
Gd	1.38	4.8	130	7.1	7
Gd	2.03	3.6	160	6.8	7
Tb	1.00	4.8	85	9.0	9
Dy	1.05	7.3	95	8.9	10
Ho	1.00	5.1	80	9.6	10
Er	1.00	4.2	65	9.5	9
Tm	1.50	5.5	45	6.0	7
Yb	0.85	4.9	20	4.4	4
Lu	1.00	6.0[b]		0	0

[a] Measured at 4.2 K in fields to 50 kG.
[b] Measured at 4.2 K in fields to 140 kG.

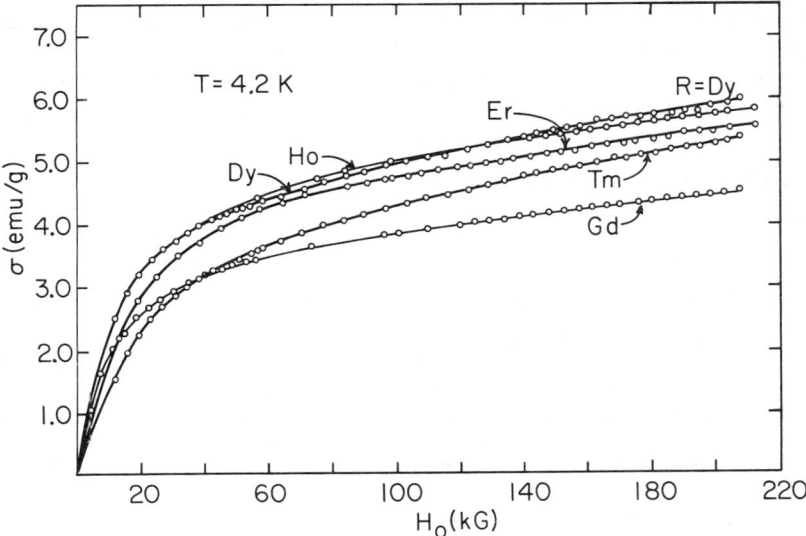

Fig. 4 — Magnetic moment, σ, vs applied field at 4.2 K for $Pd_{1-x}R_x$ alloys where R = Gd, Dy, Ho, Er or Tm and $x \simeq 0.01$. The solid lines represent the best fit of Eq. (2) to the data. The systematic error for the CFT fit for Dy (see Ref. 14) (for W = -30 mK, x = 0.75) is ~1.5% or less and would not be noticeable on the scale of this figure.

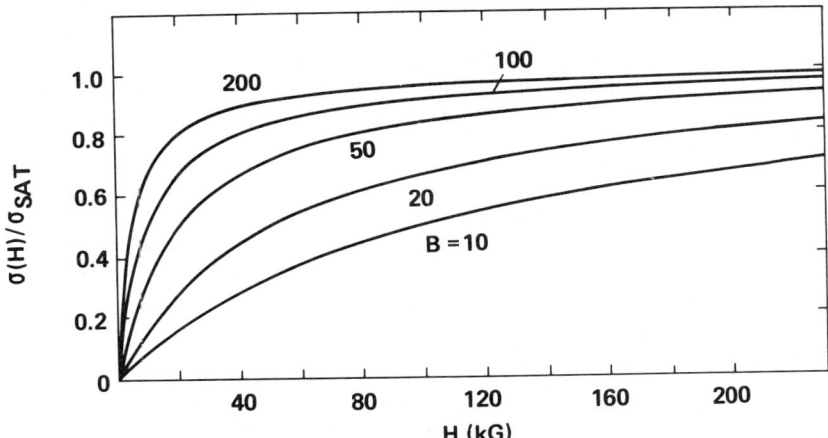

Fig. 5 — Normalized magnetic moment, $\sigma(H_o)/\sigma_{SAT}$, vs H_o where
$$\frac{\sigma(H_o)}{\sigma_{SAT}} = \frac{BH_o}{1 + B|H_o|}$$
(obtained from Eq. (2) when C = 0). The units of the parameter B are $(10^6 G)^{-1}$. When B = 100, saturation is ~95% complete at 200 kG.

field rare earth susceptibility. The parameter B is a measure of the ease of saturating the rare earth moment: large B means saturation is readily achieved. As an illustration, in Fig. 5 we plot the ratio of the RE ion magnetic moment to its saturation value vs H_o for different values of the parameter B in Eq. (2). The results of this analysis using Eq. (2) are shown in Table I for some of the $Pd_{1-x}R_x$ alloys. Within the experimental error, in most cases, the RE magnetic moment agrees with the theoretical moment expected for an R^{3+} ion. The exceptions are Pr and Ce, which we find go into solution in Pd as Pr^{4+} and Ce^{4+}. A discrepancy occurs in the $Pd_{1-x}Gd_x$ alloys. For x between 0.0068 and 0.02 the value of p_{SAT} equals that of the R^{3+} ion within experimental error (see Table I). However, for smaller x the magnetic moment of the RE ion seems to increase. This apparent increase of the RE magnetic moment is not understood at present.

In the remainder of the paper we will discuss Gd, Dy and Er in Pd (details for the other Pd(RE) alloys can be found in Ref. 14). For these three alloys we have the results of EPR measurements, the analysis of the magnetic data using Eq. (2) and the analysis of the magnetic data using the CFT Hamiltonian, Eq. (1).

Analyses of the $Pd_{1-x}Er_x$ alloys seem to be the most satisfactory. The EPR data[7] can be described equally well by the CFT analysis or the effective spin Hamiltonian approximation. The value of x = +0.465 accounts for the intensity and anisotropy of all the observed resonances. For this case, where the effective spin Hamiltonian is satisfactory, the results of the CFT analysis are extremely insensitive to the value assigned to W, as expected. A least squares fit of the CFT to the magnetic data gives the value x = +0.05 and W = -105 mK (corresponding to an overall splitting of the levels of ~55 K). As mentioned in the previous section the fit to the magnetization is weakly dependent on the values of W and x. One shallow minimum of the rms error of the fit is obtained at the values quoted above for W and x. The value of x differs appreciably from that obtained from the fit of the CFT to the EPR data (see Fig. 1). These differing values of x may be due to a combination of factors. For example, there may be concentration effects — the CFT analysis of the magnetic measurements was done for the 1% alloy, whereas the EPR was obtained on single crystal 0.1% samples. Also the uncertainty in x may be sufficiently large to produce agreement. The CFT analysis of the magnetization yields a value of p_{SAT} in good agreement with that obtained with Eq. (2); the band susceptibility is somewhat larger for the CFT analysis.

Analysis of the $Pd_{1-x}Dy_x$ alloys is less satisfactory. Well defined values of x = -0.5418 and W = -89.3 mK are obtained from the CFT fit to the most intense and anisotropic EPR line (see Fig. 2 and Refs. 7 and 16). The g-factor is strongly dependent on H_o. However, the remaining EPR lines are not reproduced well. Fitting CFT to the magnetization data yields W = -30 mK and x = +0.75, both quite different from the EPR values. Although some uncertainty in x could occur as in Er above, the difference between the two values of x (see Fig. 1) appears to be much too large. Furthermore, we expect that W would be only weakly dependent on concentration, whereas the above values of W differ by a factor of ~3. In addition, the

CFT analysis of the magnetic moment data yields a p_{SAT} for Dy^{3+} which is too small, and a band susceptibility which is too large. When we fit the magnetic moment data with Eq. (2), p_{SAT} is also too small, and the band susceptibility is also too large (compared to that of the other RE ions, see Table I).

A summary of the results of various analyses for Dy and Er is given in Table II.

Table II: Comparison of analyses of EPR and magnetic moment data with crystal field theory (CFT) and Eq. (2) for Er and Dy in Pd; χ_{band} is the band susceptibility (=C for Eq. (2)).

		Er	Dy
EPR (CFT)	W	--	-89.3 mK
	x	+0.465	-0.5418
	fit	good	good for one line only
Magnetization (CFT)	W	-105 mK	-30 mK
	x	+0.05	+0.75
	fit	good	small systematic deviations
	$p_{SAT}(\mu_B/\text{RE ion})$	9.0	7.9
	$\chi_{band}(10^{-6}/g)$	6	8.6
Magnetization (Eq. (2))	$p_{SAT}(\mu_B/\text{RE ion})$	9.5	8.9
	$\chi_{band}(10^{-6}/g)$	4.2	7.3
	fit	good	good

Finally, we consider the $Pd_{1-x}Gd_x$ alloys. The ground state of the Gd^{3+} ion is an S-state. Here we expect no crystalline field effects to first order, and a Brillouin function is predicted for the magnetization if the small fine structure[9] of the energy levels is neglected. Large and systematic deviations between the magnetic moment data and a Brillouin function are observed.[16,14] Clearly the CFT does not apply in this case.

As noted earlier, a concentration dependence of the EPR g-factor has been reported.[12,13] Recently, we have found that p_{SAT} for the Gd ion in Pd seems to increase with decreasing concentration. This is illustrated in Table I, where a systematic variation of B with concentration also is apparent. For 0.1% Gd, which is similar to the concentration used for the EPR fine structure studies,[9] we find that $p_{SAT} \simeq 10\mu_B$/Gd atom (and $g \simeq 1.8$, Ref. 9) which should be compared with the free ion value of $p_{SAT} = 7\mu_B$/Gd atom, $g = 2$ and $J = 7/2$. So far, no consistent explanation has been given for the concentration dependence of the magnetic moment data.

In conclusion, RE atoms dissolved in Pd show several general features of interest. The RE ion magnetic moment and valence state can be deduced from suitable theoretical analysis applied to high field magnetic moment data. A detailed comparison is made of the crystal field parameters derived from fits of EPR and magnetic moment data for Er and Dy in Pd. The results obtained are more consistent for Er than for Dy. In Gd, the expected Brillouin function behavior of the magnetization is not observed, and concentration dependent effects also remain unexplained.

REFERENCES

1. D. Griffiths and B.R. Coles, Phys. Rev. Letters 16, 1093 (1966).
2. L.L. Hirst, G. Williams, D. Griffiths and B.R. Coles, J. Appl. Phys. 39, 844 (1968).
3. D. Davidov, R. Orbach, L.J. Tao and E.P. Chock, Phys. Letters 34A, 379 (1971).
4. D. Davidov, R. Orbach, C. Rettori, D. Shaltiel, L.J. Tao and B. Ricks, Phys. Letters 35A, 339 (1971).
5. D. Davidov, R. Orbach, C. Rettori, L.J. Tao and E.P. Chock, Phys. Rev. Letters 28, 490 (1972).
6. R.A. Devine, J.M. Moreti, J. Ortelli, D. Shaltiel, W. Zingg and M. Peter, Solid State Comm. 10, 575 (1972).
7. R.A. Devine, W. Zingg and J.M. Moret, Solid State Comm. 11, 233 (1972).
8. M. Peter, D. Shaltiel, J.H. Wernick, H.J. Williams, J.B. Mock and R.C. Sherwood, Phys. Rev. 126, 1095 (1962).
9. R.A. Devine, D. Shaltiel, J.M. Moret, J. Ortelli, W. Zingg and M. Peter, Solid State Comm. 11, 525 (1972).
10. K.R. Lea, M.J.M. Leask and W.P. Wolf, J. Phys. Chem. Solids 23, 1381 (1962).
11. H.C. Praddaude, to be published in Phys. Letters A (1972).
12. A.M. Harris, J. Popplewell and R.S. Tebble, Proc. Phys. Soc. 85, 513 (1965).
13. H.C. Praddaude and H. Gärtner, Phys. Letters 34A, 217 (1971).
14. R.P. Guertin, H.C. Praddaude, S. Foner, E.J. McNiff, Jr. and B. Barsoumian, to be published in Phys. Rev. 1972. A discussion of the details of all the magnetic moment measurements, temperature dependent susceptibility and residual electrical resistivity is given in this paper.
15. D. Shaltiel, J.H. Wernick, H.J. Williams and M. Peter, Phys. Rev. 135, A1346 (1964).
16. H.C. Praddaude, Phys. Letters 34A, 281 (1971).

THE SPIN MAGNETISM OF BINARY ALLOYS*

H. Fukuyama[†]

Division of Engineering and Applied Physics, Harvard University
Cambridge, Massachusetts   02138

ABSTRACT

Recent developments in the theoretical treatments of the magnetic properties of the substitutional binary alloys, $A_x B_{1-x}$, are reviewed. The model is the itinerant electron system with the short range Coulomb interactions treated in Hartree-Fock approximation and the effects of the random array of atoms are taken in the coherent-potential-approximation (CPA). Various theoretical consequences are discussed and compared with experiments both in the paramagnetic and the ferromagnetic systems.

SECTION 1. INTRODUCTION

Studies of the electronic and magnetic properties of metallic alloys afford us much information about the electronic structures of solids. Especially in metals which show strong magnetism it is essential to look at their alloys since, by changing the host-impurity (or majority-minority in highly concentrated alloys) combination, we can guess the essential mechanism for the occurrence of various kinds of magnetism.

The alloy theories of magnetic metals were originated by Friedel,[1] and later Anderson,[2] Wolff,[3] Clogston[4] and Moriya[5] gave visual physical pictures to the concept of the local magnetic moments by treating one magnetic impurity in an otherwise pure system. The essential ingredients in all of these models which differ in some respects are the local resonance state at a magnetic site and the mutual Coulomb interactions between opposite spin electrons at this site. The latter is treated in Hartree-Fock approximation. The refinements of the models appeared in various directions. The most important of them might be those of Alexander and Anderson,[6] Moriya[7] and others.[8,9] Guided by the observation that the type of magnetism (ferromagnetism, antiferromagnetism, etc.) would strongly depend on local properties in the transition metals, these authors discussed a pair of magnetic atoms to find a qualitative agreement with real systems as regards the relationship between the type of magnetism and the degree of filling the d-band. It was found by Inoue and Moriya[8] that two models, due to Anderson and Wolff respectively, gave similar results to this problem. Since the other physical consequences are not expected to be so much different between Anderson's extra orbital model and Wolff's tight-binding model, we confine our consideration

---

*Supported in part by Grant No. GH-32774 of the National Science Foundation.
[†]Permanent Address: Department of Physics, Tohoku University, Sendai, Japan.

in this paper to the latter tight-binding model.

In contrast to the detailed treatment of the system with two magnetic impurities, theoretical approach to these types of alloys at finite concentrations have been based on the simplest virtual crystal approximation.[10] In this scheme electrons feel the potential energy uniform in space which is given by the sum of two kinds of potentials of each atom weighted by its concentration. Thus the shape of the density of states curve is always the same, but only the Fermi energy varies due to alloying. This approximation could not yield a local (resonance) level in the dilute limit which was the essential thing in the discussions of Anderson and others, nor could it explain the experimental facts[11] of the different values of the local magnetic moments on different kinds of atomic sites. Or, in other words, the theories at finite concentration were not in the same level of treating the scattering potential as those in the dilute limit.

Similar status was found in the theory of non-magnetic alloys, until the CPA[12] was proposed. This approximation is satisfactory, although not without defects, of course, because it reduces to the exact solution in the low concentration limit small x and smoothly interpolates between x=0 and x=1. As the nature and usefulness of the CPA gets realized,[13] it is natural to apply it to the magnetic systems. So far two different kinds of approximations are proposed. One is due to Hasegawa and Kanamori (HK)[14] and Levin, Bass and Bennemann.[15] This is also employed by the present author.[16] The other is the one due to Harris and Zuckermann.[17] The same procedures were also employed by Kato and Shimizu[18] quite recently.

The purpose of this review is to discuss these new developments in the light of the existing experiments for the better understanding of the itinerant magnetism. By the reason given in the next section, we focus our attention on the approach due to HK which is briefly sketched in Sec. 2. In Sec. 3 we give the transverse susceptibility $\chi_{+-}(Q,\omega)$ valid both in the paramagnetic and the ordered state on which almost all of the physical statements in this paper are based. Sec. 4 discusses the validity of the present approximation.

SECTION 2. PHYSICAL PICTURE OF THE PRESENT SCHEME

We consider the substitutional binary alloys represented by

$$\mathcal{H} = \sum_{i,j,\sigma} t_{ij} a^+_{i,\sigma} a_{j,\sigma} + \sum_{i,\sigma} \varepsilon^0_i n_{i,\sigma} + \sum_i U_i n_{i,\sigma} n_{i,-\sigma} \qquad (1)$$

where t_{ij} is the transfer integral which is independent of the kinds of atoms. ε^0_i and U_i are the potential energy and the mutual Coulomb interaction at the i-th site, which are assumed to take ε^0_A or ε^0_B and U_A or U_B. In Hartree-Fock approximation for U_i, the atomic energy at the i-th site will be

$$\varepsilon_{i,\sigma} = \varepsilon_i^o + U_i \langle n_{i,-\sigma} \rangle \tag{2}$$

where $\langle n_{i,-\sigma} \rangle$ is the thermal average. In the pure system where U_i and $\langle n_{i,-\sigma} \rangle$ are uniform in space this change of the atomic energy due to Coulomb interactions results only in the redefinition of the energy origin. However, in alloys $\langle n_{i,-\sigma} \rangle$ has a spatial dependence and then the second term in Eq. (2) could be important. Although the model is different from Eq. (1), this is clear in the original discussion by Anderson,[2] where not the electron number averaged over all sites but the <u>local</u> electron number at a magnetic site affected the location of the d-level and consequently the condition of the local moment formation. In reality the spatial dependence of $\langle n_{i,-\sigma} \rangle$, which is to be determined self-consistently, is very complicated. The simplest approximation might be to assume

$$\langle n_{i,-\sigma} \rangle = \begin{cases} n_{A,-\sigma} & i \in A \\ n_{B,-\sigma} & i \in B \end{cases} \tag{3}$$

and then the atomic energy levels at A and B sites are given by

$$\varepsilon_{A,\sigma} = \varepsilon_A^o + U_A n_{A,-\sigma}$$
$$\varepsilon_{B,\sigma} = \varepsilon_B^o + U_B n_{B,-\sigma} \tag{4}$$

where $n_{A,-\sigma}$ ($n_{B,-\sigma}$) is the electron number averaged over all A(B) sites of the system. Although this does not appear quite satisfactory, this obviously takes into account the above-mentioned important aspect of the problem. (If U_i becomes large, Eq. (2) no longer holds, however (see Sec. 4).)

By the simplification of Eq. (3) the problem to be solved becomes precisely the same as in the usual binary alloys[12] with one exception that the atomic levels $\varepsilon_{A,\sigma}$ and $\varepsilon_{B,\sigma}$ are to be determined self-consistently. Various parameters which are important in connection with experiments are $U_A, U_B, \delta \equiv \varepsilon_A^o - \varepsilon_B^o$, x and the total electron number in the system. The density of states of a pure system

$$\rho_o(\varepsilon) \equiv \sum_k \delta(\varepsilon - \varepsilon(k)) \tag{5}$$

where $\varepsilon(k) = \sum_i e^{ik(R_i - R_j)} t_{ij}$, is chosen so as to reproduce the reproduce the results of the band calculations.

The case of the dilute magnetic A atoms (x→0) has been discussed by several authors within this scheme. Moriya[7] considered one A atom in Wolff's model ($U_B = 0$). In this case the scattering potential at a magnetic site (0-th site) is

$$\delta_\sigma = \delta_o + U_A n_{A,-\sigma} \tag{6}$$

and $n_{A,-\sigma}$ is given in terms of Green's function

$$G_{ij}^\sigma(\varepsilon) = i \int_{-\infty}^{\infty} dt\, e^{-i\varepsilon t} <Ta_{i,\sigma}(0)a_{j,\sigma}^+(t)> \tag{7}$$

as

$$n_{A,-\sigma} = -\pi^{-1} \mathrm{Im} \int_{-\infty}^{\varepsilon_F} d\varepsilon\, G_{00}^{-\sigma}(\varepsilon) = -\pi^{-1} \mathrm{Im} \int_{-\infty}^{\varepsilon_F} d\varepsilon\, F_0[1-\delta_{-\sigma}F_0]^{-1} \tag{8}$$

where

$$F_0 = N^{-1} \sum_k (\varepsilon - \varepsilon(k) - \varepsilon_B^o + i0)^{-1}.$$

Note that G_{00}^σ can be drastically different from G_{ii} ($i \neq 0$) due to the existence of an impurity, $\delta_{-\sigma}$, at $i = 0$, and then it is essential to take $n_{A,-\sigma}$ in Eq. (6) instead of n, the average electron number of the whole system. The case of two magnetic impurities in this model was examined by Inoue and Moriya[8] who included other ingredients in the Hamiltonian. In all these approximations mutual Coulomb interactions as well as the elastic potential $\delta_o = \varepsilon_A^o - \varepsilon_B^o$ contribute to the scattering of electrons. Indeed even if $\delta_o=0$, the electrons are scattered if $U_A \neq U_B$. In the pioneering paper by Wolff and Clogston[4] on the local moment formation, this effect of the change of a potential energy at a magnetic site due to Coulomb interactions was neglected (δ_o instead of δ_σ). As regards the dynamical susceptibility, $\chi(Q,\omega)$, we know that, due to the general arguments[19] of the conservation properties, the self-consistent treatments of the Coulomb interactions are needed between the modification of the one-particle spectrum (self-energy) and the two-particle correlation function. Although there are some discussions[20] on $\chi(Q,\omega)$ in the limit of small numbers of A atoms, these authors were not interested in the second term of Eq. (4) and its resultant effects on $\chi(Q,\omega)$.

The alloys at the finite concentration represented by Eqs. (1) and (4) have been the subject of the recent interest. Hasegawa and Kanamori[14] examined, above all, the magnitude of the local magnetic moments in the various ferromagnetic alloys and obtained the nice agreement with experiments. Levin et al.[15] also showed that the present scheme could explain some of the characteristic features in the paramagnetic transition metal alloys which had not been treated before. Recently the problem of the surface state of the ferromagnetic metals was also discussed based on the same scheme.[21] Such dynamical aspects as the effects of the paramagnons, the possibility of the spin density wave states and the spin wave are analyzed in Ref. 16.

As mentioned in Sec. 1, another way was proposed to treat Eq. (1) by Harris and Zuckermann[17] and Kato and Shimizu,[18] who considered the effects of Coulomb interactions only in the form of two-particle correlation and then neglected the modification of the atomic energy (the second term in Eq. (4)). Although this scheme is somewhat complementary to the former,[16] this violates the conservation properties except if $\varepsilon_{A\sigma}=\varepsilon_{B\sigma}$. If one could include the self-energy corrections consistent with this scheme, this would be a nicer approximation to the present problem. Since no one succeeded in it for the moment and the second term in Eq. (4) seems important also from the physical point of view, we discuss the former approach (HK) in detail in this paper.

SECTION 3. $\chi(Q,\omega)$ IN THE PRESENT SCHEME

$\chi(Q,\omega)$ of the interacting electrons in the pure system has been obtained by Izuyama, Kim and Kubo[22] by use of the equation of motion method. In the present case of the alloys where two different kinds of sites are present, we could solve the problem by writing the equation of motion for $n_{A,\sigma}$ and $n_{B,\sigma}$ and by coupling them together. This is more clearly and systematically done by a diagrammatical way. Referring the readers to Ref. 16, I only write the final results which hold both in paramagnetic and ferromagnetic states.

Various quantities appearing in $\chi(Q,\omega)$ are written in terms of the one-particle thermal Green's function[23]

$$G(k,i\varepsilon_n) \equiv \int_0^B d\tau \, e^{i\varepsilon_n\tau} <Ta_k(0)a_k^+(\tau)>$$

$$= [i\varepsilon_n - \varepsilon(k) - \varepsilon_{B,\sigma} - \Sigma_\sigma(i\varepsilon_n)]^{-1} \qquad (9)$$

where $\varepsilon_n = (2n+1)\pi T$ ($T=\beta^{-1}$ is the temperature). The CPA equation for the self-energy Σ_σ is

$$\Sigma_\sigma = x\delta_\sigma[1-(\delta_\sigma-\Sigma_\sigma)F_\sigma]^{-1}, \qquad (10)$$

where

$$F_\sigma = F_\sigma(i\varepsilon_n) = N^{-1}\sum_k G(k,i\varepsilon_n). \qquad (11)$$

We take the convention that $G_\uparrow \equiv G_\uparrow(k+Q, i\varepsilon_n+i\omega_\nu)$ and $G_\downarrow \equiv G_\downarrow(k,i\varepsilon_n)$, where $\omega_\nu=2\pi\nu T$ is the external frequency in the thermal representation. Similarly $F_\uparrow \equiv F_\uparrow(i\varepsilon_n+i\omega_\nu)$ and $F_\downarrow \equiv F_\downarrow(i\varepsilon_n)$. By use of this convention, we can express $\chi(Q,\omega)$ as follows.

$$\chi(Q,\omega) = [x\chi_A(Q,i\omega_\nu) + y\chi_B(Q,i\omega_\nu)]_{i\omega_\nu \to \omega+i0^+} \qquad (12)$$

$$\chi_A(Q,i\omega_\nu) = 2\mu_B^2[(1+K_{BB})\mathcal{J}_A - K_{AB}\mathcal{J}_B][(1+K_{AA})(1+K_{BB})-K_{AB}K_{BA}]^{-1} \qquad (13)$$

and $\chi_B(Q,i\omega_\nu)$ is given by Eq. (13) by replacing suffix A by B. $K_{\mu\nu}$ and $\mathcal{S}_\mu$ are defined as follows

$$K_{AA}(Q,i\omega_\nu) = U_A x^{-1} T \sum_n [A\eta \zeta_A^2 + F_\uparrow F_\downarrow \phi_A], \tag{14}$$

$$K_{AB}(Q,i\omega_\nu) = U_B x^{-1} T \sum_n [A\eta \zeta_A \zeta_B - F_\uparrow F_\downarrow \phi_A] \tag{15}$$

$$\mathcal{S}_A(Q,i\omega_\nu) = -T \sum_n A\eta \zeta_A \tag{16}$$

where

$$A = \sum_k G_\uparrow G_\downarrow \tag{17}$$

$$\eta = [1 - A(\delta_\uparrow \Sigma_\downarrow - \delta_\downarrow \Sigma_\uparrow)(\delta_\downarrow F_\downarrow - \delta_\uparrow F_\uparrow)^{-1}]^{-1} \tag{18}$$

$$\zeta_A = (F_\uparrow \Sigma_\uparrow - F_\downarrow \Sigma_\downarrow)(\delta_\uparrow F_\uparrow - \delta_\downarrow F_\downarrow)^{-1} \tag{19}$$

$$\zeta_B = 1 - \zeta_A \tag{20}$$

$$\phi_A = (\Sigma_\uparrow \delta_\uparrow^{-1} - \Sigma_\downarrow \delta_\downarrow^{-1})(\delta_\uparrow F_\uparrow - \delta_\downarrow F_\downarrow)^{-1} \tag{21}$$

$$\phi_B = \phi_A - (\delta_\uparrow F_\uparrow - \delta_\downarrow F_\downarrow)^{-1}. \tag{22}$$

Expressions of K_{BA}, K_{BB} and $\mathcal{S}_B$ are obtained by K_{AB}, K_{AA} and $\mathcal{S}_A$ by interchanging suffix A by B and x by y respectively.

Let us examine Eq. (12) in some typical cases.

A. The paramagnetic state

(1) In the dilute limit of magnetic A atoms in Wolff's model ($U_B=0$) the criterion of the local moment formation is given by

$$1 = U_A \chi_{loc}(0) = -\left(\frac{\partial n_{A,\sigma}}{\partial n_{A,-\sigma}}\right)_{\varepsilon_F} \tag{23}$$

where

$$\chi_{loc}(\omega) = -\left(T \sum_n F_\uparrow F_\downarrow [(1-\delta_\uparrow F_\uparrow)(1-\delta_\downarrow F_\downarrow)]^{-1}\right)_{i\omega_\nu \to \omega+i0}. \tag{24}$$

In Eq. (24), F_σ corresponds to that of a pure B system. This criterion is identical to that by Moriya,[7] whereas δ_o, instead of δ_σ, appears in Wolff's discussions.[8] The corresponding equation derived in Ref. 20 does not include the factor $[(1-\delta_\uparrow F_\uparrow)(1-\delta_\downarrow F_\downarrow)]^{-1}$, whose existence shows that χ_{loc} is the local susceptibility at an impurity

$$\omega = D_M Q^2$$

$$D_M = [2\pi(n_\uparrow - n_\downarrow)]^{-1} \text{Im} \sum_k \left(\frac{\partial \varepsilon}{\partial k_x}\right)^2 \int_{-\infty}^{\varepsilon_F} d\varepsilon [G_\uparrow(k,\varepsilon+i0) - G_\downarrow(k,\varepsilon+i0)]^2. \quad (25)$$

Eq. (25) reduces to that of Ref. 22 in pure systems. Although individual excitations generally start from $\omega=0$, the spin wave is not damped (D_M is real), since in our model total spin is conserved. Since D_M, Eq. (25), is not determined solely by the properties at the Fermi energy, it behaves smoothly even if the specific heat changes abruptly. For example, in Ref. 14, the majority spin band is shown to touch the Fermi energy if $x > x_0$ ($x_0 \cong .5$) in $Ni_{1-x}Fe_x$ where γ increases proportionally to $(x-x_0)^{1/2}$ for x near x_0. From Eq. (25) we see D_M change proportionally to $(x-x_0)^{3/2}$ in this region. Thus, D_M is insensitive to the details of the system.

If $m=0$ but $m_A \neq 0$ and $m_B \neq 0$, the spin wave spectrum becomes $\omega = D_A Q$ with the real number D_A as in the case of usual antiferromagnets. This is, however, highly accidental in itinerant systems. Instead, if $m_A \neq 0$, $m_B \neq 0$, m is generally finite. We know that in the Heisenberg ferromagnets in the regular array of lattice with two sublattices having different kinds of spins there exist[30] an optical branch for the collective mode besides the spin wave of the type of Eq. (25). However, this optical branch is broadened in the disordered system. Similar situations are found in the present itinerant system. In the limit of small x, it is possible that $\chi(Q,\omega)^{-1}=0$ have two roots. One is $\omega=D_M Q^2$ and the other $\omega=\omega_{op}$ in small Q. The latter corresponds to the local excitation at the A site. ω_{op} is the solution of

$$\lim_{x \to 0} (1 - K_{AA}) = 0$$

or

$$1 = U_A \chi_{loc}(\omega_{op}), \quad (26)$$

where $\chi_{loc}(\omega)$ is defined by Eq. (24). This mode is physically the same as was discussed by Lederer[31] and Wolfram.[32] The mathematical equation for ω_{op} due to Wolfram[32] is, however, different from Eq. (26) since his basic equation is the same as that of Ref. 20. This optical mode seems different from the one discussed by Doniach and Wohlfarth[33] since they behave differently in the limit of small x.

In finite x, this branch is broadened. In experiments, however, it may happen that this mode would give visible maximum in the cross section of the neutron scattering.

SECTION 4. DISCUSSIONS

Some of the consequences of the new treatments of the magnetic

site, where electrons feel excess potential δ_σ.

(2) $\lim_{Q \to 0} \lim_{\omega \to 0} \chi(Q,\omega) \equiv \chi_0$ reduces to those of Refs. 14 and 15.

Hasegawa and Kanamori[14] evaluated the Curie temperature T_c of $Ni_{1-x}Fe_x$ by looking at the solution of $\chi_0^{-1}=0$. There they encountered the difficulty since T_c in the limit of small x does not tend to that of pure Ni. As they observed, this is due to the fact that, since the fluctuation is totally neglected in this scheme, we could not discriminate the formation of the local moment from the onset of the long range order and that T_c in this case corresponds to the temperature where the local moment is formed at the Fe site. This is one of the defects common to the simple mean field theory. On the other hand, Ref. 15 discussed the paramagnetic susceptibility of Rh-Pd, Pt-Pd, Ni-Rh and Ni-Pd. They could explain the concentration dependence in a consistent way with the estimate of the parameters by the renormalized atom theory.[24]

(3) $\chi^{-1}(Q,0)=0$ determines the instability of the paramagnetic state toward the spin density wave state with the wave vector Q. This is examined in Ref. 16 for the Wolff model ($U_B=0$) and is the extension of the investigations by Penn[25] to the disordered system. The results show that the state with finite Q is probable even in highly disordered systems and in turn suggest the existence of the spin density wave states with long range order.

(4) If the system is nearly ferromagnetic, $\chi_0^{-1} \cong 0$, there exist paramagnons.[26] For small Q and ω, $\chi(Q,\omega)$ can be written as

$$\chi(Q,\omega) = \chi_0 DQ^2 [DQ^2 - i\omega]^{-1},$$

or the diffusive mode exists[27] whose diffusion constant is very small, $D \propto \chi_0^{-1}$. The electronic specific heat, γ, is modified by such excitations.[28] For example, near the critical concentration of the alloys where the ferromagnetism occurs (e.g. $x_0 = .63$ in $Ni_x Rh_{1-x}$), γ behaves like $\gamma \propto (x_0-x)^{-1/2}$.

B. The ordered state

In this case both $m_A (\equiv n_{A\uparrow} - n_{A\downarrow})$ and m_B remain finite. They can have either sign depending on the choice of the parameters and $m \equiv xm_A + ym_B$ is also finite except in some accidental cases.

(1) The magnitude of a local spin

Although the average magnitude of magnetic moment is well represented by the Slater-Pauling curve, the moment at A or B sites could differ much from this average value. Various experiments[11] observed this difference. The calculations in Ref. 14 result in good agreement with these experiments in such various alloys as fcc $Ni_{1-x}M_x$(M=Fe, Co, Mn, Cr) and bcc $Fe_{1-x}M_x$(M=Ni, Co, Mn, Cr). These authors observe that the present treatments work well if the density of states at the Fermi energy, $\rho(\varepsilon_F)$, is small. This is easy to understand since the fluctuations might be small if $\rho(\varepsilon_F)$ is small.

(2) The spin wave[29]

If $m \neq 0$, there exists a well defined spin wave whose spectrum in small Q($Q /\!/ \hat{z}$) is given by

alloys are represented. This approach is based on the Hartree-Fock approximation (Eq. (4)) to Coulomb interactions and the CPA to disorderedness. Since the effects of the fluctuation of spins are totally neglected, the present approximation yields good agreement with experiments only when the magnitude of the magnetic moment is large in the ferromagnetic state or when the paramagnetic systems are not near to the ferromagnetic instabilities.

As is frequently said, the CPA is the single site approximation and then it is not appropriate to discuss the pair of two magnetic atoms. As the theory is being developed to treat such pairing effects[34] in the simple binary alloys, it is of interest to apply this to the present systems.

In the present paper we assumed that the shapes of the density of states of the pure A and B system are identical, which is one of the restrictions of our model. In reality they may be appreciably different. Such problems of off-diagonal disorder[35] are also being investigated intensively in simple alloys which would have direct bearings on the magnetic alloys.

Last, but not least, one would notice that there exist apparent inconsistencies of our present treatments in the sense that the Coulomb interactions are treated as weak (Hartree-Fock) first and the resulting scattering potential δ_σ is treated as possibly very strong (CPA). If CPA is essential for treating δ_σ, Coulomb interactions must be treated as strong. Although there is some hope that the orbital degeneracy of the realistic systems would guarantee Eq. (4) to some extent if we properly understand U, it obviously needs further examination to ascertain this.

On the other hand, although the problem of strong correlations is not easy, one possible way to treat U_i in Eq. (1) is to apply Hubbard's approximation.[36] As was noted by Velický et al.,[12] Hubbard's alloy analogy is the same as the CPA, if we properly translate the notations between the two. Ehrenreich and the present author[37] applied both this alloy analogy and the CPA to the present model. They found that, if δ_0 is infinite and then A sites are inaccessible to electrons, there are no ferromagnetic instabilities for B bands. As was pointed out by Harris and Lange,[37] Hubbard's simple approximation is too restrictive to ferromagnetism. Although this is also the case in the present alloy analogy, orbital degeneracy seems essential to the occurrence of ferromagnetism, too. Irrespective of such fundamental difficulties, it is clear that the scope of our understanding of the itinerant magnetism is widened by the newly developed scheme.

The author thanks Professors H. Ehrenreich and A. Luther for useful discussions. He is also grateful to Professor J. Kanamori for sending preprints prior to publication.

REFERENCES

1. J. Friedel, Nuovo Cimento 2, 287 (1958).
2. P. W. Anderson, Phys. Rev. 124, 41 (1961).
3. P. A. Wolff, Phys. Rev. 124, 1030 (1961).

4. A. M. Clogston, Phys. Rev. 125, 439 (1962).
5. T. Moriya, Prog. Theor. Phys. 34, 329 (1965) and Proceedings of the International School of Physics "Enrico Fermi", Course XXXVII, p. 206, edited by W. Marshall (Academic Press, New York, 1967).
6. S. Alexander and P. W. Anderson, Phys. Rev. 133 A, 1594 (1964).
7. T. Moriya, Prog. Theor. Phys. 33, 157 (1965).
8. M. Inoue and T. Moriya, Prog. Theor. Phys. 38, 41 (1967).
9. D. J. Kim and Y. Nagaoka, Prog. Theor. Phys. 30, 743 (1963), B. Caroli, J. Phys. Chem. Solids 28, 1427 (1967), S. H. Liu, Phys. Rev. 163, 472 (1967), D. J. Kim, Phys. Rev. B1, 3725 (1970).
10. M. Shimizu, T. Takahashi and A. Katsuki, J. Phys. Soc. Japan 17, 1740 (1962), R. D. Lowde, M. Shimizu, M. W. Stringfellow and B. H. Torrie, Phys. Rev. Letters 14, 698 (1965).
11. For example, in the case of $Ni_{1-x}Fe_x$, C. G. Shull and M. K. Wilkinson, Phys. Rev. 97, 304 (1955), M. F. Collins, R. V. Jones and R. D. Lowde, J. Phys. Soc. Japan 17 B-III, 19 (1962).
12. P. Soven, Phys. Rev. 156, 809 (1967), B. Velický, S. Kirkpatrick and H. Ehrenreich, Phys. Rev. 175, 747 (1968).
13. H. Ehrenreich, Proceedings of International Conference of Liquid Metals (Tokyo, 1972).
14. H. Hasegawa and J. Kanamori, J. Phys. Soc. Japan 31, 382 (1971) and preprints.
15. K. Levin, R. Bass and K. H. Bennemann, Phys. Rev. B6, 1865 (1972).
16. H. Fukuyama (to be published).
17. R. Harris and M. J. Zuckermann, Phys. Rev. B5, 101 (1972).
18. T. Kato and M. Shimizu, J. Phys. Soc. Japan 33, 363 (1972).
19. G. Baym and L. P. Kadanoff, Phys. Rev. 124, 287 (1961).
20. P. Lederer and D. L. Mills, Phys. Rev. Letters 20, 1036 (1968), S. Engelsberg, W. F. Brinkman and S. Doniach, Phys. Rev. Letters 20, 1040 (1968).
21. A. Luther, P. Fulde and R. E. Watson, Proc. of the 18th Annual Conference on Magnetism and Magnetic Materials (Denver, 1972).
22. T. Izuyama, D. J. Kim and R. Kubo, J. Phys. Soc. Japan 18, 1025 (1963).
23. A. A. Abrikosov, L. P. Gorkov and I. E. Dzyaloshinski, Methods of Quantum Field Theory in Statistical Physics (Prentice-Hall, Englewood Cliffs, New Jersey, 1963), Chap. 3.
24. R. E. Watson, H. Ehrenreich and L. Hodges, Phys. Rev. Letters 24, 224 (1970), L. Hodges, R. E. Watson and H. Ehrenreich, Phys. Rev. B5, 3953 (1972).
25. D. R. Penn, Phys. Rev. 142, 350 (1966).
26. N. F. Berk and J. R. Schrieffer, Phys. Rev. Letters 17, 433 (1966), S. Doniach and S. Engelsberg, Phys. Rev. Letters 17, 750 (1966).
27. P. Fulde and A. Luther, Phys. Rev. 170, 570 (1968).
28. S. Engelsberg, W. F. Brinkman and S. Doniach, Phys. Rev. Letters 20, 1040 (1968).

29. H. Fukuyama (to be published).
30. H. Kaplan, Phys. Rev. $\underline{86}$, 121 (1952).
31. P. Lederer, Thesis, A La Faculté des Sciences de L'Université de Paris, 1967 (unpublished).
32. T. Wolfram, Phys. Rev. $\underline{182}$, 573 (1969).
33. S. Doniach and E. P. Wohlfarth, Proc. Roy. Soc. (London) $\underline{A296}$, 442 (1967).
34. R. N. Aiyer, R. J. Elliott, J. A. Krumhansl and P. L. Leath, Phys. Rev. $\underline{181}$, 1006 (1969), K. F. Freed and M. H. Cohen, Phys. Rev. $\underline{B3}$, 3400 (1971), F. Cyrot-Lackmann and F. Ducastelle, Phys. Rev. Letters $\underline{27}$, 429 (1971), L. Schwartz and H. Ehrenreich, Phys. Rev. $\underline{B6}$, 2923 (1972).
35. H. Shiba, Prog. Theor. Phys. $\underline{46}$, 77 (1971), J. A. Blackman, D. M. Esterling and N. F. Berk, Phys. Rev. $\underline{B4}$, 2412 (1971), L. Schwartz, H. Krakauer and H. Fukuyama (to be published).
36. J. Hubbard, Proc. Roy. Soc. (London) $\underline{A277}$, 237 (1964).
37. H. Fukuyama and H. Ehrenreich (to be published).
38. A. B. Harris and R. V. Lange, Phys. Rev. $\underline{157}$, 295 (1967).
39. J. H. Van Vleck, Nuovo Cimento $\underline{6}$, 886 (1957).

Section 35.   Magnetic Compounds

ANOMALOUS MAGNETIC BEHAVIOR OF MnSi

Lionel M. Levinson
General Electric Corporate Research and Development
Schenectady, New York 12301

G.H. Lander
Argonne National Laboratory
Argonne, Illinois 60439

M.O. Steinitz
Department of Physics
University of Toronto
Ontario, Canada

ABSTRACT

The results of macroscopic magnetization measurements for $2.5°K < T < 900°K$ and $H < 20$ kOe, and neutron diffraction studies, on powder and single crystal MnSi are reported. The magnetization measurements are not compatible with zero-field ferromagnetism below the ordering temperature. An antiferromagnet → spin-flop → saturated paramagnet behavior typical of simple antiferromagnets for $T \ll T_N$ is observed upon increasing H at low temperatures. $\chi_\perp$ is nearly constant in the ordered state, and $\chi_{initial} = 2\chi_\perp/3$ for powdered material. However, in neither powder nor single-crystal neutron-diffraction experiments were any additional magnetic reflections observed. Below 30°K certain nuclear reflections consistent with spontaneous zero field ferromagnetism increase in intensity. The depolarization of polarized neutrons transmitted through a polycrystalline sample suggests that MnSi is neither a simple ferromagnet nor a simple antiferromagnet.

INTRODUCTION

The cubic compound MnSi is known to order magnetically at $T_o = 30°K$[1,2]. For $H > 6$ kOe the material acts like a saturated ferromagnet, but at lower fields the behavior is complex[3-5]. We report here magnetization and neutron diffraction data on powder and single crystal MnSi in fields $0 < H < 20$ kOe and temperatures $2.5°K < T < 900°K$.

The MnSi used throughout this study was prepared by fusion of 99.99% Mn and 100 ohm-cm semiconductor grade silicon in a water-cooled copper hearth. Large, high quality single crystals were produced by the Bridgman technique using quartz crucibles.

MAGNETIZATION

Figure 1 gives the sample magnetization M as a function of applied field H. The saturation moment $M_s \approx 150$ gauss ($0.4\,\mu_b$/Mn) is low and demagnetizing effects can be neglected. Experimentally, our results were independent of sample shape. The curves of Fig.1 are strongly suggestive of an antiferromagnet → spin-flop transition at H_1 (see insert) and a spin-flop → saturated paramagnet (ferromagnet) transition

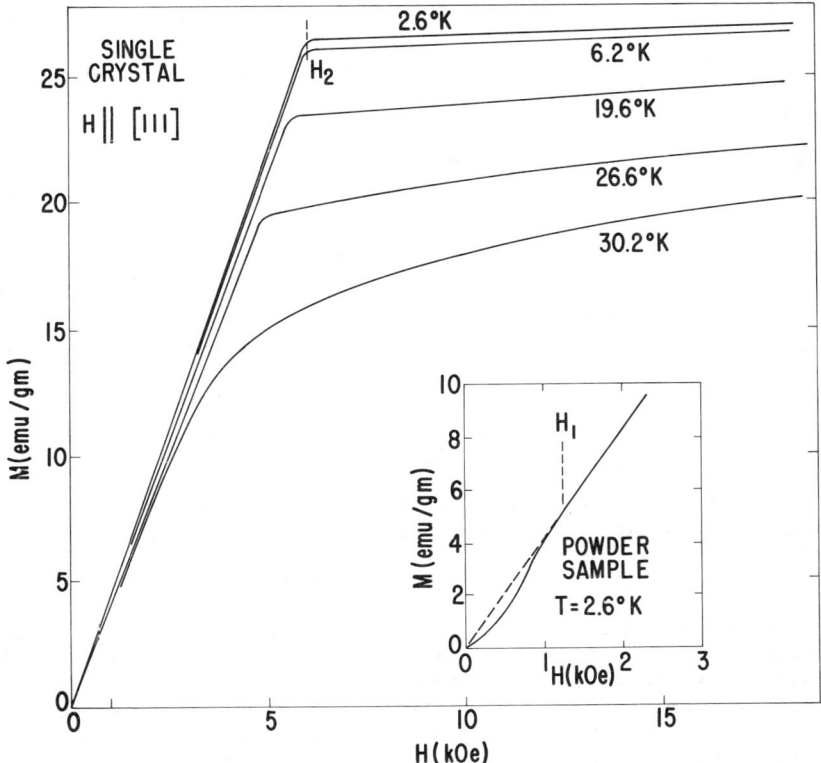

Fig.1 Magnetization M of MnSi as a function of applied field H.

at H_2. The former transition is evident for the powder sample but vanishes when the field is applied along the $\langle 111 \rangle$ direction of a single crystal. For H along $\langle 100 \rangle$ (not shown) the spin-flop transition at H_1 is present also in the single-crystal data. Above H_1, M is isotropic. Neither zero-field remanence nor any hysteresis was observed.

If we define $\chi_\perp = M/H$ for $H_1 < H < H_2$, then for the powder sample $\chi_{initial} = 2\chi_\perp/3$ as expected for an antiferromagnet. Values of $\chi_\perp$ and H_2 were the same for powders and single crystals. Note that $\chi_\perp$ is reasonably temperature independent for $T < T_o$ and that H_2 exhibits a temperature dependence characteristic[6] of the spin flop → saturated paramagnet transition.

In Fig.2 we give the temperature dependence of the magnetization measured at H = 2 and 20 kOe, and $1/\chi_{initial}$ for $T > T_o$. Notice that the curve of $1/\chi_{initial}$ intercepts the temperature axis at $T=+30°K=T_o$ which is indicative of a predominantly ferromagnetic exchange energy. The force of this argument is lessened, however, upon noting that (a) the effective moment computed from the curve of $1/\chi_{initial}$ versus T is $1.4\,\mu_B$ and not $\sim 0.4\,\mu_B$ as expected from the value of M_s, and (b) for $300°K < T < 900°K$, (not given) $1/\chi_{initial}$ versus T is no longer linear but concave downward.

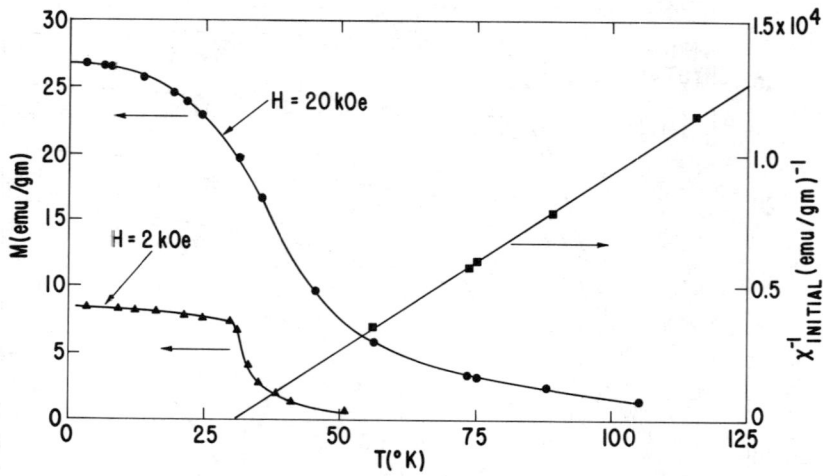

Fig.2 Magnetic moment M and inverse susceptibility $1/\chi_{initial}$ of MnSi as a function of temperature. $M(H = 2\ kG) \propto \chi_\perp$.

NEUTRON DIFFRACTION

The present neutron diffraction experiments were undertaken to determine whether MnSi is antiferromagnetic below 30°K in zero field as implied by the above magnetization results. In agreement with a previous investigation,[7] an unpolarized neutron study of a polycrystalline sample in zero field gave no evidence of antiferromagnetic ordering; any antiferromagnetic component is less than 0.25 μ_B/Mn.

A second zero-field unpolarized neutron experiment was undertaken to measure the integrated intensities of a single crystal (dimensions 1.8 x 2.8 x 5.6 mm). The observed and calculated integrated intensities are given in Table I, and are in good agreement considering the large extinction. The residual of the crystallographic refinement, based on F, is 0.037. Beyond 0.35Å^{-1} the magnetic form factor (col.7) is too small for any magnetic scattering to be observed, and these reflections are omitted from the Table. The last column gives the additional intensity expected at 5°K from a ferromagnetic moment of 0.4 μ_B/Mn. Detectable changes between the data taken at 80° and 5°K are expected <u>only</u> for the (210) reflection, and the observations confirm this. In Fig.3 the integrated intensity of the (210) reflection is plotted as a function of temperature.

Careful scans for non-nuclear peaks were made along $\langle 100 \rangle$, $\langle 110 \rangle$, $\langle 111 \rangle$ and other directions in reciprocal space. No extra reflections corresponding to a non-ferromagnetic arrangement were found at 5°K. The sensitivity of the single-crystal experiment was such that any antiferromagnetic component is less than 0.06 μ_B/Mn.

In a third neutron experiment the depolarization of polarized neutrons transmitted through a 1 cm. diameter cylinder of polycrystalline MnSi was measured as a function of temperature and applied field. The results at 5°K are shown in Fig.4. Above 30°K the depolarization is

Table I. Observed and Calculated Neutron Intensities.

The observations are an average of equivalent reflections, the standard deviations being ~3%, (or 0.2 in the case of weak reflections). The following parameters were refined: occupation of Mn site = 0.98 ± 0.02; positional parameter Mn in cubic B20 structure = 0.137 ± 0.001; positional parameter Si = 0.845 ± 0.001, overall temperature factor = $1.0 \pm 0.3 \text{Å}^2$ (this is high because of correlation with extinction); g-value for extinction[8] = $(12 \pm 3) \times 10^3$; and scale factor = 2.29 ± 0.16. The fixed values $b_{Mn}=-0.387$ and $b_{Si}=0.415 (\times 10^{-12} \text{cm})$ were used. Previously reported positional parameters[9] are $x_{Mn}=0.137 \pm 0.004$ and $x_{Si}=0.842 \pm 0.004$. The quantity y is the calculated attenuation due to the extinction.[8] The form factor is taken from Ref.7.

hkℓ	sin θ/λ	Observed Intensity 80°K	Observed Intensity 4.2°K	Calc. Nuc. Int.	y	Form Factor f(K)	Calc. Ferro. Int. 0.4 μ_B
110	0.155	113.8	114.0	118.8	0.38	0.80	0.95
111	0.190	107.9	108.1	111.4	0.40	0.75	0.79
200	0.219	9.8	9.8	11.4	0.90	0.63	0.08
210	0.245	2.3	3.2	2.0	0.99	0.57	1.76
120	0.245	123.8	121.1	111.5	0.40	0.57	0.52
211	0.269	9.4	9.6	10.3	0.91	0.52	0.57
220	0.310	2.0	2.2	2.1	0.98	0.36	0.00
221	0.329	13.9	13.7	15.1	0.87	0.33	0.02
310	0.347	22.3	22.3	23.4	0.80	0.31	0.06
130	0.347	131.8	132.2	113.4	0.39	0.31	0.10
311	0.364	20.7	20.5	21.5	0.82	0.29	0.10

negligible for all fields. From the observed depolarization at H=0 one may estimate[10] the presence of ferromagnetic regions of size ~500Å at zero field. These ferromagnetic regions would have to couple antiferromagnetically to be consistent with the magnetization measurements. Note that a simple antiferromagnet would give rise to <u>zero depolarization</u> in zero field. At low fields the observed increase in the depolarization may be interpreted in terms of an induced magnetic moment in the macroscopically antiferromagnetic MnSi. At high fields we obtained a decrease in depolarization typical of ferromagnetic powders.

CONCLUSIONS

In view of the apparent contradiction between the magnetization and coherent neutron diffraction experiments we must conclude that the low-field magnetic behavior of MnSi is not fully understood. The former results (and others[3-5]) are inconsistent with simple ferromagnetism in MnSi for $T < T_o$, while the latter experiments strongly imply that MnSi is nonetheless a simple ferromagnetic. A tentative

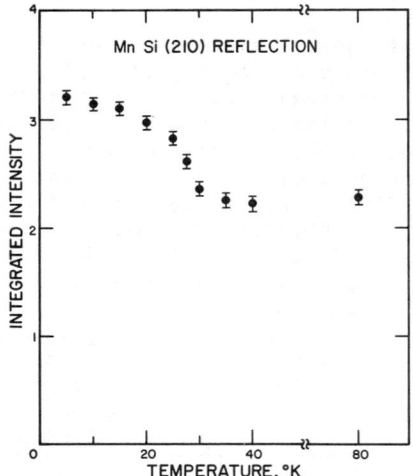 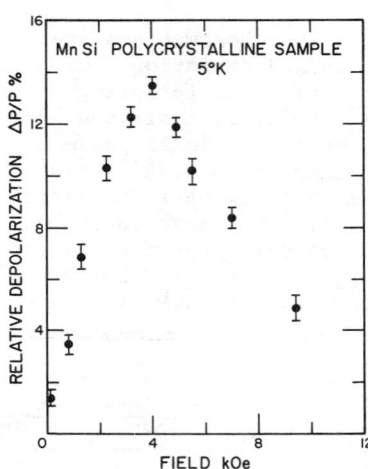

Fig.3 Variation with temperature of the integrated intensity of the (210) reflection measured with unpolarized neutrons in zero field.

Fig.4 Variation with applied field of the depolarization of polarized neutrons transmitted through a 1 cm. diameter cylinder of polycrystalline MnSi at 5°K.

model resolving this contradiction, and suggested by the depolarization measurements, is that appreciable antiferromagnetic coupling exists between microscopic (~500Å) ferromagnetic regions of MnSi.

ACKNOWLEDGMENTS

It is a pleasure to thank B.R. Cooper, I.S. Jacobs, and W.P. Wolf for their interest and comments.

REFERENCES

1. H.J. Williams, J.H. Wernick, R.C. Sherwood, and G.K. Wertheim, J. Appl. Phys. 37, 1256 (1966).
2. D. Shinoda and S. Asanabe, J. Phys. Soc. Japan 21, 555 (1966).
3. M. Kawakami and T. Hihara, J. Phys. Soc. Japan 25, 1733 (1968).
4. J.H. Condon and J.H. Wellendorf, Bull. Am. Phys. Soc. 14, 386 (1969).
5. E. Fawcett, J.P. Maita, and J.H. Wernick, Intern. J. Magnetism 1, 29 (1970).
6. K.W. Blazey and H. Rohrer, Phys. Rev. 173, 574 (1968).
7. P.J. Brown, J.B. Forsyth, and G.H. Lander, J. Appl. Phys. 39, 1331 (1970).
8. W.H. Zachariasen, Acta Cryst. 23, 558 (1967).
9. L. Pauling and A.M. Soldate, Acta. Cryst. 1, 212 (1948).
10. M. Burgy, D.J. Hughes, J.R. Wallace, R.B. Heller, and W.E. Woolf, Phys. Rev. 80, 953 (1950).

MOSSBAUER EFFECT STUDIES OF THE MAGNETIC PROPERTIES
OF THE $Bi_xLa_{1-x}FeO_3$ SYSTEM

T. P. Shaughnessy and J. H. Chen
Boston College, Chestnut Hill, Ma. 02167

ABSTRACT

Magnetic properties of the perovskite system $Bi_xLa_{1-x}FeO_3$, prepared from sintered component oxides, have been studied using the Mossbauer effect. The temperature dependences of the sublattice magnetizations were determined for this system; the Neel temperatures obtained did agree with the values from a previous bulk magnetization measurement. An antisymmetric superexchange interaction has been identified as the mechanism responsible for weak ferromagnetism in this system. Finally, the isomer shifts and the components of the electric field gradient splittings in the direction of the sublattice magnetizations are given.

INTRODUCTION

The perovskite system $Bi_xLa_{1-x}FeO_3$ has been investigated with several techniques in the past. [1,2] These techniques include X-ray, dielectric, neutron diffraction, and magnetization measurements.

Roginskaya, et al,[1] have recorded structural, dielectric and magnetic data for this system. Their X-ray analyses have indicated a rhombohedral structure for $(Bi_xLa_{1-x})FeO_3$ for x=1.0 to 0.812, and a monoclinic structure for x=0.812 to 0.0. In addition, the volume of the unit cell was shown to decrease with increasing amounts of $LaFeO_3$. Also, temperature dependent magnetizations experiments were performed. These found that the Neel temperature, T_N, varied smoothly from a high of 460°C for x=0.0 to a low of 370°C for x=1.0.

Many previous investigations have been done on polycrystalline $BiFeO_3$. X-ray[3] and neutron diffraction data[2] dielectric measurements[4], and torque magnetization[5] information have been obtained. Only recently have the results of these investigations yielded an assignment of space group and an agreement concerning electic and magnetic properties.[6,7,8,9]

Mossbauer data have been reported[10] for a polycrystalline sample of $BiFeO_3$, i.e., x=1.0. Three spectra have been recorded, all were the results of measurements done above T_N. The quadrupole splitting of $BiFeO_3$ at 404°C was 0.54 ± .05 mm/sec and decreased for the two higher temperatures. An estimate for T_N of 372°C was made.

Finally, studies[11,12] have also been made for many of the rare earth iron oxides, in polycrystalline form, including one sample corresponding to the x=0.0 of this system. Some of these investigations were specifically concerned with the origin of the weak ferromagnetism in these compounds. Eibschutz, et al,[11] combined Mossbauer measurements of iron sublattice magnetization with macroscopic magnetization data obtained by torque methods to determine the value of the canting angle between the two nearly opposed sublattices in rare earth orthoferrites. The value obtained for this quantity remained constant over a wide temperature range. On this basis, an anisotropic superexchange interaction was chosen as the source of the weak ferromagnetism in these rare earth orthoferrites.

EXPERIMENT

The Mossbauer spectrometer used for these measurements is a constant acceleration device similar to that of National Bureau of Standards design.[13]

The absorber material, $(Bi_xLa_{1-x})FeO_3$, was made by ceramic methods; the $\alpha-Fe_2O_3$ used was enriched to 60% iron-57. The final product was finely ground and a small part was subjected to Debye-Scherrer X-ray analysis. The samples for x=0.0 and x=1.0 yielded powder spectra which agreed with published values. X-ray measurements of these and of intermediate concentrations (x=.5,.6,.7) indicated no unreacted oxides. The absence of appreciable amounts of the oxide Bi_2O_3 would preclude the presence of the phase $Bi_2Fe_4O_9$. The powdered absorber material was then pressed into a disc of thickness 15 mg/cm² of perovskite material.

RESULTS AND DISCUSSION

TABLE 1
NEEL TEMPERATURES AND EFFECTIVE MAGNETIC FIELDS
(at room temperature)
$(Bi_xLa_{1-x})FeO_3$

SAMPLE X =	T_N °C	Heff in koe
0.5	400	513
0.6	375	505
0.7	358	503
1.0	342	492

TABLE 2
ISOMER SHIFTS AND QUADRUPOLE SPLITTINGS AT ROOM TEMPERATURE

SAMPLE	I.S.	Q.S.
0.5	0.598mm/sec	0.0019
0.6	0.589mm/sec	0.0035
0.7	0.591mm/sec	0.0033
1.0	0.417mm/sec	0.594

Errors: I.S. = ± 0.04mm/sec
Q.S. = ± 0.03

Curves recording the behavior of the effective magnetic field at the Fe^{57} nuclear site as a function of temperature are shown in Figure 1. At any given temperature, the internal field decreases with increasing amounts of Bi^{3+}. This is as expected from considerations of lattice cell size which increases with Bi content. For each sample the temperature where the sublattice magnetization decreases to zero is identified as T_N. The curve for each sample has been examined near this critical point and the

Neel temperatures so obtained are listed in table 1.

Also listed in the same table are the effective internal magnetic fields at room temperature for each sample. Table 2 shows the isomer shifts at room temperature relative to sodium nitroprusside. In addition the room temperature quadrupole splittings in the direction of the sublattice magnetization are given. Those obtained for samples $X = 0.5$, 0.6, and 0.7 are quite comparable to those of the rare earth orthoferrites; the value obtained for the Bi FeO_3 sample above T_N (not shown, but only slightly lower than the room temperature value) is in agreement with a previous measurement[10].

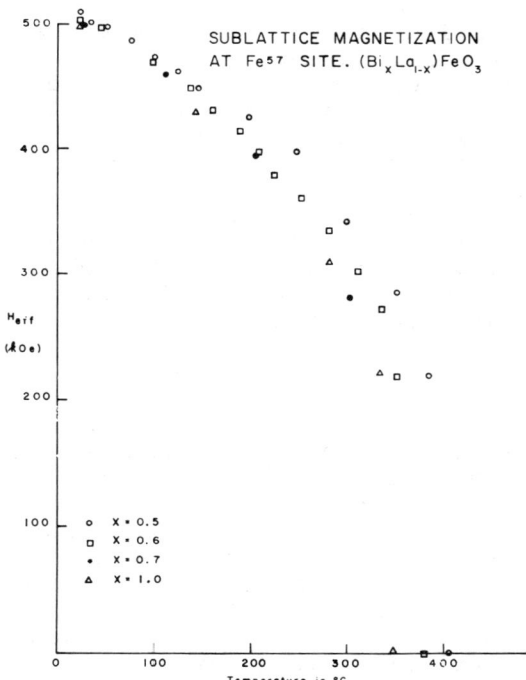

Fig. 1 Sublattice Magnetization Versus Temperature

The following relation has been observed between T_N and the superexchange bond angle for the rare earth orthoferrites

$$T_N = T_N(0) \cos \Phi$$

where Φ is the superexchange bond angle and $T_N(0)$ is the highest Neel temperature, i.e. for $LaFeO_3$[14]. This relation was proven for fourteen orthoferrites where the rare earth ion was smaller than La^{3+} and the $Fe^{3+} - O^{2-} - Fe^{3+}$ linkage angles were computed from accurate x-ray data. The same type of zigzagging octahedral structure is likely for the $(Bi_xLa_{1-x})FeO_3$ system. It was found the same monotonically decreasing behavior, i.e. a substitution of increasing amounts of the smaller Bi^{3+} ion affects T_N in the same manner as a substitution of smaller rare earth ions. Therefore it seems likely that the structural effect of the addition of larger amounts of Bi^{3+} is a lessening of the superexchange angle.

The behavior of the relative sublattice magnetization in the temperature region just below T_N is of interest. In this so called critical region the temperature dependence of the relative sublattice magnetization can be expressed as

$$\frac{\sigma_S(T)}{\sigma_S(0)} = D(1 - T/T_N)^\beta$$

where β and D are constants.

These two constants have been determined experimentally for many materials including rare earth orthoferrites. Values obtained experimentally for rare earth orthoferrites are β values ranging from 0.339 to 0.357 and D values from 1.11 to 1.17.[11] In addition a theoretical calculation by Callen and Callen has been made of the reduced magnetization of insulating ferromagnets based upon a Heisenberg model.[15] Their two spin cluster theory has yielded β = 0.33 and D = 1.222; these values apply over a relatively large range of temperatures below T_N.

The experimental values for β and D in this system were β = 0.32 and D = 1.10 for the temperature range $0.45 < T/T_N < 1.00$. The considerable temperature range away from the immediate vicinity of the phase transition over which the β = 1/3 power is applicable has been interpreted by Callen and Callen as indicating a stable phase rather than an aspect of critical behavior.

The iron ions of the orthoferrites can be described as constituting two sublattices which are strongly coupled antiferromagnetically. A small residual ferromagnetic moment has been observed over a temperature range for macroscopic samples of $Bi_xLa_{1-x}FeO_3$ lying in a direction perpendicular to that defined by the two antiferromagnetic sublattices.[1] This weak ferromagnetic moment can be explained as the result of a slight canting of one magnetic sublattice relative to the second. If M is the magnetization due to this weak ferromagnetic moment then

$$\alpha = \frac{1}{2} \frac{M}{\sigma_S}$$

where α is the canting angle between the two antiferromagnetic sublattices at zero external field. The sublattice magnetization, σ, has been measured in this experiment for some of the same samples for which macroscopic magnetizations are available. The canting angle for samples x = 0.5, 0.6 0.7 and 1.0 over a temperature range from 25°C to T_N is substantially constant over a wide range of temperatures.

There are two types of interactions which produce weak ferromagnetism; one, the antisymmetric superexchange interaction, features a temperature independent angle of cant.[16] The evidence compiled for this system indicates this type of mechanism as the origin of weak ferromagnetism. This same explanation for weak ferromagnetism has been found to be true for many of the rare earth orthoferrites.[11]

In conclusion, the above magnetic and structural information suggests that the addition of increasing amounts of Bi to a rare earth orthoferrite results in an increase in cell size and a lessening of the superexchange angle. These changes result in a lessening of the internal magnetic field and a decrease in T_N. Increasing concentrations of bismuth also reduce the ionic character of the bonding. Lastly, the structural deformity of the orthoferrite is increased by the insertion of bismuth.

REFERENCES

1. Yu E Rogenskaya, Yu N. Venevtsev, S.A. Fedulov, and G.S. Zhadanov, Soviet Phys. - Crystallography 8, 4, 490 (1964).
2. S.V. Kiselev, A.N. Kshnyakina, R.P. Ozerov, and G.S. Zhdanov, Soviet Phys. - Solid State 5, 11, 2425 (1964).
3. Yu Ya Tomashpolskii and Yu N. Venevtsev, Soviet Phys. - Crystallography 12, 1, 18, (1967)
4. G. D. Achenbach, W. J. James, and R. Gerson, J. Am. Ceramic Soc. 437 (1967).
5. C. Michel, J. Moreau, G. D. Achenbach, R. Gerson, and W. J. James, Solid State Commun, 7, 701 (1969)
6. N.N. Krainik, N. P. Khuchua, V. V. Zhdanova, and V. A. Evseev, Soviet Phys. - Solid State 8, 2, 654, (1966).
7. G. A. Smolenskii and V. M. Yudin, Soviet Phys. - Solid State 6, 12, 29, 36 (1965).
8. A. G. Tutov, Soviet Phys. - Solid State 11, 9, 2170 (1970).
9. R. J. Smith, G. D. Achenbach, R. Gerson, and W. J. James, J. Appl. Phys. 39, 1, 70 (1968).
10. V. D. Bhide and M. S. Multani, Solid State Commun. 3, 271, (1965)
11. M. Eibschutz, S. Shtrikman, and D. Treves, Phys. Rev. 156, 2, 562 (1967)
12. D. Treves, J. Appl. Phys. 36, 3, 1033 (1965).
13. F. C. Ruegg, J. J. Spijkerman, and J. R. De Voe, Rev. Sci, Instr. 36, 356 (1965).
14. D. Treves, M. Eibschutz, and P. Coppens, Phys. Letters 18, 3, 216 (1965).
15. E. Callen and H. B. Callen, J. Appl. Phys. 36, 1140 (1965).
16. J. Moriya, Phys. Rev. 120, 1, 91 (1960).

SOME MAGNETIC AND ELECTRIC PROPERTIES OF MATERIALS IN THE SYSTEM $Cu_{.5}Fe_{.5}Cr_2S_{4-x}Se_x$ (x = 0 TO x = 2)

E. M. Gyorgy, M. Robbins, P. Gibart*, W. A. Reed and F. J. Schnettler
Bell Laboratories
Murray Hill, New Jersey 07974

ABSTRACT

Curie temperatures of materials in the systems (Fe, Co or Mn) $Cr_2S_{4-x}Se_x$ decrease slowly with increasing Se content.[1] Both $FeCr_2S_4$ and $CoCr_2S_4$ exhibit negative magnetoresistance at T_C ($\sim$ 190°K and 220°K respectively). Therefore, it was decided to substitute Se for S in $Cu_{.5}Fe_{.5}Cr_2S_4$ ($T_C \sim$ 350°K) in order to lower T_C to room temperature, and to investigate the possible existence of a room temperature magnetoresistance effect. It was found that the spinel phase is formed between x = 0 and x = 2.75. Where x > 2.75 a mixture of two phases (spinel and defect NiAs) is observed. Curie temperature measurements show that T_C decreases monotonically between $\sim$ 350°K for x = 0 and 280°K for x = 2.75. At x = 1.5, T_C is $\sim$ 300°K (room temperature). Pellets of x = 1, 1.5 and 2 were hot pressed at 650°C and 25,000 lbs/in^2 to give $\sim$ 99.5% of x-ray density. Conductivity measurements show that these samples are semiconductors with a negative magnetoresistance ($\sim$ 3%) at T_C. This means that magnetic semiconductors in this system can be prepared such that T_C and the associated negative magnetoresistance effect can be obtained at room temperature.

INTRODUCTION

Many semiconducting chromium chalcogenide spinels have been found [1,4] to exhibit negative magnetoresistance effects at the magnetic ordering temperature (T_C). All of these spinels have ordering temperatures which are below room temperature where the magnetoresistance effect is of little practical importance. In a few cases spinels with $T_C \geq$ room temperature have been prepared. However, these compounds have been either metallic conductors [5] or compounds with serious stoichiometry problems [6,7] which affect their electrical properties.

In a recent investigation it was observed that the Se substitution for S, in systems of the type $M Cr_2 S_{4-x}Se_x$ (M = Co, Mn, Fe), resulted in decreasing T_C with increasing Se content. It was hypothesized that the decrease in T_C was due to decreasing strength of the A-B (tetrahedral M^{2+}-octahedral Cr^{3+}) superexchange interaction.

*Work performed while on leave from Laboratoire de Magnetisme, CNRS, Bellevue, France, on NATO Fellowship and CNRS Grant.

The spinel $Cu_{.5}Fe_{.5}Cr_2S_4$ has been reported [9] to be magnetic ($T_c \sim 340°K$) and semiconducting. Magnetoresistance measurements were not made and if such an effect did exist it would occur in the region of T_c, which is above room temperature. We, therefore, decided to study the system $Cu_{.5}Fe_{.5}Cr_2S_{4-x}Se_x$ in order to determine if the addition of Se for S would decrease T_c as in the $MCr_2S_{4-x}Se_x$ systems and if these materials would remain semiconducting, exhibiting magnetoresistance effects. In such a system T_c could be decreased to room temperature and likewise the magnetoresistance effect.

PREPARATION

All of the samples in the system $Cu_{.5}Fe_{.5}Cr_2S_{4-x}Se_x$ were prepared in polycrystalline form by reacting mixtures of the elements, in pellet form, in evacuated quartz tubes at 800°C for 48 hrs. It was observed that the spinel phase was formed in the compositional region of $x = 0$ to $x = 2.75$. Where $x > 2.75$ to $x = 4$ a mixture of spinel and defect NiAs phases were formed. Spinel unit cell parameters as a function of composition follow Vegard's law as shown in Fig. 1

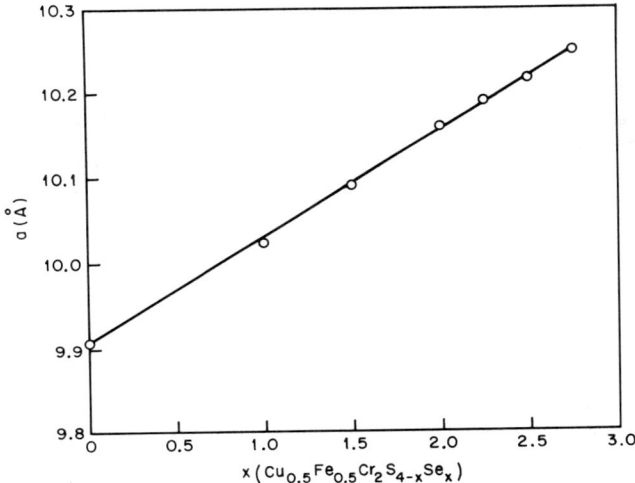

Fig. 1. Unit cell parameters vs. composition in the system $Cu_{.5}Fe_{.5}Cr_2S_{4-x}Se_x$.

High density, polycrystalline, compacts were made by hot pressing preprepared pellets at 800°C and 25,000 lbs/in² pressure. Pellets with up to 99.5% of theoretical density were obtained.

MEASUREMENTS

Curie temperatures (T_c) were obtained for all compositions at H = 5000 Oe. As shown in Table 1, T_c decreases with increasing Se substitution. 0°K magnetic moments for some of the compositions are also shown in Table 1. These values are in agreement with previously published values for $Cu_{.5}Fe_{.5}Cr_2S_4$.[9]

For electrical measurements, samples were obtained by cutting the pressured sintered pellets into rectangular parallelopipeds ($\sim 1 \times 1 \times 5$ mm). Leads, for the four terminal d.c. galvanomagnetic measurements were attached with indium solder. Both the resistivity and the magnetoresistivity (at 5 kOe) were measured as a function of temperature. Two compositions (x = 1.0 and x = 1.5) were prepared and measured as described. Both were semiconducting and both exhibited negative magnetoresistance in the region of T_c. Room temperature resistivity and the magnitude of the magnetoresistance effect are shown in Table 2. Fig. 2a and 2b show typical resistivity and magnetoresistivity curves. This type of magnetoresistance curve has been observed for ferrimagnetic spinel semiconductors [4]. Using the sample where x = 1.5, the sign of the Hall coefficient was also measured, between 77 and 300°K, and found to be negative.

Table 1
Crystallographic and Magnetic Properties of Materials in the System $Cu_{.5}Fe_{.5}Cr_2S_{4-x}Se_x$
x = 0 to x = 2.75

x	$T_c(°K)$	$\sigma_0(\mu_B)$
0	340	2.90
1.0	315	
1.5	300	2.86
2.0	290	
2.5	283	2.87
2.75	280	

Table 2
Results of Conductivity Measurements on Samples Where x = 1.0 ($Cu_{.5}Fe_{.5}Cr_2S_3Se$) and x = 1.5 ($Cu_{.5}Fe_{.5}Cr_2S_{2.5}Se_{1.5}$)

x	$T_c(°K)$	$\rho(\Omega\text{-Cm})$	$\Delta\rho/\rho\vert_{T_c}$ (H=5kOe)
1.0	315	0.03	-0.17
1.5	300	0.068	-0.026
1.5(a)	150	0.14	
1.5(b)	340	0.51	-0.05

Cut bars of the sample with x = 1.5 ($Cu_{.5}Fe_{.5}Cr_2S_{2.5}Se_{1.5}$) were annealed in flowing H_2 (700 C for 21 hrs) and in a closed tube with a mixture of 2.5/1.5 mole S to Se mixture at 700,C for 6 hours. As shown in Table 2 (1.5a) the sample annealed in H_2 exhibited a sharp decrease in T_c and a disappearance of the magnetoresistance effect. X-ray diffraction indicated that the material was still a spinel (no extra phases). Weight loss determination and chemical analysis yields an approximate composition of $Cu_{.5}Fe_{.5}Cr_2X_{3.8}$ (S + Se = 3.8 moles).

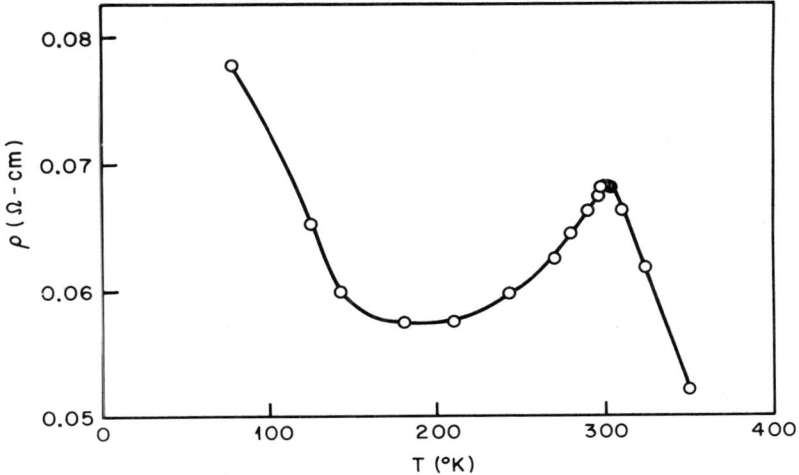

Fig. 2a. Resistivity (Ω-cm) vs. temperature for the composition $Cu_{.5}Fe_{.5}Cr_2S_{2.5}Se_{1.5}$.

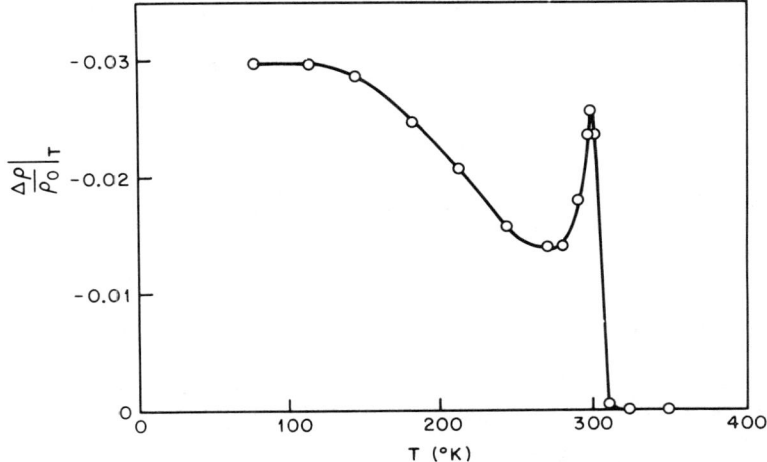

Fig. 2b. Negative magnetoresistance vs. temperature for the composition $Cu_{.5}Fe_{.5}Cr_2S_{2.5}Se_{1.5}$. $H = 5$ kOe and $\Delta\rho/\rho_0 |_T = (\rho(H)-\rho(0))/\rho(0) |_T$.

The sample annealed (closed tube) in the S-Se (1.5b in Table 2) exhibited an increase in T_c, ρ, and the magnetoresistance. The sample weight decreased as did the unit cell parameter and chemical analysis indicated that sulfur was substituted for selenium in the spinel.

DISCUSSION AND CONCLUSIONS

Most studies involving chalcogenide spinels must also take into consideration the problem of nonstoichiometry. Small anion deficiencies do not significantly affect T_c but can cause significant changes in the magnitude of the magnetic moment and in the electrical properties. For this reason, the previously described annealing experiments were carried out.

The H_2 anneal of the material $Cu_{.5}Fe_{.5}Cr_2S_{2.5}Se_{1.5}$ undoubtedly resulted in the introduction of anion vacancies. This anion deficiency could easily be compensated by the reduction of Fe^{3+} to Fe^{2+}. The spinel $Cu_{.5}Fe_{.5}Cr_2S_4$ has been shown [10] to contain Cu^+ and Fe^{3+} on the A (tetrahedral) sites, and the substitution of Fe^{2+} for Fe^{3+} would cause the observed decrease in T_c. The substitution of S for Se during the S-Se anneal of $Cu_{.5}Fe_{.5}Cr_2S_{2.5}Se_{1.5}$ indicates that the excess chalcogenide must be richer in Se, possibly to offset the lower vapor pressure of selenium as compared to sulfur. Therefore, further experiments are necessary to determine the ratios of S to Se needed to maintain any given spinel composition during annealing experiments.

As in spinels of the type $MCr_2S_{4-x}Se_x$ (M = Mn, Fe, Co) the substitution of Se for S brings about a decrease in T_c. Both of the compositions tested exhibited magnetoresistance effects at T_c. The composition $Cu_{.5}Fe_{.5}Cr_2S_{2.5}Se_{1.5}$ has T_c and a negative magnetoresistance effect ($\sim 3\%$), which occur at room temperature.

REFERENCES

1. H. W. Lehmann, Phys. Rev. 163, 488 (1967).
2. Y. Pellerin and P. Gibart, C. R. Acad. Sci. 269, serie B, 615 (1969).
3. P. F. Bongers, C. Haas, A.M.J.G. van Run and G. Zanmarchi, J. Appl. Phys. 40, 958 (1969).
4. L. Goldstein and P. Gibart, AIP Conference Proceedings #5, 883 (1972).
5. F. K. Lotgering, Proceedings of the International Conference of Magnetism, p. 533, Nothingham, England (1964).
6. M. Robbins, R. Wolfe, A. J. Kurtzig and M. A. Miksovsky, J. Appl. Phys. 41, 1086 (1970).
7. M. Robbins, P. K. Baltzer and E. Lopatin, J. Appl. Phys. 39, 662 (1968).
8. P. Gibart, M. Robbins, V. G. Lambrecht, Jr., submitted for publication, J. Phys. Chem. Solids.
9. G. Haacke and L. C. Beegle, J. Phys. Chem. Solids 28, 1699 (1967).
10. F. K. Lotgering, R. P. Van Stapele, G. H. A. M. van der steen and J. S. Van Wreringen, J. Phys. Chem. Solids 30, 799 (1968).

MAGNETIC PROPERTIES OF THE SPINEL SYSTEM
$MnCr_2S_{4-x}Se_x$ (x = 0 TO x = 2)

M. Robbins, P. Gibart*, L. M. Holmes+, R. C. Sherwood
and G. W. Hull
Bell Laboratories
Murray Hill, New Jersey 07974

ABSTRACT

In the system $MnCr_2S_{4-x}Se_x$ the spinel phase can be prepared between x = 0 to x = 2. Materials from this system are being studied with respect to their magnetic properties. $MnCr_2S_4$ exhibits strong B-B interactions, weaker A-B interactions, and at low temperatures (< 5°K) weak A-A interactions. The substitution of Se for S causes a slow decrease in T_c (T_c = 74°K for x = 0 to T_c = 56°K for x = 2), with increasing Se, which is presumably caused by weakening of the A-B interactions. It is also shown that the low temperature A-A interaction is also weakened by Se additions. Extrapolation of inverse susceptibility to $1/\chi$ = 0 yields θ's which increase with increasing Se content (θ = -27°K for $MnCr_2S_4$ to +50°K for x = 2.0). This further indicates the decrease of J_{AB} with increasing Se. Samples where x = 1.0, 1.5 and 2.0 exhibit a linear increase in magnetization with applied field up to 60 kilogauss. Using experimental data, calculated values of J_{AA}, J_{AB} and J_{BB} are shown.

INTRODUCTION

The magnetic properties of the chalcogenide spinel, $MnCr_2S_4$, have been extensively studied [1,4]. In this spinel Mn^{2+} occupies the A (tetrahedral) site and Cr^{3+} the B (octahedral) site. The variation of magnetization with temperature, for $MnCr_2S_4$, is shown in Fig. 1. At 5.5°K, $MnCr_2S_4$ was found [1,3] to exhibit a transition from Yafet-Kittel to Néel ferrimagnetism. The exchange interactions J_{AA}, J_{AB} and J_{BB} were estimated [1,2] using neutron diffraction, low temperature (T < 5.5°K) high field magnetization and T_c (Curie temperature) measurements. The magnetic properties of $MnCr_2S_4$ were ascribed to positive B-B interactions and negative A-A interactions which were nearly equal in strength to the negative A-B interaction resulting in a low temperature canting of the A site.

The theoretical 0°K magnetic moment of $MnCr_2S_4$ is 1 μ_B [2 × 3 $\mu_B(Cr^{3+})$ - 5 $\mu_B(Mn^{2+})$]. The observed moment of 1.3 μ_B results

---

*Work performed while on leave from Laboratoire de Magnetisme, CNRS, Bellevue, France, on NATO Fellowship and CNRS Grant.

+A portion of the work performed at the Laboratory for Electrophysics, The Technical University, DK-2800 Lyngby, Denmark.

from the canting of the A (Mn^{2+}) site which has a moment of $4.7\ \mu_B$ instead of $5\ \mu_B$.

In a recent investigation [5] we have found that it is possible to substitute up to 2 moles of Se for S in the system $MnCr_2S_{4-x}Se_x$ while still maintaining the spinel phase. Where $x > 2$ the materials form with the defect NiAs structure. At this time we are reporting the results of magnetic measurements on spinels in this system where $x = 0$ to $x = 2$.

MEASUREMENTS

The preparation of materials in this system and a partial phase diagram have been reported previously [5].

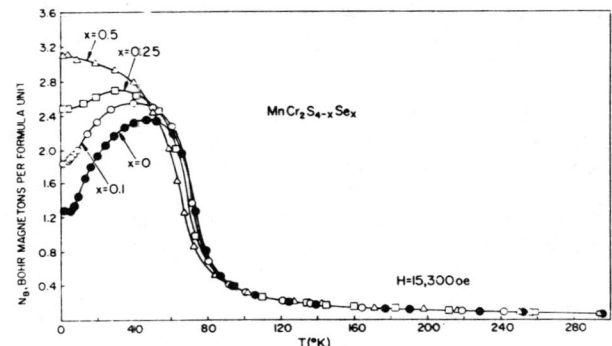

Fig. 1. Magnetization vs. temperature of materials from the system $MnCr_2S_{4-x}Se_x$ where $x = 0$ to $x = 0.5$.

Magnetization as a function of temperature at 15 kOe for materials in the region $x = 0$ to $x = 0.5$ are shown in Fig. 1. From Fig. 1 and Table 1 it can be seen that T_c decreases with increasing Se substitution. It can also be seen that the shape of the magnetization curve changes rapidly so that at $x = 0.5$ the magnetization curve resembles that of a standard ferromagnet or Neel type ferrimagnet. The magnetization curves for $x = 1, 1.5$ and 2 are similar to that of $x = 0.5$

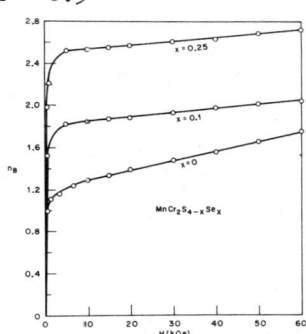 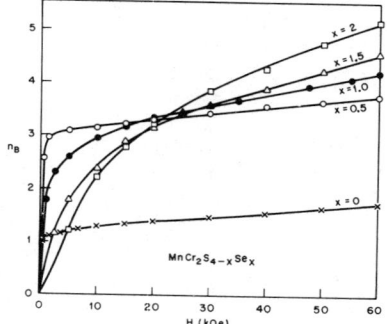

Fig. 2a. Magnetization vs. field strength at 1.5°K ($x = 0$ to $x = 0.25$).

Fig. 2b. Magnetization vs. field strength at 1.5°K ($x = 0.5$ to $x = 2$).

In Fig. 2a and b, magnetization (μ_B) as a function of external field strength (to H = 60 kOe) at 1.5 K are shown. As shown the moment of MnCr$_2$S$_4$ increases linearly with field up to 60 kOe. For x = 0.1 and 0.25 the slope decreases but for x = 0.5, 1, 1.5 and 2.0 the slope of the linear region increases with increasing x. Zero field moments, obtained by extrapolation are shown in Table 1.

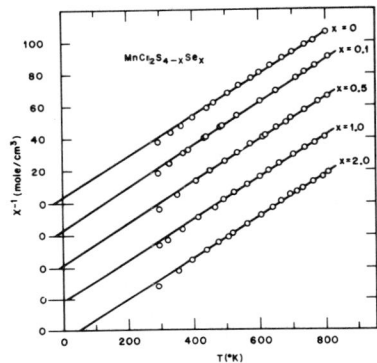

Fig. 3. Inverse susceptibility vs. temperature for some compositions in the system MnCr$_2$S$_{4-x}$Se$_x$. Zero point displacement on the susceptibility scale is included for clarity.

$1/\chi$ as a function of temperature (up to 800 °K) was measured (Fig. 3) and θ_p obtained by extrapolation of the linear region to $1/\chi = 0$. Values of C_M, calculated from the slopes, were similar (7.71 ± .12) for all compositions and in good agreement with the previously reported [2] value for MnCr$_2$S$_4$ (~ 7.7). As shown in Table 1, θ_p increases linearly from -27°K for x = 0 to + 50°K for x = 2. The value of θ_p, for MnCr$_2$S$_4$, reported by us (-27°K) is more negative than that which was reported by Lotgering [1] (-10°K).

Table I Magnetic properties of materials in the system MnCr$_2$S$_{4-x}$Se$_x$ (x = 0 to x = 2)

x	T_c(°K)	θ_p(°K)	$\sigma^+(n_B)$	$\frac{d\sigma(\mu_B)}{dH}$	J_{AA}	J_{AB}	J_{BB}(°K)
0	74	-27	1.20	.0093	-1.8	-1.92	+2.70
0.1	72	-23	1.64	.0037			
0.25	70		2.49	.0041			
0.5	68	-9	3.20	.0091	-1.8	-1.14	3.71
1.0	63	12	2.94	.0205	-0.81	-0.5	3.94
1.5	59		2.63	.0317	-0.53	-0.40	3.77
2.0	56	50	2.58	.0428	-0.39	-0.30	3.65

$^+\sigma(n_B)$ at H = 0 and T = 1.5°K.

DISCUSSION

As a result of the initial substitution of Se (x = 0 to x = 0.5) the zero field moment (σ_o) increases from 1.2 μ_B at x = 0 to 3.20 μ_B at x = 0.5. This rapid increase probably reflects a decrease in the effective A site moment with increasing x, and might be explained by an increase in the A-site canting. However, this possibility is not supported by preliminary neutron-diffraction data [6], which give no evidence of A-site canting at $x = 0.1$. The initial increase in σ_o can be explained in a collinear structure if ferromagnetic coupling predominates wherever isolated Mn-Se-Cr superexchange interactions are found in the Se-substituted material. This possibility is supported as well by the observed increase in θ_p (see Table 1), and by the strongly reduced values of $d\sigma/dH$ (as expected for a collinear structure) at low concentrations of Se.

From $x = 0.5$ to $x = 2$ the moment (σ_o) decreases ($3.2\mu_B$ to $2.5\mu_B$) with increasing x. In this compositional region the slope of magnetization with field ($d\sigma/dH$) increases with increasing x (Fig. 2b). The change in σ_o and $d\sigma/dH$ with c can be interpreted by assuming that in this compositional region the B site moment starts to cant, possibly in the form of a spiral as observed for $HgCr_2S_4$.

In the ferromagnetic chalcogenide spinels ($CdCr_2S_4$, etc.), there exist two types of B-B interactions. These are the nearest neighbor B-B interactions (6 in number) which are ferromagnetic and the more distant B-B interactions (~ 30) [7], involving the A site cation, which are antiferromagnetic. In spinels of the type MCr_2S_4 (M = Mn, Co, Fe) the next nearest neighbor interactions are insignificant with respect to the more direct, antiferromagnetic A-B (M-S-Cr) interactions. Of the three spinels (containing Mn, Fe or Co) $MnCr_2S_4$ exhibits the weakest A-B interaction (J_{AB}) as shown by a comparison of θ_p (-27°K for $MnCr_2S_4$, -290°K for $FeCr_2S_4$ and -480°K for $CoCr_2S_4$ [8]). The substitution of Se for S, in $MnCr_2S_4$, decreases the strength of the antiferromagnetic A-B interaction as indicated by T_C which decreases and θ_p which increases linearly with increasing x. Substitution of Se for S in $FeCr_2S_4$ and $CoCr_2S_4$ [5] gives a similar effect wherein T_C decreases with increasing Se substitution. With decreasing A-B interaction strength it is possible that the more distant B-B interactions become sufficiently strong to bring about the hypothesized B site canting. The increase in θ_p/T_C from -0.36 at $x = 0$ to +0.89 at $x = 2$ is consistent with a model implying canting.

Lotgering calculated J_{AA}, J_{AB} and J_{BB} in $MnCr_2S_4$ using equations (a), (b) and θ_p

$$\text{(a)} \quad \frac{d\sigma}{dH} = \frac{-\mu_B^2}{4J_{AA}}$$

$$\text{(b)} \quad \sigma_{o,o} = 2\left(1 - \frac{3}{4}\frac{J_{AB}}{J_{AA}}\right)\mu_B$$

An alternate calculation for J_{BB} involves using T_C as Plumier [4] did. The two calculations give widely different results (10.0 vs. 3.5°K). Using equations (a) and (b) to obtain J_{AA} and J_{AB} and T_C for J_{BB}, similar calculations were made for some of the compositions in this system and are shown in Table 1. For $MnCr_2S_4$, it is possible to get good agreement with the previously reported values of J_{AA}[1,9], J_{AB}[1] and J_{BB}[4].

The addition of Se for S which seems to eliminate the A site canting and possibly introduce B site canting makes the calculation of J_{AA}, J_{AB} and J_{BB} of very limited value. The trends observed, however, (e.g., the decrease in $|J_{AB}|$ and $|J_{BB}|$) are consistent with the observed data and models which are proposed.

CONCLUSIONS

The magnetic properties of $MnCr_2S_4$ can be described in terms of comparative strengths for all of the interactions (J_{AA}, J_{AB} and J_{BB}). The substitution of Se for S further complicates the system by bringing about unequal changes in J_{AA}, J_{AB} and J_{BB}.

At this time the magnetic data indicate that Se substitution initially ($0 < x < 0.5$) leads to a decrease in the resultant A site moment possibly due to a single ion effect. In the compositional region $x = 0.6$ to $x = 2$ the observed decrease in $\sigma_{o,o}$ and increase in $d\sigma/dH$ can be related to a noncollinear B site moment as observed in other sulfide spinels.

It is quite clear that additional investigation, especially neutron diffraction measurements, are needed to further elucidate the properties of this system.

ACKNOWLEDGMENTS

The authors wish to extend their thanks to Dr. J. B. Goodenough for his very constructive discussions.

REFERENCES

1. F. K. Lotgering, Phillips, Res. Rep. 11, 190 (1956); F. K. Lotgering, J. Phys. Chem. Solids 29, p. 2193 (1968).
2. N. Menyuk, K. Dwight and A. Wold, J. Appl. Phys. 36, 1688 (1965).
3. R. Plumier and M. Sougi, CR, 268B, 1549 (1969); J. Denis, Y. Allain and R. Plumier, CR, 269B, 740 (1969); J. Appl. Phys. 41, 1091 (1970); R. Plumier, R. Conte, J. Denis and M. Nauciel-Bloch, J. Phys. 32, C1-55 (1971).
4. N. Nauciel-Bloch, A. Castets and R. Plumier, Phys. Lett. 39A, 311 (1972).
5. P. Gibart, M. Robbins and V. G. Lambrecht, Jr., to be published.
6. L. M. Holmes, B. Lebech, M. Robbins, and P. Gibart (unpublished).
7. P. J. Wojtowicz, P. K. Baltzer and M. Robbins, J. Phys. Chem. Solids 28, 2423 (1967).
8. P. Gibart, J. L. Dormann and Y. Pellerin, Phys. Stat. Sol. 36, 187 (1960).
9. L. Darcy, P. K. Baltzer, and E. Lopatin, J. Appl. Phys. 39, 898 (1968).

MAGNETIC ORDERING IN $Li_{2x}Cr_{2x}Ni_{2-4x}O_2$

A. Tauber *
US Army Electronics Technology and Devices Laboratory (ECOM)
Fort Monmouth, New Jersey 07703

E. Banks
Polytechnic Institute of Brooklyn
Brooklyn, New York 11201

ABSTRACT

Magnetic ordering in the $Li_{2x}Cr_{2x}Ni_{2-4x}O_2$ system has been studied in polycrystalline compositions isotypic with α-$NaFeO_2$. The susceptibility measurements obtained between 4.2 and 900 K indicated that all compositions investigated ordered antiferromagnetically at low temperatures. The asymptotic Curie points as a function of x follow a law of the type $\theta = \theta_a(1-x)^2 + \theta_b 2x(1-x) + \theta_c x^2$, where $\theta_a = Ni^{2+}-Ni^{2+}(d^8-d^8)$ interaction, $\theta_b = Ni^{2+}-Cr^{3+}(d^8-d^3)$ interaction, and $\theta_c = Cr^{3+}-Cr^{3+}(d^3-d^3)$. These results are compared with the predictions of the Goodenough-Kanamori rules for contributions to 180° and 90° exchange.

INTRODUCTION

Compounds isotypic with the ordered rocksalt α-$NaFeO_2$ (space group $R\bar{3}m$) afford one the opportunity to probe the sign and magnitude of both 90° and 180° exchange for various $3d^n$ combinations. We have reported the nature of the exchange in the system $LiCr^{3+}_{1-x}Fe^{3+}_x O_2$.[1] Here d^3-d^3, d^5-d^5, and d^3-d^5 interactions were studied and the results compared with predictions from the Goodenough-Kanamori (GK) rules.[2,3] In the present investigation d^8-d^8, d^3-d^3, and d^3-d^8 interactions are examined in the system $Li_{2x}Cr_{2x}Ni_{2-4x}O_2$ and results compared with the expectations from G-K rules.

EXPERIMENTAL AND RESULTS

All the compositions described were prepared by solid-state reaction as reported in ref. (1). The final annealing temperature was between 1100 and 1200 K. Powder x-ray diffractometry and optical microscopy were employed to determine the presence of a single phase. Lattice parameters and susceptibility were determined as in ref. (1). Attempts to grow single crystals from lithium borate flux were unsuccessful.

Like $Li_{2x}Fe_{2x}Mg_{2x-4x}O_2$, the system $Li_{2x}Cr_{2x}Ni_{2-4x}O_2$ exhibits a transition from an ordered to disordered state near 50 mole %.[4] The c lattice parameter, Fig. 1 is essentially constant with increasing Ni^{2+} substituent while the a parameter increases. At the transition a sharper discontinuity in c is noted.

*Abstracted in part from a dissertation submitted by Arthur Tauber to the Polytechnic Institute of Brooklyn in partial fulfillment of the requirements for the degree of Doctor of Philosophy in Chemistry.

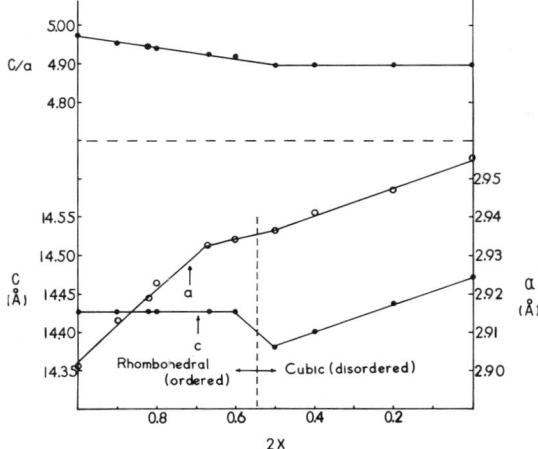

Fig. 1. Variation of lattice parameters (a,c) referred to a hexagonal unit cell and c/a as a function of 2x.

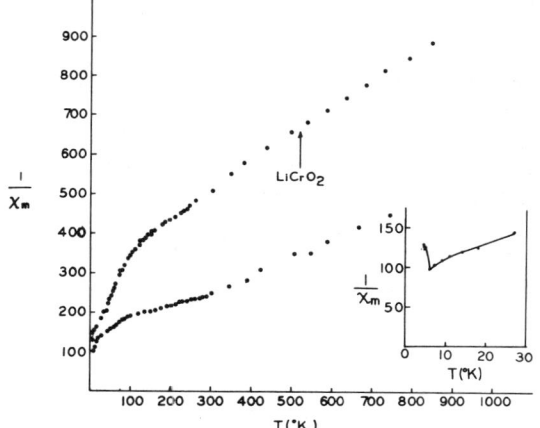

Fig. 2. Inverse molar susceptibility in moles/cm³ as a function of temperature for $Li_{0.9}Cr_{0.9}Ni_{0.2}O_2$. The insert is the low temperature region of $Li_{0.9}Cr_{0.9}Ni_{0.2}O_2$.

The susceptibility was measured for only ordered compositions with 2x=0.9, 0.84, 0.80, 0.67, and 0.60. The $1/\chi_m$ vs T curves were least squares computer fitted to a Curie-Weiss law above 400 K. Minima were observed in the $1/\chi_m$ vs T curves below 20 K, Fig. 2. These were interpreted as Néel points. The curves for 2x=0.60 and 0.67 exhibit sharp breaks in the susceptibility and essentially constant $1/\chi_m$ as a function of temperature below 200 K, Fig. 3. Such data suggest that a transition to a three-dimensional ordered antiferromagnetic state may be occurring at temperatures near 200 K.

The magnetic parameters computed from the Curie-Weiss law fit are given in Table I. The magnetic moments are compared with spin only values. These in general show a reduction in moment with increasing substitution, that is in the direction of decreasing χ. As in the $LiCr_{1-x}Fe_xO_2$ system, the reduced moment may arise from anisotropic magnetic coupling. When the asymptotic Curie points were plotted as a function of 2x, Fig. 4, we observed a parabolic dependence previously observed in $LiCr_{1-x}Fe_xO_2$. This behavior must be due to a ferromagnetic Cr^{3+}-Ni^{2+} exchange interaction. In compounds isotypic with α-$NaFeO_2$ the metal ions order in alternate metal planes

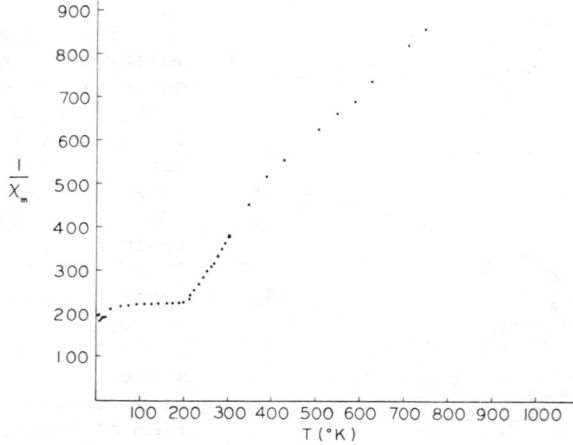

Fig. 3. Inverse molar susceptibility in moles/cm³ as a function of temperature for $Li_{0.6}Cr_{0.6}Ni_{0.8}O_2$.

Table I. Magnetic Parameters in the System $Li_{2x}Cr_{2x}Ni_{2-4x}O_2$

x	Composition $Li_{2x}Cr_{2x}Ni_{2-4x}O_2$	C_m $C_m{}^3°K/M$	μ_B exp.	$\mu_B{}^a$ calc.
0.50	$LiCrO_2$	1.58	3.57	3.87
0.45	$Li_{0.9}Cr_{0.9}Ni_{.2}O_2$	2.49	4.48	4.05
0.42	$Li_{0.84}Cr_{0.84}Ni_{0.32}O_2$	1.89	3.90	4.16
0.40	$Li_{0.80}Cr_{0.80}Ni_{.40}O_2$	1.61	3.60	4.23
0.334	$Li_{0.667}Cr_{0.667}Ni_{0.667}O_2$	1.63	3.60	4.46
0.30	$Li_{0.6}Cr_{0.6}Ni_{0.8}O_2$	1.21	3.12	4.58

a - spin only, high spin

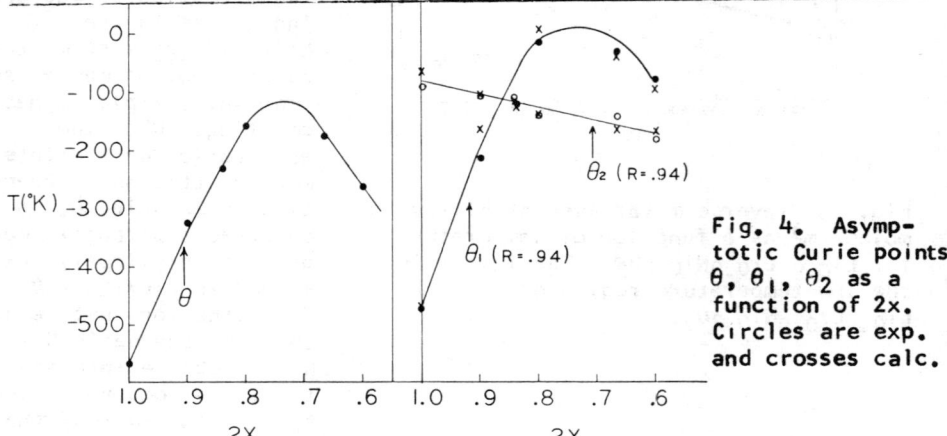

Fig. 4. Asymptotic Curie points θ, θ_1, θ_2 as a function of $2x$. Circles are exp. and crosses calc.

perpendicular to the c axis. Oxygen layers intervene between all metal layers so that interplanar order leads to the following sequence of layers: $(Ni^{2+}_{1-x} Cr^{3+}_x) - O^{2-} - (Li^+_x Ni^{2+}_{1-x}) - O^{2-} - Ni^{2+}_{1-x} Cr^{3+}_x)$. By virtue of the large interplanar separation (average is 4.8 Å for LiCrO$_2$) and the smaller intraplanar separation (average is 2.93 Å) of magnetic ions, a considerable anisotropy in the strength of the magnetic interaction can initially be expected. Within a plane the most important interactions expected would be 90° M^{3+}-O-M^{3+} superexchange, and direct M^{3+}-M^{3+} exchange and between planes ~ 180° superexchange.

Exchange parameters were computed using molecular field theory as previously described.[1] Although Néel points were obtained from susceptibility curves, J_2/k was computed from $C_m/\chi(T_n)=8J_2S(S+1)/k$ which yields a more accurate estimate.[5] The values for the exchange parameters are given in Table II. The intraplanar exchange constant J_1/k is found to be much larger than J_2/k as the concentration of Ni^{2+} is initially increased. At $2x = 0.84$ J_2/k is slightly larger than J_1/k and continues to increase with decreasing $2x$. This variation was not observed in the system $LiCr_{1-x}Fe_xO_2$ because Fe^{3+} replaced only Cr^{3+} ions and had little effect on interplanar exchange. In the system at hand, since Ni^{2+} is entering the Li^+ ion layer, interplanar exchange distances become comparable to the intraplanar and J_2/k becomes greater.

Table II. Exchange Parameters in the System $Li_{2x}Cr_{2x}Ni_{2-4x}O_2$

x	θ °K	θ$_1$ °K	θ$_2$ °K	T$_n$ °K	J$_1$/k °K	J$_2$/k °K
0.50	-570	-472	- 98	15	-33	- 6.9
0.45	-327	-217	-110	6	-10.2	- 6.1
0.42	-238	-113	-125	19	- 7.4	- 8.2
0.40	-161	- 18	-143	14	- 1.4	-11.1
0.334	-180	- 35	- 45	--	- 2.7	-11.0
0.30	-266	- 81	-185	6	- 8.3	-19.1

In the system $Li_{2x}Cr_{2x}Ni_{2-4x}O_2$ we define the following exchange interactions: θ_a=Ni^{2+}-Ni^{2+} interaction, fraction of neighbors = $(1-x)^2$; θ_b=$Cr^{3+}Ni^{2+}$ interaction, fraction of neighbors = $2x(1-x)$; θ_c=Cr^{3+}-Cr^{3+} interaction, fraction of neighbors = (x^2). The θ values were obtained by regression analysis for an equation, $\theta_n=\theta_a(1-x)^2 + \theta_b2x(1-x)+ \theta_cx^2$ first applied by Jonker[6] to perovskites; where n=1 is for intraplanar and n=2 interplanar exchange. The coefficient of determination, R was 0.94 for both θ values. The fit of the calculated θ's to the experimental values is shown in Fig. 4. The final constants obtained were,

Type of interaction	θ$_1$	θ$_2$
Ni^{2+} - Ni^{2+}	θ_a = -526 K	-47 K
Cr^{3+} - Ni^{2+}	θ_b = +589	-575
Cr^{3+} - Cr^{3+}	θ_c = -1048	-276

Goodenough and Kanamori[2,3] have formulated rules for predicting the sign and approximate strength of 180° and 90° direct and superexchange interactions for ions in octahedral coordination. For the α-NaFeO$_2$ structure this has previously been discussed for Cr^{3+} - $Cr^{3+}(d^3-d^3)$ where observation and prediction concur on antiferromagnetic for 180° and 90° exchange.[1]

The observed antiferromagnetic $Ni^{2+}-Ni^{2+}(d^8-d^8)$ interaction agrees with the prediction of Goodenough[7] for the dependence on cation separation of the exchange interaction between octahedrally coordinated high-spin cations sharing edges. The latter is consistent with the expansion of the $\underline{a}$ lattice parameter.

For $Cr^{3+}-Ni^{2+}(d^3-d^8)$ the following types of 90° superexchange are distinguished: (1) delocalized antiferromagnetic $Cr^{3+}t_{2g}$ to $Ni^{2+}e_g$ transfer; (2) delocalized antiferromagnetic $Ni^{2+}e_g$ to $Cr^{3+}t_{2g}$ transfer; (3) delocalized ferromagnetic $Ni^{2+}t_{2g}$ to $Cr^{3+}e_g$ transfer; (4) ferromagnetic correlation superexchange simultaneously via $O^{2-}-p_x$ and p_y transfer to $Cr^{3+}e_g$ and $Ni^{2+}e_g$; (5) ferromagnetic correlation superexchange via simultaneous O^{2-}-2s transfer to $Cr^{3+}e_g$, $Ni^{2+}e_g$.

For 180° superexchange the following types of interaction are predicted: (1) delocalized ferromagnetic $Ni^{2+}e_g$ transfer to $Cr^{3+}e_g$; (2) delocalized antiferromagnetic $Cr^{3+}t_{2g}$ transfer to $Ni^{2+}e_g$; (3) delocalized antiferromagnetic $Ni^{2+}e_g$ transfer to $Cr^{3+}t_{2g}$; (4) ferromagnetic correlation superexchange simultaneously via p_x or p_y transfer to $Cr^{3+}e_g$ and $Ni^{2+}e_g$; (5) ferromagnetic correlation superexchange via simultaneous O^{2-}-2s transfer to $Cr^{3+}e_g$ and $Ni^{2+}e_g$.

The intraplanar exchange for $Cr^{3+}-Ni^{2+}$ is, as expected, dominated by the stronger ferromagnetic exchanges. The interplanar exchange between these ions, due primarily to $\sim$ 180° mechanism, differs from the prediction of the G-K rules.

REFERENCES

1. A. Tauber, W. M. Moller, and E. Banks, J. Solid State Chem. $\underline{4}$, 138 (1972).
2. J. B. Goodenough, "Magnetism and the Chemical Bond," p. 180, Interscience, New York, 1963.
3. J. Kanamori, J. Phys. Chem. Solids $\underline{10}$, 87 (1959).
4. E. Kordes and J. Petzoldt, Zeit f Anorg. und Allegem. Chemie $\underline{335}$, 138 (1965).
5. J. S. Smart, Magnetism, Vol. III, Editors, G. T. Rado and H. Suhl, Ch. 2, p. 90, Academic Press, N. Y. (1963).
6. G. H. Jonker, Physica $\underline{22}$, 707 (1956).
7. Ref. 3, p. 270.

MAGNETIC LONG RANGE ORDER IN $FeCO_3$*

R. F. Altman and S. Spooner
Georgia Institute of Technology, Atlanta, Ga. 30332

D. P. Landau
University of Georgia, Athens, Ga. 30601

ABSTRACT

The sublattice magnetization of $FeCO_3$ has been obtained as a function of temperature from neutron diffraction data. Integrated magnetic Bragg intensities were collected from 4.2K to several degrees above the ordering temperature ($T_N \sim 38$ K). These measurements are compared with the results of Monte Carlo calculations based on an Ising-like $S = 1/2$ spin model having the $FeCO_3$ structure. Both nearest and next-nearest neighbor interactions are considered in the calculations. When the results of these calculations are normalized to give the observed ordering temperature, it is found that the shape of the measured sublattice magnetization curve can best be fit using an antiferromagnetic nearest neighbor interaction, and a ferromagnetic next-nearest neighbor interaction that is approximately one-quarter as large. These interaction constants are consistant with the interaction constant obtained by combining the Monte Carlo results with measured critical field and ordering temperature values.

INTRODUCTION

Among the carbonates which have the calcite structure, iron carbonate seems to provide a unique example of an antiferromagnetic system with an Ising-like ground state. The highly anisotropic crystal field environment of the iron ions, together with an unquenched orbital angular momentum which couples with the spin of the unpaired 3d electrons in the Fe^{2+} ions are thought to be responsible for this Ising character. The Ising character of the ground state was predicted in a paper by Kanamori,[1] and has since been tentatively experimentally confirmed.[2,3] Although there is some question as to the correctness of Kanamori's ground state description,[4] the results of the sublattice magnetization measurements and calculations reported in this paper add further evidence that the ground state of iron carbonate is Ising-like.

EXPERIMENTAL PROCEDURE

Integrated Bragg intensities were measured from a ground spherical crystal (5.5 mm diam.) of natural iron carbonate (7% by weight impurity content) at the Georgia Tech Neely Nuclear Research

---

*Supported by A.E.C. Contract No. AT-(40-1)-3674 and the National Science Foundation.

Center. The crystal was mounted in a cold finger helium cryostat with the 111 and 100 reflections in the scattering plane. Integrated intensities from the 100, 300, and 344 reflections were collected at temperatures from 4.2 K to 41 K. Thermal gradients in the sample were minimized by mounting the crystal inside an aluminum chamber thermally connected to the cold finger. Temperature was controlled by a resistance thermometer attached to the cold finger and the crystal temperature was measured independently by a calibrated germanium thermometer mounted close to the sample. The sample was maintained at each temperature until no drift could be detected in the sample temperature (< 0.1 K) or in the integrated intensity.

A nuclear reflection was measured after each magnetic determination to monitor instrumental drift. As a precaution against possible systematic errors, data were taken over the temperature range in a random order of temperatures and some of the first measurements were repeated at the end of the experiment. The nuclear integrated intensities indicated a variation of $\pm 1.5\%$ while the purely statistical counting error was 0.5%.

The effect of extinction was determined by comparing the intensity versus temperature profiles of the 3 reflections which diffrered greatly in counting rate because of the magnetic form factor. The normalized profiles for these reflections were different, but they could be brought into coincidence by applying an isotropic extinction correction to the measured intensities. The maximum extinction correction in the 003 reflection was 2% and the intensities from this reflection were used for the comparison with the Monte Carlo results.

MONTE CARLO CALCULATION

The behavior of 3 dimensional Ising systems has been studied extensively by a variety of techniques. Some, such as mean field theory, are straightforward but yield quantitatively unreliable results. Others, such as series expansion, are more accurate but are quite unwieldy, particularly when the next-nearest neighbor interactions are considered.

A Monte Carlo procedure described elsewhere,[5] allows an accurate calculation of thermodynamic properties over a wide temperature range. In addition, this Monte Carlo procedure allows one to include in the calculation any energy mechanisms that may be of importance in a real physical system, such as interactions with many different shells of neighbors and contributions from excited energy states.

For reasons discussed later, only first and second nearest neighbor interactions were considered in the present paper. The Hamiltonian was taken to be of the form:

$$H = K_{NN} \sum_{i,j} \sigma_{iz} \sigma_{jz} + K_{NNN} \sum_{i,k} \sigma_{iz} \sigma_{kz}$$

where the sums i,j and i,k are taken over pairs of nearest and next-nearest neighbors respectively (both six in number) and $\sigma_z = \pm 1$.

In order to correct for finite sample size effects we carried out calculations for three different lattice sizes: 4x4x12, 6x6x18 and 8x8x24. Some of our work on other lattices has indicated that bulk properties should vary inversely with the linear dimensions of the sample, viz., the maximum possible correlation length. For this reason we chose all the lattices to be of the same shape so that we could choose any of the linear dimensions of the sample as the extrapolation parameter. It was found that the finite size effects were negligable for $T/T_N < 0.9$. Edge effects were eliminated by applying periodic boundary conditions. The ordering temperature was determined from the inflection point in the energy-temperature data.

COMPARISON OF NEUTRON DATA AND MONTE CARLO RESULTS

In Figure 1, the neutron sublattice magnetization measurements are plotted with the best fitting curve obtained from the Monte Carlo calculations.[6] This curve was obtained for an interaction ratio K_{NNN}/K_{NN} of -0.4. For comparison, the spontaneous magnetization for a Brillouin function for a $S = 1/2$ model is also shown. (The $S = 1/2$ model was chosen for comparison because the ground state of Fe^{2+} ion is believed to be an exchange split doublet.) The better agreement gives an indication of the improvement of the Monte Carlo data over the molecular field results. Due to the inaccuracies in the extrapolation technique and statistical fluctuations inherent in the Monte Carlo results, the shape of the sublattice magnetization curve has not been determined exactly (only the smoothed curve is shown). When this uncertain-

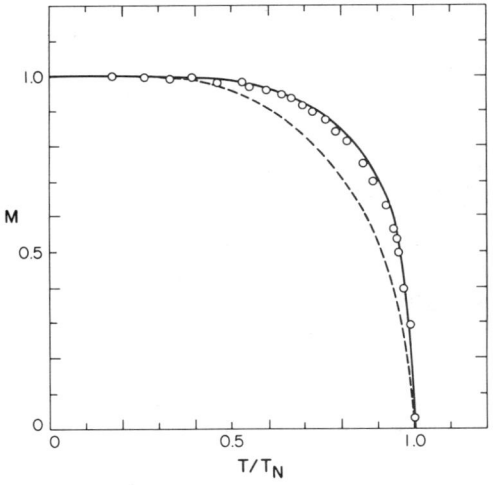

Fig. 1. Sublattice magnetization vs. temperature. Open circles-neutron data: solid line-smoothed Monte Carlo results: dashed line-molecular field curve. Neutron data error limits less than circle size.

ty is taken into account, the agreement between the neutron data and the Monte Carlo curve is reasonably good. However, the neutron data lie consistently below the calculated curve and it is unlikely that this is a purely statistical effect.

There are a number of possibilities for the source of this discrepancy. One is the impurity content of the sample. Another is that the thermal population of excited states is providing an additional mechanism for lowering the average sublattice magnetization. Although one of the crystal field levels of $FeCO_3$ has been measured,[7,8] a consistant scheme for the entire level structure has not been confirmed, and so no attempt was made in these first calculations to account for the effects of excited states. An estimate of the dipole-dipole interactions for more distant neighbors indicates they are almost two orders of magnitude smaller than the exchange interactions and so it is unlikely that their inclusion would alter our results significantly.

The Monte Carlo results can be combined with recent critical magnetic field and magnetization measurements[2] to estimate K_{NN} and K_{NNN} in another way. The magnetic measurements can be used to determine K_{NN} directly since only the shell of nearest neighbors is turned over with respect to the central ion by application of a magnetic field along the spin direction (the second nearest neighbors lie in a ferromagnetic plane with the central ion). Using the low temperature critical field and magnetization values of 145 KOe and 4.6 μ_B per Fe^{2+} ion respectively, K_{NN} has a value of +7.3K. When K_{NN} is fixed at this value and K_{NNN} is varied, T_N resulting from the Monte Carlo calculation is found to vary in the way shown in Figure 2. This graph shows that K_{NNN}/K_{NN} = -0.2 will give the observed ordering temperature of 38.5K. This number is an average of ordering temperatures reported in the literature for better specimens. (The ordering temperature of the sample used in the neutron diffraction study was measured to be 38.0K ±0.3K.) Since no allowance has been made for errors in the magnetic parameters used to calculate K_{NN}, the ratio of -.2 is in reasonable agreement with the neutron value of -.4.

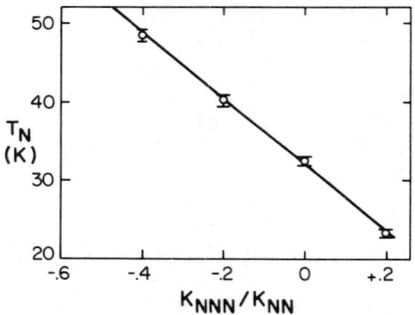

Fig. 2. Dependence of ordering temperature on ratio of next-nearest to nearest neighbor constants. K_{NN} is fixed at 7.3 K.

CONCLUSIONS

It is clear that further work remains to be done. When the properties of the low lying states of the Fe^{2+} ion have been unambiguously determined, these states will be included in the Monte Carlo calculations to see what effect they have on the sublattice magnetization and ordering temperature. Other properties, such as the susceptibility remain to be calculated, and the effects of the impurities need to be accounted for. However, the results for the neutron sublattice magnetization measurements indicate that the ground state of the Fe^{2+} ion in $FeCO_3$ is reasonably well described by a three dimensional, S = 1/2, Ising model with a ferromagnetic next nearest neighbor interaction that is about one-fourth of the nearest neighbor interaction. The comparison of the Monte Carlo results with the neutron data indicates that this technique represents a significant improvement over the techniques employed in molecular field approximations.

REFERENCES

1. J. Kanamori, Prog. Theor. Phys., 20, 890 (1958).
2. I. S. Jacobs, private communcation.
3. N. C. Koon, Thesis, Georgia Institute of Technology (1969).
4. J. S. Griffith, The Theory of Transition Metal Ions, p. 355 (Cambridge, 1964)
5. D. P. Landau, J. Appl. Phys., 42, 1284 (1971).
6. In the following reference, some data on the temperature dependence of a magnetic Bragg reflection is presented. However, the data are not sufficiently accurate to allow a meaningful comparison with our present results. R. Alikhanow, JETP 36, 1204 (1959).
7. D. E. Wrege, S. Spooner, and H. A. Gersch, AIP Conf. Proc. No. 5, 1334 (1971).
8. H. A. Mook, private communcation.

ANTIFERROMAGNETIC TO PARAMAGNETIC TRANSITION IN Fe^{2+} : MnF_2 IN EXTERNAL MAGNETIC FIELDS

C.R. Abeledo, Dept. of Chemistry, Brandeis University,
Waltham, Massachusetts 02154

R.B. Frankel and M.A. Weber,* Francis Bitter National
Laboratory, Massachusetts Institute of Technology,
Cambridge, Massachusetts 02139

A. Misetich, Gerencia de Investigaciones, Comision
National de Engergia Atomica, Buenos Aires, Argentina

ABSTRACT

Mössbauer effect measurements across the antiferromagnetic to paramagnetic phase boundary in Fe^{2+}: MnF_2 in external magnetic fields are reported. From the data, $J(Mn-Fe) = -1.7$ cm^{-1}. The phase boundary is found to vary as H_0^2, as in pure MnF_2.

MnF_2 crystallizes in a rutile structure with a tetragonal lattice. Below the Néel temperature $T_N = 67.4K$, the magnetic properties of MnF_2 are well understood in terms of an ideal, two sublattice antiferromagnet with the spins aligned along the tetragonal c-axis. The phase diagram of MnF_2 in the H-T plane has been studied by Shapira and Foner[1] and is shown in Fig. 1. At low temperature, an external magnetic field applied along the c-axis causes a first-order realignment of the sublattice magnetization from along the c-axis to the basal plane when the magnitude of the external field reaches the critical value H_{sf}. At higher temperature, specifically above 65 K, the external field causes a second order transition to the paramagnetic phase, i.e., effectively lowers the Néel temperature.

Fe^{2+} may be isomorphously incorporated into the MnF_2 lattice. The effect of the addition of iron is to increase the Néel point[2] and to increase the value of H_{sf}[3]. In a previous work, we observed the spin flop in Fe^{2+}: MnF_2 using Mössbauer spectroscopy in ^{57}Fe[4]. In this paper we report measurements of the antiferromagnetic to paramagnetic phase transition in external magnetic fields for single crystals of MnF_2 doped with 1% ^{57}Fe.

A large single crystal of ~ 1% $^{57}Fe^{2+}$ doped MnF_2 was grown from the melt by Optovac, Inc. The single crystal was oriented and a 6 mil slice was cut perpendicular to the c-axis and mounted between beryllium disks. In addition, some of the crystal was crushed and the powder was cast in lucite. Measurements were made in a conventional constant acceleration spectrometer operating in the normalized mode. The sample was placed in a cryostat which was inserted in a liquid nitrogen dewar which was in turn inserted into a superconducting solenoid operating in the persistent mode up to 85 kOe. The temperature was

*On leave from the Universidad de Chile, Casilla 5487, Santiago, Chile. Supported by the Organization of American States Fellowship.

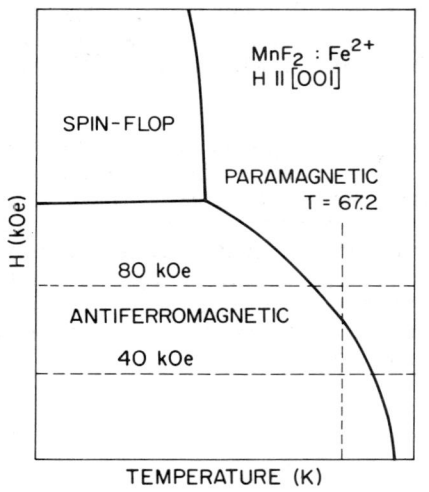

Fig. 1.

Phase diagram of pure MnF$_2$ from Ref. 1. Inclusion of Fe^{2+} raises H$_{sf}$ and increase T$_N$. The measurements reported here are across the AF-P phase boundary.

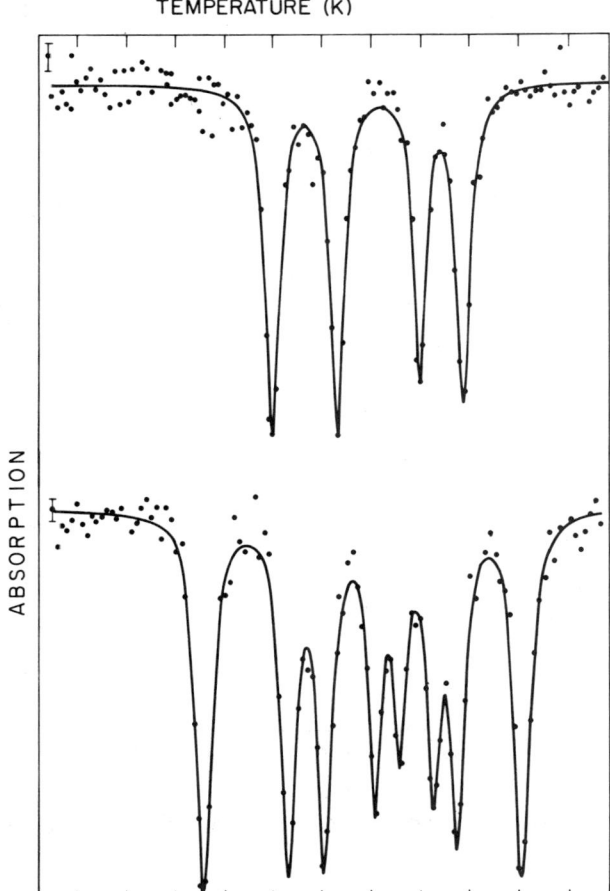

Fig. 2.

Mössbauer spectra of single crystal Fe^{2+}: MnF$_2$ with H$_O$= 80 kOe along the c-axis above and below T$_N$.

controlled by pumping the liquid nitrogen bath below the desired temperature and then heating the sample electrically. The temperature was measured using a wire wound Pt resistor and the values were corrected for the effect of the external field using the results of Neuringer et al.[5]

Because the field at the nucleus $\vec{H}_n = \vec{H}hf \pm \vec{H}_o$, the spectra of the spin up and spin down sublattices in the antiferromagnetic phase are observed independently and the complete spectrum consists typically of eight lines (Fig. 2). In the paramagnetic phase all the spins are equivalent and the spectrum consists of just four lines (the $\Delta m = 0$ lines are absent because the γ-ray propagation direction is parallel to H_o). Thus we observe the temperature dependence of the spin up and spin down sublattices independently and the transition to the paramagnetic phase in clearly delineated.

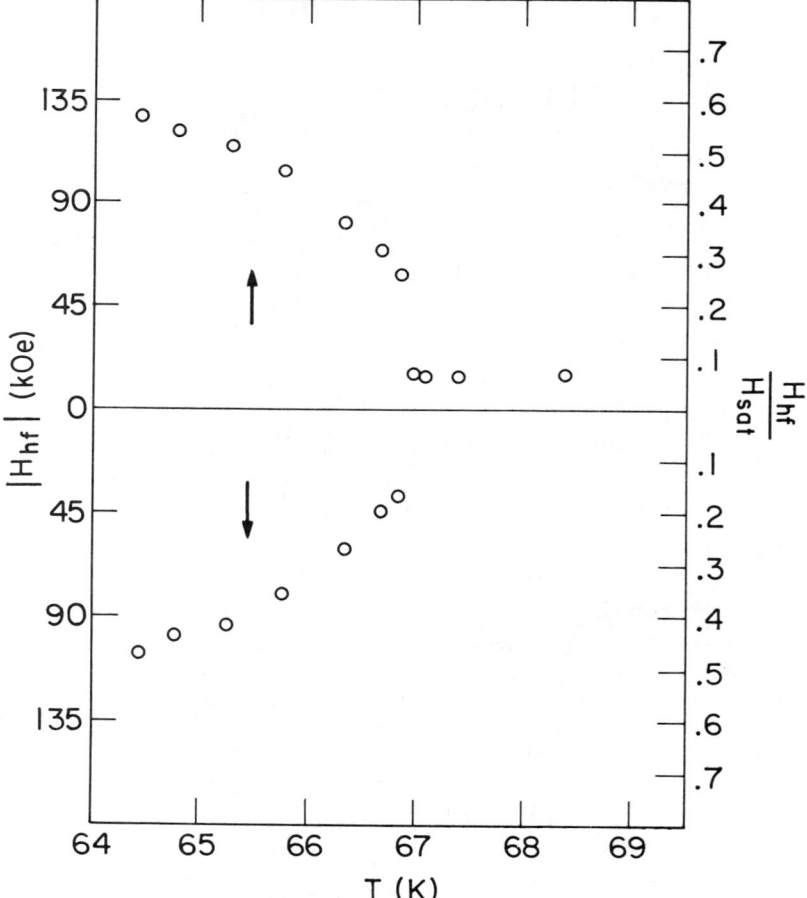

Fig. 3. H_{hf} and H_{hf}/H_{sat} plotted as a function of T for H_o=80 kOe.

The data were analyzed using a computer program of Singh and Hoy[7]. From the quadrupole splitting above T_N and powder spectra at several points below T_N we found that the quadrupole coupling parameter Q= 2.95 mm/sec. and the asymmetry parameter η= 0.4 did not change appreciably in the transition region.

In Fig. 3. we plot the hyperfine field H_{hf} in the Fe^{2+} in the spin down and spin up sublattices as a function of temperature for an applied field H_O= 80 kOe. The data have been analyzed using a molecular field approximation model wher J(Mn) and J(Fe), the Mn-Mn and Fe-Fe exchange constants respectively were chosen to give the best fit to the Néel point of pure MnF_2 and pure FeF_2. The Fe^{2+} single ion anisotropy constant D was taken from the measurements of Lowe et al;[3] D= 8.6 cm^{-1}. Using these values the best fit to the Néel point is obtained with J= -1.5 cm^{-1}, however the best fit to the magnetization above and below T_N is obtained with J= -1.7 cm^{-1}.

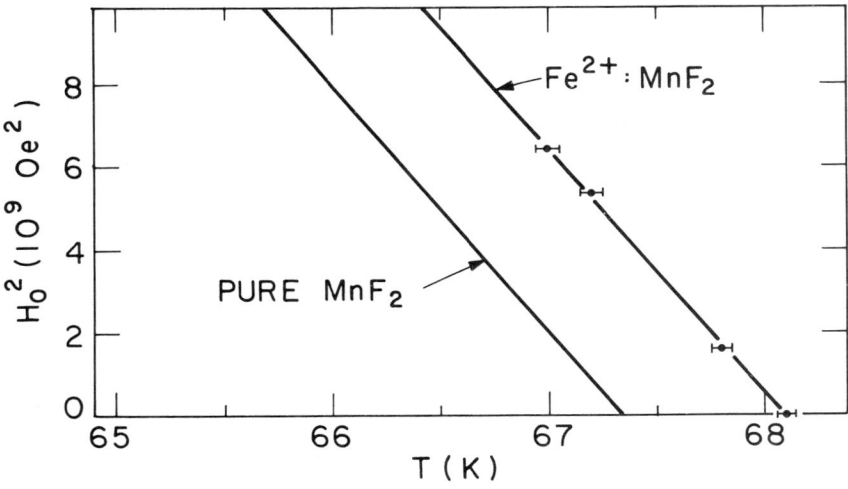

Fig. 4. H^2 plotted as a function of T for MnF_2 and $Fe^{2+}:MnF_2$

Shapira and Foner[1] and Heller[6] showed that the phase boundary in pure MnF_2 is well represented by an equation of the form $T - T_N = AH^2$, where A is a constant. In Fig. 4 we plot T_N determined from our data as a function of H_0^2, including one value obtained by holding T constant and changing H_0. The boundary is seen to vary as H_0^2 as in MnF_2 and with the same slope. This result is predicted by the M.F.A.

1. Y. Shapira and S. Foner, Phys. Rev. B1, 3083 (1970).

2. G.K. Wertheim, H.J. Guggenheim and D.N.E. Buchana, Phys. Rev. Letters 20, 1158 (1968).

3. M.A. Lowe, A. Misetich and C.R. Abeledo, Journal de Physique Colloque Cl 32, 1068 (1971).

4. C.R. Abeledo, R.B. Frankel, A. Misetich and N.A. Blum, J. Appl. Phys. 42, 1723 (1971).

5. L.J. Neuringer, A.J. Perlman, L.G. Rubin and Y. Shapira, Rev. Sci. Inst. 42, 9 (1971).

6. P. Heller, Phys. Rev. 146, 403 (1966).

7. R.P. Singh and G. Hoy, private communication.

RARE EARTH INTERCALATION AND MAGNETIC PROPERTIES OF

LAYER TYPE COMPOUNDS

G. V. Subba Rao, M. W. Shafer[*] and L. Tao[*]
IBM Research Center, Yorktown Heights, N.Y. 10598

ABSTRACT

The rare earth metals Eu and Yb and the alkaline earth Sr have been intercalated in the groups IV and VI dichalcogenides with layer type structures. The intercalations were carried out at low temperatures in liquid ammonia solutions and were shown to be free of the ferromagnetic impurities $Eu(NH_2)_2$ and $Eu(NH_3)_6$. Lattice parameter increases in the c direction of the hexagonal cell, independent of the concentration of the intercalated species, from 18.39 Å for pure 3R MoS_2 to 27.84 for the Eu intercalated material were measured. We find the intercalated species to go between every layer and ammonia is intercalated along with the metals. The composition of MoS_2 fully intercalated with Eu is $MoS_2(Eu)_{.9-1.0}(NH_3)_{1-1.5}$.

MoS_2 orders ferromagnetically at 4-5°K when intercalated with Eu in concentrations greater than $MoS_2(Eu)_{.5-.6}$. The ferromagnetic composition have paramagnetic Curie temperatures of 8-9°k and Curie constants consistent with the europium being all divalent, i.e., C_M = 7.9. When Yb and Sr are intercalated in MoS_2 superconductivity occurs at 2.8 and 5.2°K respectively.

INTRODUCTION

The transition metal dichalcogenides of groups IV, V and VI form layer type compounds where a single layer consists of metals strongly bonded between two sheets of chalcogens[1]. Adjacent layers are weakly bonded to each other, presumably by van der Waals forces, making it possible to place other ions or molecules in the gap between the layers without disturbing the structural arrangement within a given layer. This process is called intercalation. Distinct changes in the physical properties of the host layer material usually occur as a result of intercalation, presumably due to electron exchange between the intercalated species and the host.

There has been some previous work on the intercalation of magnetic ions into the gap of these structures, mainly putting in 3d metals by high temperature reactions[2-5]. In such cases structures with many three dimensional characteristics are formed.

---

*Sponsored in part by the Advanced Research Projects Agency of the Department of Defence under ARPA order No. 1588 and monitored by the U.S. Army Missile Command, Contract No. DAAH01-71-C-1313-P00002.

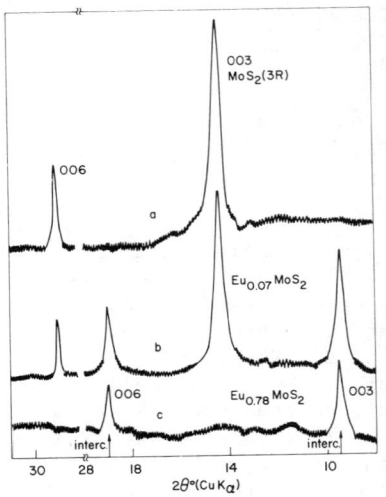

Fig. 1 Diffraction scan of (a) MoS_2(3R) and (b and c) Eu intercalated MoS_2. At low Eu concentrations both intercalated and pure phases are present (b and c).

Rudorff,[6,7] however, showed that europium, since it is soluble in liquid ammonia, can be intercalated at low temperatures in a manner similar to the alkali metals. Further, from susceptibility measurements he showed the intercalated europium to be in the divalent state.

In this paper we present further data on the intercalation and properties of europium in MoS_2, we show that ytterbium and strontium can also be intercalated, and we compare the magnetic properties of the three intercalated systems.

EXPERIMENTAL

The layer dichalcogenides were prepared as powders by the direct reaction of the elements in sealed silica tubes. The resulting powders were used as source material for crystal growth runs by the chemical vapor transport technique. Iodine was used as the carrier in most runs. The 3R polytype was obtained in all cases. The intercalations were carried out by reacting either the powders or crystal with liquid ammonia solutions of the metals i.e., Eu, Yb, or Sr. The solutions had the deep blue color characteristic of free solvated electrons in ammonia and the intercalation process was considered complete when the solution became colorless. The rate of intercalation, as determined by the time required for the solution to become colorless, depended on the usual parameters such as temperature, concentration and particle size of the dichalcogenides. However, the primary rate determining parameter for this experiment was the particle size of the material being intercalated. For example, at -50°C small crystals were not completely intercalated after 16 weeks in solution whereas fine powders ($\sim$ 3000 Å) showed complete intercalation after several days. Temperature had a lesser but still significant effect on the reaction rate. We were restricted in the temperature range in which the intercalation could be carried out because at higher temperatures i.e., room temperature, there was a tendency for the formation of europium amide $(Eu(NH_2)_2$,[8] - a ferromagnet which we could not tolerate as an impurity. Therefore, all reactions were carried out at low temperatures (-70 to -50°C) where amide formation did not occur. After the intercalation was complete the samples were washed in freshly distilled NH_3, dried in high vacuum, and then transferred to a dry box where they were removed from the reaction tube and mounted for the various examinations.

All samples were analyzed chemically for both europium and

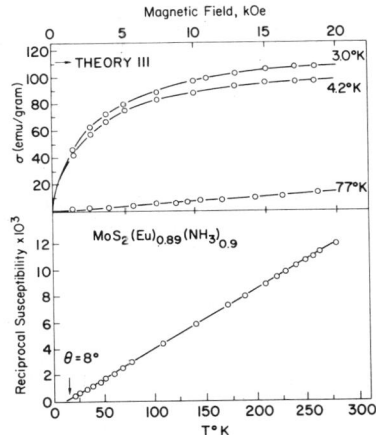

Fig. 2. Magnetic data on $MoS_2(Eu)_{.89}(NH_3)_{.9}$.

ammonia. The europium was determined by EDTA to an accuracy of ± 1% while the ammonia was done colorimetrically using Nesslers reagent. The magnetic measurements were made with a force magnetometer.

RESULTS AND DISCUSSION

The results of the intercalation from liquid ammonia solutions are summarized in Table 1. There are several significant points which warrant further emphasis. First; Ammonia is always found intercalated with europium. The ratio of Eu to NH_3 varied between ∼ 0.8 to 2.5 but the scatter in the analyses does not allow us to draw any conclusions as to the composition of the complex which is intercalated. However, in no cases could ammonia be intercalated without europium. Second; the lattice parameters in all cases are identical and independent of the concentration of the intercalated material. This means that when two layers separate to accept the intercalated species, the amount of separation is determined only by the size of the Eu^{++}, or more probably concentrations of the ammoniated europium complex. Since the low concentrations show the same spacing as the highly concentrated ones it is likely that the Eu concentration necessary to completely separate two layers (to 6.13 Å) is very low. For the low concentrations mixtures of intercalated and the pure unintercalated phase are seen in the X-ray diffraction tracings (Fig. 1b). The percent intercalated phase shows a linear increase (as determined by X-ray intensity data) as the concentration of Eu in the ammonia solution is increased and at $MoS_2(Eu)_{.58}$ only the intercalated phase is seen (Fig. 1c). We assume that at this concentration every layer has been separated by small concentrations of ammoniated europium's and further increases in the Eu content can be accounted for by the Eu concentration being increased between the previously separated layers. The lattice parameter increase in the c dimension of 9.45 Å is consistent with an intercalated species being between every layer of the 3R polytype.

Attempts were made to intercalate Eu and Sr into MoS_2 by solid state reactions at 400-800°C by reacting the metals with MoS_2 in evacuated silica tubes. Negligible amounts of Eu and about 25% Sr were intercalated. The lattice constants of these compositions were $a_0 = 3.20$ Å and $c = 3 \times 7.19$ Å. However, in all cases the corresponding metal sulfides were formed.

It is also evident from Table I that the maximum concentration of intercalated europium in MoS_2 corresponds to a composition of approximately $MoS_2(Eu)_{.95}$; to obtain this composition it was necessary to have a starting composition with a $MoS_2/Eu = .75$. In view of the fact that solutions containing less than this starting concentration always become decolorized while those containing more remained blue, it is difficult to explain why the final intercalated product was found to contain approximately only 70-80% of the added europium.

Although some of the compositions given in table I show the presence of trivalent europium, we have data to indicate this is a result of decomposition during handling and measuring. However, it is evident that europium can be intercalated to a composition of $MoS_2(Eu)\sim 1$ and that it is intercalated in the divalent states.

The magnetic data for a typical highly intercalated (with Eu) sample are shown in Fig. 2. Ferromagnetic order appears to occur at low temperatures. From paramagnetic susceptibility values in the temperature range 77-298°K a θ of 8-9°K and a molar Curie constant of 7.90 is obtained- in excellent agreement with the expected value of 7.88 for Eu^{++} with a spin value of 7/2. A ferromagnetic Curie temperature, T_c, of 4-5°K was obtained from initial permeability measurements. Several low temperature magnetization curves for this sample $MoS_2(Eu)_{.89}$ are given in Fig. 2a and it is seen that the functional dependence of the amount versus field shows some characteristics of ferromagnetic order. However, the fact that complete saturation, as evidence by the slope of the M-H curve, does not occur up to ~ 15 kOe indicates either a complex spin structure or the presence of a paramagnetic phase. If the latter were the case the data indicate it should be present in the 20-25% range - a fact which is inconsistent with the X-ray and the susceptibility data. The presence of $Eu(NH_2)_2$[8] or $Eu(NH_3)_6$[9] as ferromagnetic impurities is also ruled out by the X-ray measurements. A spin arrangement in which the europiums are coupled ferromagnetically within a sheet and adjacent sheets being coupled antiferromagnetically could account for the shape of the magnetization curve. In such an arrangement the dipolar exchange forces coupling adjacent layers are assumed to be weaker than the intra layer exchange so that only moderate fields i.e. 15-20,000 Oe are required to flip the spins to a parallel arrangement, as is seen in the magnetization curve at $\sim$ 15,000 Oe. The fact that the saturation moment is within a few per cent of theoretical for this composition (104 vs 111 emu's) shows that all Eu spins are ferromagnetically aligned.

Those samples intercalated with europium concentrations less than about $MoS_2(Eu)_{.4}$ have either zero or slightly negative paramagnetic Curie temperatures. The M versus H curve for the $MoS_2(Eu)_{.38}$ sample shows some characteristics of ferromagnetism at 2°-4.2°K but since this sample was a mixture of intercalated and pure MoS_2 we assume it contained small regions with sufficiently high Eu concentration for magnetic order to occur. It appears, that for homogenous samples, a europium concentration in the 0.4 - 0.5 Eu to one MoS_2 is necessary for ferromagnetism.

Strontium and ytterbium intercalated samples with increased lattice parameters identical to the europium ones were also prepared. No evidence of ferromagnetism was observed in these, which is good evidence showing the origin of the magnetism in the Eu sample is in the Eu layer and not the molybdenum. In fact both the Sr and Yb samples became superconducting at 5.2 and 3°K respectively. This is presumed to be a result of electrons from the intercalated species being transferred to the d band of the MoS_2.

ACKNOWLEDGMENTS

We wish to acknowledge the technical assistance of R. A. Figat and H. Lilienthal. We thank B. Olsen for the chemical analyses and A. Toxen and J. De Luca for the superconductivity measurements.

TABLE I

Starting Composition Moles MoS_2:T	Final Composition (from Analyses)	Lattice Parameter* c_o, Å	Paramagnetic Curie temperature θ, °K	Molar Curie Constant C_M
1:0	MoS_2	3 x 6.13		
1:0.2 Eu	$MoS_2(Eu)_{0.07}(NH_3)_{0.15}$	3 x 9.28	0 ± 5†	3.1
1:0.5 Eu	$MoS_2(Eu)_{0.38}(NH_3)_{0.29}$	"	0 ± 3	8.1
1:0.75 Eu	$MoS_2(Eu)_{.58}(NH_3)_{1.3}$	"	4 ± 4†	7.1
1:1.0 Eu	$MoS_2(Eu)_{.74}(NH_3)_{2.1}$	"	-†	6.0
1:1.0 Eu	$MoS_2(Eu)_{.78}(NH_3)_{1.7}$	"	9 ± 2	6.9
1:1.2 Eu	$MoS_2(Eu)_{.89}(NH_3)_{.9}$	"	8.8 ± 1	7.9
1:0.2 Yb	$MoS_2(Yb)_{0.1}(NH_3)_{.16}$	3 x 9.21	-	-
1:0.2 Sr	$MoS_2(Sr)_{.11}(NH_3)_{.4}$	3 x 9.28	-	-

*The a_o lattice parameter for all intercalated materials is 3.20 Å as compared with 3.16 Å for pure MoS_2. There were no superlattice lines observed which could be used to indicate the position of the intercalated ions.

†The presence of trivalent europium in the samples decreases the accuracy of the extrapolation from high temperature of the $1/\chi$ versus T curve because of the deviations from a straight line.

REFERENCES

1. See recent review by J. A. Wilson and A. O. Yoffe, Advan. Phys. 18, 193 (1969).
2. F. Hulligar and E. Pobitschka, J. Solid State Chem. 1, 117-119 (1970).
3. J. M. Voorhoeve-van den Berg and M. Robbins, J. Solid State Chem. 1, 134-137 (1970).
4. J. M. Voorhoeve-van den Berg and R. C. Sherwood, J. Phys. Chem. Solids 32, 167-173 (1971).
5. K. G. Verhoeven, thesis Groningen 1971.
6. W. Rüdorff, Chimia 19, 489-499 (1965).
7. W. Rüdorff and W. Ostertag, Proc. IV Rare Earth Res. Conf. Phoenix, Arizona (1965).
8. F. Hulliger, Solid State Comm. 8, 1477-1478 (1970).
9. H. Oestereicher, N. Mammano and M. J. Sienko, J. Solid State Chem. 1, 10-18 (1969).

1178 Section 36.   Lunar Magnetism

MAGNETISM AND THE HISTORY OF THE MOON

D. W. Strangway
Physics Branch, NASA, Manned Spacecraft Center, Houston, TX 77058

W. A. Gose
Lunar Science Institute, Houston, TX 77058

G. W. Pearce
Lunar Science Institute, Houston, TX 77058 and
Dept. of Physics, University of Toronto, Canada

J. G. Carnes
Lockheed Electronics Company, Houston, TX 77058

ABSTRACT

All lunar samples measured to date contain a weak but stable remanent magnetization of lunar origin. The magnetization is carried by metallic iron and is considered to be caused by cooling from above the Curie point in the presence of a magnetic field. Although at present the moon does not have a global field, the remanent magnetization of the rock samples and the presence of magnetic anomalies, both on the near and far side of the moon, imply that the moon experienced a magnetic field during some portion of its history. The field could have been generated in a liquid iron core sustaining a self-exciting dynamo, but there are some basic thermal and geochemical objections that need to be resolved.

INTRODUCTION

Only a very small number of planetary bodies have been found to have a significant magnetic field. Among these are the Earth, Jupiter and one of the satellites of Jupiter. It was therefore a surprise to most investigators to find that returned lunar samples carried a weak but quite stable remanent magnetization, in particular since early probes sent to the moon could not detect any magnetic field. Yet after examining lunar samples from five different landing sites we still have not found a sample which does not carry a measurable remanent magnetization. Furthermore, surface magnetometers revealed local magnetic fields at the Apollo 12, 14, 15 and 16 landing sites as high as 300 gammas[1] (1 gamma = 10^{-5}Oe), small values compared to the Earth's magnetic field, but nevertheless, distinctly measurable. The Apollo 15 and 16 subsatellites, orbiting the moon at a nominal height of 110 km, detected sharp magnetic anomalies up to several gammas both on the front side and on the far side of the moon. These anomalies are apparently associated with topographic and/or geologic features[2]. Reexamination of the Explorer 35 data yielded a number of magnetic irregularities where the satellite crosses the magnetic wake of the moon[3]. Coleman et al[1] noted the presence of these limb shock features from examination of Apollo 15 subsatellite magnetometer results. These features have been traced to very local magnetic anomalies on the limb of the moon. There seems to be a tendency for these anomalies to be clustered in the highlands and at low latitudes.

The remanent magnetization of the lunar samples and the presence of magnetic anomalies imply that large parts of the lunar crust must be magnetized, since the moon presently does not have a measurable global magnetic field. The data from Explorer 35 show that if the moon were to have a dipole field, the dipole moment must be less than 10^{20} emu, almost six orders of magnitude less than the dipole moment of the earth.

It is the purpose of this paper to discuss the magnetic properties of the lunar samples and the implications the magnetic data have on lunar geological processes. The last chapter will describe some thoughts about the origin of the ancient magnetic field preserved in the lunar rocks.

MAGNETIC MINERALOGY

In order to identify the ferromagnetic phases present in the lunar samples we have measured the high field magnetization (J) (3 to 10K oersted) as a function of temperature (T) in the range $4°K$ to $1100°K$. Figure 1 summarizes the full range of J - T curves above room temperature which so far have been recognized in lunar samples. In the simplest case a single Curie point of about 770°C is observed (Fig. 1a) indicating essentially pure metallic iron. The presence of metallic iron makes it necessary to heat the samples in a vacuum of better than 10^{-6} torr in order to prevent oxidation, a precaution that has proven vital for any experiment involving heating.

At the other extreme we find a curve such as that shown in Figure 1b which is the result of measuring a single grain from one of the Apollo 11 breccias. (A breccia is a rock composed of irregular fragments of various rock types held together by a fine grained soil matrix.) This grain undergoes a transition from the α phase to the γ phase at about 750°C on heating and back again at about 500-600°C on cooling. This behavior is typical of iron alloyed with a few percent of nickel. In fact, from the two temperatures indicated above, the nickel content can be estimated at about 5-8% since the nickel content controls the temperature of the α-γ phase transition. For iron with less than 20% Ni this phase transition is also a magnetic transition. This composition is typical of kamacite, a common component of meteorites, but it has also been identified in lunar igneous rocks[4].

In most of the soils and breccias which we have examined, we have seen very little of this reversible phase transition and we are led to conclude that only a fraction of the iron in any of the soils is of extralunar origin. A typical soil showing some of the effect is seen in Figure 1c and a typical soil without the effect is shown in Figure 1d.

In some of the soils and breccias there seems to be yet a third phase identifiable at the high temperatures. Both on heating and

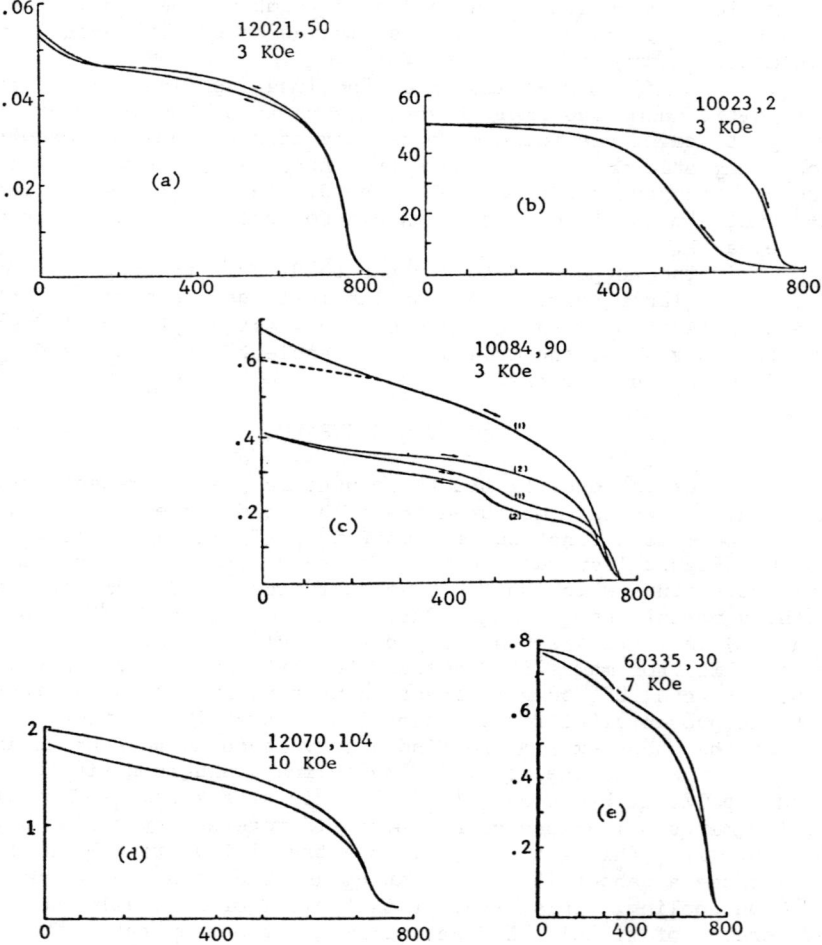

Fig. 1. Thermomagnetic curves of several lunar samples. The horizontal axis is in degrees centigrade, the vertical axis is the magnetization in emu/gm. Numbers in figures indicate the sample number and the applied field value.

cooling a small knee is seen at around 300°C (Figure 1e). We presume that this is representative of an iron phase with about 30-40% nickel which is typical of taenite, a mineral frequently found in meteorites, but taenite has also been identified in a few lunar igneous rocks[4].

Troilite (FeS) is present in many samples in about the same quantity as iron. Although troilite is antiferromagnetic with a Néel point of about 320°C, we have not observed any effect of this phase but there is one report of a possible contribution to the rotational hysteresis[19]. The lack of a magnetic signature suggests that it is almost perfectly stoichiometric since any deviation from this would lead to a small ferrimagnetic effect, a finding in agreement with microprobe analyses[48].

The most abundant phase of magnetic interest in the low temperature range is ilmenite, a significant constituent of all lunar samples. Pure stoichiometric ilmenite is antiferromagnetic with a Néel point of 57°K. The low temperature curve seen in Figure 2 clearly shows this Néel point and confirms that the ilmenite is free of ferric iron. This has also been reported by Nagata et al[5] and by Muir et al[6], the latter using Mössbauer techniques.

In addition to this there is a hint of a Néel point at around 40°K such as would be typical of pyroxene, one of the major rock-forming minerals[5]. This effect is strongly masked by the paramagnetic 1/T increase at low temperatures. There are several other phases which are known to be present in the lunar samples which have not yet been positively identified magnetically but some suggestion for their presence has been reported by Banerjee[49]. These include ulvöspinel (antiferromagnetic at 120°K) and chrome spinel (ferrimagnetic at 88°K) and their solid solution series. Undoubtedly, future studies in this area will be of importance and may give useful information on the stoichiometry of these materials.

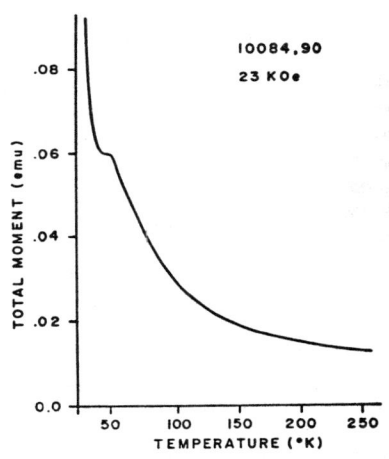

Fig. 2. Low temperature thermomagnetic curve of a soil sample.

ROOM TEMPERATURE HYSTERESIS MEASUREMENTS

The measurement of hysteresis loops is a most suitable technique to determine the concentrations of magnetic and paramagnetic components as well as the distribution of grain sizes and grain shapes of the ferromagnetic particles. The determination of the metallic iron content proved to be particularly valuable because the iron content is so small and because much of it is too fine grained to be detected by other techniques.

Magnetically determined iron concentrations are shown in Figure 3 based on our data and on data from the literature[5,7-10]. In the case of duplicate analyses the mean value has been plotted. The native iron concentration in igneous rocks is typically less than 0.1 wt% whereas the soils and breccias contain about 0.3 to 0.7 wt% Fe.

This fivefold enrichment is rather surprising and suggests either that the soils are not derived from the igneous rocks present but from some unknown parent rock or that excess iron has been added to the soil, either by the in-fall of meteoritic particles, or by some other mechanism operating on the moon. Since it is likely that we have already sampled most, if not all major rock types[11], our attention has been concentrated on the ways in which excess iron could be added to the soils. There have been only a few reports of meteoritic particles in the soil and we will shortly present magnetic evidence that also suggests that there is very little meteoritic iron in the soil.

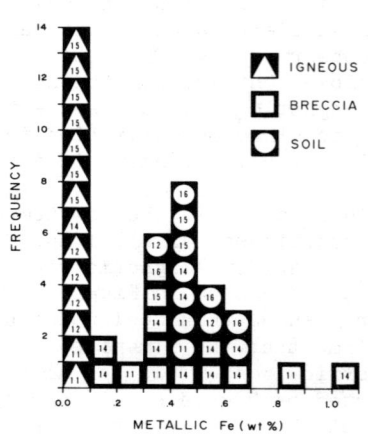

Fig. 3. Magnetically determined metallic iron concentrations. Numbers inside symbols refer to the mission number.

A series of experiments were therefore performed using simulated Apollo soils in an attempt to determine whether iron could be generated by subsolidus reduction. Pearce et al.[12] showed that simulated lunar glasses heated under reducing conditions would generate metallic iron from the iron silicates originally present. This suggests that samples which are reheated to high temperatures and then chilled as in the case of a major impact or lunar ash flow would generate excess metallic iron. The temperature at which this occurs is restricted to a range of about 700°C or greater for the range of times available in the laboratory. It would appear that successive impacts or ash flow conditions could generate glass from crystalline rocks, which on cooling would account for the excess iron seen in the soils and in the breccias which are derived from the soils.

The second surprising conclusion which can be drawn from Figure 3 is the uniformity of the metallic iron content among the soils from all landing sites, in spite of the variable nature of the igneous rocks. In light of the previously discussed mechanism for iron production it appears that the metallic iron content in the soils and breccias is more a function of the thermal history of the regolith than of the precise nature of the source rocks.

In addition to the difference in the quantity of iron present, the soils and breccias and the igneous samples also show great differences in the grain size of the iron. The ratio of saturation remanent magnetization to saturation magnetization, J_{rs}/J_s, can be used to give a rough estimate of how they differ. Iron grains that are large enough to have some sort of multidomain or other non-uniform magnetization configuration (diameter greater than about 300Å for iron) have a very small saturation remanence or J_{rs}/J_s (<.001). For thermally stable single domain particles (diameters roughly between 150Å and 300Å J_{rs}/J_s is .5 - 1.0. Particles less than 150Å in diameter are superparamagnetic and for these $J_{rs}/J_s = 0$. Thus J_{rs}/J_s, which is presented in histogram form in Figure 4 for lunar samples measured in our laboratory, gives a rough estimate of the fraction of the total metallic iron which is not multidomain or superparamagnetic. In the figure it can be seen that the soils and low grade breccias have much iron in this relatively narrow grain size range, whereas the igneous samples and the higher grade breccias have little.

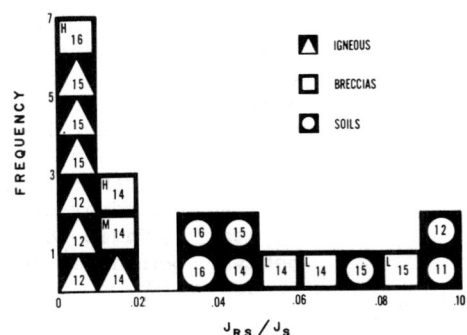

Fig. 4. Histogram of the ratio of saturation remanent magnetization J_{rs} to saturation magnetization J_s. Letters in upper left corners of breccia squares refer to approximate metamorphic grade of sample (L - low, M - medium, H - high).

TIME DEPENDENT MAGNETIZATION

Much additional information about the iron grain size distribution can be obtained by measuring the time dependence of the remanent magnetization. The experiment consists of applying a magnetic field of a few oersteds to the sample for a certain length of time and then monitoring the remanence in a zero field. This study proved to be particularly fascinating because the lunar samples contain iron particles in the size range from 20Å to about 100 microns[13], i.e., particles from the superparamagnetic range up to very large multi-domain grains.

Lunar igneous rocks in general acquire only a very weak time dependent magnetization which indicates that these rocks do not contain significant amounts of either the very small or very large iron grains. The latter statement has been verified by direct microscopic observations (to the limit of 1μm).

In the breccias, on the other hand, we found a time dependent magnetization which depends on their metamorphic grade as they change from "clods" to highly welded, recrystallized rocks[14].

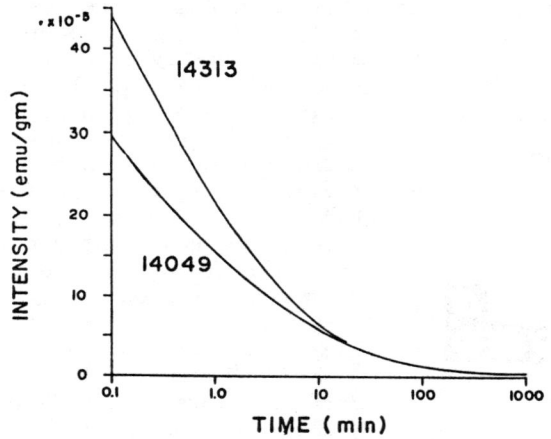

Fig. 5. Time dependent magnetization of two low metamorphic grade breccias. A 2.5 oersted field was applied for 8 minutes.

The viscous magnetization of the low grade breccias (Fig. 5) results from superparamagnetic particles. The roll-off at larger times is due to the transition from superparamagnetic to stable single domain behavior which for iron occurs at about 150Å.

The medium grade breccias exhibit a small log t dependence (Fig. 6) indicating a range of relaxation times well exceeding the experimental duration. When compared to their saturation magnetization, the viscous effect in the low grade breccias is at least a hundred times stronger than in the medium and high grade breccias. This behavior combined with a very stable remanence is expected for a grain size distribution in the range from 200Å up to about 1 micron.

A log t relationship is also found for the high grade breccias but these rocks are unstable against alternating field demagnetization. It is inferred that they contain abundant multi-domain grains above 1 micron in size.

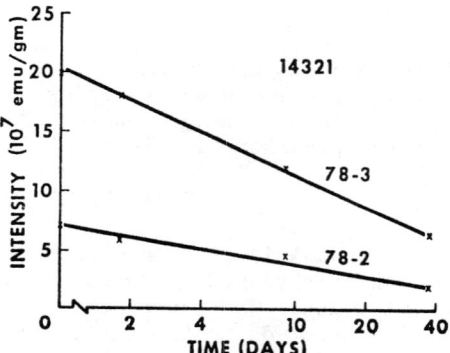

Fig. 6. Time dependent magnetization of two specimen from a medium metamorphic grade sample. The sample was exposed to the ambient magnetic field for an unknown length of time.

These inferences are completely consistent with the evidence discussed earlier for breccia formation. The spectrum of relaxation times suggests that the higher grade breccias were formed at higher temperatures so that the iron particles could grow in size, but even the low grade breccias were formed at temperatures of about 700°C or more in order that they carry a stable remanence. There are many other lines of evidence that the breccias were formed at temperatures of 700°C or more (see for example Williams[15]).

NATURE AND ORIGIN OF THE REMANENT MAGNETIZATION

For the purpose of this discussion we may classify the lunar rocks into two troups, the igneous rocks and the breccias. To date, most igneous rocks studied are mare basalts. Only one anorthositic gabbro, a rock type considered typical of the lunar highlands, has been investigated. It is expected that the Apollo 16 igneous rocks will be mainly of this type[16].

As received, the mare basalts have a remanent magnetization of the order of 10^{-6} to 10^{-4} emu/g (Fig. 7), a range reported by a number of investigators[5, 7-10, 17-21]. Upon alternating field demagnetization, however, a significant portion of the natural remanence is removed in fields typically less than 50 Oe and a second component with an intensity around 10^{-6} emu/g (Fig. 7) can be seen which can scarcely be changed in fields up to 400 oersted (Fig. 8). For a variety of reasons we believe that much of the soft component is not of lunar origin. In several cases, the directions of individual chips from the same sample were quite scattered in direction before demagnetization but after cleaning in alternating fields (AF), the directions of the magnetization were well grouped (Fig. 9). This is a strong indication that a soft random component of magnetization has been added to the samples by fields of 10 or 20 oersteds. Secondly, a sample from one of the Apollo 12 rocks was cleaned in fields of several hundred oersted and returned to the moon on Apollo 16. It remained in the Lunar Module on the outward trip and returned in the Command Module. On return it had acquired a soft magnetization which could be removed in a field of about 20 oersteds. This implied that the sample had been exposed to a DC field of 10 - 20 oersteds somewhere in the spacecraft[22]. This

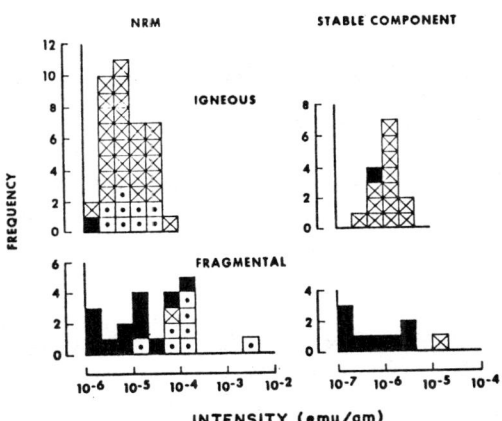

Fig. 7. Histogram of the intensity of magnetization of igneous and fragmental rocks. Dots are data from Apollo 11 samples, crosses from Apollo 12, solid symbols from Apollo 14 samples.

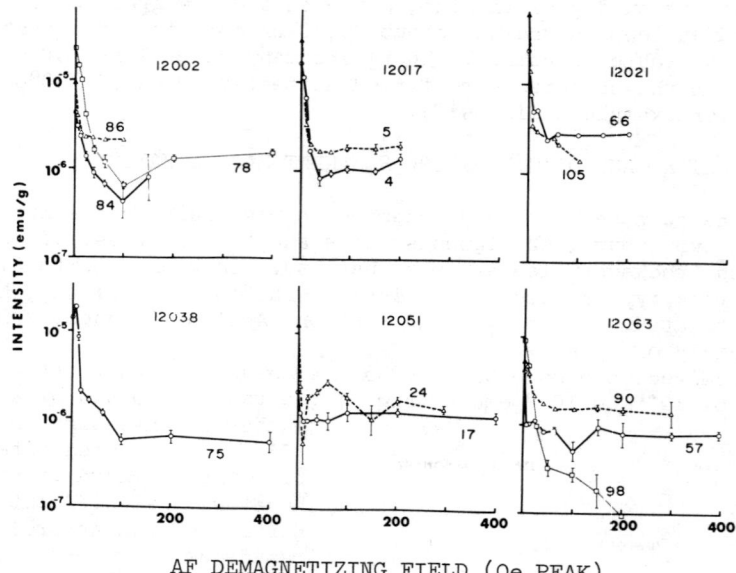

Fig. 8. Alternating field demagnetization of several igneous rocks.

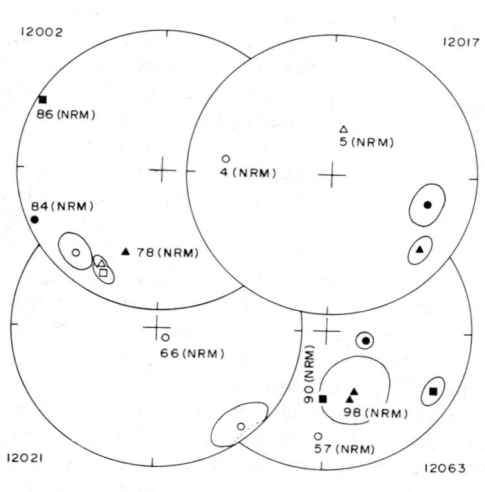

Fig. 9. Change in direction upon A.F. demagnetization of individual specimen from the same sample. Equal area projection.

effect can, moreover, be duplicated both on this sample and on many other samples by exposing them to fields of 10 - 50 oersteds (Fig. 10). We, therefore conclude that a significant portion (but not necessarily all) of the soft magnetization found in lunar samples is an isothermal remanent magnetization (IRM) of man-made origin.

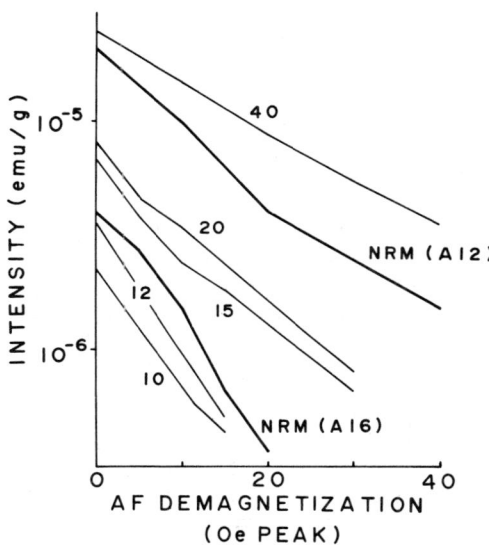

Fig. 10. A.F. demagnetization of sample 12002,78 after return from the Apollo 12 mission and after its journey aboard the Apollo 16 mission and of magnetizations acquired in fields of 10, 12, 15, 20 and 40 oersteds.

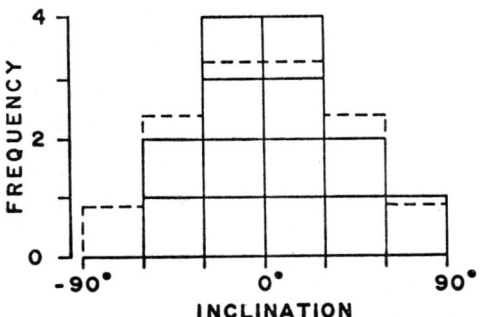

Fig. 11. Distribution of magnetic inclinations of oriented lunar samples Dashed line indicates theoretical curve for random distribution.

By contrast, the stable component of the natural remanent magnetization seems to be of definite lunar origin. For several samples it has been possible to obtain a lunar surface orientation either by γ-ray spectroscopy[23] or by micrometeorite pit counting[24]. Although these techniques only establish which side of the sample was up and which side was down, they make it possible to determine the magnetic inclination with respect to the lunar surface. If the samples were magnetized after they attained the position in which the astronauts found them, they should all be magnetized in the same direction. Alternatively, if their magnetization predates their last tumbling, their directions of magnetization should be random. The distribution of inclinations shown in Figure 11 is indeed the distribution expected for random orientation. Thus the samples are stable on a geological timescale as well.

1188

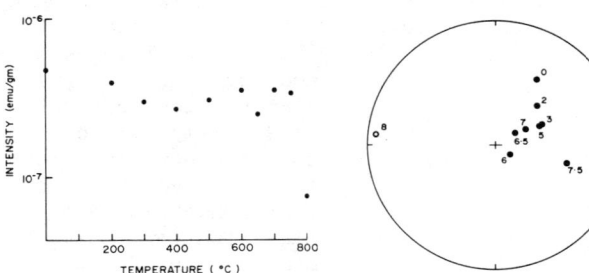

Fig. 12a. Thermal demagnetization of sample 12002.

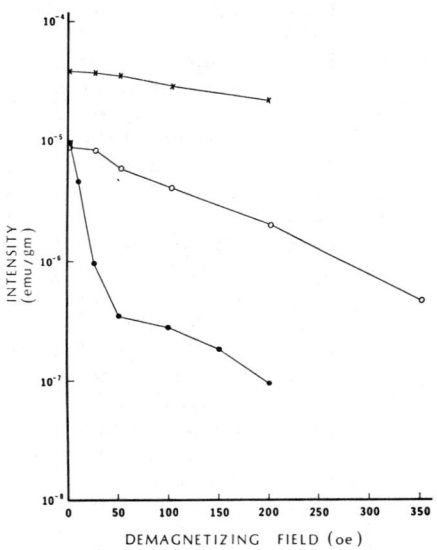

Fig. 12b. A.F. demagnetization of NRM of 12063,98 (●) and AFD of a thermoremanent magnetization of 12063,55 acquired in 0.48 Oe (x) and 0.08 Oe (o).

To establish the origin of the stable magnetization we thermally demagnetized a sample from the Apollo 12 mission after removing the soft component by alternating field demagnetization (Fig. 12a). The magnetization drops sharply between 750° and 800°C, a range which includes the Curie point of iron. When cooled from above 800°C in the presence of a weak magnetic field the sample acquires a remanence which, on AF cleaning, behaves much like the original stable remanence (Fig. 12b). The combination of an isothermal remanence and a thermoremanence can almost exactly duplicate the character of the natural remanent magnetization.

It is necessary, though, to consider other possible explanations for the origin of the stable magnetization. Shock effects may be important but in general they only modify the soft component of magnetization[25]. Moreover, the igneous rocks show little evidence of shock and it therefore appears that this mechanism is not likely to be important. Acquisition of a viscous remanent magnetization due to long exposure to the earth's magnetic field does not seem to play a major role as described in an earlier section. The growth of iron particles from the superparamagnetic grain size to the stable single domain size in the presence of a magnetic field could create a remanent magnetization but experimental evidence shows that this only happens at temperatures in excess of 600° to 700°C[12]. Therefore it appears that cooling from above the Curie point in the presence of a magnetic field - a thermoremanent magnetization (TRM) - is the most likely origin of the stable remanence of the igneous rocks.

When considering the breccias, one might expect a rather complex situation because the processes which formed the breccias are quite complex and may, in some cases, involve three or more episodes of breccia formation[26].

All the breccias which have been examined to date show a remanent magnetization which is very similar to that found in the

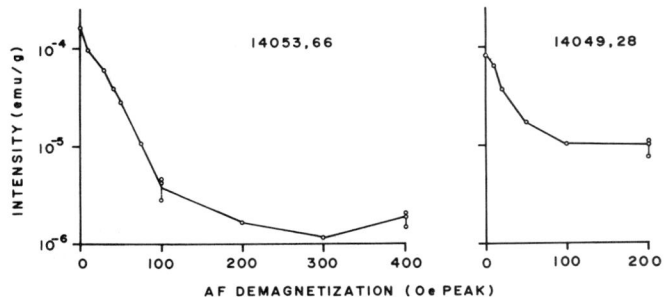

Fig. 13. Alternating field demagnetization of two breccias.

igneous rocks. In most cases they show a soft component which can be removed just as it can from the igneous rocks, and in general they have a component which is very stable. Again we need to consider the origin of this magnetization. The soft component is undoubtedly largely an IRM. The hard component may have, as before, a variety of causes, but it appears that thermoremanence is again the most likely cause. Figure 13 shows typical AF demagnetization curves for several breccias and Figure 14a shows the thermal demagnetization of one breccia sample from Apollo 15. In the same figure we also show the results of a Thellier-Thellier[27] test done to determine the intensity of the ancient field which generated the remanent magnetism. The experiment consists of heating a sample to a certain temperature,

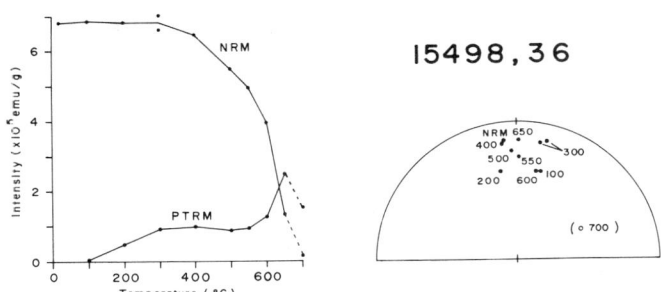

Fig. 14a. Thermal demagnetization of sample 15498,36.

and cooling in a field-free space, to determine how much remanence is lost in that particular temperature interval. It is then reheated to the same temperature and allowed to cool in a known field. The magnetization acquired in that temperature interval is then measured. The plot shown in Figure 14b is a plot of the intensity lost, versus the intensity gained in successive temperature intervals. A straight line plot can be used to derive the intensity of the ancient field. At low temperatures almost no remanence is lost so that there is no real information, but in the range from 500°C to 650°C there is a good straight-line relationship which suggests an ancient field of 2100 gammas. Above that temperature the plot deviates from a straight line probably because in spite of the good vacuum used ($<2 \times 10^{-6}$ Torr) some modification was made to the iron in the sample.

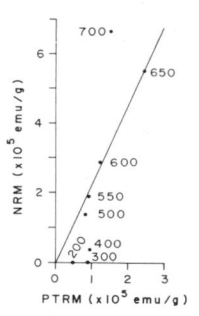

Fig. 14b. Results of the Thellier-Thellier test of sample 15498,36.

In any case, this type of evidence is strong support for thermoremanent magnetization in the lunar samples acquired by cooling from above 800°C in the presence of a weak field of a few thousand gammas.

IMPLICATIONS OF MAGNETISM TO LUNAR GEOLOGICAL PROCESSES

a) Formation of Lunar Soils and Breccias - Having reviewed the pertinent magnetic data some general conclusions can be drawn concerning geological processes on the moon. The fact that all breccias carry a stable remanent magnetization implies that their formation involved heating to about 700 or 800°C. This has been quite a surprising discovery and implies that in many ways the generation of breccias is analogous to the process which generates ash flows on earth[28]. The presence of excess iron in almost equal amounts in the soils and breccias also implies that the process creating these is a sort of homogenizing process operating over much of the lunar surface. Because the grain size of the iron increases with degree of welding, the higher grade breccias undoubtedly formed at higher temperatures. The implication, of course, is that older and more mature soils will have larger particles and a greater metallic iron content than younger soils.

The fact that only a small amount of clearly meteoritic iron-nickel is found, implies that only a very small fraction of the lunar soil is of extralunar origin.

b) Evidence of Wide-Spread Ash Flows - It is now known that magnetic anomalies are quite prevalent on the lunar surface both from surface and from subsatellite magnetometers. There is a variety of possible causes for these, but far the most probable is that the anomalies are caused by the remanent magnetization found in the samples. Figure 15, for example, shows the calculation of an anomaly for typical crater-sized features expected at the lunar

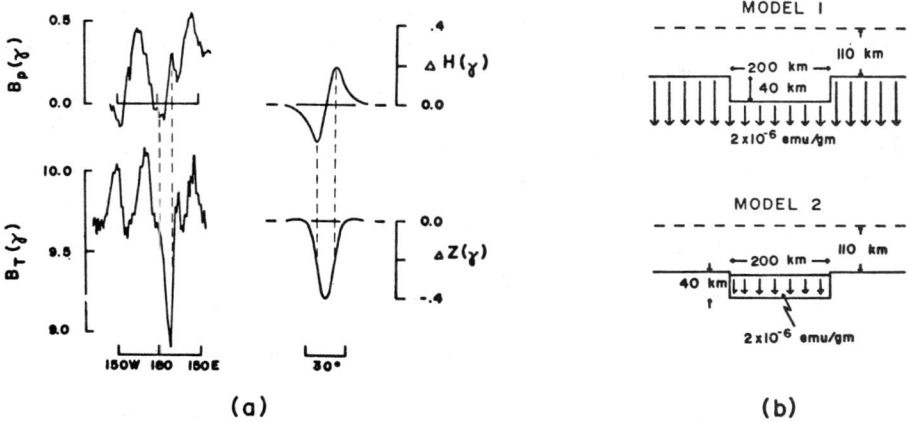

Fig. 15. Comparison of observed and computed magnetic anomaly on the lunar far side at a height of 110 km.

subsatellite height using the measured properties of the returned samples. The computed anomaly of about 0.5 to 1 gamma is just the value typical of observed magnetic anomalies, especially on the far-side of the moon, where individual features can be more clearly recognized.

At the Apollo 16 landing site, large anomalies up to 300 gammas in size were measured. These anomalies were concentrated as shown in Figure 16 - mostly down in the Cayley Plains area and up on Stone Mountain[1].

These anomalies are not random and variable, rather, they tend to have a simple pattern. The simplest explanation is that the rocks beneath the landing site are uniformly magnetized.

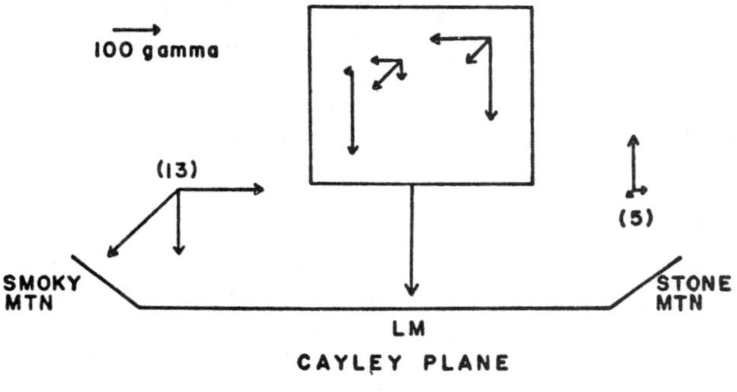

This finding carries an important implication with it. Since the Apollo 16 site is characterized by an almost monotonous series of breccias, we conclude that the breccias were emplaced at high temperatures and cooled in the presence of a magnetic field. Subsequently, surface working and some gardening took place in the uppermost few meters of the moon, but by and large a thickness of

Fig. 16. Magnetic anomalies measured at the Apollo 16 landing site.

several hundred meters of breccias is undisturbed beneath the landing site. Whether this is ultimately of impact or volcanic origin we cannot say - only that the material when it was deposited at the site was at a high temperature.

c) Environment of Magmatic Activity - The geochemists have been able to say a great deal about the partial pressure of oxygen under which the igneous processes on the moon took place. In general this is an environment low in oxygen and water. The magnetic property studies are only supportive in this context, but they confirm all other lines of evidence. The presence of metallic iron, the lack of Fe^{+++} in the ilmenite and in the ulvöspinel (it would be ferrimagnetic if not pure) and the complete lack of magnetite (with one possible exception[21]) all support the very highly reducing conditions.

THE ORIGIN OF THE MAGNETIC FIELD

One of the most intriguing aspects of the remanent magnetization preserved in the lunar rocks and reflected by the presence of the magnetic anomalies is the consideration of the origin of the magnetic field. At this time there have been extensive isotope studies done on the returned samples and it is becoming clear that all the recognizable igneous activity on the moon was restricted to the time period between 3.1 b.y. and 4.0 b.y.[29]. Considering that the moon has evolved as a planet, this rather restricted time span is a most important parameter in lunar evolution. It has also become clear that much of the impacting on the moon took place early in its history, since surfaces which contain rocks about 3.2 b.y. old are relatively fresh and uncratered[30]. From a study of the returned samples, the main igneous rocks types are mare basalts and anorthosites which are both believed to be derived from wide scale differentiation of at least the outer part of the planet[31]. In addition, the seismic experiment has shown the presence of distinct velocity contrasts at 25 and 65 km[32], below the surface. The evidence is mounting that the moon has undergone extensive differentiation.

Studies of the magnetic transient response of the moon[33,34,35] imply that the present day lunar interior temperature is perhaps as low as 800 - 1000°C. The question facing us is whether the moon could have had a fluid core early in its history which could act as a self-generating dynamo. This is a complicated problem since it requires that the moon be completely differentiated early in its history, in order to have a metallic fluid core. There are three major objections to this idea:

1. This means that the moon was hot early in its history. In this case it is difficult to make models which allow the whole moon to cool down to its present inferred, interior temperature[36,37]. The heat balance is not as severe a problem if the moon had an iron-nickel-sulfur core rather than a pure iron core[38] because the addition of sulfur and nickel lowers the melting temperature of iron from about 1700°C to as low as 940°C.

2. The mean moment of inertia implies that the moon is relatively homogeneous, and it is difficult to imagine a process that

would involve at least partial melting of the whole moon, in which the density did not increase radially with depth. Recently Gast and Guili[39] have shown that density reversals in the outer part of the moon are required to offset the influence of the high density basalts present in the mare regions. They picture these as being restricted to shallow depths however and their lunar models imply a rather uniform density-depth relation for much of the moon.

3. The moon has some inhomogeneities, as revealed from a study of its gravity field, its shape, and its moment of inertia differences. There are local features such as gravity anomalies (mascons) which require that at least the outer part of the moon has been relatively cold for the last 3 b.y.[40]. There are, however, much larger scale features represented by low order harmonics. For example, the moments of inertia differences suggest deep-seated, lateral inhomogenieties in the moon[41]. The center of figure and center of mass are offset by about two kilometers[42]. These large scale inhomogeneities are believed by some to represent the remnants of accretionary processes in the moon. The moon could not have been totally heated and differentiated enough to create a core and still preserve a relict of its early structure.

The argument about whether the deep interior was hot early in its history can not be settled in this paper, but in the final analysis we must account for the origin of the magnetic field that created the crustal magnetic anomalies and that left an imprint in the stable magnetization found in all samples - igneous and breccias - studied to date. It is tempting to suppose that the moon had a fluid core at least during the period from 4.0 to 3.2 b.y. which is the range of samples studied to date. Levy[43] has argued that the moon could not have a core large enough, and/or conductive enough, and/or spinning fast enough to generate a dynamo. This is however, a controversial point and others feel that it would be possible to have a self-generating dynamo even in a core which was only 0.2 of the moon's radius[44,45]. If this conclusion can be substantiated, it would require that the moon became hot enough before 4.0 b.y. to differentiate totally and create a core and it must have remained hot enough at least until 3.2 b.y. to retain a fluid core.

Other ideas have been proposed for the generation of the ancient magnetic field recorded in the lunar samples, but each of these also have serious drawbacks.

1. The moon was close to the earth and many of the rocks were cooled in the earth's field. This seems unlikely, because the moon would have been only 2 or 3 earth radii from the earth for almost 1 b.y.

2. The sun had a very large field which affected the moon. The required solar field, however, would need to be very much greater than any field ever observed in stars by astronomers.

3. The impact process itself generated an ionized cloud which created a field[46]. It is hard to see how such a process could last long enough for large volumes of rock to cool through the Curie point and how it could generate large volumes of uniformly magnetized crust. A variation on this is the lightning suggestion of Nagata et al[9] with the same objections.

4. There was a primordial field preserved during the accretion of the moon. The igneous activity in the outer part of the moon took place in the presence of this field and subsequently internal heating raised the temperature of the lunar interior above 750 - 800°C, the Curie point of iron and eliminated the early magnetizing field[47]. This is an appealing concept, but it does require a strong magnetic field to be present during the accretion process.

5. Other possibilities that have also been considered are variations on the core theme. It has been suggested[45] that if only the outer part of the moon were differentiated that small pockets of molten iron might collect at depths of 200 or 300 km (the raisin bran moon). Each of these pods might act in turn as sources of dynamo action. Nagata et al[5] have proposed that localized pockets of magma might behave in a similar way. These are certainly possibilities, but at the moment we have no independent evidence for their support.

Suffice it to say that future work on lunar samples will be useful in searching for variations in the strength of the ancient lunar field both in time and space. If any rocks, including breccias or glass samples can be returned that can be unequivocally dated as younger than 3 b.y., it may be possible to extend the study of the history of the lunar field over a longer time span. Continued analysis of the subsatellite magnetometer data may reveal the directions of magnetization associated with observable topographic and/or geologic features and permit a first attempt at determining the configuration of the ancient magnetizing field. Any future missions should concentrate on 1) mapping of magnetic field anomalies from orbit (preferably polar) and 2) on continuous recording of the surface fields in particular over recognized topographic and/or geologic features.

ACKNOWLEDGEMENT

We wish to thank Mr. J. A. Love of Lockheed Electronics Company for his patience in measuring the samples. Portions of this work were supported by the Lunar Science Institute which is operated by the Universities Space Research Association under Contract No. NSR-09-051-001 with the National Aeronautics and Space Administration. This paper constitutes the Lunar Science Institute contribution number 125.

REFERENCES

1. P. Dyal, C. W. Parkin, C. P. Sonett, R. L. DuBois and G. Simmons, Apollo 16 Preliminary Science Report, in press.
2. P. J. Coleman, Jr., B. R. Lichtenstein, C. T. Russell, L. R. Sharp and G. Schubert, Geochim. Cosmochim. Acta, Suppl. 3, 2271 (1972).
3. J. D. Mihalov, C. D. Sonett, J. H. Binsack and M. D. Moutsoulas, Science, 171, 892 (1971).
4. A. M. Reid, C. Meyer, Jr., R. S. Harmon and R. Brett, Earth Planet. Sci. Lett., 9, 1 (1970).

5. T. Nagata, I. Ishikawa, H. Kinoshita, M. Kono, Y. Syono and R. M. Fisher, Geochim. Cosmochim. Acta, Suppl. 1, 2325 (1970)
6. A. H. Muir, Jr., R. M. Housley, R. W. Grant, M. Abdel-Gawad and M. Blander, Science, 167, 688 (1970).
7. C. S. Grommé and R. R. Doell, Geochim. Cosmochim. Acta, Suppl. 2, 2491 (1971).
8. R. B. Hargraves and N. Dorety, Geochim. Cosmochim. Acta, Suppl. 2, 2477 (1971).
9. T. Nagata, R. M. Fisher, F. C. Schwerer, M. D. Fuller and J. R. Dunn, Geochim. Cosmochim. Acta, Suppl. 2, 2461 (1971).
10. T. Nagata, R. M. Fisher, F. C. Schwerer, M. D. Fuller and J. R. Dunn, Geochim. Cosmochim. Acta, Suppl. 3, 2423 (1972).
11. A. M. Reid, J. Warner, W. I. Ridley, D. A. Johnston, R. S. Harmon P. Jakes and R. W. Brown, Geochim. Cosmochim., Acta, Suppl. 3, 363 (1972).
12. G. W. Pearce, R. J. Williams and D. S. McKay, Earth Planet. Sci. Lett., 17, 95 (1972).
13. S. O. Agrell, J. H. Scoon, I. D. Muir, J. V. P. Long, J. D. C. McConnell, and A. Peckett, Geochim. Cosmochim. Acta, Suppl. 1, 93 (1970).
14. W. A. Gose, G. W. Pearce, D. W. Strangway and E. E. Larson, Geochim. Cosmochim. Acta, Suppl. 2, 2387 (1972).
15. R. J. Williams, Earth Planet, Sci. Lett., 16, 250 (1972).
16. Apollo 16 Sample Information Catalog (prepared by P. Butler, Jr.) NASA Manned Spacecraft Center pub. no. MSC 03210 (1972).
17. C. E. Helsley, Geochim Cosmochim. Acta, Suppl. 1, 2213 (1970).
18. C. E. Helsley, Geochim. Cosmochim. Acta, Suppl. 2, 2485 (1971).
19. S. K. Runcorn, D. W. Collinson, W. O'Reilly, M. H. Battey, A. Stephenson, J. M. Jones, A. J. Manson and D. W. Readman, Geochim. Cosmochim. Acta, Suppl. 1, 2369 (1970).
20. S. K. Runcorn, D. W. Collinson, W. O'Reilly, A. Stephenson, M. H. Battey, A. J. Manson and D. W. Readman, Proc. R. Soc. Lond., A. 325, 157 (1971).
21. D. W. Collinson, S. K. Runcorn, A. Stephenson and A. J. Manson, Geochim. Cosmochim. Acta, Suppl. 3, 2343 (1972).
22. G. W. Pearce and D. W. Strangway, Apollo 16 Preliminary Science Report, in press.
23. E. Schonfeld and G. D. O'Kelley, Second Lunar Sci. Conf., Houston, Texas, Abstracts, 125 (1971).
24. F. Hörz, D. A. Morrison and J. B. Hartung, Modern Geol., 3, 93 (1972).
25. R. Hargraves and W. Perkins, J. Geophys. Res. 74, 2576 (1969).
26. R. Grieve, G. McKay, H. Smith and D. Weill, Lunar Science - III, Lunar Science Institute, Houston, Texas, 338 (1972).
27. E. Thellier and O. Thellier, Ann. Geophys., 15, 285 (1959).
28. C. S. Ross and R. L. Smith, U. S. Geol. Survey Professional Paper 366, Gov't Printing Office, Washington, D.C.
29. D. A. Papanastassiou and G. J. Wasserburg, Earth Planet. Sci. Lett. 16, 289 (1972).
30. E. M. Shoemaker, Lunar Science - III, Lunar Science Institute, Houston, Texas, p. 696 (1972).
31. P. W. Gast, The Moon, 5, 121 (1972).

32. M. N. Toksöz, F. Press, K. Anderson, A. Dainty, G. Latham, M. Ewing, J. Dorman, D. Lammlein, Y. Nakamura, G. Sutton and F. Duennebier, The Moon, 4, 490 (1972).
33. C. P. Sonett, B. F. Smith, D. S. Colburn, G. Schubert and K. Schwartz, Geochim. Cosmochim. Acta, Suppl. 3, 2309 (1972).
34. P. Dyal, C. W. Parkin and P. Cassen, Geochim. Cosmochim. Acta, Suppl. 3, 2287 (1972).
35. G. R. Olhoeft, D. W. Strangway, H. Sharp and A. Frisillo, in preparation.
36. M. N. Toksöz, S. C. Solomon, J. W. Minear and D. H. Johnston, The Moon, 4, 190 (1972).
37. R. K. McConnell, Jr. and P. W. Gast, The Moon, 5, 41 (1972).
38. P. R. Brett, Geochim. Cosmochim. Acta, in press.
39. P. W. Gast and R. T. Guili, Earth Planet. Sci. Lett., 16, 299 (1972).
40. J. A. Hamed, The Moon, in press.
41. Z. Kopal, The Moon, 4, 28 (1972).
42. W. R. Wollenhaupt, R. K. Osburn and G. A. Ransford, The Moon, 5, 149 (1972).
43. E. H. Levy, Science, 178, 52 (1972).
44. S. K. Runcorn, Proc. Roy. Soc., A, 296, 270 (1972).
45. G. W. Pearce, D. W. Strangway and W. A. Gose, Geochim. Cosmochim. Acta, Suppl. 3, 2449 (1972).
46. R. Hide, The Moon, 4, 39 (1972).
47. H. C. Urey and S. K. Runcorn, in press.
48. B. J. Skinner, Geochim. Cosmochim. Acta, Suppl. 1, 891 (1970).
49. S. K. Banerjee, Geochim. Cosmochim. Acta, Suppl. 3, 2337 (1972).

Section 37. Exchange in Transition Metals

FERROMAGNETISM AND PHOTOEMISSION IN TRANSITION METALS

Martin C. Gutzwiller
IBM Thomas J. Watson Research Center
Yorktown Heights, New York 10598

ABSTRACT

Among the simple models of ferromagnetism, the Stoner model seems to be the most appropriate for transition metals, because it includes the metallic character from its start and allows a natural explanation of optical and photoemission data. The recent measurements of electron spin polarization in transition metals are described. In order to appreciate the difficulty of their interpretation, the recent success in calculating the photoemission yield in noble metals and its theoretical foundation is discussed. Whereas the uniform electron gas provides the information to construct an effective periodic potential in most other metals, it is necessary to invoke the narrow band (Hubbard) Hamiltonian to handle two related effects with direct bearing on ferromagnetism in transition metals: the simultaneous presence of both localized and itinerant behavior of the electrons, and the correlation in a narrow band. Some of the recent theoretical results in this area are recalled, and their possible relevance to electron spin polarization experiments is discussed.

INTRODUCTION

a) The Uncertainty Gap

The first successful model of ferromagnetism was invented by Pierre Weiss[1] in 1907 when he proposed the existence of magnetic moments which are coupled by an effective molecular field. The origin of these magnetic moments does not matter in the Weiss model. In fact, with our present understanding of solids one has a choice. Each magnetic moment is associated either with a particular atom, or with an electron in a particular Bloch state provided it belongs to a very narrow band.

As soon as we try to refine the Weiss model, however, we cannot avoid choosing between the two interpretations. If our ferromagnet is strongly ionic, or if it is a rare-earth metal, we think of localized atomic moments, and proceed from the Weiss model to the Heisenberg model of ferromagnetism. But in transition metals and their compounds of a more covalent nature, we attribute the magnetic moments to electrons in extended states.

Thus, we arrive almost immediately at the model which was invented by Stoner[2] in 1938 and has been promoted since the late 40's by Wohlfahrth. The energy of the magnetic moments is determined not only by its interaction with an external magnetic and an internal molecular field, but also by the number of states per unit interval of energy, $\nu(E)$. This feature can be explained only if we think of electrons which satisfy the exclusion principle and occupy a certain subset of available one-electron wavefunctions. In the case of a simple, metallic crystal, these wavefunctions are all of the Bloch type, i.e., they are plane waves which extend through the whole sample and are modulated by a function with the lattice periodicity.

This short description is meant to bring out one essential feature of the various models of ferromagnetism: as soon as we try to go beyond the original Weiss model, we have to face the uncertainty principle. The Heisenberg model assumes that the position of each magnetic moment is exactly determined, whereas the Stoner model relies on the wave vector k of each electron being well defined. In either case, we can include further refinements; but they do not seem to get us out of the extremes of uncertainty in which we got trapped.

That is particularly unfortunate for iron whose magnetic properties appear to be a compromise in the sense that some are well understood in terms of localized atomic moments (spin interactions at higher temperatures), and others require electrons in extended states (fractional number of Bohr magnetons per atom, Fermi surface). Our imagination and ingenuity has been quite poor in trying to cope with this intermediate case, to the point where one author[3] diagnoses the electrons in iron as schizophrenic. It is clearly our trouble that we can think only in either one of two extremes.

b) Towards a Better Theory

A lot of work has been done in the last ten years to bridge this gap, although its existence has been discussed occasionally since a much longer time. However, the Stoner model continues to enjoy great popularity, particularly for transition metals, because it is capable of incorporating many experimental results, whereas the more sophisticated theories in the middle range of uncertainty have not led as yet to any practical schemes for predicting such results.

The localized theories are excluded from the discussion because they fail such simple tests as explaining the fractional number of Bohr magnetons per atom. The itinerant theories (starting with the Stoner model) are important from a strictly opportunistic viewpoint because they can be fitted empirically so well. Also, they receive some moral support for their somewhat doubtful fitting procedures from the (sofar only) qualitative conclusions of the intermediate type theories. Without being unduly puritanical, one has to decide to what extent any particular choice of the available parameters constitutes more than a successful fit to the experimental data.

It will be argued in the remainder of this survey that the present energy band picture of transition metals relies on too many ad hoc assumptions. Although they appear to be reasonable, they seem to suffer from an essential defect which may make them useless in the future. An approximate theory may be quite useful in the initial stages of scientific inquiry, because it provides a yardstick against which the experimental results can be evaluated. But our knowledge of transition metals has passed this stage, and an approximate theory is valuable only if it can be improved in some systematic manner.

Such a systematically improvable theory has emerged in recent years to deal with the noble metals. It gives a successful interpretation of all the important experimental data (Fermi surface, optical and photoemission properties, etc.) in terms of independent,

single-electron Bloch states which require only minor
corrections due to many-body effects. The relevant
formulas will be given in section 1 because they also
form a basis for the Stoner model, and ought to be tried
out seriously for the transition metals. Their
theoretical basis will be reviewed in section 3 to show
why the treatment of transition metals requires some
new ingredients which are not needed for the noble
metals.

The photoemission experiments and, particularly,
the polarization measurements of photoemitted electrons
play a crucial role for the following reason. Ever
since the invention of the quasiparticle by Landau[4],
it has become an accepted (but unproven) truth that
low lying excitations in metals can be understood as
if we had independent, single-electron states in the
neighborhood of the Fermi surface. Therefore, our
primitive view of electrons moving through a metal in
some kind of effective periodic potential, can be tested
only by relatively large excitations such as in optical
transitions or in photoemission. The latter, of course,
yields more information because two parameters are fixed
by the experiment, the incident photon and the escaping
electron energy. In ferromagnetic substances, the
percentage of spin polarization gives a third parameter.
The results available sofar for transition metals will
be described in section 2.

Before trying to explain the electron spin
polarization results in the last section, the recent
efforts to find a consistent picture for the intermediate
range of uncertainty will be reviewed. They are
based on the narrow band (Hubbard) Hamiltonian
which can be interpreted both as breaking the present
deadlock between the two extremes of uncertainty, and
as grasping the effect of electron correlation in a
narrow band. But our understanding of the narrow band
Hamiltonian, particularly for degenerate bands and three
dimensions, is not sufficient as yet to provide the
band calculators with the kind of theoretical backing
they now have for the noble metals.

1. THE INDEPENDENT ELECTRON PICTURE

When one enumerates the important properties of iron, cobalt, and nickel, he will mention in first place that they are metallic and only in second place that they are ferromagnetic. Any theoretical explanation should, therefore, include from the very beginning our present view of metals in general. This view has changed remarkably little over the past 40 years, and we have no reason to abandon it at this time. This is why Stoner's model of ferromagnetism has to be discussed at some length.

Consider a given lattice (bcc for Fe, hcp for Co, fcc for Ni) and its Brillouin zone. To each wave vector k belongs a set of electronic energy levels, which are designated by a band index ℓ and a spin index σ. In the case of the transition metals in the first row of the periodic chart of elements, one may limit himself to a set which corresponds to the atomic 3d, 4s, and 4p levels.

The state of the metal in thermal equilibrium is described by a Fermi distribution $0 \leq n_{\ell\sigma}(\vec{k}) \leq 1$,

$$n_{\ell\sigma}(\vec{k}) = \left[\exp\left(\frac{E_{\ell\sigma}(\vec{k}) - \eta}{kT}\right) + 1\right]^{-1} \quad (1)$$

where the energy $E_{\ell\sigma}(\vec{k})$ is given self-consistently as

$$E_{\ell\sigma}(\vec{k}) = E_\ell(\vec{k}) - \mu H\sigma - I n_\sigma. \quad (2)$$

(k=Boltzmann constant, μ=Bohr magneton, H=external magnetic field). The constant I plays the role of the molecular field whose effect on the energy $E_{\ell\sigma}(\vec{k})$ is proportional to the number n_σ of electrons per lattice site with spin index σ,

$$n_\sigma = \frac{\Omega}{(2\pi)^3} \sum_\ell \int d^3k \, n_{\ell\sigma}(\vec{k}). \quad (3)$$

(Ω=volume of unit-cell). The chemical potential η is determined by making $n_\uparrow + n_\downarrow$ equal to the number n of electrons per unit cell; the magnetization m per unit-cell is given by $\mu(n_\uparrow - n_\downarrow)$.

The magnetization curve, i.e., m versus T, actually requires only the knowledge of I and the density of states per unit interval of energy,

$$\nu(E) = \frac{\Omega}{(2\pi)^3} \sum_{\ell} \int \frac{dS}{|\text{grad}_k E_\ell|} \tag{4}$$

integrated over a surface S in reciprocal space where $E_\ell(k)=E$. However, in pure metals we have a perfectly periodic lattice so that k is a good quantum number.

The full information about the dispersion $E_\ell(\vec{k})$ is necessary to explain the Fermi surface. The knowledge about the Fermi surface from de Haas-van Alphen measurements, Hall effect, and magnetoresistance has been extremely valuable[5], and certifies iron, cobalt, and nickel as real metals. Stoner's original model did not mention this possibility, but there is obviously no difficulty whatever in accommodating the Fermi surface results of the last ten years.

Further natural extensions of the Stoner model can be made to include the results of elastic neutron scattering experiments. In order to explain the magnetization distribution, $m(x)$, we associate a Bloch wave function $\psi_{\vec{k}\ell\sigma}(x)$ with each set of values $\vec{k}$, ℓ, and σ, to calculate

$$m(x) = \mu \frac{\Omega}{(2\pi)^3} \sum_\ell \int d^3k \left(n_{\ell\uparrow}(\vec{k}) |\psi_{\vec{k}\ell\uparrow}(x)|^2 - n_{\ell\downarrow}(\vec{k}) |\psi_{\vec{k}\ell\downarrow}(x)|^2 \right) \tag{5}$$

Among the important features to be realized by an appropriate choice of the ψ's are the following. The Bloch functions which are mostly responsible for the magnetization, are strongly concentrated on the lattice sites; the T_{2g} states are preferred over the E_g states for that purpose; the magnetization between the lattice sites is small, but negative[6].

The optical properties are best described in terms of the dielectric constant ε as a function of the circular frequency ω. The imaginary (absorptive) part is given in the independent electron picture by the formula

$$\varepsilon_2(\omega) = \frac{e^2}{6\pi m^2 \omega^2} \sum_{\ell j \sigma} \int d^3k \, n_{j\sigma}(\vec{k}) (1-n_{\ell\sigma}(\vec{k})) |p_{\ell j \sigma}(\vec{k})|^2$$

$$\delta(\hbar\omega - E_{\ell\sigma}(\vec{k}) + E_{j\sigma}(\vec{k})), \tag{6}$$

where the matrix element $p_{\ell j \sigma}(\vec{k})$ of the momentum operator is

$$p_{\ell j \sigma}(\vec{k}) = \frac{\hbar}{i} \int d^3x \, \psi^*_{\vec{k}\ell\sigma}(x) \frac{\partial}{\partial x} \psi_{\vec{k} j \sigma}(x) \tag{7}$$

integrated over the unit-cell (e=charge and m=mass of electron). All the wave functions are normalized to the volume of the unit cell.

The external yield of photoemitted electrons $\widetilde{D}(E,h\omega)$ is given in terms of the internal yield $D(E,h\omega)$ by

$$\widetilde{D}(E,h\omega) = \int dE' \cdot P(E,E') \cdot D(E',h\omega) \tag{8}$$

where E is the energy of the emitted electron, hω the energy of the incident photon, and

$$D(E,h\omega) = \frac{e^2}{6\pi m^2 \omega^2} \sum_{\ell j \sigma} \int d^3k \; n_{j\sigma}(\vec{k})(1-n_{\ell\sigma}(\vec{k})) |F_{\ell j \sigma}(\vec{k})|^2$$

$$\delta(h\omega - E_{\ell\sigma}(\vec{k}) + E_{j\sigma}(\vec{k})) \delta(E - E_{\ell\sigma}(\vec{k})) \tag{9}$$

The probability P(E,E') for an electron which has been excited to the energy E' inside the metal to escape with an energy E, contains all the complications of the photoemission process which were neglected in (9). In both (6) and (9), the wave vector k of the electron is conserved in the excitation process, and one speaks of a direct transition. If the summation over the spin index σ is omitted in (9), one gets the yield for each polarization separately.

The independent electron picture neglects completely whatever is included in P(E,E') such as the finite penetration depth of the incident photon, the interaction of the excited electron in band ℓ with the hole in band j, the collisions of the excited electron with other electrons as well as imperfections, and finally the transmisssion through the surface. It is indeed surprising that such an involved process should be explained by a simple formula like (9). There has been a perpetual temptation to invoke all the additional complications whenever the formula (9) seemed unable to match the experimental results.

Actually, the full force of (9) has only been exploited very recently[7]. Before that the matrix element $p_{j\ell\sigma}(k)$ was usually assumed to be independent of k and/or the indices ℓ,j, and σ. Even worse, the conservation of k was abandoned, and only the conservation of the energies was required, together with the appropriate density of states. This was called the non-direct transition approach. Such drastic simplifications were necessary because the matrix elements (7) can be calculated only once the wave functions $\psi_{\vec{k}\ell\sigma}$ are known, and the integration over $\vec{k}$ in (9) demands a very large number of sampling points in reciprocal space[8]).

These technical difficulties have not been overcome as yet for the ferromagnetic transition metals. Also, it is not quite clear how the wave functions can be obtained, because the effect of correlation on the band structure are not fully understood in this case. Therefore, the comparison with the experiments based on the Stoner model is only qualitative at the present time. Nevertheless, it is important to ask whether the formula (9) has any chance to explain the experiments, before other complications are invoked.

2. EVIDENCE FROM EXPERIMENTS

a) Experimental Support for the Stoner Model

The first quantities to be determined from experiment are the density of d-states and the effective exchange I in (2). The former is obtained quite directly from photoemission data[9] in a first rough check, ignoring the detailed structure for the time being, and the latter can then be calculated by fitting the saturation magnetization. Wohlfahrth has collected various other evidence to determine the molecular field splitting $\Delta E = (n_\uparrow - n_\downarrow)I$ in nickel and iron[10], as well as in cobalt[11]. He finds a rather remarkable consistency in the resulting figures.

The splitting ΔE can be estimated very easily if one assumes a constant density for the d-states of 2 per eV, corresponding to a total d-band width of 5eV to be shared among 10 atomic d-states. With a saturation magnetization of 2.2, 1.6, and .6, Bohr magnetrons per atom for Fe, Co, and Ni, one finds ΔE equal to 1.1, .8, and .3 eV, whereas the more detailed considerations of Wohlfahrth give 1.4+.2, 1.05+.30, and .35+.05 eV. The effective exchange $\bar{I}$ turns out to have the same value of .5 eV for all three metals in our simple computation. Actually, one should expect $I\nu(E_F) > 1$ to find ferromagnetism, whereas our estimate gives $I\nu(E_F) = 1$. The assumption of a constant density of states throughout the d-band is obviously too primitive in this respect.

The measurements of Hall-effect and magnetoresistance in Ni[5] show rather dramatically that the Fermi surface consists of several sheets, one copper-like for the majority spins, and several others both of electron and hole type without open orbits for the minority spins. Similar evidence in Fe and Co is not quite as strong, partly because the Fermi surface is even more complicated than in Ni, and partly because the experimental requirements are harder to meet. But together with the de Haas-van Alphen experiments, there can be little doubt that the Fermi surface structure in the ferromagnetic transition-metals confirms the Stoner model.

Elastic neutron diffraction provides some information about the shape of the wave functions as was noted earlier[6]. All these results are compatible with the more recent band calculations[5] in spite of the widely varying computational approaches and the uncertainties of the basic ingredients, particularly the handling of exchange and correlation effects. The magneto-optical measurements in Ni have been used to discuss the ordering of the levels near the L-point, where the majority-spin Fermi surface touches the Brillouin zone boundary[5]. With all the evidence, we have reached what may be the high-water mark for the independent electron picture in Fe, Co, and Ni at the present moment.

b) Further Experiments in Nickel

It is now time to mention the areas where the Stoner model seems to be running into trouble. There is a qualitative difference between the measured photoemission yield for Ni (without discrimination of the spin polarization) and the best calculations of formula (9) done for Ni.[12]. The former is rather smooth, without any pronounced peaks more than 1 eV below the threshold, whereas the latter shows a lot of structure at all values of E and $h\omega$. The important feature which theory and experiment have in common is the sharp rise just below the threshold, i.e., for $E < h\omega - \phi$ where ϕ is the work function, and a large peak less than 1 eV in width.

The calculations were done by Pierce and Spicer[12] who apply the interpolation scheme of Hodges, Ehrenreich, and Lang[13] to formula (9) assuming, however, a constant matrix element $p_{i\ell_0}(\vec{K})$. One may speculate whether a correction of this last deficiency will smooth out the calculated yield more than 1 ev below threshold. The calculations of Williams, Janak, and Moruzzi for Cu[7], the first and sofar only ones to include a computed (not averaged) matrix element $p_{i\ell_0}(\vec{K})$ look more smooth than the Ni computations of Pierce and Spicer; but there is still more structure than in the Ni measurements. Since the latter have been made many times and with great care, it is not clear whether we will ever get as good agreement between theory and experiment as in Cu[7].

Several efforts have been made in recent years to test specifically whether for the two spin polarizations the density of states behaves as the Stoner model predicts. Walmsley[14] tried to measure the change in work function, and hence the change in Fermi energy, of a ferromagnet upon application of a large magnetic field parallel to the surface (to avoid demagnetization). If the two spin directions have a different density of states at the Fermi level, the change in work function was assumed proportional to the ratio $[\nu_\uparrow(E_F) - \nu_\downarrow(E_F)]/[\nu_\uparrow(E_F) + \nu_\downarrow(E_F)]$. Belson[15] repeated and expanded these experiments. He found that $\nu_\uparrow(E_F)$ and $\nu_\downarrow(E_F)$ differ by a factor 1.18 at most in Fe, Co, and Ni.

Pierce and Spicer[12] measured the photoemission in Ni (without discrimination of spin polarization) as a function of temperature from $295°K$ to $678°K$, well above the Curie temperature ($631°K$). With a molecular field splitting of at least .3 eV at low temperatures, as discussed at the beginning of this section, the effect of the vanishing magnetization should have been detectable either in the rise at threshold, or in the location of the large peak immediately below, or even in the remaining structure. But nothing significant was found. The experiments were extended by Rowe and Tracy[16] to lower photon energies by covering the Ni surface with a monolayer of cesium to lower the workfuction. The location of the large peak was measured very carefully and seen to shift in a systematic manner with temperature, but the overall shift amounted to only ~45 meV.

Similar shifts have been observed in optical absorption[17] and characteristic energy loss experiments[18] which measure energy differences rather than absolute energies as

in photoemission. In the first case, the narrowing
of the main absorption peak near 5 eV with increasing
temperature correlates with the magnetization, as does
the disappearance of a small peak around 1.4 eV. In
the second case, an energy loss which is associated
with the excitation of a volume plasmon, is shifting
its peak downward linearly with increasing temperature
except near T_c where this drop becomes quite abrupt.
All the experiments suffer from a defect which tends
to reduce their importance. The molecular field
splitting ΔE is relatively small in Ni, after all, since
kT_c is a sizable fraction of the shifts $\Delta E_\uparrow$ or $\Delta E_\downarrow$ of
each spin orientation separately. There may be
compensating changes with temperature such as the volume
dependence, as was pointed out by Pierce and Spicer.
This objection would obviously not hold in Fe and Co,
but the above experiments have not been performed for
these metals.

c) Spin Polarization Measurements

The spin polarization of photoemitted electrons
in Ni was first successfully measured by Baenninger,
Busch, Campagna, and Siegmann[19], and shortly thereafter
the same group performed the measurements on Fe and
Co[20]. The main novelty in the experimental arrangement
was to make any stray magnetic field B_s outside the
sample parallel to the accelerating electric field F
so as to prevent the electrons from drifting away in
the direction $F \times B_s$. The polarization was measured
in a Mott detector of the type that had been used earlier
on other ferromagnets, e.g., Gd [21] and EuO [22]. Experiments
were also performed on amorphous thin films of Fe, Co,
and Ni where the degree of polarization was always lower
than in the crystalline sample. Finally, cesiated Co
samples were investigated[23], and the results were found
compatible with the clean samples.

There are two striking features about the results.
First, the spin polarization of the electrons is the
same as that of the majority spins in the metal, i.e.,
the magnetic moment of the photoemitted electrons lines
up preferentially with the bulk magnetization. Second,
the degree of polarization p is large compared to a
number such as $(n_\uparrow - n_\downarrow)/(n_\uparrow + n_\downarrow)$. Indeed, if we consider
only the 3d and 4s-p shell, we have $n_\uparrow - n_\downarrow = 2.2, 1.6, .6$

and $n_+ + n_- = 8, 9, 10$ for Fe, Co, Ni which gives the ratios .275, .18, .06. However, the experimental results are p=54% for Fe, 21% for Co, and 15% for Ni with an error of less than 1%.

An energy analysis of the photoemitted electrons was not made because the light source was barely above the threshold. Only in the case of the cesiated cobalt are data available which cover more than 2 eV of photon energy. In the early results on Ni the photon energy was effectively reduced by .4 eV with the help of a filter, still enough to excite electrons within "a few 100 meV below E_F".[20] But within this admittedly narrow region of incident photon energy, the degree of polarization p stayed essentially constant.

This constancy with respect to the exciting energy has been confirmed by another set of remarkable experiments. It was demonstrated by Meservey, Tedrow, and Fulde[24] that the quasiparticle spectrum in a superconducting film could be split by a large magnetic field without destroying the superconductivity provided the magnetic field was carefully aligned with the plane of the film so as to prevent any trouble with flux closure. The tunneling characteristic of such film in a junction like $Al-Al_2O_3-Ag$, i.e, conductivity dI/dV versus voltage V, shows quite clearly a double peak, both for positive and negative V, because the external magnetic field H has shifted the energy of the quasiparticles by $\pm\mu H$.

In a junction with a ferromagnetic metal, the supply of electrons near the Fermi surface is presumably different for the two spin polarizations. Therefore the double peak structure is different for negative and for positive voltage, and the lack of symmetry is simply related to the degree of polarization in the ferromagnetic metal. Tedrow and Meservey performed the experiment first on Ni,[25] and later on Fe, Co, Ni, and Gd.[26] The values for p are 44%, 34%, 11%, and 4.3%, in essential agreement with the photoemission results,[20,21] but now within 1 meV of the respective Fermi energies.

Measurements of spin polarization of field-emitted electrons have been made recently on Ni, first on polycrystalline tips by Regenfus and Hellwig[27], and then on different crystallographic faces by Gleich, Regenfus, and Sizmann[28]. The degree of polarization turns out to depend strongly on the orientation of the emitting surface, with $\langle 100 \rangle$, $\langle 110 \rangle$, and $\langle 137 \rangle$ emitting perferentially parallel to the minority spins, and only $\langle 111 \rangle$ parallel to the majority spins in the metal. The absolute values of p are all between 7 and 10% with an error of 2%. Since these electrons originate at a depth of no more than .1 eV below the Fermi level, the field emission results are drastically out of line with the previous two experiments. Of course, it is extremely valuable to have data which discriminate between various crystallographic orientations.

3. CAN THE STONER-MODEL BE JUSTIFIED

a) <u>Elementary Explanations</u>

It is natural to make first an effort which tries to fit the experimental results of the last section into the framework of the Stoner model, i.e., the formulas of section 1. This requires a set of dispersion curves $E_{\ell\sigma}(\vec{k})$ and single electron wave functions $\psi_{\vec{k}\ell\sigma}(x)$ from which the matrix elements can be calculated. These ingredients can be obtained from a number of heuristic schemes which carry various names such as tight-binding, OPW, APW, KKR, renormalized atom, and combinations thereof. They form the basis for some recent explanations of the electron spin polarization data.

Wohlfahrth[29] pointed out that the top of the d-states for the majority spins in Ni is only some 60 meV beneath the Fermi level. A resolution of a few 100 meV in the photoemission, therefore, would not detect the large difference in density of states between opposite spins near the Fermi energy. Co and Fe pose less of a problem because their Fermi levels are already inside the thicket of d-levels. Smith and Traum[30] have computed $D(E, h\omega)$ with a constant matrix element using the interpolation scheme of Hodges et. al.[13] A very intricate pattern of degrees of polarization as a function of E and $h\omega$ is obtained. Again there is obviously room for agreement with the experiments, although the numerical details are not perfect.

A more sophisticated approach is used by Politzer and Cutler[31] who calculate the field emission current

on a ⟨100⟩ face in Ni. The crucial problem is how
to fit the electron wave function outside the metal
with the Bloch states inside. It turns out that
the 4s-p states are much more likely to leak out than
the 3d states, and that the emission current for the
latter depends very strongly on $\vec{K}$ and ℓ. The result
is, therefore, a relatively small degree of polarization
which is also negative in agreement with experiment.
It should be noted that this type of argument lies still
within the limits of the independent electron picture.

As soon as the surface is taken into account (as
above), some puzzles might find an explanation because
of the subtle relation between Fermi energy E_F and surface
barrier δV.[32] If E_F is raised by the application of
a magnetic field as in Belson's experiment,[15] the electrons
spill more easily to the outside. The effective dipole
moment at the surface, i.e., the surface barrier δV,
is raised thereby. The work function $\phi = \delta V - E_F$ might
end up varying very little.[33] Pant and Rajagopal[34] have
proposed a formal treatment of this effect in a self-
consistent manner.

b) A Complete Theory

A band calculation has to explain not only the ground-
state, but the elementary excitations even if they
correspond to correlated states and/or have finite
lifetimes. No such procedure exists for ferromagnetic
metals. At least in Cu and many other metals, one can
fall back on the method of Kohn and Sham[35] which has
not been fully implemented, but provides a good conceptual
basis.

The difficulty of the problem can be appreciated
from a short description of this approach. The following
statement is first proven[36]. A one-to-one correspondence
exists between the external potential $V(x)$ in which
the electrons move (the electrostatic attraction of
the nuclei) and the resulting density distribution $n(\vec{x})$
of the electrons in the ground state, taking into account
their kinetic energy, Coulomb interaction, and the
exclusion principle. Thus, each ground state property
can be written as a functional of $n(\vec{x})$ rather than $V(\vec{x})$,
if that suits our purpose, e.g., the mass-operator $M(\vec{x}_1, \sigma_1, \vec{x}_2, \sigma_2, E)$ whose meaning will be explained later, can
be thought of as well defined once we know $n(\vec{x})$.

The first step is, therefore, the determination of the electronic density $n(\vec{x})$. This can be accomplished in the familiar form

$$n(\vec{x}) = \frac{\Omega}{(2\pi)^3} \sum_{\ell\sigma} \int d^3k \; n_{\ell\sigma}(\vec{k}) |\psi_{\vec{k}\ell\sigma}(\vec{x})|^2, \qquad (10)$$

where the wave functions ψ are solutions of an ordinary Schrodinger equation

$$-\frac{h^2}{2m} \Delta\psi + W(x)\psi = E\psi, \qquad (11)$$

whose potential is composed of two parts. First, the electrostatic potential $\Phi(\vec{x})$ which is experienced by any test-charge e (not an electron), as determined by the nuclei and the electronic density $n(x)$. Second, the exchange-correlation potential $\eta_{xc}(n(\vec{x}))$ which is a functional of $n(\vec{x})$ and which is approximated essentially by using its expression in a homogeneous electron gas. A commonly used formula for η_{xc} was proposed by Slater in 1951,[37] namely $\eta_{xc} = -3e^2 (3n/8\pi)^{1/3}$, and has been used extensively. Obviously, one has to achieve self-consistency between (10) and (11).

The second step is to determine the excitation spectrum by calculating the Green's function $G(\vec{x}_1,\sigma_1,\vec{x}_2,\sigma_2,E)$. This is done by solving Dyson's equation

$$\left(-\frac{h^2}{2m}\Delta + \Phi(\vec{x}) - E\right) G(\vec{x},\vec{x}') + \int d^3x'' M(\vec{x},\vec{x}'') G(\vec{x}'',\vec{x}') = -\delta(x-x'), \qquad (12)$$

where the spin-indices and the parameter E in G and M have not been written out for simplicity's sake. The mass-operator M contains all the effects of exchange and correlation on an electron in an excited state of energy E. Again, the dependence of M on $n(\vec{x})$ is taken from the homogeneous electron gas[38]. A practical procedure has been worked out by Hedin and Lundqvist[39].

c) Some Critical Comments

The procedures recommended by Kohn and Sham, were applied by Janak, Moruzzi, and Williams[40] to Cu, with one significant simplification in the last step. The Green's function G was not obtained, but only its main resonances as a function of E for fixed k. Generally, these resonances have a finite lifetime (non-vanishing width). Their main peak can be located by finding a wave function $\chi_{\vec{k}\ell\sigma}(x)$ from

$$\left(-\frac{\hbar^2}{2m}\Delta + \phi(\vec{x}) - E\right) \chi(\vec{x}) + \text{Re}\int d^3x'' M(\vec{x},\vec{x}'') \chi(\vec{x}'') = 0, \quad (13)$$

where only the real part of the mass-operator enters. In contrast to the ψ's which make up the electronic density $n(x)$, the wave functions χ describe the excited states. The matrix elements (7) for optical absorption and photoemission were computed from the χ's. A more careful treatment of (12) will also give the life-times: and these are currently obtained by the people who calculate low-energy electron diffraction on metals[41].

The good agreement which was found in Cu[40] depends on two essential assumptions. The first concerns the use of n_{xc} and M as suggested by the homogeneous electron gas; the second has to do with the use of the ψ's of (13) in conjunction with (7) to get the photoemitted yield (9).

The homogeneous electron gas cannot be used as a model for a ferromagnetic system, because it does not have a ferromagnetic groundstate at any density. If we use Slater's exchange in spite of this basic objection[42], we might get sensible results, but we cannot improve them in any systematic manner. The homogeneous electron gas provides a good description for the particular configuration where the two spin directions are treated in the completely symmetric manner of a filled shell. This provides a good start for the metals with open s and p shells, but not when there are partially filled d and/or f shells.

The electronic response to an external electro-magnetic field can be written as a two-particle correlation function, and the photoemission yield requires a three-particle correlation function[43], whereas equation (12) gives only the one-particle propagator. If one tries to reduce the former to the latter, he is bound to make approximations. In its widest sense, the Stoner model is good as long as one gets good agreement with experiment while using at most a one-particle propagator. On the other hand, the derivation of formula (9) from the most general treatment still presents great difficulties[44,45].

4. THE NARROW BAND PICTURE

The problem of correlation in open d-shells and its impact on the bandstructure of transition metals has been recognized for a long time. It is implicit in Slater's muffin-tin idea[46] and was discussed schematically in the case of a half-filled ferromagnetic band of electrons with one spin reversed, again by Slater.[47] The same thoughts were presented by van Vleck in terms of minimum polarity[48], where an electron in the d-bands was able to go freely from one atomic site to an adjacent one as long as the number of electrons on the old site or on the new one was still close to the average.

The electron has two possibilities for moving in the crystal. If it starts at a given site in a given localized orbital, it can go to a nearby site, or it can stay at the same site and adjust itself to the other electrons there to form an advantageous local configuration. It is important to realize that these two opportunities cannot be simultaneously included in the same Hartree-Fock type wave function. The typical band calculation sacrifices the latter opportunity to the former. As a consequence, the various local configurations at any given site occur in a well defined statistical mix. It is as if the free atom were made to occur in certain preassigned ratios over all its various states of ionization and multiplets.

This particular difficulty is again an expression of the uncertainty principle. It does not seem to be crucial in many metals with open s and p shells. But where open d shells enter the picture, it appears that the potential gains from taking advantage of the localized (on site) correlations are competitive with the gains of moving from site to site.

The most succinct formulation of this problem is found in the narrow band Hamiltonian whose simplest (non-degenerate) case is commonly called the Hubbard Hamiltonian[49]. The homogeneous electron gas quite obviously does not present this dilemma between local configurations and "steady state" motion across the whole sample with a fixed crystal momentum.

The narrow band problem for transition metals requires degenerate bands in order to make Hund's rule effective locally, and to insure a ferromagnetic ground-state without invoking nearest neighbor exchange coupling[50].

Unfortunately, this situation has not been explored sufficiently because even the non-degenerate narrow band presents formidable problems. The latter has been investigated in many different ways, particularly since it bears also on the metal-insulator transition which was first discussed by Mott[51].

The results of the narrow band theory have been used to justify particular expressions for the correlation-exchange correction in Schrodinger's equation for an independent electron[52]. Such a procedure corresponds to picking a term η_{xc} in (11) from the narrow band model rather than from the homogeneous electron gas.

Moreover, in the spirit of the Kohn-Sham approach, the single particle propagator has been taken from the Hubbard Hamiltonian to justify corrections to the basic formula (9). Anderson[53] argues that the hole occupation probability $n_{j\sigma}(\vec{k})$ should be replaced by the imaginary part of the single particle Green's function at crystal momentum $\vec{k}$, together with an additional integration over the energy of the hole. The effects of finite lifetimes, as well as correlation and exchange excitations are taken into account thereby. Doniach[54] also includes the relaxed orbital correction which shifts the main resonance for the majority spin hole because the remaining electrons are able to adjust.

Again it should be noted that the hole spectrum and the excited electron are treated by these authors as independent of each other. If we adopt their view, the photoemission experiments require only a careful study of the hole propagation in a ferromagnet, and field emission would demand only a detailed examination of the electron transmission through a surface, while the many-body corrections to the vertex (matrix element) are being ignored. Thus, the independent particle picture has not been modified all that drastically even in the more sophisticated treatments sofar.

ACKNOWLEDGEMENTS

The author has enjoyed numerous discussion with N. D. Lang, and profited from conversations with T. Di Stefano, D. Eastman, W. Grobman, J. Janak, D. Jepsen, G. Lasher, D. Paul, B. Politzer, L. Sham, P. Vashishta, S. von Molnar, and A. Williams.

REFERENCES

1. P. Weiss, Jour. Phys. <u>6</u>, 667 (1907).

2. E. C. Stoner, Proc. Roy. Soc. A <u>165</u>, 372 (1938) Cf. also F. Bloch, Z. Phys. <u>57</u>, 545 (1929).

3. W. E. Evenson, J. R. Schrieffer, and S. Q. Wang, J. of Appl. Phys. <u>41</u>, 1199 (1970).

4. L. Landau, Soviet Physics JETP <u>3</u>, 920 (1957).

5. Cf. A. P. Cracknell, Advances in Physics <u>20</u>, 1 (1971) for an exhaustive list of references.

6. H. A. Mook, Phys. Rev. <u>148</u>, 495 (1966). R. Moon, Phys. Rev. <u>136</u>, A195 (1964).

7. A. R. Williams, J. F. Janak, and V. L. Moruzzi, Phys. Rev. Letters <u>28</u>, 672 (1972).

8. E. B. Kennard, D. Koskimaki, J. T. Waber, and F. M. Mueller, in Electronic Density of States, L. H. Bennett ed., Natl. Bur. of St. Special Publication No.323, p. 795, (1971).

9. For a general survey cf. N. V. Smith, Photoemission of Properties of Metals, CRC Critical Reviews in Solid State Sciences, p. 45, March 1971.

10. E. P. Wohlfahrth, Proceedings of the Nottingham Conference on Magnetism (Institute of Physics and the Physical Society, London, 1964), p. 51.

11. E. P. Wohlfahrth, J. of Appl. Phys. <u>41</u>, 1205 (1970).

12. D. T. Pierce and W. E. Spicer, Phys. Rev. B <u>6</u>, 1787 (1972).

13. L. Hodges, H. Ehrenreich, and N. D. Lang, Phys. Rev. <u>152</u>, 505 (1966).

14. R. H. Walmsley, Phys. Rev. Letters <u>8</u>, 242 (1962).

15. H. S. Belson, J. of Appl. Phys. <u>37</u>, 1348 (1966).

16. J. E. Rowe and J. C. Tracy, Phys. Rev. Letters <u>27</u>, 799 (1971).

17. H. Shiga and G. P. Pells, J. Phys. C. Proc. Phys. Soc. London <u>2</u>, 1847 (1969).

18. B. Heimann and J. Holzl, Phys. Rev. Letters **26**, 1573 (1971).

19. U. Baenninger, G. Busch, M. Campagna, and H. C. Siegmann, Phys. Rev. Letters **25**, 585 (1970).

20. G. Busch, M. Campagna, and C. H. Siegmann, Phys. Rev. B **4**, 746 (1971).

21. G. Busch, M. Campagna, P. Cotti, and H. C. Siegmann, Phys. Rev. Letters **22**, 597 (1969).

22. G. Busch, M. Campagna, and H. C. Siegmann, Solid State Comm. **7**, 755 (1969) and J. App.. Phys. **41**, 1044 (1970).

23. G. Busch, M. Campagna, D. T. Pierce, and H. C. Siegmann Phys. Rev. Letters **28**, 611 (1972).

24. R. Meservey, P. M. Tedrow, and Peter Fulde, Phys. Rev. Letters **25**, 1270 (1970).

25. P. M. Tedrow and R. Meservey, Phys. Rev. Letters **26**, 192 (1971).

26. R. Meservey and P. M. Tedrow, Solid State Comm. **11**, 333 (1972)

27. G. Regenfus and R. Helwig, Verh. Deut. Phys. Ges. **5**, 181 (1970)

28. W. Gleich, G. Regenfus, and R. Sizmann, Phys. Rev. Letters **27**, 1066 (1971).

29. E. P. Wohlfahrth, Phys. Letters **36A**, 131 (1971).

30. N. V. Smith and M. M. Traum, Phys. Rev. Letters **27**, 1388 (197

31. B. A. Politzer and P. H. Cutler, Phys. Rev. Letters **28**, 1330 (1972).

32. Cf. the forthcoming review of N. D. Lang, The Density-Function Formalism and the Electronic Structure of Metal Surfaces, in Solid State Physics, ed. F. Seitz, D. Turnbull, and H. Ehrenreich.

33. This reasoning is due to N. D. Lang (private communication).

34. M. M. Pant and A. K. Rajagopal, Solid State Comm. **10**, 1157 (1972).

35. W. Kohn and L. J. Sham, Phys. Rev. **140A**, 1133 (1965).

36. P. Hohenberg and W. Kohn, Phys. Rev. **136B**, 864 (1964).

37. J. C. Slater, Phys. Rev. $\underline{81}$, 385 (1951).

38. L. J. Sham and W. Kohn, Phys. Rev. $\underline{145}$, 561 (1966).

39. L. Hedin and B. I. Lundqvist, J. Phys. C., Proc. Phys. Soc. London $\underline{4}$, 2064 (1971).

40. J. F. Janak, A. R. Williams, and V. L. Moruzzi, The Optical Properties of Cu (to be published).

41. D. W. Jepsen and P. M. Marcus, in Computational Methods in Band Theory, Plenum Press, New York, 1971, p. 416.

42. J. C. Slater, Phys. Rev. $\underline{165}$, 658 (1968).

43. W. L. Schaich and N. W. Ashcroft, Phys. Rev. $\underline{3B}$, 2452 (1971). This view is being disputed by G. D. Mahan, Phys. Rev. $\underline{2B}$, 4334 (1970), cf. his remarks on p. 4338.

44. D. Langreth, Phys. Rev. $\underline{3B}$, 3120 (1971).

45. H. Hermeking, Z. Physik $\underline{253}$, 379 (1972).

46. J. C. Slater, Phys. Rev. $\underline{51}$, 846 (1937).

47. J. C. Slater, Phys. Rev. $\underline{52}$, 198 (1937).

48. J. H. Van Vleck, Rev. Mod. Phys. $\underline{25}$, 220 (1953).

49. M. C. Gutzwiller, Phys. Rev. Letters $\underline{10}$, 159 (1963). J. Hubbard, Proc. Roy. Soc. (London) $\underline{A276}$, 238 (1963), J. Kanamori, Progr. Theor. Phys. $\underline{30}$, 275 (1963).

50. M. C. Gutzwiller, Phys. Rev. $\underline{134}$, A923 (1964). K. A. Chao and M. C. Gutzwiller, J. Appl. Phys. $\underline{42}$, 1420 (1971), K. A. Chao, Phys. Rev. $\underline{B4}$, 4034 (1971).

51. For a recent review, cf. S. Doniach, Adv. Phys. $\underline{20}$, 819 (1971).

52. K. J. Duff and T. P. Das, Phys. Rev. $\underline{B3}$, 192 and 2294 (1971).

53. P. W. Anderson, Phil. Mag. $\underline{24}$, 203 (1971).

54. S. Doniach, AIP Conference. Proc. Magnetism and Magnetic Materials, 1971, AIP, New York, p. 549.

k-DEPENDENT EXCHANGE AND THE DYNAMIC SUSCEPTIBILITY OF FERROMAGNETIC NICKEL*

J. F. Cooke and H. L. Davis
Solid State Division, Oak Ridge National Laboratory
Oak Ridge, Tennessee, 37830

ABSTRACT

An improved formalism is proposed for calculating the dynamic susceptibility of itinerant ferromagnets. This procedure is based on a simple extension of the random phase approximation and an interpolation method for treating wave functions and matrix elements throughout the Brillouin zone. In contrast to previous work, this formalism incorporates momentum-dependent spin-splitting of the electronic energy bands as well as multi-band effects. It is found that all parameters which appear in the theory can be determined directly from a self-consistent treatment of the ferromagnetic band structure. We have used this formalism to numerically calculate the dynamic susceptibility of ferromagnetic nickel for several "realistic" band structures which incorporate these momentum-dependent exchange effects. The spin-wave dispersion curve obtained from such calculations is found to be in excellent agreement with experiment. A comparison of our results with those obtained from the simple enhanced susceptibility theory with rigidly spin-split bands indicates that k-dependent exchange may well be an important factor in obtaining a correct description of magnetism in ferromagnetic nickel.

INTRODUCTION

Ferromagnetism in nickel has been a subject of considerable interest and controversy over the past three and one-half decades. The experimental information that has been compiled thus far seems to favor the view that the majority of unpaired or "magnetic" electrons occupy relatively narrow d-bands and that they are itinerant in character. That is, an individual electron propagates throughout the lattice, spending very little time near any particular lattice site. Thus, the motion of opposite spin electrons must be correlated in such a manner that a stable magnetic moment distribution is maintained. As a result of this correlated motion, we expect the electronic wavefunctions and energy bands to depend on the spin orientation.

For obvious reasons the model discussed above is called the itinerant model. This model is difficult to apply to physically realistic systems for several important reasons. First, there is the difficult theoretical problem of properly treating the correlation effects which are of fundamental importance in describing the

---
*Research sponsored by the U. S. Atomic Energy Commission under contract with Union Carbide Corporation.

magnetic properties of the model. Since an exact treatment in a tractable form is beyond our capabilities at present, the hope is that a reasonable approximation of the correlation problem can be found which is at least amenable to a numerical solution.

Another problem is associated with the actual solution of the equations generated by the approximate theory. Clearly these equations must be at least as complicated as present band structure equations which, in general, require computer solution at general points in the Brillouin zone (BZ). It is not sufficient, however, that a solution can be obtained at any one $\vec{k}$-point. That is, in order to check the validity of the approximate theory, and indeed the theory itself, measurable quantities must be calculated and compared with experiment. The calculation of magnetic and optical properties as well as certain Fermi surface features require the integration of terms involving electronic energies and matrix elements of various operators over all, or part, of the BZ. Thus the method of solution must be rapid enough to generate sufficient information to guarantee convergence of the required integrals in a reasonable time.

The calculation of the frequency and wave-vector dependent susceptibility, $\chi(\vec{q},\omega)$, entails all of the difficulties referred to above. It is an important quantity because it contains a considerable amount of information about the excitation spectra of magnetic systems and it can be measured directly by inelastic neutron scattering techniques. Two papers have been published recently which attempt to numerically calculate $\chi(\vec{q},\omega)$ within the random phase approximation (RPA) for ferromagnetic nickel.[1,2] Thompson's calculation was essentially based on a single-band description with parameters being chosen to give partial agreement with the neutron measurements. More recently, Lowde and Windsor (LW) proposed a generalization of the RPA result for the multi-band case and proceeded to calculate χ. Their results were found to be in good overall agreement with most of the neutron data, but in order to explain the spin-wave scattering results, they had to introduce a momentum-dependent term which could not be calculated *a priori*. In effect they adjusted the term to produce agreement with experiment.

It is the purpose of this paper to present some numerical results obtained from an improved formalism for calculating χ which is based on a simple extension of the random phase approximation of the relevant Green's function equation. Since the details of this theory have been published elsewhere,[3] we will present here only the salient features. One important aspect of the formalism is that unlike previous work it takes into account, in an approximate sense, the momentum dependence of certain screened coulomb matrix elements. This leads to an expression for χ which is more complicated than the one considered by LW and to a momentum-dependent exchange splitting of the electronic energy bands which was originally proposed by Hodges *et al.* (HEL)[4]. Calculations based on this model indicate that these momentum-dependent effects are important for nickel and that an improve description of spin-wave scattering can be obtained by including them.

The paper is divided, somewhat arbitrarily, into six sections. In Section I a brief outline of the general theory is given and a pair of equations is developed which form the basis of our approximate theory. A method of solving these equations is given in Section II. Some general features of the numerical calculation of a band structure with momentum-dependent exchange are discussed in Section III. Numerical results for the imaginary part of the transverse susceptibility are contained in Section IV. These results are compared with other work and with spin-wave scattering data in Section V. In Section VI we summarize the results that have been obtained thus far.

I. APPROXIMATE THEORY

The dynamic susceptibility, $\chi_{\mu\nu}(\vec{q},\omega)$, is a measure of the linear response of a system to a small applied magnetic field $\vec{h}(\vec{q},\omega)$ and is defined by

$$\Delta<M_\mu(\vec{q},\omega)> = \sum_{\vec{k}} \int d\omega' \sum_\nu \chi_{\mu\nu}(\vec{k}-\vec{q},\omega-\omega') h_\nu(\vec{k},\omega') , \qquad (1)$$

where $\Delta<M_\mu(\vec{q},\omega)>$ is the change in Fourier transform of the magnetic moment density upon application of the field. The symbol $<O>$ represent the thermal average of the operator O, μ and ν represent x, y, z, and $F(\vec{q},\omega)$ is the space-time Fourier transform of $F(\vec{r},t)$. It is straightforward to show that $\chi_{\mu\nu}$ may be written in the form

$$\chi_{\mu\nu}(\vec{q},\omega) = \frac{g\mu_B i}{\hbar} \int_0^\infty dt <[S^\mu(\vec{q},t), S^\nu(-\vec{q},0)]> e^{-i\omega t} , \qquad (2)$$

where $[A,B]$ is the commutator of A with B, g is the gyromagnetic ratio of the electron, μ_B is the Bohr magneton, and S^μ represents the μth component of the spin operator $\vec{S}$.

In this paper we will calculate only the transverse susceptibility, χ^{-+}. The longitudinal part, χ^{zz}, can be calculated within the framework of the theory described below but the results will not be presented here. The procedure that will be used is to relate χ^{-+} to a relevant Green's function and then to use the equation of motion coupled with the second quantization formalism to calculate it. The particular Green's function which is useful for this calculation is defined by

$$G^{-+}(\vec{q},t) = -i\theta(t)<[S^-(\vec{q},t), S^+(-\vec{q},0)]>$$

$$\theta(t) = \begin{cases} 0 & t<0 \\ 1 & t>0 \end{cases}, \qquad (3)$$

where

$$S^\pm(\vec{q},t) = S^x(\vec{q},t) \pm i S^y(\vec{q},t) . \qquad (4)$$

Then from (2), $\hbar\chi^{-+}(\vec{q},\omega)/g\mu_B$ is just the negative of the Fourier transform of $G^{-+}(\vec{q},t)$. Although we will not give the proof here, it also follows that, at low temperatures, the transverse inelastic neutron scattering cross-section is essentially proportional to the imaginary part of $\chi^{-+}(\vec{q},\omega)$.

It is useful at this point to express all quantities in terms of the second quantization formalism. In order to do this we introduce the operators $C^{\dagger}_{n\vec{k}\sigma}$ and $C_{n\vec{k}\sigma}$ which respectively create and destroy electrons in the single-particle state $B_{n\vec{k}\sigma}(\vec{r})$ where n is a band index, $\vec{k}$ is the wave-vector, and σ is the spin index. The wavefunction $B_{n\vec{k}\sigma}(\vec{r})$ is assumed to be of the form

$$B_{n\vec{k}\sigma}(\vec{r}) = \psi_{n\vec{k}\sigma}(\vec{r}) \chi_\sigma , \qquad (5)$$

where $\psi_{n\vec{k}\sigma}(\vec{r})$ is a scalar function and χ_σ is a spinor.

In terms of these quantities we can write

$$G^{-+}(\vec{q},t) = \sum_{\substack{n,m \\ \vec{k}}} (n\vec{k}|e^{-i\vec{q}\cdot\vec{r}}|m\vec{k}+\vec{q}) \hat{G}(n\vec{k},m\vec{k}+\vec{q};t) , \qquad (6)$$

where

$$\hat{G}(n\vec{k},m\vec{k}+\vec{q};t) = -i\theta(t)<[C^{\dagger}_{n\vec{k}\downarrow}(t)C_{m\vec{k}+\vec{q}\uparrow}(t), S^+(-\vec{q},0)]> . \qquad (7)$$

The matrix element of $\exp(-i\vec{q}\cdot\vec{r})$ in (6) is to be calculated with respect to the functions $\psi_{n\vec{k}\sigma}(\vec{r})$. We assume that, to a first approximation, the spin-dependence of ψ can be neglected when calculating matrix elements.

The form of the equation of motion for $\hat{G}$ depends on the commutator of the time-dependent operators in (7) with the Hamiltonian, H, where

$$H = H_{KE} + H_{ion} + H_c , \qquad (8)$$

H_{KE} is the kinetic energy operator, H_{ion} describes the electron-ion interaction, and H_c represents the electron-electron interaction resulting from the coulomb potential, U. The spin-orbit interaction which has been considered in some detail by Zornberg has been neglected in order to simplify the calculation.[5] Possible effects resulting from the inclusion of the spin-orbit interaction will be briefly discussed later on in the paper.

Since the equation of motion couples $\hat{G}$ to higher order Green's functions, some method of decoupling must be employed. In principle, the type of decoupling used should be governed by the criterion that correlation effects must be properly described. Since this criterion is not known in practice, we use the approach of examining the simplest termination first, namely, the one based on the random phase approximation. If this termination is used, it turns out that the resulting equation can be simplified considerably by choosing the $\psi_{n\vec{k}\sigma}$ to satisfy spin-polarized Hartree-Fock equations. At this stage it is clear that the RPA termination cannot be correct for nickel since energy bands calculated from Hartree-Fock equations are unsatisfactory because, for example, screening effects are neglected. A method of extending the theory beyond RPA is based on a functional derivation technique which can be used to establish an expansion in terms of matrix elements of an effective (screened) interaction, U_{eff}, instead of U. The net effect of this procedure is to modify the RPA equations by replacing U by U_{eff} in certain terms.

The resulting equations can be simplified further by introducing the complex time transform of $\hat{G}$.

$$\hat{G}(z) = \int dt \, \hat{G}(t) e^{-izt} , \qquad (9)$$

where z is a complex variable. Then we have

$$\{\hbar z - E(m\vec{k}+\vec{q}\uparrow) + E(n\vec{k}\downarrow)\}\hat{G}(n\vec{k},m\vec{k}+\vec{q};z) = \hbar(m\vec{k}+\vec{q}|e^{i\vec{q}\cdot\vec{r}}|n\vec{k}) \times$$

$$\{f_{n\vec{k}\downarrow} - f_{m\vec{k}+\vec{q}\uparrow}\} - \{f_{n\vec{k}\downarrow} - f_{m\vec{k}+\vec{q}\uparrow}\} \sum_{\substack{ij \\ \vec{p}}} (m\vec{k}+\vec{q},i\vec{p}|U_{eff}|n\vec{k},j\vec{p}+\vec{q}) \times$$

$$\hat{G}(i\vec{p},j\vec{p}+\vec{q};z) , \qquad (10)$$

$$H^{\sigma} \psi_{n\vec{k}\sigma}(\vec{r}) = E(n\vec{k}\sigma)\psi_{n\vec{k}\sigma}(\vec{r}) , \qquad (11)$$

$$H^{\sigma} = \hat{H}_{o} + \hat{H}_{c}(U) + \hat{H}^{\sigma}_{ex}(U_{eff}) , \qquad (12)$$

$$f_{n\vec{k}\sigma} = \{e^{\beta E(n\vec{k}\sigma)} - 1\}^{-1} ; \quad \beta = \frac{1}{k_B T} . \qquad (13)$$

The electronic energies and wave functions ψ are found by solving (11). The single-particle Hamiltonian, H^{σ}, is defined in terms of $\hat{H}_o$, which incorporates the kinetic energy operator and the electron-ion interaction; $\hat{H}_c(U)$, which describes the direct coulomb interaction between charge clouds and depends on matrix elements of U; and $\hat{H}_{ex}(U_{eff})$, which represents the non-local exchange interaction operator and depends on matrix elements of U_{eff}. Equation (11) reduces to the usual type of spin-polarized equation if U_{eff} is replaced by U in the exchange term.

Equations (10) and (11) represent the basic equations of this approximation theory. Correlation effects are included in a manner dictated by the termination procedure. The type of termination used also restricts the theory to low temperatures. The next step is to solve these equations.

II. A METHOD OF SOLUTION

The solution of equation (10) and (11) is obviously a non-trivial problem. First of all the equation for U_{eff} is coupled to the equation for a single-particle Green's function and the two equations must be solved simultaneously. Instead of attacking the problem in this manner we take the opposite approach, which is to assume that equation (11) is general enough to describe the band structure of nickel for some appropriate choice of U_{eff}. The idea then is to treat U_{eff} as an adjustable term to be fixed by experimental data.

Even if we assume that an appropriate U_{eff} can be found, equations (10) and (11) are still too complicated to solve numerically in a reasonable amount of computer time. The reason is that these equations contain matrix elements of both U and U_{eff}, which depend on band and momentum variables in a manner determined by the detailed nature of the function $\psi_{n\vec{k}\sigma}(\vec{r})$. Fortunately, methods have been developed in connection with solution of the band structure problem which appear to be quite useful in circumventing this, and other, problems. These methods are interpolation procedures based on an expansion of the wave functions in terms of symmetry orbitals. That is,

$$\psi_{n\vec{k}\sigma}(\vec{r}) = \sum_{\ell,\mu} a_{n\mu\sigma}(\vec{k}) \, e^{i\vec{k}\cdot\vec{R}_\ell} \, \phi_\mu^\sigma(\vec{r}-\vec{R}_\ell) , \qquad (14)$$

where the $a_{n\mu\sigma}(\vec{k})$ are expansion coefficients with n representing the band index and μ is the symmetry orbital index. The $\{\phi_\mu^\sigma\}$ are atomic-like symmetry orbitals and $\vec{R}_\ell$ is a lattice vector. For nickel we choose the first five orbitals to be d-like with t_{2g} and e_g symmetry. The remaining orbitals can be related to orthogonalized plane waves. Substitution of (14) into the band structure equation (11) leads to a matrix equation for the expansion coefficients $\{a_{n\nu\sigma}\}$ in terms of a set of parameters which arise from integrals involving the d-symmetry orbitals and OPW's. The type of approximations used to convert the band structure equation into a matrix equation determines the set of parameters to be used. In particular, different methods of approximating s-d hybridization effects lead to different interpolation formalisms.[4,5]

There are two important features of the final result. One is that it represents a way of approximately describing a given band structure in terms of a rather small set of parameters. The other feature is that once the parameters have been fixed to describe a particular band structure, the interpolation procedure provides a rapid way of determining approximate values of $a_{n\nu\sigma}(\vec{k})$ and $E(n\vec{k}\sigma)$ at arbitrary points in the BZ.

We have adopted this approach in an effort to obtain what we hope is a reasonable solution of equations (10) and (11). In particular, we have obtained the following approximate expression for the general U_{eff} and $\exp(i\vec{q}\cdot\vec{r})$ matrix elements:

$$(m\vec{k}+\vec{q}|e^{i\vec{q}\cdot\vec{r}}|n\vec{k}) \simeq F(\vec{q}) \sum_\mu a_{m\mu}(\vec{k}+\vec{q}) a_{n\mu}(\vec{k})$$

$$F(\vec{q}) = \text{magnetic form factor} \qquad (15)$$

$$(n\vec{k}_1, m\vec{k}_2|U_{eff}|i\vec{k}_3, j\vec{k}_1+\vec{k}_2-\vec{k}_3) \simeq U_{eff}^{d-d} \sum_{\mu=1}^{5} a_{n\mu}^*(\vec{k}_1) a_{m\mu}^*(\vec{k}_2) \times$$

$$a_{i\mu}(\vec{k}_3) a_{j\mu}(\vec{k}_1+\vec{k}_2-\vec{k}_3) , \qquad (16)$$

where U_{eff}^{d-d} is to be considered as an adjustable parameter. Its value will be chosen so that the band structure calculation predicts the correct spin-moment for nickel. The result given in (15) was originally derived by LW.[2] The expression in (16) takes into account only an averaged "diagonal" interaction between d-electrons and ignores s-d and s-s terms which also contribute. This means that exchange splitting between s-like states has been ignored. However, as a result of s-d hybridization effects, which are retained in the theory, there will be a non-zero moment connected with the "conduction" bands.[6]

Let us first consider the effect of the approximation given in (16) on the electronic energy bands. A close examination of the band structure model which has emerged here, as a result of the interpolation formalism coupled with the approximation given in (16), reveals that it is precisely the one proposed by HEL, with the minor exception that we have ignored the s-d exchange interaction. In particular, if equation (11) is multiplied by $\psi_{n\vec{k}\sigma}(\vec{r})$ and integrated on $\vec{r}$, then it follows from (16) that

$$E(n\vec{k}\sigma) = E_0(n,\vec{k}) + U_{eff}^{d-d} \sum_{\mu=1}^{5} |a_{n\mu}(\vec{k})|^2 f_{\mu,-\sigma}, \qquad (17)$$

$$f_{\mu,\sigma} = \sum_{n,p} |a_{n\mu}(\vec{p})|^2 f_{n\vec{p}\sigma} . \qquad (18)$$

This is the form of the electronic energy proposed by the HEL model. The most important feature of (17) is that the spin-dependent part of the energy depends on both the band index and $\vec{k}$, the wave-vector of the electron. Since this term was generated by the exchange term in (12), it can be called a momentum-dependent exchange effect. If we had ignored the band and momentum indices of the matrix element in (16) then we would have arrived at the less complicated and more familiar result that

$$E(n\vec{k}\sigma) = E_o'(n,\vec{k}) + \text{const.} \times n_{-\sigma} , \qquad (19)$$

where n_σ is the number of electrons with spin σ. The difference $E(n\vec{k}\uparrow) - E(n\vec{k}\downarrow)$ is independent of $\vec{k}$ if we use the energy expression in (19), and, therefore, we have rigid exchange splitting of the opposite-spin bands over the entire BZ. On the other hand, this difference is both band and momentum dependent if we use the energy expression in (17). In this case we have a momentum- or k-dependent exchange splitting.

Let us now turn to the problem of solving (10). As a result of the form of the approximation given in (16) the $\vec{k}$ and $\vec{p}$ dependence of the U_{eff} matrix element in (10) separates and a solution in closed form can be obtained for $\hat{G}$. Substitution of this solution into (6) gives

$$G^{-+}(\vec{q},z) \simeq \hbar |F(\vec{q})|^2 \sum_{\mu=1}^{5} \sum_{\nu=1}^{5} \sum_{\eta=1}^{5} [I+U_{eff}^{d-d}\Lambda(\vec{q},z)]^{-1}_{\mu\nu}\Lambda_{\nu\eta}(\vec{q},z)+\dots , \qquad (20)$$

where we have dropped s-d and s-s terms which are numerically small. The matrix Λ is given by

$$\Lambda_{\mu\nu}(\vec{q},z) = \sum_{\substack{n,m \\ \vec{k}}} \frac{a^*_{n\mu}(\vec{k})a_{m\mu}(\vec{k}+\vec{q})a^*_{m\nu}(\vec{k}+\vec{q})a_{n\nu}(\vec{k})\{f_{n\vec{k}\downarrow} - f_{m\vec{k}+\vec{q}\uparrow}\}}{\hbar z - E(m\vec{k}+\vec{q}\uparrow) + E(n\vec{k}\downarrow)} , \qquad (21)$$

and I is the 5 × 5 unit matrix. Equation (20) represents an improvement over previous work in that it includes both the momentum- and band-dependent effects of the U_{eff} matrix elements in a self-consistent way. These effects are incorporated into the solution in two ways. First, there is the obvious appearance of the momentum-dependent $a_{n\nu}(\vec{k})$ in the expression for Λ. Second, there is an indirect effect resulting from the k-dependent exchange term in the electronic energies. Notice also that all of the quantities appearing in (20) and (21) can be obtained from a self-consistent solution of the band structure problem. Thus, once a ferromagnetic band structure for nickel has been determined, G^{-+}, and hence χ^{-+}, is uniquely determined.

If, instead of using (16) we had assumed as before that the band- and momentum-dependence of the U_{eff} matrix element could be ignored, then by solving (10) we would have obtained in the single-band limit the enhanced susceptibility expression obtained by Izuyama et al.[7] If we retain only the $\vec{q}$ dependence of the U_{eff} matrix element appearing in (10) as a parameter, then we obtain the generalized RPA result used by Thompson[1] in the single-band limit, and by LW[2] for the multi-band case.

The form of the expression for the susceptibility then depends on whether or not the band and momentum-dependence of the U_{eff} matrix can be neglected. If it can, then we are led to the well known enhanced susceptibility expression and to a rigid splitting of the electronic bands. If not, we must consider a more complicated form for χ, like the one given in (20), and a ferromagnetic band structure model with k-dependent exchange splitting.

III. NUMERICAL RESULTS: FERROMAGNETIC ENERGY BANDS

Since the details of the calculation of the ferromagnetic energy band within the framework of the HEL model are well documented,[4,8] we will only outline briefly the basic steps of generating a "realistic" set of bands for nickel at zero temperature. First, in the interpolation approach a set of parameters must be determined which are used to describe the so-called "paramagnetic" band structure. This would have been the band structure of nickel in the event that it were not ferromagnetic. One way of determining at least initial values for these parameters is to do a traditional band structure calculation with a local potential based on a $3d^n$ configuration, where $n \simeq 9$. There are three such calculations which we have considered. Hanus' band structure corresponds to $n = 8$,[9] Mattheiss' to $n = 9$,[10] and Stocks et al. (SFW)[11] to $n = 9.4$. Once a set of parameters is chosen, a value for U_{eff}^{d-d} is then fed into a self-consistent calculation which generates the ferromagnetic band structure. The spin-moment corresponding to that value of U_{eff}^{d-d} can then be calculated. The appropriate value for U_{eff}^{d-d} is chosen to be the one that generates the experimental value of the spin-moment, which is roughly $0.55\mu_B$ at zero temperature.

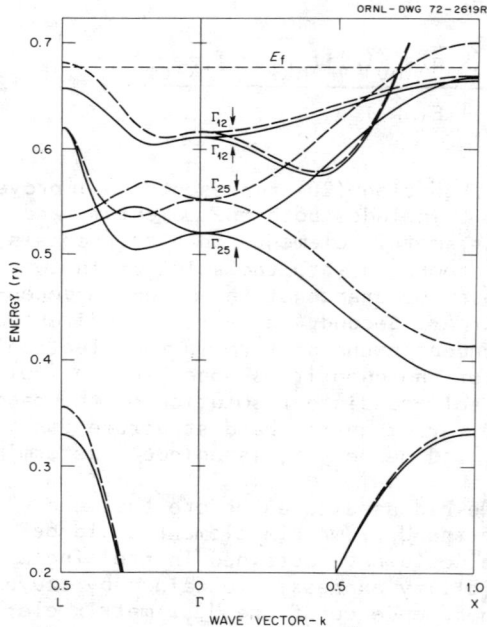

Fig. 1. Energy bands of ferromagnetic nickel with k-dependent exchange splitting. Solid curves are up-spin bands and dashed curves refer to down-spin bands. The scales along Γ to X and Γ to L are different.

An example of the type of band structure obtained in this manner is shown in Fig. 1. This particular calculation was based on the "paramagnetic" bands of SFW.[11] The ferromagnetic band structure is clearly composed of a set of narrow d-bands which cross and hybridize with a broad s-band. The solid curves represent what we have arbitrarily called up-spin bands and the dashed curves represent down-spin bands. E_f is the Fermi energy. There are several obvious features of the bands shown in Fig. 1 which follow from symmetry arguments. First, it can be shown that bands which have pure t_{2g} or e_g

symmetry along some direction are rigidly spin-split along that direction and the two symmetry types are split by different amounts. The amount of the e_g or t_{2g} splitting depends on the "number of electrons" with the e_g or t_{2g} symmetry respectively, which is automatically determined from the self-consistent solution of the band structure equations. This is clearly demonstrated by the "d-bands" along Γ to X which have either pure e_g or t_{2g} symmetry except for the e_g band which hybridizes with the s-band. When this happens, the opposite spin bands will eventually converge since we have included no s-s or s-d exchange splitting. The bands with the largest splitting have t_{2g} symmetry. Notice that at the X-point there is only one minority spin band above the Fermi energy. This agrees with de-Haas-van-Alphen measurements.[4] If we had used the rigid splitting approximation on the SFW "paramagnetic" bands and fixed the splitting parameters to give the correct spin-moment we find two minority spin bands above E_f at X, in contradiction to experiment. This incorrect feature of the rigid-band model appears to be predicted no matter what set of "paramagnetic" bands are used. This is due primarily to the fact that there is only one splitting parameter which must be relatively large in order to generate the correct spin-moment. This forces the other band in Fig. 1 (of e_g character) near E_f at X to be split by a large amount, pushing the minority spin band above E_f.

Another feature of the k-dependent band calculation is that the splitting at some arbitrary point in the BZ depends more or less on the amount of e_g (or t_{2g}) character which the wave-function at that point possesses. This is clearly demonstrated by the bands along Γ to L, where the mixing of e_g and t_{2g} symmetries increases from zero to some finite value as we move from Γ to L. The amount of the change in the spin-splitting which occurs along Γ to L is indirect evidence that the momentum-dependence of the $a_{n\nu}(\vec{k})$ is not negligible in nickel. This can and has been demonstrated explicitly by calculating the $a_{n\nu}(\vec{k})$ along various directions in the BZ.

If it had turned out that the calculation predicted little or no n or $\vec{k}$ dependence for $a_{n\nu}(\vec{k})$, then we would have been led to the approximation that the U_{eff} matrix element was a constant and thus to rigid splitting of the energy bands and the enhanced susceptibility expression for χ. However, it appears from the discussion above that the momentum and band dependence of the U_{eff} matrix element is important for nickel and that a realistic expression for χ must incorporate these effects.

IV. NUMERICAL RESULTS FOR Im $\chi^{-+}(\vec{q},\omega)$

As pointed out earlier the imaginary part of $\chi^{-+}(\vec{q},\omega)$ is proportional to the transverse inelastic neutron scattering cross-section at low temperatures. It also follows from the Greens' function formalism that the low temperature result for Im $\chi^{-+}(\vec{q},\omega)$ can be obtained by simply calculating $G^{-+}(\vec{q},\omega-i\varepsilon)$, where the limit $\varepsilon \to 0$ is implied. In this section we will present some numerical results which have been obtained for Im G^{-+} at zero temperature. Such calculations can then be compared with neutron measurements in order to test the theory.

It follows from (20) that a calculation of $\text{Im }\chi^{-+}(\vec{q},\omega)$ must begin with the evaluation of the matrix $\Lambda_{\mu\nu}(\vec{q},\omega-i\varepsilon)$. This is a complex function with the real and imaginary parts related by the Kramers-Konig relation. Thus a calculation of the imaginary part of Λ is sufficient to calculate Λ. Once Λ has been determined, it is straightforward to evaluate $[I + U_{eff}^{d=d}\Lambda]^{-1}$ and hence $\text{Im }\chi^{-+}(\vec{q},\omega)$ by using conventional computer programs designed for complex arithmetic. The calculation thus reduces to the determination of the imaginary part of $\Lambda(\vec{q},\omega-i\varepsilon)$, which is given by

$$\text{Im}\Lambda_{\mu\nu}(\vec{q},\omega-i\varepsilon) = \pi \sum_{\substack{n,m \\ \vec{k}}} W_{nm}^{\mu\nu}(\vec{k},\vec{q})\{f_{n\vec{k}\downarrow} - f_{m\vec{k}+\vec{q}\uparrow}\}\delta(\hbar\omega - E(m\vec{k}+\vec{q}\uparrow) + E(n\vec{k}\downarrow)), \quad (22)$$

$$W_{nm}^{\mu\nu}(\vec{k},\vec{q}) = a_{n\mu}^{*}(\vec{k})a_{m\mu}(\vec{k}+\vec{q})a_{m\nu}^{*}(\vec{k}+\vec{q})a_{n\nu}(\vec{k}), \quad (23)$$

where $\delta(X)$ is the delta function and at T=0

$$f_{n\vec{k}\sigma} = \begin{cases} 1 & E(n\vec{k}\sigma) < E_f \\ 0 & E(n\vec{k}\sigma) > E_f \end{cases}. \quad (24)$$

The sum of $\vec{k}$ in (22) can also be considered as an integration over the BZ.

There are basically two methods of performing the Brillouin zone sum (or integral) in (22). One is a Monte-Carlo method developed by Meuller et al.[12] The other method was developed by Gilat and Raubenheimer (GR) and involves analytic integration in small cubes.[13] We have used both methods as a cross-check on numerical results. In order to make these procedures converge we must calculate W and $E(n\vec{k}\sigma)$ at many points in the irreducible part of the BZ. As mentioned earlier, the interpolation formalism provides a procedure for rapidly calculating these quantities but we have found that it is not rapid enough to provide sufficient convergence of the integration procedures mentioned above.

In order to overcome this difficulty an additional interpolation of the energy band is performed. The quadratic interpolation scheme proposed by Meuller et al.[12] can be used to obtain an analytic representation of the electronic bands in the irreducible BZ from 916 "first principles" points which are calculated from the HEL band structure model. At present, it is not clear how to extend the quadratic interpolation procedure to the expansion coefficients, $a_{n\nu}(\vec{k})$, since these can change character quite rapidly near band crossing points. Therefore, instead of using an interpolation scheme, values for the $a_{n\nu}(\vec{k})$ are determined by storing their "first-principles" values at 916 points on a certain grid in the irreducible BZ. If we want to calculate $a_{n\nu}(\vec{k})$ at some $\vec{k}$ point, we determine the nearest grid point and call up the appropriate information.

By using these two procedures the BZ integration can be performed. It turns out that because of the relatively large number of operations that must be performed at each Monte-Carlo point the GR scheme is faster by at least an order of magnitude in performing the BZ integration. The method we use for production runs is, therefore, the GR integration scheme coupled with the quadratic interpolation of the energy bands. This is just the CLQ scheme which has been investigated and described in detail by Cooke and Wood.[14] The band sums in (22) are taken over six majority and six minority spin-bands. This includes the five d-bands and one s-band with hybridization effects.

The actual details involved in setting up the calculation so that the CLQ scheme can be used will be published elsewhere.[15] In the final analysis we obtain computer printout and plots of Im $G^{-+}(\vec{q},\omega)/|F(\vec{q})|^2$ for fixed $\vec{q}$ as a function of ω. The amount of computer time required for each $\vec{q}$ depends on the direction of $\vec{q}$. For example, it takes about 3.5 minutes on the IBM Model 360/91 per $\vec{q}$-point along [100] and about 6 minutes per $\vec{q}$-point along [111]. The time to calculate the band structure at the 916 "first principles" points and to set up the quadratic interpolation scheme for the energy band is roughly 1.5 minutes.

Most of the calculations that have been performed to date are based on ferromagnetic band structures generated from "paramagnetic" bands similar to those propsed by SFW[11] and by Zornberg.[5] These two sets of bands are quite similar to one another and are thought to provide a reasonable description of nickel's "paramagnetic" bands. If we use an interpolation procedure, such as that proposed by HEL, for describing these bands we obtain two sets of parameters which are very similar. By varying these parameters within reasonable limits we can generate a series of "paramagnetic" bands which should also be considered as reasonable for nickel. If our method of treating correlation is reasonably correct, we would expect to generate "reasonable" ferromagnetic bands from these types of "paramagnetic" bands. Instead of displaying results from a large number of calculations we have chosen to show results from a representative calculation which will be used to point out general features. These types of calculations should be considered as "demonstrative". That is, they demonstrate the type of behavior which one can obtain from the model. If this general behavior is not confirmed by experiment then we must look to more complicated models.

In Figures 2 and 3 we show results for the transverse inelastic neutron scattering cross-section, $d^2\sigma/d\omega d\Omega$, which are predicted by a calculation of Im $\chi^{-+}(\vec{q},\omega)$ based on "paramagnetic" bands which are similar to the ones proposed by SFW and Zornberg. The functions being plotted are actually histograms with a histogram box width of 2.5 meV. In both figures the relative scattering intensity is plotted versus energy for a fixed $\vec{q}$ (the momentum transfer of the neutron) which is measured in units of $2\pi/a_0$, where a_0 is the lattice constant. That is, at the X-point $|\vec{q}| = 1$. The results for each $\vec{q}$ are scaled so that the maximum is always one. We will discuss the variation of the peak height with $\vec{q}$ later on in the paper.

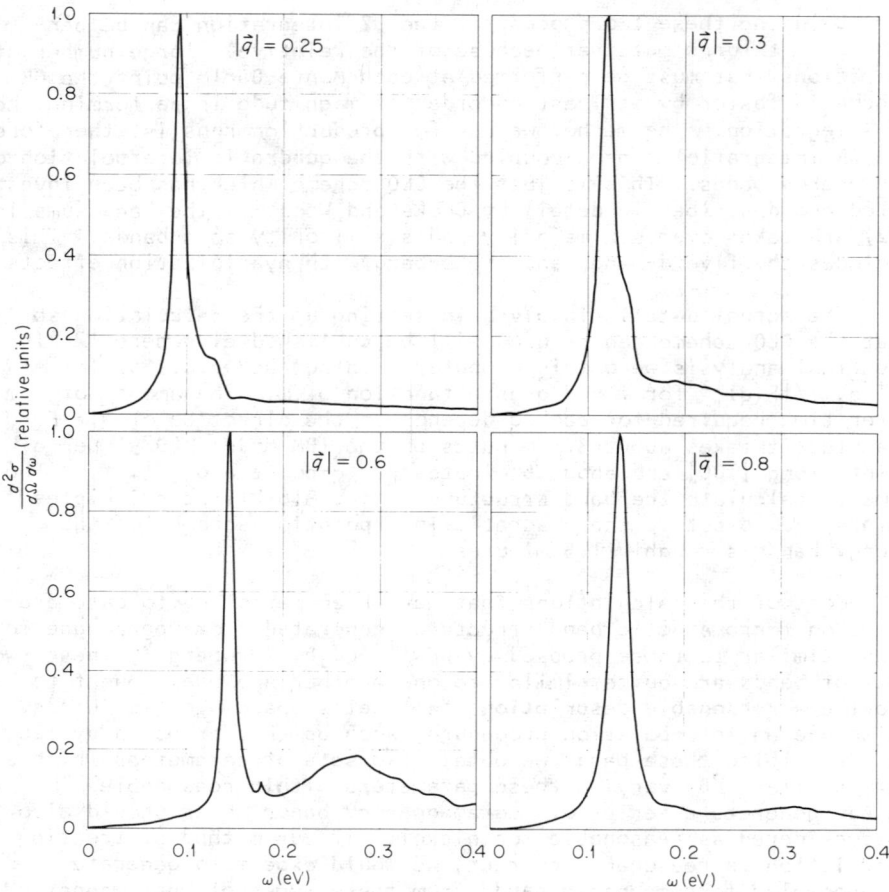

Fig. 2. Theoretical inelastic neutron scattering results for nickel for various values of neutron momentum transfer, $\vec{q}$, along [100].

The results shown in Fig. 2 are for $\vec{q}$ along [100]. The large peak in the cross-section at low $|\vec{q}|$ is, of course, the spin-wave peak. The spin-wave in the itinerant model is a bound electron-hole pair which has a net spin of one. The spin-wave dispersion relation is obtained from these calculations by locating the position of the spin-wave peak for each $\vec{q}$. The non-zero width of the spin-wave peak at low $|\vec{q}|$ is due to two effects. One is mathematical in origin and results from the histogram representation of the scattering intensity. The other, and most important, effect is connected with Stoner excitations. A Stoner excitation is defined as the excitation of an electron from a particular spin-state with momentum $\vec{k}$ below E_f into an opposite spin-state with momentum $\vec{k}+\vec{q}$ above E_f. It is useful to group these excitations by fixing $\vec{q}$ and allowing $\vec{k}$ to range over the

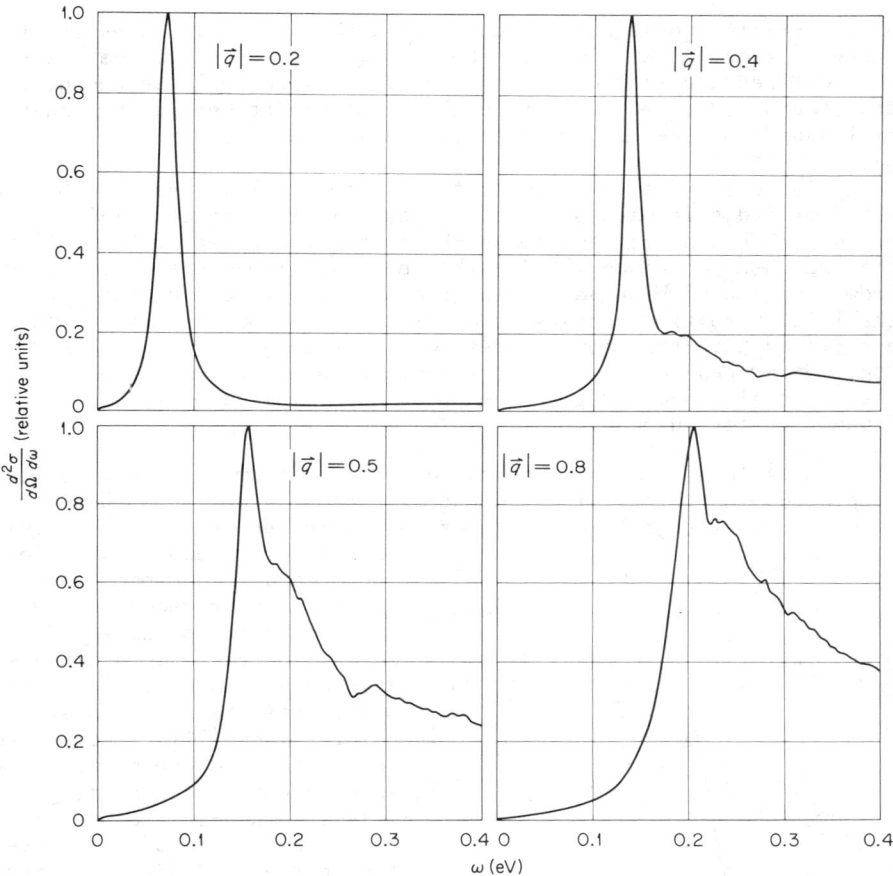

Fig. 3. Theoretical inelastic neutron scattering results for nickel for various values of neutron momentum transfer, $\vec{q}$, along [111].

BZ. In this way we can generate a density of Stoner excitations for a fixed $\vec{q}$ as a function of excitation energy. Calculations based on ferromagnetic band structures which include s-d hybridization effects reveal a non-zero density of Stoner states at all energies below a certain maximum, provided $|\vec{q}| \neq 0$.[15] That is, except for $|\vec{q}|=0$, the spin-wave is degenerate in energy with a non-zero density of Stoner excitations. As a result of this, we would expect a non-zero width of the spin-wave peak at all $|\vec{q}| \neq 0$. The well defined peaks at low $|\vec{q}|$ are the result of a low density of Stoner excitations in this region.

In the region $|\vec{q}| > 0.3$ a second broad peak begins to develop at about 200 meV. This peak could be interpreted as scattering from

Stoner excitations since it occurs in a region of high density of Stoner states.[15] Notice that for $|\vec{q}|=0.8$ the spin-wave peak is still well defined and that it has shifted to a slightly lower energy than that found for $|\vec{q}|=0.6$. Notice also that the Stoner scattering has diminished in size with respect to the spin-wave scattering.

Figure 3 shows the results for $\vec{q}$ along [111]. Again we have a well defined spin-wave peak at low $|\vec{q}|$. In the region $|\vec{q}|>0.3$ a very broad peak begins to develop which, at first, appear to be similar to the one observed along [100]. This peak is also due to scattering from a region of high density of Stoner states, but in contrast to the [100] results it occurs at a lower energy and the scattering associated with the Stoner modes continues to grow in relation to the "spin-wave" peak as $|\vec{q}|$ is increased. Even at $|\vec{q}|=0.8$ it appears that we still have a "spin-wave" peak superimposed on the broad Stoner scattering peak.

The spin-wave dispersion curve obtained from this calculation is plotted in Fig. 4. The low energy region is not shown because it appears that integration errors of less than one percent are required in order to study this region. The accuracy we expect from our integration procedure is in the range of a few percent. The solid line represents the dispersion curve along [100] and the solid dots are representative of points along [111]. Notice that the curves are approximately the same except at large energies. This is a non-trivial result since the nature of the dispersion curve in different directions depends on the detailed momentum-dependence of the energy bands and the $a_{n\nu}(\vec{k})$, which are definitely not isotropic in $\vec{k}$.

Let us now consider the behavior of the height of the spin-wave peak. As we move up the spin-wave curve in the low $|\vec{q}|$ region the peak decreases gradually while the width remains roughly constant. For $|\vec{q}|>0.3$ the behavior depends dramatically on the direction of $\vec{q}$. For $\vec{q}$ along [100] the peak height begins to decrease much faster in the region $0.3<|\vec{q}|<0.4$ than it

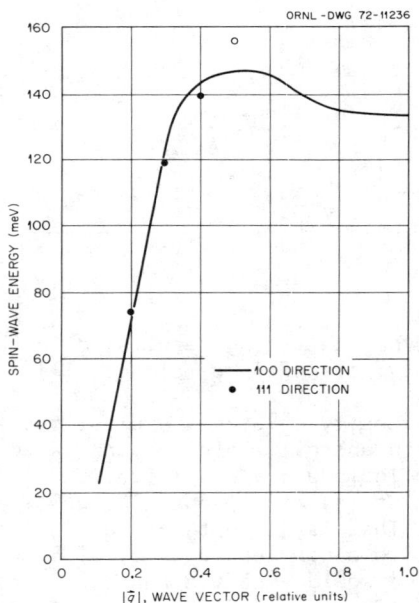

Fig. 4. Theoretical spin-wave dispersion curve for nickel. The wave-vector, $\vec{q}$, is measured in units of $2\pi/a_0$.

did in the lower $|\vec{q}|$ region. The width of the peak also broadens somewhat. For $|\vec{q}|>0.4$ the peak height increases gradually until the zone boundary is reached. At the zone boundary it is roughly an order of magnitude smaller than that found at $|\vec{q}|=0.2$ but the widths are comparable. Thus for $\vec{q}$ along [100] a reasonable spin-wave peak survives out to the zone boundary.

The behavior of the spin-wave in the [111] direction is quite different. In between $|\vec{q}|=0.3$ and 0.4 the spin-wave runs into a region of high density of Stoner states and the spin-wave peak height drops rapidly by about a factor of five. In a neutron experiment we would expect to see a well defined peak at the black dots but the "peak" located by the open circle would be washed out. Thus as far as neutron measurements are concerned, we would expect the spin-wave to disappear in the [111] direction but not in the [100].

A rough diagram of this situation is shown in Fig. 5. This diagram is intended to display only the general features of the results and, in particular, it does not give the detailed shape of the region of high density of Stoner states. The reason for this difference in behavior between the [100] and [111] directions can be attributed to the fact that the region of high density of Stoner states drops faster with $|\vec{q}|$ in the [111] direction than it does along [100]. In the [100] direction the spin-wave does not quite reach the region of high density of Stoner states and it remains a well defined magnetic excitation out to the zone boundary. It does come close enough, however, so that some scattering with respect to Stoner modes does take place. This is the reason for the rather sharp decrease in the spin-wave peak height in the region $0.3<|\vec{q}|<0.4$. For $|\vec{q}|$ above this region the energy separation of the spin-wave and Stoner regions becomes larger and the Stoner scattering gradually decreases. This is accompanied by the gradual increase in the spin-wave peak height referred to earlier. As a rule of thumb we can say that the total scattering for a fixed $\vec{q}$ is approximately independent of

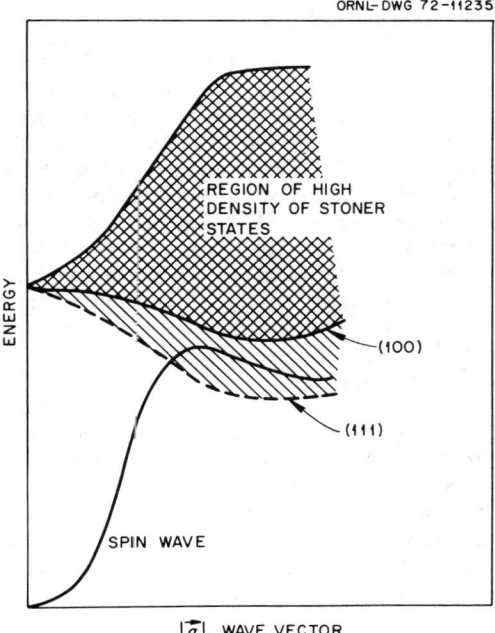

Fig. 5. Sketch of magnetic excitation spectrum for nickel. The solid line represents the bottom "edge" of the region of high density of Stoner states along [100], and the dashed line represents the "edge" along [111].

$\vec{q}$. Thus any increase in scattering from Stoner modes must be accompanied by a corresponding decrease in spin-wave scattering, and vice versa. The spin-wave along [111] runs into a region of high density of Stoner states and disappears.

The type of behavior discussed above is typical of many of the calculations which have been done. By changing slightly the parameters which control the "paramagnetic" bands a series of ferromagnetic bands can be constructed which reproduce the dispersion curve in Fig. 4 below 100 meV but which differ from one another above this energy. For example, the dispersion curve along [100] may bend over at a higher or lower energy, and in one particular case it turned out to be extremely flat after bending over. The point is that what happens in this high energy region is quite sensitive to small changes in the band structure.

V. COMPARISON WITH EXPERIMENT

Before proceeding with a comparison of our results with experiment let us first consider the predictions of the enhanced susceptibility theory. We have performed calculations similar to those reported by LW[2] based on several different "paramagnetic" band structures, and in each case we have obtained results which are consistent with their calculation. That is, we find with the rigid-splitting model that the theoretical spin-wave dispersion curve rises much too slowly with $\vec{q}$ along [100] to be in agreement with experiment. In addition, the behavior of the $\vec{q}$-dependence which is needed to force agreement between experiment and theory is similar in form and magnitude to the one they used. This suggests the argument that a non-trivial $\vec{q}$-dependent term, whose from and magnitude cannot be determined *a priori*, is necessary in order to obtain agreement between theory and experiment no matter what set of "paramagnetic" bands are used. We feel the reason for this is due in large part to the neglect of the momentum dependence of the U_{eff} matrix element.

If we turn now to the predictions of the previous section we find that the dispersion curve in Fig. 4 is in excellent agreement with some preliminary results from experiments at 4.2°K provided $\hbar\omega \leq 85$ meV.[16] These experiments, along with results from room temperature measurements which have been published,[1] also indicate that there is little, if any, directional dependence of the dispersion curve, which is also in agreement with the theory.

The agreement between theory and experiment is not as good for $\hbar\omega \geq 85$ meV. The experiments suggest that the spin-wave peak does disappear along [111] and that no large changes in the width of the peak as a function of $\vec{q}$ can be detected. These two facts are consistent with the results presented in the previous section. The energy at which the spin-wave disappears is considerably lower (< 90 meV)[1,16] than that predicted by any of the calculations. The situation in the [100] direction is worse. Although the evidence is not as conclusive, it appears that the spin-wave peak also disappears in apparent contradiction with the theory.

A possible explanation of these discrepancies in the theory can be obtained by referring to Fig. 5. Clearly if the region of high density of Stoner states could be moved down with respect to the spin-wave mode we could obtain a picture entirely consistent with experiment. This could be done by either moving the entire Stoner region or by requiring the lower limit of this region to drop faster with $\vec{q}$ in all directions. If this could be done, not only would we predict that the spin-wave peak should disappear along [111] and [100] but it would disappear in the [111] direction at a lower energy than it would along [100] which is also consistent with experiment.

In order to obtain agreement with the experiments the region of high density of Stoner states must be lowered by some 40 meV (.003ry). This is an incredibly small energy when compared to the normal scale of electronic energies. As a matter of fact, the average error in fitting one particular band structure with the HEL interpolation formalism was found to be roughly 0.007 ry.[8] This type of error could affect considerably the theoretical predictions of where the spin-wave should disappear, because this effect is directly related to energy differences between energy surfaces (one above E_f and one below E_f) which are connected by the vector $\vec{q}$. The spin-wave energy, on the other hand, would not be particularly sensitive to this type of error. Thus, it is conceivable that we could have obtained a picture similar to the one shown in Fig. 5 as a result of the interpolation formalism even though the band structure which was interpolated would have predicted agreement between theory and experiment. But, as pointed out earlier, the use of the interpolation formalism is, at present, necessary if numerical results for χ are to be obtained. It appears, therefore, that we should view our calculations of χ as being based on a model Hamiltonian, namely, the one generated by the interpolation formalism. Viewed in this manner the type of results that one would obtain with regard to the disappearance of the spin-wave might depend on the particular interpolation formalism which was adopted. It should be noted here that the errors generated by the additional quadratic interpolation of the energy bands are small compared to those produced by the interpolation formalism which was used to obtain approximate solutions of the energy band equations.

If we want to obtain better agreement between theory and experiment, we should begin by examining the possible ways of improving the calculation within the framework of our theory. These are listed below:

1. Use a different type of interpolation formalism (model Hamiltonian).

2. Use a more "realistic" set of "paramagnetic" bands.

3. Improve the treatment of correlation effects by using a "better" termination of the Green's function equation.

4. Include other interactions which may be important.

The first three points are self-explanatory. The last point was mentioned because it is possible, for example, that the spin-orbit interaction could be important in connection with the lowering of the Stoner region, since it is of the right order of magnitude and it does affect the d-bands near the X-point which contribute considerably to the susceptibility calculations. Another possibility might be the inclusion of an s-d or s-s exchange interaction. We are, at present, investigating the points listed above in an effort to improve the theory.

VI. CONCLUSIONS

An approximate method of calculating the dynamic susceptibility of nickel based on the itinerant model of magnetism has been proposed. It represents an improvement over previous work in that it incorporates momentum-dependent exchange splitting of the energy bands as well as multi-band effects. A self-consistent calculation based on this model has been presented and compared with neutron scattering experiments. Good agreement was found at energies below about 85 meV. The results above this energy were found to depend quite sensitively on the band structure and as yet only partial agreement with experiment can be obtained. The overall picture appears, however, to be reasonably good and only small changes in connection with the location of the region of high density of Stoner states are needed to obtain agreement with experiment. This may require an improvement in the theory, although it is argued that the numerical difficulties introduced by the use of an interpolation formalism to solve equations (10) and (11) could be responsible for all, or part, or the discrepancy. The results presented in this paper also indicate that as a result of the recent development of high speed computers and high flux research reactors, the neutron scattering technique can, in this particular case, be used to provide a very sensitive test of the itinerant model of magnetism.

REFERENCES

1. H. A. Mook, R. M. Nicklow, E. D. Thompson, and M. K. Wilkinson, J. Appl. Phys. **40**, 1450 (1969).

2. R. D. Lowde and C. G. Windsor, Adv. in Phys. **19**, 813 (1970).

3. J. F. Cooke, to be published in Phys. Rev. B, Jan., 1973. The Greens' function used in this reference is defined differently than the one used in this paper, but the equations of motion are identical.

4. L. Hodges, H. Ehrenreich, and N. D. Lang, Phys. Rev. **152**, 505 (1966).

5. E. I. Zornberg, Phys. Rev. B **1**, 244 (1970).

6. C. Herring, *Magnetism*, G. T. Rado and H. Suhl, eds. (Academic Press, London, 1966), Vol. 4, Chapter II.

7. T. Izuyama, D. Kim, and R. Kubo, J. Phys. Soc. Japan $\underline{18}$, 1025 (1963).

8. H. Ehrenreich and L. Hodges, <u>Methods in Computational Physics</u>, B. Alder, editor (Academic Press, New York and London, 1968).

9. J. G. Hanus, M.I.T., Solid State and Molecular Theory Group, Quarterly Progress Report, April 15, 1962, p. 25 (unpublished).

10. L. F. Mattheiss, Phys. Rev. $\underline{134}$, 192 (1964).

11. G. M. Stocks, R. W. Williams, J. S. Faulkner, Phys. Rev. B $\underline{4}$, 4390 (1971).

12. F. M. Mueller, J. W. Garland, M. H. Cohen, and K. H. Bennemann, Ann. Phys. (N.Y.) $\underline{67}$, 19 (1971).

13. G. Gilat and L. J. Raubenheimer, Phys. Rev. $\underline{144}$, 390 (1966).

14. J. F. Cooke and R. F. Wood, Phys. Rev. B $\underline{5}$, 1276 (1972).

15. J. F. Cooke and H. L. Davis, to be published.

16. H. A. Mook, private communication.

1238   Section 38.   Superexchange and Anisotropic Exchange

SUPEREXCHANGE[*]

Nai Li Huang Liu[**]

Physics Department, University of California, Riverside, CA   92502

R. Orbach[***]

Physics Department, University of California, Los Angeles, CA   90024

ABSTRACT

The theory of superexchange coupling in ionic solids is reviewed from the point of view of Keffer and Oguchi's configuration interaction, and Anderson's potential and kinetic exchange methods. Application is made to the illustrative example of right angle superexchange for d^3 ions in octahedral (weak field) environments. Identical results are exhibited for the two methods to leading order in the overlap and transfer integrals, though at first sight they appear to be different. A discussion is given of the origin of the Kanamori-Goodenough (KG) rules. Upon taking into account lack of orthogonality between cation orbitals, sign reversal occurs which is traced through both methods, and is shown to result in a net ferromagnetic interaction, in agreement with recent pair experiments for near neighbor V^{2+} ions in MgO. The systematics of right angle superexchange interactions are investigated for d^3 ions in the chalcogenides, and recent experimental work is cited in support of the predictions.

I. INTRODUCTION

This paper is intended as a critical review of current theoretical models for superexchange interactions in ionic (weakly covalent) solids. Two methods are examined in some detail: Keffer and Oguchi's[1] application of configuration interaction (CI) and Anderson's[2] potential (PE) and Kinetic (KE) exchange. The use of nonorthogonal orbitals in the former, and orthogonal orbitals in the latter, appear to cause differences between the two methods. We show in this paper that, to leading order in overlap and transfer integrals, the two methods are identical. In principle, either method can be further extended to higher accuracy. As of this writing, this has not been done because of the extreme mathematical

---

[*] A preliminary version of this paper was presented at the June, 1971 Conference on Exchange Interaction Between Ions in Crystals and Molecules, Princeton University, N.J.

[**] Supported in part by the U.S. Air Force Office of Scientific Research, Grant AFOSR 523-67, and the Univ. of Calif., Riverside, Intramural Research Grant.

[***] Supported in part by the Office of Naval Research, Contract No. N00014-69-A-0200-4032 and the National Science Foundation, Grant No. NSF GH-31973.

complexity involved.

To clarify comparisons between the two methods, and to give physical meaning to the most important of the plethora of terms which are present, we first treat direct exchange between nonorthogonal orbitals using the CI method. We then include ligand wave functions, but do not allow transfer. The transfer condition is relaxed progressively, first with respect to ligand-cation transfer, then cation-cation transfer. We delineate the relative magnitude of each contribution, and present an alternative derivation of some of the more important terms involving cation-cation overlap, giving some insight into their physical origin. We then re-calculate the superexchange using Anderson's method. We use orthogonalized linear combinations of atomic orbitals, following the procedure of Fuchikami.[3] A term by term comparison of the CI and Anderson results is presented, verifying that, to the order to which we work, the two methods are identical. Separating PE and KE, we find in the case treated explicitly in this paper (right angle [110] exchange) that the latter is not stronger than the former (indeed, the resulting exchange integral is ferromagnetic in sign). This is shown to be a consequence of McWeeny's theorem[2,4] acting on KE via a reduction of the <u>transfer</u> integral (it usually reduces PE) whereas for this geometry PE is essentially unaffected.[5] This results in a reversal of the Kanamori-Goodenough (KG) rules. The work which forms the basis of this study was performed by one of us, (N.L.H.L.), and the explicit algebraic and numeric details can be found in a series of papers[6-9] and an (unpublished) thesis.[10]

Rather than treating a variety of physical systems, we specialize to a rock salt structure, and consider the informative example of superexchange interactions between near neighbor (right angle [110]) d^3 configurations in weak octahedral fields (e.g. MgO:V^{2+}). Though this may appear restrictive, in fact we shall exhibit all of the known physically important mechanisms, and be able to demonstrate the important role played by cation-cation overlap.

In the next section (II), we derive the superexchange for MgO:V^{2+} near neighbors using the CI approach. We display the physical arguments which lead to the KG rules for right angle superexchange. We include the effects of cation-cation nonorthogonality, and exhibit their modifications of the KG rules. We compute in Section III the superexchange for the same ionic system using Anderson's method, working to lowest order in overlap and covalency. We demonstrate that, to this order, the two methods are identical, and we isolate terms in Anderson's PE and KE which correspond to equivalent terms in the CI approach. Using Anderson's physical picture, involving McWeeny's theorem, we display from yet another viewpoint the physical origin of the modified (indeed, reversed!) KG rules. We do not claim that we can execute quantitative calculations with a high degree of accuracy because of, for example, difficulty in evaluating multicenter integrals and difference between the radial distribution of e_g and t_{2g} wave functions in solid as compared to Hartree Fock ionic calculations. Rather, we shall establish physical origin of terms so that understanding (indeed, prediction) of the systematics of superexchange can be obtained. This is carried out in Section IV for the right angle superexchange between d^3 ions. Supporting evidence of our prediction for the systematics is given in recent experiments by Berger et al.[11]

II. CALCULATION OF THE EXCHANGE INTEGRAL, CI METHOD

We treat in detail the case of a weakly covalent d^3 configuration in an octahedral environment, in a [110] relationship to its neighboring d^3 cation. In a rock salt structure, this amounts to consideration of the nearest magnetic neighbor interaction. The extension to next near neighbors, or other d^n configurations, follows without conceptual difficulty. We divide our examination into a number of parts, beginning with simple and progressing to more complex models.

A. IONIC CONFIGURATION - DIRECT EXCHANGE

We begin our discussion with a single (ground) configuration. Consider the right angle array displayed in Fig. 1. For simplicity,

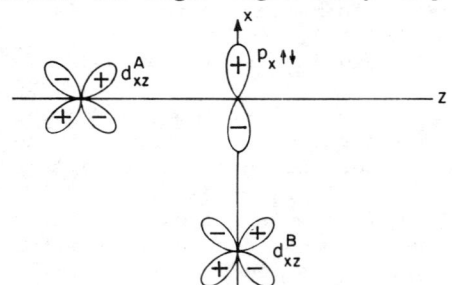

Fig. 1. A four-electron model in the ionic configuration for the near neighbor exchange in MgO:V^{2+}.

we display only the d_{xz} and p_x cation and ligand orbitals respectively. Also occupied are the d_{yz} and d_{zx} (singly), and the p_y and p_z orbitals (doubly). It is easily seen that the doubly occupied p_x orbital is orthogonal to the singly occupied d_{xz} orbital of cation B. Nor are the wave functions d_{xz}^A and d_{xz}^B orthogonal to one another. There are but four electrons in this simplified picture, so the obvious first step is the construction of the 4 x 4 Slater determinant using the d_{xz}^A, d_{xz}^B, $p_x\uparrow$ and $p_x\downarrow$ orbitals. We have suppressed the spin indices of cations A and B for the moment. The Hamiltonian for the four electron system is of the form,

$$\mathcal{H} = -\Sigma_i (\hbar^2/2m_i)\nabla_i^2 - \Sigma_{i,g}(Z_g e^2/r_{ig}) + \Sigma_{i>j} e^2/r_{ij}, \quad (1)$$

where i labels the ith electron, and g the appropriate nucleus with effective charge Z_g. The first two terms in (1) will be referred to as the one-electron part of the Hamiltonian, $\Sigma_i \mathcal{H}(i)$, the last is the two-electron part. The difference in sign between the one- and two-electron parts is crucial to the sign of the superexchange integral we shall ultimately calculate. The first order energy of (1) using the Slater determinant is complicated. To simplify our discussion, we initially ignore the p_x electrons, leaving us with a two-electron problem. The first order energy W is found from the solution of

$$\langle d_{xz}^A(1)d_{xz}^B(2)|\mathcal{H}-W|d_{xz}^A(1)d_{xz}^B(2)\rangle$$
$$\mp \langle d_{xz}^A(1)d_{xz}^B(2)|\mathcal{H}-W|d_{xz}^B(1)d_{xz}^A(2)\rangle = 0, \quad (2)$$

where the upper (lower) sign leads to the energy of the triplet (singlet), and the parentheses contain the electron designation. If d_{xz}^A and d_{xz}^B were orthogonal, the second term in (2) would reduce immediately to

$$\mp \langle d_{xz}^A(1)d_{xz}^B(2)|e^2/r_{12}|d_{xz}^B(1)d_{xz}^A(2)\rangle. \quad (3)$$

Slater[12] has proven that the matrix element in (3) is always positive, so that (2) would exhibit a lowering of the energy eigenvalue W for parallel (triplet) or ferromagnetic coupling of the spins. Forming the singlet and triplet combinations, we define an exchange energy coupling constant J

$$J = \tfrac{1}{2}(W_s - W_t), \quad (4)$$

so that the spin dependent interaction energy is written as

$$-2J(d_{xz}^A, d_{xz}^B)\vec{\mathcal{S}}_{xz}^A \cdot \vec{\mathcal{S}}_{xz}^B. \quad (5)$$

For orthogonal orbitals, following (3), we obtain

$$J(d_{xz}^A, d_{xz}^B) = \langle d_{xz}^A(1)d_{xz}^B(2)|e^2/r_{12}|d_{xz}^B(1)d_{xz}^A(2)\rangle. \quad (6)$$

If we remember that three magnetic electrons reside on each cation, A and B, we can define an overall exchange Hamiltonian by generalizing (5) to

$$\mathcal{H}_{ex} = -2\Sigma_{\alpha,\beta} J_{\alpha\beta} \vec{\mathcal{S}}_\alpha \cdot \vec{\mathcal{S}}_\beta, \quad (7)$$

where α and β label occupied spin orbitals on cation sites A and B, respectively. Then (5) is but one term of nine contained in (7). The extension of (7) to the usual form $\mathcal{H} = 2J_{AB}\vec{S}_A \cdot \vec{S}_B$ is only valid when all the $J_{\alpha\beta}$ are equal, or when $\vec{S}_A \parallel \vec{\mathcal{A}}_\alpha$ and $\vec{S}_B \parallel \vec{\mathcal{A}}_B$ for all α and β.[13] It turns out that, for all the d^3 configurations we are considering, the latter condition is fulfilled. We can, therefore, replace $\vec{\mathcal{A}}_\alpha \cdot \vec{\mathcal{A}}_\beta$ in (7) by $\frac{1}{9}\vec{S}_A \cdot \vec{S}_B$, (see Eq. (12) below) and obtain the usual form for the exchange coupling.

The interesting feature of (2) is what happens when the orbitals A and B are nonorthogonal, as in the present instance. Then, it is no longer true that the second, or exchange term, in (2) reduces to (3). Rather, contained in (2) are one-electron exchange terms of the form

$$\langle d^A_{xz}(1) d^B_{xz}(2) | \mathcal{H}(1) + \mathcal{H}(2) | d^B_{xz}(1) d^A_{xz}(2) \rangle$$
$$= 2T \langle d^A_{xz}(1) | \mathcal{H}(1) | d^B_{xz}(1) \rangle, \tag{8}$$

where $T = \langle d^B_{xz}(2) | d^A_{xz}(2) \rangle$ is the cation-cation overlap integral, and the matrix element of $\mathcal{H}(1)$ is negative, primarily due to the electron-nuclear coupling. The energy eigenvalue W is raised for parallel spins, and the stability of the singlet, or antiferromagnetic coupling of the spins, is enhanced. The difference between singlet and triplet energies is now,[14]

$$\frac{1}{2}(W_s - W_t) = [\langle d^A_{xz}(1) d^B_{xz}(2) | e^2/r_{12} | d^B_{xz}(1) d^A_{xz}(2) \rangle$$
$$+ 2T \langle d^A_{xz}(1) | \mathcal{H}(1) | d^B_{xz}(1) \rangle$$
$$- T^2 \langle d^A_{xz}(1) d^B_{xz}(2) | \mathcal{H} | d^A_{xz}(1) d^B_{xz}(2) \rangle] / (1 - T^4) \tag{9}$$
$$\equiv J(d^A_{xz}, d^B_{xz}).$$

For physical systems and interatomic distances, it is usuall found that the second (negative) term in (9) dominates the first (positive). The third term represents the difference in the coulomb energy between the singlet and the triplet state arising from the fact that the normalization factors for these states are different. The one-electron terms in the third matrix element in (9) dominant, and are negative, but are smaller in magnitude than the sume of the first two terms. Hence, J, as defined by (4), is negative (antiferromagnetic). As an example, for MgO:V^{2+} near neighbor pairs, (9) yields[9,10] a contribution of -164°K to $J(d^A_{xz}, d^B_{xz})$. This can be thought of as direct exchange, for we have omitted the role of the ligand orbitals.

To include the effect of the ligands, we return to the four-electron problem, and calculate W_s and W_t using linear combinations of determinantal wave functions constructed to be eigenstates of S_z and S^2. Direct computation of the energy difference (4) generates, in addition to (9), the following terms:

$$+4\langle p_x(1)d^A_{xz}(2)|e^2/r_{12}|p_x(1)d^B_{xz}(2)\rangle T$$
$$-2\langle p_x(1)d^B_{xz}(2)|e^2/r_{12}|p_x(1)d^B_{xz}(2)\rangle T^2$$
$$-2\langle p_x(1)d^A_{xz}(2)|e^2/r_{12}|p_x(1)d^A_{xz}(2)\rangle T^2$$
$$-2\langle d^A_{xz}(1)p_x(2)|e^2/r_{12}|p_x(1)d^B_{xz}(2)\rangle T$$
$$-2\langle d^B_{xz}(1)p_x(2)|e^2/r_{12}|d^A_{xz}(1)d^B_{xz}(2)\rangle S \qquad (10)$$
$$+\langle d^B_{xz}(1)p_x(2)|e^2/r_{12}|p_x(1)d^B_{xz}(2)\rangle S^2$$
$$-2\langle p_x(1)|\mathcal{H}(1)|d^B_{xz}(1)\rangle ST$$

where $S = \langle p_x|d^A_{xz}\rangle$. We denote (10) as the ligand ionic configuration contribution to $J(d^A_{xz},d^B_{xz})$. The largest term in (10), the first, arises from the following source: the repulsive interaction (ferromagnetic) of the overlap charge density between one cation and the delocalized part of the other (hence, the proportionality to T), with the coulomb potential of the p_x electron. We shall derive these expressions in a more physical way in Section IIE by considering directly the energy difference between the parallel and the antiparallel state. Further, we shall demonstrate in Section III that these terms, as well as the antiferromagnetic one-electron terms in (9), are contained in Anderson's KE, even though no transfer (in the CI sense) is present. This occurs because of a peculiar cancellation between part of the square of the transfer integral in the numerator and the energy transfer denominator of KE.

Explicit evaluation of (10) for MgO:V^{2+} yields a ferromagnetic contribution equal to $64°K$. By virtue of the fact that p_z plays a role equivalent to p_x, and that there are two O^{2-} ions forming right angle bonds with the V^{2+} ions, ligand dependent terms contribute a total of four times this amount, or $256°K$ (ferromagnetic), to $J(d^A_{xz},d^B_{xz})$.

So far we have maintained the ground (or ionic) configuration occupancy in our computation of J. We now relax this condition and allow for anion-cation transfer.

B. ANION-CATION TRANSFER SUPEREXCHANGE

Two equivalent anion-cation transfer configurations are exhibited in Figs. 2a and 2b. The energy "cost" for transfer is approximately 0.69 a.u. (= 18.8 eV), and comprises: (i) the difference between the unoccupied energy of a cation and the ionization energy of the ligand, (ii) the difference between Madelung energies at the cation and anion site (large positive energy), (iii) the electronic polarization energy of the surrounding ions (negative energy), and (iv) the interaction energy between the extra electron at the cation

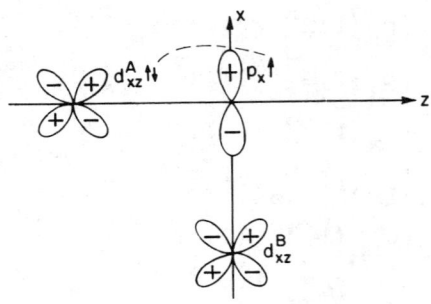

Fig. 2a. Anion→cation transfer.

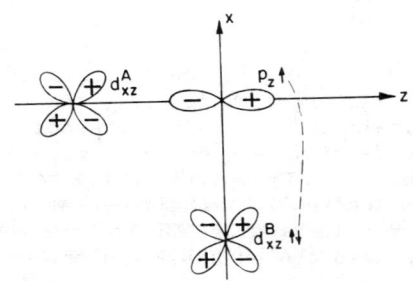

Fig. 2b. Anion→cation transfer.

site and the hole at the anion site (negative energy). The virtual character of this transfer, and the large energy cost implying rapid time variation of the virtual wave function admixture, does not allow for ionic or electronic redistributions during the transfer process, and is the reason why we have used unoccupied orbital energies and wave functions, as well as static lattice positions.

If we focus on Fig. 2a, we see that transfer of p_x is allowed by symmetry only to d^A_{xz}. If the spin of A is ↑ (defined as positive), this implies a positive spin density remaining in the p_x orbital at the ligand site, for only a p_x ↓ electron could have been transferred. The resulting spin polarized ligand now interacts through (1) with the d^B_{xz} orbital. However, p_x is orthogonal to that orbital, and only ferromagnetic coupling energies (two-electron, as in (3)) are present. The lowest energy state will be that appropriate to a d^B_{xz} ↑ spin ordering. This mechanism is an integral part of the KG rules. Yet, caution must be exercised. We shall demonstrate below that cation-cation nonorthogonality for right angle superexchange turns out to nearly cancel the effectiveness of this (ferromagnetic) mechanism.

We compute $W_s - W_t$ explicitly for anion-cation transfer, and find two kinds of terms. The first is just the KG mechanism described above: when combined with the S^2 term in (10), it yields the positive (ferromagnetic) anion-cation transfer exchange energy equal to

$$2\lambda^2 \langle p_x(1) d^B_{xz}(2) | e^2/r_{12} | d^B_{xz}(1) p_x(2) \rangle ,$$

where λ is the sum of the covalency (transfer parameter γ) and the overlap integral S. That part dependent on transfer (we have already included the overlap contribution in (10)) equals 15°K for

MgO:V^{2+}. However, cation-cation nonorthogonality generates a plethora of additional terms which, when summed, generate a negative (antiferromagnetic) anion-cation transfer exchange energy equal to $-10°$K. Thus, the net ferromagnetic strength of this process, $5°$K, is reduced in MgO:V^{2+} by two thirds from that predicted by the KG mechanism alone because of cation nonorthogonality. Taking into account the second anion in Fig. 2, a contribution equal to that from the first is generated, so that the overall contribution of the anion-cation transfer process to $J(d^A_{xz}, d^B_{xz})$ is $+10°$K.

So far we have treated only the occupied d_{xz} orbitals. However, the occupied d^A_{yz} and d^B_{xy} orbitals can couple via the p_y orbital in an anion-cation transfer process. Figure 3 exhibits the allowed process. Letting the spin of the d^A_{yz} orbital be ↑, an electron can transfer from the p_y ligand only if it is ↓, leaving a net ↑ spin density on the p_y site. However, in this instance, the spin polarized p_y orbital is <u>not</u> orthogonal to the d^B_{xy} orbital, and the one-electron terms in (1) would be expected to generate a large antiferromagnetic contribution, in consonance with the KG rules. Indeed, we calculate a contribution of $-93°$K for this process in MgO:V^{2+}. However, in this instance, cation-cation nonorthogonality acts in the opposite direction again, and actually reverses the sign of the interaction, so that an overall ferromagnetic sign results. The physical origin of this reversal will be exhibited in Sec. IIE. We compute a nonorthogonality contribution of $108°$K, leading to a net ferromagnetic interaction of $15°$K for a mechanism which the KG rules would have predicted to be antiferromagnetic. The orbitals d^A_{xy} and d^B_{yz} also generate the same contribution, so that, from anion-cation transfer,

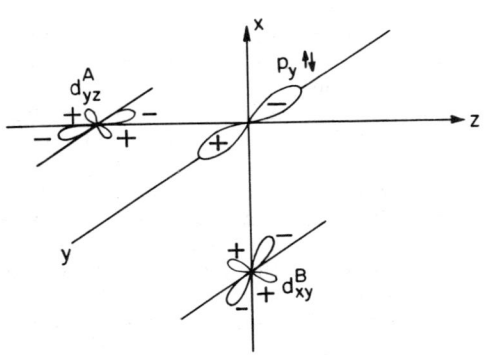

Fig. 3. A four-electron model in the ionic configuration where the ligand orbitals overlap with the cation orbitals at both sites and the cation orbitals are nonorthogonal.

$$J(d^A_{yz}, d^B_{xy}) = J(d^A_{xy}, d^B_{yz}) = 15°K .$$

This completes our study of anion-cation transfer processes.

The last important excited configuration which we shall consider is that appropriate to Fig. 4. An electron is transferred directly from the d^A_{xz} orbital to the d^B_{xz} orbital, or vice versa. This is designated as the cation-cation transfer process.

C. CATION-CATION TRANSFER SUPEREXCHANGE

This excitation process leaves unchanged the Madelung energy, but costs (1) the (large and positive) energy difference between the ionic cation configuration energies and those appropriate to the transferred states, (ii) the (negative) electronic polarization energy of the surrounding ions, and (iii) the (negative) interaction energy between the extra electron at one cation site and the hole at the other. For V^{2+} ions in MgO, the sum of these three terms is 0.47 a.u., or 12.8 eV. This is to be compared to transfer energy of 18.8 eV found for anion-cation transfer. The smaller energy denominator enhances the importance of this process relative to the anion-cation transfer process. The Pauli principle dictates that cation-cation transfer must lead to stabilization of the antiferromagnetic coupling for d^3 ions, because of the half filled character of the t_{2g} shell.

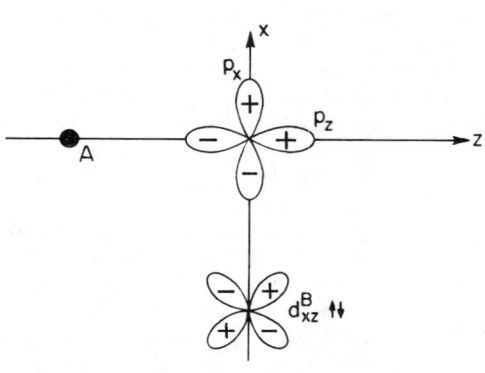

Fig. 4. Cation-cation transfer configuration.

A simple computation of the strength of this mechanism can be obtained from the 2 x 2 transfer energy determinant:

$$\begin{vmatrix} -W & b \\ b & -W + U \end{vmatrix} = 0, \qquad (11)$$

where b is the cation-cation transfer integral, and U the transfer energy. The energy of the ground (ionic) configuration is lowered by an amount $-b^2/U$ according to (11). At first glance this may appear identical to Anderson's KE result. However, as shown in Section III, the definition of the states used to calculate b and U is different in his method. Though this difference in definitions which leads to (11) and Anderson's KE are not significant in a 180° arrangement, the 90° arrangement is another matter. The large cation-cation overlap will be shown in III, to sharply reduce b in the KE expression from that value contained in the CI expression (11). This point has not always been appreciated.[15] Explicit evaluation of (11) for MgO:V^{2+} leads to $J(d_{xz}^A, d_{xz}^B) = -48°K$.

D. NET EXCHANGE COUPLING

We can combine the results of sub-sections II A-C to obtain values for the nine $J_{\alpha\beta}$, as defined in (7), and thence arrive at an overall exchange coupling. Summarizing our previous results for $J(d_{xz}^A, d_{xz}^B)$ in MgO:V^{2+},

$$J(d_{xz}^A, d_{xz}^B) = -164°K \quad \text{(direct exchange, no ligands)}$$
$$+256°K \quad \text{(ionic configuration, contribution from ligands)}$$
$$+ 10°K \quad \text{(anion-cation transfer)}$$
$$- 48°K \quad \text{(cation-cation transfer)}$$

we obtain,

$$J(d_{xz}^A, d_{xz}^B) = 54°K \quad \text{(ferromagnetic)}.$$

Returning to (7), one notes that all the electronic states in the V^{2+} ground states are parallel, so that we may write the full exchange coupling as[13]

$$\mathcal{H}_{ex} = -2\Sigma_{\alpha,\beta} J_{\alpha\beta} \vec{\lambda}_\alpha \cdot \vec{\lambda}_\beta$$
$$= -\frac{2}{9}(\Sigma_{\alpha,\beta} J_{\alpha\beta}) \vec{S}_A \cdot \vec{S}_B \quad (12)$$
$$\equiv -2J_{AB} \vec{S}_A \cdot \vec{S}_B ,$$

defining J_{AB}. Using the above value for $J(d_{xz}^A, d_{xz}^B)$, and the values for the other $J_{\alpha\beta}$ derived in previous sections of this paper, we total (12) to find,

$$J_{AB} = 9°K$$

for MgO:V^{2+} near-neighbor pairs. This number should be compared to the experimental result of Codling and Henderson,[16] who find $J_{AB} = 5.3 \pm 2.1°K$. Considering the cancellation of so many large terms, the agreement between our calculation and experiment is surprisingly good.

E. PHYSICAL ORIGIN OF THE CI CONTRIBUTIONS TO J

The purpose of this subsection is to present a simple derivation of some of the more important contributions to the superexchange found in the CI method involving cation-cation nonorthogonality: (9) and (10). For such a purpose, it is more convenient to work with parallel and antiparallel states, each described by a single determinental wave function. For antiparallel spins on the cations,

$$\psi_a = \frac{1}{\sqrt{4!}} \left| \frac{d_{xz}^A - Sp_x}{\sqrt{1-S^2}} \uparrow , p_x \uparrow , p_x \downarrow , d_{xz}^B \downarrow \right|, \quad (13)$$

with $\langle p_x | d_{xz}^B \rangle = 0$

The parallel state determinant, with orthogonalized wave functions, has the form

$$\psi_p = \frac{1}{\sqrt{4!}} \left| \frac{d_{xz}^A - Sp_x}{\sqrt{1-S^2}} \uparrow , p_x \uparrow , p_x \downarrow , \frac{d_{xz}^B - \frac{T}{1-S^2}(d_{xz}^A - Sp_x)}{\left(1 - \frac{T^2}{1-S^2}\right)^{\frac{1}{2}}} \uparrow \right|. \quad (14)$$

Calling $\varphi^A = (d_{xz}^A - Sp_x)/\sqrt{1-S^2}$

and $\varphi^B = [d_{xz}^B - \frac{T}{1-S^2}(d_{xz}^A - Sp_x)]/\sqrt{1-[T^2/(1-S^2)]}$,

all of (9) and (10) (note that no transfer has been included at this stage of the CI method) are contained in the following expression (additional terms are present below, but they are of higher order in overlap than are in (9) and (10)),

$\langle \psi_a | \mathcal{H} | \psi_a \rangle - \langle \psi_p | \mathcal{H} | \psi_p \rangle$

$= \langle d_{xz}^B | \mathcal{H}_1 | d_{xz}^B \rangle - \langle \varphi^B | \mathcal{H}_1 | \varphi^B \rangle$

$+ \langle d_{xz}^B(1)\varphi^A(2) | \frac{e^2}{r_{12}} | d_{xz}^B(1)\varphi^A(2) \rangle - \langle \varphi^B(1)\varphi^A(2) | \frac{e^2}{r_{12}} | \varphi^B(1)\varphi^A(2) \rangle$

$+ 2\langle p_x(1)d_{xz}^B(2) | \frac{e^2}{r_{12}} | p_x(1)d_{xz}^B(2) \rangle - 2\langle p_x(1)\varphi^B(2) | \frac{e^2}{r_{12}} | p_x(1)\varphi^B(2) \rangle$

$- \langle d_{xz}^B(1)p_x(2) | \frac{e^2}{r_{12}} | p_x(1)d_{xz}^B(2) \rangle + \langle \varphi^B(1)p_x(2) | \frac{e^2}{r_{12}} | p_x(1)\varphi^B(2) \rangle$

$+ \langle \varphi^B(1)\varphi^A(2) | \frac{e^2}{r_{12}} | \varphi^A(1)\varphi^B(2) \rangle$,

because we have chosen the first three orbitals in ψ_a, (13) to be the same as those in ψ_p, (14). The leading (ferromagnetic) term, $4T\langle p_x(1)d_{xz}^A(2)|e^2/r_{12}|p_x(1)d_{xz}^B(2)\rangle$, in (10) arises from the difference between the coulomb integrals $2\langle p_x(1)d_{xz}^B(2)|e^2/r_{12}|p_x(1)d_{xz}^B(2)\rangle$ and $2\langle p_x(1)\varphi^B(2)|e^2/r_{12}|p_x(1)\varphi^B(2)\rangle$. Physically, we see that the energy of the parallel state is lowered by the coulomb interaction between p_x electrons and the overlap charge density between one d orbital and the delocalized part of the other (i.e., the cross term $-2T\,d_{xz}^A d_{xz}^B$ in $\varphi^{B*}\varphi^B$). Hence a ferromagnetic interaction is obtained.

III. CALCULATION OF THE EXCHANGE INTEGRAL, ANDERSON METHOD

We compute, in this section, the exchange integral for the right angle, d^3, configuration treated in II, but using Anderson's method. The important physical point connected with his approach is the use of exact one-electron one-cluster, eigenfunctions in the Wannier sense. Each cation wave function is to be calculated in the presence of all the anions in the crystal. Then, antiparallel cation wave functions are allowed to expand from their cluster values, overlapping the neighboring cation cluster. The expansion is expressed in terms of cation-cation electron transfer, called KE because of the concomitant lowering of the kinetic energy cost of localization. Though, in principle, this method is exact in a one-electron sense, in practice such wave functions are not yet available for real calculations (but see Ellis and Freeman, Ref. 17). Thus, to date in the literature, use has been made of only orthogonalized linear combinations of atomic orbitals[3] for the computation of the exchange integral with Anderson's method (and, for that matter, for the CI method as well).

In order to compare with the CI results exhibited in II, we use the orthogonalized LCAO approximation of Ref. 3 to construct the one electron states. The wave functions have already been displayed in (13) and (14). In order to conform to the notation of Ref. 3, we rearrange these orbitals, symmetrizing the antibonding LCAO's, and allow for anion to cation transfer γ by replacing S with $\lambda = \gamma + S$. According to the three-center approximation displayed in Fig. 1, these wave functions are

$$\psi_A = \frac{1}{N_1}[d_{xz}^A - \lambda p_x - \tfrac{1}{2}T(d_{xz}^B - \lambda p_z)]; \tag{15}$$

$$\psi_B = \frac{1}{N_1}[d_{xz}^B - \lambda p_z - \tfrac{1}{2}T(d_{xz}^A - \lambda p_x)], \tag{16}$$

where N_1 is the normalization factor. The potential and kinetic exchange are given by,

$$J_{pot} = \int \psi_A^*(1) \psi_B^*(2) \frac{e^2}{r_{12}} \psi_B(1) \psi_A(2) d\tau_1 d\tau_2; \quad (17)$$

$$J_{kin} = -2b^2/U, \quad (18)$$

where the transfer integral b is defined as

$$b = \int \psi_A^* \mathcal{H} \psi_B d\tau. \quad (19)$$

$\mathcal{H}$ is the Hartree-Fock Hamiltonian, corresponding to antiparallel spin arrangement.[3] For our simple approximation of only a single d orbital at the site of a magnetic cation, we can write

$$\begin{aligned}b = &\int \psi_A^*(1) \mathcal{H}_1 \psi_B(1) d\tau_1 \\ &+ 2\int \psi_A^*(1) \psi_B(1) e^2/r_{12} [\Phi_x^*(2)\Phi_x(2) + \Phi_z^*(2)\Phi_z(2)] d\tau_1 d\tau_2 \\ &- \int \psi_A^*(1) \psi_B(2) \frac{e^2}{r_{12}} [\Phi_x^*(2)\Phi_x(1) + \Phi_z^*(2)\Phi_z(1)] d\tau_1 d\tau_2 \\ &+ \int \psi_A^*(1) \psi_B(1) \frac{e^2}{r_{12}} \psi_A^*(2) \psi_A(2) d\tau_1 d\tau_2, \end{aligned} \quad (20)$$

where Φ_x and Φ_z are the bonding orbitals defined by

$$\Phi_x = \frac{1}{N_2}(p_x + \gamma d_{xz}^A); \quad \Phi_z = \frac{1}{N_2}(p_z + \gamma d_{xz}^B).$$

Remembering that by orthogonalizing orbitals for the magnetic electrons, we introduce a term $-T d_{xz}^A/2$ in ψ_B. The presence of this term results in a reduction in the two-center coulomb interaction caused by the one-center coulomb repulsion

$$T \langle d_{xz}^A(1) d_{xz}^A(2) | e^2/r_{12} | d_{xz}^A(1) d_{xz}^A(2) \rangle / 2,$$

as is shown by substituting (15) and (16) into the last term in b. As we shall point out below, this one-center coulomb repulsion term leads to a cancellation with the leading term in the transfer energy U in the mechanism of KE, giving rise (together with PE) to results identical to CI calculation. This, in fact, is what we ought to expect for the following reason. In an antiparallel state, the orbital parts of the wave function for magnetic electrons need not be orthogonal. Orthogonalization is done only for convenience. But by doing this, one introduces a node in between the two cations, which does not really belong there. However, transfer between magnetic electron orbitals fills in the hole, and effectively we get back to the CI contribution.

The transfer energy U in (18) is given by

$$U = \langle \psi_A(1)\psi_A(2)|\frac{e^2}{r_{12}}|\psi_A(1)\psi_A(2)\rangle \qquad (21)$$
$$- \langle \psi_A(1)\psi_B(2)|\frac{e^2}{r_{12}}|\psi_A(1)\psi_B(2)\rangle,$$

with appropriate corrections caused by electronic re-arrangement (after transfer) and polarization. Substituting (15) and (16) into (21), we find

$$U = \langle d_{xz}^A(1)d_{xz}^A(2)|\frac{e^2}{r_{12}}|d_{xz}^A(1)d_{xz}^A(2)\rangle$$
$$- \langle d_{xz}^A(1)d_{xz}^B(2)|\frac{e^2}{r_{12}}|d_{xz}^A(1)d_{xz}^B(2)\rangle + O(S^2)$$
$$\equiv U_0 - U_1 + O(S^2).$$

We expand the denominator U with respect to the smallness parameter[18] U_1/U_0, as this ratio is found to be ~22% numerically. The strong ferromagnetic term, $4T\langle p_x(1)d_{xz}^A(2)|e^2/r_{12}|p_x(1)d_{xz}^B(2)\rangle$, displayed for the CI method by (10), is in fact also present in the kinetic exchange expression (18). The leading terms arising from the second and fourth integrals in (20) are found to be

$$2\langle d_{xz}^A(1)p_x(2)|e^2/r_{12}|d_{xz}^B(1)p_x(2)\rangle \text{ and}$$
$$-T\langle d_{xz}^A(1)d_{xz}^A(2)|e^2/r_{12}|d_{xz}^A(1)d_{xz}^A(2)\rangle/2,$$

respectively. When squared, the cross term between them cancels the transfer energy U_0 in the denominator, leaving the dominant CI ferromagnetic term displayed above and in (10). The presence of this ferromagnetic component, $2\langle d_{xz}^A(1)p_x(2)|e^2/r_{12}|d_{xz}^B(1)p_x(2)\rangle$ in b greatly reduces the strength of the antiferromagnetic reminder of b: $\langle d_{xz}^A|\mathcal{H}_1|d_{xz}^B\rangle - T\langle d_{xz}^B|\mathcal{H}_1|d_{xz}^B\rangle + \langle d_{xz}^A d_{xz}^B|e^2/r_{12}|d_{xz}^B d_{xz}^A\rangle - 2T\langle p_x d_{xz}^B|e^2/r_{12}|p_x d_{xz}^B\rangle$
$-T\langle d_{xz}^A d_{xz}^B|e^2/r_{12}|d_{xz}^A d_{xz}^B\rangle$, causing a substantial reduction in the magnitude of KE (which, of course, is always antiferromagnetic overall). This is an example of McWeeny's theorem. The orthogonality condition on the overlapping cation wave functions introduces a node between the cations, greatly reducing b, the transfer integral. However, McWeeny's theorem does not affect the anion to cation transfer term for right angle superexchange, which in turn dominate the ferromagnetic potential exchange. The node between the cations caused by orthogonalization between magnetic electrons thus leads to a substantial reduction of kinetic exchange, with potential exchange essentially unaffected. This is quite the opposite to the collinear case. Because of the proximity of the two magnetic ions in the right angle case, one might worry that whether the interaction between these ions could significantly alter

the wave function after transfer, thus causing self-consistency difficulty with Anderson's basis set of wave functions. However, we find that this is really not the case. The cancellation within b causes b/U to be much less unity, hence alleviating this difficulty.

Detailed calculations for KE and PE using (15) and (16) exhibit one-to-one correspondence with the leading terms in the CI results. We find CI calculation contains terms which are attributed to both KE and PE, although physical processes for superexchange in CI method look quite different from those in Anderson's method. The concept of "cation-cation" transfer gets confused when different basis sets are used. In fact, we find that the ionic contributions and anion→cation transfer terms in CI method are partly contained in KE.

IV. SYSTEMATICS

It is possible to push the analysis of Sec. II considerably further by analyzing isomorphic d^3 compounds exhibiting right angle [110] cation relationships. Examples are the chromium chalcogenide spinels[11,19] Cr^{3+}:$ZnAl_2O_4$, $CdCr_2S_4$, $CdCr_2Se_4$, $HgCr_2S_4$ and $HgCr_2Se_4$, and the compounds Cr_3X_4 (X = S, Se, or Te)[20]. In all these systems, the ferromagnetic character increases as one passes from the oxide to the sulfide to the selenide to the telluride. The analysis of Sec. II is appropriate to these materials, for the anions occupy opposite corners of squares, and the remaining sites (opposite corners) are occupied by the Cr^{3+} ($3d^3$) ions. For such systems we have demonstrated in Sec. III that orthogonalization of the magnetic electron orbitals leads to a cancellation within the transfer integral in Anderson's method. Because of this cancellation, we find it difficult to predict the systemtatics of superexchange from examining the processes of KE and PE. Instead, we adopt the CI method which is, though clumsier, more explicit. Because there exists a one-to-one correspondence between the results of CI and Anderson's method, same prediction for the systematics must obtain if one chooses to use Anderson's method.

We recapitulate the conclusions of Sec. II to discuss the variation of individual exchange terms with increasing anion size:

A.1. <u>Antiferromagnetic</u> cation-cation direct exchange, arising from the attractive interaction of the d_{xz} electron, and the effective nuclear charge of the near neighbor cation (Sec. IIA, analogue of (9)).

2. <u>Ferromagnetic</u> coupling due to the overlap charge density between one cation and the delocalized part of the other, with the coulomb potential of the neighboring anion (Sec. IIA, analogue of (10)).

B.1. <u>Ferromagnetic</u> anion-cation superexchange arising from the spin polarized ligand in an orthogonal position in relation to the other cation (Sec. IIB, first part).

2. <u>Ferromagnetic</u> (net anion-cation superexchange arising from the spin polarized ligand in a nonorthogonal position in relation to the other cation. The interaction would be expected to be antiferromagnetic according to the KG rules,

but cation-cation nonorthogonality reverses the sign (Sec. IIB, last part).

C. <u>Antiferromagnetic</u> exchange via cation-cation transfer (Sec. IIC).

As one replaces O^{2-} by the heavier (and larger) anions, the Cr-Cr near neighbor distance will of necessity increase [0.95 Å upon going from Cr:ZnAl$_2$O$_4$ (2.849 Å) to HgCr$_2$Se$_4$ (3.801 Å]. This will decrease the antiferromagnetic direct exchange (A), as well as the antiferromagnetic cation-cation transfer contribution (C). The latter follows because the cost of electron transfer U is unconnected with the Madelung potential (the only major energy which changes with changes in the lattice constant), but the transfer integral b varies roughly in proportion to the overlap integral $\langle d_{xz}^A | d_{xz}^B \rangle$. The latter diminishes sharply as the cation-cation distance increases. The ferromagnetic terms of A.2. decrease linearly with T upon going from oxide to selenide, assuming that the matrix elements remain roughly constant (the increase in ligand size compensates for the increase in lattice constant). The combined effect of A.2., A.1. and C. will then be to decrease the antiferromagnetic coupling because, the latter two, being proportional to T^2, change most rapidly.

Mechanisms B are proportional to the unpairing of the π ligand orbitals. Covalency increases upon going from the oxide to the telluride, primarily because of the reduction in Madelung potential. However, the ligand-cation overlap integrals are expected to remain roughly constant, again because of the increase in ligand size compensating for the increase in lattice constant. The ferromagnetic superexchange term B.1. is approximately given by

$$\lambda_\pi^2 \langle d_{xz}^B(1) p_x(2) | e^2/r_{12} | p_x(1) d_{xz}^B(2) \rangle, \quad (22)$$

where λ_π represents the unpairing of the ligand spin due to overlap and π-bond covalent mixing, and $\langle d_{xz}^B(1) p_x(2) | e^2/r_{12} | p_x(1) d_{xz}^B(2) \rangle$ is simply the self-energy of the overlap charge density $(d_{xz}^B)^* p_x$. Constancy of overlap implies a constancy of the latter, so that the ferromagnetic mechanism B.1. increases in proportion to λ_π^2 upon going from the oxide to selenide.

The antiferromagnetic part of B.2. is given by the product of λ_π^2 and the exchange integral J_{pd} between the nonorthogonal p_y and d_{xy}^B orbitals. The dominance of the attractive interaction between the p_y electron and the effective nuclear charge of d_{xy}^B is responsible for the antiferromagnetic character of this interaction. As the covalency increases, this effective nuclear cation charge diminishes, thereby reducing J_{pd}. We can lump mechanisms B.1 and B.2. into the (rough) expression:

$$\lambda_\pi^2 [\langle d_{xz}^B(1) p_x(2) | e^2/r_{12} | p_x(1) d_{xz}^B(2) \rangle - J_{pd}]. \quad (23)$$

Explicit computations demonstrate the dominance of the former over the latter term in the square brackets of (23). Yet the former term remains constant (constant overlap) while the latter diminishes (decreasing effective nuclear cation charge), as one passes from the oxide to the telluride, so that, taken together with increasing λ_π^2, the ferromagnetic character of B also increases. Finally, the ferromagnetic correction to the B.2. contribution will diminish as the first power of the overlap $\langle d_{yz}^A | d_{xy}^B \rangle$ while the antiferromagnetic contributions A and C diminish as the square of the overlap $\langle d_{xz}^A | d_{xz}^B \rangle$. Hence the strength of the latter interactions will diminish more rapidly than the former with increasing lattice constant and covalency, so that a relative increase in the ferromagnetic character of the exchange coupling will again result.

Evidence of the microscopic validity of these arguments can be found in an elegant paper by Berger, Budnick and Burch.[11] They exhibit a ^{53}Cr isotropic hyperfine interaction, H_{iso}, in the chromium chalcogenide spinels which varies linearly with the near neighbor exchange coupling, as shown in Fig. 5. It is known that H_{iso} decreases with increasing covalency, but increases with increasing overlap. The trend of H_{iso} with J is unmistakeable, and provides evidence of an increase of covalency, and at worst a small increase of overlap (at best, constancy) as one passes from the sulfide to the selenide. It is impressive to compare compounds with the same anions, but differing non-magnetic cations (e.g. Cd and Hg). Here, the ligand-chromium overlap must remain constant because the lattice constant does not change. The larger, more polarizeable non-magnetic cations, will reduce the transfer energy, thereby increasing the covalency. A comparison between $CdCr_2X_4$ and $HgCr_2X_4$ (where X = S or Se) demonstrates a monotonic decrease in H_{iso} associated with an increase in the near-neighbor exchange coupling. This can only be related to proportionality of the exchange coupling to covalency, as argued in this section. Similar results obtain for the series $CrCl_3$, $CrBr_3$, and CrI_3. On balance, therefore, the systematic predictions of this theory appear valid on a microscopic, as well as macroscopic basis.

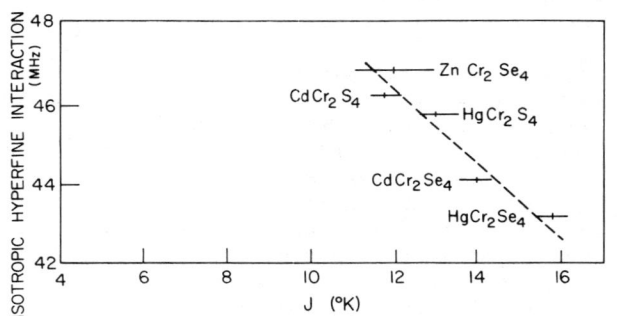

Fig. 5. Dependence of ^{53}Cr isotropic hyperfine interaction on J in chromium chalcogenides determined by Berger, Budnick and Burch.[11]

V. ACKNOWLEDGEMENTS

The authors are deeply indebted to Drs. C. Herring and P. W. Anderson who informed us of a major error in a previous version of this paper. Though the circulation of that preprint was limited, we do wish to apologize for any negative implications which may still be present concerning the validity of the Anderson method of superexchange. We also wish to thank Professor J. R. Schrieffer for a very helpful discussion of the role of orthogonality in the superexchange problem.

REFERENCES

1. F. Keffer and T. Oguchi, Phys. Rev. 109, 730 (1959).
2. P. W. Anderson, in Solid State Physics Vol. 14, ed. F. Seitz and D. Turnbull (Academic Press, New York and London), p. 99 (1963).
3. Noboku Fuchikami, J. Phys. Soc. Japan, 28, 871 (1970).
4. R. McWeeny, Proc. Roy Soc. A223, 63 (1954).
5. J. Kanamori, J. Phys. Chem. Solids 10, 87 (1959), and J. B. Goodenough, Phys. Rev. 100, 564 (1955); Phys. and Chem. Solids 6, 287 (1958).
6. N. L. Huang and R. Orbach, Phys. Rev. 154, 487 (1967).
7. N. L. Huang, Phys. Rev. 157, 378 (1967).
8. N. L. Huang and R. Orbach, in Proc. of the International Congress on Magnetism, Boston, 1967 (J. Appl. Phys. 39, pt. I, 426 (1968)).
9. N. L. Huang, Phys. Rev. 164, 636 (1967).
10. N. L. Huang, Ph.D. thesis, Department of Physics, University of California at Los Angeles (1966).
11. S. B. Berger, J. I. Budnick and T. J. Burch, Phys. Rev. 179, 272 (1969).
12. J. C. Slater, Quantum Theory of Atomic Structure Vol. I (McGraw Hill), p. 486 (1960).
13. J. H. Van Vleck, Rev. Univ. Tucuman A (Argentina) 14, 189 (1962).
14. Similar results have been derived by W. J. Carr, Phys. Rev. 92, 28 (1953); A. J. Freeman and R. E. Watson, Phys. Rev. 124, 1439 (1961); C. Herring, Magnetism IIB (ed. G. T. Rado and H. Suhl), Academic Press, 1966, p. 1.
15. D. E. Rimmer, J. Phys. C (G.B.), 2, 329 (1969).
16. A. J. B. Codling and B. Henderson, J. Phys. C (G.B.) 4, 1409 (1971).
17. D. E. Ellis and A. J. Freeman, Proc. International Congress on Magnetism, Boston (1967), published in J. Appl. Phys. 39, 424 (1968). Unrestricted Hartree-Fock calculations for two neighboring magnetic ions surrounded by ligands are to be published by these authors.
18. Using ionic Hartree-Fock wave functions, we find $U_o = 0.77$ a.u. and $U_1 = 0.17$ a.u.
19. K. W. Blazey, Solid State Comm. 4, 541 (1966), P. K. Baltzer, P. J. Wojtowicz, M. Robbins, and E. Lopatin, Phys. Rev. 151, 367 (1966).
20. A. C. Gossard, V. Jaccarino, and J. P. Remeika, Phys. Rev. Letters 7, 122 (1961), A. Narath, Phys. Rev. Letters, 7, 410 (1961); Phys. Rev. 140, A854 (1965).

MECHANISMS OF MAGNETIC ANISOTROPY IN RARE EARTH METALS

A. R. Mackintosh
AEK Research Establishment Risø, Roskilde, Denmark

ABSTRACT

A brief discussion is given of the way in which magnetic anisotropy in the rare earth metals results from a combination of the crystal field and indirect exchange interactions. The efficiency of neutron scattering for studying these interactions is emphasized.

The principal magnetic interactions in the rare earth metals are the indirect exchange coupling between the magnetic ions, mediated by the conduction electrons, and the interactions of the crystal field with the localized moments, which generally have a large effect because of the high orbital momentum of the 4f electrons and the associated anisotropy of the charge distribution. The relative effects of these two interactions differ markedly between the light and heavy rare earths, primarily because the effective exchange depends on $(g-1)J$, which is large in the latter and relatively small in the former.

In the heavy rare earths, the coupling of a given magnetic moment to the rest of the crystal may be elucidated very effectively by studying the magnetic excitations through inelastic neutron scattering[1]. For example, in the ferromagnetic metals, the single-ion anisotropy forces may be deduced from the dependence of the energy of long wavelength magnons on magnetic field and temperature. Such experiments, together with macroscopic magnetic measurements, have allowed a relatively complete characterization of the origin and magnitude of the anisotropy forces in Tb.[1,2] The axial anisotropy may be attributed to crystal fields, whereas the planar component is principally due to magnetoelastic effects. The "frozen-lattice" model is valid, so that the lattice strain cannot follow the spin-precession associated with a spin wave, and consequently the energy gap cannot be reduced to zero by a field applied in the hard direction in the plane. The total hexagonal anisotropy, as measured either macroscopically or microscopically, is smaller than that predicted from magnetoelastic effects alone. This implies either that

the crystal field contribution has the opposite sign to
that given by the point-charge model or, more likely,
that some other mechanism, such as anisotropic exchange, is contributing.

Spin wave measurements at finite wavevector may
give information on the anisotropic coupling between
the ions. We may distinguish between anisotropic interactions which mix spin and space variables, such as the
dipolar interaction or pseudo-dipolar exchange, and
interactions anisotropic in spin space which single out,
for example, the component of the bilinear exchange in
the c-direction. The former change the symmetry of the
Hamiltonian, and may be detected by the lifting of
degeneracies in the magnon dispersion relations. Such
effects, greater than those attributable to dipolar
forces alone, have been observed in Tb^1 and, in a more
pronounced form, in Dy^3.

The first indications of a wavevector-dependent
anisotropy appeared in measurements of the magnon
energies in the conical phase of Er^4, which could be
explained in terms of the conventional Hamiltonian only
if the bilinear exchange in the c-direction were an
order of magnitude greater than that in the plane.
Although the effects of the hexagonal anisotropy, which
is very large in Er, were not taken into account in the
analysis, these results are at least suggestive of a
substantial anisotropy in the two-ion coupling. Such anisotropy can readily be studied through the field-dependence of the magnon energies at finite wavevector, and
the results of such measurements for $Tb^{2,5}$ indicate
that there is indeed a significant anisotropy between
the c-axis and planar directions. This anisotropy is
however very much smaller than that proposed for Er.

In the light rare earths, the relative dominance
of the crystal fields may lead to striking anisotropies.
In particular, the crystal field ground states of the
ions at both hexagonal and cubic sites in dhcp Pr are
singlets[6], so that magnetic ordering does not apparently occur at temperatures as low as 1.5 K. The ground
state at the hexagonal sites is the $|M_J\rangle = |0\rangle$ singlet
and, if a magnetic field is applied in the plane, a
moment is developed by mixing with the $|\pm 1\rangle$ excited
doublet. On the other hand, if the field is applied
along the c-axis, no moment is initially developed on
the hexagonal sites, as confirmed by neutron diffraction. However, there is an extremely sharp first-order

transition to a metamagnetic phase in a field of about 320 kilogauss, due to the crossing of crystal field levels.[7] The magnetic excitations in Pr have been studied by inelastic neutron scattering [8], and the dispersion relations again give information about the crystal fields and exchange. Extra branches in certain directions may be an indication of anisotropic exchange [6] but the study of the light rare earths is still at a relatively early stage and extensive experimental and theoretical work remains to be performed.

REFERENCES

1. A.R.Mackintosh and H.Bjerrum Møller, in Magnetic Properties of Rare Earth Metals, edited by R.J.Elliott, Plenum Press (1972).

2. J.Jensen and J.C.G.Houmann (to be published).

3. R.M.Nicklow and N.Wakabayashi, in Neutron Inelastic Scattering, IAEA, Vienna (1972) p. 611.

4. R.M.Nicklow, N.Wakabayashi, M.K.Wilkinson and R.E.Reed, Phys.Rev.Letters 27, 334 (1971).

5. H.Bjerrum Møller, J.C.G.Houmann, J.Jensen and A.R.Mackintosh, in Neutron Inelastic Scattering, IAEA, Vienna (1972) p. 603. It should be noted that further experiments have confirmed the alternative interpretation of the results mentioned in this paper, so that Fig. 3 requires substantial modification.

6. B.D.Rainford, Magnetism and Magnetic Materials - 1971, American Institute of Physics (1972) p. 591.

7. K.A.McEwen, G.J.Cock, L.W.Roeland and A.R.Mackintosh (to be published).

8. B.D.Rainford and J.C.G.Houmann, Phys.Rev.Letters 26, 1254 (1971).

Section 39. Rare-Earth Compounds and Complexes

THE EFFECT OF COULOMBIC AND MAGNETIC DISORDER ON TRANSPORT IN MAGNETIC SEMICONDUCTORS*

S. von Molnar and F. Holtzberg
IBM Research Center, Yorktown Heights, N. Y. 10598

ABSTRACT

This paper describes transport measurements on the magnetic semiconductors $Gd_{3-x}v_xS_4$. The vacancies, v, are randomly distributed throughout the lattice and lead to fluctuating repulsive potentials and band tailings. Furthermore, since our largest measured carrier concentrations are small compared to the maximum number of vacancies ($\sim 2.3 \times 10^{21} cm^{-3}$), a rigid band model should be applicable. This is in contrast to ordinary semiconductors where the energy dependence of the density of states is generally a strong function of the dopant concentration. Recent transport measurements in Eu doped EuS, which demonstrate the applicability of a model for transport in a band tail of localized states will also be reviewed. The two systems will be compared and discussed in terms of a model first suggested by Cutler and Mott for paramagnetic $Ce_{3-x}v_xS_4$ and modified here to include magnetic interactions.

I. INTRODUCTION

Transport of charge carriers in many magnetic semiconductors is dominated by potential fluctuations of both coulombic and magnetic origin. In contrast to the coulombic case, the binding energies of states localized by spacial fluctuations in magnetic order may be both temperature and magnetic field dependent.[1,2,3]

The purpose of this paper is to review recent advances in describing and understanding the physical properties of materials in which the electrons are localized by coulombic disorder and to extend these ideas to magnetic materials. In particular we compare the properties of a magnetic semiconductor $Gd_{3-x}v_xS_4$ with its isostructural paramagnetic counterpart $Ce_{3-x}v_xS_4$. We hope to show that such a comparison is helpful in that it allows us to look at the anomalous properties of the magnetic solid in terms of the well established band tail model. We find that we can account qualitatively for our transport measurements by using this model.

For the case of materials where static potential fluctuations are invoked as the cause of disorder, e.g. amorphous or heavily doped non-magnetic semiconductors, the concentration for which defects and impurity wavefunctions overlap has been described in terms of band tailing.[4] The basic concept is that there exists a series of electronic states (See Fig. 1) below what was formerly the bottom of the unperturbed conduction band. Concomitant with this tail, there exists an energy, E_c, below which electrons are

─────────────

*Supported in part by ARPA Contract DAA H01-71-C-1313 and ONR Contract N00014-70-C-0272.

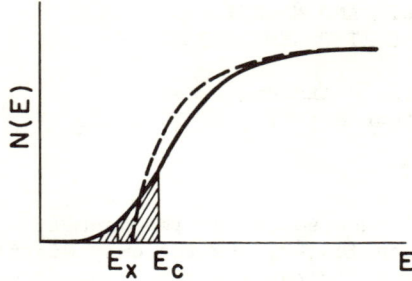

Fig. 1. Density of States, N(E) vs energy for a disordered solid. The shaded area indicates localized states and the dashed line is the density of states for the ordered material. (After Cutler and Mott, Ref. 15).

more or less immobile and have to move by hopping and above which the states are extended and the electron's movement is limited only by scattering. E_c is a consequence of quite general considerations concerning electronic localization in a disordered lattice.[5] The electron localization depends critically on the magnitude of the potential fluctuation. For example, van Vechten[6] has recently concluded on theoretical grounds that no band tail is expected for covalent amorphous semiconductors, since the covalent bonds will tend to rearrange themselves in such a way as to minimize the distortion energy and thus preserve short range order. Band tailing was not detected experimentally in strain free amorphous Si and Ge.[7] On the other hand Busch, Campagna, and Siegmann[8] were able to show from an analysis of photothreshold in photoemission experiments that in ionic materials large band tails can be produced in the disordered state.

A severe limitation in most attempts to study transport of disordered semiconductors has been the lack of information about the energy dependence of the density of states and mobility. We would like to review here the salient features of transport analysis on three compounds in which the band tail has been well studied and its occupancy is relatively well controlled.

Redfield[9] has compared experimental data on heavily doped, closely compensated GaAs with a band tail model. His samples, all of which were doped to the same carrier concentration and later compensated, provided a "rigid" band tail, whose density of states, N(E), was not expected to change substantially between samples since the net carrier concentrations were smaller than the donor concentration by at least 2 orders of magnitude.[10] It was thus possible to calculate the conductivity, using the formula

$$\sigma(T) = e \int N(E) \, f[E-E_F(T)] \mu(E) dE \qquad (1)$$

where N(E) is the density of states, $f[E-E_F(T)] = f$ is the Fermi function and $\mu(E)$ is an energy dependent mobility. The prescription is, first, to assume a form for N(E) (which, in the case of GaAs, had been calculated to be Gaussian), to calculate $E_F(T)$, the temperature dependence of the Fermi energy, by using particle conservation, and then to try several functional forms for $\mu(E)$. Redfield[9] found remarkable agreement with his data[10] by choosing an energy dependent form for μ which was also a Gaussian tail.

Thompson, et al,[11] used a similar approach to analyze their data on the magnetic semiconductor EuS. It should be pointed out that these authors were focusing their attention on the non-magnetic transport properties of this compound and consequently quenched the magnetic field dependent resistivity peak near the ordering temperature by applying a field of 32 kOe. Their results for Eu-rich samples show that the data are well described by the following model: 1) conduction occurs in a "rigid" band tail (the tail is exponential rather than Gaussian); 2) at high temperatures $\mu(E)$ is energy and temperature dependent, $\mu(E) \alpha \exp[(E_c-E)/kT]$; 3) the electrons deepest in the band tail do not contribute to the Hall effect; 4) at low temperature conduction is described by Mott's variable range hopping formula,[12] $\sigma = \sigma_o \exp(T_o/T)^{1/4}$, with T_o predicted from the high temperature data. An unexpected aspect of this work is that a rigid band tail model works as well as it does. The samples were chosen by varying the Eu/S ratio and one might expect $N(E)$ to change substantially. Apparently there is some contribution of unknown defects common to all samples.

It is obvious from the above discussion that a form for $N(E)$ must be found before one can make quantitative comparisons between theory and experiment. Furthermore, consistency between samples can only be tested if a rigid band tail is assumed. On the other hand, a qualitative description of transport behavior is possible, even when $N(E)$ is not preciseley, as long as we can be assured that the rigid band tail model is valid. A case in point is the system $Ce_{3-x}v_xS_4$, where the vacancies, v, are an intrinsic structural property of the solid. The interesting feature of this system is that one can fill the vacant sites with the same cation and thereby create charge carriers. In comparison with the large number of vacancies ($\sim 2.1 \times 10^{21}$), the carrier concentration can be varied over a significant range, presumably without seriously affecting the energy dependence of $N(E)$. This system was studied experimentally by Cutler et al.[13,14] and its transport properties were interpreted by Cutler and Mott.[15]

These compounds are only one of many rare earth Th_3P_4 type structures, and interest in them arises for at least two reasons: first of all, the $Re_{3-x}v_xX_4(X=S,Se,Te)(0<x\leq 1/3)$ compounds change continuously from metals to insulators with increasing vacancy concentration, x, leading to a variety of cooperative effects such as superconductivity and ferromagnetism;[16] secondly, the disordered distribution of vacancies at the Th sites (which are discussed below), is expected to lead to fluctuating repulsive potentials and tailing of the conduction band in which the electronic states are localized.[15]

In the following we describe some aspects of the Th_3P_4 structure and briefly review the arguments of Cutler and Mott[15] pertaining to their analysis of transport in the cerium compound. In Section II we present transport data obtained on samples of the isomorphic, but magnetic, compound $Gd_{3-x}v_xS_4$ and compare our measurements to the paramagnetic cerium system. Section III presents our interpretation in terms of the concept of localization of electron states first suggested in Ref. 15, and extended here to include magnetic interactions.

The Th_3P_4 type structure was first described by Meisel.[17] In 1949 Zachariasen[18], on the basis of x-ray powder diffraction and density data, showed that the defect Ce_2S_3 compound crystallized with the Th_3P_4 structure, in the space group $I\bar{4}3d-T_d^6$ with four molecules per unit cell. In this space group the 12-fold cationic sites are fixed while the 16-fold anionic sites are determined by a parameter, μ Both Zachariasen's and Meisel's analyses fixed the value of μ at 1/12 or 0.083. The deficit Ce_2S_3 was described by Zachariasen as having 10 2/3 Ce atoms statistically distributed over the twelve cation sites, whereas the sixteen fold anion sites were filled. On the basis of these results Zachariasen predicted that the 4/3 vacant cation sites per unit cell could be filled and therefore the structure should exist over a single phase region extending from S:Ce=1.5 to S:Ce=1.33.

The coordination polyhedra were described in detail by Krpyakevich[19] and by Holtzberg and Methfessel[20]. Single crystal x-ray analysis on related materials[21,22] showed that the parameter μ for rare earth compounds was nearer to 0.075 which displaces the anion from its symmetrical position.

Recently Carter[23] suggested the possible existence of vacancy and charge ordering in Th_3P_4 defect type structures but concluded that there is insufficient experimental evidence to confirm the existence of vacancy ordering and it is not clear that the arguments for charge ordering are conclusive. One might suspect, however, that it is unlikely to find adjacent cation vacancies and it is therefore reasonable to assume that some short range order exists. Although this would reduce the magnitude of the static potential fluctuations, the existence of a rigid band tail in Th_3P_4 structures is clearly a good assumption as long as the donor concentration is small compared to the total number of vacancies ($\sim 2.1 \times 10^{21}$ cm^{-3} in $Ce_{3-x}v_xS_4$).

The latter assumption is the basis for the analysis of transport data on $Ce_{3-x}v_xS_4$. Cutler and Mott[15] argue that when $E_F(0)$, the low temperature Fermi energy, lies below E_c, conduction is thermally activated and approaches 0, as $T \to 0$. When $E_F(0)$ is above E_c, conduction occurs in extended states and remains finite as T approaches 0. They also show, that, regardless of the mode of transport the fundamental equations for σ, the conductivity and S, the thermopower, can be written in the familiar form

$$\sigma = -\int \sigma(E)(\partial f/\partial E) dE \qquad \text{and} \qquad (2)$$

$$S\sigma = \frac{k}{e} \int \sigma(E) \frac{E-E_F}{kT} \partial f/\partial E\, dE \quad .^{15,4} \qquad (3)$$

The function $-kT\partial f/\partial E = f(1-f)$ is centered about $E_F(T)$ of width kT, where k is Boltzmann's constant. Consequently the transport will depend very sensitively on the magnitude of $|E_c-E_F|$ compared to kT. Cutler and Mott then specialize their discussion to electron concentrations and temperature ranges where E_F lies in the region of localized states and where $E_c-E_F \gg kT$. They conclude that, when the electronic current is carried by states near E_F, the thermoelectric power, S, is given by

$$S = \frac{\pi^2}{3} \frac{k^2 T}{e} \frac{d(\ln\sigma)}{dE} \qquad E=E_F \quad . \qquad (4)$$

This is the familiar metallic formula and when applied to thermally activated conduction of the form

$$\sigma = \sigma_o(E)\, e^{-W(E)/kT} \quad , \qquad (5)$$

where $W(E)$ is the mean activation energy for hops, equation 4 yields

$$S = \frac{\pi^2}{3} \frac{k}{e} \left(kT \frac{d\ln\sigma_o}{dE} - \frac{dW}{dE} \right)_{E=E_F} \quad . \qquad (6)$$

When, however, the electronic current is negligible, i.e. $\sigma(E)\to 0$, for energies below some energy E_x, (See Fig. 1) the thermoelectric power takes on a form similar to that for semiconductors, i.e.

$$S = \frac{k}{e} \left(\frac{E-E_F}{kT} + \frac{1+2kT\, d\ln\sigma/dE + \ldots}{1+kT\, d\ln\sigma/dE + \ldots} \right)_{E=E_x} \quad . \qquad (7)$$

Equation 6 predicts linear behavior in T with non-zero intercept (contrary to the case for metals). This behavior is observed in moderately doped ($\sim 5 \times 10^{18} - 10^{19}$ cm^{-3}) $Ce_{3-x}V_xS_4$. The dependence on inverse temperature, Eq. 7, is observed in considerably more insulating samples containing $5-9 \times 10^{17}$ carriers/cm^3. One particularly interesting sample (n=5.2×10^{18} cm^{-3}) obeys Eq. 6 between ~ 100 and 300°K, but changes to 1/T behavior at lower temperatures.[14] Presumably E_F changes with temperature in this sample and the quantity E_c-E_F is small enough so that above $\sim 100°$K much of the charge is transported either by hopping or in extended states (both would give a term for S linear in T). The reason we belabor this point is that similar behavior is observed in $Gd_{3-x}V_xS_4$ as shown in Fig. 4.

When $|E_c-E_F|$ and kT become comparable, as is apparently the case in EuS1I, no approximate formula is valid and a full integration over energy has to be carried out to compare theory with experiment.[9,11] In $Gd_{3-x}V_xS_4$ we suppose that $|E_c-E_F|$ is not only a function of temperature and concentration but also a function of magnetization and we shall attempt to relate the experimental values for $\rho = 1/\sigma$ and S to this concept.

II. EXPERIMENTAL RESULTS FOR $Gd_{3-x}V_xS_4$.

Single crystals of $Gd_{3-x}V_xS_4$ were grown by slowly cooling the melts (m.p. ~1800°C) contained in sealed tungsten crucibles to about 1000°C, annealing at that temperature for several hours and quenching to room temperature. Although we have not, thus far, determined the exact compositions of the materials it is clear that the Gd_2S_3 composition is insulating ($\rho > 10^6 \Omega cm$), transparent and anti-ferromagnetic and that the composition for which $x \approx 0$ the samples are metallic and ferromagnetic. Lattice parameters obtained with a Guinier focussing camera using Cu radiation are essentially constant throughout the single phase region with $a_0 = 8.375 \pm .002 Å$.

TABLE I

Sample I.D.	$n_\infty (cm^{-3})$	$\mu_{RT} (cm^2/v\text{-}sec)$	ρ_{RT} (Ω-cm)	$\Theta (°K)$
1. 24-118 (673)	†5.6 x10^{19}	1.25	9.45×10^{-2}	††-6.7 to -9.0
2. 24-60 (658)	*8.7±.8x10^{19}	2.5	2.5×10^{-2}	††8.6 to 11.7
3. 23-76 (594)	*1.6±.5x10^{20}	2.3	1.66×10^{-2}	16.1
4. 23-10	*2.5±.2x10^{20}	2.5	8.23×10^{-3}	22.2

† n_∞ is the value extrapolated to $1/T \to 0$, on the assumption that μ=constant.

* $n_\infty \equiv \lim_{\chi \to 0} (10^{-8}/e) \div (e_H/H_A)$, where χ=susceptibility, e_H is the Hall resistivity, and H_A is the applied magnetic field.

†† These values were not determined on the specific samples used in transport but represent the variations for several crystals from the same crystal growth experiment.

Table I summarizes some physical properties of the samples studied. Preliminary measurements by us on samples 3 and 4 have been published previously.[24] The single crystal material does not show an easy direction for cleavage and the magnetic measurements were performed on small samples of arbitrary shape. For the present purposes the important features of the magnetic data are: a) the high temperature susceptibility, χ, follows a Curie Weiss law (resulting in the extrapolated Θ values, b) below ~70°K, $1/\chi$ departs from linearity indicating possibly cluster formation, c) the general trend is from negative to positive Θ's, as the carrier concentration, n_∞, increases, d) in the ordered state the Gd^{3+} moments are not completely aligned and the magnetization shows a pronounced field dependence.[24] Five probe dc resistivity, Hall, and thermoelectric power measurements were made on crystals shaped as rectangular parallelopipeds, with the exception of sample 1. The latter was a thin platelet and resistivity and Hall measurements were made in the Van der Pauw configuration.

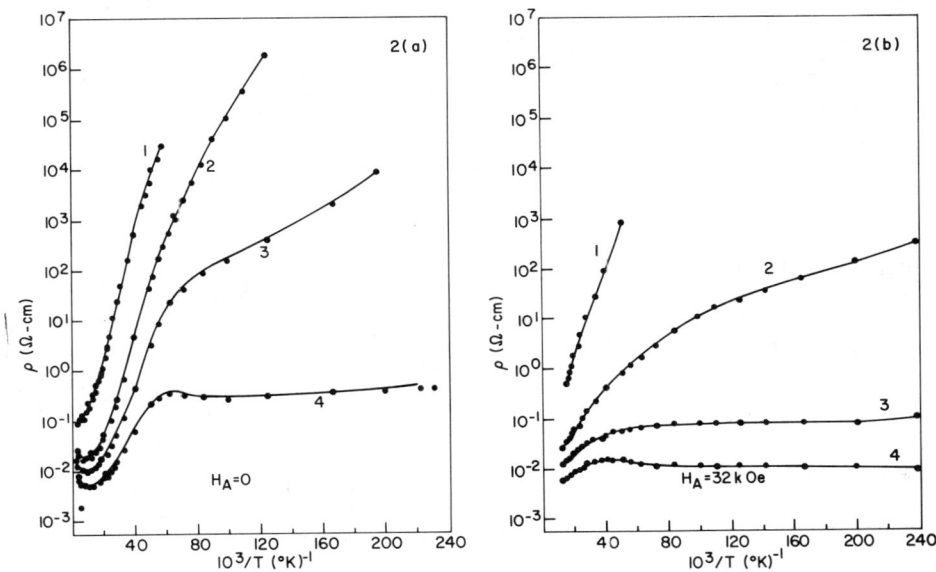

Fig. 2. Electrical resistivity, ρ, of $Gd_{3-x}V_xS_4$ for varying compositions (See Table I) as a function of temperature, T; (a) with no external magnetic field applied, (b) in the presence of an applied field of 32 kOe.

The results of measurements of the resistivity with and without an applied field of 32 kOe as a function of $10^3/T$ are shown in Figs. 2a and b. The following features should be noted: 1) Samples 2, 3, and 4 exhibit a linear decrease in ρ with decreasing temperature between 300 and $\sim$200°K (not shown) and have a resistance minimum near 100°K. The resistivity of sample 1 is activated at all temperatures. 2) Between $\sim$100°K and $\sim\theta$, $\rho(H=0)$ is activated for samples 2, 3, and 4 but the activation energy changes significantly as θ is approached. 3) Application of a magnetic field reduces ρ over the entire temperature range below $\sim$50°K in all samples. The curves for samples 2-4 strongly suggest that the increase in resistivity is directly related to the state of magnetic order (either through temperature or applied magnetic field H_A). This is obviously only partially true of sample 1, whose resistivity is activated at room temperature. Even in this case part of the activation energy below $\sim$50°K is magnetic in origin.

Figure 3 depicts the results of the thermoelectric power, S, for samples 2-4. $|S|$ decreases linearly with decreasing T, has a mini-

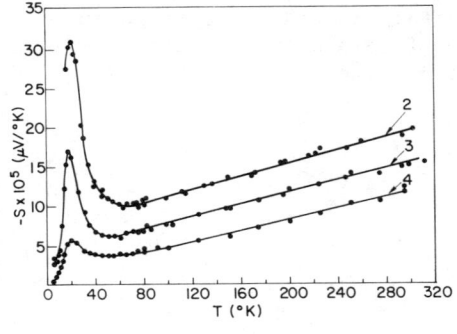
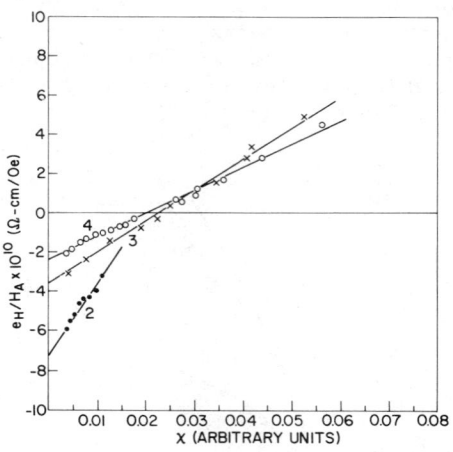

Fig. 3. Thermoelectric power, S, of $Gd_{3-x}V_xS_4$ for varying compositions (see Table I) as a function of temperature T.

Fig. 4. Hall coefficient, e_H/H_A, of $Gd_{3-x}V_xS_4$ for varying compositions (see Table I) as a function of magnetic susceptibility, χ.

mum between 40 and 80°K (depending on sample) resembling the dependence of ρ on T, rises to a maximum near but above Θ, and then decreases towards 0 as T→0. The sign of S is always <u>negative</u>.

The Hall coefficient e_H/H_A, on the other hand, is <u>negative</u> at room temperature but reverses sign (measured in samples 3 and 4) at lower temperatures (see Fig. 4). This sign reversal was unexpected and we have attempted to eliminate such spurious causes of anomalous Hall behavior as surface effects[25] and an unusually large Nernst-Ettinghausen effect. The former was tested by comparing the temperature dependent resistivities of samples with clean and lapped surfaces. No significant differences were observed. The latter was tested by comparing 210Hz and d.c. results[26] on sample 4 at ~56°K and 300°K. Both the magnitude and sign of the Hall coefficient agreed.

Shapira et al[27] have pointed out that the Hall resistivity e_H, in the NaCl type Eu-chalcogenides is best expressed as

$$e_H = R_o B + R_1 M, \qquad (8)$$

where B is the magnetic induction inside the sample, M is the magnetization, and R_o and R_1 are the normal and anomalous Hall coefficients. In the paramagnetic region Eq. 8 reduces, upon substitution of $B = H_A + (4\pi - N)M$, and $M = [\chi/(1+N\chi)]H_A = \chi^* H_A$, where N is the demagnetizing factor, to

$$\frac{e_H}{H_A} = R_o + [R_o(4\pi - N) + R_1]\chi^* . \qquad (9)$$

If it is assumed that R_o and R_1 are independent of temperature, then a plot of e_H/H_A vs $\chi*\tilde{\chi}$ should yield a straight line with R_0 as intercept. This relationship is roughly obeyed for temperatures above 77°K as can be seen in Fig. 4. The values for n_∞ quoted in Table I are derived from the magnitude of R_0 obtained in this manner. It is also obvious that, if our assumption regarding R_0 and R_1 is correct, the conduction is n-type and the change in sign of e_H/H_A is due to a large positive anomalous Hall coefficient (R_1).

III. DISCUSSION

If we restrict our discussion to the region in temperature above $\sim 100°K$, the transport behavior of the $Gd_{3-x}V_xS_4$ samples is very similar to that of the cerium isomorph. In $Ce_{3-x}V_xS_4$, Cutler et al[14] observed activated, non-metallic transport for carrier concentrations smaller than $\sim 8 \times 10^{19} cm^{-3}$. Higher concentration materials behave like metals for all temperatures and are adequately described by conventional band theory.[13] These observations are in complete agreement with our measurements on the Gd compound. The resistivity of sample 1 is thermally activated over the entire experimental temperature range (See Fig. 2a), whereas the resistivities of samples 2 through 4 initially decrease linearly with temperature as the samples are cooled from $\sim 300°K$ (This effect is indicated by the apparent resistivity minimum towards the left of the figure). The fact that $n_\infty > 8 \times 10^{19} cm^{-3}$ for samples 2-4 is in excellent agreement with Cutler and Levy's observations in $Ce_{3-x}V_xS_4$[14], although this result is possibly fortuitous, since the lattice parameter in the Ce compound is somewhat larger than in the Gd compound. It does, however, confirm our earlier remark that the low temperature behavior of samples 2-4 must be dominated by magnetic effects. The result also suggests that E_F lies near, but above E_C in our samples 2-4, whereas $E_F < E_C$ for sample 1. The latter then appears to be similar to the non-metallic EuS samples studied in detail by Thompson et al[11], and we shall concentrate further discussion on samples 2-4.

The thermoelectric power, S, is linear with an intercept which decreases with increasing carrier concentration, n. For sample 4, then, it seems reasonable to assume the unperturbed density of states indicated by the dashed line in Fig. 1 and to find E_F from the slope of S vs T. This result is readily derived from Eq. 4, if we assume $\sigma(E) = \text{const } E^x$. In this case

$$S = \frac{\pi^2}{3} \frac{k^2 Ty}{e E_F} . \qquad (10)$$

If y=1, consistent with the assumption that lattice and neutral impurity scattering dominates the resistance, $E_F \simeq .07 ev$. The density of states effective mass m*, defined as

$$m* = \frac{\hbar^2}{2 E_F} (3\pi^2 n)^{2/3} , \qquad (11)$$

is $\sim 1.6 m_e$, where m_e is the free electron mass. This value is smaller than similar calculations on $Ce_{3-x}v_xS_4$.[13] E_F cannot be determined for samples 2 and 3 from the linear part of S because $|E_F - E_c| \lesssim kT$ over much of the experimental range and current carriers from both above and below E_c contribute to S. Equation 2 thus should be evaluated numerically. This requires, as we pointed out earlier, a knowledge of the functional form for $N(E)$. Although we lack this information for $Gd_{3-x}v_xS_4$, Thompson[28] has calculated $S(T)$ for one of the EuS samples described in Ref. 11. His calculation is in good agreement with our experimental values for S on the same EuS sample.[29]

From the foregoing discussion we might expect small changes in Hall constant in samples 2 and 3, since the electron distribution above E_c is expected to change with temperature. If n, and consequently the normal Hall coefficient, R_0, is changing, the effect is completely masked in our experiments by R_1. Furthermore, measurements of R_0 in EuS exhibit surprisingly weak temperature dependences for dopant concentrations and temperatures such that $|E_c - E_F| \lesssim kT$. Guided by this result and the data of Cutler and Levy[14] we have confidence that our extrapolation procedure for obtaining n_∞ is reasonable.

Whereas the Ce and Gd compounds behave very similarly at high temperature, $Gd_{3-x}v_xS_4$ exhibits new magnetization dependent phenomena at low temperatures. In order to study these magnetic effects at temperatures below $\sim 200°K$ we have analyzed our resistivity data in terms of an activation energy, $\mathcal{E}(T)$, in a manner similar to Penney, et al[30]. We assume, on the basis of the $Ce_{3-x}v_xS_4$ results and our high temperature data, that samples 2-4 would remain metallic in the absence of magnetic interactions and define $\mathcal{E}(T)$ from the data $\sigma(T)$ by

$$\sigma(T) = \sigma_M(T) e^{-\mathcal{E}/kT} , \quad (12)$$

where $\sigma_M(T)$ is roughly of the form $A+BT$, with A and B specified by the linear high temperature behavior. $\mathcal{E}(T)$ is then computed from $\sigma(T)$ point by point for all temperatures. Our results are shown in Fig. 5.[31]

The peaks in $\mathcal{E}$ occur in the neighborhood of but slightly higher than the measured Θ values. Although the curves are not shown, $\mathcal{E}$ is substantially reduced by an applied magnetic field, i.e. when $\mathcal{E}$ is derived from the data of Fig. 2b. It follows that $\mathcal{E}$ is an activation energy for conduction which

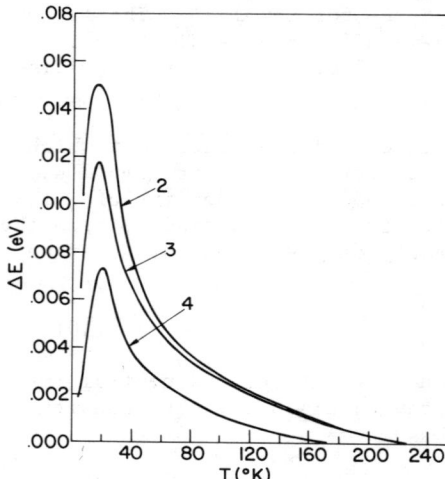

Fig. 5. Activation energy for electrical resistivity, ΔE, of $Gd_{3-x}v_xS_4$ calculated from the data (Fig. 2a).

is related to magnetic order. We have implicitly neglected effects on the mobility due to critical scattering. This is justified, since the changes in resistivity are too large to be accounted for by scattering theory and since, in our model, transport occurs predominantly by hopping near the Curie temperature Θ.

Our hypothesis for the change in ΔE, which we claim to be primarily magnetic in samples 2-4, has its origin in the idea of a magnetic polaron bound by the Coulomb field of a donor or vacancy first discussed by Kasuya[2]. Torrance has developed this concept and applied it to the insulator-metal transition in EuO[3] and the resistivity peak in EuS.[32] The cause for the localization is the exchange interaction, I_{c-f}, between the conduction electron and the localized Gd^{3+} 4f spins. The magnetic binding energy is related to I_{c-f} ($<S>_{cluster} - <S>_{lattice}$), where the term in brackets expresses the difference in average magnetization between a cluster of spins in the neighborhood of an electron and spins elsewhere in the magnetic lattice. This term encourages an electron in the coulomb field of a defect to contract forming a small magnetic spin cluster.[2,3] Furthermore, the combined effect of both magnetic and Coulomb interactions make the localized state stable over relatively wide temperature ranges.[33]

Regardless of the microscopic mechanism of localization, the bracketed term in the magnetic binding energy explains several features of ΔE observed experimentally in $Gd_{3-x}V_xS_4$. If we assume, for simplicity, that $<S>_{cluster} \simeq S$, i.e. the ferromagnetic cluster is saturated, a magnetic binding energy is possible as long as $<S>_{cluster} > <S>_{lattice}$. This situation is possible in the paramagnetic region of a magnetic material or below the magnetic ordering temperature in materials which do not order ferromagnetically. The bracketed term also explains the reduction in binding energy due to an applied field, since the latter tends to reduce the difference in average magnetization by increasing $<S>_{lattice}$. With complete ferromagnetic order, metallic conduction is predicted at low temperatures for samples which are metallic at high temperatures. In our samples, where ferromagnetic alignment is not complete[24], the bound state represents a local region of full saturation and therefore, can exist to lowest temperatures. This is observed experimentally, in that the samples do not show metallic conductivity, but a magnetic field does tend to reduce the resistivity.

The model is also capable of explaining the qualitative features of the observed thermoelectric power. Of course, the simplest assumption to explain the temperature dependence of the thermopower is that $\sigma(E) \to 0$ below the mobility edge, E_c. This means that we set $E_x = E_c$ in Eq. 7, in which case S has as its leading term $(k/e)(E_c - E_F)$. Substitution of ΔE (Fig. 5) for $E_c - E_F$ gives a peak in the thermopower slightly above Θ as observed. However, the peak value of S predicted from ΔE is approximately twice as large as found experimentally for sample 2. The discrepancy is even larger for sample 3. Eq. 7 is not valid for sample 4, since the measured S is smaller than k/e. One possible explanation for this inconsistency is to assume, as did Cutler and Mott[15], that E_x the energy below which $\sigma(E) \to 0$, is below E_c (as is indicated in Fig. 1). In this case Æ would include both a term E_x-E_F and an activation energy for hopping, W. S, on the other hand, would be dominated at low temperatures by the first term in Eq. 7, which does not include W. The hopping energy would then be the difference between Æ and S, at least for samples 2 and 3.

With the use of the bound magnetic polaron model, a natural explanation which does not involve an arbitrary cutoff in energy, is possible. If E_F is below E_c, then for temperatures such that $|E_c-E_F| \gg kT$, transport of electrons occurs by hopping near E_F with conductivity given approximately by

$$\sigma \simeq e^2 p R^2 N(E_F), \quad 34 \tag{13}$$

where R is the distance covered by each hop, and p is given by

$$p = (\nu_{ph} \exp(-2\alpha R))[\exp(-W/kT)] \tag{14}$$

Here ν_{ph} is the factor depending on the phonon spectrum, $\exp(-2\alpha R)$ is a factor depending on the overlap between localized states and W is the difference in energy between the occupied and unoccupied state. At relative high temperatures but still small enough so that only the electron distribution near E_F contributes to transport the first term in Eq. 14 remains constant and the conductivity is thermally activated with activation energy W. At lower temperatures variable range hopping occurs and both $\exp(-\alpha R)$ and W change. Mott has shown that this leads to his well known $T^{-1/4}$ law. It should be noted, however, that the physical process for both temperature ranges can be regarded as phonon assisted tunneling. The changes in the observed activation energy reflect the fact that at low temperatures only very few phonons are available to facilitate the hop. If we follow the reasoning of Thompson et al[11] we can regard the electron motion as being defined by the average barrier height of the potential well, i.e. the distance from E_F to E_c. Thompson et al find that the high temperature value for W can be used to predict α, the decay of the bound electron wave function outside the potential well, and consequently, the low temperature variable range hopping behavior. We can, therefore, identify the measured activation energy, Æ, with the distance E_c-E_F. This binding energy is dominated, at low temperatures by the magnetic term, with electrons hopping from site to site as described above.

Fritzsche[35] has pointed out that the thermopower has a simple physical meaning since it is related to the Peletier coefficient π as $S = \pi/T$. π is defined as the energy carried by the electrons per unit charge and the thermopower thus depends very sensitively on the total energy transport per carrier, either directly or by interactions with other normal modes of the system. This accounts for the phonon and magnon drag effects often encountered in metals. It also apparently accounts for the difference in activation energies observed by Emin et al.[36] in their studies of σ and S in chalcogenide glasses. These workers interpret the result that the activation energy $\mathit{Æ}_\sigma$ is larger than $\mathit{Æ}_s$ as being due to small dielectric polaron formation. The physical idea is that the excess charge carrier relaxes to a lower energy state by exchanging energy with the lattice (through small lattice distortions) forming the small polaron. In order to move, it has to overcome an energy barrier roughly equal to the polaron binding energy, W_H. If no net energy is transported in the hopping process, however, the hopping contribution to S is 0. Justification for the latter assumption has been given by e.g. Austin and Mott.[37] They point out, however, that this result is only true if the total activation energy is equally divided between the two hopping sites. If the two sites are not equivalent a term cW_H/kT enters the thermopower. c is a constant much smaller than one[37] and reflects the inequavelence of the sites.

We view the thermopower results (Fig. 4) as being analogous to the above description. The trapped magnetic polaron binding energy in $Gd_{3-x}v_xS_4$ is a distribution of energies because of static fluctuating potentials caused by the disordered vacancies. We may expect, therefore, a finite contribution to S from the term cW_H/kT. We expect, furthermore, the anomalous peak in the thermopower to be quenched in the presence of a large magnetic field because the bound state requires both the coulombic and magnetic interaction to be stable. Experiments to verify this prediction are underway and preliminary results indicate a decrease in the thermopower peak with applied magnetic field.

SUMMARY

We have studied the transport properties of samples in the more metallic concentration region of the defect $Gd_{3-x}v_xS_4$ solid solution system. Because of the large number of randomly distributed vacancies in these materials they represent transitional phases between the well ordered crystalline state and the totally disordered amorphous state. We have shown that it is possible to grow stable crystals of varying vacancy concentration under equilibrium conditions.

The transport properties of paramagnetic $Ce_{3-x}v_xS_4$ have been compared with its magnetically ordered Gd isomorph. Whereas, for concentrations above approximately $8 \times 10^{19} cm^{-3}$, the cerium compound is metallic for all temperatures, $Gd_{3-x}v_xS_4$, for similar concentrations, exhibits anomalous transport properties which we attribute to magnetic interactions. This is verified by the fact that with the application of a magnetic field, activated transport is quenched, and the conductivity approaches that of cerium sulfide.

The data has been interpreted on the basis of a band tail model in which the quantity $E_c - E_F$ is strongly affected by localized states formed by magnetic and Coulombic forces.

We gratefully acknowledge the technical assistance of R. Hamilton, S. Hanrahan and P. Lockwood and thank S. Kirkpatrick, and W. Thompson for several helpful discussions. We also thank T. Penney and J. Torrance for valuable discussion and critical reading of the manuscript.

REFERENCES

1. T. Kasuya and A. Yanase, Rev. Mod. Phys. $\underline{40}$, 684 (1968).
2. T. Kasuya, in "Proc. 10th Int. Conf. on the Physics of Semiconductors, Cambridge, Mass., 1970," S. P. Keller, J. C. Hensel and F. Stern, eds., CONF-700801 (U.S. AEC Div. of Tech. Info., Springfield, Va., 1970), p. 243.
3. J. B. Torrance, M. W. Shafer, and T. R. McGuire, Phys. Rev. Letters $\underline{29}$, 1168 (1972).
4. See, e.g. N. F. Mott and E. A. Davis, "Electronic Processes in Non-Crystalline Materials" (Clarendon Press, Oxford, 1971).
5. P. W. Anderson, Phys. Rev. $\underline{109}$ 1492 (1958).
6. J. A. van Vechten, Solid State Commun. $\underline{11}$, 7 (1972).
7. D. T. Pierce and W. E. Spicer, Phys. Rev. B$\underline{5}$, 3017 (1972); W. E. Spicer and T. M. Donovan, J. Non-Cryst. Solids $\underline{2}$, 66 (1970).
8. G. Busch, M. Campagna and H. C. Siegmann, to be published.
9. D. Redfield, J. Non-Cryst. Solids $\underline{89}$, 602 (1972).
10. D. Redfield and R. S. Crandall, "Proc. 10th Int. Conf. on the Physics of Semiconductors, Cambridge, Mass., 1970," S. P. Keller, J. C. Hensel and F. Stern, eds., CONF-700801 (U.S. AEC Div. of Tech. Info., Springfield, Va., 1970), p. 574.
11. W. A. Thompson, T. Penney, S. Kirkpatrick, and F. Holtzberg, Phys. Rev. Letters $\underline{29}$, 779 (1972).
12. Ref. 4, pg. 42.
13. M. Cutler, J. F. Leavy, and R. L. Fitzpatrick, Phys. Rev. $\underline{133}$, A1143 (1964).
14. M. Cutler and J. F. Leavy, Phys. Rev. $\underline{133}$, A1153 (1964).
15. M. Cutler and N. F. Mott, Phys. Rev. $\underline{181}$, 1336 (1969).
16. Landolt-Börnstein Tables, K. H. Hellwege and A. M. Hellwege eds., $\underline{4a}$, pp. 41-109, Springer-New York (1970).
17. K. Meisel, Z. Anorg. Chem. $\underline{240}$, 300 (1939).
18. W. H. Zachariasen, Acta Cryst. $\underline{2}$, 57 (1949).
19. P. I. Kripyakevich, Soviet Physics-Cryst. $\underline{7}$, 556 (1963).
20. F. Holtzberg and S. Methfessel, Journ. of Applied Physics $\underline{37}$, 1433 (1966).
21. F. Holtzberg, Y. Okaya, and N. Stemple, Amer. Crystallogr. Ass. Meeting, Gaithersburg, Tenn. p. 46 (1965).
22. W. L. Cox, H. Steinfink, and W. F. Bradley, Inorg. Chem. $\underline{5}$, 318 (1966).
23. F. L. Carter, J. Solid State Chem. $\underline{5}$, 300 (1972).
24. S. von Molnår, F. Holtzberg, T. R. McGuire, and T. J. A. Popma in "AIP Conference Proc. 5," C. D. Graham and J. H. Rhyne, eds. p. 869 (1972).

25. E. H. Putley, "The Hall Effect and Related Phenomena," Butterworth, London (1960).
26. Olof Lindberg, Proc. I.R.E. 40, 1414 (1952).
27. Y. Shapira and T. B. Reed, Phys. Rev. B5, 4877 (1972).
28. W. Thompson, private communication.
29. S. von Molnår, to be published.
30. T. Penney, M. W. Shafer, and J. B. Torrance, Phys. Rev. B, 5, 3669 (1972).
31. It is clear that little meaning can be attached to the magnitudes of the derived energies away from the peak near 200°K. Both the term f(1-f) and the magnetic localization energy are competing to produce an effective ΔE. But near the peak, $\Delta E >> kT$ and is a good measure of the activation energy.
32. J. Torrance, private communication.
33. Without the Coulomb interaction, the magnetic polaron in a ferromagnet is stable only very close to the magnetic ordering temperature [c.f. T. Kasuya, A. Yanase, and T. Takeda, Solid State Commun. 8, 1543 (1970)].
34. Ref. 4, Eq. 2.36.
35. H. Fritzsche, Solid State Commun. 9, 1813 (1971).
36. D. Emin, C. H. Seager, and R. K. Quinn, Phys. Rev. Letters 28, 813 (1972).
37. I. G. Austin and N. F. Mott, Advances in Physics 18, 41 (1969).

POLARIZED PHOTOELECTRONS FROM "PURE" AND DOPED EuO SINGLE CRYSTALS

K. Sattler and H.C. Siegmann
Laboratory for Solid State Physics,
Swiss Federal Institute of Technology,
8049 Zürich, Switzerland

ABSTRACT

Measurements of photoelectron spin polarization on single crystal surfaces give evidence of a paramagnetic sheet at the surface of Heisenberg ferromagnets. Doping enhances the molecular field in the surface. It is suggested that the electron escape depth is a function of surface magnetization. For $Eu_{1-x}La_xO$ (x~1%) we found $P \cong 80\%$, which is the highest electron polarization observed in a photoemission experiment so far.

INTRODUCTION

With magnetization measurements one determines the magnetization of the bulk and neglects surface effects. Until now there have been no direct measurements of the changes of the molecular field on going from the bulk to the surface. Problems related to surface magnetization have been investigated on thin films of metallic itinerant ferromagnets [1], and also on antiferromagnetic crystals of NiO [2]. The latter experiments indicate that the exchange interactions between Ni-spins in the surface are strongly reduced compared to the bulk values. Most of the existing theoretical work on surface magnetization supposes the validity of molecular field theory [3]. It is the purpose of this paper to report on a new experiment that allows one to measure the magnetization in a region very near the surface and to describe the first results.

We investigated the model Heisenberg ferromagnet EuO ($T_c = 69°K$). The single crystals were non-intentionally doped EuO as well as EuO doped with Gd, Sc, Ho and La. The dopant concentrations ranged from ~1at.% to ~4at.%. All the experiments showed the surprising result that the electron spin polarization (ESP) of photoelectrons does not saturate even if high magnetic fields (H=25kG) are applied. The results can be explained by a model which supposes a paramagnetic sheet at the surface and it is postulated that the escape depth of the photoelectrons is low due to spin disorder scattering.

EXPERIMENTAL

ESP-measurements have been made on thin films of different magnetic materials [4]. The new apparatus allows us to perform these measurements on single crystal surfaces that have been obtained by cleaving in UHV (Fig.1).

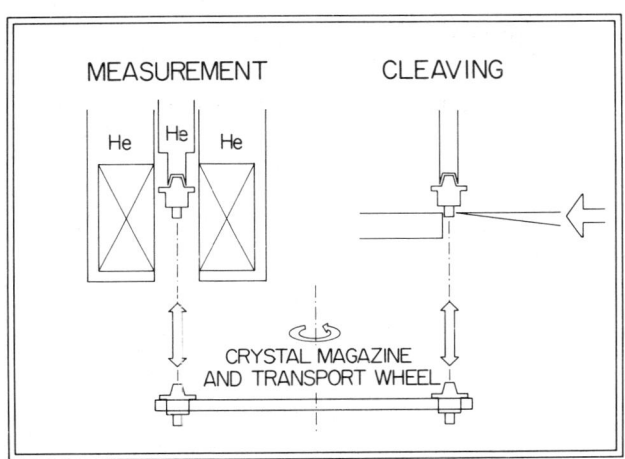

Fig.1 Preparation of single crystal Photocathodes.

22 crystals are stored in holders on a wheel. For the cleaving the crystal is moved to the cleaving chamber. For the measurement, the crystal is cooled to ~$10°K$ and moved into a homogeneous magnetic field, where a monochromatic light beam strikes the fresh (100)-surface, and photoelectrons are emitted. The photoelectrons are accelerated to 100 keV and scattered from a thin gold foil. The right/left asymmetry in this scattering is a measure of the degree of the ESP in the direction of the applied magnetic field $\vec{H}$.

RESULTS

Photoelectric magnetization curves (PMC's) of representative crystals are shown in Fig.2. The main features are (i) there is no magnetic saturation (ii) a kink occurs at 8 - 10 kG (iii) doped EuO shows higher P compared to "pure" EuO. In contrast, the ferromagnetic metal Ni [5] exhibits magnetic saturation.

In Fig.3, the dependence of ESP on photon energy at constant h = 6kG is shown. These "spectra of spin polarization" were taken with the same crystal surfaces as the PMC's in Fig.2. The ESP near photo threshold (~2eV and ~1.7eV for "pure" and doped EuO respectively) is low for the "pure" and high for the doped samples. This agrees with the well known fact that the impurities

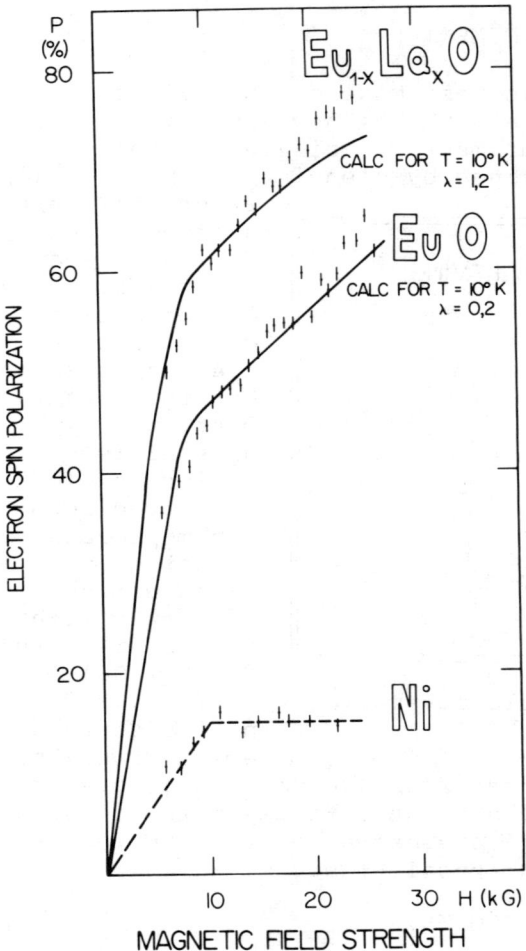

Fig.2. Photoelectric magnetization curves (PMC's) for "pure" and doped EuO, and, for comparison, for Ni (ref 5). The full lines are calculated assuming T=10°K and a molecular field constant λ, that gave the best fit to the measurements.

lie higher in energy than the 4f-states. It is surprising that the ESP of the extra electrons is higher (P=45% at $h\nu$ =3.4eV) than that of the 4f - electrons (P=40% at $h\nu$ =4.5eV).

The two spectra of doped EuO were taken 3 hrs and 5 hrs after cleaving. The close similarity shows that the surface contamination has no major influence on the results with the present vacuum conditions.

MODEL FOR INTERPRETATION

The magnetic field H_{eff} acting on a 4f-spin in the surface sheet is given by

$$H_{eff} = H_c + M_B - M_s + \lambda M_s \quad (1)$$

H_c is the applied magnetic field, M_B is the bulk magnetization, $-M_s$ is the demagnetizing field of the sheet and λM_s is the molecular field in the sheet.

Higher λ for the doped crystals indicates that the electrons introduced on doping are an effective link between bulk and surface and produce an extra molecular field in the surface. Furthermore, this explains why the polarization of the impurity electrons is higher than that of the 4f-electrons in the doped material as in Fig.3.

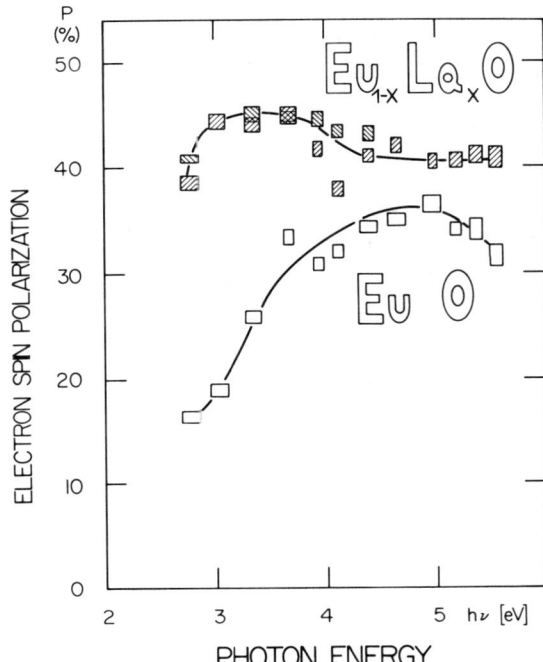

Fig.3. Spectra of spin polarization for the same crystal surfaces as Fig.2. The lower points in the spectrum of $Eu_{1-x}La_xO$ are taken 3 hrs after cleaving, the higher points 5 hrs after cleaving.

The kink occurs in the PMC's when the bulk saturates.

The demagnetizing factor of the crystals is ~1/3, and the saturation magnetization M_0=24kG, hence the bulk saturates at 1/3·24kG. The paramagnetic sheet is an effective barrier for photoelectrons from inside the bulk as long as M_s is low, because of the large spin disorder scattering in EuO [6]. Enhancement of the surface spin order at high magnetic field makes the sheet more transparent for bulk electrons, which is directly seen from the deviation of the ESP from the Brillouin function for H>20kG.

The existence of the paramagnetic sheet at an estimated T~10°K far below T_c is surprising. It leads us to the conclusion that in the europium chalcogenides the ferromagnetic part of the exchange is weaker in the surface than in the bulk.

ACKNOWLEDGEMENT

We should like to thank Prof. G. Busch for many helpful suggestions and P. Munz for important comments. This work was supported by the Schweizerischer Nationalfonds.

REFERENCES

1. U. Gradman, J. Appl.Phys. <u>40</u>, 1182 (1969).
2. T. Wolfram, R.E. Dewames, W.F. Hall and P.W. Palmberg, Surface Science, <u>28</u>, (1971) 45 - 60.
3. D.L. Mills, M.T. Béal-Monod and R.A. Weiner, Phys. Rev.B. <u>5</u>, 4637 (1972).
4. G. Busch, M. Campagna, H.C. Siegmann, J.Appl.Phys. <u>41</u>, 1044 (1970), and Phys.Rev.B, <u>4</u>, 746 (1971).
5. U. Bänninger, G. Busch, M. Campagna and H.C. Siegmann, Phys.Rev.Lett. <u>25</u>, 585 (1970).
6. C. Haas, Phys.Rev. <u>168</u>, 531 (1968).

MAGNETIC SUSCEPTIBILITY OF EXCHANGE COUPLED VAN VLECK IONS: $Sm_{1-x}La_xS$*

J. B. Torrance, F. Holtzberg and T. R. McGuire
IBM Research, Yorktown Heights, N. Y. 10598

ABSTRACT

Even though Van Vleck ions have non-magnetic ground states, exchange interactions between them can be observed by magnetic susceptibility measurements. The principles of these measurements are described as well as a technique for accounting for the paramagnetic impurity contribution to χ_M. Measurements are reported on SmS doped with La, which show that the Sm^{2+}-Sm^{2+} exchange interaction is greatly enhanced by the conduction electrons donated by the La.

INTRODUCTION

For the six 4f electrons of both Eu^{3+} and Sm^{2+}, Hund's rules indicate that L = 3 and S = 3; i.e. a ^7F term. Furthermore, the spin-orbit interaction favors an antiparallel coupling between $\vec{L}$ and $\vec{S}$, leaving the J = |L-S| = 0 state lowest; i.e. $^7F_0$. Since for such ions (Van Vleck ions) the ground state is <u>non-magnetic</u> (J=0), their magnetic properties are determined by the higher lying multiplet levels (J>0). The magnetic properties of these levels are manifested in two ways: at finite temperatures these levels are thermally populated; and, an external magnetic field admixes the J=0 and J=1 levels, giving rise to a moment in the ground state. Both effects may be examined by measuring the magnetic susceptibility of these ions. In this paper we describe how such susceptibility measurements can be used to measure the exchange interactions between Van Vleck ions; in particular, the changes in the Sm^{2+}-Sm^{2+} interactions in SmS caused by conduction electrons.

SUSCEPTIBILITY OF A VAN VLECK ION

The magnetic susceptibility of such non-magnetic ions was first calculated by Van Vleck[1], who found:

$$\chi_M = \frac{\sum_J (2J+1)\chi_J e^{-E_J/kT}}{\sum_J (2J+1) e^{-E_J/kT}} \qquad (1)$$

where, for the Jth multiplet level, E_J is the energy and χ_J is the susceptibility.[1] Assuming a simple spin-orbit coupling ($\vec{\Delta}$), we have calculated the susceptibility for various values of Δ and plotted them versus temperature in Fig. 1. In the case of no spin-orbit coupling ($\Delta=0$), the susceptibility obeys a Curie law, $\chi_M = C_M/T$, as for a usual paramagnet, with C_M = 7.5. However, Δ is typically $\sim$ 500°K for Eu^{3+} and $\sim$ 300°K for Sm^{2+} in SmS[3]. These cases are also shown in Fig. 1. Clearly, the effect of the spin-orbit coupling is

---
*Supported in part by ARPA Contract DAA-H01-71-C-1313.

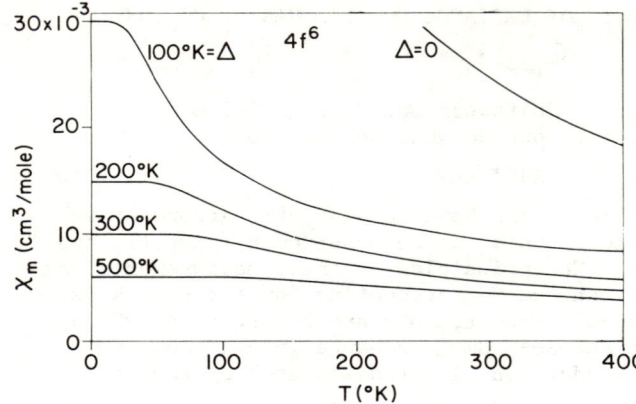

Fig. 1 The temperature dependence of the magnetic susceptibility of a $4f^6$ Van Vleck ion calculated from Eq. (1) for various values of the spin-orbit coupling, Δ.

is to dramatically reduce the susceptibility, by tending to align $\vec{L}$ antiparallel to $\vec{S}$, thus reducing the effective moment.

The temperature dependence of χ_M contains contributions due to both effects mentioned earlier: at high temperatures, the higher lying multiplet levels are thermally populated but χ_M decreases with increasing T as the Curie law susceptibility of these moments decreases. At lowest temperatures, the thermal population of the higher multiplets is negligible and χ_M is dominated by the moment admixed from J=1 by the applied magnetic field. This contribution to χ_M is the Van Vleck temperature independent susceptibility:[1,4]

$$\chi_{vv} = \chi_M(0) = 8N\beta^2/\Delta \qquad (2)$$

where Δ is more generally defined as the energy difference between the J=0 and J=1 states. Thus the low temperature value of χ_M can be used as a <u>direct</u> measure of Δ (Eq. (2) and Fig. 1).

EFFECTS DUE TO EXCHANGE

In the absence of an applied magnetic field, the exchange interactions between Sm^{2+} ions do not affect the J=0 ground state, since it has no moment. There will, however, be effects on the excited states which can be measured by the susceptibility. An early discussion of these effects is given by Bozorth and Van Vleck[2] for exchange interactions between Eu^{3+} ions. More recently, susceptibility measurements by Bucher, Narayanamurti and Jayaraman[5] revealed large differences in $\chi_M(0)$ between SmTe, SmSe, and SmS. It was suggested by Mehran, et al[6] and by Birgeneau, et al[3] that these differences were due to different Sm^{2+}-Sm^{2+} exchange interactions. Using molecular field theory, Birgeneau, Bucher, Rupp and Walsh[3] showed that the exchange interactions between Sm^{2+} ions alter the splitting, Δ, between the J=0 and J=1 states, so that Eq. (2) becomes:

$$\chi_M(0) = \frac{8N\beta^2}{\Delta_o - 8\sum_i z_i J_i} \qquad (3)$$

where the change in Δ from Δ_0 is caused by the exchange interaction J_i between z_i neighbors. For a ferromagnetic exchange, this reduction in Δ gives rise to an increase in $\chi_M(0)$. Birgeneau, et al[3] have used Eq. (3) with $\Delta_0 = 415°K$ to obtain the values for the Sm^{2+}-Sm^{2+} exchange interactions in SmTe, SmSe and SmS. In this paper we report measurements of the increase in the Sm^{2+}-Sm^{2+} exchange interaction in SmS caused by conduction electrons.

The exchange interactions between Sm^{2+} ions and <u>impurity spins</u> (eg. Eu^{2+}, Mn^{2+} and Gd^{3+}) have also been observed by Mehran, et al[6], by Birgeneau, et al[3], and by Walsh, et al[7] in SmTe, SmSe and SmS by measuring the large EPR g-shifts.

EFFECTS OF PARAMAGNETIC IMPURITIES

Since the variations in the exchange interactions that we expect to measure are small compared to Δ_0, the changes in $\chi_M(0)$ will be correspondingly small (Eq. 3) and particularly accurate measurements will be necessary. Also since the spin-orbit coupling greatly reduces χ_M (Fig. 1), a relatively small amount of paramagnetic impurities can have an important effect on our measurement of the exchange interaction. Samples of SmS generally have as much as 0.1% impurities, largely originating from the commercial Sm metal used in the synthesis. As an example, the measured χ_M data for two samples of average purity are shown in Fig. 2a: one undoped and one with nominally 3% La (La^{3+} is diamagnetic). Note the rise in χ_M at lowest temperatures which is presumably caused by paramagnetic impurities. Since the difference in $\chi_M(0)$ between these two samples

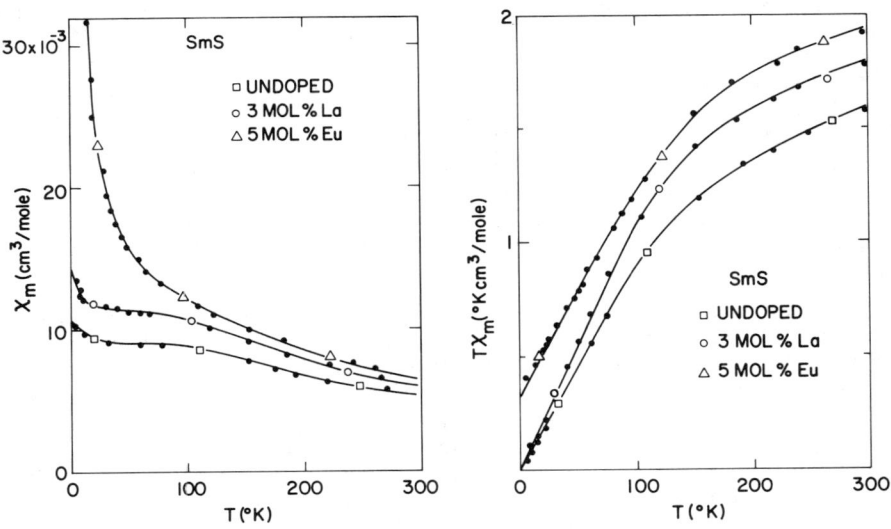

Fig. 2 (a) The susceptibility per mole Sm measured for three samples of SmS showing the paramagnetic impurity contribution at lowest temperatures. (b) a plot of $T\chi_M$ for the same three samples showing the straight line behavior below $\sim$ 75°K predicted by Eq. (4).

is nearly the largest we have found, we must be able to accurately measure smaller changes in $\chi_M(0)$ and hence we must subtract off the impurity contribution from the data. Also included in Fig. 2a is a sample of SmS intentionally doped with nominally 5% Eu. In this extreme case, the Eu^{2+} paramagnetic susceptibility completely obscures the temperature independent part of χ_M.

In order to eliminate the impurity contributions we assume that the impurities are purely paramagnetic (i.e. $\chi \sim C/T$). We then examine the product of T times the total measured susceptibility:

$$T\chi_M = 8N\beta^2 T/\Delta + C \qquad (4)$$

Thus, a plot of $T\chi_M$ versus T should give a straight line at low temperatures, with a slope related to Δ and a T=0°K intercept which measures the impurity contribution. Such a plot for each of the three samples of Fig. 2a is shown in Fig. 2b. Note that even for the 5% Eu sample we get a straight line. For that sample, the T=0 intercept converts to a Eu^{2+} concentration of 4.5% and the value of Δ obtained from the slope is 346°K, compared to $\sim$ 335°K for SmS. On the other hand, the value of Δ for the 3% La sample is 273°K and reflects the large increase in the exchange interaction caused by the conduction electrons donated by the La. The T=0°K intercepts for the two samples without intentional Eu doping indicate, for example, that if all the impurities were Eu^{2+}, there would be a 0.15% concentration.

La DOPED SmS

Since La in SmS is expected to be in the trivalent state, each La should add one conduction electron to the system, which will give rise to an indirect interaction between Sm^{2+} ions. Since La^{3+} itself is diamagnetic with about the same diamagnetic susceptibility as Sm^{2+}, the susceptibility behavior of La doped SmS should be similar to undoped SmS: any changes in χ_M (per mole Sm) will then be attributed to changes in the Sm^{2+}-Sm^{2+} exchange interaction caused by the conduction electrons. Using the techniques described above, we have obtained the sum of the exchange interactions between Sm^{2+} ions, $\Sigma z_i J_i$, for a series of samples with differing La concentrations. The results are plotted in Fig. 3, with the error bars on the data points representing a $\pm$ 1/2% error in χ_M.

Also plotted in Fig. 3 are the data points for undoped SmS obtained by Bucher, et al[5] and by Birgeneau, et al[3]. Note the variation between these two samples and our two undoped samples. This variation probably represents to a large extent real differences in the exchange interactions caused by differences in stoichiometry and/or impurity content, since we shall see that a small concentration of electrons can dramatically increase the exchange interaction. Thus, the errors associated with these effects although difficult to estimate are probably larger than the error bars in Fig. 3.

The major feature of the data in Fig. 3 is that the exchange interactions suddenly increase with increasing La concentration, so that near 3% La they are <u>almost twice</u> the value of undoped SmS. Further increases in the La **concentration** cause the exchange to

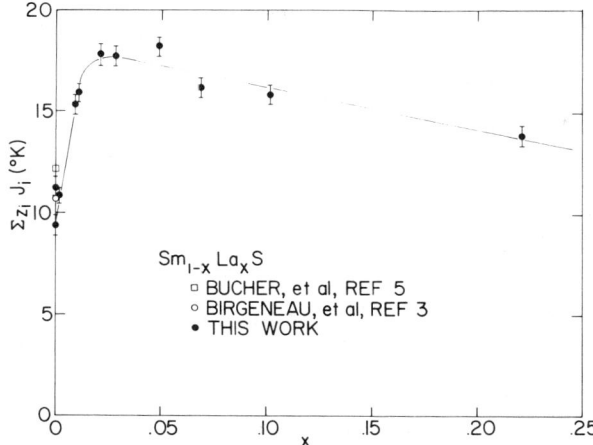

Fig. 3 The Sm^{2+}-Sm^{2+} exchange interaction in La doped SmS showing the increase in the exchange due to the conduction electrons. The error bars represent a $\pm 1/2\%$ error in the measurement of χ_M.

decrease. A discussion of the physical interpretation of these results as well as a comparison to similar behavior in the magnetic systems $Eu_{1-x}La_xS^8$ and $Eu_{1-x}Gd_xS^9$ will be the subject of a future publication.

We wish to acknowledge H. R. Lilienthal and L. J. Tao for the susceptibility measurements, P. G. Lockwood for technical assistance in preparing the samples, and J. D. Kuptsis for microprobe analyses.

REFERENCES

1. J. H. Van Vleck, Theory of Electric and Magnetic Susceptibilities (Oxford Univ. Press., London 1932) Ch. IX.
2. R. M. Bozorth and J. H. Van Vleck, Phys. Rev. 118, 1493 (1960).
3. R. J. Birgeneau, E. Bucher, L. W. Rupp, Jr., and W. M. Walsh, Jr., Phys. Rev. B 5 3412 (1972).
4. For Δ in units of °K and χ_M in units of cm^3/mole, Eq. (2) can be written $\chi_M(0) = 3.0093/\Delta$.
5. E. Bucher, V. Narayanamurti, and A. Jayaraman, J. Appl. Phys. 42 1741 (1971).
6. F. Mehran, K. W. H. Stevens, R. S. Title, and F. Holtzberg, Phys. Rev. Letters 27, 1368 (1971).
7. W. M. Walsh, Jr., L. W. Rupp, Jr., R. J. Birgeneau and L. D. Longinotti, A.I.P. Conf. Proc. 10 (1973) (in this volume).
8. T. R. McGuire and F. Holtzberg, Annual Technical Report, ARPA Contract No. DAA-H01-71-C-1313, June 1972.
9. T. R. McGuire and F. Holtzberg, A.I.P. Conf. Proc. 5, 855 (1972).

KRAMERS-KRONIG ANALYSIS AND MAGNETOREFLECTANCE OF EUROPIUM CHALCOGENIDES

G. Güntherodt and P. Wachter
Laboratorium für Festkörperphysik der ETH Zürich
8049 Zürich, Hönggerberg, Switzerland

ABSTRACT

The reflectivity of EuO, EuS and Gd-doped EuO single crystals has been analysed at 300°K in the spectral region from 250μ to 12 eV using the Kramers-Kronig relation. For EuS the optical constants have also been determined above and below the Curie temperature for photon energies from 1.5 to 5.7 eV. The imaginary part of the dielectric function has been tentatively related to recent OPW and earlier APW band structure calculations. For EuS a magnetic field-modulated magnetoreflectance has been measured and analysed by means of the Kramers-Kronig relation.

INTRODUCTION

The investigation of the optical properties of the Eu-chalcogenides has proved to be a powerful tool in determining their electronic structure and magnetic properties. Therefore the reflectivity of single crystals has been analysed at 300°K by means of the Kramers-Kronig relation. In the one-electron approximation the optical constants can be related to the interband transitions of the electrons in the solid. But a definite assignment of optical structure to interband transitions can only be done in conjunction with band structure calculations. Especially for the magnetic semiconductors it is very interesting to investigate the optical constants at low temperatures because of the effect of magnetic ordering and also of the higher resolution of the optical structure. While reflectivity measurements at low temperatures in the vacuum ultraviolet region turned out to be very difficult we have determined in the meantime the optical constants at low temperatures for photon energies up to 6 eV by a polarized light method which has been used earlier at room temperature[1].

Information about the transport properties of semiconductors is obtained by doping. The free carriers manifest themselves in the infrared as plasma reflection edges.

To gain a better resolution of the reflectivity spectra and of the involved magnetooptical transitions a modulation technique has been applied using circularly polarized light and an alternating magnetic field as modulating parameter.

OPTICAL CONSTANTS AND INTERBAND TRANSITIONS

Our earlier determination of the optical constants of Eu-chalcogenide single crystals at 300°K for photon energies from 0.5 to 6 eV[1] has been extended to the spectral range from 250μ to 12 eV. The optical constants have been obtained from a Kramers-Kronig (KK) analysis of the room temperature reflectivity of polished EuO and EuS single crystals. The reflectivity is shown in fig.1 for photon energies from 1 to 12 eV and for undoped EuO (x=0) for wavelengths up to 250μ in fig.4. For wavelengths longer than 50μ the reflectivity data of Axe[2] have been used together with his fit up to infinite wavelengths. Our infrared reflectivity up to 50μ is in good agreement with the one measured by Axe. The reflectivity spectra of Pidgeon et al.[3] on cleaved single crystals compare reasonably well with our spectra in fig.1. But our measurements show more resolved structures for photon energies above 5 eV and especially for EuS higher absolute values above 6 eV. The reflectivity spectrum of Grant et al.[4] for EuO seems not to be reliable above

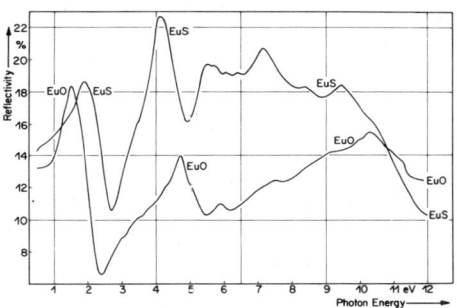

Fig.1 Reflectivity of EuO, EuS at 300°K

about 7 eV. The same refers to the whole spectrum of EuS by Mullen et al.[5] concerning the absolute values.

For the KK analysis a high energy extrapolation of the reflectivity has been assumed in form of a sequence of power laws[6]. This sequence has been chosen such as to give best agreement of the calculated optical constants with our earlier determined ones for photon energies from 0.5 to 5.8 eV[1]. This fit over such a large energy region is the main advantage of our analysis and gives a good confidence in the reliability of the optical constants above 6 eV and for wavelengths longer than 2.5 .

As a consequence it is for the first time that the imaginary part of the dielectric function of EuO and EuS at 300°K is available up to 12 eV for comparison with band structure calculations. Besides our earlier proposed energy level scheme of the Eu-chalcogenides[1], a first tentative assignment of optical structure to interband transitions has been carried out on the basis of the one-electron approximation and recent OPW band structure calculation for EuO[7]. The spectra of $\varepsilon_2(E)$ in fig.2 show similar behavior consisting of two pronounced peaks up to 5 eV with decreasing separation in energy from EuO to EuS. The main contributions to ε_2 extend for EuO up to 10 eV and for EuS up to about 7 eV. For EuO the most dominant interband transitions have been indicated as determined by the OPW calculation. In this scheme the transitions up to 6 eV are dominantly originating from the 4f-level. At about 6 eV the first "intrinsic" transition from the valence to the conduction band occurs. The crystal field splitting amounts to 3.1 eV in good agreement with the earlier determined value[1], but now is at-

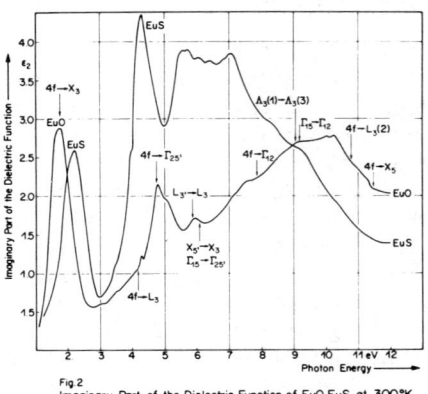

Fig. 2
Imaginary Part of the Dielectric Function of EuO, EuS at 300°K

tributed to the peaks at 4.8 and 7.9 eV in contrast to the earlier assignment[8,1]. At first glance the OPW and the APW calculation of Cho[9] seem to be similar, but there is a big difference concerning the Γ_1^--point. In the OPW calculation the Γ_1^--point is positioned above the Γ_{12}^+-point in contrast to Cho's calculation where it is situated between the X_3^- and the $\Gamma_{25'}^-$-points. Only Kasuya et al.[10] have assumed the Γ_1^--point below X_3. A theoretical calculation of $\varepsilon_2(E)$ could elucidate best the origin of the peaks at 1.8 and 2.2 eV in EuO and EuS, respectively.

The optical constants of EuS at low temperatures have been determined for photon energies from 1.5 to 5.7 eV by a polarized light method which has been used earlier at 300°K[1]. The resulting spectra of $\varepsilon_2(E)$ in fig.3 show a better resolution of the optical transitions at low temperatures with respect to 300°K. The red shift of the absorption edge and of the dominant peaks at 2.2 and 4.3 eV with decreasing temperature is obvious. The position of the absorption edge (ε_2=0 in fig.3) is not very accurate as determined by reflectivity measurements. For 5°K (T<T_c) light scattering on magnetic domains seems to have an increasing influence above 5 eV. Earlier measurements on thin evaporated films [11,1] showed in the energy region of the first maximum a structure consisting of 3 peaks. Fig.3 shows at 19 and 5°K for energies from 1.7 to 2.5 eV additional structure consisting of 6 peaks. The reason seems to be that for the first time these low temperature measurements have been performed on single crystals (in the reflectivity this fine structure is not well pronounced). The observed structure is similar to the one of EuF_2 and $Sr_{0.99}Eu_{0.01}S$[12] and can be attributed to the ^7F_J (J=0...6) multiplet of the $4f^6$ configuration[1]. The analogous structure in the maximum at 4.3 eV is difficult to assign because of overlapping transitions.

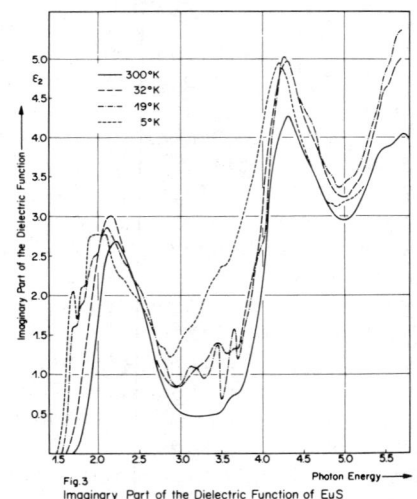

Fig. 3
Imaginary Part of the Dielectric Function of EuS

Gd - DOPED EuO

The KK analysis for undoped EuO in the infrared has given the position of the optical one-phonon absorption modes as following: LO mode at 23μ, TO mode at 55μ. These values agree well with the ones obtained by Axe[2] from a fit of the reflectivity with dispersion theory. The static dielectric constant as obtained from the KK analysis amounts to 26.5. A value of 25.6 results from the sum rule and gives a good confirmation of the accuracy of our analysis. The value reported by Axe[2] is 23.9±4.0.

The effect of doping on the infrared reflectivity is shown in fig.4. Instead of the minimum at 21μ for x=0 due to the LO mode a separate minimum appears and shifts to shorter wavelengths with increasing carrier concentration. By means of the KK analysis the long wavelength edge of this minimum could be identified as the plasma reflection edge. The condition $\varepsilon_1=0$ indicates the position of the plasma resonance at 18, 5 and 4μ for x=0.66, 1.26 and 5.6%, respectively. The reflectivity for wavelengths longer than 50μ shows an increasing "metallic" behavior as a function of x. The real part of the dielectric function changes sign and the intersection at $\varepsilon_1=0$ indicates the position of a coupled mode of plasmons with LO phonons[13].

MAGNETOREFLECTANCE

For EuS a modulation technique has been applied using circularly polarized light and an alternating magnetic field (147 cps, 200 Oe pp) as modulating parameter. Thus by changing the direction of the magnetic field the light polarization is changed from right to left circular. The change in the reflectivity is measured by phase sensitive detection in form of a difference between the reflectivity of right and left circularly polarized light ($\Delta R/R = R_+ - R_-/R$). In the region from 1.2 to 2.7 eV the change of sign can roughly be explained by taking into account the ^7F_J-multiplet as discussed above. For right circularly polarized light the transition probability dominates for the lower J-levels whereas for left circularly polarized light the inverse is true[1]. In the range from 1.5 to 2.55 eV the magnetoreflectance spectrum of fig.5 at 5°K compares qualitatively well (except the additional peak at 2.5 eV in fig.5) with the one reported by Aggarwal et al.[14] at 1.5°K. But the latter has been measured in a magnetic field of 42 KOe. Generally a complete interpretation of modulated reflectance spectra can only be

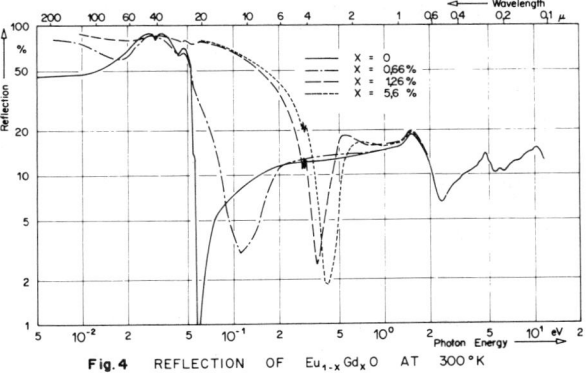

Fig.4 REFLECTION OF $Eu_{1-x}Gd_xO$ AT 300°K

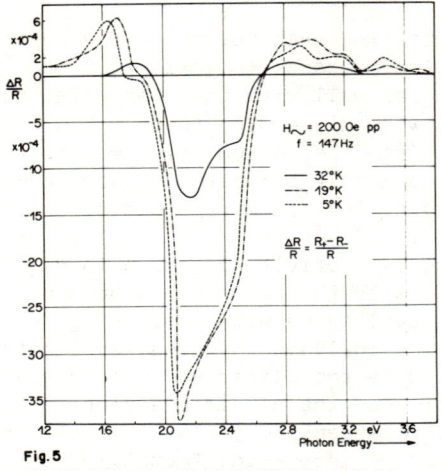

Fig. 5
MAGNETIC FIELD-MODULATED MAGNETOREFLECTANCE OF EuS

done in conjunction with the optical constants because $\Delta R/R$ is composed of a change in the real and imaginary part of the dielectric function. In a $\Delta\varepsilon_1$-dominated region "dispersive" line shapes will be preserved in the reflectance response and in a $\Delta\varepsilon_2$-dominated region "absorptive" line shapes remain. Therefore the spectra in fig.5 have been analysed by means of the KK relation to obtain the change of the phase angle $\Delta\Theta$. From $\Delta\Theta$ and $\Delta R/R$ together with the optical constants the quantities $\Delta\varepsilon_1$ and $\Delta\varepsilon_2$ could be calculated for the first time. Hence the regions near the absorption edge and around 2.55 eV turn out to be $\Delta\varepsilon_1$-dominated and the regions of 2.1 eV and above 2.8 eV are $\Delta\varepsilon_2$-dominated. In between fractional admixtures of both types exist.

The authors are obliged to Dr. P. Brüesch, BBC Baden, for measuring the reflectivity of Gd-doped EuO above 50μ. Many valuable discussions with Dr. K. Lendi are gratefully acknowledged. We also wish to thank Dr. E. Tosatti for help with the KK analysis and Mr. B. Naef for his experienced technical assistance.

REFERENCES

1. G. Güntherodt, P. Wachter and D.M. Imboden, Phys.kond.Mat. <u>12</u>, 292 (1971).
2. J.D. Axe, J. Phys. Chem. Solids <u>30</u>, 1403 (1969).
3. C.R. Pidgeon, J. Feinleib, W.J. Scouler, J.O. Dimmock, T.B. Reed, IBM J. Res. Devel. <u>14</u>, 309 (1970).
4. P.M. Grant and J.C. Suits, Appl. Phys. L. <u>14</u>,1972 (1969).
5. J. Mullen and A.W. Lawson, Phys. L. <u>24A</u>, 303 (1967).
6. E. Tosatti, private communication.
7. K. Lendi, thesis ETH Zürich, 1972, to be published.
8. S. Methfessel, F. Holtzberg and T.R. McGuire, IEEE trans. on Magn. <u>2</u>, 305 (1966).
9. S.J. Cho, Phys. Rev. <u>B1</u>, 4589 (1970).
10. T. Kasuya and A. Yanase, Rev.Mod. Phys. <u>40</u>, 684 (1968).
11. R.L. Wild, M. Shinmei and A.L. Andersen, Proc. Int. Conf. on the Semicond., Leningrad 1968, p. 1191.
12. S. Methfessel and D.C. Mattis, Handbuch der Physik XVIII, 1968, p. 476.
13. G. Güntherodt and P. Wachter, Phys. L., to be published.
14. R.L. Aggarwal and C.R. Pidgeon, Proc. of the 10th Int. Conf. on the Phys. of Semicond., 531 (1970).

Presented at the 18th Conference on Magnetism and Magnetic Materials,
Nov. 28 - Dec. 1, 1972, Denver.
To be published, A.I.P. Conf. Proc. 10 (1973)

BOUND MAGNETIC POLARONS AND THE SUSCEPTIBILITY OF EuO[*]

T. R. McGuire, J. B. Torrance, and M. W. Shafer
IBM Research, Yorktown Heights, N. Y. 10598

ABSTRACT

Magnetic susceptibility (χ) measurements at temperatures up to 700°K have been made on samples of Eu-rich EuO which show insulating, metal to insulator, and metallic conductivity. Above 200°K, χ obeys a Curie-Weiss law with a molar Curie constant of $C_M = 7.7$ for all samples. The value of θ, however, increases from 76 to 84°K with increasing Eu concentration. We interpret this increase as evidence that the electrons associated with excess Eu are contributing to the exchange interaction. The increase in θ is inconsistent with the He-like model for the metal-insulator transition but is consistent with a model of bound magnetic polarons. The constant value of C_M at high temperatures indicates that the ordering within the polarons is relatively weak. Our results are also compared to the susceptibility measurements of Menyuk, Dwight and Reed, on EuO.

INTRODUCTION

In general, the doping of Europium chalcogenides with a variety of donors can cause large increases in the Curie temperature (T_c).[1] Certain theoretical aspects of this problem concerning the enhanced exchange interactions caused by the conduction electrons have been investigated by Kasuya and Yanase.[2] The work on EuO presented here relates to the nature of the exchange interactions for relatively light doping.

For any experiments on EuO it is important to recognize that crystals of EuO exist over a relatively wide range of stoichiometry, from oxygen-rich to Eu-rich.[3] Corresponding to this range in composition, there is a wide variation in the infrared and conductivity properties, as has been shown by Oliver, et al[4] and by Shafer, Torrance and Penney.[3] The electrical conductivity, for example, of oxygen-rich EuO behaves like an insulator[3], while slightly Eu-rich samples exhibit a 13 order of magnitude metal-insulator transition just below T_c.[3-5] More heavily Eu-rich EuO appears metallic over the entire temperature range. In this paper we examine the differences in the magnetic properties between insulator (I), metal-insulator (M/I) and metallic (M) samples.

This work has been stimulated by an attempt to understand the metal-insulator (M/I) transition; in particular we hope to be able to distinguish between the two models that have been proposed for this transition in EuO: the Helium-like model[6,4] and the bound magnetic polaron model.[5] In the paramagnetic region, according to the He-like model, the two electrons in the oxygen vacancy have their spins antiparallel, and the Eu^{2+} ions surrounding the vacancy are disordered.

---
[*]Supported in part by the Advanced Research Projects Agency of the Dept. of Defense under ARPA Order #1588 and monitored by the U. S. Army Missile Command, Contract #DAAH01-71-C-1313.

In the BMP model, on the other hand, the two electron spins are parallel and they ferromagnetically align the Eu^{2+} moments around the vacancy, forming a magnetic cluster. Our investigation uses the magnetic susceptibility to study the existence and formation of these clusters.

EXPERIMENTAL PROCEDURES AND RESULTS

The samples studied were prepared and characterized as described in Ref. 3. The crystals were grown by reacting excess Eu-metal with Eu_2O_3 powder in sealed tungsten crucibles. Depending on the reaction temperature and the Eu/Eu_2O_3 ratio, the resulting crystals may be very oxygen-rich with a second phase of Eu_3O_4, or slightly oxygen-rich with Eu^{3+} present, or Eu-rich with oxygen vacancies. The approximate composition was inferred from infrared measurements.[3]

Magnetic susceptibility measurements using the Faraday method were made over the range 69 to 700°K, where the temperature was measured by a thermocouple calibrated against the susceptibility of Gd_2O_3. Above ~200°K the data closely fit a Curie-Weiss law: $\chi_M = C_M/(T-\theta)$. From a least square fit of the data, we obtained the values of C_M and θ for five representative samples shown in Table I, along with their approximate concentration. The values of T_c were determined from moment-field curves in the vicinity of T_c.

Figure 1 shows the low temperature reciprocal susceptibility data for the three types of samples. The same measurements are plotted in Fig. 2 in a different form. Space does not allow showing the high temperature data; however, it fits very closely the Curie-Weiss expression from which θ and C_M were obtained.

DISCUSSION

The critical magnetic properties of EuO have been carefully measured by Menyuk, Dwight and Reed.[7] In addition to their measurements near T_c, they have measured the high temperature susceptibility. Using a simple Curie-Weiss law in order to compare their data to ours, one obtains $\theta = 81°K$ and $C_M = 7.4$. This value of C_M is significantly lower than our value of $C_M = 7.7 \pm 0.05$, which is still lower than the theoretical $C_M = 7.87$. A possible explanation for the difference between these experimental values of C_M is that their sample may be highly oxygen-rich. If this were the case, the C_M value would be lower (due to the presence of Eu^{3+} and possibly Eu_3O_4) and the sample would

y	Type	T_c (°K)	θ (°K)	C_M
~ .000	I	69.2	76.9	7.69
~ .003	M/I	69.3	79.5	7.70
~ .003	M/I	70.6	81.0	7.64
~ .005	M	71.3	84.3	7.73
~ .005	M	---	83.0	7.73

TABLE I: Magnetic data for $Eu_{1+y}O$

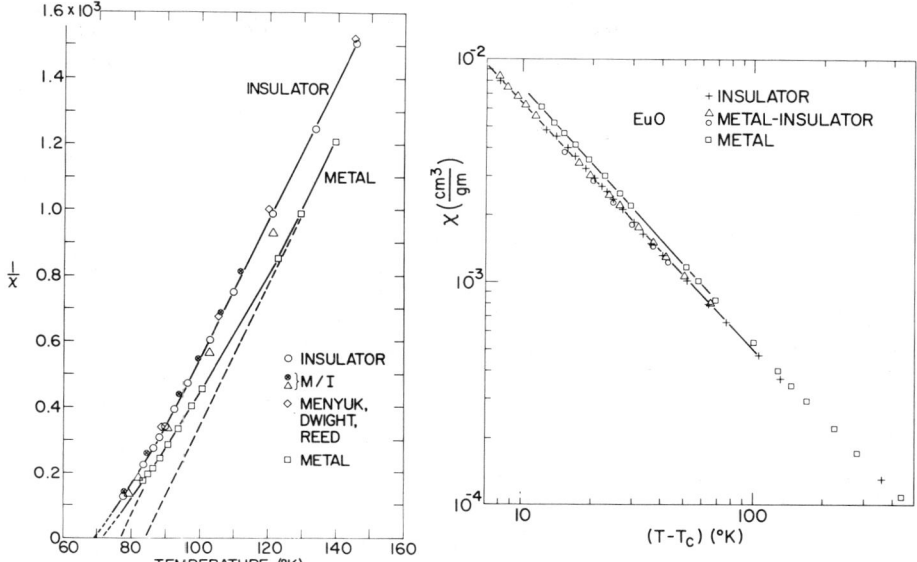

Fig. 1. Reciprocal susceptibility vs. temperature for three types of EuO. Data of Menyuk, Dwight and Reed (Ref. 7) are shown for comparison.

Fig. 2. Data of Fig. 1 with temperature normalized to T_c.

be highly insulating, as they observe. A 4% concentration of Eu^{3+} would be necessary to account for the lower value of C_M. Such a high concentration is possible according to Ref. 3. If their sample were indeed very oxygen-rich their results concerning the critical properties of EuO would likely remain perfectly valid, but their discussion concerning the magnitude of the exchange interactions might have to be reconsidered.

The difference between our value of C_M=7.7±0.05 and the theoretical C_M=7.87 cannot be explained in a similar way, because we observe the same C_M value in Eu-rich samples which have no Eu^{3+} or Eu_3O_4. We belive that the experimentally observed value of C_M is reduced due to the admixture of the 5d state, which[8] also gives rise to the nearest neighbor Eu-Eu exchange interaction.

The major feature of our data is that samples with oxygen vacancies have a <u>larger</u> magnetic susceptibility at all temperatures than samples without these vacancies. This fact is inconsistent with the simple He-like model of the insulator-metal transition, for the following reason. In this model[6,4] the two electrons associated with the oxygen vacancy have their spins aligned antiparallel. The tendency of one of these electrons to induce ordering of the Eu^{2+} spins neighboring the vacancy is nullified by the other antiparallel electron. The net result is that these Eu^{2+} spins are as disordered as if there were no vacancy electrons present. Thus, according to the simple He-like model, the susceptibility should be the same for samples with and without oxygen vacancies, contrary to what we have observed.

On the other hand, the bound magnetic polaron model[5] predicts that the spins of the two electrons associated with the oxygen vacancy will be ferromagnetically aligned. They will thus be able to induce a relatively large polarization of the Eu^{2+} spins neighboring the vacancy. The presence of these polarons (or clusters) is clearly expected to <u>increase</u> the susceptibility and, as we shall show, can account for the susceptibility behavior that is observed. At high temperatures (T>200°K) the ordering of the Eu^{2+} spins within the polaron is relatively weak (although they still exist). This temperature region has been discussed in Refs. 2 and 5, where it is shown that the effect of the polarons on the susceptibility is to <u>increase</u> the Curie-Weiss θ, as we observe (Table I). To the extent that θ may be viewed as a measure of the sum of the exchange interactions between Eu^{2+} spins, the increase in θ is due to an increase in these interactions caused by the presence of magnetic polarons.

In this paper we concentrate on studying how the polarons affect the susceptibility in the vicinity of T_c. At temperatures near, but above T_c the n Eu^{2+} spins in the polaron (cluster) do not behave independently, but act as a single, large moment with a total spin S_c for the cluster. For a concentration y of these clusters, the measured value of C_M is increased and becomes equal to $C_M' = (1-ny)C_M + yC_M S_c(S_c+1)/S(S+1)$, which can be written as:

$$\frac{\Delta C_M}{C_M} = \frac{C_M' - C_M}{C_M} = ny \left\{ \frac{S_c(S_c+1)}{nS(S+1)} - 1 \right\} \quad (1)$$

Very near T_c where the clusters are almost saturated, $S_c \sim nS$ and for .5% oxygen vacancies (with n=6), Eq. (1) predicts $\Delta C_M \sim 1$. As the temperature increases, the clusters start to disorder and the value of S_c decreases. Correspondingly, the value of ΔC_M is expected to decrease, vanishing at high temperatures.

Experimentally in the temperature region below 140°K considerable curvature in $1/\chi$ vs T is found (Fig. 1). In order to interpret these data we have plotted in Fig. 2, χ vs $(T-T_c)$, on log scales typical of analyses used to obtain critical constants. In Fig. 2 the susceptibility for the metallic sample lies higher than the samples with fewer oxy-

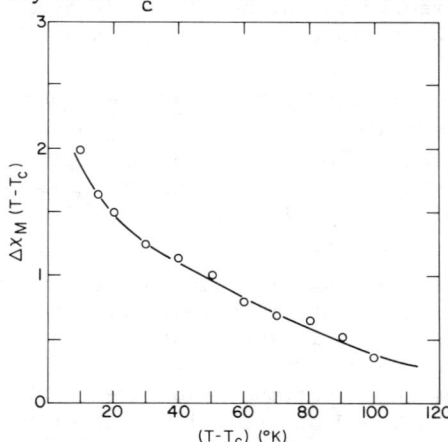

Fig. 3. Susceptibility calculated from the difference of the two data lines in Fig. 2.

gen vacancies. The additional $\Delta\chi_M$, i.e. the difference between the two lines in Fig. 2, is shown in Fig. 3 as a Curie's law plot where $\Delta C_M = \Delta\chi_M(T-T_c)$. We interpret this increase of C_M as evidence for magnetic polarons. Thus, near T_c the clusters appear to give a contribution $\Delta C_M \sim 2$, compared to our prediction above of $\Delta C_M \sim 1$. This agreement is quite reasonable considering the error associated with measuring y, the simplicity of the model and the possibility that n might be somewhat larger than 6. As we expect, the cluster contribution, ΔC_M, decreases with increasing temperature as the clusters become more disordered.

In conclusion, the magnetic susceptibility of samples with oxygen vacancies is larger than those without vacancies. This increase in χ_M is due to the presence of bound magnetic polarons (clusters), which have the effect of increasing the C_M value obtained from the low temperature susceptibility and increasing the θ value of the high temperature χ_M.

We wish to acknowledge H. R. Lilienthal and L. J. Tao for carefully making the susceptibility measurements.

REFERENCES

1. J. J. Rhyne and T. R. McGuire, IEEE Trans. on Magnetics <u>MAG-8</u>, 105 (1972). (This article reviews the effects on the magnetic properties of doping of europium compounds.)
2. T. Kasuya and A. Yanase, Rev. Mod. Phys. <u>40</u>, 684 (1968).
3. M. W. Shafer, J. B. Torrance, and T. Penney, AIP Conf. Proc. <u>5</u>, 840 (1972), and J. Phys. Chem. Solids (to be published).
4. M. R. Oliver, J. O. Dimmock, A. L. McWhorter, and T. B. Reed, Phys. Rev. <u>B5</u>, 1078 (1972).
5. J. B. Torrance, M. W. Shafer, and T. R. McGuire, Phys. Rev. Letters, <u>29</u>, 1168 (1972), T. Kasuya, <u>Proc. 10th Internat. Conf. Semicond.</u> (1970) p. 243.
6. S. von Molnar and T. Kasuya, <u>Proc. 10th Internat. Conf. Semicond.</u> (1970) p. 233.
7. N. Menyuk, K. Dwight, and T. B. Reed, Phys. Rev. <u>B3</u>, 1689 (1971).
8. T. Kasuya, IBM J. Res. Develop. <u>14</u>, 214 (1970).

COVALENCY, SPIN DENSITIES AND CORE ELECTRON BINDING ENERGY SPLITTINGS IN RARE EARTH ION COMPLEXES*

E. Byrom and D.E. Ellis
Physics Department, Northwestern University, Evanston, Illinois 60201

and

A.J. Freeman
Northwestern University and Argonne National Laboratory, Argonne, Illinois 60439

ABSTRACT

Results are reported for first principles calculations on the $(EuF_6)^{4-}$ cluster using the (spin and orbital) Unrestricted Hartree-Fock-Slater Molecular Orbital Scheme, recently developed to treat transition metal ion complexes. The effects of covalency on spin densities, neutron magnetic form factors, and exchange splitting of one-electron binding energies are described.

*Supported by the U.S. Air Force Office of Scientific Research, the National Science Foundation, and the U.S. Atomic Energy Commission.

INTRODUCTION

Although covalency (electron transfer) is now well established as playing a vital role in the observed magnetic, optical and transport properties of transition metal compounds, very little is known theoretically about covalent effects in rare earth compounds. The sharp localization of the 4f-electrons inside the 5s and 5p shells of the rare earth, and their small direct overlap onto neighbouring ions has led to the common use of the electrostatic crystal field despite its failures for the 3d transition metal ions.

Transferred hyperfine, optical and neutron diffraction measurements have required a review of the belief that covalency effects are small for the rare earths. This paper describes some results obtained by first principles calculations for the $(EuF_6)^{4-}$ cluster using the (spin and orbital) Unrestricted Hartree-Fock-Slater Molecular orbital scheme, recently developed to treat the transition metal complexes.[1]

METHOD OF CALCULATION

Details of the method for solving the HFS equations have been given elsewhere.[1] The basic features are the use of the effective Hamiltonian

$$H = \frac{-\nabla^2}{2} + V_{nuclear} + V_{coulomb} + V_{exchange} \qquad (1)$$

and the approximation of the exchange term by a local potential

$$V_{\text{exchange}, \sigma}(r) = -3\alpha \left[\frac{3}{4\pi} \rho_\sigma(r)\right]^{1/3} \quad (2)$$

where σ refers to the spin state: i.e. spin-up electrons are subject to an exchange potential derived from the density of spin-up electrons.[2] In this calculation the parameter α is set equal to 1.

One-electron HFS eigenfunctions are obtained in this method by using an LCAO (linear combination of atomic orbitals) basis set.

$$\Psi_i(r) = \sum_j A_j(r) C_{ji} \quad (3)$$

Defining an error for a state i at point r as

$$\delta_i(r) = (H - \epsilon_i) \Psi_i(r) \quad (4)$$

an optimal set of coefficients C_{ji} is determined by minimizing weighted averages of the errors over a discrete set of sample points. A secular equation is obtained, identical to the Rayleigh-Ritz equation

$$HC = SCE$$

but with the matrix elements given in discrete form.

TABLE I   One electron energies and exchange splittings (a.u.)

Principal Character	$\epsilon\uparrow$	$\epsilon\downarrow$	$\epsilon\uparrow - \epsilon\downarrow$	Subsidiary Character Mixed in
Eu 4s a1g	12.54	12.19	0.35	
4p t1u	10.30	9.95	0.35	
4d t2g	6.24	5.89$_5$	0.34$_5$	
eg	6.23	5.88$_5$	0.34$_5$	
5s a1g	1.88	1.70	0.18	F 1s, 2s, 2p
F 2s t1u, a1g, eg	1.45	1.45	0.0	Eu 4s, 4p, 4d, 4f, 5s, 5p
Eu 5p t1u	1.24	1.09	0.15	F 1s, 2s, 2p
4f t1u	1.20	0.88	0.32	F 1s, 2s, 2p; Eu 4p, 5p
t2u	1.20	0.86	0.33	F 2p
a2u	1.17	0.84	0.33	- - - -
F 2p, a1g, eg, t1g t2g, t1u, t2u	0.76	0.76	0.0	Eu 4p, 4f, 5p

The basis functions $A_i(r)$ are linear combinations of Slater type orbitals, which are basis functions of the irreducible representations of the symmetry group of the cluster. H_{ij} and S_{ij} are elements of the Hamiltonian and overlap matrices respectively. In this method the eigenvalues and eigenvectors converge rapidly as the number of sample points is increased, and preliminary studies of large molecules can be done in short computer times.

In this calculation STO's centered on both[3,4] F^- and Eu^{2+} ions, and optimized for the respective free ions, were used; giving a basis set of approximately double-zeta quality, except for the inclusion of four 4f-like STO's.

All 121 electrons of the seven-atom cluster are treated in the framework of the full HFS Hamiltonain. Matrix elements of the Hamiltonian are evaluated directly by a Diophantine (numerical) integration procedure, to overcome the problem of molecular multicenter integrals. In this calculation, 800 sample points were found adequate for a preliminary discussion of eigenvalues and spin densities. The bond length is chosen to be 5 a.u. The starting potential and charge densities are derived from a superposition of Eu and F free atom densities in the usual manner. The secular equation is diagonalized once: the calculation is not carried to self-consistency.

RESULTS

The spin density may be divided into two contributions: of the molecular orbitals with dominant Eu 4f character, only the spin-up orbitals are occupied; and this 4f spin-density polarizes the other orbitals. An indication of the atomic orbitals contributing to each MO is given in Table I. Figure 1 shows the unpaired spin-up 7t1u orbital along the [100] direction. Even at the separation of 5 a.u., which is large for europium-fluorine systems, there is some covalent mixing of fluorine atomic orbitals with the europium 4f, 5s, 5p orbitals; and thus a covalent, or overlap, contribution to the spin density around the fluorine ions.

Figure 2a shows the closed-shell-polarization contribution to the spin density. The maximum values of the closed-shell and total spin densities along the [100] direction, are respectively, 11.9 a.u. and 35.2 a.u. Most of this large closed shell contribution arises from the 4p, 4f, 5p hybridization allowed by the cubic symmetry of the cluster, and caused by the "crystal field" of the ligands at the Eu site. The 4f and 5p orbitals, in particular are nearly degenerate (Table I). For the same reason, the spin density in the region of the Eu^{2+} ion shows an angular dependence with maxima along the bond axes. Figure 2b shows the total spin density in the region of an F^- ion. The density at the nucleus is -0.002 a.u. This is the correct sign and order of magnitude, compared to Eu: CaF_2 experiments[5], but the absolute value is not considered significant.

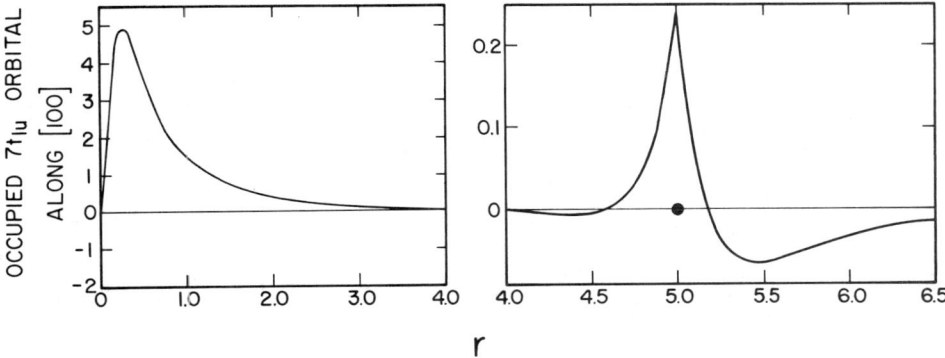

Figure 1. Occupied 7t1u orbital along [100]: (a) "4f" component; (b) region around F⁻ ion.

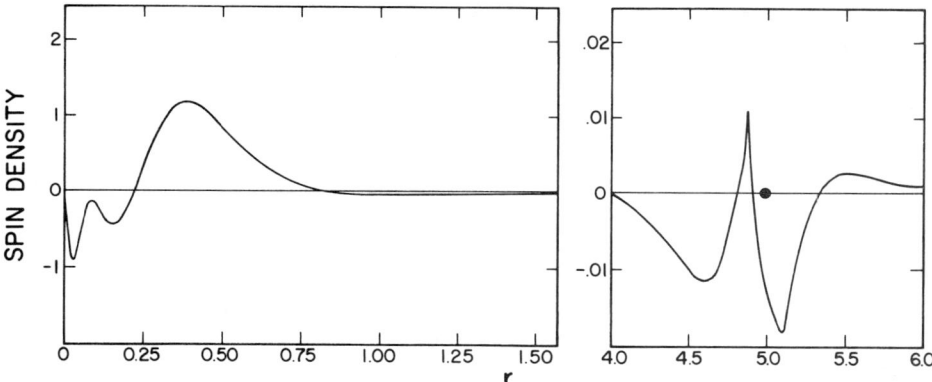

Figure 2. Spin density along [100]: (a) Closed shell contribution near Eu core; (b) total spin density around F⁻ ion.

The spherically averaged magnetic form factor $f_o(k)$ is the same as that of the free Eu^{2+} ion, to within the accuracy of the calculation. The $f_4(k)$ and $f_6(k)$ components of the form factor are small, becoming significant ($\sim 1\%$ of $f_o(k)$) in the region of $k = 5$. These non-spherical contributions are larger in the $(EuO_6)^{10-}$ cluster.[6]

The exchange splittings of one-electron energies are given in Table I. It is interesting to note that even through the individual one electron energies for spin up or spin down are different from their free ion counterparts, the exchange splittings are essentially unchanged. The effects of covalency on the 5s splittings have been found to be small, in agreement with experiment.[7] Additional splittings of the 4f levels in the cluster, due to "crystal field" effects, amount to approximately 1eV. Splittings of the fluorine levels are negligible. More exact calculations are under way for the $(EuF_6)^{4-}$ cluster as well as for the more strongly-interacting $(EuO_6^6)^{10-}$ cluster.[6]

REFERENCES

1. D.E. Ellis and T. Parameswaran, Intern. J. Quantum Chem. <u>5</u>, 443 (1971); D.E. Ellis and G.S. Painter, Phys. Rev. <u>B2</u>, 2887 (1970); T. Parameswaran and D.E. Ellis, J. Chem. Phys. (to appear).

2. J.C. Slater, Phys. Rev. <u>81</u>, 385 (1951); J.C. Slater, T.M. Wilson, and J.H. Wood, Phys. Rev. <u>179</u>, 28 (1969).

3. E. Clementi, "Tables of Atomic Functions", IBM J. Res. Develop. Suppl. <u>9</u>, 2 (1965).

4. J.V. Mallow, A.J. Freeman and P.S. Bagus, unpublished results.

5. J.M. Baker and J.P. Hurell, Proc. Phys. Soc. (London) <u>82</u>, 742 (1963).

6. Preliminary results of E. Byrom, D.E. Ellis and A.J. Freeman, unpublished.

7. J.F. Herbst, D.N. Lowy, and R.E. Watson, Phys. Rev. <u>B6</u>, 1913 (1972); A.J. Freeman, P.S. Bagus, and J.V. Mallow, Int. J. Magnetism (to appear).

FAR-INFRARED MAGNETIC RESONANCE ABSORPTION TO LOW-LYING EXCITED STATES IN DyPO4

G.A. Prinz, J.L. Lewis and R.J. Wagner
Naval Research Laboratory, Washington, D.C. 20390

ABSTRACT

$DyPO_4$ exhibits several interesting magnetic properties, such as its close approximation to an ideal three-dimensional Ising antiferromagnet, and its large magneto-electric effect. To aid in understanding such magnetic effects, we have studied the crystal field states of the lowest J manifold of Dy^{3+}, $^6H_{15/2}$. Transitions from the ground state to these low-lying levels fall in the far-infrared. Because of the unusually large ground state splitting factor ($g_\parallel \approx 20$), these transitions can be easily examined using the technique of far-infrared magnetic resonance, which offers both high intensity and high resolution in this difficult region of the spectrum. This technique employs a molecular gas laser (H_2O/D_2O) which provides a variety of fixed far-infrared frequencies. By varying an applied magnetic field (up to 90 kOe) the crystal field levels were Zeeman tuned through resonance absorption at the laser frequencies 114.4, 94.3, 90.5, 86.7 and 84.3 cm^{-1}. Both $\pi(E\|C)$ and $\sigma(E\perp C)$ spectra were obtained. The low temperature, zero field locations of the crystal field levels were found by extrapolation and correction for the dipole field. The first two excited state locations and their g-factors were determined to be 77 cm^{-1} ($g_\parallel = 0.6$) and 85 cm^{-1} ($g_\parallel = 1.1$). A complete report of this work is being prepared for publication.

RELATIVISTIC ELECTRONIC BAND STRUCTURE AND PROPERTIES OF THE HEAVIER ACTINIDES: A SECOND RARE-EARTH SERIES[*]

D. D. Koelling and A. J. Freeman
Physics Department, Northwestern University
Evanston, Illinois 60201
and Argonne National Laboratory, Argonne, Illinois 60439

ABSTRACT

The heavier actinides i.e., Am and beyond, have been studied by means of RAPW calculations and are found to differ dramatically from the lighter metals: the 5f electrons appear to be more localized spatially, overlap only weakly with their 5f neighbors and show band widths which are small relative to the Coulomb correlation energy. Hence, just as in the case of the rare-earth metals, the 5f electrons may not be described by the band model. Instead, a localized description is required in which, as in the rare-earths, Hund's rule intra-atomic exchange coupling produces localized 5f magnetic moments which can be ordered magnetically by indirect exchange interaction.

Until very recently, little was known about the electronic band structure of the actinide metals with which to interpret their magnetic properties. Quite aside from the expected theoretical complications of treating relativistically the complex crystallographic structures formed by these metals, their conduction band structures are expected to be more complicated than that of either the transition metals or the rare-earth metals. Here we report on the fcc phases of these metals. In the lighter actinides, detailed Symmetrized Relativistic APW (SRAPW) studies[1] have shown that the 5f electrons are not well localized (unlike the 4f orbitals in the rare-earth metals) so their itinerant nature makes them hybridize strongly with both the 6d and 7s bands. This strong hybridization of the broad 5f itinerant states with the very broad 6d-7s bands shown in Fig. 1 for fcc plutonium metal results in a band structure which has an overall large effective band width for the (possible) magnetic carriers. (The plane wave state can be seen starting at Γ_6^+ with the parabolic shape. The d states at Γ are the spin-orbit split t_{2g} (Γ_8^+ and Γ_7^+) and the higher e_g(Γ_8^+). All states shown with negative parity at Γ are f-states.) This large f bandwidth makes it energetically too costly to produce any magnetization (flipping of spins); i.e., in the band picture[2] the kinetic energy penalty of promoting an electron of one spin band into the opposite spin band exceeds the lowering of energy due to exchange interactions (because the effective width is just too great). This same

---

[*]Supported by the Atomic Energy Commission, the Air Force Office of Scientific Research and the National Science Foundation.

qualitative picture appears to hold for all the lighter actinide metals which show no magnetic ordering.

In the lighter actinides, the itinerant or band point of view is applicable because of the large overlap between neighboring 5f electrons resulting in their band widths exceeding the Coulomb correlation energy. By contrast, the 4f electrons in the rare-earths are highly localized, do not overlap neighbors appreciably and form a very narrow energy band in the solid whose effective width is much smaller than the Coulomb correlation energy.[3] Thus unlike the case of the 5f electrons in these actinides, the 4f electrons may not be treated as band electrons because of the large errors ($\sim$ 10-15 eV) introduced by neglect of the intra-atomic Coulomb correlation terms.

We have investigated the heavier actinide metals Am, Cm and Bk by means of the SRAPW method. The effects of varying α, the parameter multiplying the free electron exchange potential ($\rho^{1/3}$) and atomic starting configurations (relative number of occupied 5f, 6d and 7s electrons) on the resulting band structure was studied-- all in the "warped" muffin tin approximation. Since space does not permit a detailed presentation of our results, a summary will be given. Fig. 1 includes the band structure of these heavier metals obtained with $\alpha = 2/3$ (the Kohn-Sham Gaspar exchange approximation). We have found that there is a sharply increased localization of the 5f electrons in these metals with increasing atomic number, i.e., greatly decreased overlap and rapid narrowing of their band widths. (This localization is found to be much greater in the $\alpha = 1$ calculations, which we feel, however, are not as physically meaningful.) These much narrower bands do not hybridize greatly with the 6d and 7s itinerant bands. Further, since the Coulomb correlation is large relative to the effective band width, the itinerant (or band) description is no longer valid. Instead the localized description of the ionic 5f electrons (i.e., large spin-orbit and Hund's rule coupling of L, S and J--all maximized--interacting with their environment through the crystalline electric field and the exchange interactions via the conduction electrons) appears now more appropriate, as in the case of the rare earth metals. Indeed, an apt description of these metals is that they appear to form a second rare-earth series with all that implies. For magnetism we may well expect for the heavier actinide metals the diversity (and complexity) of magnetic ordering phenomena found for the rare-earth metals. It thus appears to be a new fruitful (but difficult) area for magnetic research.

One final remark should be made about the Cm band structure where the f-localization is not nearly as pronounced as for either Am or Bk. This is due to the anomalously small lattice constant reported for Cm. Had we derived an effective lattice constant from the α phase (dhcp) near neighbor distances, the resulting f-band widths would have been as small as for the Am and Bk.

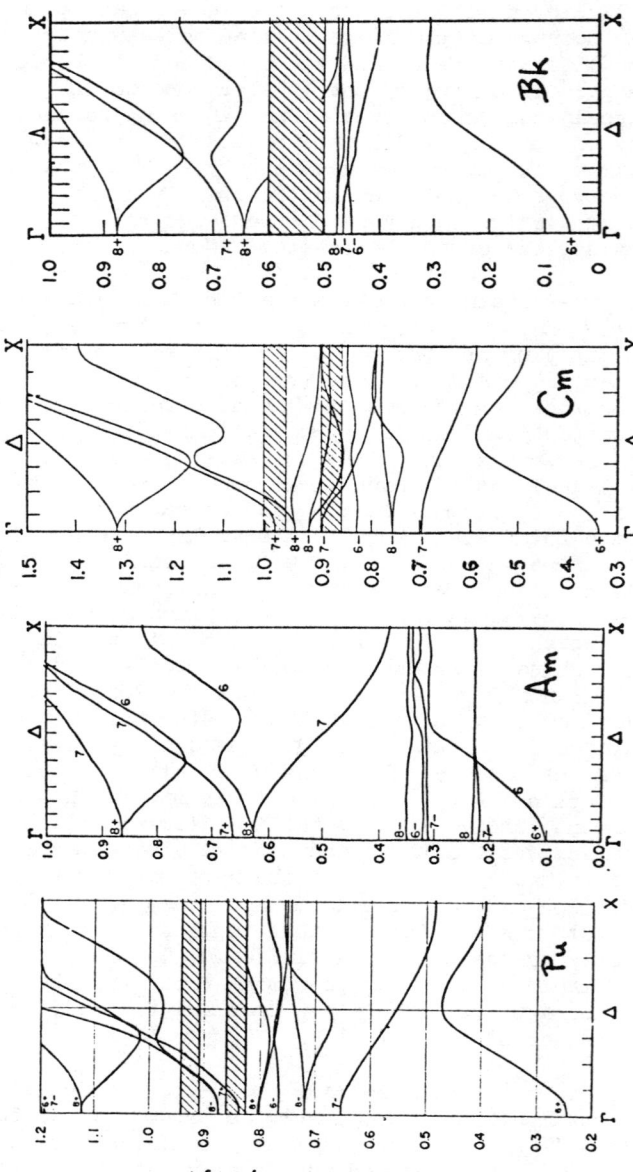

Fig. 1   Energy Bands in (100) direction for fcc Pu, Am, Cm, and Bk

REFERENCES

1. D. D. Koelling, A. J. Freeman, F. M. Mueller, Phys. Rev. $\underline{1B}$, 1318 (1970); A. J. Freeman and D. D. Koelling, Bull. Am. Phys. Soc. $\underline{14}$, 360 (1969); G. O. Arbman, D. D. Koelling, A. J. Freeman, BAPS $\underline{15}$, 344 (1970); D. D. Koelling, A. J. Freeman, G. O. Arbman, Plutonium 1970 and Other Actinides, ed. W. N. Miner, Nucl. Met. $\underline{17}$, 194 (1970); A. J. Freeman, D. D. Koelling J. de Physique (in press).

2. J. C. Slater, Phys. Rev. $\underline{49}$, 537, 931 (1936); $\underline{52}$, 198 (1937).

3. See the review chapter by A. J. Freeman in Magnetic Properties of the Rare-Earth Metals, (R. J. Elliot, ed. Plenum Press, London) 1972, p. 245.

ELECTRONIC STRUCTURE AND MAGNETIC PROPERTIES OF SCANDIUM*

Shashikala G. Das[†]
Argonne National Laboratory, Argonne, Illinois 60439

A. J. Freeman
Physics Department, Northwestern University
Evanston, Illinois 60201
and
Argonne National Laboratory, Argonne, Illinois 60439

D. D. Koelling[††]
Magnetic Theory Group, Physics Department
Northwestern University, Evanston, Illinois 60201
and

F. M. Mueller
Argonne National Laboratory, Argonne, Illinois 60439
and
Magnetic Theory Group, Physics Department
Northwestern University, Evanston, Illinois 60201

ABSTRACT

The band structure of scandium was studied on the basis of two different models: (1) an _ab initio_ calculation using the Relativistic APW method and (2) an interpolation scheme which used s, p and d type functions in the tight-binding representation. The warped muffin-tin APW potential was obtained from overlapping charge densities which were derived from the atomic configuration $3d4s^2$ and exchange interaction based on the Slater approximation. The Hamiltonian of the interpolation scheme was parameterized in terms of 16 parameters in a two-center approximation. The total density of states, the factorized s and d density of states, etc., appear to be in good agreement with the experimental magnetic susceptibility. A discussion is given of the "s" band contribution to the Knight shift and the relation of the factorized density of states belonging to the representations A_1', E' and E'' to the measured nuclear spin relaxation time and the anisotropy in the magnetic susceptibility.

INTRODUCTION

The study of the electronic structure of scandium is important not only because of fundamental interest in the clarification of the role of d electrons in the transition metals, but also because of the anisotropy observed in its various properties such as susceptibility,[1,2] spin lattice relaxation time,[2] etc.

---

*Supported by the Atomic Energy Commission, the Air Force Office of Scientific Research, and the National Science Foundation.

[†]Present address: Northwestern University, Evanston, Illinois 60201.

[††]Present address: Argonne National Laboratory, Argonne, Ill. 60439.

METHODS OF CALCULATION

The scandium band structure has been studied previously by Altman and Bradley[3] using the cellular method and by Fleming and Loucks[4] using the APW method. In our work (which we can only summarize here because of space limitations), the band structure was first calculated for scandium using the RAPW technique with the additional rotational symmetry being exploited wherever possible.[5] The potential was constructed from overlapping relativistic atomic charge densities[6] using the configuration $3d^{1.4}s^2$ and an exchange interaction treated using the full Slater $\rho^{1/3}$ term. This potential was then approximated as a warped muffin-tin potential (i.e., a potential spherically averaged inside the muffin-tin spheres of radius 3.082 au. but allowed to vary from a constant in the interstitial region). The lattice constants used were $a_o = 6.2421$ and $c_o = 9.9466$.

The energy bands obtained from the SRAPW at a few high symmetry points (20) were then parameterized in terms of interpolation scheme parameters, so as to facilitate the calculation of other quantities as the factorized density of states, the susceptibility, etc. The interpolation scheme used here is based on the LCAO scheme of Slater and Koster[7] in the sense that all the basis functions were taken in the tight-binding representation.[8] For hcp scandium with two atoms per unit cell the non-relativistic electronic structure can be adequately described by an 18 x 18 Hamiltonian matrix, corresponding to one s function, three p functions and five d functions centered on each atom of a unit cell. The one-electron wave function in our scheme is expressed as a linear combination of Bloch sums of Wannier functions as no overlap matrix was introduced. The Hamiltonian was parameterized in terms of interaction integrals[9] α, the matrix elements of the Hamiltonian between the tight-binding basis set. Here the number of independent integrals in the two-center approximation is just 16. These parameters were then further determined by using the Hellman-Feynman technique described by Connolly.[8] We used the quadratic interpolation scheme technique (QUAD) of Mueller, et al.,[10] to calculate the total and factorized density of states. The rms deviation of the interpolation-scheme fit to the RAPW eigenvalues at 21 high symmetry points is of the order of three millirydbergs.

RESULTS

Figure 1 presents the total density of states and the d and s-density of states. The Fermi energy was obtained by integrating the total density of states curve to yield three electrons per atom. The $N(E_F)$ is found to differ from the specific heat value.[11] We have also calculated the factorized density of states belonging to the representation A_1', E_1' and E''. From the observed anisotropy of the spin-lattice relaxation rate $R = (T_1 T)^{-1}$, Ross, et al.,[2] have derived the fractional admixture coefficients at E_F. We find that the p-character at the Fermi surface is rather large (14.8%) contrary to the usual assumptions. This is justified in the NMR analysis by the small hyperfine field due to p electrons relative to that of the s-

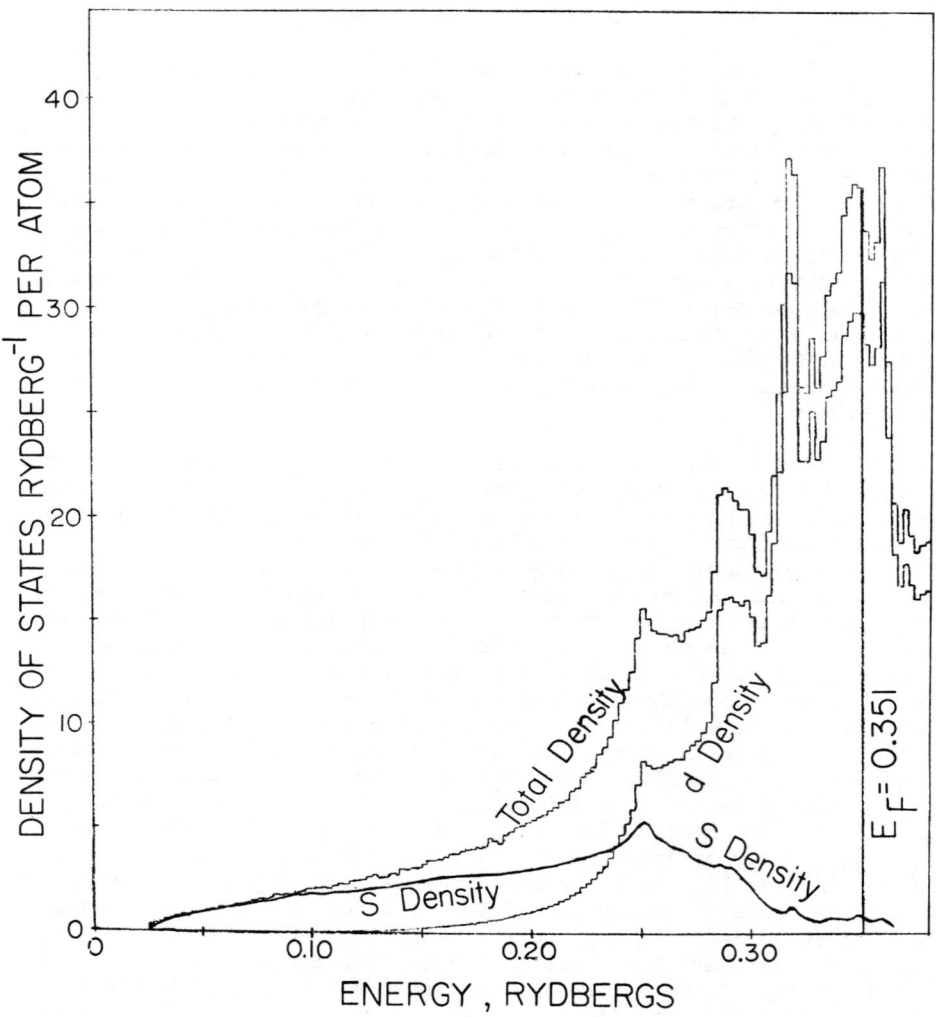

electrons. Thus accordingly we have calculated the fractional s character ρ_s, obtained from the Knight shift measurements and relaxation time measurements, from the relation

$$\rho_i = N_i(E_F)/[N_s(E_F) + N_d(E_F)] \qquad (1)$$

$N_s(E_F)$ and $N_d(E_F)$ being the s and d density of states at Fermi energy. According to Ross, et al.,[2] and Altman and Bradley,[3] the major contribution to the d-density of states at E_F comes from the functions belonging to A_1' symmetry, whereas we find the contribution from the functions belonging to E' and E'' to be quite large. The anisotropy of the susceptibility measurements can also be explained (assuming that the main contribution comes from the Van Vleck orbital susceptibility) by knowing the factorized density of states belonging to A_1', E' and E'' symmetry.[2] Our present work shows a great sensitivity to the precise details of the calculation and we could not determine the sign of the quantity $\chi_a - \chi_c$ to our satisfaction.

Table I. Comparison of the factorized density of states.

	Altman & Bradley[3]	Fleming & Loucks[4]	Present Calc.	Exp.[2]
Total Density of States (Ry^{-1}/atom)	17.6	31.0	35.6±0.8	38-57
Fractional[a] s character ρ_s	--	--	1.9%	2.4-7.3%
Fractional[a] d character ρ_d	--	--	97±3.9%	85-97%

[a] For definition, see Eq. (1) of text.

It might be noted that even the experiments reported so far do not agree on the sign of this quantity.[2] Finally, we found that the fractional character of the functions belonging to the symmetries E' and E'' is also a very sensitive function of the Fermi energy. While Ross, et al.,[2] have found that the fractional character belonging to E'' is consistently larger than that belonging to E', we do not observe this property in our calculations until we move away from the Fermi energy by small amounts (20 mRy).

Further work is in progress to refine the potential and other factors and to determine such quantities as the Van Vleck susceptibility.

REFERENCES

1. V. I. Chechernikov, I. Pop, O. P. Naumkin, and V. F. Terekhova, Zh. Eksperim. Soviety Phys. JETP 17, 265 (1963).

2. J. W. Ross, F. Y. Fradin, L. L. Isaacs, and D. J. Lam. Phys. Rev. 183, 645 (1969).

3. S. L. Altman and G. J. Bradley, Proc. Phys. Soc. 92, 764 (1967).

4. G. S. Fleming and T. L. Loucks, Phys. Rev. 173, 685 (1968).

5. D. D. Koelling, Phys. Rev. 188, 1049 (1969).

6. D. Liberman, J. T. Waber, and D. T. Cromer, Phys. Rev. 137, 127 (1965).

7. J. C. Slater and G. F. Koster, Phys. Rev. 94, 1498 (1954).

8. J. W. D. Connolly, Electronic Density of States Conference, Gaithersburg, Maryland, 1969, NBS Special Publication 323.

9. Maria Maisek, Phys. Rev. 107, 92 (1957).

10. F. M. Mueller, J. W. Garland, M. H. Cohen, and K. H. Bennemann, Annals of Physics 67, 19 (1971).

11. G. S. Knapp and R. W. Jones, Phys. Rev. B6, 1761 (1972).

SPIN POLARIZED ENERGY BAND STRUCTURE, SPIN DENSITIES AND THE NEUTRON MAGNETIC FORM FACTOR OF GADOLINIUM METAL[*]

B. N. Harmon
Physics Department, Northwestern University, Evanston, Illinois 60201

and

A. J. Freeman
Physics Department, Northwestern University, Evanston, Illinois 60201
and Argonne National Laboratory, Argonne, Illinois 60439

ABSTRACT

Conduction electron spin densities and their contribution to the diffuse part of the neutron magnetic form factor of Gd metal, measured by Moon et al., have been determined by means of spin polarized APW calculations. Although our spin density is similar to that of Moon et al., we are unable to reproduce their large region of negative spin density. Thus, our theoretical form factor is considerably lower than experiment at the first (hkℓ) reflection. Possible reasons for this difference are discussed here.

INTRODUCTION

Moon, Koehler, Cable, and Child[1] have measured a very precise neutron magnetic form factor for Gd which they separated into local and diffuse components (identified with a 4f local moment and a conduction electron spin density respectively). The 4f density was found to be significantly expanded relative to the Freeman-Watson[2] non-relativistic Hartree-Fock results for Gd^{3+}. Recently, the local form factor was found to agree with the predictions of a fully relativistic (Dirac-Fock) solution[3] for atomic Gd. From their diffuse form factor, derived by subtracting off their local contribution, Moon et al., obtain projections of the corresponding moment density assumed to be due to the conduction electrons. The projections show large negative spin density columns running in the z direction and located in the region of the unit cell farthest from the atomic sites. Such large negative densities (equivalent to about one third the peak in a single atomic 5d electron density - see Fig. 1) are unexpected from an atomic viewpoint for a system with a net positive spin of $0.5\mu_B$/atom. To study these features we have carried out a spin polarized APW band calculation and have obtained a conduction electron spin density.

METHOD OF CALCULATION

For the APW calculations a spin up and spin down crystal potential was created from superimposed spin polarized Hartree-Fock-Slater atomic charge densities (Fig. 1) using the full Slater exchange and a

[*]Supported by the U. S. Air Force Office of Scientific Research, the National Science Foundation and the U. S. Atomic Energy Commission.

starting configuration $[4f^7_\uparrow 5d^{0.5}_\uparrow 6s^1_\uparrow 5d^{0.5}_\downarrow 6s^1_\downarrow]$. The two spin bands were found to vary slightly from each other and from the paramagnetic bands. The net moment obtained was $0.74\mu_B$/atom rather than the experimental $0.55\mu_B$/atom, indicating the approximate (non self-consistent) nature of the potential or more probably too strong an exchange parameter. Since the bands are similar we adjusted Fermi levels less than 0.01 Ry to get the correct moment instead of using a smaller value of the exchange parameter. This means that the shifting of the 5d radial distributions of Fig. 1 will be slightly overemphasized in our charge densities.

RESULTS AND DISCUSSIONS

The spin up Fermi level was found to be in a region of low density between two peaks, while the spin down level lies on the second peak in the density of states (Fig. 2). An interesting result of the muffin tin calculation which helps to explain the distribution of the spin density is the character of the wavefunctions found in these two peaks. An average wavefunction in the first peak has its maximum density directed away from the atomic sites towards the empty part of the cell - the c site in an fcc stacking arrangement. Because the negative spin radial functions are shifted away from the atomic sites (cf Fig. 1), and the first peak in the density of states is completely populated for both spins, the negative spin density at the c site is not completely unexpected. An average wavefunction in the second peak has its maximum in the x, y plane of atoms, in a direction which points between the atoms located at $\pm 1/4c$ in the z direction. (It appears that wavefunctions from both these peaks could be described as antibonding.) It should also be noted that this character implies a three fold symmetry about the z-axis which is found to give large terms proportional to Y_{33} for the spin density and Y_{53} for the total charge density. These terms are not usually considered in crystal field theory because they contribute in second order only, but since we find them to be a factor of three larger than the even ℓ terms, they may be important.[5] However, since we are finding that these odd ℓ terms are quite sensitive to the small changes in the potential caused by including the warped muffin tin terms, it is really too early to say these terms will persist when a better potential is used.

Our calculated spin density in the basal plane is shown as a contour mapping in Fig. 3. In this plot we have only used the expansion of the density outside the APW spheres so that values inside the shaded regions are not strictly valid. The main features of these results are seen in the similar shape of the spin density contours to the experimental result (Fig. 9 of Ref. 1) and the fact that our negative spin region is much smaller and down in magnitude by a factor of 25. This factor of 25 is roughly equivalent to removing all spin-up electrons in the outside region using the calculated densities.

We have done our calculation in the non-relativistic regime. This is justified for the energy bands because, except for more band crossings allowed by symmetry, the non-relativistic bands for Gd[6] are similar to the relativistic bands.[7] The Fermi surfaces for both cases are very similar, and except for a broader s band density in the relativistic case, both density of states curves are alike and show the familiar two peaks common to hcp transition metals. The concept of a spin density is treated easily non-relativistically since spin is a good quantum number. Also we are primarily interested in the outer regions of the cell where direct relativistic effects are small. It should be mentioned, however, that indirect relativistic effects, such as are the key to the agreement obtained for the local form factor,[3] may also play a role here.

Since the relative magnitude of the experimental diffuse spin density in the basal plane is completely determined by the first reflection in the form factor, any non-spin density mechanism which contributes to this reflection would bring experiment and theory closer together. An orbital contribution from unquenched 5d electrons is a possibility, as is additional hybridization of the 4f electrons with the other electrons. We should also note that an important improvement in the calculation of the spin density would be to go to self-consistency, using a warped muffin tin potential in the interstitial region. This would help to build up more negative spin in the outer region of the cell. It is also possible that exchange will have to be treated more exactly. We believe, however, that the most important reason for the difference is that the local form factor should not be treated as an atomic entity, but rather effects of the crystal environment upon the 4f, 5s, 5p "core" states must be included--especially for low scattering angles. We are presently studying the sensitivity of the calculated spin density to some of these improvements.

We are greatly indebted to Dr. D. D. Koelling for invaluable assistance.

REFERENCES

1. R. M. Moon, W. C. Koehler, J. W. Cable, and H. R. Child, Phys. Rev. B5, 997 (1972).
2. A. J. Freeman and R. E. Watson, Phys. Rev. 127, 2058 (1962).
3. H. L. Davis and J. F. Cooke, Magnetism and Magnetic Materials, ed. by C. D. Graham, Jr., J. J. Rhyne (A.I.P., New York, 1971), p. 1441; A. J. Freeman and J. P. Desclaux, Int. J. Magnetism 3 (1972) to appear.
4. D. D. Koelling, J. Phys. Chem. Solids 33, 1935 (1972). For a discussion of wavefunction properties and convergence see B. N. Harmon, D. D. Koelling, and A. J. Freeman (to be published).
5. D. J. Newman, Advances in Physics 20, 197 (1971).
6. J. O. Dimmock and A. J. Freeman, Phys. Rev. Letters 13, 750 (1964).
7. S. C. Keeton and T. L. Loucks, Phys. Rev. 168, 672 (1968).

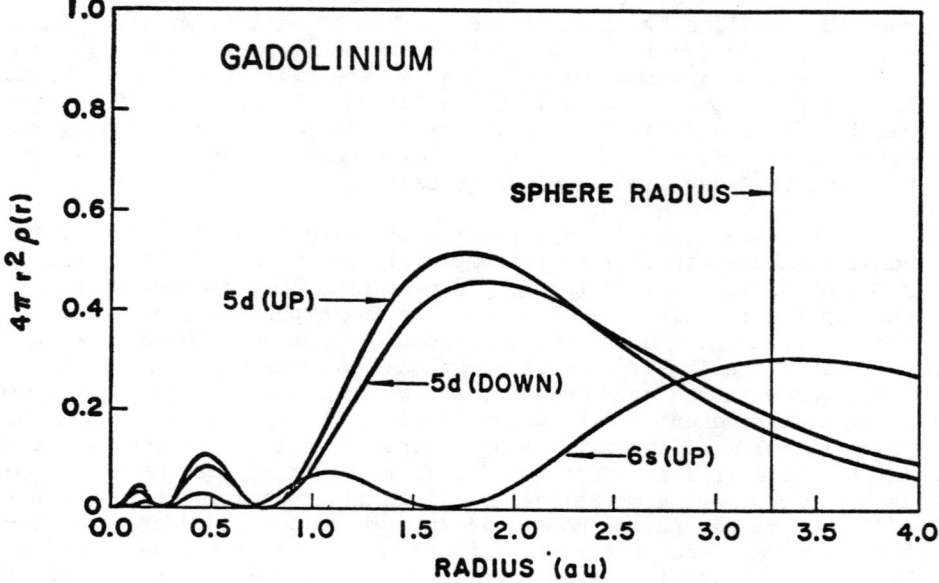

Fig. 1. Atomic, spin polarized, radial densities.

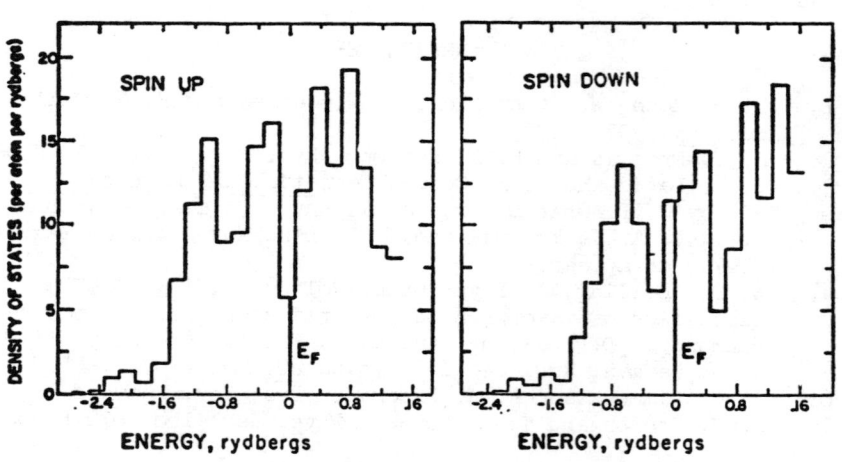

Fig. 2. Density of states for spin up and spin down bands.

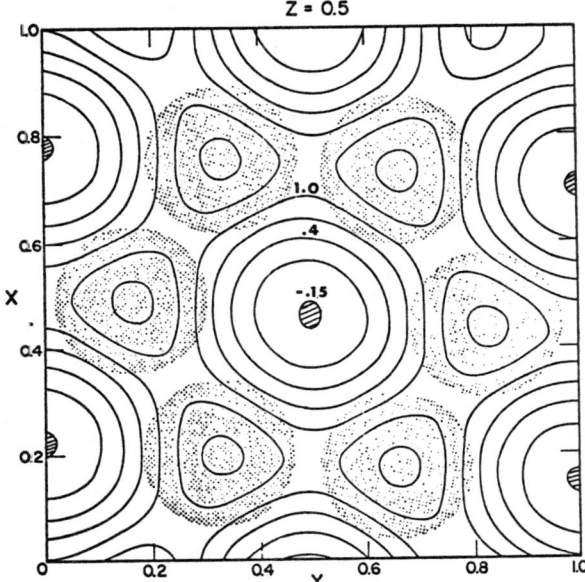

Fig. 3. Projection in the basal plane showing the calculated spin density. The shaded regions are inside the APW spheres. The contours are in units of .01 μ_B/A^3.

MAGNETIC SCATTERING AMPLITUDES OF SAMARIUM*

R. M. Moon and W. C. Koehler
Solid State Division, Oak Ridge National Laboratory
Oak Ridge, Tennessee 37830

ABSTRACT

Approximate ground states for the cubic and hexagonal sites in metallic Sm are calculated. The cubic ground state is adjusted to give rough agreement with observed neutron scattering amplitudes for high angle reflections. Large deviations occur at low scattering angles which are attributed to conduction electron polarization. The ratio of conduction spin polarization to 4f spin polarization is 0.15.

INTRODUCTION

We have recently determined the magnetic structure of metallic Sm through neutron diffraction experiments.[1,2] In this paper we attempt to deduce characteristics of the ground state of metallic Sm by interpreting the observed magnetic scattering amplitudes.

The crystal structure[3] of Sm has a large unit cell of nine layers along the $\vec{c}$ axis. Two-thirds of the sites have point symmetry $3m(C_{3v})$, with a near-neighbor coordination like that of a distorted HCP structure. The rest of the sites have point symmetry $\overline{3}m(D_{3d})$, with near-neighbors arranged as in a distorted FCC structure. The hexagonal sites order antiferromagnetically at 106°K with a structure which results in zero exchange field at the cubic sites, which then order independently at 14°K. The cubic-site structure has a magnetic cell which is four times larger than the chemical cell along both $\vec{a}$ and $\vec{c}$, resulting in magnetic Bragg peaks at very small scattering vectors where conduction electron polarization should be evident.

The combination of large unit cells and weak magnetic intensities resulted in a difficult experimental situation. A large cell produces a high density of Bragg spots in reciprocal space and a high incidence of simultaneous reflections. The effects of simultaneous reflections can be minimized by reducing the size of the sample but we were forced to use a fairly large crystal because the magnetic intensities were very weak. We estimate that all the magnetic amplitudes are subject to an error of about ±15% from simultaneous reflections. In addition there were serious background and resolution problems for many of the peaks. One should be cautious in ascribing significance to all details of the data.

Our general approach is to use standard theoretical approximations to identify the important parameters in the 4f electronic ground state, then vary these selected parameters to give rough agreement with the neutron amplitudes. The theoretical task is to

---

*Research sponsored by the U. S. Atomic Energy Commission under contract with the Union Carbide Corporation.

calculate the effects of the crystal field and exchange on the free-ion $^6H_{5/2}$ ground state multiplet, taking into account a possible admixture from the $^6H_{7/2}$ multiplet which lies at about 1500°K.

THEORY

The basic model assumes localized 4f electrons in a crystal field of appropriate symmetry with an exchange interaction between different sites via polarization of the conduction electrons. Expanding the crystal field in spherical harmonics and making the molecular field approximation for the exchange term, the effective single-ion Hamiltonian for the 4f electrons becomes

$$H = V_2^0 + V_4^0 + V_4^3 + V_6^0 + V_6^3 + V_6^6 + 2\beta H_e S_z. \qquad (1)$$

The magnitude and sign of the individual V_n^m terms depend on the site symmetry and the polar axis is along $\vec{c}$. We have assumed that the exchange field is also along $\vec{c}$, as indicated experimentally. It can be shown that the ground states we obtain are consistent with this assumption. We diagonalize that portion of the energy matrix involving states in the lowest lying multiplet ($J = 5/2$) and then use perturbation theory to estimate the admixture of states from the $J = 7/2$ multiplet. The problem simplifies greatly when only the $J = 5/2$ states are considered because then all matrix elements involving crystal field terms of the sixth order are zero.

In evaluating the crystal field terms we used the point charge model with $+3|e|$ on each lattice site and the lattice sums carried to convergence. We have used the values of $<r^n>$ tabulated by Lewis[4] based on relativistic Dirac-Slater wave functions. We do not expect accurate results from such a calculation but hopefully it should identify the important parameters and give the form of the approximate ground state. The exchange term was evaluated in the usual way by expanding $<S_z>$ in powers of H_e, assuming $H_e = -\lambda <S_z>$ and determining λ at the Néel point. The low-temperature ground state was then determined by a self-consistent requirement on $<S_z>$.

For the cubic sites, considering only the $J = 5/2$ multiplet, the ground state has the form

$$\psi_G = \cos\alpha |\tfrac{5}{2}, \tfrac{5}{2}> - \sin\alpha |\tfrac{5}{2}, -\tfrac{1}{2}>. \qquad (2)$$

We estimate $\sin\alpha = -0.306$, giving a 4f moment of 0.634 μ_B, while the saturation moment for the free ion is 0.714 μ_B. Our calculations show the $J = 7/2$ admixture to be quite small, so we adopt Eq. (2) as our approximate wave function.

For the hexagonal sites the point charge calculation indicates that the trigonal terms in Eq. (1) are very small and can be neglected. Within the $J = 5/2$ manifold, all non-vanishing matrix elements are diagonal and the predicted ground state is $|5/2, 5/2>$. In this case the much larger exchange term produces a small admixture of the higher multiplet and we calculate an approximate ground state of

$$\psi_G = 0.996 \ |\tfrac{5}{2}, \tfrac{5}{2}\rangle - 0.084 \ |\tfrac{7}{2}, \tfrac{5}{2}\rangle. \tag{3}$$

This admixture causes an increase in the expectation value of L_z and an equal decrease in S_z, resulting in a net decrease in the total moment. We obtain a moment of 0.564 μ_B for the state given by Eq. (3).

The calculation of elastic neutron scattering amplitudes involves the evaluation of

$$(\vec{\mu}f) \equiv \langle\psi_G|\hat{K} \times [\vec{M}(\vec{K}) \times \hat{K}]|\psi_G\rangle/\sin\beta, \tag{4}$$

where $\vec{K}$ is the scattering vector, $\vec{M}(\vec{K})$ is the Fourier inversion of the magnetic moment density and β is the angle between $\vec{K}$ and $\vec{M}(0)$. A general method for making this calculation has been given by Lovesey and Rimmer.[5] A great reduction in the complexity of the calculations is achieved by making the dipole approximation, which for our cases becomes

$$\mu f = \langle\psi_G|(\langle j_0\rangle + \langle j_2\rangle)L_z + 2\langle j_0\rangle S_z|\psi_G\rangle, \tag{5}$$

where

$$\langle j_n\rangle = \int \rho(r) \ j_n(Kr) dr. \tag{6}$$

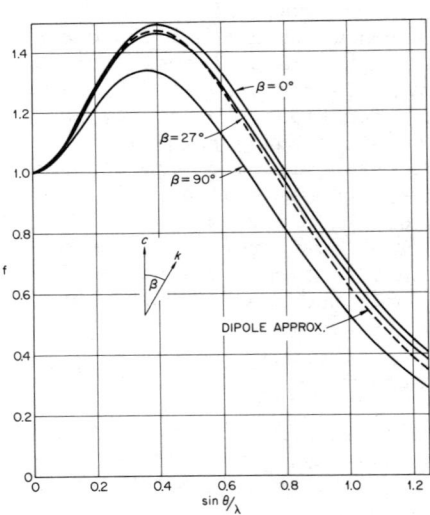

Fig. 1. Comparison of the dipole approximation with exact calculation of magnetic form factors for Sm^{3+}, $J_z = 5/2$.

Here $\rho(r)$ is the ground state radial charge density and $j_n(Kr)$ is the spherical Bessel function of order n. In making this approximation we lose all information on the asphericity and noncollinearity of the magnetic moment distribution. A comparison between the dipole approximation and the exact calculation for various values of β for the $|5/2,5/2\rangle$ state of Sm^{3+} is shown in Fig. 1. The exact calculation is by Davis[6] using relativistic Hartree-Fock-Slater wave functions. The amplitudes have been normalized to unity by dividing by the net moment. Note that for $\sin\theta/\lambda$ from 0.3 to 0.7 the dipole approximation is very close to the exact calculation for $\beta = 27°$ and is greater than the exact calculation for $\beta = 90°$ by 8 to 15 percent. Since our approximate ground states are both predominantly $|5/2,5/2\rangle$, we should expect a similar spread in the observed values when plotted against $\sin\theta/\lambda$. A complete analysis seems unwarranted because of the approximate nature of our data. We will use the dipole approximation and attempt to adjust the ground-state wave function so that the calculated scattering amplitude gives an upper bound to the data for β varying between 30° and 90°.

COMPARISON OF THEORY AND EXPERIMENT

The magnetic intensities were reduced following standard procedures and placed on an absolute basis by comparison with nuclear intensities. The magnetic amplitudes for the cubic sites, obtained from the observed magnetic intensities, are shown in Fig. 2. The corresponding data for the hexagonal sites are not really worthy of analysis. All that can be said is that the level of scattering is about the same as in the cubic case, which is consistent with our approximate wave functions.

The expected scatter in the data of Fig. 2 is evident and there is a dramatic decrease in the amplitudes for small values of $\sin\theta/\lambda$. We expect any evidence of conduction electron polarization to show up at small scattering angles, so we will temporarily neglect the data inside of $\sin\theta/\lambda = 0.2$ in comparing with the calculation for the 4f electrons. For wave functions of the form given by Eq. (2) the dipole approximation simplifies to

$$(\mu f) = g\langle\psi_G|J_z|\psi_G\rangle (\langle j_0\rangle + 6\langle j_2\rangle), \qquad (7)$$

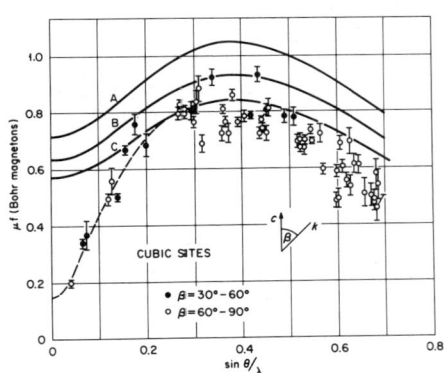

Fig. 2. Magnetic scattering amplitudes for the cubic sites in metallic Sm compared with various calculations (see text).

with $g = 2/7$. Changing the admixture of the state with $J_z = -1/2$ does not change the shape of the calculated scattering function but does change the scale factor $g\langle J_z\rangle$. Curve A is drawn for $\sin\alpha = 0$ in Eq. (2) and curve B for $\sin\alpha = -0.306$, corresponding to our calculated ground state. We believe that curve C, drawn for $g\langle J_z\rangle = 0.571$, or $\sin\alpha = -0.409$, represents a better fit to the data based on the expected behavior of the dipole approximation as discussed in the previous section. We can get a larger admixture of the state with $J_z = -1/2$ by strengthening the V_4^3 crystal field term, or weakening the terms V_2^0 or V_4^0. The changes in crystal field parameters required to go from curve B to C are not excessive in view of the known shortcomings in a metallic point charge calculation. If curve C represents our best estimate of the 4f electronic contribution to the magnetic scattering, then the large deviations at small $\sin\theta/\lambda$, where the dipole approximation should be very good, must be ascribed to polarization of the conduction electrons. We make the crude assumption that the total scattering amplitude is given by a 4f contribution corresponding to curve C, plus a conduction contribution made up of a linear combination of 5d and 6s terms,

$$\mu f = (\mu f)_{4f} + \mu_c [\alpha <j_o>_{5d} + (1-\alpha)<j_o>_{6s}]. \qquad (8)$$

The dashed curve in Fig. 2 is a least-squares fit of Eq. (8) to the low-angle observations, with μ_c and α as adjustable parameters. The $<j_o>$ functions were calculated by Davis[6] for a $(4f)^5(5d)^1(6s)^2$ configuration using relativistic Hartree-Fock-Slater wave functions for the free atom. The fitting procedure gave $\mu_c = -0.42 \pm 0.05$ μ_B and $\alpha = 0.77 \pm 0.24$. A fit of about equal quality was obtained using free-atom form factors but with Wigner-Seitz normalization. The use of free-atom form factors to represent the conduction electrons is certainly questionable, but it seems to give a surprisingly good fit to the data and is probably a reasonable method of extrapolating the data to $\sin\theta/\lambda = 0$ to obtain the total moment. The rough ratio of 3/1 for 5d to 6s spin may indicate a high density of 5d states at the Fermi surface. The apparent free-atom behavior of the conduction electron polarization in Sm is quite different from that deduced from neutron studies on Gd.[7]

We believe that the total moment per atom for the cubic sites is (0.15 ± 0.05) μ_B and is made up of the following parts: a 4f orbital moment of 3.42 μ_B, a 4f spin moment of -2.85 μ_B and a conduction spin moment of -0.42 μ_B. This gives a ratio of conduction spin to 4f spin of 0.15, as compared to a similar ratio in Gd of 0.08. If this conduction polarization persists in the paramagnetic region, it will have a major effect on the paramagnetic susceptibility, as discussed by Stewart.[8] Unfortunately, his analysis of the Sm susceptibility data must be reconsidered because he assumed all sites to be paramagnetic above 15°K.

ACKNOWLEDGMENT

We are indebted to H. L. Davis of this Laboratory for calculating the various $<j_n>$ functions and for help in evaluating the point-charge crystal field coefficients.

REFERENCES

1. W. C. Koehler, R. M. Moon, J. W. Cable and H. R. Child, AIP Conference Proc. No. 5, 1434 (1972).
2. W. C. Koehler and R. M. Moon (to be published).
3. A. H. Daane, R. E. Rundle, H. G. Smith and F. H. Spedding, Acta Cryst. 7, 532 (1954).
4. W. Burton Lewis, Proceedings of the Colloque Ampere, "Magnetic Resonances and Related Phenomena," Bucharest, 1970.
5. S. W. Lovesey and D. E. Rimmer, Rept. Prog. Phys. 32, 333 (1969).
6. H. L. Davis (private communication).
7. R. M. Moon, W. C. Koehler, J. W. Cable and H. R. Child, Phys. Rev. B 5, 997 (1972).
8. A. M. Stewart, Phys. Rev. B 6, 1985 (1972).

NEUTRON SCATTERING STUDIES OF ^{147}Pm[*]

W. C. Koehler, R. M. Moon, and H. R. Child
Solid State Division, Oak Ridge National Laboratory
Oak Ridge, Tennessee 37830

ABSTRACT

Neutron scattering experiments have been carried out on a polycrystalline rod of ^{147}Pm at temperatures ranging from 320°K to 7.5°K. The double hexagonal close-packed structure characteristic of Pr and Nd was observed over this entire temperature range. Evidence for magnetic ordering in ^{147}Pm was sought in a variety of neutron scattering experiments. From conventional measurements it appears that any ordered moment must be less than about 0.4 μ_B. Neutron depolarization measurements indicate a small ferromagnetic moment at low temperatures and a transition temperature of about 98°K. These observations are not consistent with point charge model crystal field calculations which predict a doublet ground state for ions on both hexagonal and cubic sites.

INTRODUCTION

In recent years the magnetic properties of a number of the light rare earth metals have been intensively studied both by classical macroscopic methods and by neutron scattering techniques. The element promethium, falling between Nd and Sm in the periodic table, has been little studied by any means primarily because it is not a naturally occurring element. However, the isotope ^{147}Pm is found, in relatively copious amounts, as a by-product of fission in nuclear reactors. It has a relatively long half-life for easily shielded β-emission, and a not unreasonable neutron capture cross section.

Highly pure metal can be obtained from the Target Development Center of the Oak Ridge National Laboratory in which methods have been developed and facilities built for the production, purification, and fabrication of gram quantities of elemental ^{147}Pm. Finally, since the crystal structure of ^{147}Pm appeared to be the relatively simple dhcp structure[1] characteristic of Nd rather than the more complex structure characteristic of Sm, we were encouraged to carry out a number of neutron scattering experiments on a polycrystalline specimen in an attempt to characterize this material.

SAMPLE PROPERTIES

A reduction-distillation technique was used to produce high purity ^{147}Pm according to the reaction

$$2 \text{ Pm}_2\text{O}_3 + 3 \text{ Th} \xrightarrow{1600°C} 3 \text{ ThO}_2 + 4 \text{ Pm}\uparrow \qquad (1)$$

---

[*]Research sponsored by the U. S. Atomic Energy Commission under contract with the Union Carbide Corporation.

Prior to the collection of Pm in a hemispherical quartz dome just above the reactor still, the reactor temperature was held at 1400°C for 30 minutes which was sufficient to reduce and distill off the ^{147}Sm daughter impurity in the oxide. The metallic ^{147}Pm was subsequently arc melted and drop cast to form a cylinder 0.635 cm in diameter and 2.54 cm long. The arc melting and casting process removed any ^{147}Sm decay product formed since the distillation; thus "time zero" for ^{147}Sm growth in the diffraction specimen could be accurately known. (With a half-life of 2.62 years, the β decay produces about 2.2% ^{147}Sm per month. The data to be reported here were collected one month after the casting of the specimen, and are thus for an alloy containing about 2% ^{147}Sm.)

The diffraction specimen was placed in an aluminum capsule, sealed with an epoxy cement, and loaded into one of the cryostats at HB-1 of the HFIR. The β activity was effectively shielded by the aluminum capsule and the Bremsstrahlung radiation level (greater than 100 rem at a distance of 1.5") was under 100 mrem at 15 feet from the instrument. A total number of 5159 curies was in the sample. The thermal power of the specimen was 1.85 watts and this presented something of a challenge to obtaining low sample temperatures. Nevertheless, we were able to reach a temperature, as measured by a sensor located close to the specimen, of 7.5°K.

A transmission experiment on our specimen yielded a value for the total neutron cross section σ_T = 110 ± 2b at a neutron energy of 0.07 eV in good agreement with recent measurements of the total cross section of ^{147}Pm in the thermal region.[2]

A neutron diffraction pattern taken at room temperature showed lines characteristic of the double hexagonal close-packed structure with a ≈ 3.65 Å, c ≅ 11.65 Å in accord with the room temperature x-ray diffraction data.[1] Patterns taken at various lower temperatures show that this structure is retained to 7.5°K. From the intensities observed in a pattern taken at 97°K, calibrated in the usual way against a standard scatterer, the coherent scattering amplitude and cross section for the alloy were found to be b = 1.26 ± 0.04 x 10^{-12} cm and σ_{coh} = 20.0 ± 1.3 barns.

According to Hund's rules the free ion ground state for the $4f^4$ configuration of Pm^{+3} is $^5I_4$(L = 6, S = 2, J = 4) for which the fully ordered moment gJ is 3/5·4 = 2.4 μ_B. The degeneracy of this free ion ground state can be lifted by the crystal field and Pm^{+3} with an even number of electrons can have a singlet, non-magnetic ground state in both the hexagonal and cubic sites of the dhcp structure.

RESULTS

We turn now to results of the several neutron scattering experiments which we have carried out in an effort to investigate the magnetic properties of promethium.

1. <u>Search for Antiferromagnetic Ordering</u>. In the absence, to our knowledge, of magnetic susceptibility or of other physical data indicating a magnetic ordering transition, we first made diffraction patterns at room temperature, and at a low temperature, 7.5°K, and

looked for additional reflections of magnetic origin. Patterns were made as well at 97°K. Except for slight changes in the intensities of the nuclear reflections, these patterns were not significantly different from each other, and this suggests that the ordered moment in any antiferromagnetic structure of Pm must be very small or zero.

If we assume, for definiteness, that the magnetic structure of Pm is the same as that of polycrystalline Pr[3] which in turn is supposed to be similar to that of Nd,[4] and if only one set of sites is ordered, then from the sensitivity of the data we conclude that the ordered moment must be smaller than about 0.3 μ_B.

2. <u>Search for an Ordered Structure of Any Type</u>. When a transition from the paramagnetic state to a magnetically ordered state occurs it is sometimes possible to assess the magnitude of the moment in the ordered state from the difference in magnetic diffuse scattering observed at temperatures well above and well below the transition without a detailed knowledge of the structure. The method gives reliable results when the neutron energy is very high compared to the overall splitting of the ground state. This is probably only approximately true for Pm with an incident neutron energy of 70 meV. Assuming the method to be applicable, a background difference which is just significant would correspond to an ordered moment of 0.6 μ_B.

3. <u>Search for a Ferromagnetically Ordered State</u>. A very sensitive probe for the detection of small ferromagnetic moments is a beam of polarized neutrons. When such a beam is transmitted through an unmagnetized ferromagnet, the random direction of the magnetic domains produces depolarization of the beam. According to Halpern and Holstein[5] this depolarization depends upon the thickness of the sample, d, the mean domain size $\bar{\Delta}$, and the spontaneous induction, B, in the domains according to the relation

$$P/P_o = D = \exp - (\gamma_n^2 B^2 \bar{\Delta} d / 3v^2) \qquad (2)$$

where γ_n is the gyromagnetic ratio and v the velocity of the neutron and P and P_o are the beam polarizations with and without the specimen. Neutron depolarization measurements on Ni[6] near its Curie point have shown that the variation of $(B^2\bar{\Delta})^{\frac{1}{2}}$ with temperature follows the bulk magnetization data, and experiments on Dy[7] showed that in zero applied field the magnetic domains were very small, between 10^{-6} and 5×10^{-5} cm.

In order to check for a weak ferromagnetic moment in the ^{147}Pm specimen we made measurements of the polarization ratio R_T of the neutrons transmitted through the specimen as a function of temperature. The quantity R_T is simply the ratio of the intensity of neutrons detected when the flipper is off to those counted when the flipper is activated; that is, the ratio of neutrons in the + spin state to those in the - spin state, and it is directly related to the polarization of the transmitted beam $R_T = (1 + P)/(1 - P)$. As shown in Fig. 1, the polarization ratio increases with temperature becoming constant above about 98°K. Such behavior is typical of a ferromagnet with a Curie point of 98°K.

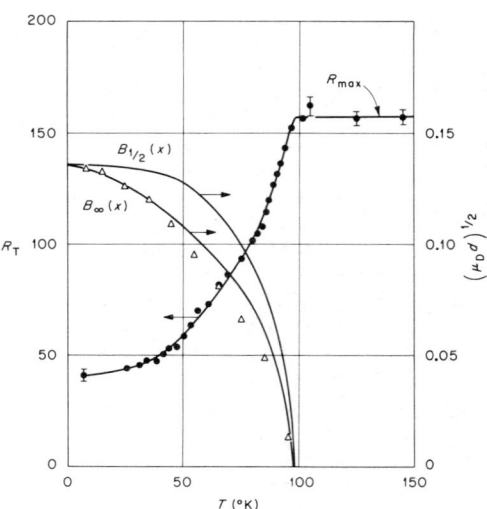

Fig. 1. Variation of R_T and $(\mu_D D)^{1/2}$ with temperature. For comparison with the $(\mu_D d)^{1/2}$ data are shown Brillouin functions for $J = 1/2$ and $J = \infty$.

When the instrumental corrections are small, as is the case when R is over 100, one may express P_o, P_f, and P_A, the efficiencies of the polarizer, flipper, and analyzer as $1-2\delta_o$, $1-2\delta_f$, and $1-2\delta_A$, respectively. It is easy to show that the inverse flipping ratio with the sample out (actually sample in but above 98°K) is approximately $1/R_{max} \cong \delta_o + \delta_f + \delta_A$. Below 98°K there is depolarization of the beam and the sample can be characterized by a small linear depolarization coefficient μ_D. Then
$D = P/P_{max} = 1 - 2\mu_D d$ and
$1/R_T \cong \delta_o + \delta_f + \delta_A + \mu_D d$.
Thus the difference between the inverse polarization ratio at temperature T and the maximum is directly a measure of the depolarization.

At the lowest temperature the observed value of $\mu_D d = 0.01806$. If we assume all atoms have the same moment, μ, expressed in Bohr magnetons we find from Eq. (2) $\mu^2 \bar{\Delta} = 5.7 \times 10^{-6}$. With a mean domain size of 10^{-4} cm, say, this would imply a moment of 0.24 μ_B.

Having obtained evidence for weak ferromagnetism in the specimen from the depolarization experiment, we measured the intensity of the strong (102) reflection as a function of temperature in order to detect a possible magnetic contribution to this reflection. An estimated Debye-Waller correction factor was applied to the intensities after which it appeared that a ratio of low temperature to high temperature intensities of 1.003 would not be significant. If all atoms carry the same moment this result means that a moment $\mu \lesssim 0.4$ μ_B would not have been detected.

The ideal way to measure the magnetic contributions to Bragg reflections is to measure the flipping ratio of a series of Bragg peaks. Such a measurement is best carried out with a magnetic field strong enough to saturate the specimen directed normal to the scattering vector. At present we are not equipped to apply a vertical field to a specimen at very low temperatures so we attempted to measure the flipping ratio of the (102) reflection by applying a horizontal field at an angle to the scattering vector and making use of a small normal component of the sample magnetization. We estimate that a moment per atom $\lesssim 0.3$ μ_B would not have been detected.

SUMMARY AND DISCUSSION

None of the diffraction experiments that we have carried out on ^{147}Pm has shown any evidence for magnetic ordering, which suggests that the crystal-field ground state may be a singlet. If so, the existence of a small net moment, as suggested by the depolarization experiments, can be understood as arising from exchange induced admixture of an excited level into the ground state. Quantitatively, however, it is difficult to reconcile a low moment with such a high transition temperature. In addition, point-charge crystal field calculations show that the ground state should be a doublet with a moment near to the saturated free-ion value for both sets of sites. At present, we are far from a satisfactory explanation of our experimental observations.

As we have mentioned, our sample contains a few atomic percent ^{147}Sm, which is present in the double hexagonal structure. The susceptibility of pure dhcp Sm^8 shows a sharp maximum characteristic of an antiferromagnet at 27°K and a kink of undetermined origin between 100° and 120°K. While it is not likely that a ferromagnetic transition at 98°K can be attributed to a 2% ^{147}Sm impurity, it is disturbing that we have no direct diffraction evidence for ferromagnetism. We propose, in the near future, to repeat the flipping ratio experiments with the aid of a 60 kOe split coil superconducting magnet.

ACKNOWLEDGMENT

The authors are grateful to E. H. Kobisk and H. L. Adair, Target Development Center, Oak Ridge National Laboratory, for preparing and encapsulating the specimen.

REFERENCES

1. P. G. Pallmer and T. D. Chikalla, J. Less Common Metals 24, 233 (1971).
2. G. J. Kirouac, H. M. Eiland, R. E. Slovacek, C. A. Conrad, and K. W. Seemann, *Neutron Cross Sections and Technology*, D. T. Goldman, Ed., National Bureau of Standards, Washington, D. C., 1968, p. 687.
3. J. W. Cable, R. M. Moon, W. C. Koehler, and E. O. Wollan, Phys. Rev. Letters 12, 553 (1964).
4. R. M. Moon, J. W. Cable, and W. C. Koehler, J. Appl. Phys. 35, 1041 (1964).
5. O. Halpern and T. Holstein, Phys. Rev. 59, 960 (1941).
6. H. K. Bakker, M. Th. Rekveldt and J. J. Van Loef, Phys. Letters 27A, 69 (1968).
7. E. Löffler and H. Rauch, J. Phys. Chem. Solids 30, 2175 (1969).
8. A. Jayaraman and R. C. Sherwood, Phys. Rev. 134A, 691 (1964).

IMPLICATIONS OF THE MAGNETIC PROPERTIES OF PURE α-CERIUM FOR ITS ELECTRONIC STRUCTURE

A.J.T. Grimberg and C.J. Schinkel
Natuurkundig Laboratorium der Universiteit van Amsterdam,
Valckenierstraat 65, Amsterdam, The Netherlands

Recently we have reported the magnetization as a function of magnetic field and temperature of α-cerium samples containing a few percents of β-phase. The correction for the β contribution as given in [1] can be made more sophisticated [2] and the resulting α-phase magnetic properties are analyzed in terms of itinerant electron paramagnetism in two models. The first is a one band Stoner model. Combination of the low temperature specific heat [3], the temperature dependence and the field dependence of the susceptibility leads to $dn(E_F)/dE = \pm(120\pm20)eV^{-2}$ per atom and $d^2n(E_F)/dE = (55\pm20) \times 10^2 eV^{-3}$ per atom, where $n(E)$ is the density of states. In the second model a distinction is made between Coqblin type electrons in hybridized 4f states and conduction electrons. Then for the width of the Lorentzian 4f-peak in the density of states and the distance of the center above the Fermi energy we derive: $(0.04\pm0.02)eV$ and $(0.4\pm0.2)eV$, respectively.

The details of the analysis will be published elsewhere [4].

1. A.J.T. Grimberg, A.P.L.M. Zandee and C.J. Schinkel, Solid State Comm. **11**, 1579, (1972).
2. A.J.T. Grimberg and C.J. Schinkel, to be published in Solid State Comm.
3. N.T. Panousis and K.A. Gschneidner, Solid State Comm., **8**, 1779, (1970).
4. A.J.T. Grimberg, Thesis University of Amsterdam, to be published.

NMR IN CeSn$_x$In$_{3-x}$ ALLOYS*

L. B. Welsh
Northwestern University, Evanston, Illinois 60201

J. B. Darby, Jr.
Argonne National Laboratory, Argonne, Illinois 60439

ABSTRACT

A study of the Sn^{119} NMR Knight shifts and spin-lattice relaxation time has been made in the CeSn$_x$In$_{3-x}$ alloy system for $0.6 \leq x \leq 3.0$ to determine the variation of the magnetic behavior of the Ce f-electron with x. For CeSn$_3$ the Sn NMR properties are qualitatively similar to those of LaSn$_3$ which is consistent with the Ce f-electron occupying a virtual bound state. The NMR properties of the alloys are consistent with a sharp increase in the f-electron susceptibility with decreasing x down to x=1.8 and suggest that a local moment picture for the Ce f-electron may be roughly correct for $0.6 \leq x \leq 1.8$.

INTRODUCTION

From magnetic susceptibility[1-4] (χ) and specific heat[2,5] (γ) measurements it is known that the magnetic behavior of the Ce f-electron in CeIn$_3$ and CeSn$_3$ is quite different. While CeIn$_3$ orders antiferromagnetically (AFM) at 11°K, for CeSn$_3$, χ has a Curie-Weiss behavior above 250°K, a peak near 200°K, and appears to decrease at low temperature with no indication of magnetic ordering. Indeed, several studies have indicated that the f-electron in CeSn$_3$ is non-magnetic in the Friedel-Anderson sense.[2,6] In this paper we report the results of a low temperature Sn^{119} nuclear magnetic resonance (NMR) study of the increasing localization of the Ce f-electron in the CeSn$_x$In$_{3-x}$ alloy system as the electron/atom ratio (e/a) decreases. We find that the Sn^{119} Knight shifts and spin-lattice relaxation rate are consistent with a sharp increase in the f-electron susceptibility at low temperature as x decreases from 3.0 to 1.8, and that from x=1.8 to 0.6 the NMR properties of these materials show a considerably smaller variation.

NMR RESULTS

The alloys used in this study were made from 99.9% Ce, 99.9999% Sn, and 99.999% pure In arc melted in a high purity He-Ar atmosphere and homogenized five days at 800°C. After the lattice parameter was determined by X-ray analysis, powders of 250 mesh were prepared and annealed at 400°C for 2½ hr. The NMR measurements were made using mostly pulsed NMR techniques, enhancing the Sn spin echo either using boxcar integration or using a Fabritek 1072 instrument computer.

The Sn NMR properties were studied from 1.55 to 4.2°K and from 4.5 to 19.5 MHz and are summarized in Table I where the results at 1.55°K and 8 MHz are listed. The Sn Knight shift was measured relative to the proton resonance in glycerine. The values of the isotropic Knight shift, K_{iso}, the anisotropic Knight shift, $3K_{ax}$, and the

TABLE I

Sn^{119} NMR properties in the $CeSn_xIn_{3-x}$ alloys at $1.55°K$ and 8 MHz. Numbers in parentheses indicate uncertainty in preceeding digit.

x	0.6	1.0	1.4	1.8	2.2	2.6	3.0
$\Delta H_{\frac{1}{2}}$ (Oe)	45(5)	37(2)	33(2)	40(2)	19.0(5)	10.4(3)	6.8(1)
K_{iso} (%)	2.56(10)	2.55(5)	2.88(5)	3.10(5)	1.73(2)	1.36(1)	1.030(5)
$3K_{ax}$ (%)	1.3(2)	1.34(5)	1.27(5)	1.50(5)	0.75(4)	0.39(2)	0.145(3)
$T_1T \times 10^2$ (sec °K)	------	2.15(10)	1.78(8)	1.30(6)	3.15(15)	4.25(15)	4.80(10)
K^2T_1T/S	------	7.4	7.9	6.5	4.9	4.1	2.70

half linewidth at half amplitude, $\Delta H_{\frac{1}{2}}$, were determined by comparison of the measured lineshapes with computer generated lines assuming an intrinsic Gaussian lineshape. No field dependence was observed for K_{iso} or $3K_{ax}$. With the exception of $CeSn_3$, which will be discussed below, K_{iso}, $3K_{ax}$, and $\Delta H_{\frac{1}{2}}$ were independent of temperature.

The Sn Knight shifts and linewidth listed in Table I show the same qualitative dependence on x. They increase rapidly as x decreases from 3.0 to 1.8 with K_{iso}, $3K_{ax}$, and $\Delta H_{\frac{1}{2}}$ increasing by factors of 3, 10, and 6 respectively. For $x<1.8$, these quantities are relatively constant with values much larger than observed in $CeSn_3$ and only a little smaller than at x=1.8. The large increase in the linewidth as x decreases probably results from a combination of inhomogeneous Knight shift broadening and demagnetization effects, both of which are proportional to χ.

The Sn spin-lattice relaxation time, T_1, was measured using a $180°-90°-180°$ pulse sequence and is listed in Table I as T_1T. The recovery of the Sn NMR signal was exponential in all cases. No magnetic field or temperature dependence was observed for T_1T. The behavior of T_1T with x is qualitatively similar to that of the Knight shifts and linewidth. As x decreases from 3.0 to 1.8, T_1T decreases by a factor of 3.5 and increases somewhat for x below 1.8.

For $CeSn_3$, the Knight shift and spin-lattice relaxation time were also measured at 77°K. While $3K_{ax}$ and T_1T are temperature independent from 1.55 to 77°K, K_{iso} has values of 1.175, 1.060, and 1.030% at 77, 4.2 and 1.55°K respectively. Thus contrary to the result of Borsa et al.[3] that K_{iso} is constant at K_{iso}=1.24% from 77 to ~180°K, we find a value at 77°K below their peak value and a further decrease of K_{iso} at 4.2°K. Thus the low temperature behavior of K_{iso} appears to correlate with a decrease of χ at low temperatures observed by Cooper et al.[2]

DISCUSSION

The susceptibility of the $CeSn_xIn_{3-x}$ alloys used in the NMR study has been measured from 100 to 300K and for several alloys between x=0.6 and 1.5, χ has been measured for several temperatures between 5.7 and 100K.[7] As x decreases, χ at 5.7°K increases by roughly a factor of 6 for $0.6 \leq x \leq 1.5$ relative to χ in $CeSn_3$ which is consistent with the variation of the Sn NMR properties with x observed at $4.2°K$. The susceptibility appears to follow a Curie-Weiss law with $\theta \approx 100K$ over most of the temperature range for $0.6 \leq x \leq 1.8$. For the most In rich alloys the Curie-Weiss behavior extends to $5.7K$.[7] Neither the susceptibility nor the NMR results indicate magnetic ordering occurs for $0.6 \leq x \leq 3.0$. It is interesting that the magnitude of χ at $5.7K$ for the In rich alloys is slightly larger than that of $CeIn_3$ (although it is smaller than observed in $CePb_3$[1]) which would appear to be consistent with the existence of a Ce f-electron local moment in these alloys, if such is the f-electron state in $CeIn_3$. However, the θ values for these alloys are about twice as large as measured for $CeIn_3$. Buschow et al[4] have pointed out that the large θ values determined from χ in the temperature range $100 \leq T \leq 300K$ result at least in part from crystal field effects and that the low temperature value of θ differs substantially from the θ values determined in the temperature range $100 \leq T \leq 300K$.

With the increase in the low temperature susceptibility with decreasing x described above, a qualitative understanding of the Sn NMR properties of the $CeSn_xIn_{3-x}$ alloys can be obtained by comparing the Sn Knight shift, spin-lattice relaxation time and the Korringa product[8] K^2T_1T/S (where $S=(\hbar/4\pi k_B)(\gamma_e/\gamma_n)^2$) with the values for a non-magnetic isomorphic system such as the $LaSn_xIn_{3-x}$ alloys?[9] For the Ce alloys the Korringa product is listed in Table I. For the La alloys K_{iso}, T_1T, and the Korringa product are temperature independent and vary from 0.64%, 0.033 $sec°K$, and 0.64 in $LaSn_3$ to 0.65%, 0.090 $sec°K$, and 2.0 at $x=1.8$ respectively.[9] For $x<1.8$ these quantities slowly approach the values for $LaSn_3$ as x decreases. The most obvious result of such a comparison is that the Sn NMR properties of $CeSn_3$ are similar to those of the La alloys. This is rather surprising in view of the factor of five increase in γ from $LaSn_3$ to $CeSn_3$ which indicates a large contribution from the Ce f-electron to the local density of states at the Ce site. However, as x decreases in the Ce alloys, K_{iso}, $3K_{ax}$, and T_1T clearly deviate from the values found in the La alloys. From the large Knight shifts and the fact that the Korringa product is much larger than unity for $x \leq 1.8$, it is clear that as x decreases these alloys behave more like rare earth compounds where the f-electron susceptibility results either from local moment behavior or from strong spin fluctuation effects. Inspection of Table I and a comparison of these values with those of the La alloys suggests that with decreasing e/a ratio, the Ce alloys change from a nearly transition metal-like behavior at $CeSn_3$, with the Ce f-electron occupying a virtual bound state in the Friedel-Anderson sense, to a more localized Ce f-electron state for alloys in the vicinity of $x=1.8$. This picture of the electronic state of the Ce f-electron in $CeSn_3$ is in agreement with the conclusions of Harris et al[6] and Cooper et al.[2]

A more quantitative analysis can be given by considering the form of the isotropic Knight shift at the nonmagnetic site[10] which is

$$K_{iso} = K_o + \alpha \chi_f, \qquad (1)$$

where χ_f is the susceptibility of the f electron and α is a coupling constant. K_o arises from the conduction electron bands and is typically estimated from the Knight shift of a non-magnetic isomorph. For the case of an f-electron local moment the Knight shift is written as

$$K_{iso} = K_o + N(0)H_{hfs}\Gamma(g_J-1)\chi_f/(g_J N \mu_B), \qquad (2)$$

where $-\Gamma \bar{s} \cdot \bar{\sigma}$ is taken as the form of the localized f-electron conduction electron interaction, $N(0)$ is the density of states per spin direction, H_{hfs} is the hyperfine field, g_J is the Lande g-factor. For the Sn rich Ce alloys, the problem of determining the correct division of K_{iso} between the two terms of Eq. 1 is difficult in view of the lack of accurate χ data at low temperatures. In $CeSn_3$, K and χ appear to have the same qualitative variation with temperature but insufficient data exists to determine if K and χ really scale. In addition, the large Pauli susceptibility of $CeSn_3$,[6] and the possibility of strong interband mixing make the procedure of determining K_o by extrapolating χ to zero suspect since χ and χ_f may differ, although this method gives a reasonable value of $K_o \approx 0.55\%$ using the $CeSn_3$ data of Borsa et al.[3] K_o can also be estimated from a comparison of the NMR properties of $LaSn_3$ and $CeSn_3$. Since K_{iso} and T_1T for $LaSn_3$ both result primarily from the s-contact hyperfine interaction,[9] and the ratio of the T_1T values for $CeSn_3$ and $LaSn_3$ is 1.45, then K_o for $CeSn_3$ can be estimated using the Korringa relation assuming T_1T for $CeSn_3$ is determined by the s-contact interaction. Thus K for $CeSn_3$ is roughly 20% less than that of $LaSn_3$. This gives $K_o \approx 0.60\%$ for $CeSn_3$ in reasonable agreement with the value of K_o obtained by extrapolating χ to zero suggesting that χ and χ_f do not differ appreciably. Assuming that K_o for x=1.8 is the same in the Ce and La alloys, the f-electron contribution to K_{iso} increases by about a factor of 6 from x=3.0 to 1.8 which is consistent with the increases observed in $3K_{ax}$ and $\Delta H_{\frac{1}{2}}$ and χ. Rough values of α in Eq. 1 can be obtained by comparing the concentration dependence of K_{iso} with that of the extrapolated value of χ at 4.2K from the data[7] for[8] $100 \leq T \leq 300K$. Determining α from $\Delta K/\Delta \chi$ at 4.2K as x decreases (and taking K_o as roughly concentration independent) indicates α is larger (~ 5 mole/emu) near x=1.8 than in $CeSn_3$ (~ 2.5 mole/emu).

While a great deal of information exists on Knight shifts for the non-magnetic site in rare earth compounds, few relevant spin-lattice relaxation time studies have been reported and several of these were actually on actinide systems. Fradin and co-workers have examined the problem theoretically both for the spin fluctuation contributions of the f-electrons[12] and the f-electron local moment contribution.[11] Since the susceptibility and NMR data suggest that the local moment picture may be roughly correct for the Ce f-electron for $x \leq 1.8$, it is interesting to see if the local moment relaxation mechanism gives a reasonable value for $(T_{1f}T)^{-1}$ where T_{1f} is the f-electron contribution to the relaxation time at the non-magnetic site. From Fradin[11] the relaxation rate is of the form

$$(T_{1f}T)^{-1} \simeq C \sum_{\bar{q}} (\Gamma(\bar{q}))^2 \chi_f(\bar{q}) \qquad (3)$$

where C is a constant and $\bar{q}$ is the change in the conduction electron $\bar{k}$ vector. The behavior of T_{1f} when compared with K and χ will depend on whether ferromagnetic (FM) or antiferromagnetic (AFM) ordering is favored. For the FM case T_{1f}, K_{iso}, and χ all depend on $\chi_f(\bar{q}=0)$ while for the AFM case $\chi(\bar{q} \neq 0)$ will increase faster than $\chi(\bar{q}=0)$ so that the use of the static susceptibility will underestimate the f-electron contribution to T_{1f}.

A rough estimate can be made of the contribution of Eq. 3 to T_{1f} writing $(T_{1f}T)^{-1}=\eta\theta^{-1}$ where the evalution of η is discussed by Fradin.[11] Using the Sn density of states estimated from the La alloys[9] gives a value of $(T_{1f}T)^{-1}$ less than an order of magnitude smaller than the measured value for the x=1.8 alloy if $\Gamma(\bar{q})$ and $\chi_e(\bar{q})$ equal $\Gamma(0)$ and $\chi_e(0)$ respectively. Since $CeIn_3$ is AFM ordered at low temperature, the $\bar{q}\neq 0$ values of Γ and χ_e will probably be enhanced over the $\bar{q}=0$ values bringing the contribution to $(T_{1f}T)^{-1}$ from Eq. 3 into somewhat better agreement with the measured value. At present through the lack of detailed low temperature susceptibility data prevents a detailed comparison of the bulk and NMR properties of the $CeSn_xIn_{3-x}$ alloy system in order to determine if local moment behavior or strong spin fluctuation effects can account for the Sn spin-lattice relaxation rate. However, the rouch features of these data indicate that as In is substituted for Sn, the Ce f-electron becomes more localized suggesting that a local moment picture of the f-electron may be roughly correct for the In rich alloys.

REFERENCES

*Supported by AFOSR, and NSF through N. U. Materials Research Center, and in part by U. S. AEC.

1. T. Tsuchida and W. E. Wallace, J. Chem. Phys. $\underline{43}$, 3811 (1965).
2. J. R. Cooper, C. Rizzuto, and G. Olcese, J. Phys. $\underline{32}$, C1-1136 (1971).
3. F. Borsa, R. G. Barnes, and R. A. Reese, Phys. Stat. Sol. $\underline{19}$, 359 (1967).
4. K. H. J. Buschow, H. W. de Wijn, and A. M. van Diepen, J. Chem. Phys. $\underline{50}$, 137 (1969).
5. A. M. van Diepen, R. S. Craig, and W. E. Wallace, J. Phys. Chem. Solids $\underline{32}$, 1867 (1971).
6. I. R. Harris, M. Norman, and W. E. Gardner, J. Less-Common Metals $\underline{29}$, 299 (1972).
7. A. T. Aldred, private communication.
8. R. J. Korringa, Physica $\underline{16}$, 601 (1950).
9. L. B. Welsh, A. M. Toxen, R. J. Gambino, Phys. Rev. $\underline{B4}$, 2921 (1971); Ibid. B6, 1677 (1972); and Proceedings of LT13, 1972. A. M. Toxen, R. J. Gambino, L. B. Welsh, to be published.
10. E. D. Jones, Phys. Rev. $\underline{180}$, 455 (1969).
11. F. Y. Fradin, J. Phys. Chem. Solids $\underline{31}$, 2715 (1970).
12. F. Y. Fradin, M. B. Brodsky, and A. J. Arko, AIP Conf. Proc. $\underline{10}$.

HALL AND MAGNETORESISTANCE EFFECTS IN Yb THROUGH THE FCC-HCP TRANSFORMATION

C.M. Hurd and J.E.A. Alderson
National Research Council of Canada, Ottawa K1A 0R9

ABSTRACT

The temperature dependence of the Hall and transverse magnetoresistance effects has been measured in a polycrystalline sample of Yb (RRR = 17) showing the fcc-hcp transformation. The data cover the temperature range 1.8-420°K and field range up to 26 kOe, and they indicate that the behaviour of the Hall effect through the transition arises primarily from the structural change and not from the associated magnetic transition. However, the entire interpretation is handicapped by the present lack of detailed knowledge of the Fermi surface of hcp Yb.

INTRODUCTION

Recent work[1] has established a first-order polymorphic transformation (fcc-hcp) at 1 atm in unstrained bulk Yb of sufficient purity. The transformation occupies typically the range 150-250°K upon cooling and 320-370°K upon heating. The higher temperature phase is fcc and paramagnetic, corresponding to about 0.8% Yb^{3+} ions in the Yb^{2+} matrix, while the lower temperature phase is hcp and diamagnetic — indicating that practically all the Yb^{3+} are presumably changed to Yb^{2+} during the structural rearrangement. We describe below what seems to be the first study of the temperature dependence of the Hall effect in samples showing this transformation. Earlier results by Anderson et al.[2] in the range 80-300°K showed no evidence of the transformation, while more recent results[3] are restricted to 4.2 and 300°K.

HALL COEFFICIENT AND ELECTRICAL RESISTIVITY

These results are shown in Fig. 1 for a polycrystalline sample which, after light-cold rolling and cutting to the desired shape, was annealed in vacuum at 673°K for 30 mins before measurement. The RRR of the resulting sample (referred to the hcp phase) was 17. (Specific experimental details are available in a report published elsewhere[4].)

The resistivity data confirm the gross features observed previously[1]: with increasing temperature we find an impurity-dominated range extending to about 6°K. This

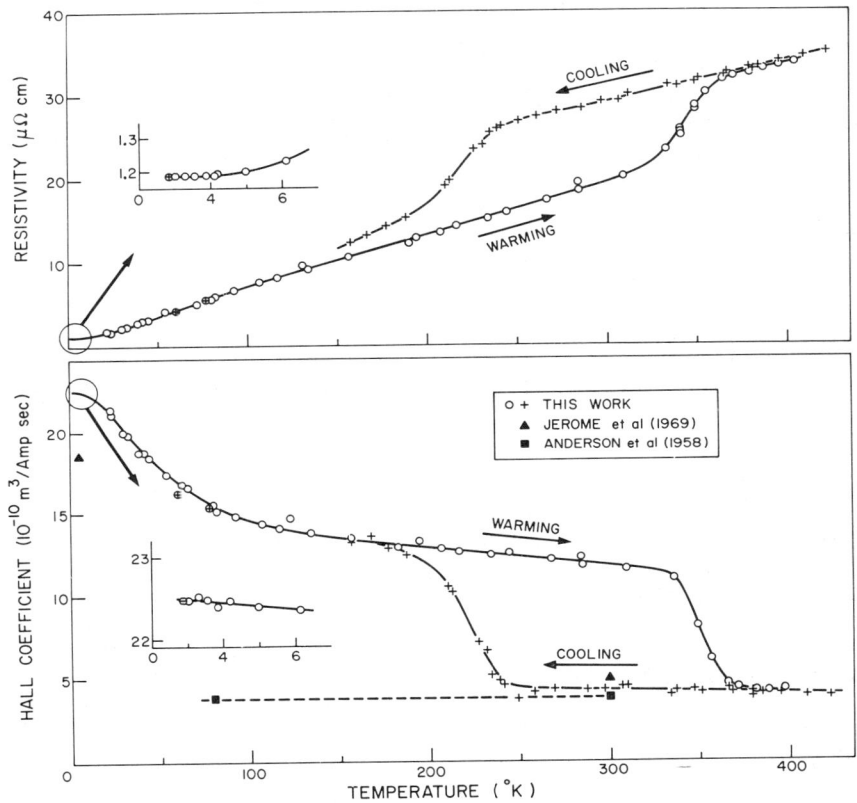

Fig. 1 Hall and resistivity data

is followed by an intermediate interval (6-20°K), with subsequently a linear temperature dependence in the hcp phase extending from about 20 to 320°K. Upon cooling from above 400°K, a linear dependence is observed in the fcc phase down to about 240°K.

Turning to the temperature dependence of the Hall coefficient, R_H, the data of Fig. 1 (obtained at 20 kOe) can be divided into three regions: (a) the impurity-dominated region, which we saw above extends to about 6°K, (b) the range ~6-150°K in which increasing temperature produces a progressive decrease in R_H, and (c) the range ~150-400°K occupied by the transformation.

In the region (c) both a magnetic and a structural change (with an associated valency change) is involved[1]. Either would be expected separately to alter the Hall effect; the structural/valency change will modify the relative numbers of hole and electron-like carriers while a magnetic→nonmagnetic transition would remove the spin scattering mechanism known[4] to be a contributor to R_H. We turn to the field dependence of R_H to judge whether either of these two influences is predominantly responsible for the behaviour observed in Fig. 1. For if the change in magnetism is the primary influence, we would expect magnetic scattering to be an important contributor to the Hall effect, and so R_H should show a marked field dependence. We find, in fact, that the Hall voltage shows essentially the same linear field dependence up to 26 kOe at 300°K in either the fcc (paramagnetic) or hcp (diamagnetic) phase, suggesting that spin scattering is not an important contributor. Therefore, the observed behaviour of R_H probably arises primarily from the structural and valency changes.

Space does not permit discussion of the behaviour in region (b), which in any case must inevitably be inconclusive because of the lack of all but rudimentary knowledge of the Fermi surface of the hcp phase[5]. Briefly (further details given in ref. 4), the monotonic variation is reminiscent of that seen in other metals where it is attributed to changes in the effective occupancy of different bands resulting from the anisotropy of the phonon scattering. The progressive freezing out of Umklapp processes between different regions of the Fermi surface is an example of one possible contributing effect.

TRANSVERSE MAGNETORESISTANCE

Data obtained under isothermal and isomagnetic conditions are shown in a Kohler plot in Fig. 2. The quadratic dependence characteristic of the low-field regime[5] is evident for both the hcp and fcc phases, but displaced along the ordinate. Such a shift can be caused[6] by a change in the anisotropy of either the band structure or the dominant electron scattering mechanism. However, it is impossible to disentangle these contributions in the present context, again, largely because of the poor knowledge of the band structure of the hcp phase.

What is interesting in Fig. 2 is the shift toward a more linear field dependence shown by the Kohler plot for higher values of $H/\rho(H=0)$. This is characteristic of the onset of the intermediate-field condition[6], where some groups of electrons are able to make several complete revolutions of their cyclotron orbit during their lifetime. It is surprising that this is possible with the modest combinations of magnetic

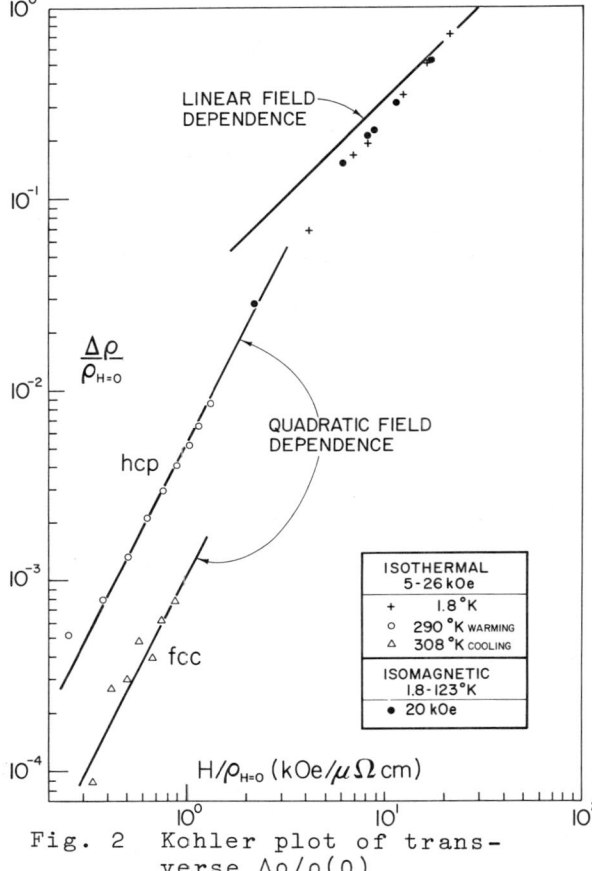

Fig. 2   Kohler plot of transverse $\Delta\rho/\rho(0)$

field strength and temperature available to us. For example, a simple free-electron calculation gives $\omega\tau = 0.04$ for hcp Yb at 2°K in an applied field of 20 kOe, and yet Fig. 2 shows the clear tendency away from the quadratic behaviour observed in such conditions. Note the superposition in this region of the isomagnetic and isothermal data; we therefore cannot dismiss the increasing deviation from quadratic behaviour as due, for example, to changes with temperature in the sample's multiphase inhomogeneity.

In our opinion the data represent the onset of high-field conditions for some of the more mobile electrons on particular parts of the Fermi surface[5], and they underline the inapplicability of simple free-electron theory to a metal such as Yb. An analogous situation exists in the group-1B metals; intermediate-field effects are observed in Cu and Au samples up to ∼40°K in fields of 15 kOe[7]. Here again the corresponding $\omega\tau$ values seem unreasonably small, but we should bear in mind that such a quantity calculated from free-electron theory gives an idealised and averaged value for all electrons contributing to the conduction. It does not necessarily reflect the dynamical condition of particular bands of cyclotron orbits which contribute to the galvanomagnetic effects and which can relatively easily be rendered to the high-field condition[7].

REFERENCES

1. E. Bucher, P.H. Schmidt, A. Jayaraman, K. Andres, J.P. Maita, K. Nassau and P.D. Dernier, Phys. Rev. B2, 3911 (1970).

2. G.S. Anderson, S. Legvold and F.H. Spedding, Phys. Rev. 111, 1257 (1958).

3. D. Jerome, M. Rieux and J.C. Achard, Propriétés Physiques des Solides sous Pression (Editions du CNRS, Paris, 1970), p. 157.

4. J.E.A. Alderson and C.M. Hurd, Solid State Commun. (In press).

5. O. Jepson and O.K. Anderson, Solid State Commun. 9, 1763 (1971).

6. C.M. Hurd and J.E.A. Alderson, J. Phys. Chem. Solids 32, 2075 (1971); Phys. Rev. B6, 1894 (1972).

7. C.M. Hurd and J.E.A. Alderson, Phys. Rev. B (In press for Feb. 1973).

Section 41.   Ferrites   1335

RECENT ADVANCEMENT IN THE FIELD OF HIGH FREQUENCY FERRITES

Mitsuo Sugimoto
Institute of Physical and Chemical Research
Wako-city, Saitama Prefecture, Japan

ABSTRACT

A comprehensive review is given of the technological progress concerning high frequency ferrites which has taken place recently in fundamental research, manufacturing techniques, electromagnetic features and trends in applications. The many problems awaiting solution and the prediction of the status in the future will be discussed.

1. INTRODUCTORY REMARKS

The advancement of high frequency ferrites was initiated by the work done by Snoek[1], who found that, associated with excellent properties in the high frequency range, manganese zinc ferrites and nickel zinc ferrites provide a family of magnetic materials useful to radio and TV sets as well as carrier telephony as cores of inductors, transformers and so forth. Immediately thereafter, a number of research workers and engineers became involved in basic studies, manufacturing improvement, measurements and applications of ferrites. In the following years, through fifties, new ferrites continued to be discovered, including a number having different crystalline structures or new properties. Among the noteworthy are barium ferrite[2], a magnetically hard material; manganese-magnesium ferrite[3] for memory application; Ferroxplana type ferrites[4] and garnet type ferrites[5], both for microwave use; the acicular powders[6] of CrO_2 for magnetic recording; and ortho-ferrites[7], which have been of recent concern in bubble domain technology. Since the end of this exciting period no new ferrites have been found as if sources for them have been exhausted.

During the above period of time, however, high frequency ferrites have seen remarkable improvements in performance and ever-increasing annual production rates. These technological developments in high frequency ferrites have resulted from achievements in basic studies as well as by the progress in manufacturing technique.

In this paper, the author will present a review of the recent advances in the field of high frequency ferrites, while citing the problems with which it is faced, with particular attention paid to the manufacturing process, electromagnetic performance and application aspects.

2. ADVANCES IN BASIC RESEARCH

It results from full use having been made of achievements in basic studies on physics and chemistry of ferrites that both the quantitative and the qualitative advancements have been achieved as we see to date in the field of high frequency ferrite.

<u>Physical Research</u>: Nearly definite conclusions have been reached concerning crystal structure, magnetic ordering and saturation magnetization, the special preference of metallic ions for interstitial sites in the spinel lattice, crystal anisotropy, magnetostriction, the high frequency magnetization mechanism, and the frequency dependence of permeability.

For interpreting the mechanisms of the disaccommodation in manganese zinc ferrites containing excess iron, a number of models[8] have been proposed. According to the currently accepted interpretation, the disaccommodation at room temperature, which is important to practical use, is closely related to the vacancy concentration.

<u>Chemical Research</u>: In chemical research, a great number of problems remain to be solved. With respect to phase equilibria, phase diagrams have been reported concerning the ternary systems Cu-Fe-O[9], Ni-Fe-O[10], Co-Fe-O[11], Mg-Fe-O[12], and Zn-Fe-O[13] containing the ferrite regions. Nevertheless, no detailed investigation for Mn-Fe-O system has as yet been reported, that is useful in manufacturing the manganese zinc ferrites finding the most wide-spread use today. This is because the manganese ferrite is sensitive to the partial pressure of oxygen in the atmosphere. Information for the Mn-Fe-O system as determind by Schwerdtfeger and Muan[14] is available. This, however, is not comprehensive enough to be of practical use. Consequently, so far as the preparation of manganese ferrite is concerned, we must refer to the study[15] on the solid state reaction of the raw materials at elevated temperature.

As indicated in Fig. 1[10], the single phase region of ferrite spinel in the ternary system Fe_3O_4-$NiFe_2O_4$-Fe_2O_3 is very narrow. The oxygen isobaric surface intersecting the spinel region is bent toward the two phase region containing excess Fe_2O_3 at a lower temperature and toward that containing excess nickel oxide at a higher temperature. The manufacture of the ferrites, therefore, requires a knowledge of the effect of the atmospheric oxygen pressure upon the resulting ferrite. In connection with this, research by Blank[16] and by Slick[17] are worthy of note.

Most of the high frequency ferrites finding current practical use are solid solutions with zinc ferrite. The evaporation of zinc in the sintering process[18] is a critical problem. Nevertheless no comprehensive report has so far been published concerning this aspect. Incidentally, the author would like to add that a report[19] is available on the loss of lithium during the sintering process.

In the preparation of single crystals, the phase equilibria data for the molten ferrite is very useful. Studies which have been published cover the Ni-Fe-O[20], Mn-Fe-O[21] and MgO-FeO-Fe_2O_3[22] systems, all are concerned with the liquid phase. A study has also

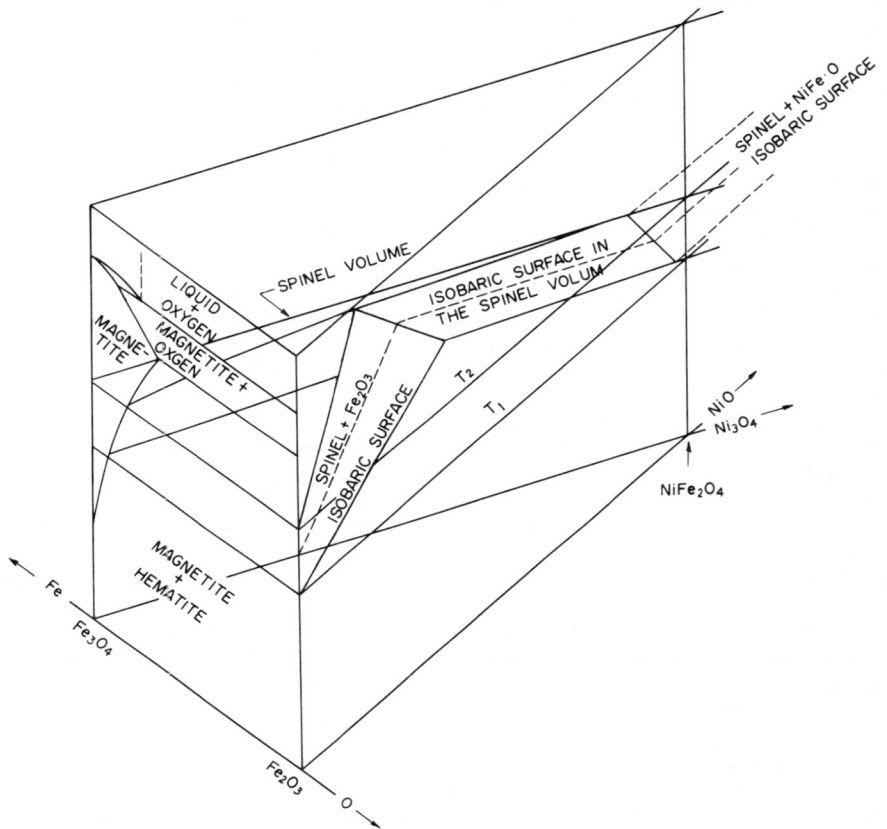

Fig. 1. Spinel single phase region between temperatures T_1 and T_2 in the ternary system of Fe_3O_4-$NiFe_2O_4$-Fe_2O_3.

been conducted on the evaporation of zinc from the liquid phase in the $MnFe_2O_4$-$ZnFe_2O_4$ system[23].

Concerning the sintering mechanism, Kuczynski[24] has reported that the sintering is dominated by the interdiffusion of metallic ions in the chemical potential gradient with the capillary force as a subordinate.

The grain size as well as the quantity, the size and the configuration of the pores in the microstructure serve as the information enabling the progress of sintering to be probed. Besides, they exert a substantial effect on the physical properties[25]. Stuijts[26] states that, a slight deficiency of Fe_2O_3 from the stoichiometric composition leads to a material with large grains and small pores, that an excess of Fe_2O_3 causes the pores to be increased in size and in quantity, and that the aforementioned observations are consistent with an enhanced concentration of cation vacancies playing an important role in the mass transport mechanism. A study has also been conducted on the equilibrium shape of pores created in the course of sintering[27].

3. ADVANCEMENT IN THE MANUFACTURING PROCESS

Of great importance to high frequency ferrites are high permeability, high flux density, low loss, low disaccommodation and good homogeneity of products. These properties are apparently highly sensitive to the composition and the manufacturing process. Current mass-production facilities for ferrites such as automatic presses, continuous sintering furnaces, and testing devices have been enhanced in performance and regulation[28,29]. The introduction of these methods has enabled products to be made with excellent performance and high homogeneity. Further, machining techniques have been improved for lapping, grinding and cutting the sintered ferrites. For example, threaded cores of high quality are now mass-produced for radio and TV sets. A process in which ferrite raw materials are mixed with a binder, rolled, and shaped to the desired configuration by punching, is finding increased use.

The extension of the application field has occasionally resulted in the traditional ceramic technologies failing to yield the products fulfilling the requirements or found difficulties in manufacturing, because of required high quality and complicated shapes. In order to alleviate these difficulties new ferrite manufacturing techniques have been introduced such as hot-pressing, slip-casting, the mass-production techniques for large single-crystalline ferrites, and arc-plasma spraying.

Hot-Pressing: In general, as-sintered high frequency ferrites contain pores of around a few percent. In order to minimize the porosity and thus enhance the density, to obtain a high permeability and or a high flux density ferrite, the hot-pressing process has been introduced. Of vital importance to hot-pressing of ferrites, is a die material that has a small thermal expansion coefficient from room through the sintering temperature, and that is sufficiently unreactive chemically to the ferrite. To date either dies made of silicon nitride[30] or of pure alumina sintered to high density[31] are finding frequent use. However, these materials are liable to chemically react with the ferrites, so that the alumina powders are stuffed between the die wall and the ferrite. Many reports[30,31,32] are available describing the relationship between the condition of hot-pressing and the properties of the product. Continuous hot-pressing facility[31] have also been described.

To date, the hot-pressing process serves to mass-produce ferrites with porosity of 0.1% or less and initial permeabilities of 30,000 or higher. This process, however, has difficulties including the unavailability of a die enabling the ferrite powders to be hot-pressed in direct contact with it, the resultant necessity of re-machining of the hot-pressed sample, the difficulty of hot-pressing to complicated shapes, and the susceptibility to cracking or stripping of a small grains taking place at the grain boundary in machining. Another process[33] is finding practical use for making high density ferrites. In this process, the ferrite is initially fired with pressure reduced to 10^{-3} through 10^{-4} Torr. at temperature between 1,200 and 1,300°C, and subsequently re-sintered in an equilibrium atmosphere. This process yields ferrites with an initial permeability of around 20,000 and a high anisotropy constant K_1 of negative polarity, and noticeably, with reduced magnetostriction in the [111] direction. In this process, the ferrite is liable to suffer from tiny cracks in the course of re-sintering.

Slip-Casting[34]: The slip-casting process as refered to here has been developed specifically for mass-producing items of either complicated shape or large size. In perfoming a slip-casting, a ferrite is mixed with a peptizator that disperses the powder, a lubricant that makes it easy to shape and prevents the powders from adhering the die, and the binder that gives sufficient strength to the compact specimen. This measure has been taken because ferrites do not have plasticity. Needless to say, the cited additives must be selected with care. The slip-casting process enables a compact sample to be formed which is uniform in density throughout the body and under minimum stress. Furthermore, as shrinkage during the sintering process is uniform, the change in dimension of the sintered ferrites will be very small and the density will be high. Figure 2 illustrates the wide angle deflection core for colour TV sets as manufactured by the slip-casting process.

Fig. 2. Large size colour TV deflection yoke core made by the slip-casting process. Outside diameter is 15 cm.

Table 1 compares in performance Mn-Mg-Ni-Zn ferrite as shaped

Table 1. Comparison of properties of the sample by the slip-casting process with those by the oil press process for the firing temperature of 1,210°C.

	Initial permeability	Mechanical strength	Density (g/cm^3)	Resistivity (Ω-cm)
Oil press	570	140	4.3	10^{10}
Slip-casting	840	180	4.6	10^{10}

by the traditional oil press with that as shaped by the slip-casting process, demonstrating that the permeability, the strength and the density of the latter are improved by about 10%, 30% and 8%, respectively, in comparison with those of the former.

Growing of Large Single Crystal: Specifically for VTR head

material having a high resistivity against wear in contact with magnetic tape, a production technology has been developed for preparing large single crystals of spinel ferrite. Such crystals are also finding applications in manufacturing cores for pulse transformers or cores for an practical miniaturized inductor. Of a number of processes available for growing single crystals[35], the Bridgman method is most frequently adopted at the present. By means of this process single crystal manganese zinc ferrite is mass-produced with diameters up to 60 mm and lengths up to about 150 mm[36].

Figure 3 illustrates the inner structure of a Bridgman type electric furnace. The interior of the furnace central tube, where the crystal is grown, is filled with pure oxygen gas, while the exterior of the central tube, where the heater is applied, is filled with pure nitrogen gas. Provisions are made for gas pressures to be automatically balanced between the interior and the exterior, so that the central tube is not subjected to pressure.

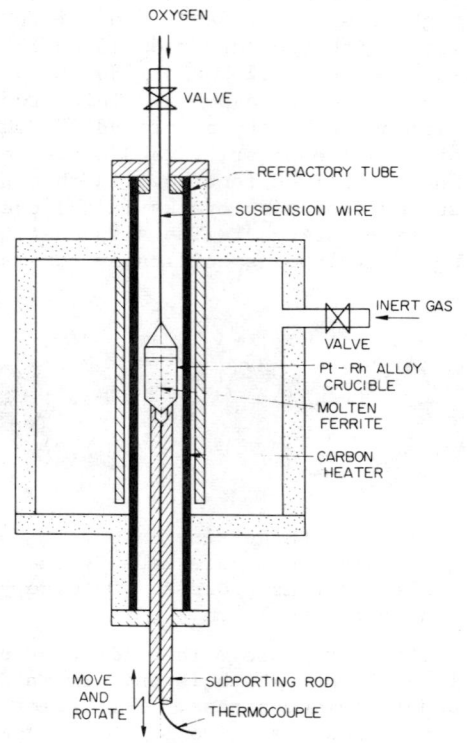

Fig. 3. Inner structure of the Bridgman type electric furnace for growing large single crystal manganese zinc ferrite.

In operation, the platinum-rhodium crucible, with diameter of 60 mm and length of 180 mm, loaded with the as-sintered

Table 2. Properties of single crystal manganese zinc ferrites.

Initial permeability	Saturation magnetization (G.)	Coercive force(Oe.)	Curie temp. (°C)	Resistivity (Ω-cm)	Vicker's hardness
5,000	3,600	0.05	210	1	700
10,000	3,800	0.05	180	3	700
13,000	3,400	0.05	230	3	700
28,000	3,300	0.05	160	30	700

manganese zinc ferrite block weighing totally about 2.5 Kg is placed inside the electric furnace and pre-heated to 1,640°C, whereby the ferrite block is fused. In fusing the ferrite, the crucible is required to be heated gradually from the bottom. Otherwise the fused ferrite will erupt outside the crucible. After holding for a few hours in an oxygen atmosphere at a pressure of 2 to 3 atm., the fused ferrite is cooled at a rate around 5 to 10°C/min.

At the present time single crystal manganese zinc ferrites are being produced with properties as listed in Table 2. Initial permeabilities at frequencies of 1 MHz, 8 MHz and 10 MHz are 2,000, 800 and 500, respectively. Close precaution must be exercised against the dissociation of Fe_2O_3, the evaporation of zinc and the inhomogeneous distribution of component cations, all given rise in the course of crystal growth.

Arc-Plasma Spraying[37]: Need for a miniaturized inductor has renewed interest in ferrite films. The arc-plasma spraying process, as refered to here, enables a dense ferrite film to be prepared. As illustrated in Fig. 4[38], a torch has been developed allowing the ferrites to be sprayed in an oxidizing atmosphere. The miniature inductor is produced in the manner described below. As shown in Fig. 5a, copper is sprayed onto the substrate to form half sections of a coil. This is followed by the spraying of ferrite to form the core as illustrated in Fig. 5b. After this the copper is sprayed again to form the complete thin film inductor as illustrated in Fig. 5c. The ferrite films produced by this method are very useful for radio-wave absorbers and so forth. It is added here that a radio frequency sputtering method has recently been reported for preparing ferrite films[39].

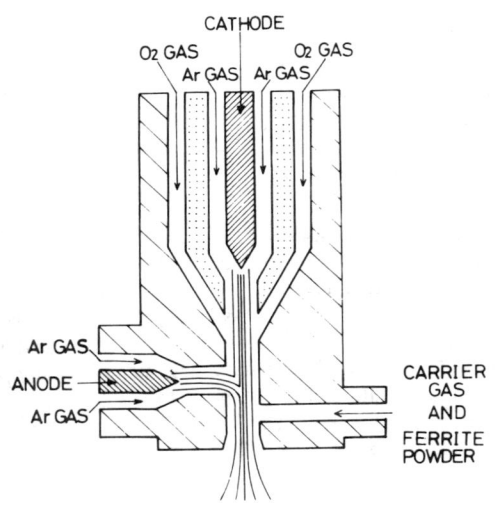

Fig. 4. Sketch of an experimental arc-plasma torch which is capable of spraying the ferrites in the jet flame containing the oxygen of 90% or more.

Miscellaneous: As a recent trend in the manufacturing of high frequency ferrites, fine particle raw materials are used. They are produced by a wet method that yields particles with sizes ranging from 0.05 to 1.0 μ[40]. The same method yields spinel ferrite powders with sizes much smaller than this[41]. The freeze-drying process[42] is used to produce raw oxides, and there is also a process in which

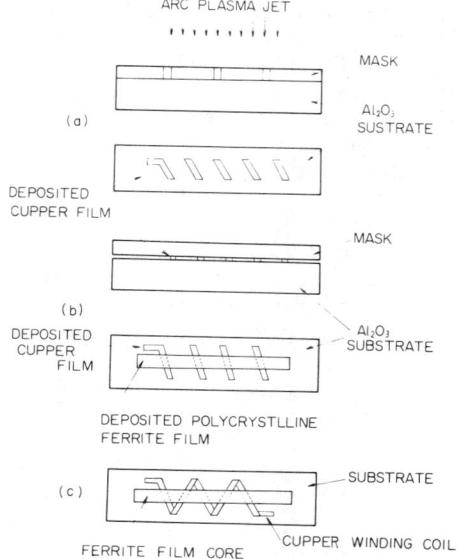

Fig. 5. Illustrative example of process for fabricating the miniaturized inductor by means of the arc-plasma spraying process.

a little amount of alkalimetal sulfates is added to prepare the ferrite at a lower temperature[43]. Furthermore, the gas bonding technique[44] wherein manganese zinc ferrite powders are produced.

4. ADVANCEMENT IN PHYSICAL AND ELECTROMAGNETIC FEATURES

In association with the advancement in manufacturing technology, high frequency ferrites have seen remarkable improvements in performance, enabling the cores to be reduced in size. In connection with this, comprehensive reports have been published on the effect of processing parameters upon the property of manganese zinc ferrite[45], nickel zinc ferrite[46] and lithium ferrite[47,48]. Studies have been also reported on the effects of the particle size and shape of the raw material, Fe_2O_3[49,50], and the mixing of the raw material[51] upon the spinel formation and the properties of sintered ferrites. Acicular or coarse particles of Fe_2O_3 reduce the rate of spinel formation, giving rise to discontinuous grain growth which results in an inhomogeneous microstructure.

Following are descriptions of the advancements in the high frequency ferrites in reference to the major properties including the peameability, the disaccommodation factor, the loss factor, the temperature coefficient and the flux density.

Permeability: About a decade ago, it was for the first time reported that a manganese zinc ferrite with initial permeability of 10,000 was obtained experimentally, which however, had Curie temperature as low as 40°C[52]. A few years later a ferrite was presented with initial permeability of 20,000[53]. Currently, ferrites are being mass-produced with Curie temperatures ranging from 100 to 130°C, and ferrites with initial permeability of 40,000[54] have successfully been prepared experimentally with Curie temperatures ranging from 95 to 130°C[55]. In general practice the ferrites with high permeability and high flux density are prepared by resorting to pure raw materials, high pressure forming as well as high temperature sintering, all contributing to enhancing the uniformity of grain distribution and lowering porosity[56]. Recently, the hot-pressing process and the vacuum-sintering process have enabled high

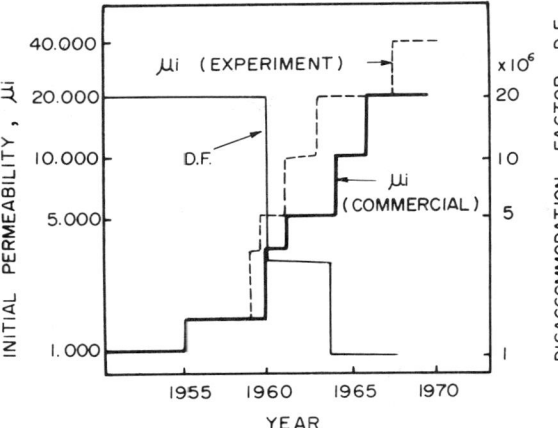

Fig. 6. Chronological progress of manganese zinc ferrite in terms of the initial permeability and the disaccommodation.

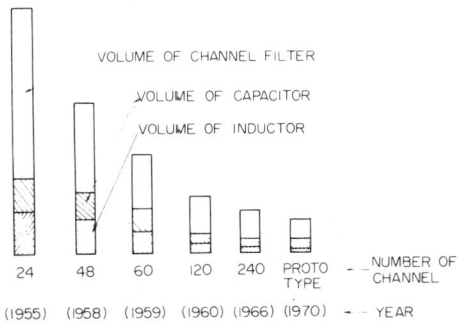

Fig. 7. Chronological progress toward reduction in size of the communication pot cores.

permeability ferrites to be prepared with considerable ease. Figure 6[34] illustrates the chronological innovation of the high frequency ferrites with specific reference made to initial permeability.

Permeability enhancement in the manganese zinc ferrite has resulted in inductor cores being reduced in size, as shown in Fig. 7[57] which illustrates the chronological trend toward the size reduction in the pot-cores.

Disaccommodation Factor: It is of great importance in practical applications to reduce the disaccommodation around room temperature. The disaccommodation is highly susceptible to the Fe_2O_3 content and the partial pressure of oxygen in the sintering atmosphere. Noteworthy is also the fact that an appropriate addition of CaO[58,59], SiO_2[58], TiO_2[59], or ZrO_2[59] contributes to the reduction of disaccommodation. Hence, in order to obtain ferrites having a low disaccommodation and a high stability, the above three parameters must be closely controlled. The advancement of the disaccommodation factor is shown in Fig. 6.

Magnetic Loss: At the present time, the low loss ferrites are obtained through the appropriate choice of composition, structure control and the addition of proper additives. In particular the loss factor given as $\tan\delta/\mu_i$ is improved by the appropriate additives, the effect of the addition of $CaO-SiO_2$, $CaO-Ta_2O_5$ or $CaO-Nb_2O_3$ being prominent in manganese zinc ferrites[34]. The recent achievement in terms of $\tan\delta/\mu_i$ is 0.8×10^{-6} at 100 KHz.

Addition of a small amount of TiO_2 results in the eddy current loss of manganese zinc ferrite being reduced down to 5×10^{-6} at 1 MHz and 1×10^{-6} at 100 KHz[60]. On the other hand, addition of a small amount of cobalt results in the magnetic loss of nickel zinc ferrite being reduced around the FM frequency band. This may be ascribed to the domain walls stabilization which caused by the ordering of Co^{2+} ions over the available octahedral sits[61], so that the magnetization arises mainly from the domain rotation[62].

Iron-deficient nickel zinc ferrites exhibit a low loss even when subjected to an intense disturbing field or to a polarizing field, thereby lending themselves to output transformers in RF power amplifiers of RF tuners for accelerating cavities of the proton synchrotron[63].

Temperature Coefficient: Temperature coefficient is determined by relative positions of the first and the second peak of the μ-T curve which result from the Hopkinson's effect and from the reduction of the anisotropy constant, K, respectively. In order to position the quoted peaks as desired, a close control of composition is indispensable. This is evident from findings such as an error of 0.02 mol% in composition causes the second peak to be shifted by 1°C on the μ-T curve[34].

Addition of a small amount of TiO_2 or ZrO_2 reduces the temperature coefficient. On the basis of this fact, a ferrite has recently been developed having the initial permeability of 1,500 to 3,000, and $tan\delta/\mu_i$ of around 2×10^{-6} at 100 KHz with temperature coefficient of null over 0 to 40°C. Addition of the Ferroxplana type ferrite, Co_2Y, to the amount of about 2 wt.% causes the temperature coefficient of permeability of nickel zinc ferrite to be reduced approximately to 10^{-7} [64].

Table 3. Chronological progress in the magnetic properties of ferrite core for TV flyback transformers.

Year	Initial permeability	Flux density at 15 Oe.(G.)	Coercive force(Oe.)	Curie temp.(°C)	$tan\delta/\mu_i$	
1950	400	2,000	0.5	90	1	10^{-4}
1955	1,100	3,800	0.15	160	2	10^{-5}
1960	1,400	4,300	0.21	210	2	10^{-5}
1965	2,500	4,800	0.13	200	1	10^{-5}
1972	3,000	4,800	0.10	200	1	10^{-5}

In connection with the subject of interest, a report has been published on a method enabling the temperature change of the inductance of coil to be measured automatically to a high accuracy, thereby permitting the temperature dependences of ferrite cores to be measured to a high accuracy[65].

Flux Density: The ferrous zinc ferrite of the composition, $Fe^{2+}_{1-\delta}Zn_\delta Fe_2O_4$[66] has a saturation magnetization as high as 6,750 G. at room temperature for δ ranging from 0.2 to 0.4. The author feels that it might be a potential candidate for a TV flyback transformer core material in the coming age. Table 3 shows the chronological progress of the magnetic properties of ferrite core for the TV flyback transformers.

5. NEW APPLICATION ASPECTS

So far, high frequency ferrites have been finding wide-spread application for antenna cores and IF transformers for radio sets, flyback transformers, deflection yokes, IF transformers, pincusion transformer, horizontal width regulators and sound discriminators for TV sets, filter inductors for communication equipment, and pulse transformers for computers.

The prime new field of application under current development where particularly great demands are expected to emerge is that of the magnetic recording head using ferrite cores[67]. The operational frequencies being high, all the VTR heads are expected to be fabricated from ferrite cores either of the hot-pressed or of the single crystal type. A great demand is also expected to arise for instrument recording heads. Recently, ferrite heads have begun to find application also in the audio frequency region in place of the permalloy alloy head in current usage.

The second major application is found in the field of broadcasting much of which is expected to be conducted in the future in the VHF band using frequency modulation in place of the so-called BC band used today. This should cause the demand for portable FM radio receivers to be increased, resulting in the need for high sensitivity ferrite antennas being enhanced in place of whip antennas. Reports have already been published on experimental compact ferrite antennas[68] fabricated from ferrite as well as on experimental miniaturized tuners[69] made of ferrite.

Besides, ferrites are finding increasing use for the flux keepers on IC substrates, ferrite cores for pocketable telephones and so forth. Applications based on the Curie temperature effect, wherein the ferrite loses the magnetic property when the working temperature reaches a prescribed value, are finding the ever-increasing applications. These applications include the temperature controls for constant-temperature ovens or air conditioners, sensors for anti-fire facilities such as fire alarm and anti-fire shutters, and sensors for precenting excessive cooling or heating.

In addition to application as electronic materials, ferrites recently found specific applications as an insoluble electrode for electrolysis, and as a contrast material[70] for the X-ray diagnosis of the digestive organs.

6. PREDICTION OF FUTURE OF HIGH FREQUENCY FERRITES

The future will also see the improvement of high frequency

ferrites in performances resulting from the advancements in academic researches and manufacturing technologies. The extent to which these improvements will be achieved is very difficult to predict at the present time, much as no one could foresee a decade ago such tremendous improvement of the performance in terms of the initial permeability or the disaccommodation factor as well as such a vast amount of productions, as seen at the present time. As initially indicated, however, structurally new ferrites appear to be hard to find and develop. Nevertheless, in sight are the development of new application fields, the improvement of manufacturing technologies, and the resultant performance improvement as well as the production rise, allowing the conclusion to be reached that the high frequency ferrites are certain to have a bright future.

ACKNOWLEDGEMENTS

The author wishes to express his sincere thanks to Prof. Dr. N. Ogasawara, Tokyo Metropolitan University, for his numerous helpfull discussions. Thanks are also due to Dr. T. Hiraga and Mr. S. M. Tominaga of TDK Electronics Co., Ltd. for their permission to refer to some of their works.

REFERENCES

(1) J. L. Snoek, New Developments in Ferromagnetic Materials. (Elserier Publ. Co. N.Y., 1947).
(2) J. J. Went, G. W. Rathenau, E. W. Gorter, G. W. Van Oosterhout, Philips Tech. Rev. $\underline{13}$, 194 ($^{1951}/_{1952}$).
(3) E. Albers-Schoenberg, J. Appl. Phys. $\underline{25}$, 152 (1954).
(4) G. H. Jonker, H. P. J. Wijn, P. B. Braun, Philips Tech Rev. $\underline{18}$, 145 ($^{1956}/_{1957}$).
(5) F. Bertaut, F. Forrat, Compt Rend. $\underline{242}$, 382 (1956).
(6) P. A. Arther, Jr., M. T. Swoboda, Jap. Pat. No.236633 (1957).
(7) M. L. Keith, R. Roy, Amer. Mineral, $\underline{39}$, 1 (1954).
S. Geller, M. A. Gilleo, J. Chem. Phys.$\underline{24}$, 1236, 1239 (1956).
S. Geller, E. A. Wood, Acta. Cryst. $\underline{9}$, 563 (1956).
(8) A. Marais, T. Merceron, C. R. Acad. Sci. Paris, $\underline{249}$, 2511 (1959); $\underline{256}$, 2560 (1963).
K. Ohta, J. Phys. Soc. Japan, $\underline{16}$, 250 (1961).
S. Iida, J. Phys. Soc. Japan, $\underline{17}$, 123 (1962).
A. Yanase, J. Phys. Soc. Japan, $\underline{17}$, 1005 (1962).
A. Braginski, T. Merceron, J. Phys. Soc. Japan, $\underline{17}$, 1611 (1962).
T. Okada, T. Yamadaya, S. Miyahara, J. Phys. Soc. Japan, $\underline{17}$, 1799 (1962).
S. Krupicka, J. Phys. Soc. Japan, $\underline{17SB-1}$, 304 (1962); Czech. J. Phys. $\underline{B-14}$, 29 (1964).
F. J. Schnettler, E. M. Gyorgy, J. Appl. Phys. $\underline{35}$, 330 (1964).
(9) A. M. M. Gadalla, J. White, Trans. Brit. Cer. Soc. $\underline{65}$, 1 (1966).
T. Yamaguchi, Proc. Fac. Eng. Keio Univ. Japan, $\underline{19}$, 36 (1966).

(10) A. E. Paladino, Jr., J. Am. Cer. Soc. $\underline{42}$, 168 (1959).
A. E. Paladino, Jr., Signal Corps. Contract, DA-36-039-SC-74987, Unclassified report (1959).
(11) J. Smiltens, J. Am. Chem. Soc. $\underline{79}$, 4881 (1957).
B. D. Roiter, A. E. Paladino, Jr., J. Am. Cer. Soc. $\underline{45}$, 128 (1962).
(12) D. Woodhouse, J. White, Trans. Brit. Cer. Soc. $\underline{54}$, 333 (1955).
D. L. Fresh, Proc. IRE, $\underline{44}$, 1303 (1956).
A. E. Paladino, Jr., J. Am. Cer. Soc. $\underline{43}$, 183 (1960).
T. Katsura, S. Kimura, Bull. Chem. Soc. Japan, $\underline{38}$, 1664 (1965).
D. H. Speidel, J. Am. Cer. Soc. $\underline{50}$, 243 (1967).
(13) T. Yamaguchi, T. Takei, Sci. papers Inst. Phys. Chem. Res. $\underline{53}$, 207 (1959).
(14) K. Schwerdfeger, A. Muan, Trans. MS-AIME. $\underline{239}$, 1114 (1967).
(15) A. Bergstein, M. Rozsival, M. Mikulas, Collection Czechoslov. Chem. Commun. $\underline{24}$, 885 (1959).
V. Montoro, Gazy. Chim. Ital. $\underline{70}$, 145 (1940).
(16) J. M. Blank, J. Appl. Phys. $\underline{32}$, 378S (1961).
(17) P. I. Slick, H. Basseches, IEEE Trans. $\underline{MAG-2}$, 603 (1966).
P. I. Slick, Proc. of the Int. Conf. on Ferrites, Kyoto, Japan, Univ. of Tokyo Publ. Co. (1970) p. 81.
(18) J. M. Brownlow, J. Appl. Phys, $\underline{29}$, 373 (1958).
T. Yamaguchi, Sci. papers Inst. Phys. Chem. Res. $\underline{54}$, 124 (1960).
(19) A. J. Pointon, R. C. Saull, J. Am. Cer. Soc. $\underline{52}$, 157 (1969).
(20) M. W. Shafer, J. Phys. Chem. $\underline{65}$, 2055 (1961).
(21) M. W. Shafer, IBM. J. Res. Develop. $\underline{2}$, 193 (1958).
(22) B. Phillips, A. Muan, J. Am. Cer. Soc. $\underline{45}$, 588 (1962).
B. Phillips, S. Somiya, A. Muan, J. Am. Cer. Soc. $\underline{44}$, 167 (1961).
(23) M. Sugimoto, J. Appl. Phys. Japan. $\underline{5}$, 557 (1966); Proc. of the Int. Conf. on Ferrites, Kyoto, Japan, Univ. of Tokyo Publ. Co. (1970) p. 318.
(24) G. C. Kuczynski, Proc. of the Int. Conf. on Ferrites, Kyoto, Japan, Univ. of Tokyo Publ. Co. (1970) p. 87.
(25) M. Paulus, Proc. of the Int. Conf. on Ferrites, Kyoto, Japan, Univ. of Tokyo Publ. Co. (1970) p. 114.
(26) A. L. Stuijts, Proc. of the Int. Conf. on Ferrites, Kyoto, Japan, Univ. of Tokyo Publ. Co. (1970) p. 108.
(27) T. Takei, T. Yoshida, T. Yamaguchi, T. Yodogawa, A. Okamoto, T. Hibiya, M. Kamoshita, Proc. of the Int. Conf. on Ferrites, Kyoto, Japan, Univ. of Tokyo Publ. Co. (1970) p. 125.
(28) U. H. Banga, W. Mesman, Philips Tech. Rev. $\underline{27}$, 337 (1966).
(29) M. A. Strivens, Proc. of the Int. Conf. on Ferrites, Kyoto, Japan, Univ. of Tokyo Publ. Co. (1970) p. 249.
(30) A. Ikeda, M. Satomi, H. Chiba, E. Hirota, Proc. of the Int. Conf. on Ferrites, Kyoto, Japan, Univ. of Tokyo Publ. Co. (1970) p. 337.
(31) G. J. Oudemans, Philips Tech. Rev. $\underline{29}$, 45 (1968).
(32) A. L. Stuijts, Science of Ceramics. $\underline{5}$, 335 (1970).
(33) Y. Shichijo, E. Takama, Proc. of the Int. Conf. on Ferrites,

(33) Kyoto, Japan, Univ. of Tokyo Publ. Co. (1970) p. 210.
(34) T. Hiraga, Proc. of the Int. Conf. on Ferrites, Kyoto, Japan, Univ. of Tokyo Publ. Co. (1970) p. 179.
(35) F. W. Harrison, R. F. Pearson, K. Tweedale, Philips Tech. Rev. **28**, 135 (1967).
(36) S. Kobayashi, I. Yamagishi, R. Ishi, M. Sugimoto, Proc. of the Int. Conf. on Ferrites, Kyoto, Japan, Univ. of Tokyo Publ. Co. (1970) p. 326.
(37) D. H. Harris, R. J. Janowiecki, C. E. Semler, M. C. Willson, J. T. Cheng, J. Appl. Phys. **41**, 1348 (1970).
(38) M. Sugimoto, H. Tateno, S. Kojima, T. Ichimiya, presented at the 1972 InterMag. Conf, Kyoto, Japan.
(39) R. Metselaar, P. Rem, Czech. J. Phys. **B21**, 558 (1971).
(40) T. Takada, M. Kiyama, Proc. of the Int. Conf. on Ferrites, Kyoto, Japan, Univ. of Tokyo Publ. Co. (1970) p. 69.
(41) T. Sato, C. Kuroda, M. Saito, M. Sugihara, Proc. of the Int. Conf. on Ferrites, Kyoto, Japan, Univ. of Tokyo Publ. Co. (1970) p. 72.
(42) F. J. Schnettler, D. W. Johnson, Proc. of the Int. Conf. on Ferrites, Kyoto, Japan, Univ. of Tokyo Publ. Co. (1970) p. 121.
(43) D. G. Wickham, Proc. of the Int. Conf. on Ferrites, Kyoto, Japan, Univ. of Tokyo Publ. Co. (1970) p. 105.
(44) A. Sawaoka, S. Saito, Proc. of the Int. Conf. on Ferrites, Kyoto, Japan, Univ. of Tokyo Publ. Co. (1970) p. 102.
(45) S. Natansohn, D. H. Baird, J. Am. Cer. Soc. **52**, 127 (1969).
(46) D. R. Secrist, H. L. Turk, J. Am. Cer. Soc. **53**, 683 (1970).
(47) D. H. Ridgley, H. Lessoff, J. D. Childress, J. Am. Cer. Soc. **53**, 304 (1970).
(48) M. Amemiya, Proc. of the Int. Conf. on Ferrites, Kyoto, Japan, Univ. of Tokyo Publ. Co. (1970) p. 154.
(49) J. H. Magee, V. Morton, R. D. Fisher, I. J. Lowe, Proc. of the Int. Conf. on Ferrites, Kyoto, Japan, Univ. of Tokyo Publ. Co. (1970) p. 217.
(50) T. Akashi, I. Sugano, T. Okuda, T. Tsuji, Proc. of the Int. Conf. on Ferrites, Kyoto, Japan, Univ. of Tokyo Publ. Co. (1970) p. 96.
(51) G. Chol, J. P. Aubaile, Proc. of the Int. Conf. on Ferrites, Kyoto, Japan, Univ. of Tokyo Publ. Co. (1970) p. 243.
M. A. Strivens, G. Chol, Proc. of the Int. Conf. on Ferrites, Kyoto, Japan, Univ. of Tokyo Publ. Co. (1970) p. 239.
(52) E. Röss, E. Moser, Z. angew. Phys. **Bd13**, 247 (1961).
(53) E. Röss, I. Hanke, E. Moser, Z. angew. Phys. **17**, 504 (1964).
Y. Shichijo, G. Asano, E. Takama, J. Appl. Phys. **35**, 1646 (1964).
(54) E. Röss, Electronic Components Bulletin. **1**, 138 (1966).
A. Beer, J. Schwarz, IEEE Trans. **MAG-2**, 470 (1966).
(55) T. Akashi, Meeting of the Japan Soc. of Powder Metallurgy, Sept. (1962).
T. Hiraga, Meeting of the Japan Soc. of Powder Metallurgy, Sept. (1962).

(56) E. Röss, Proc. of the Int. Conf. on Ferrites, Kyoto, Japan, Univ. of Tokyo Publ. Co. (1970) p. 203.
(57) T. Akashi, I. Sugano, Y. Kenmoku, Y. Shinma, T. Tsuji, Proc. of the Int. Conf. on Ferrites, Kyoto, Japan, Univ. of Tokyo Publ. Co. (1970) p. 183.
(58) T. Akashi, Trans. Japan Inst. Metals. $\underline{2}$, 171 (1961).
(59) T. G. W. Stijntjes, J. Klerk, C. J. M. Rooymans, A. Broese Van Groenou, R. F. Pearson, J. E. Knowles, P. Rankin, Proc. of the Int. Conf. on Ferrites, Kyoto, Japan, Univ. of Tokyo Publ. Co. (1970) p. 191.
T. G. W. Stijntjes, A. Broese Van Groenou, R. F. Pearson, J. E. Knowles, P. Rankin, ibid. p. 194.
(60) E. Röss, I. Hanke, Phys. Stat. Sol.(a) $\underline{2}$, K185 (1970).
(61) J. G. M. de Lau, A. L. Stuijts, Philips Res. Rept. $\underline{21}$, 104 (1966).
(62) I. Mikami, Proc. of the Int. Conf. on Ferrites, Kyoto, Japan, Univ. of Tokyo Publ. Co. (1970) p. 221.
(63) H. Yokoyama, Y. Hirose, S. Chiba, Proc. of the Int. Conf. on Ferrites, Kyoto, Japan, Univ. of Tokyo Publ. Co. (1970) p. 233.
(64) A. Arai, T. Ido, Proc. of the Int. Conf. on Ferrites, Kyoto, Japan, Univ. of Tokyo Publ. Co. (1970) p. 225.
(65) C. V. Newcomb, E. C. Snelling, Philips Tech. Rev. $\underline{28}$, 184 (1967).
(66) A. L. Stuijts, D. Veeneman, A. Broese Van Groenou, Proc. of the Int. Conf. on Ferrites, Kyoto, Japan, Univ. of Tokyo Publ. Co. (1970) p. 236.
(67) H. Sugaya, IEEE Trans. $\underline{\text{MAG-4}}$, 295 (1968).
R. D. Fisher, J. H. Magee, I. J. Lowe, V. Morton, Proc. of the Int. Conf. on Ferrites, Kyoto, Japan, Univ. of Tokyo Publ. Co. (1970) p. 340.
H. Abe, T. Iwasawa, S. Ohtsuki, ibid. p. 343.
E. Hirota, T. Mihara, A. Ikeda, H. Chiba, IEEE Trans. $\underline{\text{MAG-7}}$, 337 (1971).
(68) D. Mitchell, K. G. Van Wynen, Bell System Tech. Journal, $\underline{40}$, 1239 (1961).
G. Schiefer, Philips Tech. Rev. $\underline{24}$, 332 ($^{1962}/_{1963}$).
A. E. Kerwin, L. H. Steiff, Bell System Tech. Journal, $\underline{42}$, 527 (1963).
(69) C. G. Sontheimer, U.S. Pat. No.2973431 (1961).
M. Sugimoto, N. Matsuoka, U.S. Pat. No.3430175 (1969)
Deutsches Pat. No.1466123 (1971).
(70) E. H. Frei, E. Gunders, M. Pajewsky, W. J. Alkan, J. Eshchar, J. Appl. Phys. $\underline{39}$, 999 (1968).
M. Sugimoto, T. Watari, N. Watanabe, M. Tobe, H. Takizawa, presented at the 1972 InterMag. Conf, Kyoto, Japan.

Section 42. Transition-Metal
Alloys and Compounds

ORIGIN OF MÖSSBAUER LINEWIDTH IN STAINLESS STEEL*

R. C. Reno[†] and L. J. Swartzendruber
National Bureau of Standards, Washington, D.C. 20234

ABSTRACT

The hyperfine interactions which broaden the Mössbauer effect (ME) spectrum in austenitic stainless steel have been investigated with the aid of ^{57}Fe time-differential perturbed angular correlations (TDPAC). The TDPAC measurements reveal a distribution of electric field gradients at the ^{57}Fe nuclei with a mean value corresponding to a Mössbauer splitting of 0.14 mm/s. This splitting is not sufficient to explain the total line width and isomer shifts are invoked to account for the remaining width.

INTRODUCTION

For ME spectra with unresolved or just barely resolved structure, it is not possible to establish the broadening mechanism (i.e., magnetic, quadrupole, isomer shift, or some combination of these) on the basis of the ME spectra alone. We have recently shown that the application of time-differential perturbed angular correlations (TDPAC) to the study of such cases can *unambiguously* determine the origin of the hyperfine interaction when combined with existing Mössbauer data[1]. In Ref. 1 the technique was applied to the barely resolved ME doublet characteristic of dilute Fe in Cu-Ni alloys. Here similar results are presented on the unresolved broadening found in the ME spectra of austenitic stainless steels. In contrast to the Cu-Ni-Fe case, a significant contribution to the linewidth can be attributed to a distribution of isomer shifts.

MÖSSBAUER EFFECT IN STAINLESS STEEL

The Fe moment in fcc (γ) stainless steel is much smaller than that found for Fe in the bcc (α) phase. In addition, stainless steel exhibits antiferromagnetic behavior at sufficiently low temperatures[2]. Several Mössbauer experiments have been performed on austenitic (γ-phase) stainless steels[3-5] in the hope of understanding the hyperfine interactions involving the Fe nuclei both below and well above the Néel temperature. Although there are slight differences, depending in part on alloy concentration, all room temperature ME spectra appear to be well described by a single broadened line, roughly Lorentzian in shape, with a linewidth of 0.40-0.50 mm/s. Since lifetime broadening alone (0.194 mm/s) accounts for almost half the total stainless steel linewidth, it is not possible to resolve structure in the line which is due to hyperfine interactions. It is therefore impossible to attribute correctly the broadening to any

---

* Work supported in part by the Office of Saline Water
[†] NRC-NBS Postdoctoral Research Associate

one of the three most common hyperfine mechanisms: isomer shift distributions, electric quadrupole effects, or magnetic hyperfine interactions. Such ambiguity is particularly troublesome when studies of the antiferromagnetic-paramagnetic transition are carried out as in Refs. 3 and 4. These experiments showed that as the temperature is lowered from room temperature the ME linewidth remains constant down to a certain temperature, whereupon it then begins to broaden but still maintains the appearance of a single line. This effect was interpreted as due to the onset of antiferromagnetism and the appearance of a magnetic hyperfine field, the value of which was obtained by fitting an unresolved six-line spectrum to the ME spectrum. This procedure may be invalid if additional temperature-dependent interactions are present, such as those proposed to explain the ME spectra of Au-Fe alloys near their ordering temperatures[6].

In this paper, the contributions from the various possible hyperfine interactions in stainless steel are determined by a comparison of ME and TDPAC data taken on the same source. Knowledge of the amount of each type of hyperfine interaction present at room temperature should make possible more reliable analysis of low temperature data. In addition, the experiment described here could itself be performed near the phase transition in order to study directly the effect of temperature on the various hyperfine interactions.

RESULTS AND CONCLUSIONS

A source of (^{57}Co) 310 stainless steel was prepared by evaporating approximately 5 µCi of ^{57}Co onto a piece of 310 stainless steel, 0.0025 cm thick, and by diffusing in a Hydrogen atmosphere at 950°C for 10 minutes. The source was placed in a three-counter TDPAC spectrometer and a spectrum was accumulated for 14 days. The reduced data are shown in Fig. 1a. The solid curve indicates that

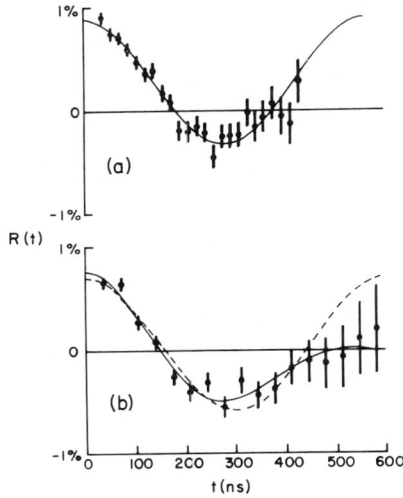

FIGURE 1. Room temperature TDPAC spectra of ^{57}Co in stainless steel. The ordinate, R(t), defined in Ref. 1, is a measure of the perturbation to the time-dependent coincidence rate caused by a hyperfine interaction.
a) 5 µCi ^{57}Co source. The solid line is a fit to a single cosine which would result from a single-valued electric quadrupole interaction.
b) 2.5 µCi ^{57}Co source. The dashed line is a fit to a single cosine and the solid line is a fit to a Gaussian distribution of cosines.

the data can be fit by a single cosine of period 575 ns. However, the limited time range of 450 ns makes it difficult to distinguish between a pure cosine or a distribution of cosines with periods peaked near 575 ns. In order to resolve this question, the source was cut in half to reduce accidental coincidences (thereby extending the time range by ~150 ns) and a second 14 day run was taken. The data appear in Fig. 1b with a fit to both a single cosine and a Gaussian distribution of cosines. The comparison shows that the distribution of cosines gives a better fit.

The 5 µCi source was also placed in a Mössbauer spectrometer. The spectrum using a single-line potassium ferrocyanide absorber, is shown in Fig. 2 and has the appearance of a singlet with a linewidth (FWHM) of 0.44 ± 0.01 mm/s. Since the same absorber

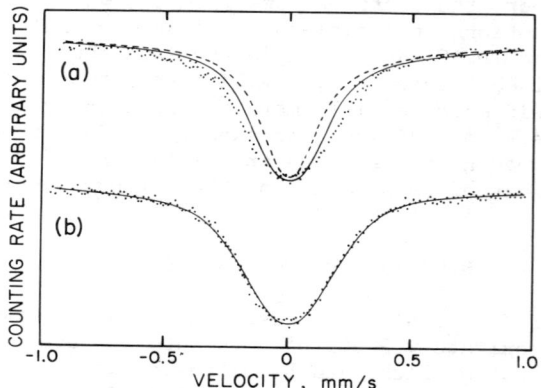

FIGURE 2. Room temperature Mössbauer spectrum of 5 µCi ^{57}Co in stainless steel. Absorber was potassium ferrocyanide at room temperature.

a) The dashed curve shows a Lorentzian line with the same width obtained using the potassium ferrocyanide absorber and a (^{57}Co)Pd source. The solid curve shows a theoretical spectrum generated assuming the same absorber and the EFG distribution deduced from the TDPAC data in Fig. 1.

b) Data and fit (solid line) to a model incorporating the TDPAC-deduced EFG distribution and an additional distribution of isomer shifts.

(The centroid of the stainless steel spectrum was taken as zero isomer shift).

when used with a (^{57}Co)PD source gives a linewidth of 0.26 ± 0.01 mm/s, hyperfine interactions must account for the additional linewidth. The mechanisms responsible for this broadening can be deduced as follows:

1) The sole source of Mössbauer line broadening cannot be isomer shift distributions. If such were the case, a null TDPAC spectrum would have been observed since TDPAC is not sensitive to isomer shifts. The TDPAC spectrum must result from either electric quadrupole or magnetic dipole interactions.

2) The TDPAC spectra were fit assuming that the oscillations were due to magnetic hyperfine fields. As in Ref. 1, although the TDPAC data can be fit reasonably well by even a single magnetic hyperfine field, the value of the field deduced (13.5 kG) would predict a Mössbauer spectrum 0.20 mm/s broader than actually observed. It is therefore possible to rule out magnetic hyperfine fields as a major source of broadening.

3) The TDPAC spectrum must be due primarily to electric quadrupole interactions and the Gaussian distribution of cosines must be interpreted as due to a Gaussian distribution of electric field gradients (EFG's) at the ^{57}Fe nuclei. In particular, the TDPAC data were least square fit by a distribution of EFG's characterized by:

$$p(\Delta E) \propto \exp[-(\Delta E - \Delta E_o)^2/2\sigma_E^2]$$

where $p(\Delta E)$ is the probability of occurrence of an EFG giving a ME doublet with a splitting of ΔE. Values of the parameters ΔE_o and σ_E so derived are listed in Table I. (The Gaussian form used here is only a model; the precise shape of the distribution cannot be deduced from the data.) It should be noted at this point that a model incorporating a mixture of EFG's and magnetic hyperfine interactions is also consistent with both ME and TDPAC data provided the magnetic contribution is less than about 3 kG. Larger magnetic admixtures result in highly asymmetric ME spectra, inconsistent with the stainless steel spectrum.

4) The distribution of EFG's deduced from the TDPAC data would result in a ME spectrum shown by the solid line in Fig. 2a. The alternative model of EFG's with small admixtures of magnetic interactions would result in an almost identical spectrum. Since the linewidth in either case is only 0.36 mm/s, it is necessary to conclude that a distribution of isomer shifts is also present to account for the remaining 0.08 mm/s of linewidth. The solid curve in Fig. 2b is a fit to the Mössbauer data assuming that the quadrupole spectrum shown by the solid line in Fig. 2a is distributed with isomer shifts given by:

$$p(IS) \propto [-IS^2/2\sigma_{IS}^2]$$

where $p(IS)$ is the probability that the quadrupole spectrum described by the parameters ΔE_o and σ_E (Table I) has an isomer shift IS. The value of σ_{IS} so derived is also listed in Table I.

The amplitude of the PAC spectrum near t=0 is in good agreement ($\pm$ 10%) with the amplitude expected if all ^{57}Fe nuclei experienced EFG's characterized by the parameters ΔE_o and σ_E given in Table I. This indicates that the Mössbauer spectrum of stainless steel consists of a superposition of quadrupole doublets rather than a mixture of doublets and isomer-shifted singlets.

The existence of electric field gradients and isomer shifts in stainless steels is not surprising given that the material is a disordered alloy containing appreciable amounts of Ni and Cr. What does seem a bit surprising, however, is the lack of correlation between the isomer shift spread and the EFG spread when the stainless steel measurements are compared to our previous Cu-Ni measurements[1]. In the latter case we found no detectable isomer shift contribution although the EFG's were larger and more spread out.

The fact that the Mössbauer spectrum can not, by itself, establish the existence of EFG's in stainless steel merely illustrates the limitation of the technique when applied to materials experiencing weak hyperfine interactions. It is clear that TDPAC can provide useful additional information in such cases.

The authors wish to thank Drs. L. H. Bennett, M. Kuriyama, and D. D. Hoppes for useful discussions and comments. The technical assistance of Mr. R. D. Robbins is also gratefully acknowledged.

TABLE I

Parameters derived from fits to ME and TDPAC data

	ΔE_o (mm/s)	σ_E (mm/s)	σ_{IS} (mm/s)
TDPAC	0.14 ± 0.008	0.029 ± 0.007	------
ME	0.14*	0.029*	0.10 ± 0.01

* Values constrained in ME least squares fit

REFERENCES

1. R. C. Reno and L. J. Swartzendruber, Phys. Rev. Letters 29, 712 (1972).
2. E. I. Kondorskii and V. L. Sedov, Sov. Phys. JETP 8, 1104 (1959)
3. U. Gonser, C. J. Meechan, A. H. Muir and H. Weidersich, J. Appl. Phys. 34, 2373 (1963).
4. L. D. Flansburg and N. Hershkowitz, J. Appl. Phys. 41, 4082 (1970).
5. B. P. Srivastava H. N. K. Sarma and D. L. Bhattacharya, phys. stat. sol. (a) 10, K117 (1972).
6. M. S. Ridout, J. Phys. C. 2, 1258 (1969).

PRESSURE VARIATION OF THE CURIE TEMPERATURE AND SPONTANEOUS MAGNETIZATION IN Fe_2P AND $Fe_2P_{0.9}As_{0.1}$ [*]

J. B. Goodenough, J. A. Kafalas, K. Dwight, and N. Menyuk
Lincoln Laboratory, M.I.T., Lexington, Mass. 02173

A. Catalano
Dept. of Chemistry, Brown University, Providence, R.I. 02912

ABSTRACT

The transition-metal pnictides $(M_{1-y}M'_y)P_{1-x}As_x$ exhibit structural relationships and magnetic properties that indicate the presence of filled valence bands, empty conduction bands, and partially filled 3d bands active in metal-metal bonding. In many cases they support spontaneous magnetism, thereby offering the opportunity to study itinerant-electron magnetism as a function of 3d bandwidth and occupancy. In particular, the hexagonal system $Fe_2P_{1-x}As_x$ is ferromagnetic, but for $x < 0.33$ its spontaneous moment at $T = 0K$ is reduced from the $\mu_0 = 3.0\mu_B$/molecule predicted for itinerant, spin-only ferromagnetism. We investigated the pressure dependence of T_c and μ_0 to 11 kbar. In Fe_2P, the relation between ΔT_c [°C] and P [kbar] is: $P = -0.252(\Delta T_c) - 0.0012(\Delta T_c)^2$. In $Fe_2P_{0.9}As_{0.1}$, it is: $P = -0.71(\Delta T_c) - 0.0017(\Delta T_c)^2$. Pressure did not change significantly the value of μ at 58K, but it promoted a remarkably exchange-enhanced susceptibility above T_c. We interpret these results to mean that the reduced moment in Fe_2P is not due to conduction-band overlap of the Fermi energy, but to a 3d bandwidth that is just narrow enough to support spontaneous ferromagnetism. A critical pressure $P_c \approx 13$ kbar is estimated for a ferromagnetic-to-metamagnetic transition.

INTRODUCTION

Transition metals (M) and their alloys are characterized by narrow 3d bands that are overlapped by a broad 4s band. In the pnictides M_2X, introduction of the X atoms splits the broad bands into a valence band and a conduction band that are separated by a finite energy gap. If the Fermi energy E_F falls in this gap, then the number z_d of 3d electrons per M atom can be inferred from the formal valence X^{3-}. The absence of X-X pairing, even where M = Ni, and an average formal valence of only 1.5+ at the M atoms, indicate that the broad valence bands are filled. In order to establish that the broad conduction band is empty, it is useful to have a theoretical prediction for the magnitudes of the atomic moments as a function of z_d for an itinerant-electron ferromagnet.

---

[*]This work was sponsored by the Department of the Air Force.

Where M-M bonding via 3d electrons contributes a binding energy that is small compared to intraatomic-exchange stabilization, there the atomic moments at the M atoms can be deduced from crystal-field theory, and the Weiss molecular fields are given by superexchange (or double-exchange) perturbation theories. Where this condition is not fulfilled, an itinerant-electron model of ferromagnetism is generally employed. In this model, both intraatomic and interatomic exchange interactions contribute to the Weiss molecular field. As discussed more fully elsewhere,[1] the celebrated Slater-Pauling curve for the spontaneous magnetization vs electron/atom ratio of the ferromagnetic transition metals and their alloys can be successfully rationalized if it is assumed that the binding energy, being stronger than any magnetic energy associated with the Weiss molecular field, keeps the bonding orbitals occupied. With this assumption, the maximum ferromagnetic moment per M atom contributed by unpaired electron spins in a ν-fold degenerate band is $(\nu/2)\mu_B$, which occurs where the bands are one-quarter or three-quarters filled. The ferromagnetic moment falls off linearly to zero for empty, half-filled, and full bands. However, antiferromagnetism may be associated with a half-filled band, and the atomic moments to be associated with antiferromagnetic order are not predictable from these simple considerations. Finally, since any orbital contribution to the atomic moment is relatively small for itinerant electrons, it follows that fivefold-degenerate 3d bands ($\nu = 5$) should exhibit a maximum atomic moment u_A (max) $\approx 2.5\mu_B$ at $z_d = 7.5$ and that $du_A/dz_d = +1\mu_B$ for $5 < z_d \leqslant 7.5$, $du_A/dz_d = -1\mu_B$ for $7.5 \leqslant z_d \leqslant 10$. In the system $Fe_2P_{1-x}As_x$, filled valence and empty conduction bands would leave a $z_d = 6.5$, and the predicted spin-only molecular moment for ferromagnetic coupling and $\nu = 5$ would be $u_o = 3.0\mu_B$, corresponding to an average $1.5u_B$ per Fe atom.

Pure samples of Fe_2P exhibit a $\mu_o = 2.20\mu_B$, noticeably reduced from the predicted spin-only value of $3.0\mu_B$.[2,3] Such a reduction implies either a smaller z_d (due to an overlapping conduction band) or a 3d band that is too broad for the Weiss molecular field to empty all the antibonding states of antiparallel spin. Substitution of As for P increases the M-M separation, thereby reducing the widths of the 3d bands. It also increases the covalent mixing, thereby raising the bottom of the conduction band relative to the 3d bands. Therefore, if the theory outlined above is applicable, then u_o should increase with x in $Fe_2P_{1-x}As_x$ until it saturates at a $\mu_o = 3.0\mu_B$. This critical experiment has been performed,[2,3] and indeed μ_o saturates at $3.0\mu_B$ for $x \geqslant 0.33$. In order to distinguish whether the reduced moment in Fe_2P is due to conduction-band overlap or to broad 3d bands, we have investigated the pressure dependence of several magnetic properties.

EXPERIMENTAL

The Fe_2P and $Fe_2P_{0.9}As_{0.1}$ samples studied were prepared by direct combination of the elements and were the same as those used for the atmospheric-pressure study.[2] High-pressure measurements were made by

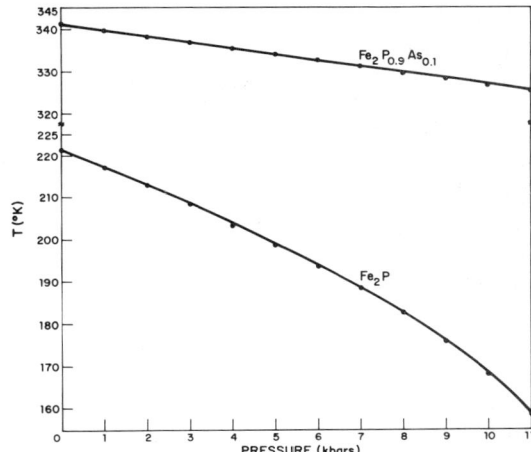

Fig. 1. Pressure dependence of the Curie temperature in Fe_2P and $Fe_2P_{0.9}As_{0.1}$.

using a gas generator with a vibrating-coil magnetometer, as described previously.[4] The Curie-temperatures T_C were obtained at low (100 Oe) fields, and plots of T_C vs hydrostatic pressure P are shown in Fig. 1. The points are experimental, and the solid lines are the analytic functions (with P in kbar and ΔT_C in °C)

$$P = -0.252(\Delta T_C) - 0.0012(\Delta T_C)^2 \text{ for } Fe_2P \tag{1}$$

$$P = -0.71(\Delta T_C) - 0.0017(\Delta T_C)^2 \text{ for } Fe_2P_{0.9}As_{0.1} \tag{2}$$

Extrapolation of eq. (1) to the extremum defined by $\partial P/\partial (\Delta T_C) = 0$ gives a critical pressure $P_c = 13.3$ kbar above which the ground state of Fe_2P should no longer be ferromagnetic.

Surprisingly, there was no significant change with pressure in the magnetization per molecule, μ, at 58K. Examination of μ vs the applied field H at T = 58K showed no appreciable change on passing from 1 atm to 10 kbar pressure. However, measurement of μ vs T/T_C for Fe_2P gave strikingly different results for 1 atm and 10 kbar. At H = 10 kOe, the inflection in μ vs T occurs at $T \approx 1.07\ T_C$ at P = 1 atm, but at $T \approx 1.15\ T_C$ at P = 10 kbar. Furthermore, an extraordinarily large exchange enhancement of the susceptibility, which extends to temperatures well above T_C, is markedly greater at 10 kbar than at 1 atm. At P = 10 kbar and $T = 1.02\ T_C$, a plot of μ vs H is extremely nonlinear, resembling the initial magnetization curve of a ferromagnet at $T < T_C$.

DISCUSSION

If the moment of Fe_2P were reduced from $\mu_0 = 3.0\mu_B$ because of conduction-band overlap of E_F, pressure should change z_d, and hence μ_0, more dramatically than T_c. Therefore, we conclude that the moment of Fe_2P is reduced because the molecular fields are not strong enough to empty all the antibonding states of antiparallel spin.

The existence of antiparallel-spin electrons in the ground state would create a ferromagnetic spin-density wave in the magnetically ordered phase (antiparallel-spin excited electrons create spin waves), and a spin-density wave reflects a long-range antiferromagnetic component to the interatomic-exchange interactions. Since longer M-M separations decrease the width of the 3d bands, the relative importance of this antiferromagnetic component must decrease with increasing As concentration x, which would account for the sharp rise with x in T_c (from 221 to 443K)[2] over the interval $0 \leq x \leq 0.33$. On the other hand, pressure would increase the antiferromagnetic component, and the critical pressure P_c presumably marks a transition from a ferromagnetic spin-density wave to a metamagnetic state.

The remarkable susceptibility above T_c in Fe_2P, and its enhancement by pressure, would seem to indicate that T_c is suppressed by pressure more rapidly than is the paramagnetic Curie temperature θ. Suppression of T_c relative to θ by weak, long-range antiferromagnetic interactions has been observed[5] in the metamagnetic thiospinel $Zn[Cr_2]S_4$, which contains localized 3d electrons. In Fe_2P, the ferromagnetic short-range order above T_c must be exceptional and appears to extend well above T_c, although the magnetic interactions are three-dimensional. This behavior is quite different from that found in $CoS_{2-x}Se_x$, where the ferromagnetic moment is also reduced because the bandwidth is too large.[6] The $CoS_{2-x}Se_x$ 3d bands are broadened with x, and the ferromagnetic-to-metamagnetic transition is marked by a reduction in θ that makes $\theta < T_c$.[7]

Within the molecular-field approximation, the paramagnetic Curie temperature θ is given by

$$\theta = (2/3k)S(S+1)\sum_v z_{uv} J_{uv} \tag{3}$$

where z_{uv} is the number of v atoms near-neighbor to a u atom. If the interatomic exchange energy falls off more rapidly than linearly with decreasing atomic separation, then

$$J_{uv} \approx \sum_\ell J^o_{uv\ell} \sum_j \left[1 + \beta^j_{uv}\epsilon_j - \frac{1}{2}(\gamma^j_{uv}\epsilon_i)^2 + \cdots \right] \tag{4}$$

The $J^o_{uv\ell}$ are components of the uv exchange interaction in the unstrained sample (P = 0) and $\beta^j_{uv} > 0$ because an $\epsilon_j < 0$ increases the 3d bandwidth, thereby lowering J_{uv}. The strain at equilibrium is given by[4]

$$\epsilon_j = \sum_i K_{ji} \left[\sum_{u,v} (\partial J_{uv}/\partial\epsilon_i) \vec{S}_u \cdot \vec{S}_v - P + T\sum_k \alpha_k c_{ki} \right] \tag{5}$$

where K_{ji} = cofactor $c_{ij}/$(determinant c_{ij}), $\vec{S}_u$ and $\vec{S}_v$ are the thermodynamic expectation values of the spins at sites u and v, α_k is a thermal-expansion coefficient, and the c_{ki} are elastic constants. Since μ_o appears to be relatively independent of pressure, and hence of ϵ_j, the spin quantum number S in eq. (3) is assumed constant. Furthermore, since $|S| = 0$ at $T > T_c$, it follows that for any temperature $T > T_c$

$$\frac{\Delta\theta}{\theta_o} = -P_1^{-1}P - \frac{1}{2}P_2^{-2}P^2 \qquad (6)$$

where $\Delta\theta = (\theta - \theta_o)$ and θ_o is the value of θ at $P = 0$. If the influence of thermal expansion is neglected, the parameters are $P_1^{-1} \equiv \sum_j A_j$ and $P_2^{-2} = \sum_j \lambda_j^2 A_j^2$, which contain $\lambda_j \equiv \gamma_{uv}^j/\beta_{uv}^j$, $A_j \equiv \sum_i \beta_{uv}^j K_{ji}$, $B_j \equiv \sum_i \beta_{uv}^j K_{ji} \sum_k \alpha_k c_{ki}$. The remarkable susceptibility above T_c in Fe_2P indicates that

$$\theta = T_c(1 + a + p^{-1}P + \cdots) \text{ or } \Delta\theta \approx (1+a)\Delta T_c + T_c p^{-1}P \qquad (7)$$

So long as $\lambda_j^2 \Delta\theta/\theta_o \ll 1$ remains valid, substitution of eq. (7) into eq. (6) gives

$$P = -Q_1 \Delta T_c - Q_1^2 Q_2 (\Delta T_c)^2 \qquad (8)$$

where $Q_1 \equiv (1+a)[(\theta_o/P_1) + (T_c/p)]^{-1}$ and $Q_2 \equiv (\theta_o/2P_2^2)[(\theta_o/P_1) + (T_c/p)]^{-1}$. Comparison of eq. (8) with eqs. (1) and (2) shows that eq. (4) has the correct form and that $Q_1 Q_2 \sim 3 \times 10^{-3} [K]^{-1} \sim \theta_o^{-1}$. Therefore $(P_1/P_2) \sim 1$, or $\lambda_j \sim 1$. If all the constants but θ_o and T_c in $Q_1 Q_2$ are the same for Fe_2P and $Fe_2P_{0.9}As_{0.1}$, the ratio of the respective θ_o are $252 \times 1.7/710 \times 1.2 \approx 1/2$. The measured Curie temperatures at 1 atm are $T_c = 221K$ and $341K$, respectively, which demonstrates the essential self-consistency of the analysis. In fact, the small discrepancy can be qualitatively accounted for by the observation that the pressure sensitivity of T_c, and hence p, is larger in Fe_2P.

REFERENCES

1. J. B. Goodenough, Progress in Solid State Chemistry, Vol. 5, H. Reiss, ed. (Pergamon Press, 1972) Chap. IV; Proceedings of the Winter School in Solid State Chemistry, C. N. R. Rao, ed. (Plenum Press, New York) in press.
2. A. Catalano, R. J. Arnott, and A. Wold, J. Solid State Chem. (in press).
3. A. Roger, Thesis, Univ. of Paris, Orsay (1970).
4. N. Menyuk, J. A. Kafalas, K. Dwight, and J. B. Goodenough, Phys. Rev. 177, 942 (1969).
5. F. K. Lotgering, Proc. Int. Conf. Magnetism, Nottingham 1964, (Inst. Phys. and Phys. Soc., London) p. 533.
6. J. B. Goodenough, J. Solid State Chem. 3, 26 (1971) and its references.
7. Unlike $CoS_{2-x}Se_x$, crystal-field effects influence the magnetic interactions in Fe_2P and may introduce some antiferromagnetic near-neighbor interactions.

MAGNETIC MOMENT DISTRIBUTION IN ORDERED AND DISORDERED FeCo

S. Spooner
Georgia Institute of Technology, Atlanta, Georgia 30332

J. W. Cable
Solid State Division, Oak Ridge National Laboratory
Oak Ridge, Tennessee 37830

ABSTRACT

The magnetic moment distribution of ordered and disordered alloys of FeCo has been determined from single crystals by using polarized neutrons. The asymmetry of the spin density of average atomic moment in both ordered and disordered alloys corresponds to a 52% E_g character which is similar to the 53% E_g character of pure iron. However, in the ordered alloy the spin density of the cobalt moment corresponds to a 63% E_g character and the iron moment has a 45% E_g character. The electron asymmetry of the ordered alloy moments is interpreted qualitatively in terms of a rigid band model using the iron band structure of Wakoh and Yamashita. The effect of ordering is to increase the average local moment by $3\pm1\%$. The local moments in the ordered alloy are $3.2\mu_B$ for iron and $2.4\mu_B$ for cobalt. The non-local moment is $-0.2\mu_B$ per atom for both alloys. Thus, within the accuracy of these experiments, the increase in the bulk magnetization upon ordering can be accounted for by the change in the local moments.

Research sponsored jointly by Georgia Institute of Technology and by the U. S. Atomic Energy Commission under contract with the Union Carbide Corporation.

HYPERFINE FIELDS AT Sn^{119} NUCLEI IN ORDERED Fe-Co AND γ-Fe-Mn

G.P. Huffman and G.R. Dunmyre
U.S. Steel Corp. Res. Lab., MS-98, Monroeville, Pa. 15146

ABSTRACT

Mössbauer results for the temperature dependence of Sn^{119} hyperfine fields in ordered FeCo and in a random antiferromagnetic $Fe_{0.65}Mn_{0.35}$ alloy are presented. In FeCo, Sn atoms apparently strongly prefer to enter the Co site where they exhibit a large hyperfine field (-268 kG at 0°K) which decreases considerably faster with temperature than does the host magnetization. The average Sn^{119} hyperfine field in Fe-Mn is also large ($\sim$175 kG at 0°K) and decreases in an approximately linear fashion with T^2.

INTRODUCTION

The hyperfine fields at dilute Sn^{119} nuclei in the 3d transition metals and alloys show unusual variations in both value and temperature dependence[1-6] and several models have been proposed to explain the experimental results.[2-5] Results for Sn^{119} hyperfine fields in ordered alloys are of particular interest since the Sn impurity can occupy sites with only a few distinct nearest neighbor (nn) configurations.[6] Ordered FeCo has the CsCl structure[7] and should thus provide only two simple nn environments for Sn. It is also of interest to examine the values and temperature dependence of Sn^{119} hyperfine fields in simple antiferromagnetic systems, most previous work having involved ferromagnetic hosts. In this paper, we present the results of Mössbauer measurements of the temperature dependence of the Sn^{119} hyperfine field in ordered FeCo and in the random antiferromagnetic alloy, $Fe_{0.65}Mn_{0.35}$.

EXPERIMENTAL PROCEDURE

Alloys of FeCo and $Fe_{0.65}Mn_{0.35}$ containing 1 at.%Sn (enriched to 87% Sn^{119}) were prepared by arc-melting. Nominal purity of the starting material was better than 99.99 in all cases. The alloys were sealed in quartz in a helium atmosphere, annealed for 3 days at 850°C, and quenched into ice water. The Fe-Mn sample was then cold-rolled and hand polished to a thickness of 1.5 mils after which it was reannealed for 3 hours at 800°C in helium and quenched in ice water. The FeCo sample was crushed to a powder finer than 200 mesh and annealed in helium for 1 week at 480°C, followed by 2 weeks at 450°C, followed by 4 weeks at 400°C and quenched. The Mössbauer spectrometer and temperature variation and control equipment used in this work have been described elsewhere.[1,6] The source was 20 mCi of Sn^{119} in $BaSnO_3$.

RESULTS

The Fe and Co atoms in ordered FeCo have moments of approximately 3.05 and 1.8μ_B, respectively.[8] Thus, one would expect the Sn^{119} spectra to show two distinct hyperfine patterns arising from Sn atoms with 8Fe or 8Co nn. Instead, as shown in Fig.(1), one sees a single set of slightly broadened magnetic peaks and a weak central "doublet" pattern. The magnetic peaks exhibit a field of -268 kG at 0°K (field sign measured at room temperature only) and a room temperature isomer shift of $-.18 \pm .05$ mm/sec with respect to natural Sn. Both the large magnitude of the field and the fact that Sn is more soluble in Fe than in Co indicate that the magnetic peaks probably arise from Sn in Co sites. The broadness of the peaks ($\Gamma \sim 2$ mm/sec) is not unexpected, since the annealing temperature would indicate a long range order parameter of about .95, in which case 81.7, 16.8 and 1.5% of the Sn atoms would have 0, 1 and 2 Co nn, respectively. As in a number of previous cases,[1-6] the temperature dependence of the hyperfine field, shown in Fig.(2), differs considerably from that of the host magnetization, indicated by the dashed line. We have suggested that such behavior can be explained by the existence of a small parasitic polarization on the Sn impurity.[3,4] The size of the Fe moment would seem to indicate a local state density at Co sites with the spin up band full and the spin down band partially filled, which condition is favorable for the formation of such polarization.[4] A local-moment molecular field model described previously[3] gives values of about -162 and -106 kG for the conduction electron and core polarization fields at the Sn

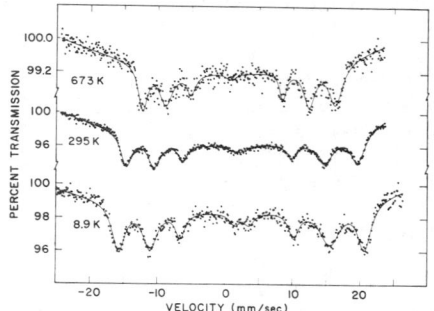

Fig.1. Typical spectra of Sn^{119} in ordered FeCo. For all spectra in this report, the absorber temperature is indicated, the zero of velocity is the position of the $BaSnO_3$ line and the solid curves are computer fits.

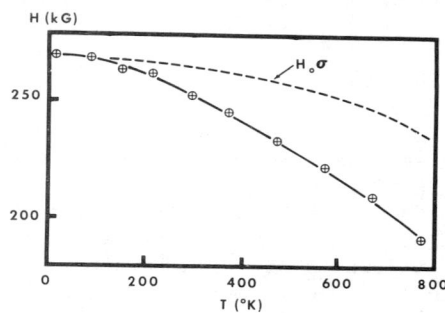

Fig.2. Temperature dependence of the hyperfine field for Sn^{119} in Co sites.

nucleus at T = 0. If the parasitic polarization is carried primarily by p electrons, the local Sn moment would be about $0.35\mu_B$.[9]

To try to determine the origin of the doublet, a second FeCo sample was annealed for 15 days at 600°C and quenched. As seen in Fig.(3), this increased the doublet intensity by a factor of about 5. Two interpretations are possible: (1) the doublet is a true quadrupole doublet with a splitting of 1.53 mm/sec and arises from a Sn rich precipitate; or (2) the solubility of Sn in the psuedo "bcc Co lattice" is greatly increased at 600°C, and the "doublet" is the magnetic spectrum arising from Sn atoms in Fe sites. Although the shape of the doublet is more typical of quadrupole than of magnetic splitting, a preliminary scanning electron microscopy investigation failed to identify any Sn rich particles. A measurement of the dependence of the doublet splitting on temperature and applied field should resolve this question.[10]

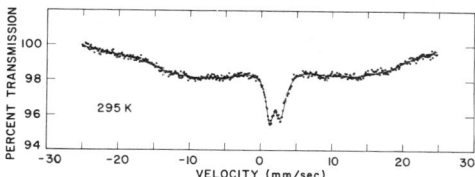

Fig.3. Sn^{119} spectrum of an FeCo sample annealed 15 days at 600°C. The broadness of the magnetic peaks reflects the decreased order at this temperature.

Typical spectra of the $Fe_{0.65}Mn_{0.35}$ (1 at.%Sn^{119}) alloy are shown in Fig.(4). The most probable spin structures of γ-Fe-Mn alloys, which have been touted as possible prototypes of γ-Fe, are discussed in Refs.(11) and (12). While the temperature dependence of the susceptibility indicates that a collective electron description of the antiferromagnetism in these alloys is appropriate, the size of the average moment at 0°K ($1.9\mu_B$), the small Fe^{57} hyperfine fields (30-50 kG), and the increase of T_N with Mn concentration[12] all indicate that a considerably larger moment resides on Mn than on Fe atoms. Since μ_{Fe} is probably $\lesssim 0.5\mu_B$, the relatively large Sn^{119} hyperfine field (∿175 kG at 0°K) probably arises primarily from conduction electron polarization produced by large nn Mn moments. A proper analysis of these rather complicated spectra, using probability distributions which consider the number, type and spin direction of nn moments to the Sn impurity is in progress. For the present, we merely indicate

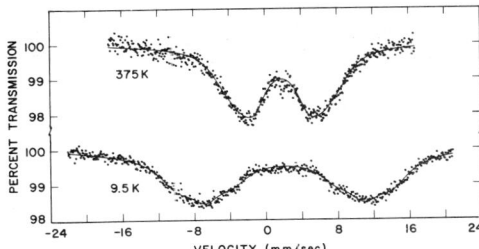

Fig.4. Typical spectra of Sn^{119} in $Fe_{0.65}Mn_{0.35}$.

the temperature dependence of the average Sn^{119} hyperfine field by showing the temperature dependence of the full width of the spectra at half maximum (FWHM) in Fig.(5). This parameter decreases linearly with T^2, indicating that the Sn hyperfine field decreases with temperature in proportion to the sublattice magnetization.

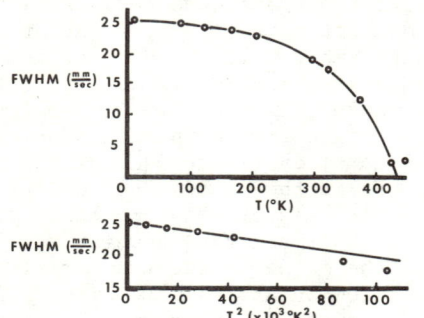

Fig.5. Temperature dependence of the Sn^{119} spectral width for $Fe_{0.65}Mn_{0.35}$.

REFERENCES

1. G.P. Huffman, F.C. Schwerer and G.R. Dunmyre, J.Appl.Phys. 40, 1487 (1969).
2. T.E. Cranshaw, J.Appl.Phys. 40, 1481 (1969).
3. G.P. Huffman and G.R. Dunmyre, J.Appl.Phys. 41, 1323 (1970).
4. G.P. Huffman and G.R. Dunmyre, J.Appl.Phys. 42, 1613 (1971).
5. N.N. Delyagin and E.N. Kornienko, Sov.Phys.-JETP 32, 832(1971); A.E. Balabanov and N.N. Delyagin, Sov.Phys.,-JETP 30, 1054 (1970).
6. G.P. Huffman and G.R. Dunmyre, AIP Conf.Proc. No.5, Magnetism and Magnetic Materials, 1971, p.544 (A.I.P., N.Y. 1972).
7. C.G. Shull and S. Siegel, Phys.Rev. 75, 1008 (1949).
8. D.I. Bardos, J.Appl.Phys. 40, 1371 (1969); M.F. Collins and J.B. Forsyth, Phil.Mag. 8, 401 (1963).
9. A.M. Clogston and V. Jaccarino, Phys.Rev. 121, 1357 (1961).
10. If the doublet turns out to be a Sn rich phase, an alternative explanation for the existence of only one set of magnetic peaks might be that the FeCo order is not CsCl but some other structure in which all sites are equivalent. One can think of several such structures, but the ones which would give the superlattice reflections of Ref.(7) are rather unusual.
11. J.S. Kouvel and J.S. Kasper, J.Phys.Chem.Solids 24, 529 (1962).
12. H. Umebayaski and Y. Ishikawa, J.Phys.Soc. Japan 21, 1281, (1966).

MAGNETIZATION AND ELECTRICAL RESISTIVITY OF NEAR EQUI-ATOMIC FeRh
ALLOYS IN FIELDS UP TO 370 kOe

C.J. Schinkel and R. Hartog
Natuurkundig Laboratorium der Universiteit van Amsterdam, The Netherlands

ABSTRACT

The magnetization curves and electrical resistivities of $Fe_{50+x}Rh_{50-x}$, with x=0, 0.2, 0.5 and 2, taken at 4.2 K in magnetic fields up to 370 kG, are reported. It is shown that the saturation magnetization corresponds to 5 μ_B per formula unit. The role of excess iron atoms can be characterized by the building up of giant moments in the antiferromagnetic matrix, their magnitude being 135 μ_B per excess iron atom. The influence of a magnetic field on the resistivity is extremely large.

INTRODUCTION

Many more or less contradictory results about the properties of near equi-atomic FeRh alloys can be found in the literature[1]. Despite the extensive collection of experimental data a full understanding of the antiferro-ferromagnetic transition is not accomplished. In an attempt to clarify this situation, and in particular the influence of high magnetic fields on the transition, we recently started a systematic study of the magnetization and electrical resistivity in magnetic fields up to 400 kOe. In this work we will present and discuss the data taken at 4.2 K on alloys of the composition $Fe_{50+x}Rh_{50-x}$ with x=0, 0.2, 0.5 and 2, respectively.

EXPERIMENTAL

The starting materials were iron (rods, Johnson Matthey, Specpure) and rhodium (sponge, Johnson Matthey, Specpure). The rhodium sponge was first arc-melted in order to get bulk metal. Then weighed amounts of iron and rhodium were arc-melted together in an argon atmosphere, the weight losses being less than 0,1%. Finally the button shaped ingots (ca 4.5 gram each) were annealed in vacuum sealed silica capsules at 1050°C for 2 days. Samples of the required dimensions for the various measurements were cut by means of spark-cutting machine. The high fields were generated by the high field installation in our institute[2], these fields can be kept constant during 0.1 sec. or can be varied in time, in a controlled linear way. The samples are immersed in liquid helium. The magnetizations are measured in an inductive coil system and for the resistance measurements use was made of the apparatus described by Chang[3].

RESULTS

The magetization curves of the alloys at 4.2 K are shown in figu-

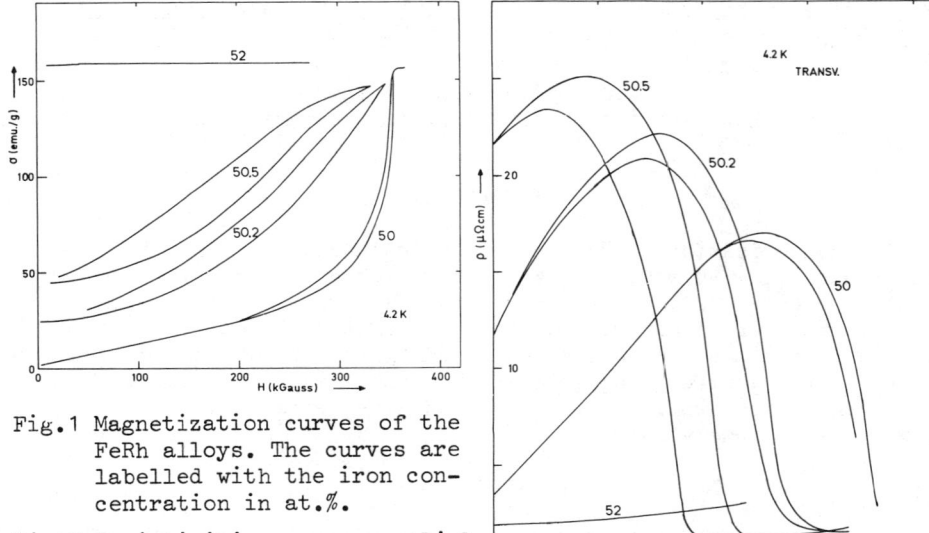

Fig.1 Magnetization curves of the FeRh alloys. The curves are labelled with the iron concentration in at.%.

Fig.2 Resistivities versus applied magnetic field.

re 1. Although the curves for $Fe_{50.2}Rh_{49.8}$ and $Fe_{50.5}Rh_{49.5}$ seem to have a large intercept with the magnetization axis no appreciable remanence was measured. The $Fe_{52}Rh_{48}$ alloy has a remanence of about 0.1 of its saturation magnetization. The resistivities as a function of applied transverse magnetic field at 4.2 K are given in figure 2. In the stoichiometric compound we find a large difference between the longitudinal and the transverse magnetoresistance, whereas in the other alloys the difference was much smaller. Magnetization and resistivity measurements as a function of temperature showed a small shift of the transition temperature with increasing iron content up to 50.5 at.%. Furthermore, the magnetization curve at 77 K of the stoichiometric compound had a broader hysteresis than at 4.2 K but the field at which the magnetization saturated was unchanged.

DISCUSSION AND CONCLUSIONS

Although the work described here is by no means a complete systematic investigation yet and although we can not give a consistent explanation of the results at the moment, some interesting conclusions can be drawn.

First about the saturation magnetization: the experimental value of 156 emu./gram for $Fe_{50}Rh_{50}$ corresponds to about 5 μ_B per formula unit, a very high value in view of the 4 μ_B as determined by neutron diffraction experiments at room temperature, and in view of the usual behaviour of iron as well as rhodium atoms in metals.

Secondly, the effect of excess iron atoms in the antiferromagnetic matrix can be characterized by a giant moment inducement of about

135 μ_B per excess iron atom, as determined from the intercept of the magnetization curve of $Fe_{50.2}Rh_{49.8}$ with the σ-axis. This localized character is still more evidenced by the value of the field needed for saturation, which is as high as for the stoichiometric alloy. The magnetization of $Fe_{52}Rh_{48}$ is practically constant with respect to the applied magnetic field H for H > 10 kGauss, thus it seems that 2% excess iron atoms are enough to convert all of the antiferromagnetic matrix into a ferromagnet.

The resistivity curves shown in figure 2 show first, that the resistivity in zero applied field at 4.2 K is determined mainly by magnetic scattering. A second striking feature is the very high positive magnetoresistance in the antiferromagnetic phase, in particular when compared with the ferromagnetic phase, indicating that the transition from antiferromagnetism to ferromagnetism involves a drastic change in electronic structure.

Finally the rapid decrease of the resistivity requires attention, and in particular the relatively low fields at which it happens. When one combines magnetization and resistivity curves it follows that for all of the alloys the resistivity falls down at the "ferromagnetic value" for a magnetization of about 100 emu./gram, which is significantly lower than the saturation magnetization and still in the hysteresis region. An admittedly somewhat naive explanation might be that at 100 emu./gram the highly resistive antiferromagnetic parts of the sample are short-circuited by the much less resistive ferromagnetic part.

REFERENCES

1. See the references of J.A. Ricodeau and D. Melville, J. Phys.F. Metal.Phys. 2, 337 (1972).
2. L.W. Roeland, F.A. Muller and R. Gersdorf, Coll.Int.C.N.R.S. 166, 175 (1967).
3. K.H. Chang, thesis Universiteit van Amsterdam (1972).

ANOMALOUS MAGNETIC PROPERTIES OF MICTOMAGNETIC Fe-Al ALLOYS*

G. P. Huffman

U.S. Steel Corp. Res. Lab., MS-98, Monroeville, Pa. 15146

ABSTRACT

We report Fe^{57} Mössbauer results for CsCl and DO_3 ordered $Fe_{1-x}Al_x$ alloys (x = .23 to .33) and a molecular field theory which explains all magnetic properties of the Fe-Al system reasonably well. At low temperatures, the hyperfine fields decrease as $T^{3/2}$ for $x \lesssim .27$ and as T^2 for $x \gtrsim .33$.[1] For $.27 \lesssim x \lesssim .33$, the fields fall rapidly from their maximum values at T = 0 to sharp minima at intermediate temperatures (T_{min} = 140°K for x = .3), rise to secondary maxima, then decrease again. The theory assumes a strong d electron exchange interaction which aligns nearest neighbor spins parallel, while RKKY interactions between separated spins are predominantly antiferromagnetic (a.f.). By random chance, sizeable regions are built up in which all spins are parallel. For $.27 \lesssim x \lesssim .33$, the mean radii of these regions are ~160 to 50°A and one has a system of superparamagnetic (s.p.) "particles" separated by narrow regions in which a.f. RKKY coupling is dominant. Above the "Néel points" of the a.f. regions, but below the s.p. blocking temperatures, the particles behave as small single ferromagnetic domains. For $60 \lesssim T \lesssim 200°K$, the spins in the a.f. regions become ordered and exert exchange anisotropy fields ~20 to 200 gauss on the particles, which can either rotate particle moments directly or lower energy barriers sufficiently to allow thermally excited rotation. This causes the particle moments to flip rapidly, leading to hyperfine field minima, zero remanence and s.p. behavior. Further cooling increases the magnetization and magnetocrystalline anisotropy energy of the a.f. regions, which locks the particle moments and give rise to large hyperfine fields and broad hysteresis loops which are shifted if the system has been field cooled. For $x \lesssim .27$, the parallel spin regions are large enough to behave as normal ferromagnetic domains while for $x \gtrsim .33$, they are small enough that their blocking temperatures are below the Néel temperatures of the a.f. regions and a.f. behavior dominates.

*(A somewhat more detailed account of this work will appear in the Proceedings of the International Symposium on Amorphous Magnetism, held at Wayne State University, August 1972 [to be published by Plenum Press]).

1. G.P. Huffman, J.Appl.Phys. **42**, 1606 (1971).

LOCAL INTERACTIONS AND SPIN TRANSFER MECHANISMS IN THE HEUSLER-TYPE ALLOYS $Pd_{1+x}MnSb$ AND $Pd_2MnSb_{0.9}Sn_{0.1}$

L. J. Swartzendruber
National Bureau of Standards, Washington, D.C. 20234

B. J. Evans
University of Michigan, Ann Arbor, Michigan 48104

ABSTRACT

The magnetic hyperfine fields at ^{121}Sb, $H_{eff}(Sb)$, in the Heusler-type alloys $Pd_{1+x}MnSb$ are found to be strongly dependent on x, varying from about 600 kG for x=1 to about 300 kG for x=0. As x is decreased from unity, the single unique value observed in Pd_2MnSb is replaced by a distribution in $H_{eff}(Sb)$, with an average value which decreases in a regular manner with the number of Pd vacancies. In $Pd_2MnSb_{0.9}Sn_{0.1}$, $H_{eff}(Sn)$ is found to be similar in magnitude to $H_{eff}(Sn)$ in Pd_2MnSn. Both these results indicate that local spin transfer mechanisms are important in determining the magnitude and sign of H_{eff} at the Sb site in Heusler-type alloys.

INTRODUCTION

Ordered intermetallic compounds with the composition X_2MnZ are generally referred to as Heusler alloys. The (often incomplete) ordering is of the Cu_2MnAl type ($L2_1$) and, for example, X is Cu, Ni, Pd, or Co and Z is Al, In, Sn or Sb. Most of these alloys are ferromagnetic, with exceptions such as Pd_2MnIn which is antiferromagnetic.[1] A related series of compounds with the composition X MnZ exhibit order of the MgAgAs (Cl_b) type and are also mostly ferromagnetic. The crystal structures are illustrated in Fig. 1. Measured magnetic moments range between 3.5 and 4.5 μ_B per formula unit and are (with exceptions[2] when X is Co) localized on the Mn site[3]. In these alloys the Mn-Mn separations are large and indirect exchange interactions are an important factor in determining their magnetic properties[4].

In a previous study[5] we have measured $H_{eff}(Sb)$ in $Ni_{1+x}MnSb$ for $1 \geq x \geq 0$. These alloys exhibit a continuous transition between the $L2_1$ and Cl_b structures[6]. The saturation value of $H_{eff}(Sb)$ is approximately +300 kG, with a relatively small dependence on x. Here we present results on the isoelectronic alloys $Pd_{1+x}MnSb$. In contrast to the $Ni_{1+x}MnSb$ alloys, $H_{eff}(Sb)$ in $Pd_{1+x}MnSb$ has a considerable dependence on x. The Sb and Sn hyperfine fields in $Pd_2MnSb_{0.9}Sn_{0.1}$ were also measured. While the presence of the Sn decreases the average magnitude of the Sb hyperfine field, the average Sn field has about the same magnitude as that found at the Sn site in Pd_2MnSn.

EXPERIMENTAL

Alloys of $Pd_{1+x}MnSb$ for x=1.00, 0.75, 0.50 and 0, and an alloy of $Pd_2MnSb_{0.9}Sn_{0.1}$ were prepared by arc melting weighed quantities in a gettered argon atmosphere. The ingots were homogenized at

920 K for four days and water quenched. A metallographic examination followed. Predominantly single phase alloys with less than 2% of unidentified second phases were revealed in each case. The ingots were crushed to a powder and transmission Mössbauer spectra obtained on samples containing approximately 350 mg of the powder embedded in a plastic disc 2.5 cm in diameter. Sources were (^{121}Sn)BaSnO$_3$ for the ^{121}Sb spectra and (^{119}*Sn)BaSnO$_3$ for the ^{119}Sn spectra. X-ray measurements, performed on the crushed powder, revealed no extraneous phases. The measured lattice constants are listed in Table I. Crystallographic order, as evidenced from the resolution of the α_1,α_2 doublet of the (224) x-ray peak, increased as x decreased.

RESULTS

For x=0 and x=1, satisfactory least squares fits could be obtained to the Pd$_{1+x}$MnSb spectra using a single-valued magnetic hyperfine field pattern. For intermediate values of x, a hyperfine field distribution is evident. These spectra were fitted using two single-valued magnetic hyperfine field patterns. Such fits are probably not unique. However they give as satisfactory a representation of the distribution as is justified by the scatter in the data, and give accurate values for the average hyperfine field values. They can also be used to prove that, for intermediate values of x, the spectra are not simply the sum of the Pd$_2$MnSb and PdMnSb spectra. A similar fit was used to represent the ^{121}Sb spectrum from the Pd$_2$MnSb$_{0.9}$Sn$_{0.1}$. The ^{119}Sn spectrum from the same sample was fitted to a single valued hyperfine field pattern and thus only an average hyperfine field value was obtained. Table I lists results for the ^{121}Sb spectra. Table II compares average hyperfine fields for a number of Heusler-type alloys. Table II also lists Curie temperatures. (In all cases, the hyperfine field at 100K is within 10% of its value extrapolated to 0 K.) The ^{121}Sb spectra and least squares fits are shown in Fig. 1. Some of the least squares fits show a slight assymetry which is due to a quadrupole splitting used as a free parameter in the fitting procedure. In no case did the quadrupole splitting exceed 0.2 mm/sec.

DISCUSSION

The Sb hyperfine fields in Ni$_2$MnSb, NiMnSb and PdMnSb are all approximately 300 kG. In contrast, Sb in Pd$_2$MnSb has a hyperfine field nearly twice as large[9]. This large field value is confirmed here in another sample and has also been measured using the spin echo technique[10]. For intermediate values of x in Pd$_{1+x}$MnSb, the ^{121}Sb ME spectra are not simply superpositions of spectra for Pd$_2$MnSb and PdMnSb. The introduction of Pd vacancies into the Pd$_2$MnSb lattice, or the substitution of Sn for Sb, tends to decrease H$_{eff}$(Sb).

It has recently been recognized[5,8,11,12] that the virtual bound state-spin polarization model of Caroli and Blandin[13] fails to give a complete description of the hyperfine fields in Heusler alloys. This failure is most evident for the positive hyperfine fields observed in a number of alloys, e.g. at Sn in Ni$_2$MnSn, for which the model predicts a negative hyperfine field.[2] Considering the Heusler alloys as β-phase electron compounds[14] with an electron-

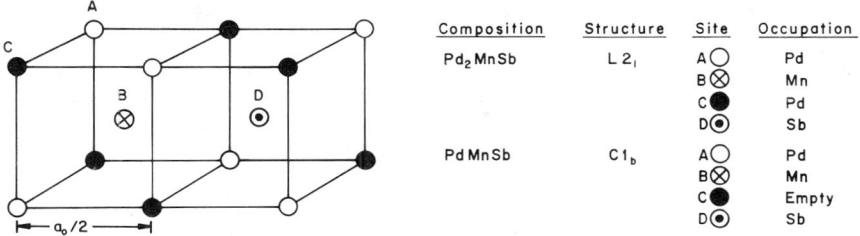

FIGURE 1. Schematic illustration of the $Pd_2MnSb(L2_1)$ and $PdMnSb(Clb)$ structures.

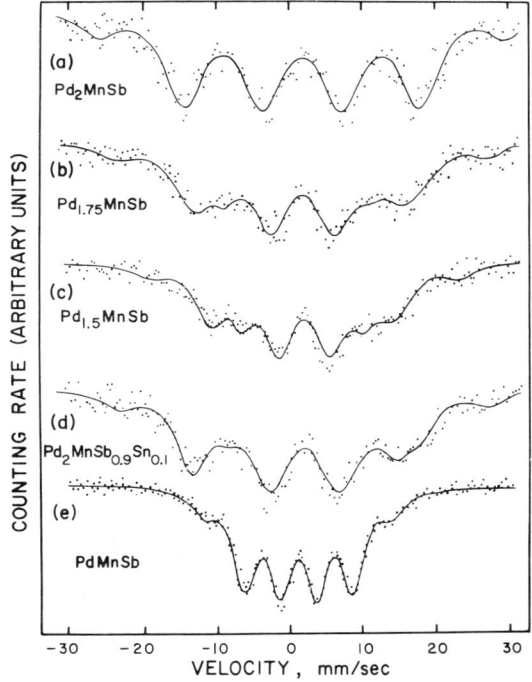

FIGURE 2. ^{121}Sb Mössbauer effect spectra for several Heusler-type alloys at 100K. The solid line is a least square fit to the data points as described in the text.

to-atom ratio near 3/2, the Caroli-Blandin theory predicts (within the customary free-electron picture) a negative hyperfine field (for such atoms as Sn, Sb, and In) at the D-site for every alloy. The existence of positive hyperfine fields indicates that mechanisms other than the type of conduction electron polarization described by Caroli and Blandin must be contributing to the observed hyperfine field. The results obtained here on $Pd_{1+x}MnSb$ suggest that local interactions involving the Pd atoms play a considerable role in transferring the hyperfine field from a Mn to an Sb atom. These local interactions could be responsible for the very large hyperfine field found for Sb in Pd_2MnSb. The sign of $H_{eff}(Sb)$ in PdMnSb and Pd_2MnSb has not been determined, although the variation of $H_{eff}(Sb)$ with x in $Pd_{1+x}MnSb$ indicates that the sign is the same in PdMnSb and Pd_2MnSb.

TABLE I

Parameters obtained in least squares fits to the 100 K ^{121}Sb Mössbauer Spectra of several Pd-based Heusler-type alloys. Two magnetic hyperfine field patterns with identical linewidths and isomer shifts were assumed. Isomer shift is with respect to an InSb absorber at 100 K. Lattice constants determined by x-ray analysis are also listed.

	Ha kG (±5)	Hb kG (±5)	Area Fraction of pattern b (±0.05)	FWHM mm/sec (±0.2)	Isomer Shift mm/sec (±0.2)	χ^2 per point	Lattice Constant nm
Pd_2MnSb	585	–	0	4.9	1.2	1.26	0.6428
$Pd_{1.75}MnSb$	526	367	0.42	4.3	1.1	1.11	0.6381
$Pd_{1.5}MnSb$	443	299	0.49	3.6	0.9	1.19	0.6338
PdMnSb	–	307	1.0	3.5	0.7	1.14	0.6231
$Pd_2MnSb_{0.9}Sn_{0.1}$	526	336	0.29	4.4	1.2	1.06	0.6416

TABLE II

Average Z-ion hyperfine fields, H_{eff}^{avg}, at 100 K in several X_2MnZ and XMnZ Heusler-type alloys. Signs of the fields and Curie temperatures are indicated where known.

Alloy	H_{eff}^{avg} kG(±5)	Reference	T_c K	Reference
Pd_2MnSb	585	[7]	247	[1]
PdMnSb	305	[7]	500	[15]
NiMnSb	+307	[5]	720	[16]
Ni_2MnSb	+291	[5]	360	[17]
Ni_2MnSn	+ 87	[8]	344	[17]
Pd_2MnSn	− 35	[8]	189	[1]
$Pd_2MnSb_{0.9}Sn_{0.1}$	471(Sb)	[7]		
	37(Sn)			

ACKNOWLEDGMENTS

The authors wish to thank Dr. L. H. Bennett and Dr. R. E. Watson for useful discussion, R. D. Robbins for technical assistance, D. P. Fickle for sample preparation, and C. H. Brady for metallographic examination.

REFERENCES

1. P. J. Webster and R. S. Tebble, Phil. Mag. **16**, 347 (1967).
2. J. P. Felcher, J. W. Cable and M. K. Wilkinson, J. Phys. Chem. Solids **24**, 1663 (1963); P. J. Webster, Phil. Mag. 16, 347 (1967).
3. P. J. Webster, J. Phys. Chem. Solids **32**, 1221 (1971).
4. C. Zener and R. R. Heikes, Rev. Mod. Phys. **25**, 191 (1953).
5. L. J. Swartzendruber and B. J. Evans, AIP Conf. Proc. **5**, 539 (1972).
6. L. Castelliz, Mh. Chem. **82**, 1059 (1951).
7. This work. Figures represent the average hyperfine field.
8. D. J. W. Geldart, C. C. M. Campbell, P. J. Pothier and W. Leiper, Can. J. Phys. **50**, 206 (1972).
9. L. J. Swartzendruber and B. J. Evans, Phys. Letters **38A**, 511 (1972).
10. S. K. Malik, R. Vijayaraghavan, Le Dang Khoi and P. Veillet, Phys. Letters **40A**, 161 (1972). Structure is apparent in the spin echo spectrum due to Pd-Sb disorder. However, H_{eff}(Sb) is nearly the same on either site, resulting in only a slight broadening of the ME spectrum. The slight differences between the present value for H_{eff}(Sb), the average spin-echo frequency, and previous ME results [ref. 9], probably reflects small differences in the amount of disorder in different samples.
11. J. M. Williams, J. de Physique **32**, C1-790 (1971).
12. R. Segnan, W. A. Ferrando, D. Sweger, and P. J. Webster, J. de Physique **32**, C1-792 (1971).
13. B. Caroli and A. Blandin, J. Phys. Chem. Solids **27**, 503 (1966).
14. B. R. Coles, W. Hume-Rothery and H. P. Myers, Proc. Roy. Soc. (London) **196A**, 125 (1949).
15. K. Endo, J. Phys. Soc. Japan 29, 643 (1970).
16. L. Castelliz, Mh. Chem. **82**, 1059 (1951).
17. P. J. Webster, Contemp. Phys. **10**, 559 (1969).

SPIN-DEPENDENT TRANSPORT PROPERTIES OF SnTe-MnTe SYSTEMS

A. Ghazali, M. Escorne, H. Rodot and P. Leroux-Hugon
Laboratoire de Physique des Solides, CNRS, 92 Bellevue, France

ABSTRACT

Resistivity and Hall effect have been studied in monocrystalline samples of ferromagnetic semiconducting alloys $Sn_{1-x}Mn_xTe$, with $0 < x < 15$ at. %, at low temperature. Contrary to a previous claim, these materials exhibit a resistance maximum around the magnetic ordering point ; the peak position varies with Mn concentration x as does the magnetic ordering point itself. This maximum, due to critical scattering, is suppressed by application of a moderate magnetic field. From these data, we obtain an estimate of the s-d like exchange coupling constant. Hall resistivity curves show an ordinary and an anomalous[10] contribution. The anomalous Hall resistivity, ascribed to the skew scattering of spin-polarized current carriers, is compared with a theoretical estimate.

SnTe, a highly degenerate p-type semiconductor provides a suitable matrix to dilute magnetic ions, in order to study the carrier-magnetic moment interaction in magnetic semiconductors. Ferromagnetic ordering has been reported to occur in SnTe-MnTe solid solutions[1,2] ; the ordering temperature depends on the Mn concentration[1] but also on the carrier concentration[2] as expected if the coupling between magnetic moment were (partially) due to their indirect interaction throught the carriers. Resistivity[3] and anomalous Hall effect[1] measurements have also been reported. We have undertaken a more detailed study of the transport properties of this material in order to find out how the scattering of the carriers is modified by their interaction with magnetic moments.

Monocrystalline samples of $(Sn_{1-x}Mn_x)_{.97}Te$ alloys have been cut in ingots grown by a Bridgman technique. By electron microprobe analysis, we have checked the homogeneity of the solid solution and measured the local Mn concentration. Transport measurements have been performed on an a.c. bridge using a lock-in detector. Some magnetic measurements have also been carried out on the same samples using a vibrating sample magnetometer.

Some resistivity measurements are plotted on figure 1. One first notices that increasing the Mn concentration increases the resistivity. Second, at the lowest temperature, where the resistivity of the pure matrix does no longer vary, a slight increase of the alloy resistivity with increasing temperature is apparent. Should the Mn concentration increase, the temperature range of the resistivity variation shifts toward higher temperatures as the magnetic ordering point itself does. Finally some samples exhibit a resistivity maximum around the ordering point ; this maximum we attribute to the critical scattering of carriers by spin fluctuations[4]. This interpretation is supported by the fact that a moderate magnetic field (which suppresses the divergency of the spin

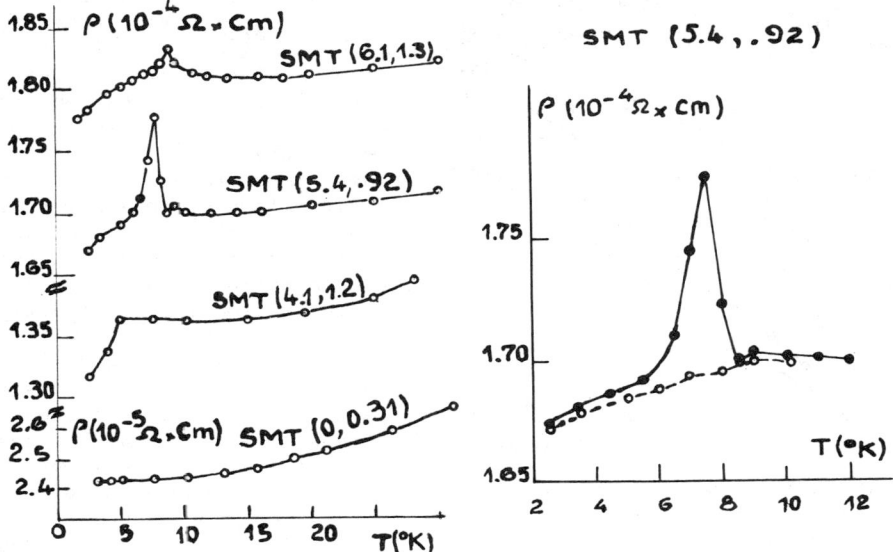

Fig. 1. Resistivity versus temperature for different alloys. Samples are hereafter denoted as SMT (x, y) : x - Mn concentration in at. % , y - apparent carrier concentration in unit of $10^{21} cm^{-3}$.

Fig. 2. Influence of the magnetic field on the critical behaviour of the resistivity - full line, H = 0 - broken line H = 1kOe.

fluctuation range) does suppress the resistivity maximum (Fig. 2).

In discussing these measurements, one has to take into account the complex band structure of SnTe[5] which we suppose unmodified by the addition of Mn. One can then calculate the carrier concentration dependence of the mobility in the SnTe matrix, assuming[6] the scattering to be due to short range scatterers, with a concentration N_L equal to the carrier concentration and specified by a potential W_L which we shall suppose constant within the volume Ω of the unit cell and vanishing elsewhere, in such a way that the relaxation time τ_L is :

$$\frac{1}{\tau_L} = \frac{m^* k_F}{\pi \hbar^3} N_L (W_L \Omega)^2$$

To estimate the mobility in SnTe - MnTe alloy we shall add to the previous mechanism the scattering associated with Mn ions which is specified by a (spin independent) potential W_M and an exchange term under the usual form :

$$- \sum_n J \vec{s} \cdot \vec{S}_n \qquad (1)$$

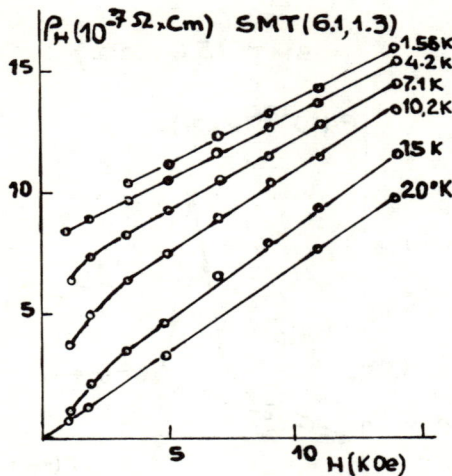

Fig. 3. Hall resistivity versus magnetic field at different temperatures for SMT (6.1,1.3).

with different Mn concentrations ; the figure 5 gives the magnetization curve of the same samples at 4.2°K. These data clearly show that an anomalous component is superimposed on the ordinary Hall voltage. An extrapolation to low field allows us to separate the ordinary Hall effect from which we deduce an (apparent) carrier concentration ; this carrier concentration is not directly modified by the addition of Mn.

If we ascribe the anomalous Hall effect to the skew scattering of the current carriers, then a straightforward extension of a theory developped by two of the authors[9] allows us to estimate the associated Hall angle. This theory emphasizes the role of the mixing of spin components in the wave functions, mixing which is due to the interband spin-orbit coupling ; for the sake of simplicity we have used an estimation of the mixing coefficients which is valid for the standard band structure of InSb but with parameters (E_g = 0,27eV, Δ = 0,47eV) suitable for SnTe. Two scattering mechanisms then contribute to the skew scattering : the first is the scattering by the

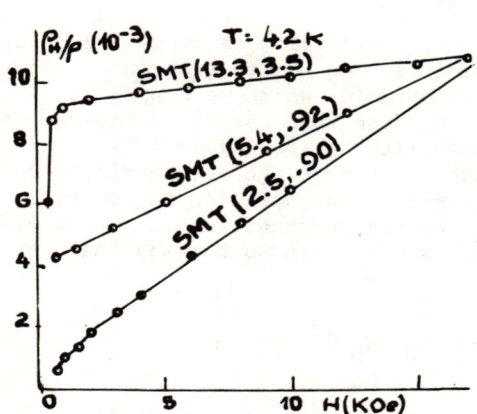

Fig. 4. Hall angle versus magnetic field at 4.2°K for different samples.

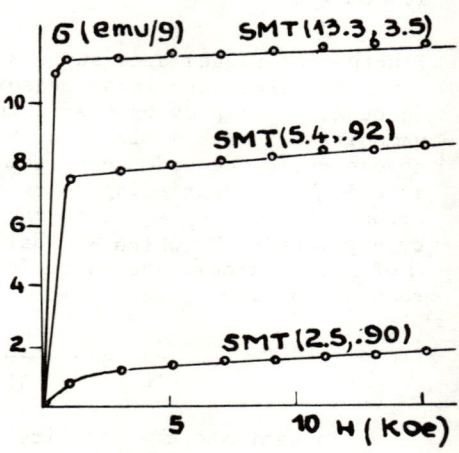

Fig. 5. Magnetization versus magnetic field at 4.2°K for different samples.

where $\vec{S}_n$ is the localized spin carried by the Mn ion, $\vec{s}$ the carrier spin and J is a short range potential. J and W_M are defined as W_L is. The corresponding relaxation time is then :

$$\frac{1}{\tau_M} = \frac{m^* k_F}{\pi \hbar^3} N_M \left\{ (W_M \Omega)^2 + \left(\frac{J\Omega}{2}\right)^2 [<s^2> - <s>^2] \right\}$$

Comparing the mobility for $T > T_c$ ($<s>^2 = 0$) and $T \sim 0$ ($<s>^2 = S^2$), one may extract the (small) spin dependent contribution to the resistivity. This calculation shows that the value of the potential W_M associated with Mn ions is close to the one of native defects W_L which fact seems to indicate that the Mn ions are not resonant scatterers. One may also get an estimate of the exchange constant $J = 0.6$ eV. This value corresponds to the one which is commonly used in the s-d exchange scattering problem in magnetic semiconductors[7]. It also roughly agrees with the estimate given by a simple molecular field calculation based upon the experimental value of the ordering temperature : in a rigid (parabolic) band approximation, assuming a ferromagnetic ordering, the internal energy per unit volume as a function of reduced magnetization σ, is[8] :

$$\epsilon_{int} = \epsilon^o_{int} + \frac{6}{5} n \epsilon^o_F - \frac{3}{2} n N_M^2 S^2 \frac{(J\Omega)^2}{4\epsilon^o_F} \sigma^2$$

where 2n is the carrier concentration.

On the other band, the entropy for the spin system is :

$$\mathscr{S}_M = N_M k_B \left[\ln(2S+1) - \frac{3}{2} \frac{S}{S+1} \sigma^2 \right]$$

the magnetization dependent part of the electronic entropy being negligeable in front of the spin one. We may then write down the free energie expansion :

$$\mathscr{F} = \mathscr{F}^o - \frac{3}{2} n N_M^2 S^2 \frac{(J\Omega)^2}{4\epsilon^o_F} \sigma^2 + \frac{3}{2} k_B T N_M \frac{S}{S+1} \sigma^2$$

The σ^2 term vanishes for the ordering temperature :

$$T_c = \frac{1}{k_B} \frac{(J\Omega)^2}{4\epsilon^o_F} n N_M S(S+1)$$

As a numerical example, let us take SMT (5.4, .92), with $T_c = 7.5$ K, we find $J = 0.4$ eV.

This result seems to indicate the indirect exchange alone[8] may account for the magnetic properties.

We have plotted on the figure 3, the Hall resistivity of a sample as a function of applied magnetic field at different temperatures and, on the figure 4, the Hall angle at 4.2°K for samples

local potentials of either native or Mn defects ; the second involves a combination of the spin-independent part and of the exchange part of the scattering associated with Mn ions. The anomalous Hall angle is given, at the lowest temperature, by :

$$\frac{\rho_H}{\rho} = \frac{m^* k_F b}{6\pi \hbar^2} \left[N_L (W_L\Omega)^2 + N_M (W_M\Omega)^2 + N_M \left(\frac{J\Omega}{2}\right)^2 s \right]^{-1}$$

$$\times \left\{ \frac{\Delta n}{n} (2\sqrt{2}c - b) \left[N_L (W_L\Omega)^3 + N_M (W_M\Omega)^3 \right] \right.$$

$$\left. + (2\sqrt{2}c + b) N_M (W_M\Omega)^2 \left(\frac{J\Omega}{2}\right) s \right\}$$

where b and c are the mixing coefficient and $\Delta n/n$ refers to the carriers polarization associated with the exchange term (1). With the quoted values of the parameters, the estimates agree with experimental data ; let us take the sample SMT (5.4, .92) as an example, then $(\rho_H/\rho)_{exp} = 4.7 \times 10^{-3}$ and $(\rho_H/\rho)_{calc.} = 7.6 \times 10^{-3}$.

In summary, we have shown that the transport properties of ferromagnetic semiconducting alloys SnTe-MnTe include a spin-dependent scattering mechanism which may be accounted for in terms of s-d like exchange coupling between the localized magnetic moments and the current carriers. We would like to emphasize that we have got in this way a fair estimate of the anomalous Hall angle.

REFERENCES

1. J. Cohen, A. Globa, P. Mollard, H. Rodot and M. Rodot, J. de Phys. (Paris), 29, C4 - 143 (1968).
2. M. Mathur, D. Deis, C. Jones, A. Patterson, W. Carr Jr and R.C. Miller, J. Appl. Phys., 41 - 1005 (1970).
3. M. Mathur, D. Deis, C. Jones, A. Patterson and W. Carr Jr, J. Appl. Phys. 42, 1693 (1971).
4. P. de Gennes and J. Friedel, J. Phys. Chem. Sol., 4, 71 - (1958).
5. see R. Allgaier and B. Houston, Phys. Rev. B, 5, 2186 (1972).
6. Y. Kanai, R. Nii and N. Watanabe, J. Appl. Phys. suppl. vol. 32, 2146 (1961).
7. C. Haas, Phys. Rev. 168, 531 (1968).
8. K. Yosida, Phys. Rev., 106, 893 (1957).
9. P. Leroux-Hugon and A. Ghazali, J. Phys. C (Sol. Stat. Phys.), 5, 1072 (1972).
10. By anomalous Hall effect, we mean the magnetization dependent part of the total Hall effect which is also referred to as extraordinary or spontaneous.

MAGNETIC STRUCTURE OF DO_{19} TYPE COMPUNDS

G.J. Zimmer and E. Kren
Central Research Institute for Physics
Budapest 114, P.O.B. 49, Hungary

ABSTRACT

All of the magnetic structures possible in a DO_{19} lattice with magnetic and crystallographic cells of the same dimension are compiled. Two new models, related by a rotation of 90° of magnetic moments, are proposed for the magnetic structure of Mn_3X (X=Ga, Ge, Sn) compounds. As essential feature of these models is the orthorhombic magnetic symmetry. The presence of the weak ferromagnetism and of the hyperfine field at the tin site can be easily interpreted by a simple picture of the interactions present. The neutron diffraction experiments are also fully explained. The only interaction which is able to stabilize the proposed structure is an antisymmetric exchange of proper sign and magnitude, clearly indicating the importance of Dzyaloshinski-Moriya type interaction in these metallic compounds.

INTRODUCTION

The unit cell of the hexagonal DO_{19} lattice (space group $P6_3mmc$) contains two M_3X formula units, with M and X atoms at 6h and 2d positions, respectively. The projection of 4 unit cells on a basal plane is shown in Figure 1.

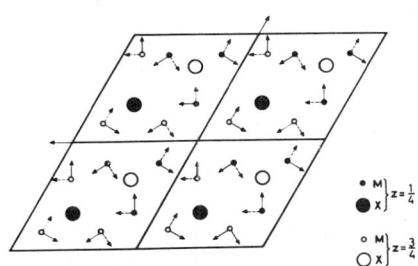

Fig.1. The projection of 4 unit cells on a basal plane showing the proposed magnetic structures.

Among the magnetically ordered representatives of DO_{19} both ferromagnetic (Fe_3Sn) and weakly ferromagnetic (Mn_3X, X=Ga^2,Ge^3, Sn^3) examples are known. The ferromagnetism of Mn_3Sn is particularly weak and disappears entirely on cooling to low temperatures.[3,4] There is a quite substantial hyperfine field at Sn nuclei (45 kOe at 0 K)[5] which is only slightly affected by the disappearance of net magnetization.

Neutron diffraction measurements on Mn_3X powder

samples has revealed[2-4,6] that the crystallographic and magnetic unit cells are of identical dimensions (except in the low-temperature phase of Mn_3Sn, which we shall not deal with here); the X-atoms do not carry a detectable magnetic moment, while the Mn moments are coplanar, those of Mn atoms related by an inversion parallel, and the structure is triangular antiferromagnetic, i.e. the magnetic moments of Mn atoms in the same basal plane are at an angle of approx. 120°. The plane of the magnetic moments was initially found to be perpendicular[3] to the basal plane, but later, in a series of samples, parallel[2,6] to it.

The triangular models proposed by Kádár and Krén[6] are shown in Table I. as structures 2 and 3. While these structures were observed in $Mn_3RhN_{0.20}$ and in $Mn_3PtN_{0.25}$ respectively, the diffraction intensities for Mn_3X could only be explained, according to them, either by the simultaneous presence of both structures (i.e. the triangle rotated by 45°) or by an admixture in equal proportions of grains displaying one or the other triangular structure.

There are, however, serious objections against these mixed models. One would hardly expect to find a physical mechanism that not only accounts for a 45° rotation or for a systematic variation of the easy and hard directions (which would permit admixing) but which is also equally effective in samples of different compositions. Further, there is no obvious reason why the triangle should distort to produce weak ferromagnetism, since the magnetic moments all point in an easy direction. Moreover, in both suggested models the hyperfine field at the tin sites can only be due to such distortion and thus would have to reflect changes of net magnetization brought about by heat treatment or by the transition to the low-temperature phase. No such effect was observed.[8] One must conclude, therefore, that neither of the models proposed can account for all the observed properties of the Mn_3X compounds.

POSSIBLE MAGNETIC STRUCTURES

Assuming that the crystallographic and magnetic unit cells are of identical size, it is easy to single out all the non-equivalent magnetic structures which transform according to irreducible representations of the space group $P6_3mmc$ (D_{6h}^4) and give nonvanishing axial vectors (magnetic moments or hyperfine fields) for the 6h (M) or 2d (X) sites. Table I. is a compilation of these structures.
The moment directions for only three M and one X atom on each basal plane are indicated because the direction of moments of corresponding atoms can only be parallel

Table I. The basic magnetic structures possible in DO_{19} compounds

	REPRESENTATION	STRUCTURE	ORIENTATION	MAGN. SPACE GROUP
1	Γ_2^+ (Γ_4^-)		∥C	*$P6_3\,mm'c'$ ($P6_3'm\,m'c$)
2	Γ_3^+ (Γ_1^-)		⊥C	$P6_3'\,m'm'c$ ($P6_3\,m'm'c'$)
3	Γ_4^+ (Γ_2^-)		⊥C	$P6_3'\,m'm\,c'$ ($P6_3\,m'm\,c$)
4	Γ_5^+ (Γ_6^-)		⊥C	*$P\,m'm\,a'$ ($P\,m'm\,a$)
5				
6			⊥C	*$P\,m'm'a$ ($P\,m'm'a'$)
7				
8	Γ_6^+ (Γ_5^-)		∥C	(*)$P\,m\,m'a'$ ($P\,m\,m'a$)
9				$P\,m\,m\,a$ ($P\,m\,m'a$)

or antiparallel. For the latter case the symbol of the representation and of the magnetic space group is given in parenthesis. The MSGs marked by an asterisk permit bulk magnetization to occur. The absence of an arrow by an atom signifies that no axial vector is permitted to reside on that particular atomic site in the structure.

All of the possible magnetic structures can be described by a linear combination of the types contained in Table. Landau's theory of second order phase transformations, however, tells us that only structures belonging to a single irreducible representation may appear in a given transition. It is therefore impossible to have structures 2 and 3 in combination with each other, or with structures permitting both bulk magnetization as well as a hyperfine field at the X site (structures 1, 4 and 6).

Combination of structures 4 and 5 or 6 and 7, on the other hand, is allowed. By investigating the 2nd and 6th order terms in the power series expansion of thermodynamic potential as a function of the magnitude of the sublattice magnetization, it can be shown that, in the case of a simultaneous appearance of 4 with 5 or

6 with 7, one of the two structures must strongly dominate, while no essentially new structure can be obtained by combining structures 4+5 with 6+7. Structures 8 and 9 are not allowed to appear in combination with any of the other structures.

DISCUSSION

It is proposed that the magnetic structure of Mn_3X compounds is actually either structure 5 or structure 7 with some admixture of structures 4 or 6, respectively. The two models we propose, shown in Fig. 1, are in fact connected by a 90° rotation of all magnetic moments.

One of Kádár and Krén's models (structure 2) is compared in Fig. 2. with one of the new models (structure 7) to show their connection. The angle between the directions of adjacent Mn moments is the same for both, as required for a predominance of antiferromagnetic exchange; the sole difference is that the moment directions for two of the three Mn atoms are interchanged. For symmetric exchange purposes structures two and seven are obviously equivalent. Structure 2 will certainly be favoured by anisotropy forces, since the moment directions for all Mn atoms are parallel to local symmetry directions, while in structure 7 this is true for only a third of them. The symmetry of the lattice also permits a Dzyaloshinski-Moriya type interaction to take place with a $\underline{D}$ vector parallel to the c axis. Depending on the sign of $\underline{D}$, this interaction may favour either structure 2 or 7, since the vector product of the sublattice magnetizations must obviously be of opposite sign for the two cases. Thus, if antisymmetric exchange is strong enough and of proper sign, it will stabilize structure 7 against structure 2.

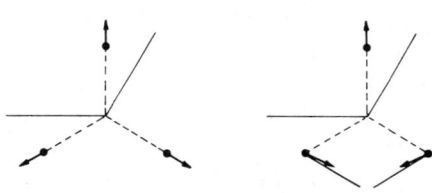

Fig. 2. Comparison of an early and a new model (structure 2 and 7)

In the proposed new model (structure 7) a net magnetization is, of course, permitted on symmetry grounds, but the physical reason for its appearance is also apparent from Fig. 2. The anisotropy forces try to align the moments parallel to the easy direction against the exchange interactions. An equilibrium will be established with angles somewhat different from 120°, so cancellation of the sublattice magnetizations will not be complete.

In structure 7 the hyperfine field at the X site is

also possible on symmetry grounds, but in addition it is easy to show directly (e.g. by assuming a simple dipole-dipole interaction) that this undistorted, completely antiferromagnetic structure really can give a large hyperfine field. Since it is not the weak ferromagnetism which is responsible for the major part of the hyperfine field at the tin sites, it is not surprising that no change is detected on altering the weak moment.

The interpretation of the neutron results is more complicated. Disregarding the ambiguity in the sign of magnetic moments, three different type 7 structures can be established by selecting alternate pairs of Mn atoms to interchange their moments. (The structures thus obtained are simply those permitted by the Landau theory for combinations of structure 5 with 7.) The detected neutron intensities will be an average of these. If the sample does not have a preferred direction, it is easy to show that these averages give exactly the same intensities for the observed reflections as the $45°$ triangular model of Kádár and Krén, i.e. the intensities that were observed experimentally. No decision between strucutes 5 and 7 can be made, because the averages are the same for both. Should the sample have a preferred direction, it should be possible to single out the structure that actually occurs, though a magnetic field would not be very effective for this purpose, because the net moment involved is very small and the magnetic susceptibility is very nearly isotropic.

REFERENCES

1. C. Jannin, P. Lecocq and A. Michel, C.R. Acad. Sc. Paris 257, 1906 (1963).
2. E. Krén and G. Kádár, Solid State Commun. 8, 1953 (1970)
3. J.S. Kouvel and J.S. Kasper, Proc. Internat. Conf.on Magnetism, Nottingham 1964. (The Institute of Physics and the Physical Society, London, 1965). p.169.
4. G.J. Zimmer and E. Krén, Magnetism and Magnetic Materials, 1971. (AIP Conference Proceedings No.5. Ed. C.D. Graham, Jr. and J.J. Rhyne. American Inst. of Physics, New York, 1972.) p.513.
5. L. Meyer-Shützmeister, R.S. Preston and S.S. Hanna, Phys. Rev. 122, 1717 (1961).
6. G. Kádár and E. Krén, Internat. J. Magnetism, 1, 143 (1971)
7. E. Krén, G. Kádár, M. Barberon and R. Fruchart, Internat. J. Magnetism, 1, 341 (1971).
8. L. Cser, private communication.

HYPERFINE FIELD SPECTRA OF MAGNETITE (Fe_3O_4)

Mark Rubinstein and George H. Stauss
Naval Research Laboratory, Washington, D.C. 20390

Frank J. Bruni
Oak Ridge National Laboratory, Oak Ridge, Tenn. 37830

ABSTRACT

The hyperfine field spectra of magnetite (Fe_3O_4) have been investigated at 4.2°K (below the Verwey transition temperature T_V) and at 300°K (above T_V). The low temperature spectrum is composed of at least eight inequivalent "ferrous-like" B-site lines, eight "ferric-like" B-site lines, and eight ferric A-site lines. The high temperature B-site spectrum shows structure caused by anisotropic hyperfine fields. The small anisotropy found suggests either a d-band width greater than the B-site trigonal splitting or electron hopping between B sites with energy levels severely lifetime broadened.

INTRODUCTION

On lowering through about 120°K magnetite undergoes a first-order phase transition from a high-conductivity to a low-conductivity state and changes crystal symmetry from cubic spinel to a less symmetric structure. There is now abundant evidence from recent neutron- and electron diffraction studies that the crystal structure below T_V, once thought to be orthorhombic, is actually less symmetric, and that the simple ordering of electronic charges on B-site ions postulated by Verwey is not correct. Moreover, although it is generally accepted that the phase transition is driven by the Coulomb repulsion between electrons on the B-sites, there is little agreement concerning the details of the mechanism. Recent theories of the Verwey transition predict the existence of either fractional charges[1] or charge density waves[2] below T_V; however, there is at present no theory complete enough to explain why the low-temperature phase of magnetite is as complex as the nuclear resonance studies discussed below reveal it to be.

This work is an extension of a previous study on the hyperfine spectra of magnetite.[3] More careful measurements have enabled us to improve the resolution and reliability of the low temperature spectrum and to separate its ferrous- and ferric-like components. Room temperature spectra have also been obtained from powdered samples showing good low temperature spectra.

In addition to several natural single crystals, the present study utilized synthetic single crystals. These were prepared by mixing iron and Fe_2O_3 powders in a sealed platinum crucible which was supported in a sealed tantalum container. The container was heated slowly to 1000°C, held at 1000°C for one hour, slowly heated to 1675°C, held for one-half hour, and cooled at 35°C/hr to 820°C. Aside from a sharper spectrum from the synthetic crystals, no differences were found between these and the best natural crystals.

LOW TEMPERATURE PHASE

Fig. 1 shows ^{57}Fe hyperfine field spectra of single crystal magnetite at 4.2°K with zero external magnetic field. The spectra were obtained with a variable frequency spin-echo spectrometer by measuring the amplitude of the nuclear echo signal as a function of frequency. The A- and B-site spins were found to be anti-aligned, and the A-site lines were identified by their upward frequency shift upon application of external magnetic fields in the easy [001] direction.[3] The A-site spectrum consists entirely of the strong sharp features between 69 and 71 MHz overlying part of the B-site spectrum. The natural crystal spectrum can be seen to be broadened in its individual B-site features in comparison with the synthetic crystal spectrum, and to spread more into the gap between B-site spectral segments. This behavior is typical of poorer quality samples; in extreme cases the gap is completely closed.

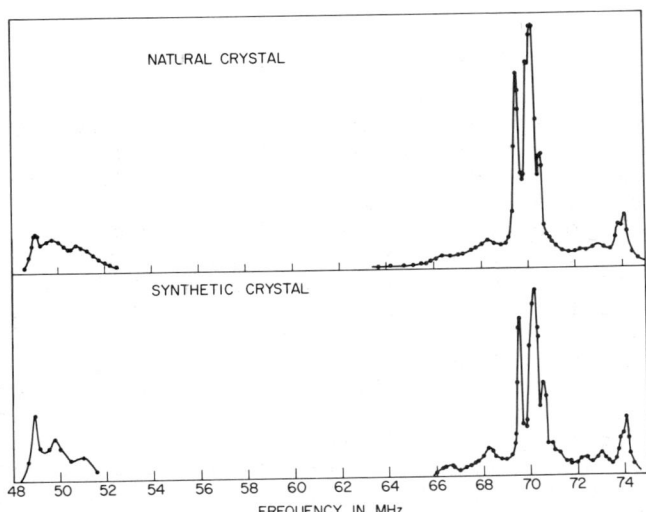

Fig. 1 4.2°K NMR spectra of ^{57}Fe in single crystal magnetite. The natural crystal was obtained from the Smithsonian Institution.

Verwey's original model for the low-temperature phase consisted of Fe^{2+} and Fe^{3+} B-site ions in alternate (001) planes. If such a picture were adequate, only two B-site resonances and one A resonance would be observed. Instead, we find that the resonance spectrum can be resolved into at least 24 lines, eight "nearly ferrous" and eight "nearly ferric" B-site resonances, as well as eight ferric resonances from A-sites whose role in the transition is evidently passive.

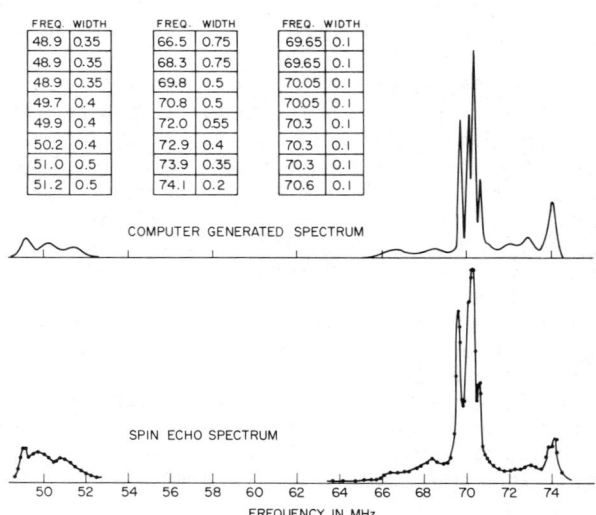

Fig. 2. 4.2°K zero field spectrum of single crystal (lower); and computer generated spectrum (upper) composed of 24 Gaussian lines. Line positions and widths are tabulated (left to right) for "Fe^{2+}" - B, "Fe^{3+}" - B, and Fe^{3+} - A sites.

FREQ.	WIDTH	FREQ.	WIDTH	FREQ.	WIDTH
48.9	0.35	66.5	0.75	69.65	0.1
48.9	0.35	68.3	0.75	69.65	0.1
48.9	0.35	69.8	0.5	70.05	0.1
49.7	0.4	70.8	0.5	70.05	0.1
49.9	0.4	72.0	0.55	70.3	0.1
50.2	0.4	72.9	0.4	70.3	0.1
51.0	0.5	73.9	0.35	70.3	0.1
51.2	0.5	74.1	0.2	70.6	0.1

In Fig. 2 the spin-echo spectrum is shown in the lower half of the picture; above it is a computed spectrum obtained by summing 24 Gaussian lines of variable width and position but of equal area. The widths and central frequencies, chosen to agree best with the experimental spectrum, are tabulated in the figure. The intensity at each point in the computed spectrum has been divided by f^3 (f is the resonance frequency) to match the variation of the experimental echo amplitude with frequency due to the Boltzmann distribution, Faraday's law of induction, and enhancement effects. When these corrections are taken into account, it appears that the B-site spectrum is divided into two main parts of equal intensity, corresponding to Verwey's separation of ferrous and ferric ions. Additional structure is also present, however, which further divides each main group into a minimum of eight lines. The low temperature unit cell contains at least eight Fe_3O_4 molecular units.

To account for this reproducible but complex structure, we find the concept of charge density waves[2] most adequate. At least three sinusoidal waves of charge density on the B sites, including one of double period along the magnetization direction, are needed to cause the observed splitting of the A sites. The charge fluctuation amplitudes need not alter the Verwey order by more than a few percent if orbital hyperfine field contributions are significant. The gross behavior of the B-site spectrum can be accounted for in this way with plausible ad hoc assumptions concerning the influence of crystal fields on the change in hyperfine field with charge density, but a detailed development is difficult.

HIGH TEMPERATURE PHASE

The ^{57}Fe NMR echo spectrum of magnetite at 300°K, well above T_V, is shown in Fig. 3. Above T_V, magnetite has the cubic spinel structure, with two crystallographically equivalent A sites and four equivalent B sites per unit cell. When the sample magnetization is introduced, however, the B sites become <u>magnetically</u> inequivalent and the anisotropic hyperfine fields at the trigonal B sites will split the B-site resonance line into two, with an intensity ratio of 3:1. This splitting is observed (Fig. 3) and corresponds to 7.5 kOe at 300°K.

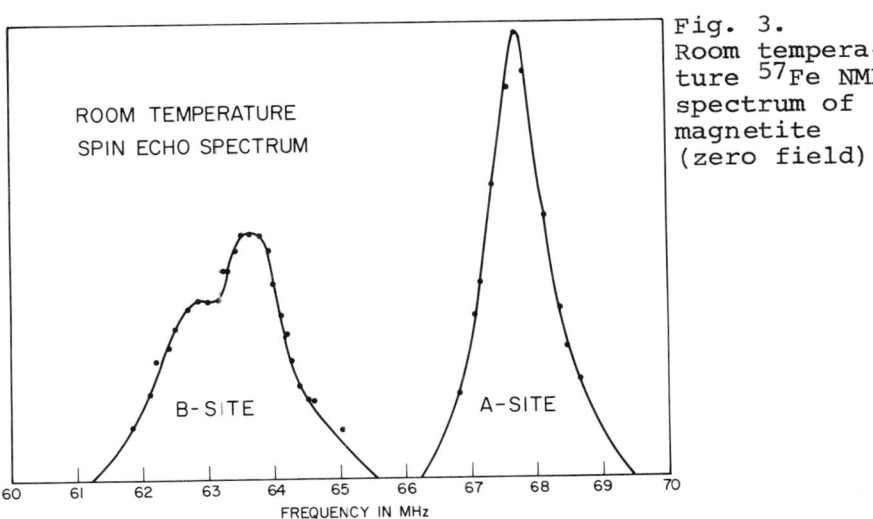

Fig. 3. Room temperature ^{57}Fe NMR spectrum of magnetite (zero field).

Above T_V, electronic hopping between B-site ions is sufficiently rapid that each B-site wave function is 50% ferrous and 50% ferric in character. The ferrous wave function would be expected to dominate the hyperfine field anisotropy at a trigonal B site through the spin-dipolar interaction. However, typical values[4] lead to anisotropic hyperfine fields of about 200 kOe for wave functions, at spinel B sites, which are 50% Fe^{2+}. This is over an order of magnitude greater than the 7.5 kOe observed here.

If one invokes a band model of magnetite, this discrepancy can be explained by using a d-electron bandwidth W which is greater than the single-ion trigonal splitting Δ. We then expect the anisotropic hyperfine interactions to be reduced by about Δ/W, since the "hopping" or banding interactions would tend to equalize the populations of the three t_{2g} orbitals. To give the low hyperfine anisotropy in magnetite, the bandwidth must be an order of magnitude larger than the trigonal splitting, whereas band theories of magnetite have so far assumed just the opposite.[1,2] Alternatively, a localized model with hopping type mobility must have its energy levels lifetime broadened to a similar degree.

1. J. R. Cullen and E. Callen, Phys. Rev. Letters <u>26</u>, 236 (1971).
2. J. B. Sokoloff, Phys. Rev. <u>5</u>, 4496 (1972).
3. M. Rubinstein and D. W. Forester, Solid State Commun. <u>9</u>, 1675 (1971).
4. F. Hartmann-Boutron and P. Imbert, J. Appl. Phys. <u>39</u>, 775 (1968).

ELECTRICAL PROPERTIES OF PURE AND SUBSTITUTED MAGNETITE FROM 4.2 TO 300K

C. Constantin
Institute of Physics Bd. Pacii 222, Bucharest, Romania

M. Rosenberg
University of Bucharest, Bucharest, Romania

ABSTRACT

Electrical properties of the compositional series $Fe_{3-x}Me_xO_4$ with Me = Ni, Co, Mn were investigated over the temperature range 4.2-300K. The results obtained were interpreted in terms of a hopping of localized electrons below the Verwey point.

INTRODUCTION

Several models have been proposed to explain the semiconducting behavior of magnetite below the Verwey temperature T_V ($\approx$ 119K). Cullen and Callen[1] have suggested an itinerant-electron model, while most other descriptions, the most recent by Chakraverty[2] and Camphausen,[3] are based on a localized-electron model. Mott[4] has proposed that the transition represents a Wigner crystallization, itinerant-electron behavior above T_V becoming localized-electron behavior below T_V, and Goodenough[5] has suggested the formation of Fe-Fe molecular-orbital states that are randomly ordered above T_V. Buchenau[6] has considered the possibility of an ordering of these pair states below T_V.

In order to obtain further information about the character of the down-spin (mobile) electrons at the B-cation subarray in magnetite below T_V, we have investigated the temperature dependence of the electrical conductivity σ and the Seebeck coefficient θ of pure and substituted magnetite below room temperature. Preliminary results have been given elsewhere.[7] Measurements of the temperature dependence of θ have also been reported by others.[8-10]

EXPERIMENTAL TECHNIQUE AND RESULTS

Polycrystalline samples in the compositional series $Fe_{3-x}Me_xO_4$, with $0 \leqslant x < 1$ for Me = Ni and $0 < x \leqslant 0.1$ for Me = Co or Mn, were prepared from oxidic powders obtained, in all but two cases, by heating at 500° C an appropriate mixture of coprecipitated oxalates. The oxidic powders were isostatically pressed at 2000 kgf/cm^2, sintered at 1480° C, and cooled under a protective atmosphere of argon mixed with purified nitrogen.

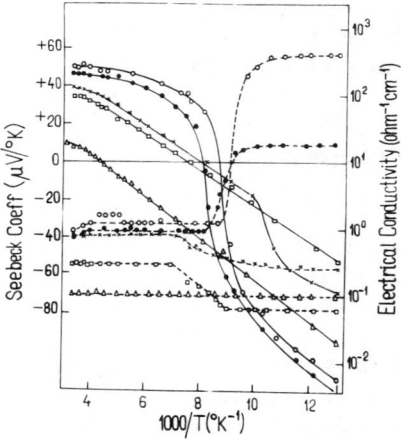

Fig. 1. Electrical conductivity (-) and Seebeck coefficient (...) vs. T for $Fe_{3-x}Ni_xO_4$: o x = 0; ● x = 0.01; x x = 0.02; □ x = 0.2; △ x = 0.4.

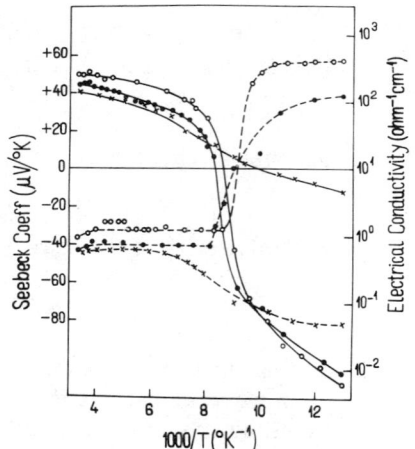

Fig. 2. Electrical conductivity (-) and Seebeck coefficient (...) vs. T for $Fe_{3-x}Co_xO_4$: o x = 0; ● x = 0.01; x x = 0.1.

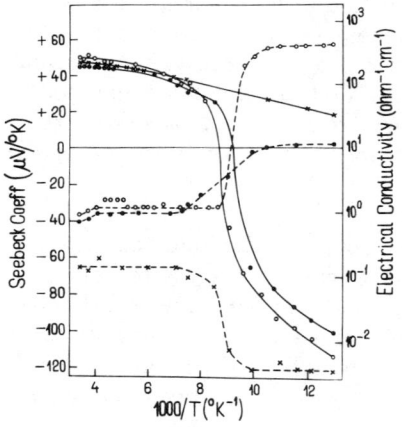

Fig. 3. Electrical conductivity (-) and Seebeck coefficient (...) vs. T for $Fe_{3-x}Mn_xO_4$; o x = 0; ● x = 0.01; x x = 0.1.

X-ray, chemical, and metallographic analysis have shown that all the samples were single-phase, cubic spinels at room temperature and had compositions very close to the nominal ones.

The d.c. conductivity, as well as the a.c. conductivity at 53.5 and 83 MHz, between 4.2 and 300K and the Seebeck coefficient in the temperature range 77 < T < 300K were all measured with standard techniques in especially designed cryostats. The results for σ and θ in the interval 77 < T < 300K are shown in Figs. 1-3, and the low-temperature conductivity data are shown in Fig. 4. The d.c. conductivity data are in good agreement with those obtained by previous investigators.[11-14] The slope of the conductivity curve for magnetite, $-d(\ln\sigma)/d\beta$ with $\beta \equiv 1/kT$, is plotted in Fig. 5. For each composition, σ increases with frequency below a critical temperature, but decreases with temperature above it. We assume that the decrease of σ with frequency is due to the skin effect.

DISCUSSION

From Figs. 1-3 it is seen that the Seebeck coefficients θ are essentially temperature-independent below T_V for all x if Me = Ni or Mn and that θ decreases with x, in the interval $0 \leq x \leq 0.01$, where the conductivity shows an identifiable T_V below which long-range electron ordering occurs. For x = 0.02, there is no sharply defined T_V, and θ remains negative for all temperatures. For Me = Co, the variation of θ with x and T is only slightly different.

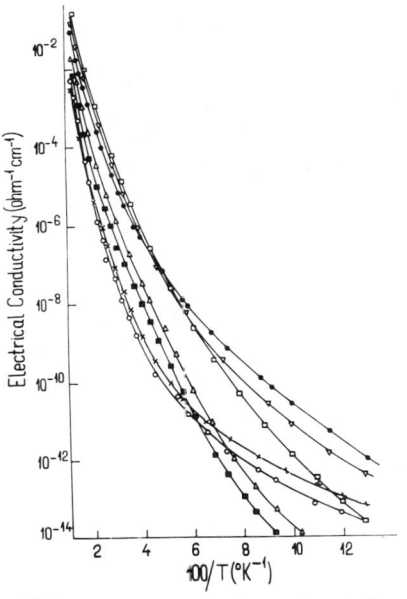

Fig. 4. Electrical conductivity vs. T for $Fe_{3-x}Ni_xO_4$: ○ x = 0; × x = 0.01; ● x = 0.02; ∇ x = 0.1; □ x = 0.2; △ x = 0.4; ■ x = 0.63.

A temperature-independent Seebeck coefficient is not compatible with the itinerant-electron model of Cullen and Callen,[1] but it is compatible with a localized-electron regime having a constant number of charge carriers of one type. In this latter case, the Seebeck coefficient is given by[15]

$$\theta_\pm = \pm \frac{k}{e} \{\ln[(1-n_\pm)/n_\pm] + \alpha_\pm\}$$

where $n_\pm$ is the number of charge carriers per available polaron site and $\alpha_\pm$ is a transport term, which is generally negligible for small polarons in oxides.

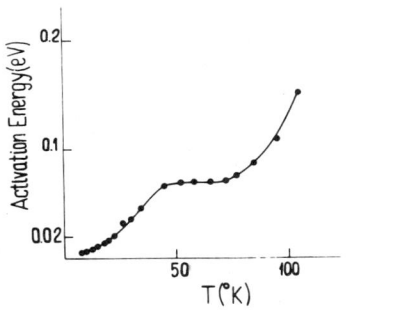

Fig. 5. Energy ≡ $-d(\ln\sigma)/d\beta$ vs. T for magnetite; $\beta \equiv 1/kT$.

According to the localized-electron model, first proposed by Verwey,[16] below T_V the Fe^{2+} and Fe^{3+} ions of $Fe^{3+}[Fe^{3+}Fe^{2+}]O_4$ order respectively into [110] and [1̄10] rows of the octahedral-site subarray. A θ < 0 above T_V and a θ > 0 below T_V for Fe_3O_4 then implies the existence of a few cation vacancies randomly distributed over this subarray, so that the number of mobile electrons is less than half the number of octahedral iron atoms and below T_V there are a few Fe^{3+} ions -- or small-polaron holes -- on the [110] rows of Fe^{2+} ions. In this model, the addition of randomly distributed

Me^{2+} ions would increase the number of holes on the [110] rows of Fe^{2+} ions below T_V, thereby decreasing the magnitude of $\theta_+ > 0$. Conversely, above T_V the number of mobile electrons per octahedral-site iron atom is reduced, so $\theta_- < 0$ above T_V increases in magnitude. However, for $0.02 \leq x \leq 0.2$ the nature of the electronic ordering appears to change, and the Seebeck coefficient remains negative for all temperatures, only increasing in magnitude below T_V. This observation implies a noncooperative trapping of electrons at identifiable Fe^{2+} sites, thereby decreasing the number of mobile electrons per octahedral-site iron atom. For $x > 0.2$, there appears to be no electronic ordering at low temperatures.

In conclusion, investigation of the electrical properties of pure and substituted magnetite below 300K favor a localized-electron model below T_V. The mobile electrons at the octahedral-site array of iron ions in Fe_3O_4 may be trapped as small polarons at identifiable Fe^{2+} ions or at Fe-Fe pairs. The evolution with x of the temperature dependence of the Seebeck coefficient θ in the systems $Fe_{3-x}Me_xO_4$, Me = Mn, Co or Ni, can be qualitatively understood in terms of a small-polaron model below T_V and a change from long-range, cooperative ordering in the compositional range $0 \leq x \leq 0.01$ to noncooperative, short-range ordering in the compositional range $0.02 \leq x \leq 0.2$.

The authors would like to thank J. B. Goodenough for suggesting changes in the manuscript.

REFERENCES

1. J. R. Cullen and E. Callen, Solid State Comm. 9, 1041 (1971).
2. B. K. Chakraverty, J. Phys. (Conf. Int. Mag.) C1-1112 (1970).
3. D. L. Camphausen, Preprint (1972).
4. N. F. Mott, Adv. Phys. 13, 116 (1967).
5. J. B. Goodenough, Prog. Solid State Chem. 5, 509 (1971).
6. U. Buchenau, Solid State Comm. 11, 1287 (1972).
7. C. Constantin and M. Rosenberg, Solid State Comm. 9, 675 (1971).
8. F. J. Morin and T. H. Geballe, Phys. Rev. 99, 467 (1959).
9. A. A. Samokhvalov and A. G. Rustanov, Sov. Phys. Solid State 7, 961 (1965).
10. B. A. Griffiths, D. E. Elwell and R. Parker, Phil. Mag. 175, 163 (1970).
11. D. J. Epstein, Thesis, M.I.T. (1954).
12. J. M. Burkey and R. Poplowski, J. Appl. Phys. 39, 5813 (1968).
13. C. A. Domenicali, Phys. Rev. 78, 458 (1950).
14. N. Miyata, J. Phys. Soc. Japan 16, 206 (1961).
15. G. H. Jonker and S. van Houten, Halbleiterprobleme 6, 118 (1961).
16. E. J. W. Verwey and E. L. Heilmann, J. Chem. Phys. 15, 174 (1947); E. J. W. Verwey and P. W. Hayman, Physica 8, 979 (1941).

DOPANT EFFECTS UPON THE VERWEY TRANSITION IN Fe_3O_4*

James J. Bartel and Edgar F. Westrum, Jr.
The University of Michigan, Ann Arbor, Michigan 48104

ABSTRACT

Additional adiabatic heat capacity calorimetry has confirmed the bifurcated nature of the Verwey transition in pure and doped Fe_3O_4 samples. Two sharp peaks occur at 113.3 and 118.8 K for a 99.99% pure ceramic sample. Data on a hydrothermal sample confirmed the presence of both peaks. However, the presence of 0.2 weight per cent of Mn(II) raised them by 3.7 and 4.1 K, respectively. Large single-crystals of naturally occurring magnetite revealed only a broad transition with a high temperature shoulder. The total transitional enthalpy for the pure ceramic sample was 155 cal mol^{-1} or a value about 10 $\pm$ 1% higher than that for the two doped samples. Observation of the positive shift of transition temperatures for the Mn(II)-doped (hydrothermal) sample appears contradictory not only to the trend reported in a recent study (in which magnetic sensing was used) but in particular to that observed for a Zn(II)-doped sample. Calorimetric sensing of the transition temperature appears to be more reproducible than either resistance or magnetic sensing and to have other significant advantages.

INTRODUCTION

The earlier report[1] of two heat capacity peaks in synthetic, ceramic Fe_3O_4 in the vicinity of the Verwey transition was sufficiently radical to require confirmation. This was provided by studies on Mn(II)-doped hydrothermally prepared Fe_3O_4 and natural magnetite crystals (one of which contained Ti^{3+}). The effect of dopants [(e.g., Ti(IV), Mn(II)], which do not contribute electrons to the conduction band of the host on the multiple-ordering phenomena is an initiatory aspect of a highly-coordinated, systematic, interdisciplinary investigation of this extensively studied but poorly understood transition. Mössbauer data have confirmed that at least for Mn(II) doping no variation occurs in the apparent electron concentration.

RESULTS

The (summed) enthalpies of transitions, ΔH_t, obtained by direct calorimetric measurements for each of the samples to be considered are presented in Table I. For consistency, the results of Millar were regraphed and the net enthalpy calculated, as were all others, by interpolating the essentially lattice heat capacity through the transition region. Subtraction of the lattice

---

*Supported by the U.S. Atomic Energy Commission and the National Science Foundation.

Table I. Properties of the Verwey Transition

Sample	$\Sigma\Delta H_t$ cal mol^{-1}	Transition T_l, K	T^* T_h, K	Ref.
Ceramic Fe_3O_4	155	113.3	118.9	1
Mn(II)-doped Fe_3O_4	139	117.0	123.0	-
Two magnetite crystals [#]	141	117.3	(119)	-
Magnetite	108	114.2		3

*T_l and T_h refer to the lower and higher transition temperatures

[#]Titanium content of one crystal (about 50% sample) = 0.2 wt. percent

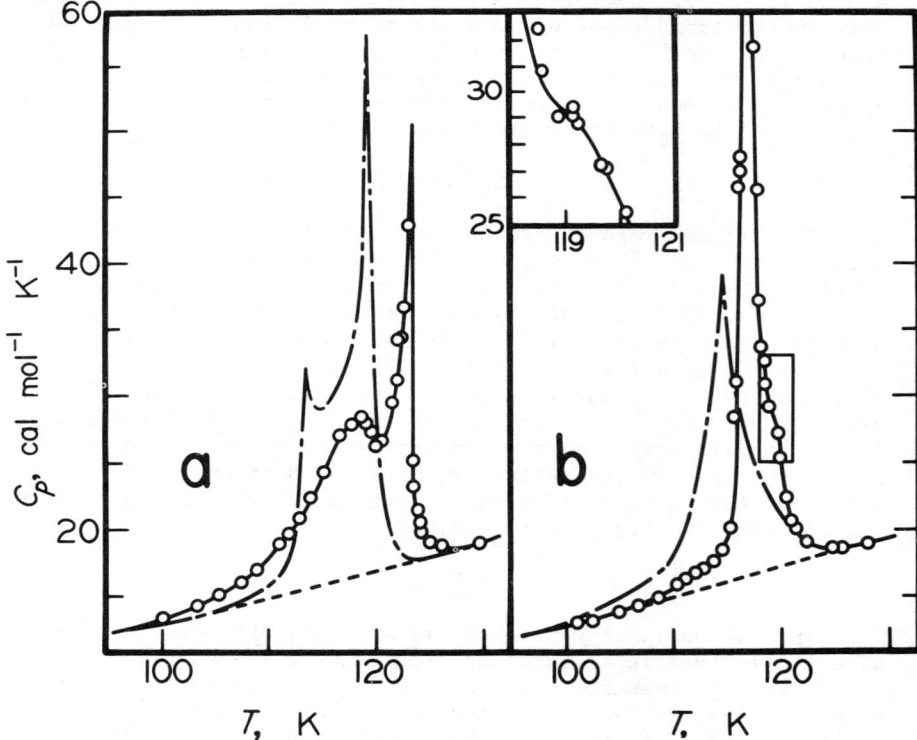

Fig. 1a. ——— ceramic Fe_3O_4, Westrum and Grønvold[1], —o— Mn(II)-doped Fe_3O_4, ---- interpolated lattice heat capacity. 1b ——— magnetite, Millar, —o— two natural magnetite crystals, ----- interpolated lattice heat capacity.

Fig. 2. The effect of cooling rate and the extent of cooling upon the shape and enthalpy increment near the T_h peak for the Mn(II)-doped (hydrothermal) Fe_3O_4. ○ cooled through the transition to 50 K over 96 hours. □ cooled just through the T_h peak at 0.1 K per hour. △ cooled just through the T_h peak at 14 K/per

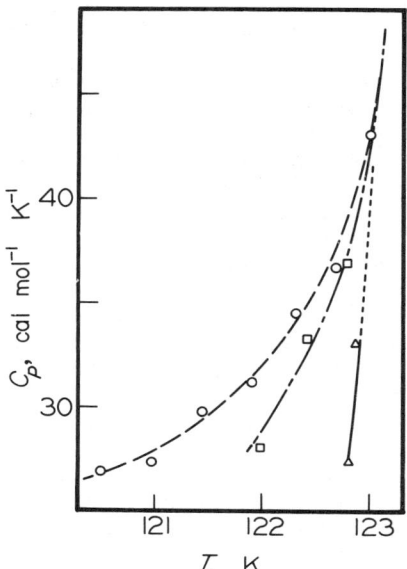

contribution from the total enthalpy yielded the $\Sigma\Delta Ht$'s recorded in Table I. The heat capacity peaks of the magnetite crystals and of Millar's magnetite are presented in Fig. 1a. Those for the hydrothermal sample and the (purer) ceramic sample, previously reported[1], are presented in Fig. 1b.

That variation of the cooling rates affects the apparent heat capacity in the transition regions and that sufficiently slow cooling is mandatory to obtain repoducible results is evident from Fig. 2. We found that the hydrothermal sample had to be cooled to at least 90 K at an average rate of heat removal of 550 cal/(mol hr) to accurately reproduce the transitional heat capacities. However, regardless of the rate of cooling, the temperature of the heat capacity maximum, determined by monitoring the cooling rate, was always the same as that on heating.

DISCUSSION

The six experimental series through the transition region of a single sample yielded about 80 experimental heat capacities, consistent to about ± 0.02%. Temperatures are relatively accurate to a mK and on an absolute basis to 40mK referred to the International Temperature Scale. Compared to resistivity and magnetic[4] measurements in which thermocouples are employed as the temperature standard, adiabatic calorimetric techniques are an order of magnitude greater in sensitivity and relatively independent of contaminating magnetic material.

The variation in observed ΔHt's, the shifts in the transition temperatures, and the effect of the cooling and heating rate upon the form of the transition have several important implications.

After a typical energy input to the sample producing a temperature increment of 0.5 - 3.0 K, 0.7 to 1.5 hours were required to achieve thermal equilibration in the transition region. The longer equilibrium times were associated with the larger $\Delta T's$. The implication is that a heat-transition region are being investigated.

The shift of the Mn(II)-doped (hydrothermal) sample peaks by 2.0 and 4.2 K towards higher temperatures than those for the ceramic sample is contrary to the results of Miyahara[4] who found by magnetometer measurements that Mg(II)-, Zn(II)-, and Ga(II)-dopants (which like Mn(II)-dopant all have closed shell structures) significantly lowered the Verwey transition temperature. However his measurements were dynamic in contrast to our equilibrium measurements and even at high dopants he failed to observe an increase in lattice parameter. Our Mn(II)-doped sample had a lattice parameter of 8.399 ± 0.001 Å in comparison with a value of 8.396 ± 0.001 Å for the ceramic sample[1].

Thermophysical measurements on the two natural crystals do not offer positive verification of the temperature elevation effect since the peaks overlap, but the maxima are above the temperature of T_1 for the ceramic sample.

However, our observations are consistent with Wigner[5] crystallization since the Mn(II) would decrease carrier density and as a consequence the temperature of the transition would be expected to increase.

The Mössbauer evidence[2] together with the thermophysical evidence for the bifurcated transition found in this study support the multiple-ordering theory[6] for the Verwey transition itself. That the two transitions are occasioned by different mechanisms is evident for the differences between the heat capacity maxima vary from sample to sample. Although Millar's data only show one anomaly, the temperature increments are too large to rule out the possibility of a second, due to superposition of the two peaks, as was nearly the case with the two natural crystals reported here.

ΔHt appears to decrease with enhanced dopant concentration. The low ΔHt observed on Millar's sample argues for a high impurity content on this basis; hence an entire peak may have been quenched in his sample. Studies of impure (Kiruna)magnetite show no evidence of *any* thermal anomaly in the region of the Verwey transition.[1]

In conclusion, the bifurcated nature of the Verwey transition has been confirmed and its sensitivity to low levels of impurities established. Increasing the temperature of transition and lowering the enthalpy of the transitions are two manifestations of Ti(II)- and Mn(II)-dopant inclusion. Although the transition was shown to be sensitive to too rapid cooling, experimental conditions necessary to achieve reproducible results are formulated (below).

EXPERIMENTAL

The hydrothermal sample was kindly supplied by Mr. John L. Haas Jr. of the U. S. Geological Survey. Analysis showed 0.2 mass per cent Mn, 0.02 mass per cent Ni and less than 0.02 for the sum of all other impurities. Heat capacities of a 212 g. sample were measured in a gold-plated, copper calorimeter in the Mark III adiabatic cryostat. Similarly two natural crystals (47 g. total mass) supplied by Professor B. J. Evans of the University of Michigan were studied in a similar calorimeter. Non-destructive analysis showed that a Ti impurity was present on the same level as the Mn in the hydrothermal sample; microprobe analysis revealed a second phase (probably hematite) in one of the two natural crystals. The impurity level of the hydrothermal sample corresponds to the formula:

$$[Fe(III)_{0.992} Mn(II)_{0.008}]_{Td} [Fe(II)Fe(III)]_{Oh} O_4$$

(Td = tetradhedral sites, Oh = octahedral sites.)

Heat capacity measurements were made from 50-350 K for both samples. Temperatures were measured by a capsule-type, platinum, resistance thermometer (A-3) calibrated by NBS against the International Temperature Scale above 90 K. The general procedure was to cool the sample gradually to 50 K (over a twenty hour period). Exploratory data with 10 K temperature increments were made to roughly locate the transition maxima; subsequent passes with increments of 3 K and 1 K were made to evaluate the details of the transitions. In the peak region increments as small as 0.03 K were used. Finally the transition region was traversed several times in a single enthalpy increment to determine the total enthalpies of the transitions. This value agreed with the summation of the individual increments through the transition regions with + 0.5%. The hydrothermal sample was subjected to varied cooling rates to determine if this affected the heat capacity curve. In separate experiments, the heat capacities of the calorimeters were determined over the range 5-350 K. These data plus adjustment for the masses of helium, grease, and solder were used to arrive at a net molal heat capacity. The molecular weight of Fe_3O_4 was taken as 231.55.

REFERENCES

1. E. F. Westrum, Jr. and F. Grønvold, J. Chem. Thermodynamics $\underline{1}$, 543 (1969).
2. B. J. Evans and E. F. Westrum, Jr., Phys. Rev. B $\underline{5}$, 3791 (1972).
3. R. W. Millar, J. Amer. Chem. Soc. $\underline{51}$, 215 (1929).
4. Y. Miyahara, J. Phys. Soc. Japan, $\underline{32}$, 629 (1972).
5. E. Wigner, Trans. Faraday Soc., $\underline{34}$, 678 (1938).
6. J. R. Cullen and E. Callen, Solid State Commun. $\underline{9}$, 1041 (1971). and references therein.

ELECTRICAL CONDUCTIVITY AND HYPERFINE INTERACTIONS IN $M_xFe_{3-x}O_4$ ABOVE AND BELOW T_N

B. J. Evans
The University of Michigan, Ann Arbor, Mich. 48104

ABSTRACT

For $T<T_N$ the results of the present study of the ^{57}Fe Mössbauer spectra of Zn- and Ni-substituted Fe_3O_4 are in qualitative agreement with those of recent investigations, supporting a band model for the conduction mechanism in Fe_3O_4. Asymmetric line shapes are observed, however, for the B site pattern of the substituted samples. Attempts to simulate the asymmetry using a binomial probability distribution and linear hyperfine field perturbations are only marginally successful. The effects of Cd substitution are considerably greater than those of Ni and Zn. A localized conduction mechanism appears to set in at lower x values for Cd. These results are consistent with existing ideas concerning conduction mechanisms in transition metal oxides. The band-like character of the conduction electrons does not appear to be diminished above T_N, and the isomer shifts indicate a close approach to the +3 oxidation state for all Fe ions. A quadrupole doublet and a singlet pattern with similar isomer shifts may account for the spectra above T_N.

INTRODUCTION

Recently, there have been several ^{57}Fe Mössbauer studies of substituted Fe_3O_4 below T_N.[1-3] Similar conclusions supporting a band model for the conduction mechanism in pure and substituted Fe_3O_4 were reached in each case. No measurements have been reported for $T>T_N$. It has been suggested that the delocalization of the conduction electrons should be enhanced by the $Fe^{2+}(B)-Fe^{3+}(B)$ double-exchange interaction.[4,5] Measurements above T_N are therefore of some significance, and here we report the results of such measurements for $Ni_xFe_{3-x}O_4$, $0<x<0.4$, and $Zn_xFe_{3-x}O_4$, $x = 0.2$ and 0.8. As the transition from the collective to the localized electron state is expected to depend upon the Fe-Fe separation and the covalence of the Fe-O bonds,[5] measurements have also been made on $Cd_xFe_{3-x}O_4$, which differs from $Zn_xFe_{3-x}O_4$ mainly in these two respects. In each of these systems the substituted ions have strong site preferences, with the Cd and Zn ions occupying the tetrahedral(A) sites and the Ni ions occupying the octahedral(B) sites. Hence, only Fe ions occupy the B sites in $Cd_xFe_{3-x}O_4$ and $Zn_xFe_{3-x}O_4$ with nominal $Fe^{2+}(B)/Fe^{3}(B)$ and $Fe(B)/Fe(A)$ ratios of $(1-x)/(1+x)$ and $(1-x)/2$, respectively. In $Ni_xFe_{3-x}O_4$ the values for $Fe^{2+}(B)/Fe^{3+}(B)$ and $Fe(B)/Fe(A)$ are $(1-x)$ and $1/(2-x)$, respectively. Because of the different modes of substitution, T_N increases with increasing x for $Ni_xFe_{3-x}O_4$ and decreases with increasing x for Zn and Cd substitution.

In the main, our results and conclusions for $T<T_N$ agree with

those of the previous studies. However, consideration of the B site
linewidths and line shapes reveals differences in the influence of
the different ions. The effect of Cd is considerably greater than
that of either Zn or Ni in agreement with the ideas of Goodenough on
conduction mechanisms in transition metal oxides.[5] For $T>T_N$ the A
and B site patterns are not resolved and isomer shifts are quite
similar, tending toward a value characteristic of the +3 oxidation
state. In keeping with the larger perturbation of Cd for $T<T_N$, a
weak absorption, due possibly to incipient Fe^{2+}, is observed for
$Cd_{.4}Fe_{2.6}O_4$.

EXPERIMENTAL

The samples were prepared by mixing previously characterized
Fe_3O_4,[7] $ZnFe_2O_4$, $CdFe_2O_4$,[6] spectroscopic grade NiO, and Fe_2O_3 in the
desired proportions and repeatedly grinding and firing in evacuated
silica tubes at 1300 K until a single phase spinel was obtained. The
spectra were obtained with an electromechanical transducer and a 1024
channel analyzer. A vacuum furnace (10^{-5} torr) was used for the high
temperature measurements. 99.999% pure Al foil was used for the
radiation shields, and no spectra were obtained from materials in the
path of the γ-ray other than the sample. The Co^{57}/Cu source was at
298 K for all measurements.

RESULTS

The ^{57}Fe Mössbauer spectra of $Cd_{.2}Fe_{2.8}O_4$ at 298 K and of Fe_3O_4
at 900 K are given in Figs. 1 and 2; the spectra of the other materials studied were similar to these, especially for $T>T_N$. The results of fitting two six line patterns to the spectra at 298 K and
a single line pattern to the spectra above T_N, using Lorentzian lines
in each case, are given in Tables I and II. Each of these fits led
to a χ^2 of ~2 per data point. The asymmetric lines in the spectrum
of Fig. 1 are reminiscent of those observed in partially inverse,
insulating spinel ferrites,[8] and the spectra obtained below T_N were
also fitted assuming a binomial probability distribution and a
linear relation between the change in $H_{eff}(B)$ and the number of non-
Fe neighbors. The fits improved and the asymmetry in the lines was
qualitatively reproduced; the weak structure in the experimental
data was more diffuse, however, than that resulting from the fits to
the model. The decrease in $H_{eff}(B)$ was 12 kOe per Zn and 32 kOe per
Cd A site neighbor. As shown in Fig. 2 and also from the large χ^2
values, the single line fit is inadequate in accounting for the
details of the spectra above T_N. A symmetric (quadrupole) doublet
and a single line pattern may be used to fit these data. For such
fits χ^2 decreased to ~1 per data point. The parameters of the single line pattern were nearly the same as those in Table II. The
isomer shift and splitting of the doublet for Fe_3O_4 were +0.1 mm/s
and 0.29 mm/s, respectively; the values of these parameters for the
other samples differed only slightly from those of Fe_3O_4. An additional, weak absorption at ~1.5 mm/s was observed in $Cd_{.4}Fe_{2.6}O_4$.

TABLE I

Parameters obtained in least squares fits of two six line patterns to the ^{57}Fe Mössbauer spectra of $M_xFe_{3-x}O_4$ at 298 K. a_B/a_A is the area ratio of the B to A site patterns.

Sample	$H_{eff}(A)$ kOe (± 2)[a]	$H_{eff}(B)$ kOe (± 2)	$\delta_B - \delta_A$ mm/s (± 8)	a_B/a_A (± 5)	$\Gamma_1(B)$ mm/s (± 3)	$\Gamma_1(A)$ mm/s (± 3)
Fe_3O_4	493	461	0.391	1.85	0.37	0.30
$Ni_{.1}Fe_{2.9}O_4$	490	458	0.373	1.70	0.72	0.33
$Ni_{.2}Fe_{2.8}O_4$	490	461	0.371	1.43	0.88	0.38
$Ni_{.4}Fe_{2.6}O_4$	492	470	0.393	1.24	1.25	0.50
$Zn_{.2}Fe_{2.8}O_4$	488	454	0.393	2.69	0.76	0.30
$Cd_{.2}Fe_{2.8}O_4$	481	430	0.337	3.00	1.37	0.36
$Cd_{.4}Fe_{2.6}O_4$	478	425	0.367	3.74	1.63	0.43

[a]Indicates error in last digit of tabulated values

TABLE II

Parameters obtained in least squares fits of a single line pattern to the ^{57}Fe Mössbauer spectra of $M_xFe_{3-x}O_4$ at 900 K. Line positions are uncorrected for the second-order Doppler shift and are given with respect to an Fe absorber at 298 K.

Sample	Line Position mm/s (± 6)[a]	Linewidth(Γ) mm/s (± 1)
Fe_3O_4	0.124	0.56
$Ni_{.1}Fe_{2.9}O_4$	0.123	0.61
$Ni_{.2}Fe_{2.8}O_4$	0.109	0.69
$Ni_{.4}Fe_{2.6}O_4$	0.069	0.64
$Zn_{.2}Fe_{2.8}O_4$	0.134	0.62
$Zn_{.8}Fe_{2.2}O_4$	0.068	0.63
$Cd_{.2}Fe_{2.8}O_4$	0.124	0.89
$Cd_{.4}Fe_{2.6}O_4$	0.105	0.73

[a]Indicates error in last digit of tabulated values

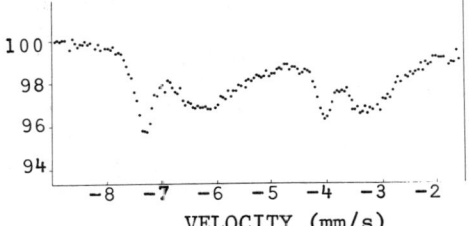

Fig. 1. Lines 1A, 1B, 2A, and 2B of the ^{57}Fe Mössbauer spectrum of $Cd_{.2}Fe_{2.8}O_4$ at 298 K.

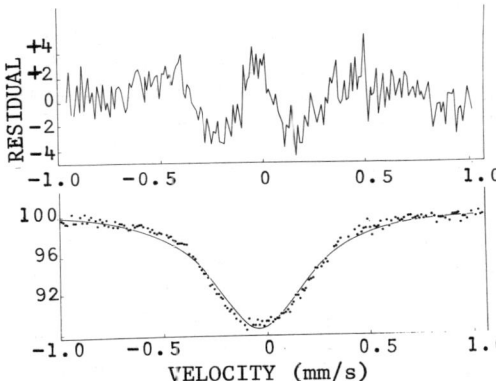

Fig. 2. ^{57}Fe Mössbauer spectrum of Fe_3O_4 at 900 K. Bottom: Least squares fit (solid line) of a single line pattern to experimental data (dots). Top: Residual of single line fit showing presence of doublet pattern.

DISCUSSION

Qualitatively, the results of the present study agree with previous measurements on $Ni_xFe_{3-x}O_4$ and $Zn_xFe_{3-x}O_4$ and support a band model for the conduction mechanism in pure and slightly impure Fe_3O_4. In addition, the relative covalence of the substituted ion is found to be an important factor in determining the strength of its perturbation on the conduction mechanism. The deviations of the B/A area ratios in Table I from the site occupancy ratios are mainly artifacts of fitting Lorentzian line shapes to the B pattern; these deviations, however, provide us with the clues to the differences in the influences of the different substituted ions. For $Ni_xFe_{3-x}O_4$ and for x>.1, the experimental area ratios are less than the site occupancy ratios, with the difference resulting from changes in the magnetic exchange interactions and charge transfer processes which tends to increase $H_{eff}(B)$. The B site pattern is asymmetric and overlaps strongly the A site pattern; the fitting of a single Lorentzian line about the <u>apparent</u> center of gravity of the B pattern results in a low estimate of the its total area. A detailed interpretation of very similar data for $Ni_xFe_{3-x}O_4$ has been reported.[3] The differences between the B/A area ratios and the site occupancy ratios are evidence for incipient Fe^{3+} ions and the onset of localized hopping conduction for .4<x<.6 . The B/A area ratios of

$Zn_xFe_{3-x}O_4$ and $Cd_xFe_{3-x}O_4$, in contrast to $Ni_xFe_{3-x}O_4$, are greater than the site occupancy ratios. Again, the lines of the B pattern are inhomogeneously broadened but in a manner opposite to that in $Ni_xFe_{3-x}O_4$, resulting in an overestimate of the contribution of the tails of the lines of the B pattern to the A pattern. There is no evidence for Fe^{3+}(B) ion in these systems, incipient or otherwise, for x values up to 0.4 and possibly as high as 0.6.[1] Both the simple six-line pattern and local-molecular-field-like fit(which leads to inhomogeneously broadened line envelopes) indicate a weak influence of Zn on the conduction mechanism. Similarly, either fitting technique leads to the conclusion that Cd perturbs the B pattern to a much greater extent than in insulating ferrites.[6] The stronger perturbations by Cd arise from its greater influence on the electrical conduction mechanism. The increase in the Fe(B)-Fe(B) and Fe(B)-O internuclear separations and the attendant increase in the ionicity of the Fe(B) bonds occasioned by Cd substitution[6] is expected to lead to a greater tendency toward localized hopping conduction, as observed. No evidence for pair-wise hopping conduction is found for either $Zn_xFe_{3-x}O_4$ or $Cd_xFe_{3-x}O_4$.

While double-exchange interactions below T_N may enhance the band character of the conduction electrons, the spectra above T_N indicate that facile electron transfer is also possible in its absence. Indeed, the isomer shifts above T_N are consistent with an increase in the delocalization of the conduction electrons: At 900 K the isomer shifts of an A and a B site Fe^{3+} ion and a B site Fe^{2+} ion are expected to be ~0 mm/s, ~0.1 mm/s, and ~1 mm/s, respectively.[6,9] The measured value of ~0.1 mm/s is somewhat less than the expected value of ~0.4 mm/s for a +2.5 oxidation state. The fitting of the spectra above T_N to the singlet and doublet patterns is perhaps non-unique, and other combinations of patterns and line shapes are possible. Nonetheless, the conclusions of this study as to the degree of electron delocalization above and below T_N and the influence of different kinds of substituted ions are not expected to be altered significantly by other modes of fitting the data.

REFERENCES

1. D. C. Dobson, J. W. Linnett, and M. M. Rahman, J. Phys. Chem. Solids 31, 2727 (1970).
2. I. Bunget, C. Nistor, and M. Rosenberg, J. Phys. 32, C1-274, (1971).
3. J. W. Linnett and M. M. Rahman, J. Phys. Chem. Solids 33, 1465 (1972).
4. A. Rosencwaig, Can. J. Phys. 47, 2309 (1969).
5. J. B. Goodenough, Mat. Res. Bull. 9, 967 (1971).
6. B. J. Evans, S. S. Hafner, and H.-P. Weber, J. Chem. Phys. 55, 5282 (1971).
7. B. J. Evans and E. F. Westrum, Phys. Rev. B5, 3791 (1972).
8. B. J. Evans and L. J. Swartzendruber, J. Appl. Phys. 42, 1628 (1971).
9. D. P. Johnson, Solid State Commun. 7, 1785 (1969).

MAGNONS IN NICKEL SULPHIDE

M. T. Hutchings, R. D. Lowde, D. H. Saunderson, M. W. Stringfellow
and C. G. Windsor
Atomic Energy Research Establishment, Harwell, England.

ABSTRACT

The antiferromagnon dispersion function of the lower-temperature-range, nickel-arsenide-structure phase of NiS has been determined at various temperatures in a specimen exhibiting a spin of about 0.4 by neutron scattering techniques, and fitting parameters derived. If the observed value of the spin be used with an effective equivalent Heisenberg-model antiferromagnet, the nearest-neighbour and next-nearest-neighbour coupling constants are found to be −100 and 40 meV respectively.

- - - - - - - - - - - - - - - - - -

By presenting this paper it is hoped to make a brief mention in the mainstream literature on magnetism of our measurements on spin waves in hexagonal NiS, hitherto published only in a rather wider connexion in the proceedings of a specialised conference on neutron inelastic scattering [1]. Our investigation is described below only in general terms; for a fuller discussion the reader is referred to our earlier account [1].

Hexagonal NiS exhibits a semiconductor–to–metal transition; at 264°K its electrical conductivity alters abruptly by a factor of at least forty [2]. In the low-temperature, less-conductive phase the material is antiferromagnetic with magnetic moments of the order $\sim 1.7 \mu_B$; however, above the transition the moment appears negligibly small [2,3].

We have studied the neutron scattering from a single crystal kindly supplied by Dr. S. E. R. Hiscocks. From the Bragg reflexions we conclude that in the low-temperature phase the magnetic moment on nickel sites in our specimen was only $0.8 \pm 0.3 \mu_B$. The conductivity transition, moreover, appeared to take place gradually over the temperature range 100-200°K. It is likely that these anomalies result from the existence of a concentration gradient across the crystal amounting to a nickel deficiency varying from 2.7 to 1.4% of the nickel atoms [1,2].

Our inelastic-scattering data were fitted to a cross-section formula derived by convoluting the resolution function of the instrument with the mosaic spread distribution of the crystal and a sharp spin-wave dispersion law of the form $\sqrt{(c^2 + d^2 q^2)}$. (d could be different in different crystallographic directions). We are able to say that $c = 12.6 \pm 0.2$ meV; d for the [100] direction is 440 ± 35 meV Å; the data indicated no detectable renormalisation of these values over the whole range of temperature up to the transition. An attempt was made to measure d along [001], but our resolution function proved to be too broad in that direction to allow any significant result to be obtained.

The data are insufficient to fit the spin-wave dispersion to a Heisenberg model taking proper account of the interactions between the different types of neighbour pair; however, if we take a simple Hamiltonian

$$H = -\sum_{i<j} 2J_{ij}\underline{S}_i \cdot \underline{S}_j - D\sum_i S_i^{z\,2} \qquad (1)$$

and employ merely $J_1 = J_2$ for the nearest-neighbour coupling parameter in the a_1 and a_2 directions and J_3 for that in the a_3 direction, it can be shown that (neglecting D in comparison with either J)

$$c^2 = -8S(2S-1)J_3 D, \qquad (2)$$

$$d_{100}^2 = -24S^2 J_1 J_3 a_1^2, \qquad (3)$$

$$d_{001}^2 = 4S^2 J_3^2 a_3^2, \qquad (4)$$

where $a_1 = 3.45$ Å and $a_3 = 5.37$ Å are the appropriate unit cell dimensions at absolute zero. Thus the spin wave dispersion at small q would actually be isotropic in NiS if

$$J_3 = -2.5 J_1. \qquad (5)$$

Our inability to measure d_{001} has prevented us from estimating J_3 directly from (4), but our determination of d_{100} gives the geometric mean of J_1 and J_3 on this model; in fact we obtain from (3)

$$S\sqrt{(-J_1 J_3)} = 26 \pm 2 \text{ meV}. \qquad (6)$$

Presumably, J_1 is to be taken much smaller than $26/S$ and $-J_3$ much larger. (With our reduced spin of about 0.4 the spin wave isotropy condition would obtain for values in the region ~ 40 and ~ 100 respectively). With S apparently always less than 1, this is an unusually high number; the figure of 26 meV is itself several times greater than is commonly observed for coupling parameters in covalent materials, and is substantially above the value 19 meV for the antiferromagnetic nearest-neighbour coupling constant in NiO [4].

White and Mott [5] have pointed out that the low $\chi_\perp$ observed in NiS also corresponds to a large J. If we regard the quantity labelled $\chi_\perp - \chi_\parallel$ in figure 1 of Townsend et al [6] as the magnitude to take for this $\chi_\perp$ (the observed $\chi_\parallel$ might well indicate a contribution from Van Vleck temperature-independent susceptibility) we obtain a minimum figure of 2×10^{-6} emu gm^{-1}; and from

$$\chi_\perp = -g^2 \mu_B^2 / 2ZS(S+1)J_3 \qquad (7)$$

with $Z = 2$, $S = 1$, a $-J_3$ in the region of 100 meV is deduced. If 100 is indeed thus an upper limit for $-J_3$, it would give, with our $\sqrt{(J_1 J_3)}$, a lower limit for J_1 between 7 and 40 meV depending on the philosophy adopted about spin in this material.

It is difficult to analyse the situation further, since the available data on NiS are still palpably insufficient to pin down the magnetic parameters convincingly. However, it is clear that the coupling constants in this material are quite exceptionally large, the preferred values of J_1 and $-J_3$ being of the order of 40 and 100 meV respectively. In general the emergence of such big numbers from a Heisenberg-model analysis is an indication that we have an itinerant-electron system; and in this respect our results fall into line with the discussion of White and Mott.

REFERENCES

1. G. A. Briggs et al. in: Proceedings of the fifth International Symposium on Neutron Inelastic Scattering, Grenoble 1972, IAEA, Vienna, 1972.
2. J. T. Sparks and T. Komoto, J. Appl. Phys. <u>34</u>, 1191 (1963) and Rev. Mod.Phys. <u>40</u>, 752 (1968).
3. D. B. McWhan, M. Marezio, J. P. Remeika and P. D. Dernier. In course of publication.
4. M. T. Hutchings and E. J. Samuelsen, Phys. Rev. B<u>6</u>, 3447 (1972).
5. R. M. White and N. F. Mott, Phil. Mag. <u>24</u>, 845 (1971).
6. M. G. Townsend et al., J. Phys. C<u>4</u>, 598 (1971).

DISORDERED BOND MODEL FOR $V_{1-x}Cr_xO_2$

T. M. Rice
Bell Laboratories, Murray Hill, N. J. 07974

ABSTRACT

A model is proposed for the M_3- phase of $V_{1-x}Cr_xO_2$, in which all V atoms are locally bonded but the arrangement of the bonds on half the V sites is disordered. In the average structure, seen by classical X-ray methods, these V atoms appear to be unbonded. It is further proposed that at M_3-M_2 these V sites become localized V^{4+} atoms. Brief comments are made on other VO_2 alloy systems.

INTRODUCTION

While it has been well established that the metal-insulator transition in VO_2 arises from the pairing of V ions in the low temperature monoclinic phase,[1,2] the nature of the driving force has been the subject of considerable controversy. On the one hand, Berglund and Guggenheim,[3] Paul[4] and Hearn[5] have interpreted the transition as being driven by the lattice distortion and the electronic changes as due to the consequent rearrangement of the band structure. On the other hand Zinamon and Mott,[6] Rice, McWhan and Brinkman,[7] and Lederer et al.[8] have argued that in VO_2 the d-electrons are close to localization as V^{4+} ions and that the intra-atomic correlations are important. In Ref. 7, it was argued that it would be difficult to interpret the magnetic and electrical properties of $V_{1-x}Ti_xO_2$ with other than a local point of view. Detailed studies of the $V_{1-x}Nb_xO_2$ system have been carried out by the Orsay and Bordeaux groups[8-10] and also support this point of view. They find that at low temperatures each Nb atom breaks a V-V bond and forms a V^{3+}-Nb^{5+} complex. At high temperatures a continuous transition within the rutile structure is observed from the metallic state to a localized insulating phase at $x \approx 0.12$. Accurate X-ray crystal structure determinations[11] below room temperature on the system $V_{1-x}Cr_xO_2$ reveal a complex phase diagram as shown in Fig. 1 with two new monoclinic insulating phases M_2 and M_3. In both phases classical X-ray methods give the same structure in which one half the V atoms are bonded while the other half are not. The structure is illustrated in Fig. 2 by the large solid dots. Such a structure cannot simply be explained on a localized point of view since, in the lower temperature M_3 phase, the small susceptibility is inconsistent with local moments on the unpaired V sites. Hence, the dilemma that, while the overall properties of the mixed $V_{1-x}M_xO_2$ systems are in favor of a localized point of view, the insulating nature of the low temperature M_3 phase would appear to support a band picture.

In this paper a model is proposed for the M_3-phase in which all V atoms are bonded but a disordered arrangement of the bonds leads

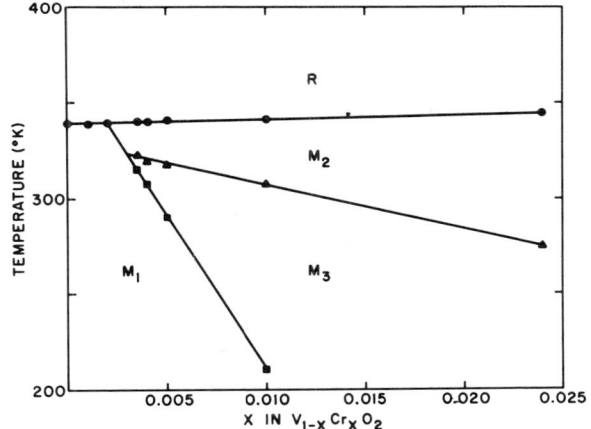

Fig. 1 The Phase Diagram of $V_{1-x}Cr_xO_2$ found by Marezio et al.[11]

to an average structure in which half the atoms appear to be unbonded.

THE M_3-PHASE OF $V_{1-x}Cr_xO_2$

In VO_2 there are two strong forces. One is the bonding attraction between near neighbor V atoms. The other is the requirement to maintain the balance of ionic charges. Marezio et al[12] have pointed out that this latter requirement accounts for the twisting of the V-atoms. The V-atoms move off the c-axis in order to shorten the distance to the oxygen atoms situated between V atoms which have moved apart along the adjacent c-axis. This effect is illustrated in the arrangement of V atoms shown in Fig. 2. The rutile phase may be split into two interpenetrating V sublattices with alternate V chains in Fig. 2 belonging to the first and second sublattices respectively. The two sublattices interact via the O atoms. The bonding in pairs on one sublattice causes the twisting on the other sublattice. Thus, if the bonding is random on one sublattice, it will cause random twisting on the other. Such a model is shown in Fig. 2 where a (110) plane of the rutile lattice is shown. On the first sublattice, the bonding is fixed as illustrated by the solid lines, while on the second the bonding is random, with successive V chains being paired at random, either as shown by the - - - - lines, or the ___ . ___ . ___ lines. The twisting is fixed on the second sublattice. The random bonding on the second sublattice will induce a random twisting on the first sublattice in the (1$\bar{1}$0) plane i.e. perpendicular to the plane of the paper. While it is clear from Fig. 2 that the bonding on the second sublattice can be random in a (110) plane between each chain of V atoms in the second sublattice, the same is not true in a (1$\bar{1}$0) plane. In the latter case if the bonding is not regular there will be some V atoms which are forced to compensate two oxygen atoms

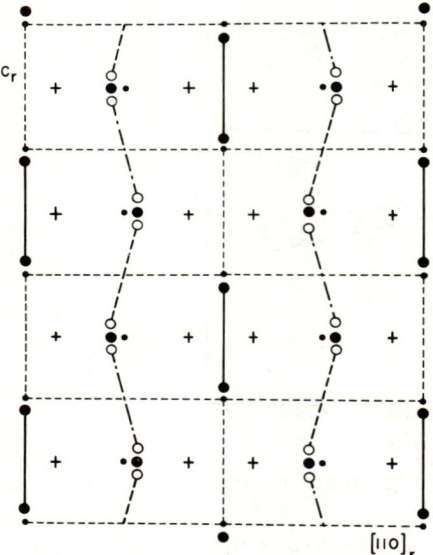

Fig. 2. The atomic positions in a rutile [110] plane. The O atoms are denoted by +. The average V atomic positions by large solid dots. The two possible bonding arrangements are denoted by the ------- and ———.———.——— lines.

simultaneously, violating the second restriction discussed above. Thus one expects the bonding arrangement to be regular in each $(1\bar{1}0)$ plane but disordered between successive $(1\bar{1}0)$ planes. This planar disorder will be averaged in a classical X-ray pattern. It is clear from Fig. 2 that the average positions correspond to the large solid dots which are seen experimentally.

Disorder manifests itself in classical X-ray methods as anomalously large thermal parameters. The authors of Ref. 11 have recently reexamined the Bragg reflections in the M_3-phase of a 0.5% Cr crystal. They found that the r.m.s. average displacements of the V atoms on the first and second sublattices were 0.095 Å and 0.094 Å respectively. These values are 27% larger than those found in the M_2-phase of the 2.4% crystal at the same temperature.

Such disordered structures give rise to X-ray diffuse scattering. The planar disorder proposed here should appear as rods of diffuse scattering[13] in one of the rutile [110] directions. The observation of such diffuse scattering would give conclusive evidence for the model.

THE M_2-PHASE OF $V_{1-x}Cr_xO_2$

The M_3–M_2 phase transition is marked by a rise in susceptibility.[11] After subtracting the contribution of the Cr^{3+} moments, the remaining susceptibility is $\approx 3.5 \times 10^{-4}$ and must be attributed to the V d-electrons. This is quite a large value. If we assume that the classical X-ray refinement gives the correct structure for the M_2-phase and that the unbonded V atoms are localized then we should expect a Curie-Weiss law for the susceptibility $\chi = C/(T+\theta)$.

The M_2-phase exists only over a narrow temperature range so it is not possible to determine the temperature dependence. However, if we substitute at T = 300 K assuming one half the V sites are localized V^{4+} atoms we obtain a value of $\theta \approx 200°K$. In the M_2-phase the chains of localized V^{4+} are surrounded by four chains of bonded V atoms which are only very weakly paramagnetic. Therefore, the magnetic character of the M_2-phase will be essentially 1-dimensional chains of localized spins with only very weak interaction between the chains. The rough estimate of $\theta \approx 200$ K will correspond to an intrachain exchange coupling of the same size.

The classical X-ray methods give only normal values for the thermal parameters in the M_2-phase thus supporting the view that the correct local structure is obtained from such refinements.

The $M_3 - M_2$ phase transition therefore is simply a transition from bonded (but disordered) V atoms in M_3 to localized unbonded V atoms in M_2 on the second sublattice. The greater entropy of the localized atoms favors the M_2 phase at higher temperature. Finally at the M_2-R transition all bonding breaks down and the metallic phase ensues. There are several possibilities for the way in which the bonds on the second sublattice break apart with increasing temperature in the M_3-phase. They may break abruptly at the M_3-M_2 phase transition or weaken gradually with an expansion of the short V-V distance such as occurs in V_4O_7.[14-15] A general discussion of the stability and properties of disordered bond phases has been given recently by Anderson.[16]

CONCLUSIONS

A model for the phase diagram for the $V_{1-x}Cr_xO_2$ system based on a local point of view is proposed. The considerations put forward here should also apply to other alloys of VO_2. Preliminary X-ray measurements by the authors of Ref. 9[17] on the $V_{1-x}Nb_xO_2$ system below room temperature indicate that near x = 0.1 the insulating phase is also rutile while monoclinic phases are stabilized at smaller x. If one extends the model put forward here for the M_3-phase to incorporate disordered bonds on both sublattices simultaneously then the average structure will be rutile although locally the V atoms are all bonded. The definitive test of the models proposed here would be the observation of the associated diffuse scattering.

ACKNOWLEDGMENTS

This work was begun during a stimulating stay at the Faculte des Sciences, Orsay. The author wishes to acknowledge many illuminating conversations with Drs. H. Launois, D. B. McWhan, M. Marezio, P. W. Anderson and W. F. Brinkman.

REFERENCES

1. A. Magneli and G. Andersson, Acta Chem. Scand. $\underline{9}$, 1378 (1955).
2. J. B. Goodenough, Magnetism and the Chemical Bond (Interscience Publishers, N.Y. 1963).
3. C. N. Berglund and H. J. Guggenheim, Phys. Rev. $\underline{185}$, 1022 (1969).
4. W. Paul, Mat. Res. Bull. $\underline{5}$, 691 (1970).
5. C. J. Hearn, Phys. Letters $\underline{38A}$, 447 (1972) and to be published.
6. Z. Zinamon and N. F. Mott, Phil. Mag. $\underline{21}$, 881 (1970).
7. T. M. Rice, D. B. McWhan and W. F. Brinkman, Proc. 10th Int. Conf. on Semiconductors (U.S.A.E.C. 1970), p. 293.
8. P. Lederer, H. Launois, J. P. Pouget, A. Casalot and G. Villeneuve, J. Phys. Chem. Solids $\underline{33}$, 1969 (1972).
9. G. Villeneuve, A. Bordet, A. Casalot, J. P. Pouget, H. Launois and P. Lederer, J. Phys. Chem. Solids $\underline{33}$, 1953 (1972).
10. J. P. Pouget, P. Lederer, D. S. Schreiber, H. Launois, D. Wohlleben, A. Casalot and G. Villeneuve, J. Phys. Chem. Solids $\underline{33}$, 1961 (1972).
11. M. Marezio, D. B. McWhan, J. P. Remeika and P. D. Dernier, Phys. Rev. $\underline{B5}$, 2541 (1972).
12. M. Marezio, D. B. McWhan, P. D. Dernier and J. P. Remeika, J. Sol. State Chemistry (in press).
13. A. Guinier, X-Ray Diffraction (W. H. Freeman, 1963) Chap. 7.
14. M. Marezio, D. B. McWhan, P. D. Dernier and J. P. Remeika, Phys. Rev. Letters $\underline{28}$, 1390 (1972).
15. A. C. Gossard and J. P. Remeika, Bull. Am. Phys. Soc. $\underline{17}$, 359 (1972).
16. P. W. Anderson (to be published).
17. H. Launois and G. Villeneuve (private communication).

NMR STUDIES ON ANTIFERROMAGNETIC STATE OF V_2O_3-Cr_2O_3

H. Yasuoka,* K. Motoya and Y. Nakamura
Department of Metal Science and Technology
Kyoto University, Kyoto 606, Japan

J. P. Remeika
Bell Telephone Laboratories, Murray Hill, N. J. 07974

ABSTRACT

Magnetic properties of the antiferromagnetic state of the V_2O_3-Cr_2O_3 mixed system have been investigated with a pulsed NMR technique. For pure V_2O_3, the hyperfine field studies lead to the conclusion that the magnetic moment of 1.2 μ_B/V-atom is mainly due to the spin moment which lies at 71° to the hexagonal c-axis. There was no evidence of spin-flop in field less than 60 kOe. In the mixed system, $(V_{1-x}Cr_x)_2O_3$, the V^{51} NMR frequency was not changed, while the intensity fell off more rapidly with increasing x than expected from a simple dilution.

INTRODUCTION

The existence of antiferromagnetic (AMF) long range order in V_2O_3 is by now well established below the transformation temperature from the corundum metallic phase to a monoclinic insulating phase (150-160 K).[1-3] The polarization analysis of neutron diffraction had indicated that ferromagnetic coupling exist within (110) hexagonal planes in the monoclinic phase and antiferromagnetic coupling between layers.[3] The zero-temperature atomic moment was also determined to be 1.2±0.1 μ_B/V-atom which lies at approximately 71° to the hexagonal c-axis. This surprisingly small moment is to be compared with the ordered spin only moment of 2.0 μ_B. In the $(V_{1-x}Cr_x)_2O_3$ system, the temperature-composition phase diagram was obtained from both magnetic susceptibility[4] and neutron diffraction[5] measurements.

---

*Visiting scientist at Bell Telephone Laboratories. Experiments were done at Kyoto University and manuscript done at present address, B.T.L., Murray Hill, N. J. 07974.

They show very interesting magnetic transitions accompanied with metal-nonmetal transitions. The microscopic properties were studied by V^{51} NMR in the paramagnetic metallic and insulating phases of this system.[6] Our attention here is focused on clarifying the magnetic and electronic properties in the AFM state.

EXPERIMENTAL

Most of the experiments in V_2O_3 were performed by using single crystals grown by the molten salt technique.[7] A continuous series of solid solution V_2O_3-Cr_2O_3 was prepared by a ceramic method which has been described earlier.[8] NMR measurements at temperatures up to 77 K were made by a standard high power pulsed apparatus. A superconducting solenoid was used to examine the field dependence of the resonance frequency up to 60 kOe in the stacked single crystals. The integrated amplitude of the spin-echo signal was stored in a boxcar integrator and recorded as a function of external field.

RESULTS AND DISCUSSIONS

(A) $\underline{V^{51} \text{ NMR in Pure } V_2O_3}$

i) Hyperfine Field

The spin-echo spectrum in V_2O_3 at zero external field has been interpreted as a manifestation of the combined effect of the Suhl-Nakamura interaction with the quadrupole interaction.[9] The frequency for resonance extrapolated to 0 K was 208.1±0.1 MHz which corresponds to a hyperfine field of 185.9±0.1 kOe. Although the sign of the hyperfine field could not be determined experimentally, it could be inferred to be negative, since the orbital contribution to the hyperfine field should be the same direction as the core-polarization field for ions having less than half-filled d-shells. Comparing this field to the known hyperfine coupling constant due to the spin part, $H_{spin}^{hf} \approx -140$ kOe/μ_B,[1] it is concluded that the saturation moment of 1.2 μ_B is mainly due to the spin moment. The reduction of the moment from 2.0 μ_B could thus not be explained by the orbital contribution and the reason for the moment reduction remains still in question.

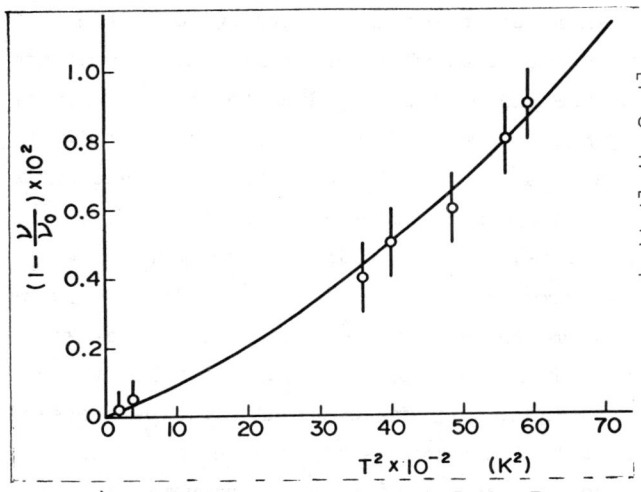

Fig. 1
The fractional change of the V^{51} resonance frequency in V_2O_3. The solid curve is a least-squares fit to the spin-wave theory.

ii) Temperature Dependence of the Resonance Frequency

In Fig. 1, the fractional change of the resonance frequency which generally resembles that of the sublattice magnetization, is plotted against the square of temperature. The solid curve indicates a best fit of the data to a spin wave prediction for a two sublattice antiferromagnet. Within the limit of small spin wave energy gap, the temperature dependence of the sublattice magnetization can be expressed as, $\Delta M/M_o = \alpha T^2 + \beta T^4 + \ldots$ From the least-squares fit, we found the coefficients α and β to be 7.83×10^{-7} $(K)^{1/2}$ and 1.14×10^{-10} $(K)^{1/4}$, respectively. Using these coefficients, the Néel temperature was estimated to be about 300 K. This is in agreement with $T_N \approx 285$ K obtained from the neutron scattering measurement.[3]

iii) External Field Dependence of the Resonance Frequency

Using stacked single crystals with the hexagonal c-axis being the same, the external field dependence of the resonance frequency was measured for both the field parallel and perpendicular to the c-axis. Since we could not set the crystals to a particular crystallographic axis within the c-plane, the observed spectra for $H_o \perp$ c-axis showed an asymmetrical shape. Analyzing these spectra by a random distribution of the magnetic moment in the c-plane, we found the extreme points which correspond to the minimum angle between the moment and the external field. Taking these points, together with the case of $H_0 \parallel$ c-axis where the spectrum was symmetrical, the frequency-field diagram was obtained as shown in Fig. 2. The dashed lines in Fig. 2 indicate the slopes corresponding to the γ-value for free V^{51} nucleus (11.193 MHz/10 kOe) in the two sublattice antiferromagnet. The solid curves were calculated taking into account the angle between the magnetic moment and the c-axis to be 71°, which was determined by the neutron scattering measurement.[3] The agreement between the experimental points and the calculated curves clearly indicates that the moment is indeed tilted from the c-axis by 71°. Two interesting features should be noted here. First, no substantial orbital shifts induced by the external field were observed. This is consistent with the previous discussion of the zero field hyperfine field, where we concluded that the magnetic moment of 1.2 μ_B is mainly the "spin only" value and the hyperfine field is due to the core-polarization induced by this moment. Secondly, no evidence of spin-flop was observed in fields up to 60 kOe. This clearly means that the anisotropy energy, even within the c-plane, must be quite large. Since the magnetic moment is tilted away from the c-axis and the orbital angular momentum seems to be almost quenched, the origin of the large anisotropy is interesting to investigate further.

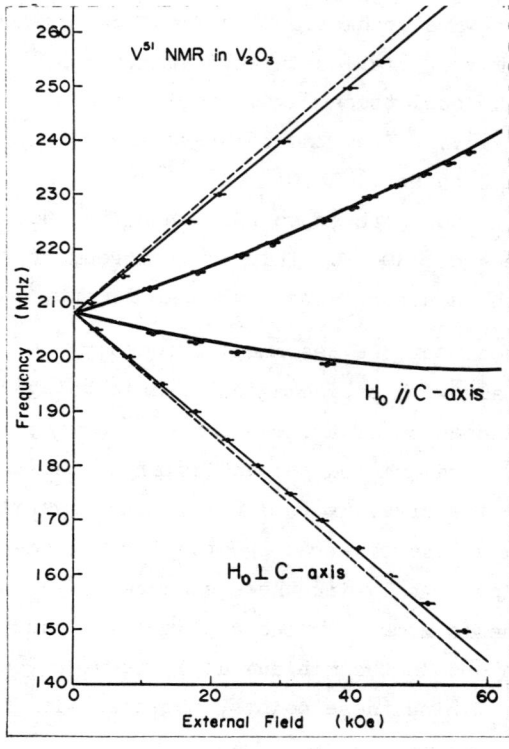

Fig. 2

The resonance frequencies against the external field parallel and perpendicular to the hexagonal c-axis at 4.2 K.

(B) V^{51} and Cr^{53} NMR in $(V_{1-x}Cr_x)_2O_3$

In order to further understand the magnetic state in V_2O_3, where the V atoms are in a rather peculiar ionic state, the V^{51} and Cr^{53} NMR were observed in the mixed system of V_2O_3 and the purely ionic insulator Cr_2O_3. For $x \leq 0.1$, the V^{51} NMR frequency does not change with increasing x, while the integrated intensity falls off more rapidly with increasing x than expected from a simple dilution. A crude analysis of the intensity measurement suggests that the magnetic state of a V atom having a Cr atom in the 13 near-neighbor metal sites will be drastically modified from that in pure V_2O_3 and presumably behave nearly a V^{3+} free ion. On the C_2O_3 rich samples, both free-induction and spin-echo signals of Cr^{53} nuclei were observed at about 70.5 MHz. The spin-echo envelope decay and the free-induction decay were both strongly modulated by the quadrupole interaction. For decreasing x from x = 1.0, no remarkable reduction of the Cr^{53} NMR intensity was observed. Detailed investigations are now in progress including the nuclear relaxation studies.

ACKNOWLEDGMENT

We wish to thank A. C. Gossard for many discussions and for critical reading of the manuscript.

REFERENCES

1) E. D. Jones, Phys. Rev. <u>137</u>, A978 (1965).
2) T. Shinjo and K. Kosuge, J. Phys. Soc. Japan <u>21</u>, 2622 (1966).
3) R. M. Moon, Phys. Rev. Lett. <u>25</u>, 527 (1970).
4) A. Menth and J. P. Remeika, Phys. Rev. <u>B2</u>, 3756 (1970).
5) A. F. Feid, T. M. Sabine and D. A. Wheeler, J. Sol. St. Chem. <u>4</u>, 400 (1972).
6) A. C. Gossard, A. Menth, W. W. Warren and J. P. Remeika, Phys. Rev. <u>B3</u>, 3993 (1971).
   M. Rubinstein, Phys. Rev. <u>B2</u>, 4731 (1970).
7) D. B. McWhan and J. P. Remeika, Phys. Rev. <u>B2</u>, 3734 (1970).
8) D. B. McWhan, A. Menth and J. P. Remeika, J. de Physique <u>23</u>, C1-1079 (1971).
9) H. Yasuoka, H. Nishihara, Y. Nakamura and J. P. Remeika, Phys. Lett. <u>37A</u>, 299 (1971).

STRONG MAGNETIC FIELD EFFECT ON MAGNETOELECTRIC PROPERTIES OF Cr_2O_3 SINGLE CRYSTAL

K. Daido, K. Hoshikawa and C. Uemura
Musashino Electrical Communication Laboratory
Nippon Telegraph and Telephone Public Corporation
Musashino-shi, Tokyo, Japan

Antiferromagnetic Cr_2O_3 shows interesting magnetoelectric behavior. Astrov[1] and Rado et al.[2] have pointed two antiferromagnetic spin states to explain the variation of sign and magnitude of the measured magnetoelectric susceptibility α. These spin states are controlled by cooling through the Neel temperature (T_N=307°K) in the presence of D.C. magnetic field in proper direction[3].

Our detailed investigations on the Cr_2O_3 single crystals grown by flux evapolation method[4] were reconfirmed the above results. For the explanation of these phenomena it is necessary to introduce the existence of some directionality in crystal in spite of no evidence from the so far stated magnetic symmetry. The following experiments have been done concerned with the nature of the directionality.

The strong magnetic field was applied on a crystal along c-axis below T_N, and then α was measured. As the crystal stayed in antiferromagnetic region, the value of α had no change. After heating above T_N followed by magnetic field cooling, however, change of α was observed which closely related with the direction and magnitude of the strong field. The variation of α with the strong field on typical three specimens are shown in the figure. Moreover it was found that the sign and value of α changed by the strong field, which was measured after reheating and magnetic field cooling, varied with time and brought back to the initial state. A similar effect was observed when the strong field was applied above T_N.

From the above experimental results, the directionality in Cr_2O_3 crystal seems to be reversed by the strong magnetic field and to be existed below and above T_N. We suppose that this directionality is originated from a magnetically aranged defect.

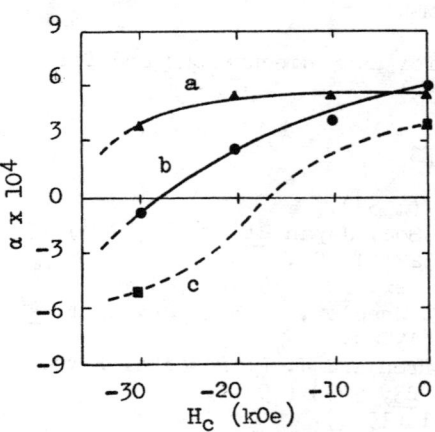

Fig. α vs strong magnetic field strength under constant magnetic field cooling condition.
(T=293°K)

1. D.N.Astrov, Soviet Physics-JETP, 13, (1961), 729
2. G.T.Rado and V.J.Folen, Phys. Rev. Letters, 7, (1961), 310
3. T.J Martin and J.C.Anderson, Phys. Letters, 15, (1964), 109
4. R.E.Barks and D.M.Roy, Proc. of I.C.C.G.-'66, (1967), 497

MAGNETOELECTRIC EFFECTS IN TbPO$_4$

G. T. Rado and J. M. Ferrari
Naval Research Laboratory, Washington, D.C. 20390

ABSTRACT

It is shown experimentally that the antiferromagnet TbPO$_4$ exhibits a linear magnetoelectric (ME) effect and that its ME susceptibility tensor $\vec{\alpha}$ has a component α_{aa} whose largest absolute value is considerably larger than any component of $\vec{\alpha}$ previously reported for any material. Here a denotes one of the axes (a, a', c) of the tetragonal crystal structure existing above the Néel temperature. The electrically as well as the magnetically induced ME effect was used to measure the temperature (T) dependence of $|\alpha_{aa}|$ on a single crystal which had been annealed magnetoelectrically (with an appropriate electric field and a necessarily "small" magnetic field) so as to maximize $|\alpha_{aa}|$. The largest value of $|\alpha_{aa}|$ is $|\alpha_{aa}|_{max} = 11 \times 10^{-3}$ Gaussian units and occurs at T = 1.92 ± 0.01°K. Using the previously proposed atomic mechanism[1] of the ME effect in rare earth crystals and the fragmentary spectroscopic data presently available, we compare this value of $|\alpha_{aa}|_{max}$ with the value $|\alpha_{aa}|_{max} = 1.2 \times 10^{-3}$ Gaussian units measured[1] in ME-annealed DyPO$_4$. While we are still investigating the symmetry of magnetoelectrically annealed TbPO$_4$, we have concluded from suitable ME experiments that the symmetry of non-annealed TbPO$_4$ in the antiferromagnetic state is lower than tetragonal. For sufficiently large absolute values of a biasing magnetic field H oriented parallel to either a, a', or c, the dependence of α_{aa} on H reveals abrupt magnetic transitions, hysteresis, and switching effects. A full account of this work will be published elsewhere.

[1] G. T. Rado, Phys. Rev. Letters 23, 644 (1969); 23, 946 (E) (1969); Solid State Commun. 8, 1349 (1970).

Section 44. Magneto-Optic Memory
and Magnetic Recording

FARADAY ROTATION AND OPTICAL ABSORPTION OF EPITAXIAL FILMS OF $Y_{3-x}Bi_xFe_5O_{12}$

S. Wittekoek, J.M. Robertson, T.J.A. Popma, P.F. Bongers
Philips Research Laboratories, Eindhoven, The Netherlands

ABSTRACT

Thin films of $Y_{3-x}Bi_xFe_5O_{12}$ ($x \leq 0.3$) have been grown on $Gd_3Ga_5O_{12}$ and $Sm_3Ga_5O_{12}$ substrates by liquid phase epitaxy. Faraday rotation and optical absorption measurements between 0.7 and 0.4 micron are reported. Near 0.5 the FR has a large negative value of -3 degrees/micron for x = 0.26 as compared with +0.3 degrees/micron for YIG. Both the optical absorption and the figure of merit (rotation/absorption) are higher for the Bi containing garnet films. The promising properties of these materials for magneto-optic applications are discussed.

INTRODUCTION

It has been reported that the Faraday rotation of garnets of composition $Ca_{3-x}Bi_xFe_{3.5+0.5x}V_{1.5-0.5x}O_{12}$ (CaVBIG) is unusually high[1,2] and cannot be explained by the phenomenological theory developed by Crossley et al.[3], and successfully applied to many ferri-oxides.[3,4]

Analysis of the Faraday rotation measurements at wavelengths down to 0.6 by Lacklison et al.[5] and A. Akselrad[6] have indicated that an anomalous contribution to the rotation is associated with the presence of Bi^{3+} ions. Investigations of the magneto-optic Kerr effect by Wittekoek et al.[7] have shown that the origin of this Faraday rotation is a magneto-optically active, electronic transition near $\lambda = 0.47\mu$ which is only present in CaVBIG and not in the reference compound YIG. The transition was attributed to a charge transfer of an electron from oxygen to Fe^{3+} at tetrahedral sites for which the magneto-optic activity is enhanced by the presence of Bi.

In order to investigate the Faraday rotation associated with this transition we have prepared thin films of the Bi-containing garnets $Y_{3-x}Bi_xFe_5O_{12}$ and measured the magneto-optical properties in the visible range of the spectrum. Of special interest in this range is the wavelength 5145 Å of the Argon laser, which is a likely light source for magneto-optic devices.

FILM PREPARATION AND ANALYSIS

Thin epitaxial films of composition $Y_{3-x}Bi_xFe_5O_{12}$ were grown on [111] oriented substrates of $Gd_3Ga_5O_{12}$ (GGG) and $Sm_3Ga_5O_{12}$ (SmGG) by the method of liquid phase epitaxy, using a dipping procedure similar to that described by Levinstein et al.[8] The starting composition of the melt was PbO 256 g, B_2O_3 6.45 g, Y_2O_3 2.615 g, Fe_2O_3 16.98 g and Bi_2O_3 18.00 g (all the chemicals were of 4 N or greater purity). The dipping temperature was varied between 960 and 700°C. In subsequent runs the Bi_2O_3 content of the melt was increased. The viscosity of the melt was found to increase with Bi_2O_3 content and with decreasing temperature. To overcome this problem the mole ratio of PbO/B_2O_3 was decreased from 12 to 6 and resulted in lower liquidus temperatures being obtainable and better quality films. As shown in Figure 1, the bismuth content in the film increased with decrease in growth temperature.

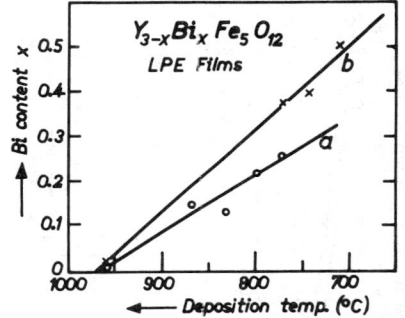

Fig. 1. Bi content versus deposition temperature for 18 g Bi_2O_3 (a) and 28 g Bi_2O_3 (b) in flux.

The composition of the films was determined by electron probe analysis. The Bi content was measured using the BiM_α and BiM_β intensities and compared with standards prepared from polycrystalline Bi-YIG samples. The highest Bi content that could be realised was $x = 0.5$. The films with high Bi content contained also appreciable amounts of Pb from the flux. The optical studies reported here have been limited to films with $x \leq 0.3$, in which the weight percentage of Pb was below 2%.

OPTICAL MEASUREMENTS

The <u>Faraday rotation</u> (FR) of several Bi containing films was measured in the spectral range between 0.7 and 0.48 μ, and for some submicron films down to 0.38μ. Since we employed a polarization modulation technique the accuracy of the rotation measurements was better than 0.01° but the specific rotations in degree/cm that are reported here are accurate within 10% only due to uncertainties about the exact thickness of the films. The rotations have been corrected for the contributions to the FR from the relatively thick (100-200μ) substrates of GGG and SmGG. As a function of the externally applied field H_0 the rotation of the films reached a maximum value for $H_o \approx$ 2 kOe which roughly equals the saturation magnetization $4\pi M_s$ perpendicular to the film plane.

In Fig. 2 the effect of bismuth doping upon the spectral dependence of the Faraday rotation is shown. The positive FR of pure YIG decreases when small amounts of Bi are incorporated and for large amounts of Bi the FR reaches a large negative value which peaks at $\lambda = 0.5\mu$.

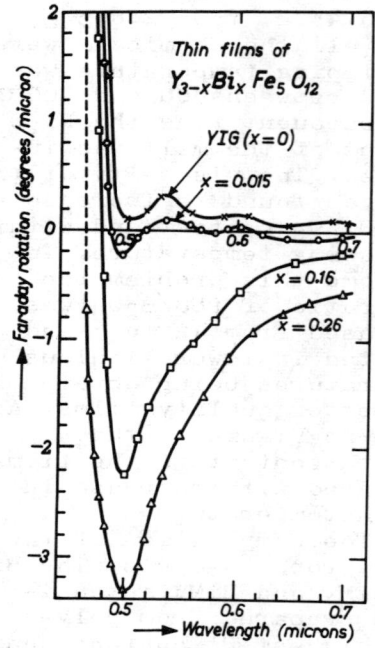 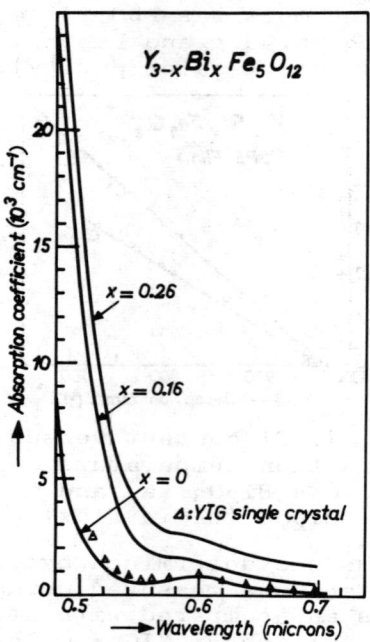

Fig. 2. Faraday rotation of LPE films of $Y_{3-x}Bi_xFe_5O_{12}$.

Fig. 3. Optical absorption of films of Fig. 2.

At shorter wavelengths the FR decreases and shows a sharp zero crossing near 0.47μ (for $x = 0.26$). In the wavelength region below that shown in Fig. 2, the FR has large positive values with a maximum value near $\lambda = 0.43\mu$ for both Bi doped and undoped YIG.

Between 0.7 and 0.55μ the FR of a pure YIG film was checked against that of a thin (23μ) single crystal grown from the flux and very good agreement was found.

The spectral dependence of the <u>optical absorption coefficient</u> of the films is shown in Fig. 3. The result has been derived from an analysis of transmission and reflection measurements. A conclusion from Fig. 3 is that the LPE films which contain bismuth have higher absorption coefficients than that of pure YIG. At the present stage of the work it is not justified to conclude from Fig. 3 that the Bi doping is solely responsible for the increased absorption. As mentioned above,

the LPE films which contain bismuth also contain Pb.
Further experiments to distinguish the contributions to
the optical absorption from both Bi and Pb are in
progress.

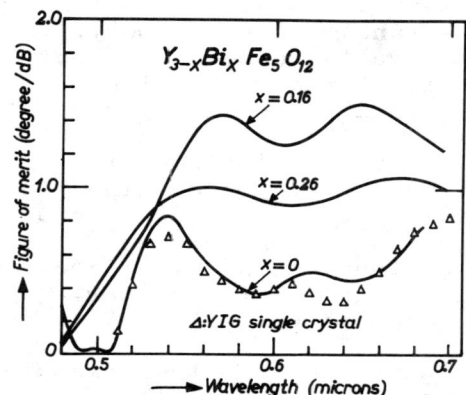

Fig. 4. Figure of merit
(rotation/absorption) for
LPE films with Bi.

In Fig. 4 we have plotted the magneto-optical <u>figures of merit F</u> (degrees/dB) for three LPE films and a YIG single crystal. For the bismuth doped films the figures of merit in the long wavelength region ($\lambda > 0.65\mu$) are lower than those reported for CaVBIG.[1,4] This can be traced back to the relatively high absorption coefficients of our films in this spectral region.

It is evident from Fig. 4 that the Bi doped films have higher figures of merit then pure YIG over the whole spectral region studied. However the increase of F obtained by Bi doping is strongly dependent on the wavelength chosen. For the wavelength of the Ar ion laser (5145 Å) the various optical properties of the films have been summarized in the following table.

TABLE I

Properties of L.P.E. films of $Y_{3-x}Bi_xFe_5O_{12}$. Results of analysis and magneto-optical measurements at 5145 Å.

Bi content	weight %			thickness	absorption $[10^3 \text{ cm}^{-1}]$	Faraday rotation $[10^3 \text{ °/cm}]$	Fig. of merit $[\text{deg/dB}]$
	Bi	Pb	Pt				
x=0	0.0	0.01	0.01	2x3 μ	1.8	2.55	0.33
x=0.16	4.0	1.5	1.5	2x0.8 μ	8.3	21.0	0.61
x=0.26	6.5	2.0	1.5	2x1.7 μ	9.8	27.0	0.64

DISCUSSION

It can be concluded that by means of liquid phase epitaxy single crystal layers of $Y_{3-x}Bi_xFe_5O_{12}$ can be grown which have superior magneto-optical properties in the visible spectral region, compared with existing garnet materials. The most important effect of the bismuth doping is the strong increase of the Faraday rotation: about a factor of ten at 5145 Å. The optical absorption coefficient and the figure of merit are both higher than for other garnets (YIG, GdIG etc.). Therefore these layers are very promising candidates for magneto-optical memory applications: The high absorption coefficient allows the use of thinner films which permits faster thermal writing and facilitates the structuring of layers, necessary to obtain spots for information storage such as a random access memory.[9] These thin layers will still have favourable read-out properties by virtue of the high Faraday rotation. For these storage applications it will be necessary to tailor the magnetic properties of the garnet material to obtain low magnetization and a compensation or Curie point near room temperature. This could be obtained in systems like $Gd_{3-x}Bi_xFe_{5-y}Ga_yO_{12}$ or $Gd_{3-x}Bi_xFe_{5-y}In_yO_{12}$. We expect that the absorption coefficient of the films will be lower when the lead content can be minimized, resulting in garnet films with high figures of merit which are useful for various magneto-optical applications.

A tentative model explaining the effect of bismuth doping is given in ref. 7, where it is suggested that the combined effect of mixing of oxygen 2p- and bismuth 6p-orbitals and the high spin-orbit coupling of the Bi-orbitals causes a strong increase in the magneto-optical rotation of a charge transfer transition near 0.47μ, occurring within the tetrahedral iron-oxygen complex. It can be expected that higher magneto-optical rotations can be realised when more bismuth can be incorporated in the garnets.

We wish to thank Ms A. Wijma and Mr. M. van Hout for skilful assistance with the experiments.

REFERENCES

1. C.F. Buhrer, J. Appl. Phys. **40**, 4500, (1969).
2. R.M. Lambert, J. Phys. D. **4**, 139 (1971).
3. W.A. Crossley, R.W. Cooper, J.L. Page and R.P. van Stapele, Phys. Rev. **181**, 896 (1969).
4. P.F. Bongers and G. Zanmarchi, Solid State Commun. **6**, 291 (1968).
5. D.E. Lacklison, H.I. Ralph and G.B. Scott, Solid State Commun., (1972) to be published.
6. A. Akselrad, Proceedings of the 17th Conference on Magnetism and Magnetic Materials, AIP Proceedings **5**, 249, (1972).
7. S. Wittekoek and D.E. Lacklison, Phys. Rev. Lett. **28**, 740 (1972).
8. H.J. Levinstein, S. Licht, R.W. Landore and S.L. Blank, Appl. Phys. Lett. **19**, 486 (1971).
9. J.P. Krumme, J. Verweel, Appl. Phys. Lett. **20**, 451 (1972).

MAGNETIC PROPERTIES AND CURIE-POINT WRITING IN EVAPORATED MnAlGe FILMS

K. Y. Ahn
IBM Thomas J. Watson Research Center, Yorktown Heights, New York 10598

E. Sawatsky and B. R. Brown
IBM Research Laboratory, San Jose, California 95114

ABSTRACT

Magnetic and structural properties of vacuum-evaporated MnAlGe films are presented and magneto-optical data storage with this material is discussed. X-ray analyses of films prepared by simultaneous evaporation and sequential layering show polycrystalline MnAlGe with bulk lattice constants. The optical absorption is featureless in the visible and near infrared regions with a typical absorption coefficient of 4.7×10^5/cm at 0.6328 μm. The Faraday rotation in the visible spectrum is also flat and is $\sim 1 \times 10^5$ deg/cm. H_c is typically 9 kOe in 1000Å thick films. Thermo-magnetic Curie-point writing was achieved at 25°C with a 2 μm×20 μm GaAs laser beam, and the written bits were read out with the same laser at a reduced intensity.

INTRODUCTION

The preparation of MnAlGe films by d-c getter sputtering, and some optical and magnetic properties, together with the feasibility of Curie-point writing were reported recently[1]. In this paper we examine and compare physical properties of MnAlGe films prepared by vacuum evaporation, and demonstrate their application in room-temperature beam-addressable data storage.

FILM FABRICATION AND STRUCTURE

In order to maintain a low background pressure during evaporation, a uhv system was used. It consists of a stainless-steel chamber, integrally mounted ion pumps, titanium sublimation pumps, and liquid-nitrogen traps. The total pumping speeds for some of the gaseous species are: $> 1.5 \times 10^5$ ℓ/s for H_2O, 2.5×10^5 ℓ/s for H_2, and $> 1.5 \times 10^4$ ℓ/s for N_2. High-purity materials (99.99% or better) were heated in a crucible and two electron-beam guns, for outgassing, prior to evaporation. Films were deposited onto a variety of substrates by simultaneous evaporation from three sources onto substrates heated to 300°C, or by sequential deposition of multiple layers (Al-Mn-Ge-Al-Mn-Ge) onto substrates held at 100°C followed by an annealing in vacuum at 300°C for 30 min. Both methods yielded homogeneous films with similar structure and magnetic properties. The single-crystal substrates of sapphire and spinel were chemically cleaned[2].

The background pressure in the bell jar rose from 4.2×10^{-10} Torr before evaporation to the following pressures during evaporation:

6×10^{-9} Torr for Mn, 2.5×10^{-8} Torr for Al, 3×10^{-8} Torr for Ge, and 8×10^{-8} Torr for coevaporation from three sources. Residual gas spectrum determined by a quadripole analyzer shows the following four major species during evaporation in descending order of amplitude: mass-to-charge ratio of 2 (H_2), 18 (H_2O), 44 (CO_2), and 28 (CO or N_2). Typical deposition rate for the simultaneous evaporation was 3 to 5 Å/s for a source-to-substrate distance of 30 cm.

The crystal structure of films was analyzed by three different x-ray diffraction techniques. In addition to the Bragg-Brentano diffraction and pole figures, glancing-angle pictures were also taken using a Debye-Scherrer powder camera at 10° angle between the incoming collimated beam and the film surface.

In films deposited on fused quartz and (111) spinel, there are powder lines corresponding to the bulk MnAlGe with tetragonal structure. However, the 002 and 003 reflections are missing, indicating that either the 002 and 003 reflections are weak or there is fibre texture in the film. Glancing-angle pictures show all lines of MnAlGe in films deposited on Al_2O_3 (c-axis $\perp$ surface) with structure in some of the lines. The exceptionally strong 112 reflection indicates a probable 112 preferred orientation perpendicular to the film plane. Additional data from Bragg-Brentano diffraction and 112 pole figures also confirm such preferred orientation.

MAGNETIC AND MAGNETO-OPTICAL PROPERTIES

Variations in the optical properties of MnAlGe with regard to crystallographic orientation are not known at this time. The absorption measurements presented below are for polycrystalline films with preferred orientation. The wavelength dependence of the optical absorption (at room temperature) and the polar Faraday rotation (at 100°C to reduce H_c) for MnAlGe on fused quartz substrates are shown in Fig. 1. The absorption was measured with a Cary 14 spectrophotometer and is not corrected for reflectivity.

Except for a gradual increase in absorption in the near uv, the absorption is quite featureless over the wavelength region covered in Figure 1. Since the optical density in the films used for this work was quite high even

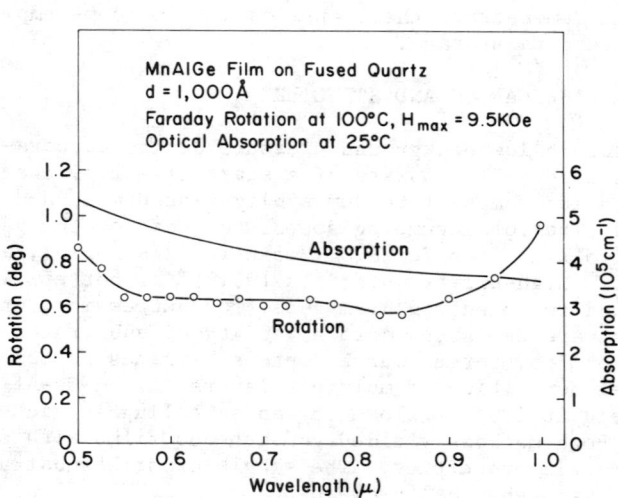

Fig. 1. Faraday rotation and optical absorption as a function of wavelength for evaporated MnAlGe film on fused quartz.

in the near ir (∼ 1.5), no interference fringes were observed in the absorption spectrum. The absorption coefficient of these evaporated, stoichiometric films is similar to that reported for d-c getter sputtered stoichiometric films[1]. RF-sputtered, nonstoichiometric films have recently been shown to have similar absorption properties[4]. The high optical density in MnAlGe is consistent with its metallic character.

The Faraday rotation was obtained from hysteresis loop measurements in an applied magnetic field limited to 9.5 kOe. Since this field is insufficient to saturate films at room temperature, the wavelength dependence of the Faraday rotation was determined at 100°C.

The Faraday rotation is essentially constant between 0.6 μm and 0.9 μm and increases slightly for wavelengths outside this region. The reasons for this variation in Faraday rotation is not understood at this time, especially since there is no structure in the absorption spectrum over the same wavelength region. The saturation rotation at room temperature for films (∼ 1,000 Å thick) on different substrates is summarized in Table I for a wavelength of 0.6328 μm together with room-temperature coercivities.

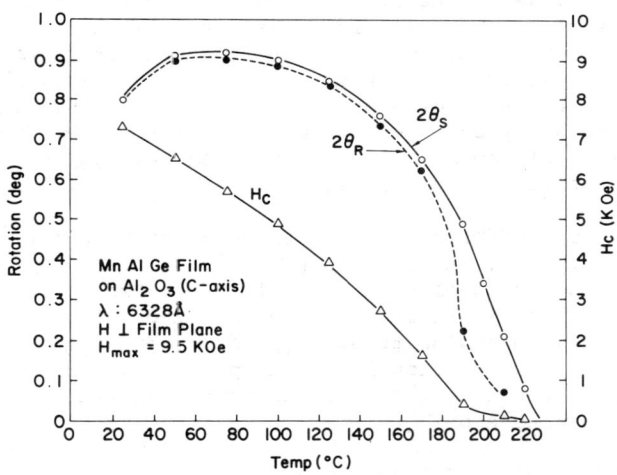

Fig. 2. Temperature dependence of the Faraday rotation and H_c for MnAlGe film on c-axis oriented Al_2O_3. The external field was applied normal to the film plane.

A typical temperature dependence of the Faraday rotation and coercive force for MnAlGe films on a c-axis oriented sapphire substrate is shown in Fig. 2. The apparent decrease in rotation near room termperature arises from incomplete saturation in the 9.5 kOe applied field, and the observed hysteresis loops near room temperature were in fact minor loops. Hence, the H_c values at room temperature (measured in another apparatus with $H \geq 15$ kOe) are higher as shown in Table I.

The temperature dependence of the Faraday rotation closely resembles that of magnetization of a typical ferromagnet, while H_c decreases almost linearly with increasing temperature. The Curie temperature T_c for this film is extrapolated to be about 230°C (Figure 2). The small difference between T_c for this film and the bulk MnAlGe (245°C) is probably due to a slight nonstoichiometry in the film. Similar results were seen in r-f sputtered films[4]. The remanent rotation Θ_R approaches the saturation rotation Θ_S, also shown in Fig. 2, and the ratio, Θ_R/Θ_S, remains almost unity up to $\sim 170°C$.

Table I Comparison of Faraday Rotation and H_c
At 25°C, $\lambda = 0.6328 \mu$, $H_{max} = 15$ kOe

Substrate	H_c (kOe)	$2\Theta_S$ (deg)
Fused Quartz	9.5	1.0
Al_2O_3 (C)	8.7	1.1
Spinel (111)	9.0	0.9
MgO (111)	9.3	1.0
Mica	10.3	1.4

Although the fundamental properties of the evaporated films are similar to those of sputtered films, the evaporated films have considerably higher coercivities. Typically, H_c is 9 kOe for 1,000 Å thick films, and it is relatively independent of substrate materials and methods of film deposition. The difference between the coercivities in evaporated films and the lower values in sputtered films (for example, 3.5 kOe at 300°C substrate temperature) are not well understood at this time. Optical microscopy with polarized light and electron transmission microscopy show the evaporated films to be similar to the sputtered films of low H_c. Comparison with the high H_c sputtered films is not possible as details of structure were not reported

Differences in H_c values between films prepared by evaporation and sputtering may be attributed, at least in part, to subtle differences in crystallization and residual strain arising from two different deposition techniques. The increased H_c is advantageous for applications in high-density recording to overcome demagnetizing fields from small bit size. No additional heating power or magnetic field is necessary when Curie-point writing is employed because of the strong temperature dependence of H_c.

DYNAMIC WRITE AND READ TESTS

A dynamic test of the capability for Curie-point writing and Faraday-effect readout was performed on the MnAlGe samples using low-temperature GaAs lasers. The sample was spun at room temperature in a carrier disk (76mm dia) at 6,000 rpm to achieve a tangential velocity of 20m/s. At the sample, the light beam is an image of the laser junction with dimension (to the half power points) of 2 μm x 20 μm. The beam is normally incident with the short dimension parallel to the sample motion, resulting in an exposure time of ~ 100 nsec for each sample point as it passes through the beam.

Curie-point writing is achieved by pulsing the laser at high power while applying a magnetic bias field. The field opposes the magnetization of the film, and is supplied by an electromagnet coaxial to the optical system. An applied field of 500 Oe was found to be adequate for writing.

Faraday-effect readout was obtained using a lower-power CW laser output of 20 mW. Polarization of the transmitted light was monitored with a differential detector system[5]. Figure 3 shows an example of the readout signal where each major division of the horizontal scale corresponds to 40um of sample. Each major vertical division corresponds to 0.45° of rotation of the detected light polarization. Writing pulse length was 0.5 μs for an effective data rate of 2×10^6 bits/s. The size of the written spot, considering beam size and sample motion, is 12 μm x 20 μm. Peak beam intensity used for writing was $7 mW/\mu m^2$.

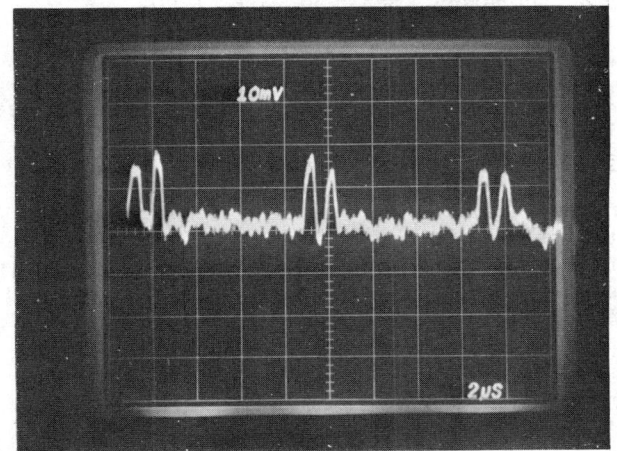

Fig. 3. Readout signal from dynamic write/read tests.

It can be estimated from the trace that, with a 2 μm x 20 μm beam, the MnAlGe film produces a peak-to-peak signal to rms medium noise ratio of about 13:1. In this experiment, the detector bandwidth is 5MHz and the observed p-p signal to rms detector system shot and electronic noise is about 20:1.

ACKNOWLEDGMENT

It is a pleasure to thank J. Angilello for X-ray analyses and W. W. Molzen for substrate preparations.

REFERENCES

1. R. C. Sherwood, E. A. Nesbitt, J. H. Wernick, D. D. Bacon, A. J. Kurtzig, and R. Wolfe, J. Appl. Phys. 42, 1704 (1971).
2. A. Reisman, M. Berkenblit, J. J. Cuomo, and S. A. Chan, J. Electrochem. Soc. 10, 1653 (1971).
3. W. Rühl, Zeitschrift für Physik 138, 121 (1954).
4. K. Lee, E. Sawatzky, and J. C. Suits, to be published.
5. B. R. Brown, IBM J. Res. & Dev. 16, 19 (1972).

MAGNETIC AND MAGNETO-OPTICAL PROPERTIES OF SPUTTERED MnGaGe FILMS

E. Sawatzky and G. Bryan Street
IBM Research Laboratory, San Jose, Ca. 95114

ABSTRACT

The new compound manganese gallium germanide has recently been shown[1] to be a uniaxial ferromagnet with a Curie temperature of 185°C. The easy axis of magnetization coincides with the tetragonal crystallographic c-axis. Thin films of this compound have been prepared by sputtering on a variety of substrates. Nearly completely c-axis oriented films with square hysteresis loops showing essentially 100% remanance normal to the film plane are obtained at a deposition temperature of 200°C followed by annealing at 400°C. The Curie temperature depends on the film composition. Curie temperatures as low as 120°C have been observed in non-stoichiometric films. These films have a room temperature coercivity of 950 oersteds compared to $\geq$ 2000 oersteds for stoichiometric films. Both the optical absorption and polar Faraday rotation measured for MnGaGe are structureless in the visible and near infrared. At a wavelength of 6328 Å the absorption coefficient α (corrected for reflectivity) is 5.8×10^5 cm^{-1} and the room temperature polar Faraday rotation F is 80,000 deg/cm. The reflectivity R at normal incidence is 0.47. MnGaGe undergoes no phase changes before melting peritectically at 616°C and the films are stable to oxidation at temperatures well above the Curie point. The low Curie temperature combined with high remanance normal to the film plane make this material interesting for magneto-optic memory application.

REFERENCES

1. G. Bryan Street, E. Sawatzky and K. Lee, accepted for publication in J. Appl. Phys., Jan. 1973.

MAGNETIC PROPERTIES OF THIN FILMS IN THE Mn-Ga-Ge SYSTEM

Kenneth Lee and J. C. Suits
IBM Research Laboratory, San Jose, California 95114

ABSTRACT

Thin films in the Mn-Ga-Ge system have been prepared by thermal vaporization and their magnetic and magneto-optical properties have been studied as a function of chemical composition. It has been found that the tetragonal MnGaGe structure exists over a wide range of compositions with corresponding large variations in properties. The Curie temperature of films with the tetragonal structure ranges from 120° to 250°C, and was found to increase with increasing Ga content and decrease with increasing Ge content. Films of Mn-Ge and Mn-Ga were ferromagnetic and did not show the tetragonal structure.

INTRODUCTION

The ferromagnetic compound MnAlGe was discovered in 1961 and was found to exhibit large magneto-crystalline anisotropy.[1] More recently thin films of MnAlGe were reported to have potential for magneto-optical memory applications.[2] Subsequently, MnGaGe was discovered and found to have somewhat similar properties.[3] The Curie temperature of MnAlGe is 245°C and of MnGaGe is 185°C. The crystal structure of MnGaGe is tetragonal and is identical to MnAlGe.

This paper presents the results of an investigation into the properties of thin films in the Mn-Ga-Ge ternary system. The tetragonal crystal structure is observed to exist over a wide range of chemical composition. Corresponding to the large allowed variations in composition, the compound shows large variation in properties. The dependence upon composition of various properties pertinent to an optical memory application, such as Curie temperature and Faraday rotation, will be described here. Films with either no Ge or no Ga show different crystal structures and properties and will also be discussed.

PREPARATION

The thin films were prepared by the simultaneous thermal vaporization of each of the three constituent elements. The base pressure in the vaporization chamber was 1×10^{-8} Torr prior to deposition and was typically 2×10^{-6} Torr during deposition. Polished fused quartz substrates were preheated to approximately 100°C and the deposition rate was 6 Å/sec. The compositional variation was accomplished by preparing two series of films, each of which was prepared by varying either the Ga or the Ge content relative to the two remaining constituents. Electron microprobe analysis was used to verify sample compositions. These compositions are listed in Table I and are in atomic proportions such that the sum equals 3, except for MnGe and MnGa, where the sum equals 2.

All films were amorphous as deposited. These amorphous films were non-magnetic. Subsequent to deposition the films were annealed and crystallization occurred (determined in a microscope hot stage) generally in the range 290-500°C. The crystallization temperatures listed in Table I show that a large deficiency in either Ga or Ge produces high crystallization temperatures. If these compositions have a defect structure, then the high crystallization temperature may be due to the high energy associated with the large number of vacant lattice sites.

Except for film 2 listed in Table I, all films containing the three constituents crystallized in the tetragonal structure isomorphic with MnAlGe. Those films closest to stoichiometry were predominantly oriented with their (112) axis normal to the film plane.

MAGNETIC PROPERTIES

The magnetic properties were monitored by measuring the polar Faraday rotation at 5000 Å as a function of temperature. The Curie temperature T_c, the coercivity H_c, the remanence, and the specific polar Faraday rotation θ_F are given in Table I. The θ_F for many of these compositions is lower than expected and may in part be due to uncrystallized regions in the films. The temperature dependence of the specific polar Faraday rotation is shown in Figure 1 for the samples with the tetragonal crystal structure.

In the first series of films the elemental component showing the greatest variation is germanium. Here, T_c decreases with increasing Ge content, and θ_F has its largest value at the stoichiometric composition ($x = y = z \overset{\sim}{=} 1$). In the second series, gallium shows the largest variation, and both T_c and θ_F increase with increasing gallium content. The thin films with the MnAlGe crystal structure did not show any significant variations in the lattice constants and yet T_c shows a considerable variation, which may be due to a change in the electron density in the conduction band. This variation in T_c allows an important flexibility in an optical memory application.

Film 2 was not magnetic after crystallizing at 430°C and the X-ray spectrum was not identifiable. Film 1 contained no Ge and the X-rays are consistent with the η-phase of MnGa. It was not possible to identify film 6 (no Ga) with any of the six known magnetic phases of the Mn-Ge system, which suggests that there may be an additional magnetic compound in this group.

ACKNOWLEDGMENTS

The authors wish to thank W. Parrish for the X-ray analysis, D. F. Kyser for the compositional analysis, E. Sawatzky and G. B. Street for illuminating discussions, and S. Lawrence and G. Guthmiller for their able technical assistance.

REFERENCES

1. J. H. Wernick, S. E. Haszko, W. J. Romanow, J. Appl. Phys., 32, 2495 (1961).
2. R. C. Sherwood, E. A. Nesbitt, J. H. Wernick, D. D. Bacon, A. J. Kurtzig, and R. Wolfe, J. Appl. Phys., 42, 1704 (1971).
3. E. Sawatzky and G. B. Street, this conference; G. B. Street, E. Sawatzky and K. Lee, J. Appl. Phys. (to be published).

Table I. Crystallographic, Magnetic and Magneto-Optical Properties of $Mn_x Ga_y Ge_z$ Thin Films

No.	Film x	y	z	Structure	T_c (°C)	H_c^* (kOe)	Remanence* (%)	θ_F^* (× 10⁴ deg/cm)	Cryst. Temp. (°C)
1)	1.0	1.0	0.0	η-MnGa	290	4.0	35	9.8	340
2)	1.26	1.43	0.31	undetermined	---	---	---	---	430
3)	1.11	1.23	0.66	MnAlGe	250	4.8	75	3.6	310
4)	0.95	0.98	1.07	MnAlGe	200	3.5	77	5.0	300
5)	0.88	0.87	1.25	MnAlGe	195	4.1	92	3.6	290
6)	1.0	0.0	1.0	undetermined	110	1.0	15	17.0	300
7)	1.34	0.46	1.20	MnAlGe	120	1.2	27	0.8	> 500
8)	1.13	0.90	0.97	MnAlGe	200	3.6	100	1.3	340
9)	1.09	1.07	0.84	MnAlGe	240	4.9	84	4.1	310
10)	0.93	1.39	0.68	MnAlGe	245	3.2	89	5.3	280

* at room temperature

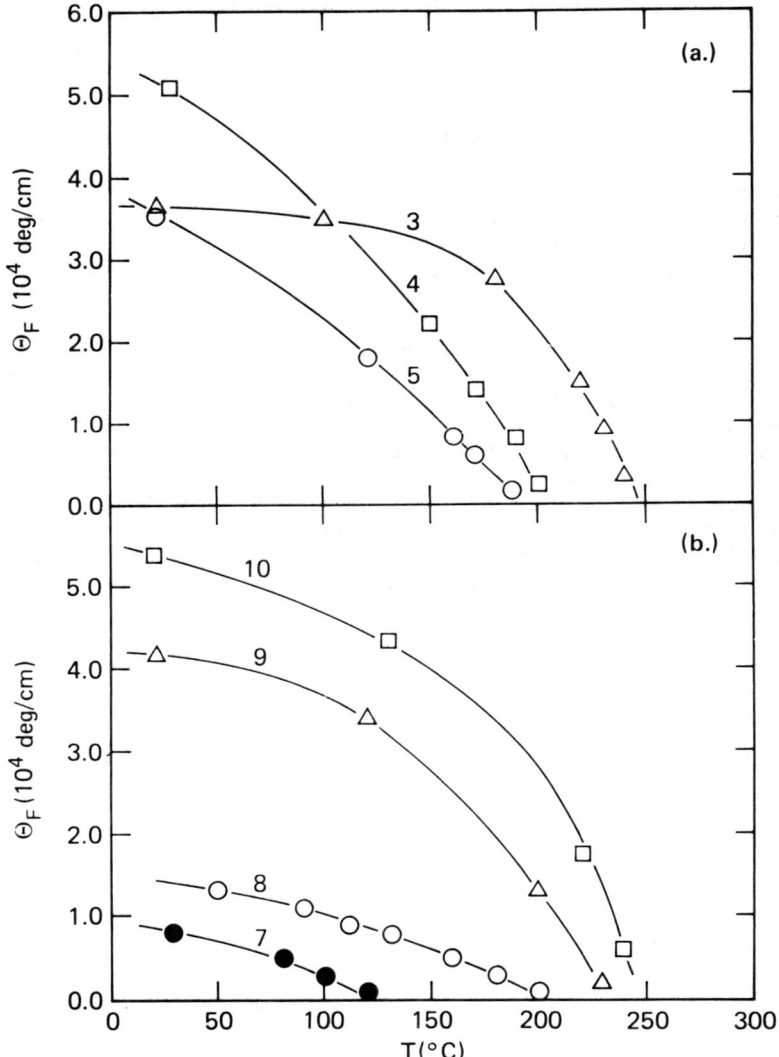

Figure 1. Temperature dependence of specific polar Faraday rotation for films with different compositions. Numbers on curves refer to compositions listed in Table I. (a) Major variation in Ge content. (b) Major variation in Ga content.

DIRECT COMPARISON OF THERMAL AND MAGNETIC PROFILES IN "CURIE POINT" WRITING ON MnGaGe FILMS

Harold Wieder and Robert A. Burn
IBM Research Laboratory, San Jose, Ca. 95114

ABSTRACT

Measurements have been made of both the real-time temperature profile and the steady state magnetization profile achieved in a film of MnGaGe as a result of "Curie point" writing with a GaAs laser pulse. A HeNe laser probe having a half-power diameter of 1 µm was used to measure both profiles. Temperature was measured during and after application of the heat pulse by relating the time-dependent transmission change of the heated area of the film to the temperature-calibrated shift of the absorption edge of MnGaGe. Magnetization information was obtained by measuring the Faraday rotation within the written area after the film had cooled to ambient. A comparison of the two profiles indicates that: (1) magnetization reversal occurs at temperatures well below the Curie point, even in the absence of an external bias field, and (2) the critical temperature required for reversal increases with an increase in the size of the heated area. Both results can be explained qualitatively by local demagnetization effects.

MAGNETIC FIELD EFFECT IN THERMOMAGNETIC RECORDING

N. Minnaja and P. L. Boschetti
Honeywell Information Systems, Italia, Pregnana Milanese, Italy

D. Chen
Honeywell Corporate Research Center, Bloomington, Minn. 55420

ABSTRACT

In a particulate medium with very high uniaxial anisotropy constant, the Ising model in the molecular field approximation is adequate to analyze the external field effect in thermomagnetic recording. The statistical properties of such a system can be obtained by examining the free energy G of a single grain. As the medium cools from above the Curie temperature T_c under the influence of an external field H, the two minima in G near T_c are separated by ΔG which increases with H. If ΔG is large compared to the thermal energy, the spins will be essentially aligned with the field. The calculated room temperature remanent magnetization agrees well with measured results obtained in thin films of MnBi of 500Å thickness, as a function of the applied field during cooling from T_c.

INTRODUCTION

The phenomena involved in thermomagnetic recording have been recently reviewed by Berkowitz and Meiklejohn[1]. Aim of the present paper is to provide a unified theory of two of the mechanisms described in the quoted paper, namely Superparamagnetic-Stable Single Domain Transitions, and Curie Writing, in the case of particulate media with high anisotropy, and to support it with evidence of thermoremanent magnetization in MnBi. The theory of thermoremanent magnetization has been outlined many years ago by Neel[2,3] and recently correlated to the theory of anhysteretic magnetization by Jeap[4]; we will put it in a slightly different form for sake of consistency with the derivation of coercivity versus temperature near the Curie point.

THEORY

We will consider a particulate ferromagnetic medium consisting of an assembly of single-domain grains, exhibiting a very high anisotropic ferromagnetic coupling among the spins. We will describe each grain by means of the Ising model in the molecular field approximation, choosing as privileged axis the easy axis, assuming a Weiss' field parallel to that axis and proportional to the net magnetization, and considering only external fields parallel to the same axis.

The statistical properties of the grain assembly are best described by the free energy of the single grain

$$G = -\frac{1}{2} CM^2 - MH - TS$$

with M magnetic moment, CM the Weiss molecular field, H external field, T absolute temperature, S entropy. For quantities M and H, which are in principle vectors, we understand with the symbols the components along one well established sense of the easy axis. In the Ising model the entropy is given by[5]

$$S = \frac{k}{2m} \left[(M_s + M) \ln \frac{2M_s}{M_s + M} + (M_s - M) \ln \frac{2M_s}{M_s - M} \right]$$

where k is the Boltzmann constant, m the elementary magnetic moment, and M_s the saturation magnetic moment of the grain. If we define x as

$$x = \frac{M}{M_s}, (-1 \leq x \leq 1)$$

and if we restrict our analysis to the region near the Curie temperature T_c ($= \frac{mM_sC}{k}$), where we are interested only in $x \ll 1$, we can write[6]

$$G \simeq \frac{kT_c M_s}{2m} \left[\frac{T}{6T_c} x^4 + (\frac{T}{T_c} - 1) x^2 \right] - M_s H x$$

up to the 4th order in x.

An analysis of the behavior of G vs x shows that for each value of H, there exists a temperature T^*, such that G exhibits only one minimum for $T \geq T^*$ and two minima for $T < T^*$; it is easy to prove that T^* is related to H by

$$|H| = \frac{2}{3} \frac{kT^*}{m} (\frac{T_c}{T^*} - 1)^{3/2}$$

Typical plots of G vs. x are sketched in Fig. 1. Notice that $T^* \leq T_c$ and the equal sign holds only for H = 0.

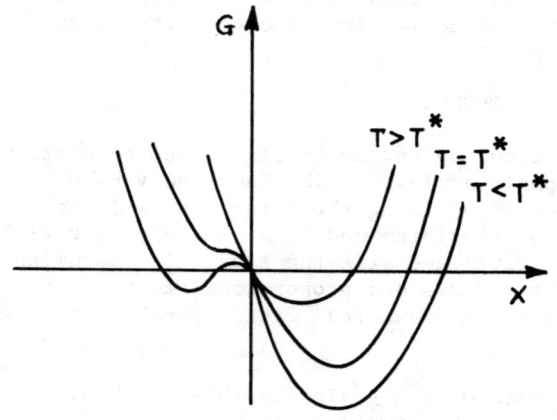

Fig. 1. Sketch of free energy G as a function of x.

Let us consider an assembly of grains cooling down from a temperature higher than T_c. For $T > T^*$ we can easily assume a complete equilibrium to be reached for the distribution of the grains at each value of x. When the temperature falls beyond that value, the redistribution between the regions of x near to the two minima is hindered by the free energy barrier. Let us assume with Neel[3] that the relaxation time t for one grain is related to the height, ΔG, of this barrier, by

$$t = \frac{1}{c} \exp\left(\frac{\Delta G}{kT}\right)$$

and with Jeap[7] that the frequency-dimensional constant c is 10^9 s^{-1} and moreover[4] that the situation is frozen when t exceeds 10^2 s. This leads us to define a blocking temperature[3,8], $T_B = \Delta G/25k$. For fields in the range of interest,

$$\Delta G \simeq \frac{3}{4} kTN \left(\frac{T_c}{T} - 1\right)^2$$

where $N = M_s/m$ is the number of elementary moments in each grain, and we find that T_B is related to the grain size by

$$\frac{T_c}{T_B} - 1 = \frac{10}{\sqrt{3N}}$$

At T_B the assembly of grains can be considered as consisting of two subassemblies, each with distribution centered around one minimum of G versus x. The difference in the two minima of G is

$$G_{m1} - G_{m2} = 2 \left(\frac{300}{N}\right)^{1/4} M_s H$$

At lower temperature no grain succeeds in transforming from one subassembly to the other, so that each subassembly evolves independently, the only allowed evolution being the adjustment of its average magnetic moment to the temperature according to the displacement of the corresponding minimum. At low temperatures the absolute value of the magnetic moment is M_s, and therefore the net magnetization of the assembly, which is referred to as thermoremanent magnetization, M_r, results to be

$$M_r = M_s \tanh \sqrt[4]{\frac{300}{N}} \frac{M_s H}{kT_c} = M_s \tanh\left(\frac{H}{H_o}\right)$$

where H_o is the equivalent field.

EXPERIMENTAL RESULTS

Thin films of MnBi were used to evaluate this analysis. This material medium possesses extremely high uniaxial crystalline anisotropy. The samples were prepared by vacuum vapor deposition technique[9] to cause the c-axis, which is the magnetic easy axis, to be oriented perpendicular to the film plane. The samples were first heated to above the Curie temperature, and then cooled back to room temperature in vacuum under the influence of a uniform applied field, normal to the film plane. Magneto-optic effect was used to monitor the remanent magnetization. The results for samples of 500Å thick prepared on glass substrate is shown in Fig. 2. Similar results measured by Unger[10] on samples with mica substrate are also given. It is to be noted that these data fit the hyperbolic tangent function well. The equivalent field value for the two sets of results are 450 Oe and 330 Oe respectively. The corresponding number of spins per grain is approximately 10^5 for our sample on glass substrate.

REFERENCES

1. A. E. Berkowitz and W. K. Meiklejohn, AIP Conf. Proc. $\underline{5}$, 764 (1971).
2. L. Neel, Ann. Geophysics $\underline{5}$, 99 (1949).
3. L. Neel, Adv. Phys. $\underline{4}$, 191 (1955).
4. W. F. Jeap, AIP Conf. Proc. $\underline{5}$, 786 (1971).
5. R. Brout, **Magnetism**, edited by G. T. Rado and K. Suhl, vol. IIA, p. 43, Eq. (2.8) with Stirling's approximation, (New York, 1963).
6. An original discussion of the analytical technique adapted here can be found in L. D. Landau and E. J. Lifshitz, <u>Statistical Physics</u>, Chapter 14, Pergamon Press, London (1958).
7. W. J. Jeap, J. Appl. Phys. $\underline{40}$, 1297 (1969).
8. C. P. Bean, J. Appl. Phys. $\underline{26}$, 1381 (1955).
9. D. Chen, J. Appl. Phys. $\underline{42}$, 3625 (1971).
10. W. K. Unger, Int. J. Mag. $\underline{3}$, 43 (1972).

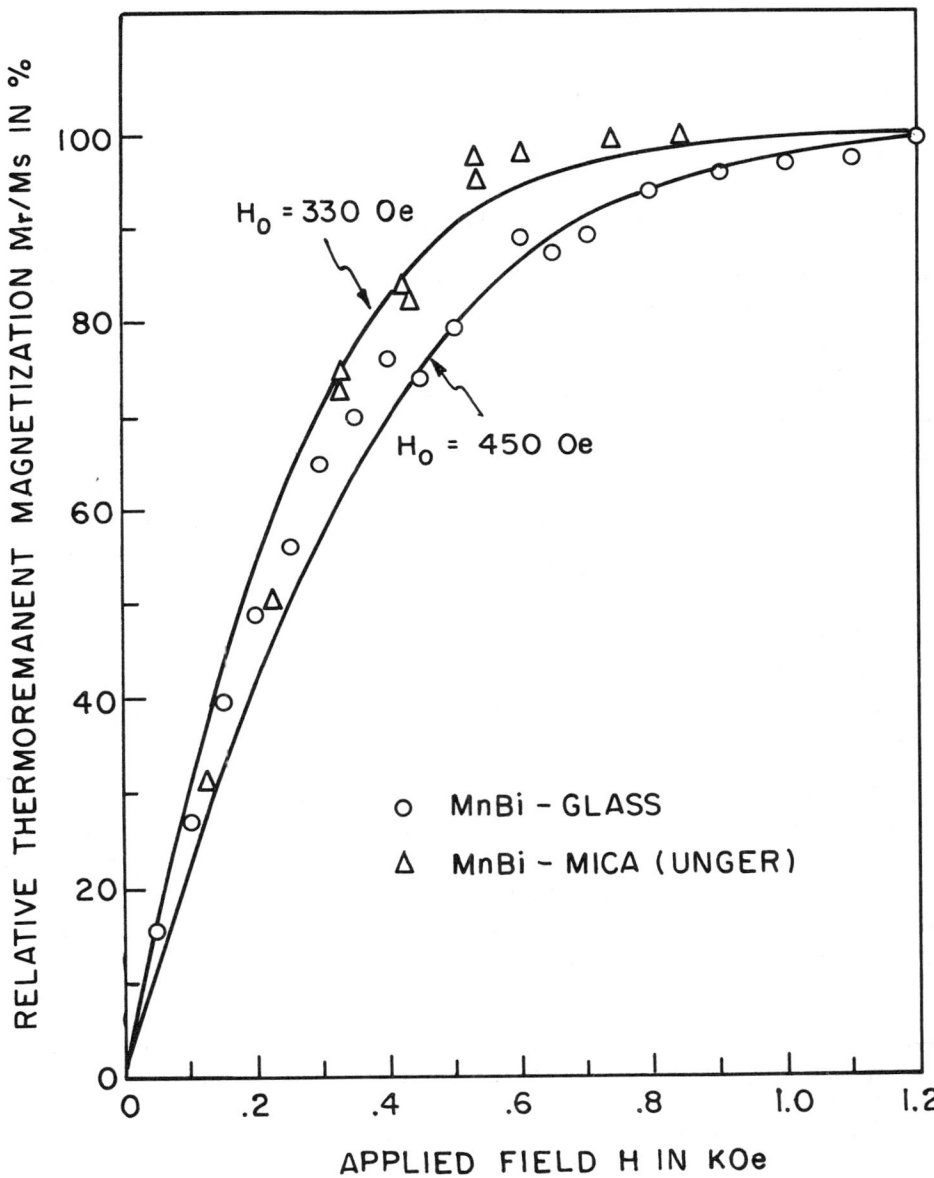

Fig. 2. Room temperature thermoremanent magnetization after heating to above the Curie temperature as a function of the applied field.

IMAGING EFFECTS IN ANISOTROPIC CONTACT DUPLICATION

H. N. Bertram
Ampex Corporation
Redwood City, CA 94063

ABSTRACT

The contact duplication efficiency is calculated for the anhysteretic printing of a prerecorded master tape onto an anisotropic copy tape in the presence of a permeable transfer head. Complete solutions are given for the cases of the transfer head either on the copy or master tape side. At short wavelengths, where transfer head imaging is ineffective, the solutions are identical and yield, for the special case of an isotropic copy tape, a copy magnetization which is an anhysterized image of the master. At long wavelengths, with a large permeability, the efficiency in the first case varies linearly with wave number (6 dB per octave); it depends only on the vertical component of the anhysteretic susceptibility, since the transfer head images out the longitudinal component of the field from the master tape. In the second case, both components of the master tape fields are imaged out which results in a quadratic wave number variation (12 dB per octave).

The contact printing process of magnetic tape consists of recording an erased copy tape via the fringing magnetic field of a prerecorded master. While the tapes are in contact a large high frequency bias field is applied which anhysterizes the copy. The problem has been treated theoretically for various special cases (e.g., ref.1-3). Recently[4], the effect of an infinitely permeable transfer head has been considered for special cases of the anhysteretic susceptibility. We present here a two-dimensional, linear calculation of the contact duplication process in which a transfer head of finite permeability and a copy tape possessing a general, diagonal anhysteretic susceptibility tensor are included.

The tape configurations are shown in Fig. 1 for the two cases. The tape thicknesses labeled are the magnetic coating thicknesses; the non-magnetic backing is not shown and the effect of the gap in the transfer head is not considered. The assumptions for the calculation follow ref. 1 with the exception that a two-dimensional field is considered. The master tape is taken to be prerecorded with a magnetization whose magnitude is uniform in the y direction but varies sinusoidally in the x direction. It is assumed that the anhysteretic process yields a copy tape magnetization which is linearly related via a diagonal anhysteretic susceptibility tensor to the net field at each point in the tape. The net field is comprised of the field from the master tape plus the demagnetization field created by the copy tape magnetization pattern and the associated image fields. The resulting integral equation is readily solved for both configuration 1 and 2. The transfer efficiency is just the ratio of the output voltages of copy to master when individually reproduced by a playback system. If playback spacing losses are neglected, the efficiency is simply the ratio of the respective peak magnetic field magnitudes (either component) at the surface of the tapes.

The contact duplication efficiencies η_1, η_2 for the two configurations are given in Table 1. λ is the recorded wavelength and χ_x, χ_y are respectively the x, y components of the diagonal anhysteretic susceptibility tensor. In the table the short and long wavelength limits of the efficiencies are given. At short wavelengths the transfer efficiencies are identical for the two configurations since the tapes are not influenced by the permeability of the transfer head. The reason is that the spatial extension of fields perpendicular to a sinusoidal magnetization source is roughly $\lambda/3$. It is interesting that for the case of an isotropic copy tape ($\chi_x = \chi_y$) the resulting short wavelength efficiency reflects a magnetization pattern which is an anhysterized image of the master.

At long wavelengths the transfer head permeability becomes effective. For configuration 1 the efficiency varies linearly with wave number k. Since a linear k dependence entails a simple transfer process, the wavelength dependent part of the demagnetization fields vanishes. The effect of a highly permeable transfer head is to image out the x component of the field from the master while doubling the y component. Thus, in addition to the k proportionality, the origin of the various terms in the long wavelength limit of η_1 is clear: the "$4\pi d$" translates magnetization to flux, the "$2\chi_y$" results from the presence of only the y component of the recording field doubled by transfer head imaging, the "$(1 + 4\pi\chi_y)^{-1}$" is a shearing term due to shape demagnetization of a thin tape, and the "1/2" occurs because the derivative of the tape flux is equal to twice the field at the tape surface.

In Fig. 2, curves of duplication efficiency (in decibels) versus the dimensionless parameter kd for s = o are shown. Curve (a) is for a typical well-oriented tape ($\chi_x = 2.5$, $\chi_y = .5$) whereas curve (b) represents an unoriented or isotropic tape ($\chi_x = \chi_y = 1.2$). For unit permeability of the transfer head both tapes have the best long wavelength output since the x component of the anhysteretic susceptibility is effective; in fact, the outputs are proportional to χ_x at these wavelengths since the y component is sheared by demagnetization. The efficiency is reduced as the permeability increases in a) and b) since the x component becomes imaged out. In fact, at large permeabilities the long wavelength efficiencies are identical since vertical demagnetization renders the susceptibility essentially independent of χ_y.

It is interesting to follow curve (a) for $\mu = 1000$ over the range of frequencies plotted. As the frequency is reduced from the short wavelength limit, the efficiency begins to fall towards a quadratic variation, as in case 2, since $\chi_x >> \chi_y$. In fact, if χ_y were identically zero, the variation would become and remain at "12 dB per octave." However, eventually the x component contribution falls enough that the small linear (6 dB per octave) y contribution dominates.

In configuration 2, the effect of the transfer head is to image out both components of the field from the master tape. Thus, at long wavelength and large permeabilities, the transfer efficiency (η_2) varies quadratically with wave number for any degree of anisotropy of the anhysteretic susceptibility tensor. As the expression in Table 1 for η_2 (k → o) indicates, the long wavelength efficiency increases with the degree of anhysteretic anisotropy.

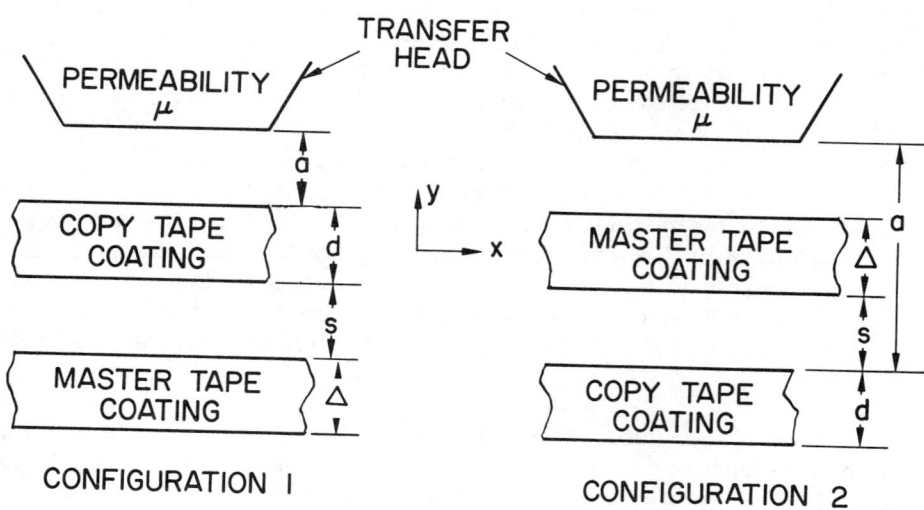

Fig. 1 Tape Configurations Considered in Efficiency Calculation

Table I Contact Printing Efficiencies

COMPLETE EXPRESSIONS

$$\eta_1 = \frac{e^{-ks}}{D}\left\{(Q-T)e^{2\alpha kd} - Q + TQ^2 - T(Q^2-1)e^{(\alpha-1)kd}\right\}$$

$$\eta_2 = \frac{e^{-ks}}{D}\left\{Q(e^{2\alpha kd} - 1)(1 - T e^{k(2s+\Delta)})\right\}$$

$$T = \frac{\mu-1}{\mu+1}e^{-2ka} \qquad \alpha = \sqrt{\frac{1+4\pi X_x}{1+4\pi X_y}} \equiv \sqrt{\frac{\mu_x}{\mu_y}} \qquad k = \frac{2\pi}{\lambda}$$

$$Q = \frac{\sqrt{\mu_x \mu_y}+1}{\sqrt{\mu_x \mu_y}-1} \qquad D = Q^2 e^{2\alpha kd} - 1 + TQ(1-e^{2\alpha kd})$$

LIMITS FOR SIMPLE CASES

Short wavelength $k \to \infty$

$$\eta_1 = \eta_2 = \frac{\sqrt{\mu_x \mu_y}-1}{\sqrt{\mu_x \mu_y}+1} e^{-ks} \qquad \text{Independent of } \mu$$

Long wavelength $k \to 0$

$$\eta_1 = \frac{1}{2} \cdot 4\pi kd \cdot \left\{\frac{2}{\mu+1} \cdot X_x + \frac{2\mu}{\mu+1} \cdot \frac{X_y}{1+4\pi X_y}\right\}$$

$$\eta_2 = \frac{1}{2} \cdot 4\pi kd \cdot \left\{X_x + \frac{X_y}{1+4\pi X_y}\right\} \cdot \left\{\frac{2}{\mu+1} + \frac{\mu-1}{\mu+1} k(2a-2s-\Delta)\right\}$$

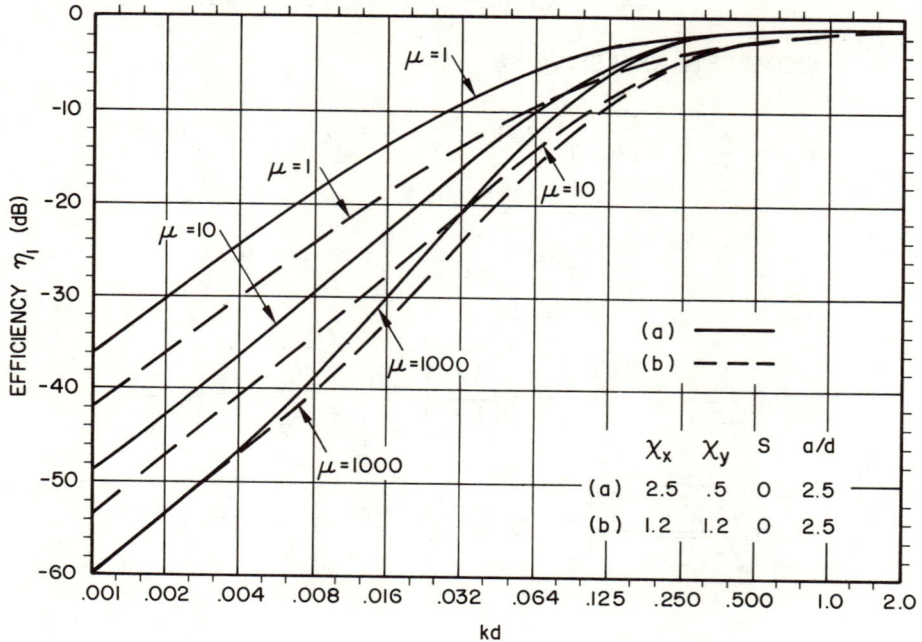

Fig. 2 Duplication efficiency for configuration 1 versus the parameter "wavenumber times copy tape coating thickness" for an anisotropic (a) and isotropic (b) copy tape.

Although the duplication efficiencies have been presented for both configurations, the discussion has centered around that in which the copy tape is closest to the transfer head. This is the most efficient configuration since it is desirable not to erase or degrade the master tape during the process. In thermoremanent duplication, if permeable material is nearby or used for the heating element, the equations apply with a trivial modification.[1] This results since anhysteresis and thermoremance are virtually identical physical processes.

REFERENCES

1. J.C. Mallinson et al, IEEE Trans. Magn. MAG-7, 524 (1971)
2. F. Koboyashi et al, IEEE Trans. Magn. MAG-7, 528, (1971)
3. D.L.A. Tjaden et al, IEEE Trans. Magn. MAG-7, 532 (1971)
4. J. Hokkyo et al, IEEE Trans. Magn. MAG-8, 397 (1972)

NUMERICAL ANALYSIS OF A MAGNETORESISTIVE TRANSDUCER FOR MAGNETIC RECORDING APPLICATIONS

R. L. Anderson, C. H. Bajorek and D. A. Thompson
IBM T. J. Watson Research Center
Yorktown Heights, N. Y. 10598

ABSTRACT

The response of the vertical magnetoresistive head of Hunt[1] is analyzed by a numerical relaxation technique which explicitly includes contributions of material anisotropy, demagnetization, exchange and bias fields. A graphical representation of the magnetization within the head reveals curling details near the edges of the magnetic films. Magnetoresistance vs. field curves show the gradual approach to saturation characteristic of real stripes. The resolving ability of the head is evaluated as a function of its magnetic properties, geometry and separation from the recording medium.

INTRODUCTION

Hunt has described a novel thin film magnetoresistive transducer for magnetic recording applications.[1] His predictions of the expected output response and comparison with experimental results were based on a simplified model which does not rigorously consider anisotropy, shape, demagnetizing, exchange, and bias field contributions. No expression for resolution was given

In this paper we describe the response of a vertical transducer as depicted in Fig. 1. For numerical examples, the material properties of 81% Ni-Fe evaporated permalloy films are used.

MATHEMATICAL MODEL

Figure 1 is a schematical representation of the vertical head geometry. It is situated a distance h above the medium, H_x is the vertical component of the sensed field, M is the saturation magnetization of the transducer and I is the sense current. Consistent with practical dimensions (t < 500Å, w < 100 microns and head length >> w) the actual head is approximated with a one dimensional model. All

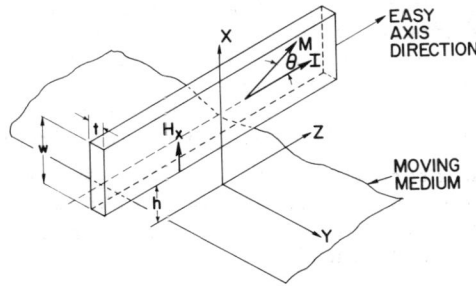

Fig. 1 The vertical head geometry.

quantities of interest except H_x are functions of x only. The applied field H_x may also depend on y, the relative distance between a transistion in the medium and the head. It is also assumed that the magnetization rotates coherently in a single domain configuration: The presence of domain walls and their discontinuous motion is neglected. Such domain wall switching is actually observed in practice, but it typically negligibly perturbs the behavior (Barkhousen jumps) expected on the basis of the single domain model.

Under the above assumpitons and including material uniaxial anisotropy, demagnetizing, exchange and applied field contributions, the total torque density T acting on M is given in MSKA units by

$$\frac{T(x)}{\mu_o M H_k} = \frac{1}{2} \sin 2\theta(x) - [\frac{H_d(x)}{H_k} + \frac{H_x(x,y)}{H_k}]\cos\theta(x) - \frac{2A}{\mu_o M H_k}\frac{\partial^2 \theta(x)}{\partial x^2} \quad (1)$$

where $H_d(x) = -\frac{tM}{2\pi} \int_h^{h+w} \frac{\frac{d\theta(\xi)}{d\xi} \cos\theta(\xi)}{x-\xi} d\xi$, (2)

$H_k = 2K/\mu_o M = 3$ oe $= 240$ amp/meter $=$ uniaxial anisotropy field,

$\mu_o = 4\pi \times 10^{-7}$, $A = 10^{-11}$ Joule/m $=$ exchange constant and $\mu_o M = 1$ w/m^2. The solution $\theta(x)$ for which $T = 0$ is then used to compute:

$$\Delta R/\Delta R_{max} = \frac{1}{w} \int_h^{h+w} \cos^2 \theta(x) dx \quad (3)$$

Equation 3 follows from the phenomenological expression $\rho = \rho_o + \Delta\rho$, where $\Delta\rho = \Delta\rho_{max} \cos^2\theta$. The fact that $\Delta\rho_{max}/\rho$ is typically less than 5% allows the use of this expression rather than the more complicated exact one. The output voltage from the transducer is $I\Delta R$.

NUMERICAL COMPUTATION

The computation proceeds as follows: For any applied field choose an arbitrary initial function θ_1. Calculate H_d by using five point polynomial interpolation and integrating the right side of Eq.2 using Simpson's Rule. The endpoint contributions are handled with delta functions. The discontinuity at $x = \xi$ is avoided with Cauchy's Principal Value Integral. The torque is then calculated using Eq. 1. The angle θ is iteratively relaxed using the relation $\theta_{i+1} = \theta_i - C T_i$, where C is typically 10^{-5} until T_f reaches a negligible value. The final θ_f values are then used to compute $\Delta R/\Delta R_{max}$ by integrating Eq. 3 using Simpson's Rule.

The calculation was programmed in APL on an IBM 360/91 system. A grid size of w/20 was normally used, w/200 changed the results by less than 1%. The iteration was stopped when $T_f < 10^{-2} \mu_o M H_k$ at all points within the stripe. This corresponds to a less than 0.1% error in ΔR. Convergence was verified by observing the asymptotic approach to θ_f from the positive and negative directions.

To ascertain the validity of the model and numerical procedure, we first considered the uniform field excitation which is most amenable to experimental verification. We then proceeded to model the real application with an isolated line change transition moved past the vertical head. The head was biased with a uniform field about the "linear operating point" $\Delta R/\Delta R_{max} = 0.5$. This leads to bipolar outputs about the most sensitive point of the R vs. H response curve. The "resolution" of the latter head was then conveniently defined as the half amplitude output pulse width.

UNIFORM FIELD RESPONSE COMPARED WITH EXPERIMENT

The curves of Fig. 2 display the predicted and experimental $\Delta R/\Delta R_{max}$ vs. uniform field response for three transducer widths. The applied field is conveniently normalized by $H_k + H_d'$ where H_d' is defined as tM/w. Measurements were made at 60 hz. The sense currents and voltages were applied and recorded using four bar probe techniques. The details for sample preparation, properties and characterization are described elsewhere.[2]

The following should be emphasized: First is the gradual approach to saturation of the curves corresponding to the stripes of finite width. This is characteristic of the actual stripes and directly associated with curling details of the magnetization near the edges of the stripes (see Fig. 3). Second, as expected from

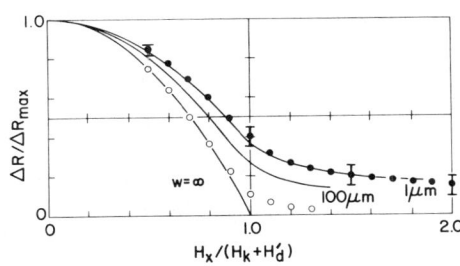

Fig. 2. Solid Curves: numerical prediction of $\Delta R/\Delta R_{max}$ vs. normalized hard axis uniform field excitation with width w as a parameter. Open o: experimental R vs. H response for a 200Å thick, 3 cm dia. film. Solid ●: experimental R vs. H response at three points for 90 and 197Å thick, 3 to 100 micron wide stripes (total of sixteen stripes). Although the data points fall on a single normalized curve, the scale factor $H_k + H_d'$ varies by more than a factor of 20 between samples.

Equations 1 and 2, the curves do not coincide. In the limit $w \to \infty$, the switching of M is governed by H_k whereas for w = 1 micron the switching is predominantly affected by demagnetizing effects. Finally, the predicted behavior is in good agreement with experiment. The indicated discrepancies are in most part due to inaccuracies in the experimentally determined dimensions and composition of the films.

The distribution of the magnetization across the stripe width in a uniform field is displayed in Fig. 3. Note that θ is lower everywhere for the narrower of the two cases. Saturation of the stripe requires fields well in excess of $H_k + H_d'$.

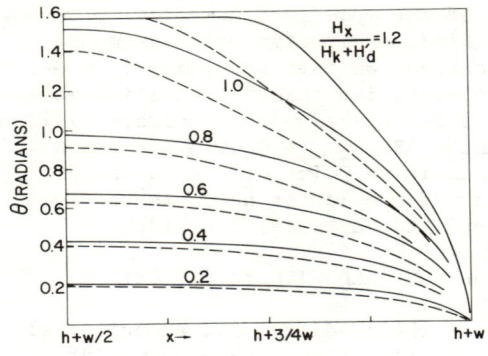

Fig.3. Distribution of θ along the strip width w with normalized uniform applied field as a parameter. The solid and dashed curves correspond to 100 and 1 micron wide stripes respectively.

Exchange represents a less than 1% effect on the predictions for stripes 1 micron wide. Exchange was therefore ignored in subsequent calculations.

NON UNIFORM RESPONSE AND RESOLVING ABILITY

The field excitation applicable to recording is modeled by an isolated transition moved past a head biased with a uniform external field about its linear operating point, $\Delta R/\Delta R_{max} = 0.5$. The relevant component of the line charge field is $H_x(x,y) = Qx/(x^2+y^2)$. The magnitude of Q is conveniently expressed as $Q = \pm CMt$, a fraction of the saturation flux for the magnetoresistor. The constant C is then adjusted to give realistic values ($H_x < 500$ oe) of the field within the stripe. The field is also chosen so that $0.25 < \Delta R/\Delta R_{max} < 0.75$. This excursion is consistent with acceptable linearity as well as output signal amplitude.

Figure 4 shows the distribution of the magnetization along the stripe width, with normalized head to transition distance as a parameter, for both positive (sense field polarity = bias field

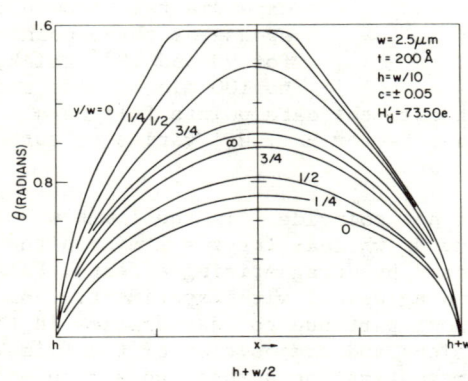

Fig. 4. Distribution of along the stripe width w, with normalized transition to head distance as a parameter, for both positive and negative transitions.

polarity) and negative transitions. The total ΔR excursion is approximately $0.4 \Delta R_{max}$ centered about $0.5 \Delta R_{max}$. The value of $C = 0.05$ was also chosen so that the sum of bias and sense fields does not reverse polarity anywhere within the stripe. A significant increase of C above this value would drive the device beyond the $\Delta R/\Delta R_{max} = 1$ point. In this case there is a noticeable asymmetry in the distribution of M along the stripe width.

The resolution of the head is deduced from the half amplitude widths of the ΔR vs y pulses obtained by integrating distributions like those of Fig. 4. The resolution is defined as b/w where b is the half amplitude width. To obtain the most optimistic case we considered small sense field excitations: $.45 < \Delta R/\Delta R_{max} < .55$. The resolution is represented by the solid curve of Fig. 5. This prediction should be valid for transducers with similar dimensions: $t < 500Å$ and $w < 25$ microns. The dashed line corresponds to the half amplitude width of the applied field component at the middle of the stripe ($x = h + w/2$). This approximation is expected to be accurate for $h \gg w$.

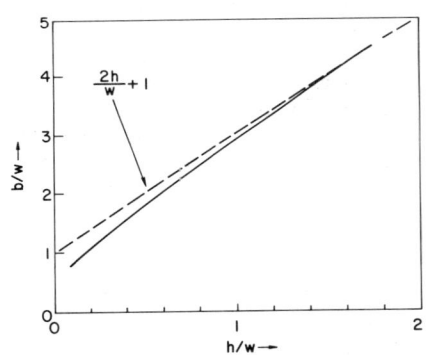

Fig. 5 "Resolution" b/w vs. normalized flying height based on small signal excitation of a 200Å thick, 2.5 micron wide stripe.

SUMMARY AND CONCLUSIONS

The response of the vertical magnetoresistive head of Hunt was analyzed by a numerical relaxation technique which explicitly includes contributions of material anisotropy, demagnetizing, exchange and bias fields. The results for a uniform field excitation were compared with and found in very good agreement with experimental results. An isolated line charge transition was then used to model the response to, and resolution in, an actual recording application. The numerical resolution prediction is in reasonably close agreement with the half amplitude field width acting at the center of the stripe. The graphical representations of the angular distribution of M across the stripe width reveal the curling details for an actual stripe.

REFERENCES

1. R. P. Hunt, IEEE Trans. on Mag., Mag-7, 150, 1971.

2. S. Krongelb, Paper E.4, presented at AIME Conf., Boston, Mass., Aug. 1972.

THE RELATIONSHIP BETWEEN MAGNETIC PROPERTY OF ELECTROLESS PLATED Co-P VIDEOTAPE AND MICROSTRUCTURES

M. INATSU and K. ITO

NHK Technical Res. Labs., 1-10-11 Kinuta, Setagaya, Tokyo, Japan

T. TERANISHI

NHK Broadcasting Science Res. Labs., 1-10-11 Kinuta, Setagaya, Tokyo, Japan

ABSTRACT

By the X-ray diffraction method were investigated the microstructures of electroless plated Co-P films which were prepared under varying conditions of plating bath. The X-ray diffraction lines of these films indicate only the hexagonal cobalt phases. From the analysis of peak breadths it was found that grain sizes of Co crystals are within the limit of single magnetic domain and more than the critical size of superparamagnetism. Integrated intensities of diffraction lines indicate that films in which magnetic easy c-axes of Co crystals are preferably oriented in the surface exhibit high coercivities (coercive forces). These suggest crystal-line anisotropy plays an important role in the magnetic properties of Co-P films under a certain condition of plating bath.

INTRODUCTION

About 1500Å thickness Co-P film, which is deposited on polyethylene telephtalate tape by electroless plating, is promised as high density magnetic video recording medium because of its high coercive force value and saturation magnetization.[1,2,3] In general, electroless plating Co-P film may exhibit various magnetic properties depending on the experimental conditions of plating bath.[4] For example, the magnetic properties are influenced by Co ion concentration, sodium hypophosphite concentration, boric acid concentration and PH of the plating bath. Until now it has not yet been clarified by what structural condition (for example, film thickness,[4,5] crystal orientation,[4] phosphorus concentration[4] and grain size[6]) the magnetic properties are mainly influenced. In the present experiments, four series of Co-P film samples were made in various bath conditions, in order to investigate the relationship between film microstructure and its magnetic properties.

PREPARATION OF SAMPLES

Substrate — Polyethylene telephtalate substrate were precleaned and its surface were etched by a hot sodium hydroxide solution and treated by $Sn\ Cl_2$ for sensitization and by $Pd\ Cl_2$ for activation.

Co-P Film on substrate — The caustic alkaline citrate bath was used to deposit Co-P on the substrate. To promote the buffer action and complex forming boric acid was used. The standard conditions of the bath is shown in Table I, and series of samples An, Bn, Cn and Dn were obtained by varing the one of these bath conditions as shown in the Table II. In these specimens the phosphorus contents will exceed about 3%.

Table I. Composition and condition of the plating bath

Composition	Concentration (M/L)
$CoCl_2 \cdot 6H_2O$	0.04
$H_3C_6H_5O_7 \cdot H_2O$	0.09
NH_4Cl	0.2
H_3BO_3	0.5
$NaH_2PO_2 \cdot H_2O$	0.05

PH : 8.0 at 25°C
(PH adjusted with NaOH)
Bath temperature : 80°C

Table II. Plating conditions of four series of film samples

A Series		B Series				
A_1	A_2	B_1	B_2	B_3	B_4	B_5
0	0.5	0.05	0.04	0.03	0.025	0.02
Boric acid concentration (M/L)		$CoCl_2 \cdot 6H_2O$ concentration (M/L)				

C Series				D Series	
C_1	C_2	C_3	C_4	D_1	D_2
0.5	1.25	2.0	3.0	7.0	8.0
$NaH_2PO_2H_2O$ /$CoCl_2 \cdot 6H_2O$ (Molar ratio)				PH	

STRUCTURE and MAGNETIC PROPERTIES

Magnetic measurement - The magnetic properties of the Co-P film are measured by an ac-type B-H curve tracer with maximum driving field of 2000 Oe.

Thickness measurement - The thickness of the plated films were measured by Hi Kocaur Electric Thickness Tester which uses a de-plating method.

X-ray diffraction measurement - Only a single sheet of the Co-P film is exposed to the $CoK\alpha$ radiation in the diffractometer. Because the diffracted beam is weak, counting of intensity of diffraction lines is continued for 400 seconds at every $0.02°$ diffraction angles by using a counting rate computer. Under this condition, background intensity is about 350,000 counts at each angle.

Below the diffraction angle of $25°$ all Co-P films show three diffraction lines, (100), (002) and (101). These are coincident with those of hexagonal α-Co and any other lines cannot be observed. From half value widths of (100) and (002) lines approximate values of the gain size are obtained by using Scherrer's formula. These grain size values and ratios of integrated intensities of two lines (100) and (002) are shown in Table III. together with film thicknesses and remanent magnetizations.

Table III. Replationship between Magnetic Properties and Microstructure of Co-P Films

Samples	Corcive Force (Oe)	Squarness Ratio	Particle Size (Å)		$\frac{I(100)}{I(002)}$	Film Thickness (Å)	Br (Gauss)
			(002)	(100)			
A_1	450	0.64	221	73.6	3.00	1651	5450
A_2	760	0.86	82	48.2	3.86	1050	5700
B_1	730	0.86	126	48.4	3.64	790	10100
B_2	760	0.86	82	48.2	3.86	1000	8000
B_3	550	0.82	118	43.2	2.31	1100	9100
B_4	410	0.80	128	30.9	1.37	1050	5700
B_5	230	0.80	110	22.0	0.97	1530	3900
C_1	200	0.73	149	25.7	0.95	838	9550
C_2	760	0.86	82	48.2	3.86	1050	5700
C_3	590	0.77	119	57.3	3.59	1778	7860
C_4	70	0.88	-	-	-	1981	7560
D_1	340	0.39	278	47.0	2.70	1300	3100
D_2	760	0.86	82	48.2	3.80	1050	5700

The sizes of Co-particles deposited on these films are not greater than 200Å and not smaller than 50Å. For Co the critical grain size above which the grain has the multi-domain structure is about 330Å. Also the critical value for the superparamagnetism is about 20Å. All experimental values of the grain size are within in these limits. Therefore the magnetizing process of these Co-P films will be the magnetization rotation within the domain. From Table III we cannot find close relations between coercive forces and film thicknesses within each series of samples.

Intensity ratio, $I(100)/I(002)$, exhibits the orientation of crystallites. The large value of this ratio shows that c-axes of Co crystallite (namely, the easy magnetization direction) are relatively on the surface plane of the film compared to the samll value. In this case the magnetizing curve has a great squareness and a large coercivity.

This close relation between a large intensity ratio, $I(100)/I(002)$, and a large coercive force holds good in our experimental data (Table III) within a certain series of specimens which contain above 3% phosphor, while Miksic's specimens contain less than 3% phosphor.[4]

This suggests that the coercive force of a electroless plating Co-P film is mainly controlled by the crystalline anisotropy, when one of the plating conditions is varied.

REFERENCE

1. K. YOKOYAMA, K. ITO, et al, NHK Lab. Note 133 (1970)

2. K. YOKOYAMA, K. ITO, et al, 1971 Intermag Conference 27.4.

3. K. YOKOYAMA, K. ITO, NHK Lab. Note, 155 (1972)

4. M.G. MIKSIC, et al, J. Electrochem Soc., 113, p.360 (1966)

5. V. Morton and R.D. Fisher ; J. Electrochem Soc., 116, p.188 (1969) ; R.D. Fisher and E.W. Jones, presented at the 1972 INTERMAG Conf., Kyoto.

6. M. Aspland, G.A. Jones and B.K. Middleton

    IEEE Trans. Magn., vol. MAG-7, March 1971, p.215

EVALUATION OF Ni-Zn FERRITE FILMS FROM MAGNETIZATION-TEMPERATURE CURVES

O.S. Lutes and R.L. Kooyer
Honeywell Corporate Research Center, Bloomington, Minn. 55420

ABSTRACT

A useful method of evaluating the composition of Ni-Zn ferrite films, and optimizing their fabrication, has been developed using magnetization-temperature (σ-T) curves. σ-T data were obtained in the 25-400°C range by a torquemeter technique. The method represents the film data as composites of bulk ferrite curves for compositions $Ni_{1-\delta}Zn_\delta Fe_2O_4$, and yields the average Zn content, $\bar{\delta}$, of the film. The method was used to evaluate 7-8 μm films of nominal composition $\delta=0.64$ produced by nitrate deposition. For non-optimum processes the data indicated inhomogeneity and Zn loss. The Zn loss was reduced and films having $\bar{\delta}=0.63$ were produced by using a relatively high temperature in the initial heating stage between layer depositions.

INTRODUCTION

Potential uses of spinel ferrite films in recording or memory applications would probably depend on achieving low magnetocrystalline anisotropy and magnetostriction through careful control of composition. The fabrication of such films departs from the well-established procedures used in the processing of bulk ferrites. For high Zn-content films, in particular, it is of interest to know whether the composition can be controlled during thermal treatment to yield films having magnetic properties similar to bulk material. Only a little attention has been directed to this point. Ni-Zn ferrite films prepared by flash evaporation and oxidation showed pronounced Zn loss during processing, based on lattice constant evaluation of composition.[1] Other studies showed the effect of annealing on magnetization and permeability in Ni-Zn ferrite films produced by vacuum arc discharge.[2] We report here a method of evaluating polycrystalline Ni-Zn ferrite films by means of magnetization temperature curves, and a study of the effect of thermal process variables on composition in films of nominal composition $Ni_{.36}Zn_{.64}Fe_2O_4$.

EXPERIMENTAL

All films used in this study were deposited on alumina substrates, using the nitrate decomposition method.[3] A film is built up by applying successive coatings, each application being followed by oven baking at temperature T_B. The duration of this baking is about one minute. After the required thickness is reached, the film is fired in air for about one hour at temperature T_F. Thicknesses were determined by weighing. Films in this study were 7 to 8 μm thick and about 1 cm² in area. H_c following firing was typically about 25 Oe. The magnetization-temperature measurements were made using a torquemeter apparatus, which included convenient provisions for temperature control, large saturating fields, and continuous X-Y recording. The sample was mounted on a quartz rod and centered inside a tube furnace between magnet pole pieces. The temperature was measured with a chromel alumel thermocouple. The film was oriented with its plane vertical and at an angle ϕ to the magnetic field H. Under these conditions the magnetization will assume a direction θ intermediate between the film plane and the field direction. M_s and θ were determined with sufficient accuracy from the torque, L, by means of an iterative calculation using the following simultaneous equations:

$$L = HM_s V \sin(\phi-\theta) \qquad (1)$$

$$\sin\theta\cos\theta = (H/4\pi M_s)\sin(\phi-\theta) \qquad (2)$$

where V is the film volume. Eq. 2 is an approximation of the more general relation, which includes a perpendicular anisotropy term.[4] Separate experiments indicated this term did not significantly alter our M_s results.

Magnetization data are reported in the following section by weight rather than volume to facilitate comparison with bulk results. For this purpose σ (gauss-cm³/g) is given by M_s/ρ, with ρ taken as 4.7 g/cm³.

RESULTS

σ-T curves were obtained on films of nominal composition $Ni_{.36}Zn_{.64}Fe_2O_4$ differing systematically in the temperatures of baking and firing. Fig. 1a shows smoothed σ-T curves obtained by varying the firing temperature following 400°C baking. As a means of comparison, a σ-T curve is shown for bulk ferrite of approximately this composition, taken from Smit and Wijn.[5] All experimental curves show a relatively low σ at 0°C and a relatively high Curie temperature, both characteristic of a lower Zn content than nominal.[5] All films of this series showed an appreciable αFe_2O_3 content of 15-20 wt. per cent as determined from x-ray diffraction following the firing step.

Fig. 1b shows the results of varying the baking temperature at a constant firing temperature of 1000°C. Here it is seen that higher baking temperature results in a pronounced rise in σ (0) and lowering of Curie temperature. Further 1100°C (3 hr.) annealing of the film baked at 1050°C resulted in an optimum σ-T curve approximating the bulk behavior. Continued annealing at 1100°C (7 hr.) and 1200°C (1 hr.) resulted in eventual degradation of the σ-T curve. No films baked at $T_B \geq 600°C$ showed αFe_2O_3 diffraction lines following baking, firing, or subsequent annealing.

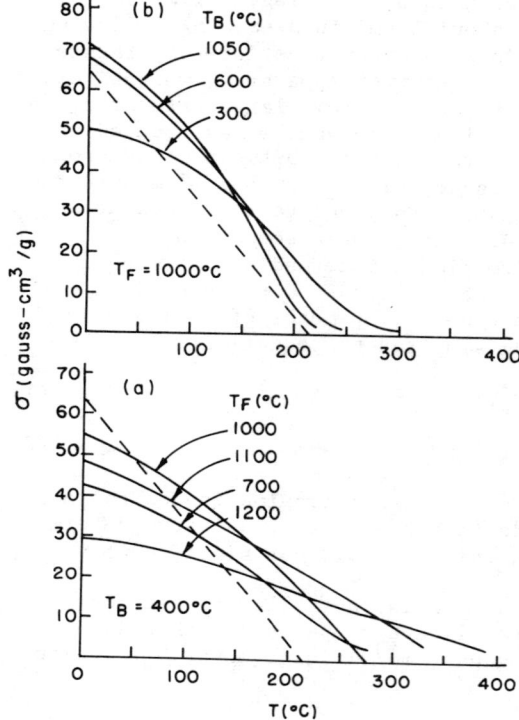

Fig. 1. Effect of thermal process variables on σ-T curves for nominal $Ni_{.36}Zn_{.64}Fe_2O_4$ ferrite. Solid curves are smoothed data. Dashed curve is approximate bulk curve for this composition. T_B is temperature of heat treatment between layer applications. T_F is firing temperature following deposition.

The σ-T results for the optimum and over-annealed film condition are shown in Fig. 2, together with fitted curves giving a measure of their Ni/Zn ratio. In each case the data are fitted to analytical expressions which are composites of bulk Ni-Zn ferrite σ-T curves taken from Smit and Wijn[5] and shown in the upper insert. In terms of the general formula $Ni_{1-\delta}Zn_\delta Fe_2O_4$, the curves of the insert are identified as follows:

$$\sigma_A(\delta=0.65), \sigma_B(\delta=0.50), \sigma_C(\delta=0.20), \text{ and } \sigma_D(\delta=0).$$

The general expression for the fitted curves is:

$$\sigma = \alpha_A \sigma_A + \alpha_B \sigma_B + \alpha_C \sigma_C + \alpha_D \sigma_D \qquad (3)$$

while the average value of the composition parameter is given by:

$$\bar{\delta} = (\alpha_A \delta_A + \alpha_B \delta_B + \alpha_C \delta_C + \alpha_D \delta_D)/(\alpha_A + \alpha_B + \alpha_C + \alpha_D) \qquad (4)$$

The results of Fig. 2 may now be described as follows: For the near-optimum heat treatment, $\bar{\delta} = 0.63$, compared to the nominal 0.64. For the excessive annealing, $\bar{\delta} = 0.42$, indicating a pronounced composition change. X-ray diffraction results on this

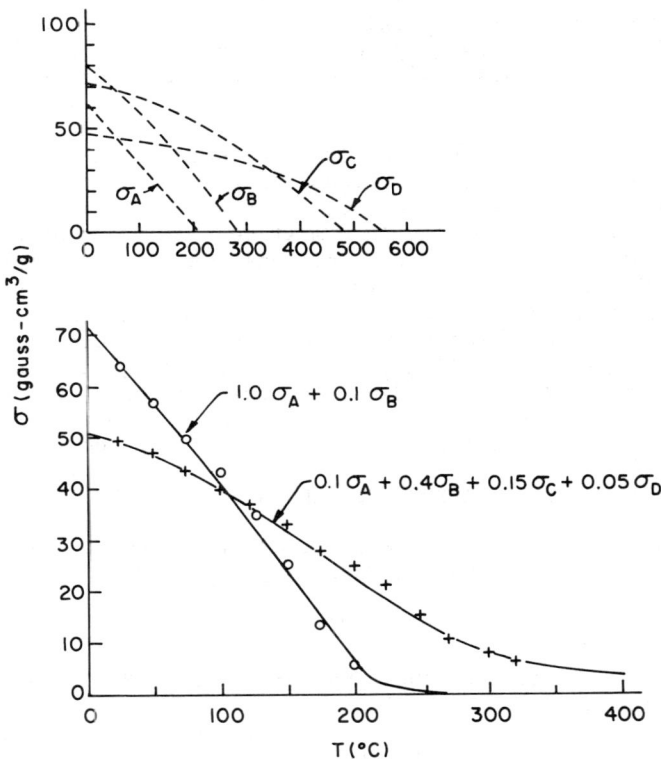

Fig. 2. Evaluation of film composition from σ-T curves. Main part of figure shows data for nominal $Ni_{.36}Zn_{.64}Fe_2O_4$ film following near-optimum heating (circles) and over-annealing (+ symbols). Expressions for calculated curves are derived from approximate bulk curves (upper insert) having different δ in formula $Ni_{1-\delta}Zn_\delta Fe_2O_4$. $\sigma_A: \delta=0.65$; $\sigma_B: \delta=0.50; \sigma_C: \delta=0.20; \sigma_D: \delta=0$.

sample revealed a reaction between film and substrate, yielding compositions such as $FeAl_2O_4$, but no detectable Fe_2O_3. It should be noted that the representation of Eq. 3 in terms of only Ni-Zn ferrites is probably not complete in this case, and is intended to give only an approximate measure of deviation from nominal composition.

DISCUSSION

The above results show that except for particular process conditions the Ni-Zn ferrite films were composites or mixtures of a range of spinel compositions. The σ-T curves of Fig. 1a indicate that for low T_B no final air firing condition can bring about the desired spinel composition. With higher T_B, on the other hand, a subsequent firing program may be found to give a film having σ-T characteristics very close to that expected for the nominal Ni/Zn ratio, as shown by the σ-T curves of Fig. 1b and Fig. 2.

An account of the above results consistent with the σ-T and x-ray diffraction data is as follows. After low temperature baking, films contain a mixture of NiO, ZnO, and Fe_2O_3. The subsequent firing procedure then results in a mixture of spinel phase of different compositions plus some unreacted oxides, as supported by the higher Curie temperatures, lower σ values, and presence of αFe_2O_3. Films baked at higher T_B, on the other hand, undergo more complete conversion to the spinel phase following each layer deposition. The final firing then serves to homogenize the phase to the desired Ni/Zn ratio. Hence, no oxide phase is found. Prolonged annealing at higher temperatures leads to degradation of the σ-T curve through a film-substrate reaction process.

ACKNOWLEDGEMENTS

C.A. Knudson furnished the x-ray diffraction results. A.E. Seitz helped with the film deposition procedures.

REFERENCES

1. H. Nosé, M. Hashimoto, and R. Kimura, Trans. Nat. Res. Inst. Metals (Japan) 11, 1 (1969).
2. M. Naoe and S. Yamanaka, Jap. J. Appl. Phys. 6, 1029 (1967).
3. W. Wade, T. Collins, W. Malinofsky, and W. Skudera, J. Appl. Phys. 34, 1219 (1963).
4. R. Baron and R.W. Hoffman, J. Appl. Phys. 41, 1623 (1970).
5. J. Smit and H.P.J. Wijn, Ferrites, (John Wiley and Sons, New York, 1959), Chap. VIII, p. 158.

Section 45. Organic Materials 1459

SOME MAGNETIC PROPERTIES OF LIQUID CRYSTALS

J. P. HURAULT

Laboratoires d'Electronique et de Physique Appliquée
3, avenue Descartes, 94450 Limeil-Brévannes, France

ABSTRACT

The constitutive molecules of liquid crystals are rod shaped and are oriented according to certain types of long range order which depend on the nature of the mesophase. Accordingly, the dielectric, conductive, diffusive, optical and magnetic properties of these substances are strongly anisotropic. If $\chi_{//}$ and $\chi_{\perp}$ denote the magnetic susceptibilities measured respectively parallel and perpendicular to the molecular axis, $\chi_{//} - \chi_{\perp}$ is generally positive and the molecules tend to align parallel to an applied magnetic field H. However, this effect is counteracted by the elastic forces which tend to restore the system in its initial configuration.

Thus, depending on the boundary conditions and on the initial configuration, a magnetic field may induce various types of distorsions on a liquid crystalline sample. As liquid crystals are optically active, these distorsions can be detected quite sensitively. We shall review here some of the most remarkable magnetic properties of nematic and cholesteric mesophases. We shall also report recent observations about the spectacular behavior of cholesteric substances submitted to a magnetic field applied parallel to the helical axis.

INTRODUCTION

The constitutive molecules of the mesomorphic substances named liquid crystals are rod-shaped. In a molecular crystal, these molecules would be fixed in a regular, three dimensional array. On the contrary, liquid crystals exhibit only long range orientational order.

Discovered in 1888 by Reinitzer, they were divided in three classes by G. Friedel in 1922[1]. In smectics, the molecules are arranged side by side in a series of layers (Fig. 1a). Different subclasses of smectics can be distinguished, depending of the type of molecular order within the smectic planes. For example, in smectics A and C, the molecules are randomly distributed within each layer and make a given angle with the normal to the layers : the angle is finite in smectics C, null in smectics A. In smectics B the molecules would possess long range positional order within each layer.

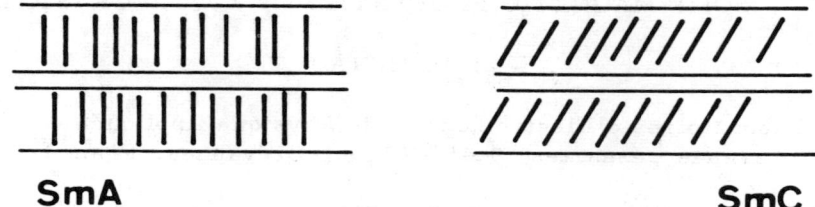

Fig 1a   Smectics

Fig 1b   Nematics

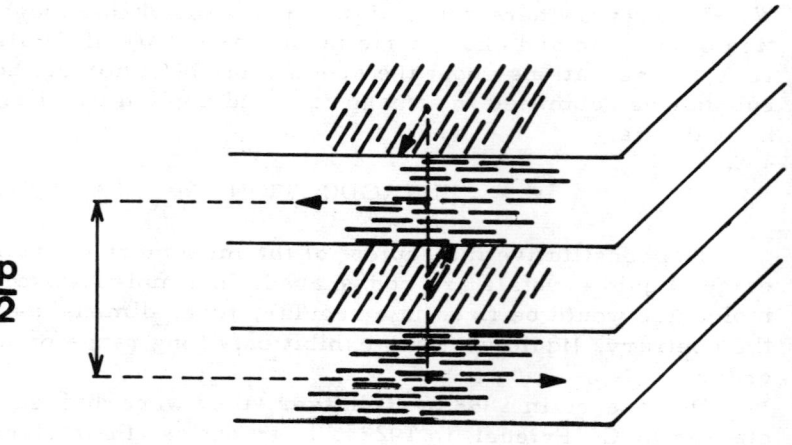

Fig 1c   Cholesterics

Fig 1   Schematic molecular description of some liquid crystals

In nematics, the molecules are randomly distributed and possess an average orientation along a prefereed axis[2] (Fig 1b)

In cholesterics, the molecules possess a nematic ordering within a given plane, but the molecular axis rotates uniformly when one moves up perpendicularly to the planes. The periodicity of this helical arrangement is characterized by the pitch P (Fig. 1c). In conventional cholesterol esters, P lies within the visible range ($P \simeq 5000$ Å). However, P may assume much larger values in nematic-cholesteric mixtures.

Due to their structure, liquid crystals possess quite remarkable properties : i) they behave mechanically like viscous fluids ; ii) they are strongly anisotropic substances.

i) in smectics, the bonds between two adjacent layers are not sufficient to prevent the layers from sliding along each other. In nematics, the molecules can move relatively freely along each other as long as their molecular axis remains parallel to the direction of alignment.

ii) As the individual molecular susceptibilities are anisotropic and as a liquid crystal exhibits a long range orientational order, the whole crystal behaves as an anisotropic medium as a whole.

- Optical properties.[3] Most liquid crystals are birefringent. Smectics A and nematics are positive uniaxial. On the contrary, cholesterics are negative uniaxial materials, the optical axis being parallel to the helical axis. Of all liquid crystals, cholesterics possess the most remarkable optical properties[4] : optical activity (the plane of polarization of a light beam transmitted along the helical axis can rotate at a rate of 10^4 degree/mm), circular dichroïsm (an unpolarized light beam is transmitted and reflected along two polarized beams), and Bragg reflection (if the wavelength of the light coincides with the pitch).

All these properties have made liquid crystals very attractive for display applications : an external mechanical, thermal, electrical or magnetic coupling may induce modifications of the texture, which affects drastically the optical properties.

- Dielectric properties. For example, in nematics, we can define a dielectric anisotropy ε_a, $\varepsilon_a = \varepsilon_{//} - \varepsilon_\perp$, $\varepsilon_{//}$ being the dielectric constant measured with a field parallel to the direction of alignment. ε_a may be either positive or negative.[5] In nematic MBBA, $|\varepsilon_a|/\varepsilon_{//} \simeq 10\%$. Thus, if changes of textures can be observed optically, they can be detected using dielectric capacitive measurements as well.

- Transport properties. (Molecular diffusion and ionic conductivity). In smectics, a foreign molecule moves more easily parallel to the layers than perpendicular to them. In nematics, this motion is

generally easier parallel the direction of alignment.[6] The anisotropy of conductivity in nematics is at the origin of the famous dynamic scattering effect[7], which focussed interest and dollars on liquid crystals.
- Magnetic properties. Liquid crystals are also anisotropic diamagnetic materials[8], as was pointed out in nematics as soon as 1929 by Foëx[9]. Typically, the averaged diamagnetic susceptibility of nematics is 5.10^{-7}. As for the dielectric anisotropy, one can define a magnetic anisotropy $\chi_a = \chi_{\parallel} - \chi_{\perp}$. In nematics, χ_a is generally <u>positive</u> (for MBBA, $\chi_a \simeq 10^{-7}$). On the contrary, in conventionnal cholesterics (cholesterol esters) the molecular magnetic anisotropy is generally <u>negative</u> and much lower in magnitude than in nematics ($|\chi_a| \simeq 10^{-9}$). However, we shall consider only, in what follows, cholesteric substances obtained by mixing nematics with small amount of cholesterol esters : we shall thus deal only with positive molecular χ_a.

Whan an external magnetic field is applied to such a substance, the molecules tend to align parallel to the direction of the field, in order to minimize the magnetic contribution to the Gibbs energy. In some sense, the same effect can be expected in the case of liquid crystals with positive dielectric anisotropy submitted to an external electric field. However, applying an electric field may give rise to important secondary effects (charge injection at the electrodes, formation of space charges) which add to the aligning effects. The fundamental advantage of applying a magnetic field lies in the fact that similar complications are not present.

It is not our pretention to give here an extensive review of the abundant work devoted to the magnetic properties of liquid crystals. We prefer to illustrate this very rich domain of research by some examples. Thus we shall not consider the magnetic properties of smectics, which have not been as extensitvely studied as the magnetic properties of nematics and cholesterics. We refer the reader to ref.10, for the very interesting flow properties of nematics in the presence of an applied magnetic field.

STATIC MAGNETIC PROPERTIES OF NEMATICS AND CHOLESTERICS: THE RULE OF THE GAME

At equilibrium, the total torque exerted on the molecules must be null. In the case of a constant applied magnetic field, two components only of this torque need be considered : a magnetic one, which tends to align the molecules in the direction of H, and an elastic one, which tends to restore the molecules in their initial configuration (imposed by the boundary conditions).

More precisely, in nematics and cholesterics, the texture of the system is described by a unit vector $\underline{n}(\underline{r})$, the director, parallel to the local alignment. The total Gibbs energy, G, is a functional of $\underline{n}(\underline{r})$ and can be written as a sum of two terms

$$G = G_{el} + G_{magn} \quad (1)$$

with G_{magn}, the magnetic contribution given by

$$G_{magn} = -\frac{1}{2} \int d^3r \, \chi_a (\underline{H} \cdot \underline{n})^2 \quad (2)$$

G_{el}, the elastic contribution, is yielded by Frank's[11] expression for long wavelength distorsions

$$G_{el} = \int d^3r \left[K_1 (\text{div} \, \underline{n})^2 + K_2 (\underline{n} \cdot \text{curl} \, \underline{n} + q_o)^2 + K_3 (\underline{n} \wedge \text{curl} \, \underline{n})^2 \right] \quad (3)$$

In (3), K_1, K_2 and K_3 are Frank's elastic constants for splay, twist and bend respectively (typically, $K_i \simeq 10^{-6}$ erg/cm) ; q_o, for a right-handed cholesteric, is worth $2\pi/P$ and is null for nematics ($P \to \infty$).
If $\underline{n}$ is defined by the two Euler angles θ and φ, the stability conditions are simply written as

$$\begin{cases} \dfrac{\delta G}{\delta \theta}(\underline{r}) = 0 \\[6pt] \dfrac{\delta G}{\delta \varphi}(\underline{r}) = 0 \\[6pt] + \text{ boundary conditions} \end{cases} \quad (4)$$

In the above equations, the derivatives are functional derivatives. The boundary conditions may assume different forms. For example, the molecules may be forced to align along a certain direction at the glass-liquid interfaces : this occurs if the glass is rubbed along a given direction[12], the molecules tending then to align parallel to that direction. Another treatment of the glass, for example rinsing with acid, forces the molecules to be normal to the glass plane.

After having outlined the formal simplicity of the theoretical analysis, we turn to the description of some specific examples.

A NEMATIC SLAB IN A MAGNETIC FIELD : THE FREEDERICKS TRANSITION

A nematic film, of thickness L ; between two parallel walls previously rubbed along two parallel directions, is aligned along the rubbing direction. It has been pointed out long ago that this configuration is unstable under a sufficiently high magnetic field H applied along the Oz direction [the Oz axis being normal to the walls, the origin being taken in the middle of the sample][1,3] The so-called Freedericks transition is a second order transition : the nematic texture is stable up to a field H_F where an infinitesimal distorsion begins to appear : the molecules are tilted, in a plane Oxz (Ox being the rubbing direction) by angle θ which depends only upon z .[14]

The magnitude of H_F can be found approximately through the following argument ; if the K_i in (3) are set equal, equation (1) can be written as

$$G = \frac{1}{2}\left[\int d^3r\, K\left[(\nabla n_x)^2 + (\nabla n_y)^2 + (\nabla n_z)^2\right] - \chi_a (\underline{n}\cdot\underline{H})^2\right] \quad (5)$$

In the present case, to the lowest order in θ, G takes the form

$$G = \frac{1}{2}\int d^3r\, K\left[\left(\frac{\partial \theta}{\partial z}\right)^2 - \chi_a \theta^2\right] \quad (6)$$

From (4), the minimization condition can be written as

$$K\frac{\partial^2 \theta}{\partial z^2} + \chi_a \theta = 0 \quad (7)$$

Relation (7) defines a "magnetic coherence length" ξ_H ($\xi_H = \frac{1}{H}\sqrt{K/\chi_a}$) over which a distorsion can be relaxed. For a distorsion to appear under increasing H, ξ_H must be at most equal to L/2. Thus, the critical field H_F is approximately defined by the condition $\xi_H \simeq L/2$. In fact, the exact value for H_F is

$$H_F = \frac{\pi}{L}\sqrt{\frac{K}{\chi_a}} \quad (8)$$

For $L = 100 \, \mu m$, $K = 10^{-6}$, $\chi_a = 10^{-7}$, $H_F \simeq 10^3$ oe. For $H > H_F$, the amplitude of the tilt angle increases rapidly as $(H - H_F)^{\frac{1}{2}}$, and for $H \gg H_F$, the system can be roughly described as follows : the molecules are nearly aligned parallel to the field in a thickness L centered at the middle of the sample and are relaxed towards the rubbing direction in the two regions adjacent to the walls.

The Freedericks transition can be sensitively detected between two crossed polarizers : if the sample is transparent for $H = 0$, the intensity of the transmitted light begins to decrease at $H = H_F$. However, the sample has not a uniform aspect (Fig. 2). This comes from the fact that, under the applied field, the molecules can rotate by two different ways ($+\theta$ and $-\theta$). This degeneracy gives rise to adjacent domains separated by walls. The static and dynamic properties of these walls have been studied recently by L. Léger [15], together with their dynamic behavior when the field is tilted to lift the degeneracy.

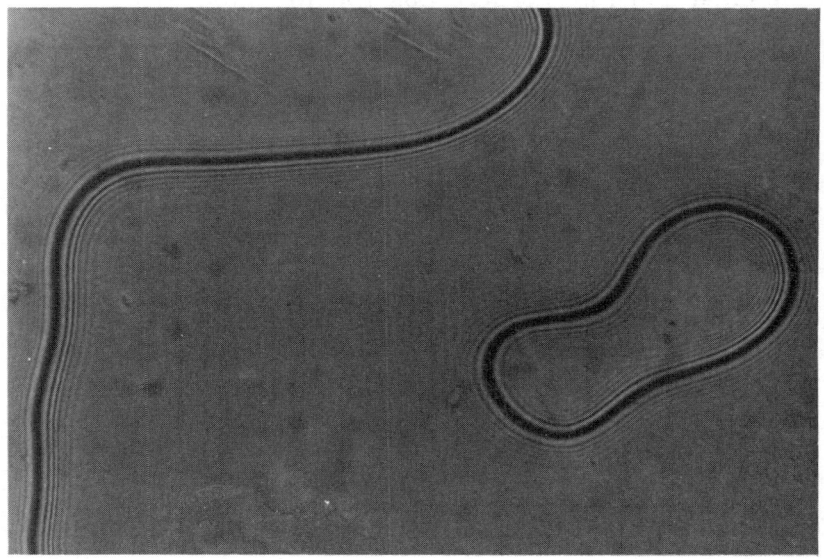

Fig. 2 : wall separating two tilted regions ($+\theta$ and $-\theta$) in nematic MBBA. Thickness : $100 \, \mu m$ - $H \simeq 1.2 \, H_F$.
(L. Leger: private communication).

MAGNETIC DISTORSIONS IN CHOLESTERICS

I - H normal to the helical axis : untwisting of a cholesteric

A nematic-cholesteric mixture ($\chi_a > 0$), of pitch P along Oz, is submitted to a field H applied perpendicular to the helical axis. When $H \to \infty$, we guess that the cholesteric will be completely unwound to the nematic state. It has been shown by de Gennes and Meyer [16], that a complete untwisting is achieved at a field H_U, H_U being given by

$$H_U = \frac{\pi^2}{P}\sqrt{\frac{K_2}{\chi_a}} \qquad (9)$$

For $H < H_U$, the spatial period increases with H (Fig. 3). These predictions have been quantitatively verified by Durand et al. and by Meyer [17].

Let us here point out that this agreement between theory and experiment has represented an important step in the understanding of the elastic and magnetic properties of cholesterics.

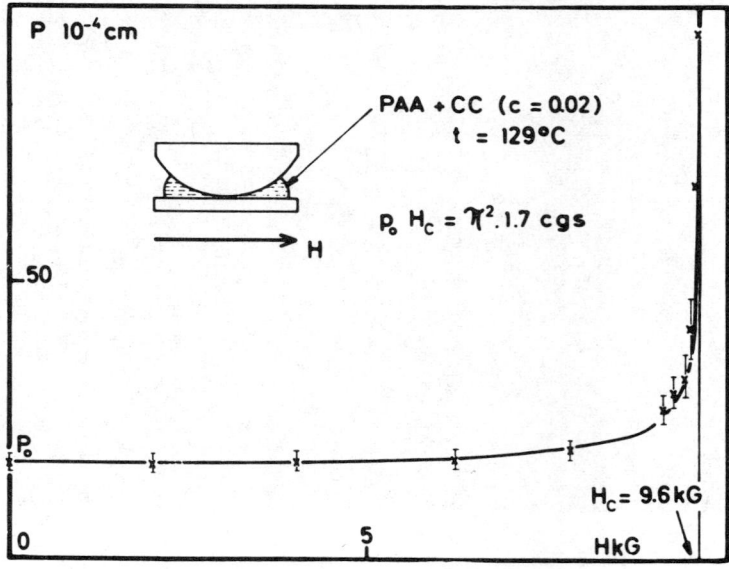

Fig. 3 : Variation of the spatial period with the applied magnetic field H. The theoretical curve is deduced from Ref. 16 (From Durand et al, Ref 17).

II - H parallel to the helical axis

We come now to a configuration which has been studied very recently. The geometry we shall consider from now on is as follows : the sample is prepared in the so-called planar texture ; i. e. is sandwiched between two semi-transparent electrodes spaced by a distance L.

At rest, the helical axis is perpendicular to the electrodes. The magnetic(or electric) field is applied parallel to the helical axis. Such a configuration gives rise to quite remarkable phenomena. Fig. 4 shows how a sample, which was initially transparent, can be distorted by the action of a low frequency electric field.

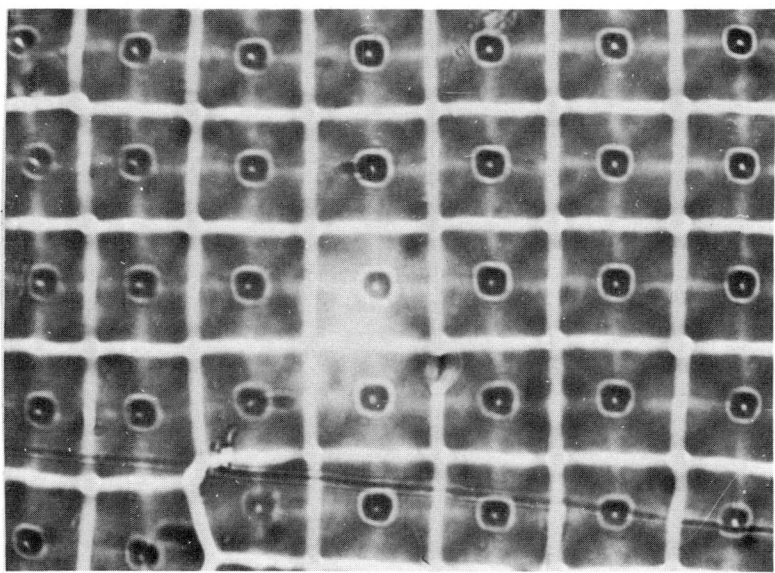

Fig. 4 : Square periodic pattern observed at threshold under low frequency electric field applied parallel to the helical axis. The material is a mixture of nematic MBBA and cholesteric cholesterol nonanoate : L = 105 μm, $P \simeq 12$ μm, $V_{eff} \simeq 10$ volts (F. Rondelez and H. Arnould, private communication).

Experimentalists were first interested in the electrical properties of conventional cholesterics. In Ref. 18, it was noticed that a planar structure (the cholesteric planes are parallel to the walls) would be unstable under a sufficiently high electric field : the sample, from transparent, turns to milky white. Moreover, if the electric excitation is switched off, the cholesteric still scatters light and comes back to its initial configuration only after a long time ($\simeq$ several hours). This is the so-called storage effect.

Others authors were primarily interested in the phenomena which occur at the instability threshold. This is the case in ref. 14, where complicated periodic patterns were observed. In 1970, Helfrich predicts that a periodic bending mode of the cholesteric planes could be nucleated at fields much lower than the unwinding field [20]: if L is the thickness of the planar texture, and if L $\gg$ P, the threshold field is proportional to $(PL)^{-1/2}$, the periodicity of the distorsion behaving like $(PL)^{1/2}$. These predictions seemed to agree roughly with the experimental results of ref. 19.

However, a real breakthrough in the understanding of these effects was achieved by the beautiful experiments by Rondelez and Arnould and by Gerritsma and van Zanten [21].

This result was obtained because
- these authors succeeded in preparing good monocrystalline samples,
- they used nematic-cholesteric mixtures, with high P values (P $\sim$ 10 μm), which made the observation at the microscope much more precise, and which yielded lower thresholds than conventional cholesterics. Fig. 4 represents the square periodic pattern observed at threshold under an electric field applied parallel to the helical axis.

However, as previously mentioned, electric fields give often rise to complicated phenomena such as charge injection and space charge formation. A much cleaner experimental situation could thus be met under a magnetic field.

Using the formalism we have just presented, the precise stability conditions of the planar structure were derived in ref. 22, first under magnetic fields and also under a.c. electric fields.

For L $\gg$ P, the resulting threshold field H_H is such that

$$H_H^2 = \frac{2\pi^2}{\chi_a} \frac{\sqrt{6K_2 K_3}}{PL} \qquad (10)$$

At H_H, the period of the bending distorsion, Λ, is given by

$$\Lambda^2 = \frac{3}{2} \sqrt{\frac{K_3}{K_2}} \, PL \qquad (11)$$

The corresponding experiments have been performed by Scheffer and by Rondelez and Hulin[23]. The nematic-cholesteric mixture (6 μm < P < 80 μm), prepared in the planar texture, was sandwiched between two semi-transparent electrodes. The onset of the distorsion was detected optically and also by the capacitive method.

The pattern observed at threshold is very similar to the one of Fig. 4, obtained under an electric field, the period Λ being the same, for a given sample, than under a low frequency a.c. field, in agreement with ref. 22. Moreover, the agreement between formulae 10 and 11 and with experiment is better than 10 %. A convincing proof of this statement is given in table 1, which shows the experimental determinations of the product $H_H \Lambda$ (from (10) and (11), H_H is independent of L and P).

EXPERIMENTAL $H_H \cdot \Lambda$

e(μm) \ P(μm)	23	13.2	6.5
50	19.5	20.4	19.0
75	21.3	19.4	20.8
100	22.0	18.2	19
175	17.8	19.4	17
250	20.2	18.9	20.2

$$H_H \cdot \Lambda = 19.5 \pm 1.5 \quad \text{gauss.cm}$$

$$\Rightarrow K_{33} = 7.45 \pm 1.1 \quad 10^{-7} \text{ dyne}$$

Table 1 : Experimental values of the products $H_H \Lambda$. From (10) and (11), $H_H \Lambda$ must not depend on P or L. (F. Rondelez and J. P. Hulin : private communication).

But this is not the end of the story. Fig. 5 represents the behavior of the sample capacitance versus H.

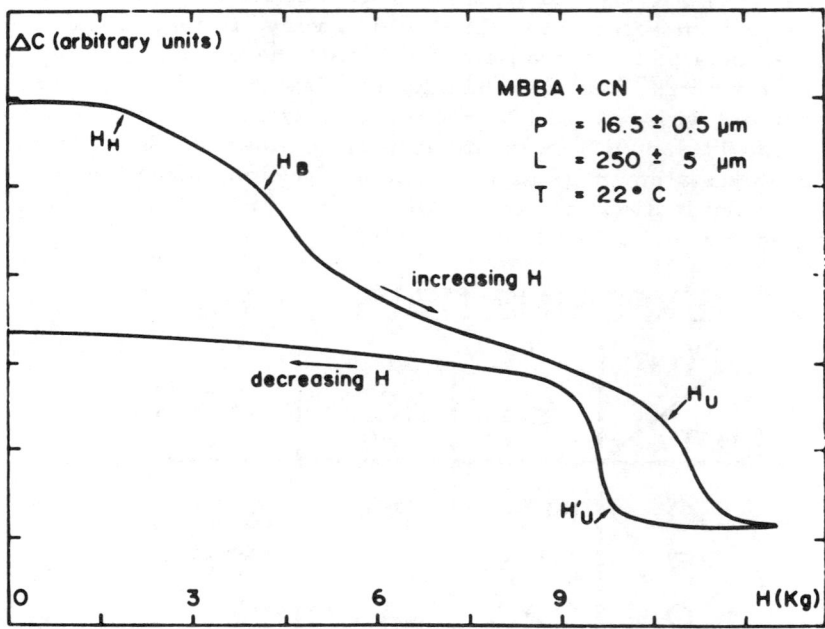

Fig. 5 : Behavior of the capacitance of a typical sample versus H (from Ref. 23).

We note first the onset of the square periodic distorsion at H_H. Then, this distorsion swells and, at $H = H_B$, an abrupt change of slope in the C vs H curve is observed : at H_B, we detect optically that a finger print pattern (Fig. 6) has replaced the square pattern : the periodicity between the dark lines in Fig. 6 is equal to half the pitch, from which we deduce that the cholesteric planes are now seen by the edge.

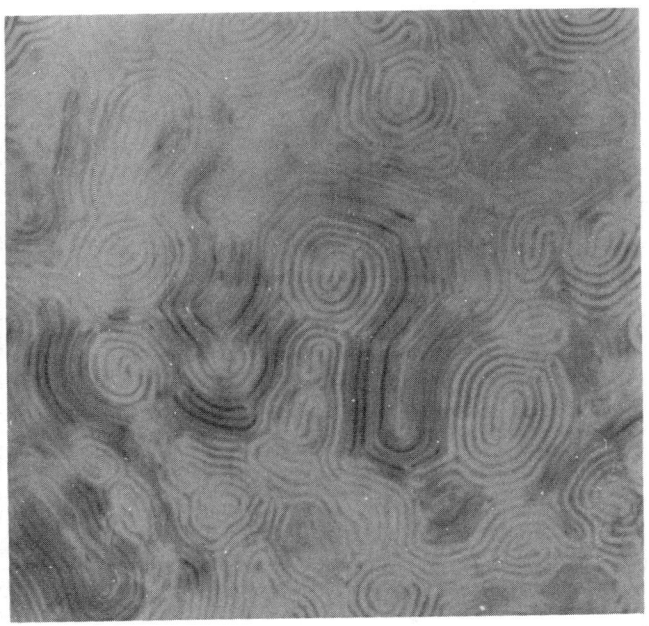

Fig. 6 : "Finger print" pattern observed at H_H. Note that the cholesteric planes have been tilted by 90°. L = 250 μm
P = 16 μm H≃1500 œ.

At H_B, the cholesteric planes have been tilted by 90°. Increasing H further on, a supplementary abrupt change of the C versus H curve is found at a field H_U : at H_U, the sample appears uniform. In the microscope, the distance between the lines is observed to increase up to H_U : H_U must thus correspond to the unwinding field calculated in ref. 16, and this is indeed what is found.

Now, if H is decreased from H_U, the capacitive behavior shows up a very strong hysteresis, and, when H is switched off to zero, the optical appearance of the sample corresponds to the "finger print" texture. From this result, we can deduce that the "finger print" texture is responsible for Heilmeier's storage mode.

A more detailed study of the life time τ of the "finger print" texture has been undertaken by Hulin[24]. It appears that τ is an exponential function of the L/P ratio (Fig. 7). Changing L/P from 2 to 15 results is changing τ from 1 min to several days !

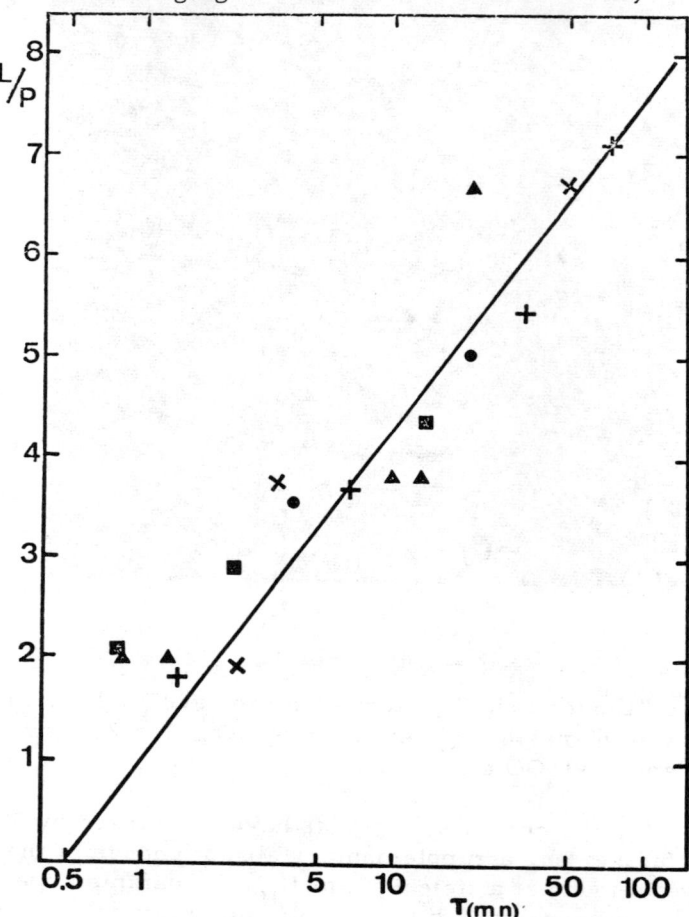

Fig. 7 : Experimental behavior of the life time τ of the storage mode versus the L/P ratio (from Ref. 24).

Finally, let us outline that the distorted pattern observed when $L \sim P$ differs strongly from the one seen for $L \gg P$. Then lines are observed instead of squares, which correspond to a direct tilting of the cholesteric planes. Other amusing effects are also detected, which shall be reported soon[25]. Finally, when $L \ll P$ (case of the twisted nematic obtained by rubbing the walls in two different directions), capacitive observations evidence a reversible tilting of the molecules at a field which increases with the twist angle[26].

CONCLUSIONS

Our description has been limited to certain magnetic properties of the mesophases. We hope we have succeded in evidencing the very rich and fascinating nature of the phenomena that can be observed. It appears that a magnetic field is an ideal tool for the investigation of the elastic and structural behavior of mesophases. However, one must be prudent about the perspectives of using the magnetic properties of mesophases for display applications. This is simply due to the fact that high field values ($H \simeq 1 k\text{œ}$) are needed to induce structural change. On the other hand, the physical principles underlying a given display can be hopefully understood using a magnetic excitation. Finally, those high field values might be considerably lowered if the magnetic coupling between the mesophase and the field could be increased : this is what might be achieved in suspensions of magnetic particles in nematic or cholesteric phases[27].

ACKNOWLEDGEMENTS

It is a pleasure to thank here P.G. de Gennes, F. Rondelez, E. Dubois Violette, L. Leger, O. Parodi and A. Rapini for frequent discussions on the magnetic properties of liquid crystals.

REFERENCES

1. G. Friedel, Annales de Physique, 9ème série, t. XVIII, 273 (1922).
2. The reader might be mislead by fig. 1, where for example in Fig. 1b, nematics are described as if a complete molecular alignment was achieved. In fact, this is an oversimplified description and Fig. 1 is basically schematic. Indeed, it must be kept in mind that thermal fluctuations give rise to important short range order orientational disorder. See for example Groupe d'Etude des Cristaux Liquides (Orsay), J. Chem. Phys. 51, 816 (1969).
3. see for example I.G. Chystyakov, Sov. Phys. Uspekhi 9, 551 (1967).
4. See for example H. de Vries, Acta Cryst. 4, 219 (1951).
5. See for example M. Schadt, J. Chem. Phys. 56, 1494 (1972). and references therein.
6. For diffusive properties, see for example : W Franklin, Mol. Cryst. and Liq. Cryst. 14, 227 (1971) ; R. Blinc et al : Mol. Cryst. and Liq. Cryst. 14, 97 (1971).
   For conductive properties, see for example T. Svedberg, Ann. Physik 44, 1121 (1914) ; R.P. Twitchell, E.F. Carr, J. Chem. Phys. 46, 2765 (1967)
7. G.H. Heilmeier, L.A. Zanoni, L.A. Barton, Proc. IEEE 56, 1162 (1968).
8. For recent experimental determinations of χ_a in nematics, see for example H. Gasparoux, J. Prost, J. de Physique 32, 953 (1971) and references therein.
9. G. Foëx, Journal de Physique et le radium 10, 960 (1929).
10. See for example : Ch Gähwiller, Phys Lett. 36A, 311 (1971) ; W. Helfrich, J. Chem. Phys. 50, 100 (1969).
11. F.C. Frank, Disc. Faraday Soc. 29, 883 (1958).
12. D.W. Berremann, Phys. Rev. Letters 28, 1683 (1972) and references therein.
13. V. Freedericks, V. Zolina, Trans. Faraday Soc. 29, 919 (1933).
14. For more recent work on the Freederick transition, see A. Rapini, Thèse de 3ème cycle, Orsay (1970) unpublished ; A. Rapini, M. Papoular, P. Pincus, C.R.A.S. Paris 267, série B, 120 (1968).
15. L. Léger, Solid State Comm., in press.
16. P.G. de Gennes, Solid State Comm., 6, 163 (1968).

17. G. Durand, L. Léger, F. Rondelez and M. Veyssié, Phys. Rev. Letters 22, 227 (1969).
    R.B. Meyer, Appl. Phys Letters 14, 208 (1969)
18. G. Heilmeier, J. Goldmacher, Phys Letters 13, 132 (1968).
19. C.J. Gerritsma, P. van Zanten, Molecular Cryst. and Liq. Cryst. 15, 257 (1971).
20. W. Helfrich, Appl. Phys. Lett. 17, 531 (1970).
21. F. Rondelez, H. Arnould, C.R. Acad. Sci. Paris 273 B, 549 (1971) ;
    C.J. Gerritsma, P van Zanten, Phys. Lett 37 A, 47 (1971).
22. J.P. Hurault, J. Chem. Phys., in press.
23. T.J. Scheffer, Phys. Rev. Letters 28, 593 (1972) ;
    F. Rondelez, J.P. Hulin, Solid State Comm. 10, 1009 (1972)
24. J.P. Hulin, Appl. Phys. Lett., in press.
25. A. Hervet, J.P. Hulin, F. Rondelez, private communication.
26. C.J. Gerritsma, W.H. de Jeu, P. van Zanten, Phys. Letters 36A, 389 (1971)
27. F. Brochard, P.G. de Gennes, J. de Phys. 31, 691 (1970).

MAGNETIC PROPERTIES OF CONDUCTING ORGANIC SALTS

A.J. Heeger and A.F. Garito
University of Pennsylvania, Philadelphia, Pa. 19104

ABSTRACT

The physics of organic charge transfer salts is reviewed with emphasis on the highly conducting compounds based on TCNQ. Specific heat studies of a family of these salts show systematic variation of the Debye temperature, indicating that charge transfer in such systems is in some cases stabilized by the Madelung energy, but that polarization contributions are important. Requirements for achieving a metallic state in organic salts are discussed in terms of crystal structure, Coulomb interaction, and cation polarizability. Magnetic properties of (NMP)(TCNQ) are discussed in the general context of the metal-insulator transition. Measurements of the transfer integral t, the on-site Coulomb repulsion U, and the exchange interaction J, provide a numerical check on the relation $J = 2t^2/U$, expected when the AF exchange arises from virtual charge transfer. In the low temperature AF regime, spin dynamics have been studied with NMR techniques. Results indicate that for spin $\frac{1}{2}$ systems in 1-d, spin wave excitations are to be treated as Fermions. The role of disorder is reviewed. The existence of metallic behavior in a symmetric cation salt, together with earlier studies, rules out cation disorder as the source of high conductivity.

INTRODUCTION

The class of solids known as organic charge transfer salts has been of considerable interest in recent years. This interest arises from the novelty of the systems themselves as well as from the fact that the flat planar molecules involved lead to anisotropic structures and therefore to pseudo one-dimensional electronic properties. More generally, the existence of a class of materials made up of organic molecules as fundamental units allows the possibility of utilizing the flexibility of organic chemistry to design and synthesize molecules with particular characteristics. The over-all goal would be to create solid state materials with a desired bulk property by the preliminary design of the constituent molecules.

The organic charge transfer salts are composed of a donor molecule, D, and an acceptor molecule, A.[1] In contrast to conventional molecular crystals (that is, crystals made up of neutral organic molecules bonded by van der Waals forces), these salts have unpaired electrons on the acceptor or donor or both as a result of the basic charge transfer reaction:

$$DA \rightarrow D^+A^- . \tag{1}$$

In the simplest cases, the energy required for charge transfer is given by:

$$\Delta E_{ct} = (I_D - A_A) - (E_M + E_{ex} + E_{pol}) \tag{2}$$

where the neutral DA system is taken as the zero of energy. I_D is the donor ionization potential and A_A is the electron affinity of the acceptor. The final three terms are of electrostatic origin: E_M is the total Madelung energy arising from the sum over the long range Coulomb forces in the ionic D^+A^- solid, E_{ex} is the exchange contribution, and E_{pol} represents the interaction of the ionic charges with the induced dipoles arising from the molecular polarizability. E_M and E_{ex} are well-known; E_{pol} may be important in systems containing molecules with large π-electron polarizabilities.

When $\Delta E_{ct} \ll 0$, the resulting system will be nearly fully ionic with unpaired electrons outside a filled core. Generally in such molecules, the affinity level corresponds to the lowest unfilled π-level of the neutral acceptor molecule so that one is dealing primarily with the $2p_z$ wavefunctions of C and N.

Much of the work done in this area has utilized the excellent acceptor tetracyanoquinodimethan (TCNQ),[1] the structure of which is shown below:

The molecular physics of this system has been discussed in detail elsewhere.[2] We note here briefly only the most important features. The relatively large electron affinity (~2eV) results from the cyanide groups at the ends of the molecule. Consequently, the excess charge density associated with the (TCNQ)⁻ anion resides primarily on the ends of the molecules where the one-electron attraction is greatest. In the event of an ionic fluctuation with two excess electrons on a single TCNQ molecule (as would be the case in a metallic state or near the metal-insulator transition),[2] there is a clear tendency for the two electrons to localize at opposite ends of the molecule and correlate to stay apart in order to reduce their mutual Coulomb repulsion.[3] This Heitler-London correlation, together with the polarizability of the cation, makes the resultant effective Coulomb repulsion sufficiently small that the metallic state can be achieved in some of the TCNQ compounds. The crucial role of the cation polarizability has been demonstrated in our earlier work.[2]

Both simple (1:1) and complex (2:1) salts exist with a wide variety of donors.[1,4] The flexibility of organic chemistry quickly leads to a wide class of materials even with the single TCNQ acceptor, and a variety of possible acceptors with special properties can be envisioned. Since $|\Delta E_{ct}|$ may be relatively small, multiple crystal phases relatively close in energy are not unexpected. As a result, the detailed crystal structure is difficult to control, and even small changes in cation structure can lead to gross changes in crystal structure and thus in physical properties.

LOW TEMPERATURE HEAT CAPACITY

In spite of their apparent complexity, these systems represent a relatively simple class of ionic solids. Only a single narrow electron energy band is involved (the other molecular levels being several volts higher in energy) and tight binding theory is an excellent approximation. Some insight into the relative simplicity of these solids can be obtained from studies of the low temperature heat capacity.

The low temperature specific heats of several of the TCNQ salts have been measured,[5] including $Li^+(TCNQ)^-$, $Rb^+(TCNQ)^-$, $Cs^+(TCNQ)^-$, $(TMPD)^+(TCNQ)^-$, $(NEP)^+(TCNQ)^-$, $(NMP)^+(TCNQ)^-$ and $Q^+(TCNQ)_2^-$. The first five compounds are insulators, whereas the NMP and quinolinium salts are metallic at room temperature. X-ray studies of these compounds show linear chains of $(TCNQ)^-$ anions separated by similar chains of cations.[6] An exception is the $(TMPD)^+(TCNQ)^-$ system which crystallizes in chains of alternating $D^+A^-D^+A^-$ ... stacking.[7]

The anisotropic structures suggest a pseudo one-dimensionality in the electronic and magnetic properties of these salts which actually has been observed experimentally.[2] The lattice dynamics, however, are expected to be less anisotropic due to the importance of the ionic bonding.

The specific heat was measured from 1.7 to 4.2 K. The data on all the above salts follow the form $C = \alpha T + \beta T^3$, which was found by minizing the standard deviation. The T^3 term is the Debye lattice contribution, whereas the linear term is of electronic or magnetic origin. In fact, only the $(NMP)^+(TCNQ)^-$ salt has a sizeable linear term.[2] In all other cases the linear term is zero (to within experimental accuracy) except possibly for the system $Q^+(TCNQ)_2^-$ where a very small linear term is inferred from the least-squares fit to the data.[5] The small coefficient, α, for $Q^+(TCNQ)_2^-$ is more than an order of magnitude less than that observed for the

NMP salt.

The Debye temperatures for the compounds studied are given in Fig. 1. The values for θ_D are relatively large and suggest strong ionic bonding. One can qualitatively compare the observed trends with expectations based on the simple Debye theory wherein $\theta_D \propto (K/M)^{\frac{1}{2}}$, with M the molecular mass of the chemical unit cell and K the average force constant. The binding energy in an ionic crystal arises from the Coulomb force between the ionic charges. The force constant K is therefore $K \simeq 1/r_{av}^3$, where r_{av}^3 is the volume of the chemical unit cell. The value of r_{av}^3 is obtained using crystallographic data when available.[6,7] For $Cs^+(TCNQ)^-$ we have used the same value of r_{av}^3 as for the K, Li, and Rb salts. Since $(NEP)^+(TCNQ)^-$ has nearly the same density and molecular weight as $(NMP)^+(TCNQ)^-$, we have used the same r_{av}^3 for both.[6]

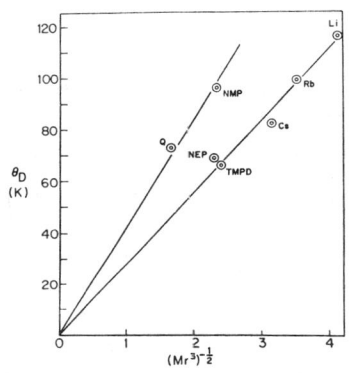

Fig. 1. θ_D vs. $(Mr_{av}^3)^{-\frac{1}{2}}$ for the TCNQ salt.

Fig. 1 shows a plot of θ_D vs. $(Mr_{av}^3)^{-\frac{1}{2}}$ in arbitrary units. There is good agreement between this simple theory and experiment for all the salts, with the notable exception of $Q^+(TCNQ)_2^-$ and $(NMP)^+(TCNQ)^-$ which show significantly larger θ_D values and appear to fall on a separate line. For example, the T^3 term in the NMP salt is nearly three times smaller than that of the NEP salt.

Madelung calculations have recently been carried out for several of the TCNQ salts.[8] Of particular interest for comparison are the systems $(TMPD)^+(TCNQ)^-$ and $(NMP)^+(TCNQ)^-$. For the TMPD salt, Metzger finds a Madelung contribution which is sufficient (when the exchange part is added[9]) to account for the stability of the c. t. state.[8] Using the standard Debye theory for ionic crystals with Metzger's Madelung constant yields $\theta_D \sim 60$ K in good agreement with experiment. The situation for the NMP salt is far less clear. The Madelung energy is nearly zero and the exchange contribution is less than 0.1eV.[9] Thus, the overall stability is in question whereas the Debye θ as given in Fig. 1 is actually larger than that of the TMPD salt. The obvious suggestion is that the polarization contribution arising from the large cation polarizability is dominant. However, detailed calculations as well as systematic studies with different cations are required before the role of the cation polarizability on charge transfer stability can be fully clarified.

THE METALLIC STATE

The achievement of a metallic state in an organic solid is all the more remarkable since we are used to thinking of molecular solids as weakly interacting collections of molecules. There are three basic requirements:

1) The existence of unpaired electrons;
2) A uniform crystal structure such that, in absence of electron-electron interactions, the electronic band structure of the system would be metallic;
3) Relatively weak electron-electron repulsive interactions.

The first requirement needs little comment. Without unpaired electrons, the system would remain insulating until the band widths exceeded the intramolecular level splittings, $\Delta E_{\pi-\pi^*}$. Since this would require intermolecular lattice spacings comparable with those between atoms within the molecule, such a state could only be achieved at very high pressures or if a true polymerization occurred. Were this the case, the resulting metallic state would be analogous to that in the alkaline earth metals. The existence of unpaired electrons in the charge-transfer salt greatly facilitates the achievement of the metallic state, for in the absence of interactions any weak intermolecular transfer integral, t, leads to formation of energy bands and delocalization of the electronic wave functions. The observation made above that $t/\Delta E_{\pi-\pi^*} << 1$ simply means that the band structure may be calculated in the tight-binding approximation with high accuracy. However, it is clear that such a band structure need not be metallic. For example, in a one-dimensional system, any weak dimerization of the structure into pairs leads to an energy gap at the Fermi surface (the Peierls instability) and a resulting semiconducting state; hence requirement (2), the need for a uniform structure.

The role of the Coulomb interaction in the metal-insulator transition has been discussed in detail elsewhere.[2,10] Basically, the Coulomb interaction tends to cause correlation of the electrons so as to stay apart. For the case of a half-filled band system, in the limit of large Coulomb repulsion, the electrons localize one per site to form a magnetic Mott-Hubbard insulator. This tendency toward localization competes with the transfer integral which tends to lower the energy of the system by delocalizing the electronic wave functions and forming energy bands. Hubbard has attempted to describe the relevant physics in terms of the model Hamiltonian[11]

$$\mathcal{H} = -t \sum_{(i,j)} c_{i\sigma}^{+} c_{j\sigma} + U \sum_{i} n_{i\uparrow} n_{i\downarrow} \quad (3)$$

where t is the near neighbor transfer integral and U represents the Coulomb repulsion between two electrons on the same site. The detailed solutions of the Hubbard model as applied to the TCNQ salts have been discussed elsewhere. Qualitatively, $U/t \gg 1$ corresponds to the insulating limit and $U/t \ll 1$ to the metallic limit. The metal-insulator boundary as given qualitatively by the Hartree-Fock approximation occurs at $U/4t = 1$. In these semi-quantitative terms, requirement (3), the need for weak Coulomb repulsion, is clear.

Requirements (1) and (3) are relatively easy to achieve and control. As indicated above, a wide class of charge transfer salts exist and flexibility in choice and design of both the cation and the anion are possible. Because of the structure of the TCNQ molecule (and other related systems) the bare Coulomb repulsion is considerably reduced from the atomic value by the Heitler-London correlation described above. The final step in achieving the reduced Coulomb interaction is via the cation polarizability. Since this polarizability arises from virtual excitation of the π-π^* singlet excitons in the cation chain, the problem can be reduced to that of interacting excitonic polarons.[12] Detailed calculations indicate the existence of an indirect attractive interaction via these virtual excitons, resulting in a total effective interaction which may be written as[12]

$$U_{eff} = U - 2E_B \qquad (4)$$

where $E_B = \Gamma^2/\hbar\omega_0$ is the excitonic polaron binding energy arising from the coupling Γ of the electronic system to the cation polarizability expressed in terms of cation excitons at energy $\hbar\omega_0$. This indirect attractive interaction may indeed by large, and under favorable circumstances can reduce the net interaction to the point where the metallic state is accessible.

Requirement (2) is more difficult. Since the overall energy $|\Delta E_{ct}|$ is relatively small, and since isotropic Coulomb forces are involved, multiple crystal phases are likely and several examples have been observed experimentally.[9,13] Generally, however, the crystal structures consist of stacks of molecules with separations between molecules within a chain of 3.2 - 3.5 Å and with the individual chains widely separated. As far as the electronic properties are concerned, the obvious anisotropy of such a structure is enhanced by the directionality of the π-electron wave functions with the result that interchain coupling can be negligible. Two basic kinds of structures have been observed. In the first, the anions stack in linear chains separated by parallel cation chains. The anion stacking may be uniform as in (NMP)(TCNQ) or dimerized as in $Rb^+(TCNQ)^-$. The uniform anion chains are <u>potentially</u> metallic; the dimerized systems are <u>necessarily</u> semi-conducting. The second kind of structure can be described as chains of the form $D^+A^-D^+A^-\ldots$ where each cation has nearest neighbor anions along the chain, and vice versa, as is the case in (TMPD)(TCNQ). The latter type of structure is unlikely to be metallic, as an accidental near degeneracy of the D^+ and A^- levels would be required. Obviously, the structure is not easy to control, although experimentally without exception the uniform structures are found with flat planar organic cations.

One final comment on structure is in order. It has been noted that in several of the highly conducting salts, there is disorder in the cation chains. Although it has been shown that this disorder does not dominate the electronic properties,[14] it may act to stabilize the desirable uniform stacking in anion chains[15] by removing the logarithmic singularity in the dielectric constant through residual disorder. Although speculative, this possibility deserves further study in the context of gaining some control on the lattice structure.

THE METAL-INSULATOR TRANSITION IN (NMP)(TCNQ)

The system (NMP)(TCNQ) is perhaps the most thoroughly studied charge transfer salt. Through experimental studies of magnetic susceptibility,[2] low temperature specific heat,[2] transport properties,[16] electron spin resonance,[2] and nuclear spin-lattice relaxation,[17,18] it has been shown that this system undergoes a gradual transition from magnetic Mott insulator to a highly correlated metal as the temperature is increased above 200 K. Using the exact solutions for the ground state and low lying excitations of the Hubbard model to analyze the low temperature data, it was possible to obtain values for the transfer integral t and the Coulomb interaction U.[2] Analysis of the higher temperature data using approximate solutions yielded similar values for t and U.[17] The overall numerical agreement and consistency of the results imply that the system (NMP)(TCNQ) is a nearly ideal experimental example of the 1-d Hubbard model. In the following paragraphs we briefly review the experimental results obtained on this compound.

A schematic phase diagram appropriate to (NMP)(TCNQ) is shown in Fig. 2. As a function of temperature one sees three regions: antiferromagnetic insulator, paramagnetic insulator, and metal. Pressure experiments are underway; however, the information on the phase diagram included in Fig. 2 is restricted to one atmosphere.

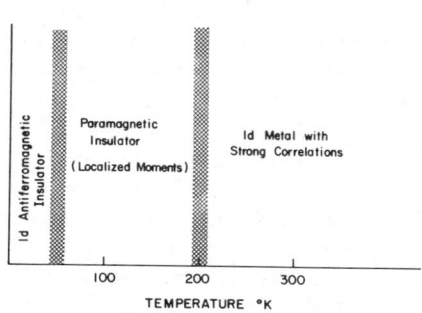

Fig. 2. Schematic phase diagram for (NMP)(TCNQ).

X-ray studies[19] at 160 K show no evidence of a distortion in the anion chains. That is, both above and below the metal-insulator transition in (NMP)(TCNQ), the (TCNQ)⁻ anions stack in a <u>uniform</u> chain with each molecule equidistant from its nearest neighbors. This direct experimental result is in agreement with the low temperature antiferromagnetic properties described below, which also imply a uniform structure. The absence of a doubled unit cell (along the chain axis) definitively rules out a Peierls instability as the source of the insulating phase. The (NMP)(TCNQ) system thus represents a unique example of a Mott-Hubbard metal-insulator transition driven by Coulomb correlations <u>without</u> an accompanying lattice distortion.

At low temperatures the specific heat and magnetic susceptibility are in agreement with the predictions of the Hubbard Hamiltonian for a 1-d system. There is a linear term (coefficient α_m) in the specific heat; the susceptibility approaches a constant value, $\chi(0)$ (aside from a small residual impurity contribution); and the ratio $\alpha_m/\chi(0)$ is in quantitative agreement with theory.[2] Using as input data the coefficient of the linear term in the specific heat, the extrapolated zero temperature susceptibility, and the magnitude of the Hubbard gap as obtained from resistivity measurements, the low temperature solutions to the 1-d Hubbard model yield $t = 0.02$ eV, $U = 0.14$ eV with the ratio $U/t \simeq 7$ being greater than one, but relatively close to the qualitative M-I boundary as indicated by the Hartree approximation.

In the limit $U/t \gg 1$, there is an additional energy parameter associated with the Hubbard Hamiltonian since virtual charge transfer onto neighboring sites provides an overall lowering of the energy. Since the exclusion principle restricts such virtual transfer to opposite spin pairs, this is equivalent to an antiferromagnetic exchange interaction. In this limit the Hubbard model becomes identical to the Heisenberg exchange Hamiltonian ($\mathcal{H} = J \sum_{i,j} S_i S_j$) with

$$J = 2t^2/U . \qquad (5)$$

As indicated in the phase diagram, below 50 K (NMP)(TCNQ) has the properties of a 1-d antiferromagnet. Although not truly in the strong coupling limit, the ratio U/t as given above is sufficiently large that the Heisenberg approximation should be reasonable at low temperatures. We therefore wish to compare the exchange interaction J obtained from the low temperature measurements with the independently determined values of t and U to check the relation given in Eq. 5.

The experimental evidence for the AF ground state comes from a variety of independent measurements. The spin susceptibility follows a Curie-Weiss law, $\chi = C/(T+\theta)$, with $\theta = 60$ K, rounding off toward a constant value below 40 K. Second, the low temperature specific heat shows a term linear in T as expected from AF spin wave excitations in 1-d. Third, nuclear relaxation studies at low temperatures are consistent with predictions for a spin $\frac{1}{2}$ AF in one dimension. Finally, the ESR linewidth shows a broad maximum at 60 K, suggesting the onset of the AF state. Each of these can be used to obtain a value for the exchange interaction J.

The coefficient of the specific heat term was evaluated by using the numerical results of Bonner and Fisher,[20] $\alpha_m = (0.7/J)Nk_B^2$. The low temperature nuclear relaxation studies yield a value for J through the relation[17] $T_1^{-1} = A^2/3.28\hbar J$. Relaxation studies in the paramagnetic-insulator give an independent measure of J through the well-known relation for exchange coupled disordered moments, $T_1^{-1} = A^2/\hbar J$.[21] A quantitative theory of the ESR linewidth does not exist. However, one intuitively expects the maximum to arise from a slowing down of the spin fluctuations in the vicinity of the magnetic transition. The final estimate comes from the Curie-Weiss θ obtained from the susceptibility. Molecular field theory yields (for

Table I. Values of the exchange integral from independent experimental measurements.

Measurement	Exchange Integral
$J = 2t^2/U$	66 K
Specific heat coefficient (Bonner-Fisher) $\alpha_m = (0.7/J)Nk_B^2$	75 K
Nuclear relaxation (AF) $T_1^{-1} = A^2/3.28\hbar J$	68 K
Nuclear relaxation (para-insulator) $T_1^{-1} = A^2/\hbar J$	60 K
ESR linewidth maximum $J \sim k_B T_{max}$	60 K
Susceptibility (Curie-Weiss law) $J = 2k_B\theta$; $\theta = 60$ K	120 K

$S=\frac{1}{2}$) $J = 2k_B\theta$.

The results of these various measurements of J are given in Table I, together with the value of $J = 66$ K obtained from Eq. 5. The agreement is excellent with the single exception of the value obtained from the high temperature susceptibility. The latter number is, however, one of the least reliable, for the applicability of a Heisenberg model is expected to be limited to temperatures well below the metal-insulator transition. Independent evidence that a simple molecular field theory is inadequate for describing the Curie-Weiss regime is found in the fact that the Curie constant, C, indicates reduced moments characteristic of the intermediate coupling regime. To summarize, the low temperature measurements of J are in good numerical agreement with Eq. 5. To our knowledge, this is the first time that it has been possible to quantitatively check the well-known result $J = 2t^2/U$ through independent measurements of t, U, and J.

The low temperature T_1 measurements (in the AF regime) are particularly interesting in that they imply that for spin $\frac{1}{2}$ in 1-d, the spin wave excitations are to be treated as Fermions.[18] Using the Bulaevskii[22] Fermion representation, the Heisenberg Hamiltonian takes the form

$$\mathcal{H} = 2Jp \sum_k (\cos ka - 1) q_k^+ q_k \qquad (6)$$

where q_k^+ and q_k are spinless Fermions, and $p \simeq 1 + 2/\pi$. The total Fermion number is not conserved. A straightforward Golden Rule calculation then yields[18]

$$\frac{1}{T_1} = \frac{2\pi}{\hbar} A^2 \rho(0) = \frac{A^2}{2\hbar Jp} \qquad (7)$$

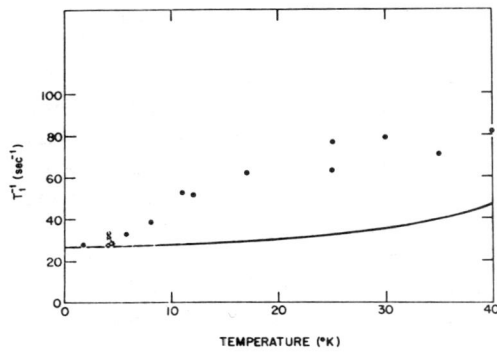

Fig. 3. Proton relaxation rate in low temperature AF phase of (NMP)(TCNQ). The solid curve is calculated using the Fermion representation.

As indicated above, this expression is in good agreement with the low temperature T_1 data using $J=2t^2/U$ as shown in Fig. 3. On the other hand, making the conventional assumption that spin waves can be treated as Bosons at low temperature, a similar calculation leads to a result off in magnitude by a factor of order 10^4 and yields an incorrect temperature dependence. The T_1 measurements thus provide very detailed information on the nature of the excitations in the $S=\frac{1}{2}$ antiferromagnet in 1-d. The somewhat surprising result that the excitations are Fermion-like evidently arises from the simultaneous restrictions of both 1-d and spin $\frac{1}{2}$. Under these circumstances, the kinematic effects are not negligible (in contrast to the well-known situation in 3-d).[23]

The paramagnetic insulating phase is characterized by uncorrelated exchange coupled local moments as indicated by the Curie-Weiss susceptibility and the proton T_1 values. Additional insight into the nature of the metal-insulator transition can be obtained from a comparison of the susceptibility and conductivity data.[16] The magnetic susceptibility data shows that local moments persist in the insulating state essentially without change in magnitude right up to 200 K,[2] whereas the apparent Hubbard gap as obtained from the temperature dependence of the conductivity begins to decrease at much lower temperatures.[16] This comparison emphasizes the important role of fluctuations in the metal-insulator transition. At intermediate temperatures, the Hubbard model can be viewed as a disordered system in which the random potentials experienced by a given electron arise from the electron-electron interaction as a result of the finite number of single particle excitations and the magnetic spin disorder. On general grounds one expects this fluctuation disorder to be reflected in a smearing of the density of states with band tailing into the Hubbard gap. This kind of self-consistent change in the density of states has been identified theoretically in the functional integral approach to the Hubbard model of Kimball and Schrieffer[24] as well as Pleischke's[25] improved theory of the Falicov-Kimball two-band model.[25]

The metallic state at higher temperatures has been characterized through measurements of the susceptibility, nuclear spin relaxation, and transport properties. The room temperature conductivity is high [$\geq 380\,(\Omega\,\text{cm})^{-1}$] and decreases with increasing temperature as would be expected for a metal. The thermoelectric power is negative with a temperature derivative consistent with a narrow band metal with $t \simeq 0.05\,\text{eV}$. More detailed information has been obtained from quantitative analysis of the magnetic susceptibility and T_1 data. The fact that little enhancement is observed in the Pauli susceptibility[2] (when compared with that expected for a tight-binding 1-d band structure using the value of t obtained at low temperature) indicates that the metallic state is highly correlated to minimize double occupancy. An RPA theory of susceptibility and T_1 for an interacting electron system with tight-binding band structure in 1-d was derived and used to analyze the data.[17] Using χ_p and T_1 as input data, the following values for t and U were obtained: $t = 0.05\,\text{eV}$ and $U = 0.18\,\text{eV}$. These values are quite close to those inferred from the low temperature data; however, a factor of two increase in t on going into the metallic state is suggested. Although the results of the RPA analysis are expected to be only semi-quantitative, the larger value of t is also required to explain the temperature independence of the Pauli susceptibility. Thus, the apparent change in t suggests that the transition from insulator to metal is accompanied by a change in effective bandwidth as a result of the change in the many-electron wave function. Even though only a factor of two, such a change is not expected within the simple Hubbard model.

Although the RPA theory is instructive and should give semi-quantitative results well away from the metal-insulator transition, it clearly breaks down near T_{MI}. It is instructive to look at the problem from the point of view of the correlation time, τ_c, associated with the lifetime of a given spin configuration on a site. One expects this to vary smoothly from $\tau_c \sim a/v_F = \hbar/2t$ in the metallic state to $\tau_c \sim \hbar/(2t^2/U)$ in the paramagnetic insulating state.[2] The RPA theory gives the correct result at high temperature as expected but diverges logarithmically as $T \to T_{MI}$. The source of the difficulty is intuitively clear in that RPA always overestimates the tendency toward collective behavior. Fluctuations must be included especially in a 1-d system. Hone and Pincus[26] have recently approached the problem from a different point of view. Their theory essentially involves a high temperature expansion of the Hubbard model in the limit $U/t \gg 1$, taking U into account to all orders. They are able to write an expression for T_1 of the form

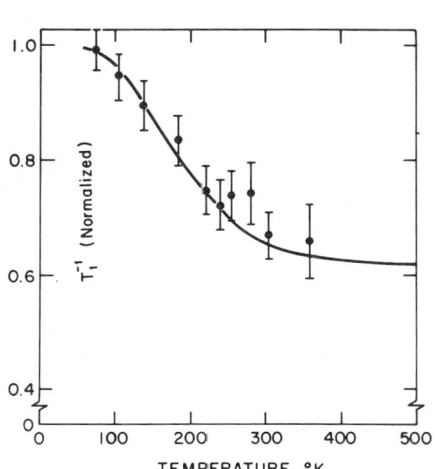

Fig. 4. Normalized temperature dependence of the proton relaxation rate in (NMP)(TCNQ) through the M-I transition. The solid curve is calculated from the theory of Hone and Pincus. The low temperature data are shown in Fig. 3.

$$T_1^{-1} = \left(\frac{A}{\hbar}\right)^2 \tau_c \qquad (8)$$

where τ_c varies smoothly between the two limits given above. A detailed comparison of their theory with the experimental results is shown by the solid curve in Fig. 4. The normalized functional dependence on temperature is excellent using $U/t \simeq 4$ and $U = 0.17$ eV; i.e., values consistent with those obtained from the RPA analysis. The absolute value is, however, off by a factor of about four. The discrepancy in magnitude is not understood in detail. It may have to do with the assumption of a Gaussian time dependence for the correlation function or with the high temperature expansion itself, which treats the electrons as nondegenerate and takes account of only leading terms in t/U. For $U/t \sim 4$, the system is sufficiently close to the metal-insulator boundary that one can perhaps expect only the qualitative features to be given correctly.

To summarize, the experimental results appear to be both qualitatively and in most respects quantitatively in agreement with the predictions of the Hubbard model throughout the entire temperature range. The major discrepancy is the apparent factor of two change in the transfer integral on going from low to high temperature. However, even this may be the result of using approximate solutions in the data analysis. In conclusion, the Mott-Hubbard transition has been established for (NMP)(TCNQ), but this is by no means the general rule for this class of solids, as demonstrated by the characteristic lack of localized magnetic moments at low temperatures in other related systems.

THE ROLE OF DISORDER

The question of the role of cation disorder on the electronic properties of the highly conducting TCNQ salts has been a subject of considerable discussion. Bloch, Weisman and Varma[27] have asserted that the random potential from the disordered asymmetric cations in the highly conducting TCNQ salts causes a localization of electronic functions and completely dominates the electronic properties. In their view, the high conductivity arises not from a metal, but by classical diffusion of electrons through the localized states. From this point of view the magnetism in (NMP)(TCNQ) at low temperatures[18] (Curie constant implying $\sim 1 \mu_B$ per molecule) would require a localization length of order one lattice constant.

The experimental situation was reviewed in the context of the disorder model, with the conclusion that disorder does not play a dominant role.[14] Conductivity, nuclear relaxation, and magnetic susceptibility were each shown to be qualitatively and quantitatively in disagreement with the disorder theory. Moreover, the basic validity of the variable range hopping theory in 1-d has been seriously questioned.[28,29]

Several new experimental facts have emerged which have relevance to this question. First of all, it has been possible to establish a cause-effect relationship between sample purity and curvature in the plot of $\log \sigma$ vs. T^{-1}.[16] High purity samples of (NMP)(TCNQ) show straight line behavior (i.e., exponential) whereas less pure samples show curvature. One must therefore conclude that the curvature (the $T^{-\frac{1}{2}}$ behavior?) was the result of impurities and not in any way intrinsic. Secondly, the detailed studies of the low temperature AF regime described briefly above indicate a spin $\frac{1}{2}$ AF in one dimension, with no evidence of disorder and with parameters numerically consistent with those obtained at high temperatures using other techniques.[18] Finally, and perhaps most important, recent work carried out in several laboratories[30-32] has demonstrated metallic behavior for systems with a <u>symmetric</u> cation, TTF (tetrathiofulvalene), the structure of which is shown below:

$$\begin{array}{c} C-S \\ \parallel \quad \diagdown \\ C-S \diagup \end{array} C=C \begin{array}{c} \diagup S-C \\ \quad \parallel \\ \diagdown S-C \end{array}$$

The room temperature conductivity of the 1:1 TCNQ salt is in the range of several hundred, i.e., comparable with (NMP)(TCNQ). This removes any remaining uncertainty on the existence of a metallic state in these compounds and unambiguously demonstrates that the high conductivity does not arise from diffusion through states localized by asymmetric cation disorder.

THE FUTURE

Before considering future prospects of this novel area of solid state research, a summary of several initial achievements appears appropriate: 1) the synthesis of well-characterized organic metals; 2) the experimental realization of a system well-approximated by the 1-d Hubbard model; 3) the discovery of a metal-insulator transition of the Mott-Hubbard type driven by electron-electron Coulomb correlation without an accompanying lattice distortion; 4) the demonstration of the role of intra-molecular Heitler-London correlation in reducing on-site Coulomb repulsion; 5) the experimental confirmation of the important role of π-polarizability in achieving highly conducting compounds; 6) the theoretical treatment of small excitonic polarons and their role in the M-I transition; 7) the discovery of a 1-d spin $\frac{1}{2}$ uniform antiferromagnet; 8) the experimental confirmation of the well-known result, $J = 2t^2/U$ arising from virtual charge transfer by independent measurements of t, U, and J; 9) the experimental demonstration that spin wave excitations in 1-d spin $\frac{1}{2}$ systems are to be treated as Fermions and not Bosons; 10) the clarification of the subtleties of 1-d systems, particularly in demonstrating that the lack of long range order and the general role of disorder have been overemphasized, and that well-correlated excitations as well as reasonably well-defined phase boundaries do exist.

Continued research in this area promises to be exciting. Although the Mott transition has been established for the (NMP)(TCNQ) salt, this is by no means the general rule. The magnetic properties of many other systems show no evidence of localized moments at low temperatures. Whether such systems at low temperatures are interrupted strand metals (as seems possible for the 2:1 salts or $\frac{1}{4}$ filled band cases), or whether they undergo lattice distortions to dimerized non-magnetic insulators [as seems likely for (TTF)(TCNQ)] is a subject which will require detailed experimental study. It is, however, quite clear that in such narrow band systems the stability of the metallic state is marginal, and some very clever molecular design will be required to maintain the metallic state to very low temperatures.

REFERENCES

1. For a review of the early work on the highly conducting charge transfer salts see O. H. LeBlanc, Jr. in <u>Physics and Chemistry of the Organic Solid State</u> edited by M. M. Labes, D. Fox, and A. Weissburger (Interscience, New York 1967) p. 123.

2. A. J. Epstein, S. Etemad, A. F. Garito, and A. J. Heeger, Physical Review B <u>5</u>, 952 (1972).

3. Extension of earlier solution studies an MO calculations on $TCNQ^-$ and $TCNQ^=$ recently reported by H. T. Jonkman and J. Kommandeur [Chem. Phys. Letters <u>15</u>, 496 (1972)] provide further evidence for the importance of intramolecular Heitler-London correlation occurring in these anions.

4. I. F. Shchegolev, Physica Status Solidi <u>12</u>, (in press).

5. S. Etemad, A. F. Garito, and A. J. Heeger, Physics Letters <u>40A</u>, 45 (1972).

6. C. J. Fritchie, Jr., Acta. Cryst. <u>20</u>, 892 (1966); A. Hoekstra, T. Spoelder, and A. Vos, Acta Cryst. B <u>28</u>, 14 (1972). See also reference 1 for other information on structures.

7. A. W. Hanson, Acta. Cryst. <u>19</u>, 610 (1965).

8. R. M. Metzzer, J. Chem. Phys. <u>57</u>, 1870 (1972); <u>57</u>, 1876 (1972); <u>57</u>, 2218 (1972).

9. Z. G. Soos and A. J. Silverstein, Molec. Physics <u>23</u>, 775 (1972).

10. For a brief review see I. G. Austin and N. F. Mott, Science 168, 71 (1970).

11. J. Hubbard, Proc. Roy. Soc. (London) A276, 238 (1963); A277, 237 (1963); A281, 401 (1964).

12. P. M. Chaiken, A. F. Garito, and A. J. Heeger, Phys. Rev. B 5, 4966 (1972); a second paper has been submitted to J. Chem. Phys.

13. L. B. Coleman, S. K. Khanna, A. F. Garito, and A. J. Heeger, Physics Letters, (in press).

14. E. Ehrenfreund, S. Etemad, L. B. Coleman, E. F. Rybaczewski, A. F. Garito, and A. J. Heeger, Phys. Rev. Letters 29, 269 (1972).

15. This point first came up in a conversation with A. N. Bloch.

16. L. B. Coleman, J. A. Cohen, A. F. Garito, and A. J. Heeger, Phys. Rev., (in press).

17. E. Ehrenfreund, E. F. Rybaczewski, A. F. Garito, and A. J. Heeger, Phys. Rev. Letters 28, 873 (1972).

18. E. Ehrenfreund, E. F. Rybaczewski, A. F. Garito, A. J. Heeger and P. Pincus, Phys. Rev. B, [in press (Jan., 1973)].

19. F. Holtzberg, private communication. The author wishes to thank Dr. Holtzberg for performing the low temperature x-ray measurement.

20. J. C. Bonner and M. E. Fisher, Phys. Rev. 135, A640 (1964).

21. V. Jaccarino, in Magnetism, edited by G. Rado and H. Suhl (Academic, New York, 1965).

22. L. N. Bulaevskii, JETP 16, 685 (1963); 17, 684 (1963).

23. F. J. Dyson, Phys. Rev. 102, 1217 (1956).

24. J. C. Kimball and J. R. Schrieffer, Phys. Rev., (in press).

25. M. Plischke, Phys. Rev. Letters 28, 361 (1972); R. Ramirez, L. M. Falicov, and J. C. Kimball, Phys. Rev. B 2, 3383 (1970).

26. D. Hone and P. Pincus, Solid State Comm., (in press).

27. A. N. Bloch, R. B. Weisman, and C. M. Varma, Phys. Rev. Letters, $\underline{28}$, 753 (1972).

28. W. Brenig, G. H. Dohler, and H. Heyzenau, Phys. Letters, $\underline{39A}$, 175 (1972).

29. J. Kurkijarvi, preprint.

30. F. Wudl, D. Wobschall, and E. J. Hufnagel, J. Amer. Chem. Soc., $\underline{94}$, 672 (1972); F. Wudl, G. M. Smith, E. J. Hufnagel, Chem. Comm. 1453 (1970).

31. J. H. Perlstein, to be published. We thank Dr. Perlstein for sending a preprint of his work.

32. A. J. Sandman, L. B. Coleman, A. F. Garito, and A. J. Heeger, to be published.

HIGH TEMPERATURE THERMODYNAMICS OF THE STRONGLY CORRELATED HUBBARD MODEL AT ARBITRARY ELECTRON DENSITY

G. Beni and P. Pincus
University of California, Los Angeles, Ca. 90024

D. W. Hone
University of California, Santa Barbara, Ca. 93105

ABSTRACT

We consider the Hubbard model for electron correlations in solids in the narrow band regime where the intra-atomic Coulomb repulsion is large compared to the band width. A high temperature perturbation expansion in the band width is performed to lowest order for the grand partition function. This procedure is carried out for both one dimensional systems, of interest for the TCNQ charge transfer salts, and for a three dimensional cubic lattice. Both the specific heat and magnetic susceptibility are computed as a function of temperature and electron density.

ELECTRON TRANSPORT AND MAGNETIC PROPERTIES OF NEW HIGHLY CONDUCTING TCNQ COMPLEXES

Jerome H. Perlstein, John P. Ferraris,
Vernon V. Walatka, Jr. and Dwaine O. Cowan
Department of Chemistry, The Johns Hopkins University
Baltimore, Maryland 21218

George A. Candela
National Bureau of Standards, Washington, D. C. 20234

ABSTRACT

Single crystals of the 1:1 complexes tetrathiafulvalinium tetracyanoquinodimethane (TTF-TCNQ) and tetrathianaphthacinium tetracyanoquinodimethane (TTN-TCNQ) have been synthesized and the electron transport properties and magnetic susceptibility have been measured from 2°K to room temperature. For TTN-TCNQ, σ at room temperature is $1\Omega^{-1}$ cm^{-1}. For TTF-TCNQ, σ at room temperature along the long axis ($\sigma_{//}$) is in the range 192-652Ω^{-1} cm^{-1} depending on sample whereas perpendicular to the long axis $\sigma_{\perp}$ is $1\Omega^{-1}$ cm^{-1}. The conductivity remains metallic down to 66°K in both directions whereupon a continuous metal to insulator transition occurs. The activation energy in the insulating state is 0.0062 ev. The transition is associated with a small hysteresis between the heating and cooling curves suggesting a possible structural change. In the metallic region, $\rho_{//}$ follows a T^2 dependence whereas $\rho_{\perp}$ follows a T^{+1} behavior. The magnetic susceptibility is diamagnetic below 20°K becoming increasingly more paramagnetic with increasing T even in the metallic region. It is suggested that spin disorder scattering may account for the anomalous temperature dependence of $\rho_{//}$.

INTRODUCTION

The semiconductor to "metal" transitions which occur in numerous organic materials involving the tetracyanoquinodimethane molecule (TCNQ) have been the subject of a number of experimental[1-4] studies. These transitions have ranged in temperature from 170°K for acridinium+(TCNQ)$\bar{}_2$ to 240°K for quinolinium(TCNQ)$\bar{}_2$.[1-3] The interpretation of the transition has so far taken two directions. One suggestion is that the transition is of the Mott-Hubbard type[4] in which the electron correlation energy gap decreases continually to zero and with the size of the gap depending strongly on the polarizability of the cation.[5] The other suggestion is that the random orientation of the asymmetric cations in these structures produces a random potential at the TCNQ sites which localizes all the electronic states in the one-dimensional band.[1,6] In this model the semiconductor to "metal" transition occurs in a region where the mobility activation energy W is of order kT so that preexponential factors dominate for W<kT.

These two models for highly conducting organic solids suggested to us that it should be possible to produce a purely metallic ground state in an organic solid by (a) reducing the electron correlation energy utilizing more polarizable donor cations, and (b) using donors which are more symmetric to eliminate the possibility of disorder. Several electron donor molecules containing sulfur suggested themselves as good candidates:

tetrathiafulvalene (TTF)   tetrathianaphthacene (TTN)

both belonging to the high symmetry point group D_{2h}. Moreover, cations of these donors contain a single unpaired spin so that interactions between cations themselves in the solid state can also produce highly conducting materials.[7,8]

Below we present some electron transport and magnetic properties of the complexes formed between these donors and TCNQ.

EXPERIMENTAL

1:1 Complexes of the above donors with TCNQ were prepared as previously described[13] and single crystal electron transport properties measured as described earlier.[2] Magnetic susceptibility of powdered TTF-TCNQ was measured using a Faraday balance.

TTF-TCNQ crystals grow as thin rectangular plates. At room temperature the conductivity parallel to the long axis[13] $\sigma_{//}$, is in the range 192-652Ω^{-1} cm^{-1}. Perpendicular to the long axis $\sigma_{\perp}$ = 1Ω^{-1} cm^{-1}. For TTN-TCNQ $\sigma_{//}$ = 1Ω^{-1} cm^{-1}.

Figure 1 shows the temperature dependence of $\sigma_{//}$ for TTF-TCNQ from 4°K to 300°K. The complex remains metallic down to 66°K with a slight hysteresis between the heating and cooling curves (see insert). Figures 2 and 3 show the temperature dependence of the resistivity $\rho_{//}$ and $\rho_{\perp}$ in the metallic region. $\rho_{//}$ follows a T^2 dependence although with different slopes for different samples whereas $\rho_{\perp}$ follows a T^{+1} behavior.

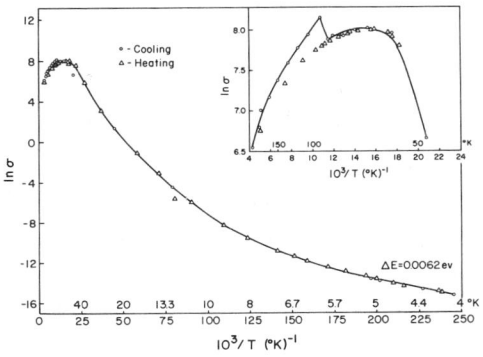

Fig. 1. Natural logarithm of $\sigma(\Omega^{-1}$ cm$^{-1})$ vs. $10^3/T$ for TTF-TCNQ 4°K-300°K
Insert: High temperature region expanded.

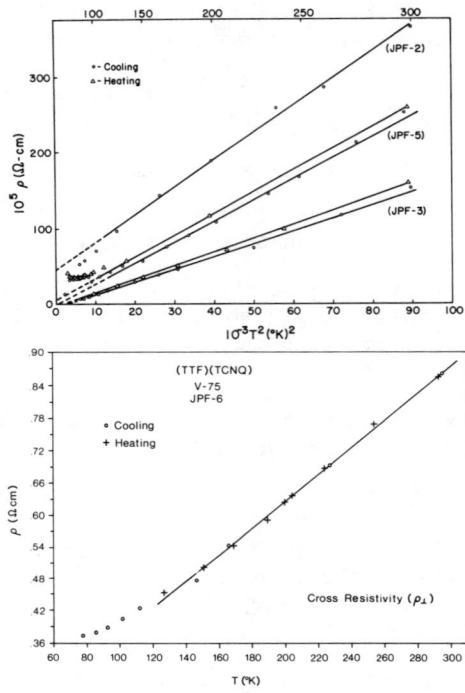

Fig. 2. Resistivity along long axis ($\rho_\parallel$) vs. T^2 for three crystals of TTF-TCNQ.

Fig. 3. Resistivity perpendicular to long axis ($\rho_\perp$) vs. T for TTF-TCNQ.

Figure 4 shows $\chi(T)$ for TTF-TCNQ from 2°K to 359°K, uncorrected. No data was collected in the dashed region.

χ is diamagnetic below 20°K with a minimum as shown suggesting the presence of about 0.07% paramagnetic impurities. Above 20°K, χ becomes paramagnetic increasing monotomically up to 359°K.

DISCUSSION OF RESULTS

TCNQ⁻ in many complexes stacks pancake style with the spacing between ions depending on the cation present. In the high conductivity complexes the TCNQ anions are equally spaced[9] whereas in the low conductivity complexes, dimers, trimers, and even tetramers appear to form. In most cases the cations are diamagnetic with van der Waals spacing between cations. The cations thus contribute to the transport properties only insofar as they effect the band structure and scattering mechanisms along the TCNQ chain.

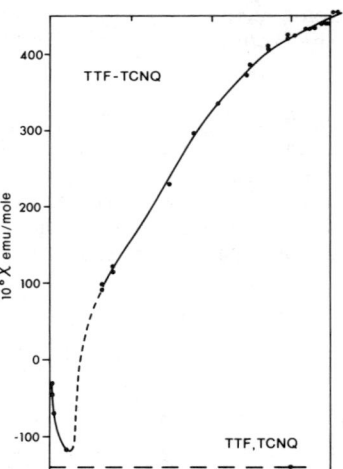

Fig. 4. $\chi(T)$ for TTF-TCNQ uncorrected. Dashed straight line is diamagnetism of TTF° plus TCNQ°.

For TTF and TTN cation radical salts, however (e.g. TTF^+Cl^- and TTN^+Cl^-) the band structure is determined by the interactions between cation radicals. Bandwidths in these salts should be wider than in TCNQ salts since the wavefunctions which contribute to the cation radical band involve sulfur $3p\pi$ orbitals which have a greater radial extent than the carbon $2p\pi$ orbitals of TCNQ.

The diamagnetism and the insulating nature of the ground state of TTF-TCNQ originally suggested to us[13] that a two electron transfer from TTF to TCNQ occurs to give $(TTF^{++})(TCNQ^=)$. The filled TCNQ band would be insulating and the TTF^{++} band empty. Excitation from the filled $TCNQ^=$ band to the empty TTF^{++} band would account for the semiconductor to metal transition in a natural way since this is equivalent to back charge transfer according to the reaction

$$TTF^{++} + TCNQ^= \rightarrow TTF^+ + TCNQ^-$$

with its concomitant decreasing Madelung energy and hence decreasing band gap.

This model however does not explain why the magnetic susceptibility above the insulator-metal transition (66°K) is not of the Pauli type nor why $\rho_{//}(T)$ is anomalous and not typical for phonon scattering. The magnetic properties in the metallic state are most unusual for TCNQ salts. It is tempting to suggest that the single particle excitations that produce the metallic state arise predominantly from the wider TTF^+ band whereas the magnetic excitations arise predominantly from $TCNQ^=$ interactions.

The magnetic properties are more typical of the lower conductivity TCNQ salts although attempts to fit[14] the data to a singlet-triplet model have not been successful. The magnetic ground state could consist of antiferromagnetically coupled $TCNQ^-$ anions as recently suggested[4] with a gradual disordering with increasing T. Alternatively a Peierls distortion of a half-filled $TCNQ^-$ band might account for the susceptibility data in the metallic region. The magnetic susceptibility would then have the form

$$\chi = \frac{4N_o\beta^2 e^{-\Delta/2kT}}{3kT} + \chi_p \quad (1)$$

where the $\frac{4}{3}$ accounts for both the Curie paramagnetism and diamagnetism of the thermally excited electron and holes, Δ is the band gap, χ_p

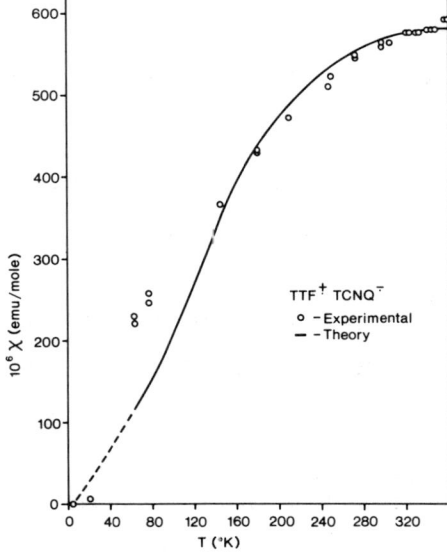

Fig. 5. $\chi(T)$ for TTF^+TCNQ^- corrected for core diamagnetism and 0.07% impurity term. Solid line is eq. (1) with Δ = 0.065 ev and χ_p = 100×10^{-6} emu/mole.

contains the Pauli-Peierls term for the TTF⁺ metallic excitations as well as any high frequency terms; assuming zero bandwidth, N_o is Avagodro's number. Fig. 5 shows a fit of this expression to the data after correction for the diamagnetism of TTF and TCNQ (-138 x 10^{-6} emu/mole) with Δ = 0.065 ev and χ_p = +100 x 10^{-6} emu/mole. The model predicts a maximum in χ at $T = \frac{\Delta p}{2K} = 378°K$. The model is not expected to hold near 66°K and below since additional gaps in the TTF⁺ density of states open up at this temperature.

The resistivity for this model in the metallic region would then contain not only a phonon contribution but a T^2 spin disorder scattering term[15] as well:

$$\rho_{//,\perp} = AT + BT^2 \quad C \qquad (2)$$

The data suggest that the spin disorder scattering term is more effective for the high conductivity direction ($\rho_{//}$) than for the low conducitivity direction ($\rho_{\perp}$).

ACKNOWLEDGEMENTS

Support of this work by the National Science Foundation and the Petroleum Research Fund administered by the American Chemical Society is acknowledged. Jerome H. Perlstein would also like to acknowledge to Dr. Peter Y. Johnson the many provocative discussions had on organo-sulfur chemistry.

REFERENCES

1. J. H. Perlstein, M. J. Minot and V. Walatka, Jr., Mat. Res. Bull. 7, 309 (1972).
2. V. Walatka, Jr. and J. H. Perlstein, Mol. Cryst. and Liq. Cryst. 15, 269 (1971).
3. L. I. Buravov, D. N. Fedutin, and F. I. Shchegolev, Sov. Phys. JETP 32, 612 (1971).
4. A. J. Epstein, S. Etamad, A. F. Garito, and A. J. Heeger, Phys. Rev. B5, 952 (1972).
5. O. H. LeBlanc, Jr., J. Chem. Phys. 42, 4307 (1965).
6. A. N. Bloch, R. B. Weisman, and C. M. Varma. Phys. Rev. Lett. 28, 753 (1972).
7. E. A. Perez, Albuerne, H. Johnson, Jr., and D. J. Trevoy, J. Chem. Phys. 55, 1547 (1971).
8. F. Wudl, D. Wobschall, and E. J. Hufnagel, J. Amer. Chem. Soc. 94, 670 (1972).
9. C. J. Fritchie, Jr., Acta Cryst. 20, 892 (1966).
10. P. Goldstein, K. Seff, and K. N. Trueblood, Acta Crys. B24, 778 (1968).
11. C. J. Fritchie, Jr., and P. Arthur, Jr., Acta Cryst. 21, 139 (1966).
12. T. Sundaresan and S. C. Wallwork, Acta Cryst. B28, 1163 (1972).
13. J. Ferraris, D. O. Cowan, V. Walatka, Jr., and J. H. Perlstein, J. Amer. Chem. Soc. (in press).
14. R. G. Kepler, J. Chem. Phys. 39, 3528 (1963).
15. D. L. Mills and P. Lederer, Phys. Rev. 165, 837 (1967).

DELOCALIZED TRIPLET EXCITATIONS IN AN ANTIFERROMAGNETICALLY COUPLED UNIFORM 1-D SYSTEM

S. Etemad and E. Ehrenfreund
University of Pennsylvania, Philadelphia, Pa. 19104

ABSTRACT

Spin susceptibility and ESR linewidth measurements are reported for (TMPD)(TCNQ), a uniformly spaced one-dimensional spin $\frac{1}{2}$ system. Analysis of the susceptibility in terms of the Bulaevskii Fermion approximation shows the magnetic excitations are delocalized triplet excitons propagating in a band. Due to exciton-exciton scattering, the long time behavior of these excitations is dominated by diffusion with interchain hopping acting as a cut-off.

INTRODUCTION

Many TCNQ radical ion salts form one-dimensional magnetic insulators[1] whose excitations may be described by the generalized Heisenberg Hamiltonian

$$\mathcal{H} = J \sum_{i}^{N/2} (\vec{S}_{2i} \cdot \vec{S}_{2i+1} + \gamma \vec{S}_{2i} \cdot \vec{S}_{2i-1}) \quad 0 \leq \gamma \leq 1 \quad (1)$$

where $\gamma = J'/J$ is the alternation parameter describing the degree of non-uniformity of the exchange coupling between sites along the chain. For $\gamma = 0$, each dimer, $\vec{S}_{2i}$ and $\vec{S}_{2i+1}$, has a singlet ground state and a localized triplet excited state populated with probability

$$\rho(T) = 3 \left[\exp(J/T) + 3\right]^{-1} \simeq 3 \left[\exp(-J/T)\right] . \quad (2)$$

For $\gamma \neq 0$, the interpair coupling leads to formation of a triplet exciton band. In the limit $\gamma = 1$, Eq. 1 is the uniform 1-d Heisenberg model which leads to spin wave excitations with a gapless energy spectrum.

Among the strongest π-electron donors and acceptors are the TMPD and TCNQ molecules respectively (Fig. 1). In the solid state (TMPD)(TCNQ) forms an ionic crystal with weakly coupled chains of alternating but <u>uniformly</u> spaced (TMPD)$^+$ and (TCNQ)$^-$ ions.[2] Furthermore, (TMPD)$^+$ and (TCNQ)$^-$ are magnetic open shell ions with spin $\frac{1}{2}$. Due to the crystal structure and the anisotropy of the $2p_z$ wave functions, the intrachain coupling is suffi-

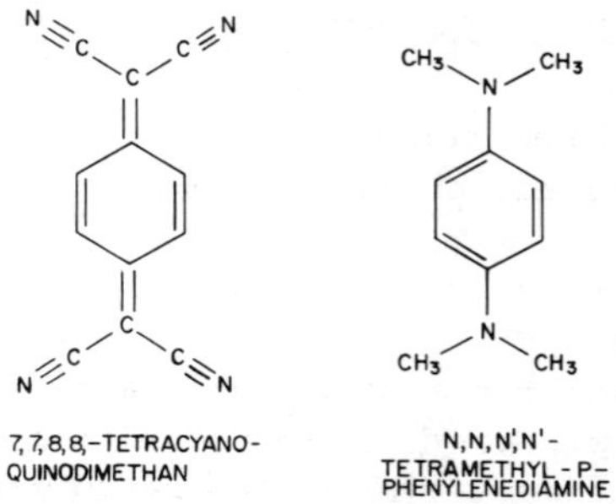

7,7,8,8,-TETRACYANO-
QUINODIMETHAN
N,N,N',N'-
TETRAMETHYL-P-
PHENYLENEDIAMINE

Fig. 1. Molecular structure of TCNQ and TMPD.

ciently greater than the interchain coupling that one can view $(TMPD)^{+}$ $(TCNQ)^{-}$ as a collection of unpaired spins arranged in a uniform chain with each spin interacting with its neighbors according to Eq. 1. However, contrary to expectations implied by this uniform chain structure, the magnetic excitations of (TMPD)(TCNQ) at all temperatures correspond to a band of triplet excitons with a finite energy gap. We therefore report measurements, on high purity (TMPD) (TCNQ), of the spin susceptibility and ESR linewidth.

EXPERIMENTAL RESULTS

The spin susceptibility χ of (TMPD)(TCNQ) as determined directly using an integrating low frequency (~30MHz) technique[3] is plotted as $\log \chi T$ vs. T^{-1} in Fig. 2. The spectrum consists of a single, frequency independent and strongly exchange narrowed Lorentzian line located at g=2 with width of order 0.2 gauss. Neither fine structure nor hyperfine structure is observed.[4] Above 210K, the temperature dependence of χ is consistent with previously reported measurements[4] and indicates the presence of triplet excitations with an effective gap of 810K.

Due to the high purity of the samples, we have been able to extend the measurements to approximately 150K. As shown in Fig. 2, starting at 210K the system undergoes a gradual transition, the existence of which is confirmed by observation of a broad anomaly in heat capacity.[5] Below 180K, the data indicate triplet excitons with an energy gap of about 1300K.[5]

The temperature dependence of the peak to peak

(TMPD) (TCNQ)
Δ=810K
a=.3
Δ=1300K
a=1
χT(10^-28 erg K/G^2)
1000
500
200
100
50
20
10
4
5
6
7
1000/T(K)

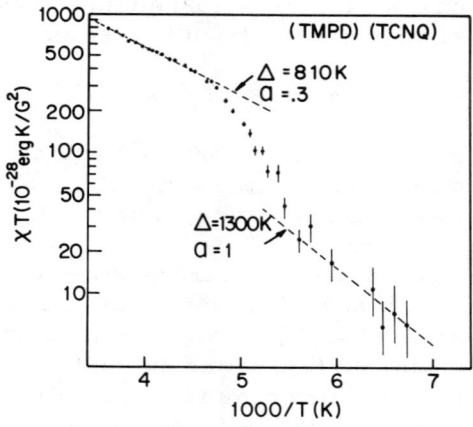

Fig. 2. Temperature dependence of spin susceptibility χ.

linewidth (ΔH) is given in Fig. 3. The most important characteristic of the high temperature regime is the exponential decrease ($\Delta H \propto \exp -\Delta/T$) with activation energy $\Delta = 430\,K$. The data are reproducible from sample to sample and are in agreement with earlier measurements.[4] Since the activation energy is approximately half the value of the singlet-triplet gap as determined from susceptibility, the linewidth (for $T>210\,K$) can be represented as

$$\Delta H_{exp} = 0.8 \sqrt{\rho} \text{ (in gauss)} \qquad (3)$$

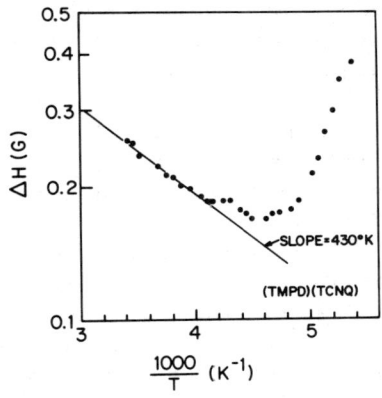

Fig. 3. Temperature dependence of ESR linewidth ΔH.

where ρ is the measured triplet exciton density. Below $180\,K$, the linewidth increases rapidly with decreasing T. However, the magnitude in this regime varied from sample to sample with the steepest slope observed in the purest samples. Thus, contrary to the arguments of Hoffman and Hughes,[4] the low temperature widths are not intrinsic.[6]

DISCUSSION

There have been many attempts to solve Eq. 1 for all values of γ. The two end points $\gamma=0$ and $\gamma=1$ are reasonably well understood, whereas for intermediate values the situation is less clear.[5] In a series of papers, Bulaevskii[7] developed a theory for the alternating Heisenberg chain for arbitrary γ. The significant feature of this approach is that in the limit $\gamma=1$, it reproduces the Bonner and Fisher[8] numerical calculations for the specific heat with reasonable accuracy, and the excitations go smoothly to zero-gap spin waves. The Bulaevskii expression for the susceptibility applied in the range $0.02 \leq T/2J \leq 0.25$ is[7]

$$\chi = [g^2 \mu_B^2 S(S+1)/k_B T] \, a(\gamma) \, \exp[-J\Delta(\gamma)/T] \qquad (4)$$

where the prefactor $a(\gamma)$ and the gap parameter $\Delta(\gamma)$ arise from the finite exciton bandwidth. In the limit $\gamma=0$, $a(\gamma)$ and $\Delta(\gamma)$ equal unity, and one has isolated triplet excitons as described by Eq. 2. For $\gamma \neq 0$, $a(\gamma)$, $\Delta(\gamma)$, and the exciton bandwidth must be evaluated numerically.[5]

A least squares fit of Eq. 4 through the susceptibility data for $T>210\,K$ yields a prefactor $a(\gamma)=0.3$. The implied alternation parameter is $\gamma=0.65$, corresponding to a triplet exciton band of width 1300 K centered at 1400 K above the singlet level. We note once again that the existence of a finite energy gap to a band of triplet excitons is contrary to expectations based on the observed uniform crystal structure. Although the data below 180 K are less extensive, a similar analysis yields $a(\gamma) \approx 1$, corresponding to localized triplet excitons in a narrow band centered at about 1300 K.

The broad exciton bandwidth inferred from the prefactor analysis implies that the triplet excitons in (TMPD)(TCNQ) above 210 K may be regarded as delocalized over the D^+A^- chains. At relatively low temperatures (i.e. $k_B T < J$), the thermal concentration of excitons is analogous to a dilute paramagnetic system with a fractional spin of $\frac{1}{2}\rho(T)$ per site. As a result, the linewidth can be dominated by the unresolved dipolar interaction. Soos[9] has argued on this basis that the Van Vleck rigid lattice second moment, $\Delta\omega_D^2$, should be reduced by the exciton density ρ; i.e., $\Delta\omega_D^2 = \rho \Delta\omega_0^2$. Similarly, since the probability of finding an exciton on a given site is $\rho(T)$, the effective exchange coupling responsible for exciton-exciton exchange scattering is $J_{ex} \approx \frac{1}{4} J\rho$.[9]

There has been recent interest in the question of exchange narrowed linewidths in 1-d magnetic systems, for the introduction of spin diffusion into the relaxation formalism leads to interesting and unique behavior.[10] Since the diffusional nature of long time behavior of the spin correlation function can be argued on general grounds,[11] the existence of frequency independent Lorentzian lines would appear to be a puzzle. To explain such observations, Soos[12] has suggested that in addition to spin diffusion, one must consider the finite lifetime of an exciton on a particular 1-d chain.

To explain the linewidth behavior of (TMPD)(TCNQ), we extend Soos' argument to the case of a delocalized exciton band. The propagating character of excitons from the point of view of band theory would at first sight appear to be in contradiction to a diffusion approach. This is, however, not the case, for the long time spin correlation function will be properly given by diffusion theory as a result of the multiple exciton-exciton exchange scattering processes among the $N\rho(T)$ excitons present at a given temperature. Again, such a system can be viewed simply as a dilute paramagnetic system with spin $\frac{1}{2}\rho$ per site. The analogy with the concentrated paramagnetic system immediately suggests a correlation function, $\psi(t)$, varying as $t^{-\frac{1}{2}}$ at long times.[10] The above argument neglects interchain hopping or other lifetime effects which, when included, lead to $\psi(t) \sim t^{-\frac{1}{2}} \exp(-\omega_r t)$[12] where ω_r^{-1} is the lifetime. For $\psi(t)$ of this form, the resulting linewidth can be expressed as[12]

$$\Delta H = \Delta\omega_D^2 (2\omega_r J_{ex}/\hbar)^{-\frac{1}{2}} = \Delta\omega_0^2 (\tfrac{1}{2}\omega_r J)^{-\frac{1}{2}} \rho^{\frac{1}{2}} \qquad (5)$$

The functional dependence on exciton density agrees with the experimental result (Eq. 3). To estimate the magnitude, we initially consider only the interchain dipolar contribution to ω_r (neglecting interchain exchange contributions). Assuming a dumbbell-shaped spin density distribution for the (TMPD)$^+$ and (TCNQ)$^-$ ions, we estimate $\omega_r \simeq 72$ gauss and $\Delta\omega_0^2 \simeq 4 \times 10^4$ (gauss)^2. Taking $J = 1400$ K as inferred from the susceptibility data gives (Eq. 5)

$$\Delta H_{theory} = 1.2 \, \rho^{\frac{1}{2}} \quad \text{(in gauss)} . \qquad (6)$$

The agreement with experiment is quite reasonable since a further reduction of Eq. 6 is expected from interchain exchange coupling. To bring Eq. 6 and Eq. 3 into agreement would require only that $J_{inter}/J_{intra} \simeq 1 \times 10^{-5}$.

These experimental studies of (TMPD)(TCNQ) thus indicate that the magnetic excitations are triplet excitons with a large energy gap even though the chain structure is uniform. This paradox is not understood and remains as one of the most interesting aspects of the problem.

We are indebted to Prof. A. J. Heeger for discussions leading to the understanding of triplet excitons, and to Profs. Heeger and A. F. Garito for the assistance which made this work possible.

REFERENCES

1. Nordio P. L., Soos Z. G., and McConnell H. M., Ann. Rev. Phys. Chem. 17, 237 (1966).
2. Hanson A. W., Acta Cryst. 19, 610 (1965).
3. Epstein A. J., Etemad S., Garito A. F., and Heeger A. J., Phys. Rev. B5, 952 (1972).
4. Hoffman B. M., and Hughes R. C., J. Chem. Phys. 52, 4011 (1970).
5. Etemad S., Ph. D. Thesis, University of Pennsylvania, 1972.
6. Hoffman and Hughes attribute the low temperature linewidth to hyperfine effects and argue that $\Delta H \propto 1/\rho$. This is incorrect. The hyperfine interaction is motionally narrowed to a negligible and temperature independent value $A^2/\gamma J$ whenever $A/\gamma J \ll 1$.
7. Bulaevskii L. N., Sov. Phys. Solid State 11, 921 (1969).
8. Bonner J. C., Fisher M. E., Phys. Rev. 135, A640 (1964).
9. Soos Z. G., J. Chem. Phys. 46, 4284 (1967).
10. Hone D., AIP Conf. Proc. Mag.-Mag. Mat. (Chicago, 1971), p. 413.
11. Gulley J. E., Hone D., Scalapino D. J., and Silbernagel B. G., Phys. Rev. B1, 1020 (1970).
12. Soos Z. G., J. Chem. Phys. 44, 1729 (1966).

STATIC PROPERTIES OF THE HALF-FILLED BAND HUBBARD MODEL*

D. Cabib and T. A. Kaplan
Physics Department, Michigan State University
East Lansing, Michigan 48823

ABSTRACT

We summarize some recent results, the details of which will be published elsewhere, and add some further discussion. Our results came mainly from a calculation of static thermal properties of the half-filled band Hubbard model for a ring of N=4 atoms. They resolved serious discrepancies between some calculations which have appeared. For weak interactions, a new kind of smooth magnetic transition was found at low temperature. For strong interactions, properties are approximately independent of N when the grand canonical ensemble is used, enabling contact to be made with recent experimental work on (NMP)(TCNQ); the comparison suggests strongly that the Hubbard model is seriously deficient as a means of description of these experiments. We show here that the new magnetic transition and the related peculiar low-T, small-interaction behavior of the specific heat do not occur for chains, and that an explanation given in the literature for this result is incorrect, due to the fact that it is based on wrong conductivity calculations.

Exact calculations on the 4 atom ring Hubbard model[1] had resolved serious confusion existing in the literature on the subject[2,3]. More specifically the qualitative behavior of the specific heat agrees with that found by Heinig and Monecke[3], (even though the quantitative results are appreciably different) and disagrees with Shiba and Pincus[2] (for weak to intermediate interactions). The results obtained with the grand canonical ensemble[1] do not differ qualitatively from those obtained with the canonical ensemble; but the use of the GCE is important for the extrapolation to a large number of atoms, N, because in this case when the ratio of hopping integral to Coulomb repulsion, $b/U \to 0$, any intensive parameter is independent of N (we take U and b > 0).

The specific heat has three peaks for $0 < U/b \lesssim 6$ and two peaks for $U/b \gtrsim 6$ both in the CE and GCE. The results obtained for the spin-spin correlation function, defined as $L_n = \langle s_{iz} s_{i+n,z} \rangle$ where s_{iz} is the z-component of spin at site i, are very illuminating on the physical significance of these specific heat peaks: the various anomalies in L_n were shown to occur always at temperatures very close to the ones at which one or another specific heat peak occurs. Since the large U/b region had been essentially understood previously[4,5] and since we agree qualitatively with the existing results[2,3,4,5] in this region, we will focus for the moment on the small U/b region. Here the specific heat has three peaks at temperatures $T_I < T_{II} < T_{III}$. (T_I and $T_{II} \to 0$ as $U/b \to 0$.) We found that $-L_1$ decreases rapidly near T_I, approaching a constant different from 0, and again near T_{III}, approaching 0. L_2 decreases from a constant value to 0 near T_{II}. From this picture we cannot characterize T_I as being similar to a Néel temperature (as HM do) because the first and second-neighbor correlations L_1 and L_2 do not have anomalies at the same temperature. This type of transition had not been found previously. HM argued, by extrapolation, that the low-T specific heat peaks will occur for large N, but we showed^1 that this is wrong. Let us summarize here the reason. First of all this phenomenon occurs only when N/2 is even; in fact in this case the one-particle energy levels for U=0 are given by

$$E_k = -2b \cos k \qquad k = 0, \pm\frac{2\pi}{N}, \pm\frac{4\pi}{N}, \ldots, \pi$$

and the N-electron ground state (half-filled band) fills completely the states with $k=0, \pm 2\pi/N, \ldots, (2\pi/N)\cdot(N/4 - 1)$ leaving 2 more electrons the possibility of occupying the four states $k=\pm\pi/2$ ($s_z=\pm 1/2$) which all correspond to the same one-electron energy. This gives rise to a six-fold degeneracy of the ground state which is removed when a small U is turned on, explaining$_6$ the existence of the low-T peaks in the spcific heat per atom. This degeneracy will remain six-fold for all even N/2 and its effect will therefore become negligible when $N\to\infty$, the height in the above peaks decreasing as 1/N. One is naturally led to ask what the situation is for a chain. In this case the hopping integral b is taken to be the same for every nearest neighbor pair of sites, the end-sites having only one-sided hopping. The one-electron energies are

$$E_k = -2b \cos k, \qquad k = \frac{\pi\ell}{N+1}, \quad \ell = 1, 2, \ldots, N$$

Therefore, when the number of electrons is even, the ground state is non-degenerate. Hence, when U increases from zero, no appreciable low-T peak in C will occur in contrast to the behavior discussed above for the ring. By appreciable we mean $\int (C/T)dT$ integrated over the peak is $\simeq k_B/N$ and by low-T we mean the peak location, $T_0 \ll b/N$ and $T_0 \to 0$ as $U \to 0$ (we have small N in mind). In other words, we cannot expect to find in this case the new type of smooth magnetic transition which was found for the ring.

HM^7 also stated that the low-lying specific heat peaks should not occur for the chain, but gave an incorrect argument. Namely they said that the vanishing of the conductivity σ for the chain implied the vanishing of the peaks in C, presumably because they had established a causal relation between anomalies in σ with those same low-T peaks in C. A reason for the incorrectness of this argument is that their calculation of σ is seriously in error, as we now show.

HM's results imply that $\sigma(\omega=0) \neq 0$ at zero temperature for any U/b. A calculation from Kubo's formula for the conductivity yields:

$$\sigma(\omega) = \frac{\pi}{Z} \sum_{nm}' |j_{nm}|^2 \frac{e^{-\beta \tilde{E}_m} - e^{-\beta \tilde{E}_n}}{\tilde{E}_{nm}} \delta(\omega + \tilde{E}_{nm})$$

$$+ \frac{\pi}{Z} \beta \sum_n e^{-\beta \tilde{E}_n} [|j_{nn}|^2 + \sum_{\substack{m \neq n \\ m \text{ degenerate} \\ \text{with } n}} |j_{nm}|^2] \delta(\omega)$$

Z is the partition function. Σ' is extended to states such that $E_n \neq E_m$ (E_n is the energy eigenvalue corresponding to the n-th energy eigenstate in zero electric field), $\tilde{E}_k = E_k - \mu n_k$ (n_k=number of electrons in the state corresponding to E_k), j_{nm} is the matrix element of the current operator,

$$\underline{j} = -ie \, \Sigma b_{ij} (\underline{R}_i - \underline{R}_j) \, c_{i\sigma}^\dagger c_{j\sigma} \, .$$

Clearly for a small system (with discrete energies) only the sum involving the square brackets contributes to $\sigma(0)$. States with an even number of electrons can always be chosen to give a zero contribution to the sum involving the first term in square brackets. (For an even number of particles the time reversal operator θ and H can be simultaneously diagonalized and since j changes sign under time reversal, $j_{nn}=0$ in this basis.) Since in our case the ground state has an even number of particles, this sum must vanish at zero T. Hence the only contribution to $\sigma(0)$ at zero T is the second sum in square brackets. For it to be different from zero at zero temperature it is necessary that the ground state be degenerate. Explicit calculations for four atoms showed that the ground state is non-degenerate for finite U and hence the HM result is incorrect. The error can be traced to the paper by Monecke[9], upon which the conductivity calculations for 4 atoms are based.

Since the 4-atom calculations for U/b>>1 do not give special results whose significance is restricted to small systems, and the GCE for b=0 does not depend on N, we[1] tried to compare zero magnetic field susceptibility calculations with the experimental measurements by Epstein et.al.[10] (see fig. 1). To this end we discuss the extrapolation to large N.

First of all, it is known that for large enough U/b and kT<<U, the Hubbard model approaches the Heisenberg model $-\Sigma_{nm} J_{nm} \underline{S}_n \cdot \underline{S}_m$, with $J_{nm} = 2b_{nm}^2/U$; for the proposed value[10] U/b=8 and small N we find behavior very similar to that of the Heisenberg model[11]; therefore we expect similar behavior for U/b=8 and increasing N. The similarity can be described in terms of the following important features: the existence of a minimum in χ^{-1}, its location, and the behavior in the Curie-Weiss region. As seen in fig. 1 the exact value[12] of χ^{-1} at T=0 (shown as x) lies above

the minimum for N=4 by about 50%; this suggests strongly that the minimum will persist when N→∞, as it does in the Heisenberg model,[13] because the height at the minimum appears to be relatively insensitive to N (it changes by ~15% in the Hubbard model from N=2 to 4[14], and by about 7% in the Heisenberg model[11] from N=4 to ∞). The location $T_o \simeq 2b^2/U$ (N≈60°K for the experimental values of U and b) only varies about 10% from 2 to 4 atoms in the Hubbard model[14] and about 20% from 4 to ∞ in the Heisenberg model[11]. Finally, the theoretical 4-atom result for χ above ~60°K approaches closely to the Heisenberg-model Curie-Weiss behavior (with a moment of one Bohr magneton and a θ~-60°K as seen in Fig.1). The use of the 4-atom χ as an approximation to the many-atom χ in this higher-T

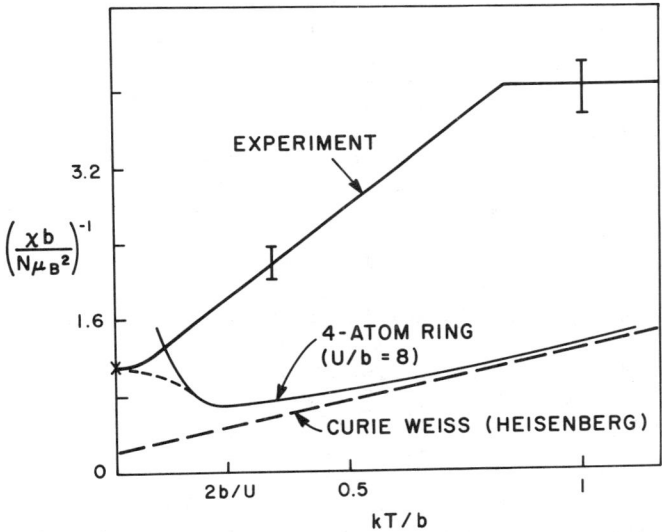

Fig. 1 Inverse susceptibility vs temperature. The exact value at T=0 for the infinite chain is shown by x.

region seems very reliable because in this region the results of Bonner and Fisher are very insensitive to N, in the Hubbard model χ^{-1} changes by only about 12% from 2 to 4 atoms[14], and the correlation L_n is short-ranged for N=4 (see Fig. la, ref. 1). Thus we are led to suggest that the curve χ^{-1} vs T for N→∞ will be closely approximated by the dotted line in Fig. 1 for kT/b<.17 and by the 4-atom curve for larger kT/b.

The experimental curve is seen to differ radically from this theoretical curve. The reduction in the moment from $1\mu_B$ was noted by Epstein et.al.[10], who gave the not implausible argument that it might be expected because b/U is large enough to give appreciable mixing of ionic states into the singly-occupied states; however, the fact that our calculations show no such reduction forces us to conclude that such an argument is highly questionable.

In fact the mixing effect discussed certainly occurs for systems with small N, but as seen in Fig. 1, it is negligible for these purposes.

We conclude that drastic changes in the Hubbard model are needed to explain the so-called [10] "metal-insulator transition" at high T and the low-T antiferromagnetic behavior.

We thank R.A. Bari for useful discussions and Dr. S.D. Mahanti for his encouragement.

References

1. D. Cabib and T.A. Kaplan, to appear in Phys. Rev. B1 (March 1973).
2. H. Shiba and Pincus, Phys. Rev. **B5**, 1966 (1972).
3. K.H. Heinig and J. Monecke, Phys. Stat. Sol. (b) **49**, K139 and K141 (1972).
4. T.A. Kaplan and R.A. Bari, J. Applied Phys. **41**, 875 (1970); and Proc. 10th Internat'l Conf. on Physics of Semiconductors, ed. by S.P. Keller, J.C. Heusel, F. Stern, CONF-700801 (U.S. AEC Div. of Tech. Information, Springfield, Va., 1970) p. 301.
5. R.A. Bari and T.A. Kaplan, Phys. Rev.B Dec.1972, to appear.
6. In the case of 4 atoms it is possible to see, directly or by perturbation theory, that the splitting between the ground state (singlet) and the first excited state (triplet) is of order U^2, whereas the next two excited states (singlets) differ from the ground state (and remain degenerate) in first order in U. This agrees with the small-U behavior of the specific heat peaks at low T displayed in figure 2 of the previous work.[1]
7. H. Heinig and J. Monecke, Phys. Stat. Sol. (b), **50**, K117 (1972).
8. K. Kubo, J. Phys. Soc. Japan **31**, 30 (1971).
9. J. Monecke, Phys. Stat. Sol. (b), **51**, 369 (1972).
10. A.J. Epstein, S. Etemad, A.F. Garito, A.J. Heeger, Phys. Rev. **B5**, 952 (1972).
11. J. Bonner and M. Fisher, Phys. Rev. **135**, A640 (1964).
12. M. Takahashi, Prog. Theor. Phys. (Kyoto) **43**, 1619 (1970).
13. The maximum in χ found[11] for the Heisenberg model is quite pronounced even in the limit of large N: $[\chi(T_o)-\chi(o)]/\chi(o) \simeq 2/5$.
14. Because the number of nearest neighbors changes from 1 to 2 when N changes from 2 to >2, we expect larger changes in behavior in going from say N=2 to 4 than from N=4 to 6 or 6 to 8, etc.

* Supported in part by NSF grants GU 2648 and GH 34565.

MAGNETIC RESONANCE STUDIES IN LiTCNQ

S.K. Khanna, E. Ehrenfreund, E.F. Rybaczewski, and S. Etemad
University of Pennsylvania, Philadelphia, Pa. 19104

ABSTRACT

Measurements of the spin susceptibility, low temperature specific heat, and ESR linewidths of high purity Li(TCNQ) are reported. The qualitative behavior and magnetic properties of Li(TCNQ) are found to be the same as those of the general class of structurally dimered one-dimensional spin $\frac{1}{2}$ alkali salts. The observed magnetic excitations are treated as triplet excitons delocalized in a band.

INTRODUCTION

The simple alkali charge transfer salts of the strong electron acceptor tetracyanoquinodimethan (TCNQ) are magnetic semiconductors that exhibit magnetic susceptibilities which vary exponentially with temperature, typical of triplet exciton behavior.[1] The primary interest in these properties arises from the fact these simple salts appear to have strongly dimerized linear structures. The crystal structure of Rb(TCNQ),[2] for example, consists of one-dimensional dimerized chains of open-shell TCNQ⁻ anions separated by chains of alkali ions. The complete transfer of a single electron onto each TCNQ site has been confirmed recently for K(TCNQ) by NQR measurements of the ^{14}N coupling constant.[3] General attention has been focused on these salts in terms of possibly associating the origin of the dimerization with the well-known Peierls instability and identifying the observed magnetic excitations with those of the alternating Heisenberg Hamiltonian

$$\mathcal{H} = J \sum_i (S_{2i} \cdot S_{2i+1} + \gamma S_{2i} \cdot S_{2i-1}) \qquad 0 \leq \gamma \leq 1 \qquad (1)$$

where the localized triplet excitations associated with the AF exchange interaction J are delocalized into triplet exciton bands by J', and $\gamma = J'/J$ is the alternation parameter.

Recently Pincus and co-workers[4] have studied the general question of lattice instabilities of linear chains in the presence of strong electron correlations. They calculated the thermodynamics of an antiferromagnetic linear chain with elastically coupled spins in the X-Y model and in the Hartree-Fock approximation to the Heisenberg Hamiltonian. Their results show that such systems exhibit an inherent tendency to dimerize. This is equivalent to the Adler-Brooks treatment for electrons in narrow bands.

EXPERIMENTAL RESULTS

Considerable care was taken in preparing pure Li(TCNQ) samples. Li(TCNQ) was prepared as glistening red microcrystals from twice gradient sublimed TCNQ and ultrapure anhydrous LiI (Alfa Inorganics) under inert handling conditions as described earlier.[5] The twice sublimed TCNQ showed less than a few ppm metal atom concentration by atomic emission and absorption, and liquid chromatograms (DuPont) showed no detectable traces of organic contaminent. Analysis: Calculated for $C_{12}H_4N_4Li$: C, 68.26; H, 1.91; N, 26.55; Li, 3.29; Found : C, 68.32; H, 1.91; N, 26.59; and Li, 3.27.

High and low field ESR measurements showed a single exchange narrowed Lorentzian line. The line was frequency independent and located at g = 2, with no observed fine or hyperfine splitting. The temperature dependence of the spin susceptibility of Li(TCNQ) is given in Fig. 1. The data were obtained by integrating the ESR curves by the Schumaker-Slichter technique described earlier.[5] Below 225 K, a linear dependence is observed in the plot of log χT vs. T^{-1}, showing the data are of the form

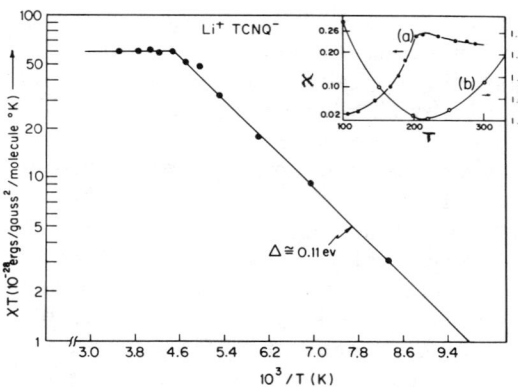

Fig. 1. Temperature dependence of spin susceptibility χ for Li(TCNQ). Insert: χ is in units of 10^{-28} erg/G^2/molecule. (a) Present work, (b) Ref. 1 [note scale change].

$$\chi = \frac{C}{T} e^{-\Delta/T} = \frac{C}{T} \rho(\Delta, T). \quad (2)$$

This form corresponds to triplet excitations of density $\rho(\Delta, T)$ separated from a singlet ground state by the energy gap Δ. From the slope of the exponential behavior below 225 K, Li(TCNQ) is found to have a singlet-triplet gap of 1300 K and a relatively small exciton density of 10^{-3} molecule^{-1} at 225 K. At lower temperatures, the spin resonance signal continues to decrease exponentially with temperature to the point where the amplitude drops below the detection limit of the measuring equipment; no signals are observed at any temperatures below 140 K, implying high sample purity.

The observed triplet exciton behavior is consistent with measurements of the low temperature specific heat on Li(TCNQ). One-dimensional antiferromagnetic spinwaves in a uniform chain system are predicted to contribute a linear term to the low temperature[5] specific heat, whereas no contribution is expected from magnetic excitations having an associated gap $\Delta \gg k_B T$. The low temperature data shown in Fig. 2 for Li(TCNQ) were computer fitted to the relation $C_p = \alpha T + \beta T^3$, where the linear term is of electronic or magnetic origin and the T^3 term is the Debye lattice contribution. By minimizing the standard deviation, the numerical analysis of the data yields a value of α of 0 ± 10 (kerg/mole-K) as expected for the triplet excitations observed in Li(TCNQ). The corresponding value of β is 16.4 (kerg/mole-K^3) and θ_D is 107 K. The θ_D value of several 1:1 alkali (TCNQ) salts was shown to follow a well-defined linear relationship in terms of the simple Debye theory of ionic binding.[5] The crystal binding in Li(TCNQ) is effectively the same as in the other alkali (TCNQ) salts.

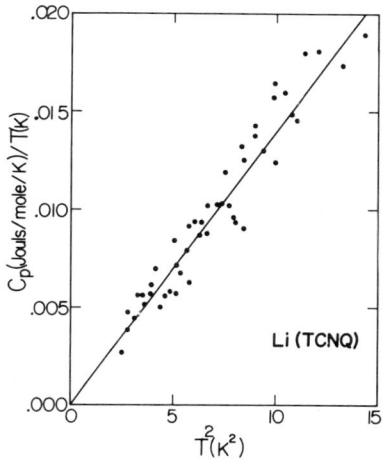

Fig. 2. Low temperature specific heat for Li (TCNQ).

At 225 K there is an abrupt change in $\chi(T)$ (Fig. 1). The fact that the exciton density, $\rho(T)$, remains constant with increasing temperature above 225 K implies that the singlet-triplet gap becomes temperature dependent, increasing almost directly in proportion to T. This is particularly noticeable in the plot of $\log \chi T$ vs. T^{-1} above 225 K. Considering the small number of magnetic excitations ($\rho \sim 10^{-3}$) present at 225 K, the associated magnetic entropy is so small that the transition in all likelihood is not magnetic in origin but rather is driven by a lattice structural phase change. The sharpness of the transition as observed in $\chi(T)$ suggests the transition is first order. The high temperature susceptibility reflects the new lattice structural phase, and presumably the observed temperature dependence of the gap Δ is the result of this triplet exciton-lattice phonon coupling.

The temperature dependence of the spin susceptibility is of the same general form as observed by Vegter et al.[1] for other simple alkali (TCNQ) salts. However, comparison of the present data with the previous results of Vegter et al.[1] on Li(TCNQ) reveals considerable disagreement as shown in Fig. 1. At lowest temperatures, the earlier data[1] show a Curie-like upturn, often observed in organic charge transfer salts. This behavior is generally ascribed to extrinsic paramagnetic impurities, which in this case would correspond to nearly 4%. This behavior is completely absent in the present data. Subtraction of the impurity contribution from the earlier data[1] suggests triplet exciton behavior at low temperatures, with no evidence of a transition occurring at higher temperatures to a temperature dependent gap region as observed in Fig. 1. The effect of the paramagnetic impurities present in the previous samples[1] is evidently to actively suppress and qualitatively change the properties and magnetic behavior of Li(TCNQ). We conclude, therefore, that the present samples are of higher quality and as such show intrinsic behavior. Further evidence that this is the case is the general agreement between the pure Li(TCNQ) data and that from all the other simple alkali salts where phase transitions of a similar nature have been observed.[1]

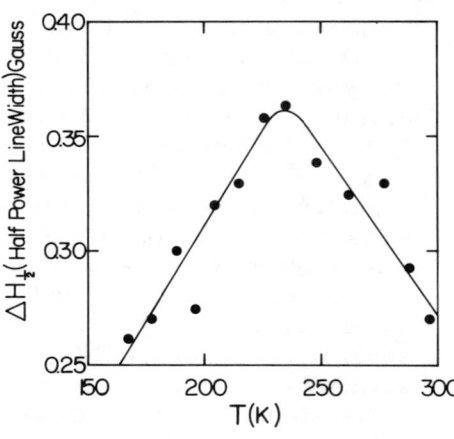

Fig. 3. Temperature dependence of the ESR linewidth for Li(TCNQ).

The temperature dependence of the ESR linewidth for Li(TCNQ) is given in Fig. 3. A sharp anomaly in the linewidth occurs at 225 K, which further shows that a phase transition does take place as observed in $\chi(T)$.

DISCUSSION

In general, the magnetic semiconducting TCNQ salts behave as Mott-Hubbard insulators, illustrated by the fact that the semiconducting gap ($2E_G$) is much larger than the magnetic excitation gap (Δ). For Li(TCNQ), $2E_G$ is approximately 7400K[6] whereas Δ is 1300K. The excitations should be viewed, therefore, not as single-particle excitations across a simple semiconducting gap, but as spin excitations of a system described by the strong coupling limit of the Hubbard Hamiltonian

$$\mathcal{H} = t \sum_{i,\sigma}^{N/2} (c^+_{2i\sigma} c_{2i+1\sigma} + c^+_{2i+1\sigma} c_{2i\sigma}) + t' \sum_{i,\sigma}^{N/2} (c^+_{2i-1\sigma} c_{2i\sigma} + c^+_{2i\sigma} c_{2i-1\sigma}) + U \sum_i n_{i\uparrow} n_{i\downarrow} \quad (3)$$

where $t(t')$ is the intra(inter)-dimer hopping integral; $c_{i\sigma}(c^+_{i\sigma})$ the creation (destruction) operator of the electron on site i with spin σ; $n_{i\uparrow}$ the corresponding number operator; and the effective U is the difference ($U_0 - U_1$) between on-site (U_0) and off-site (U_1) Coulomb repulsions.

The value 0.9 eV for U has been experimentally determined from the optical charge transfer band in K(TCNQ),[7] in general agreement with SCFMO results (U = 0.5 - 1eV).[7] The agreement indicates that effects such as cation polarizability[5] or intramolecular π-core electron screening play little or no role in reducing U.

Typical values for the transfer integrals t and t' in charge transfer TCNQ salts are approximately 0.1eV. Thus, since U is larger than t and t' in these systems, the magnetic excitations in the alkali TCNQ salts may be approximated at temperatures low compared to the single-particle excitation gap by the alternating Heisenberg Hamiltonian (Eq. 1). Bulaevskii[8] has developed an approximate treatment of Eq. 1, resulting in the susceptibility expression

$$\chi = [g^2 \mu_B^2 S(S+1)/k_B T] \, a(\gamma) \, \exp[-J\Delta(\gamma)/T] \qquad (4)$$

where $0.02 \leq T/2J \leq 0.25$, and the prefactor $a(\gamma)$ and the gap parameter $\Delta(\gamma)$ result from the finite width of the exciton band. For Li(TCNQ) below 225K, $J\Delta(\gamma)$ is 1300K, and the measured $a(\gamma) = 0.65$ corresponds to $\gamma = 0.35$. The magnitude of γ places Li(TCNQ) in the strongly alternating regime as expected from a dimerized crystal structure. The width of the triplet exciton band is 600K centered about 1500K above the singlet ground state. Since the magnetic properties of Li(TCNQ) follow the general form of behavior of the other simple alkali TCNQ salts, analysis of their magnetic properties as outlined here should yield similar results.

We are deeply indebted to Profs. A.F. Garito and A.J. Heeger for the assistance which made this work possible.

REFERENCES

1. J.G. Vegter, P.I. Kuindersma and J. Kommandeur, Proc. Conf. on Low Mobilities, Eilat, Israel (1971); J.G. Vegter, T. Hibma and J. Kommandeur, Chem. Phys. Letters 3, 427 (1969).
2. A. Hoekstra, T. Spoelder and A. Vos, Acta Cryst. B28, 14 (1972).
3. J. Murgich and S. Pissanetzky, to be published.
4. P.A. Pincus, Solid State Comm. 9, 1971 (1971); P.A. Pincus and G. Beni, J. Chem. Phys., to be published.
5. A.J. Heeger and A.F. Garito (this Proceedings) and references therein.
6. W.J. Siemons, P.E. Bierstedt and R.G. Kepler, J. Chem. Phys. 39, 3523 (1963).
7. S. Hiroma, H. Kuroda and H. Akamatu, Bull. Chem. Soc. Japan 44, 9 (1971).
8. L.N. Bulaevskii, Sov. Phys. Solid State 11, 921 (1969).

FERROMAGNETISM OF Fe(PHENANTHROLINE)Cl$_2$ AND RELATED ORGANO−METALLIC COMPOUNDS

W. M. Reiff[*]
Northeastern University, Boston, Massachusetts 02115

S. Foner
Francis Bitter National Magnet Laboratory,[†] Massachusetts
Institute of Technology, Cambridge, Massachusetts 02139

Examples of ferromagnetic order among transition metal organo-metallic compounds are rare. Here we report observations of long range ferromagnetic order in powders of Fe(phenanthroline)Cl$_2$ and some related compounds. Magnetic susceptibility data of Fe(phenanthroline)Cl$_2$ from 4.2 to 300 K fit to a Curie-Weiss law yield a paramagnetic Curie temperature $\theta = +12 \pm 4$ K, and a Curie-Weiss constant $C = 1.24 \times 10^{-2}$ emu/g corresponding to an effective moment of 5.4 μ_B consistent with quintet Fe II. Mössbauer data for Fe^{57} shows an isomer shift ($\delta(300\,K) = +1.04$ mm/sec, $\Delta E = 1.45$ mm/sec) relative to Fe metal and an effective internal field of 116 kG at 4.2 K. Magnetic moment data yields $T_c = 8 \pm 2$ K. For $T \leq 4.2$ K, magnetic saturation is not achieved at 48 kG. Results for the related compounds Fe(2,2'-bipyridine)Cl$_2$, Fe(5,5'-di-CH$_3$-2,2'-bipyridine)Cl$_2$ and Fe(4,4'-di-CH$_3$-2,2'-bipyridine)Cl$_2$ will also be presented. The field dependence of the Mössbauer spectrum and magnetic moment of Fe(2,2'-bipyridine)Cl$_2$ is also consistent with long range ferromagnetic order. X-ray and optical spectra show that all the above compounds are polymeric. The possibilities of increasing T_c in such organo-metallic compounds will also be discussed. The details of this work will be published elsewhere.

[*] Supported by the Petroleum Research Fund.
[†] Supported by the National Science Foundation.

TORQUE MEASUREMENTS ON LIQUID CRYSTALS IN A ROTATING MAGNETIC FIELD

P. J. Flanders,
LRSM, University of Pennsylvania, Philadelphia, Pennsylvania 19104

S. Shtrikman*
Department of Electronics, Weizmann Institute
of Science, Rehovot, Israel

ABSTRACT

The torque exerted on nematic liquid crystals due to a rotating magnetic field is investigated. Measurements of this torque on EBBA, MBBA and a 1:1 mixture of these as a function of temperature, magnetic field and frequency of rotation are reported. Comparison with theory shows reasonable agreement and yields the magnetic susceptibility anisotropy and the twist viscosity as a function of temperature for these liquid crystals.

INTRODUCTION

The behaviour of liquid crystals in a rotating magnetic field was first studied by Zvetkov.[1] In this work Zvetkov develops a theory to relate the torque L experienced by a nematic to H the magnetic field and ω the frequency of rotation. He reports also measurements of $<L>$, the time average torque, on PAA which are in reasonable agreement with his calculations. Recently Prost and Gasparoux[2,3] repeated the measurement of Zvetkov on PAA and extended them to MBBA. They have also improved considerably the theory developed by Zvetkov by basing it on the modern hydrodynamics of liquid crystals and getting explicit formulae for the torque, its time dependence and time average. By comparing their theoretical and experimental results they derive the temperature dependence of γ_1, the twist viscosity, and $\Delta\chi$, the magnetic anisotropic susceptibility, in PAA and MBBA.

In the present study we use the procedure developed by Prost and Gasparoux to measure the temperature dependence of γ_1 and $\Delta\chi$ in EBBA, and a 1:1 EBBA MBBA mixture. We also report our MBBA results for comparison. Our measurements were done with a rotating electromagnet allowing the use of fields up to 10 kOe. The instantaneous rather than a time averaged torque was recorded using an automatic torquemeter[4] in search of a pulsating component in the torque. Samples were held in gelatine capsules 0.5 cm in diameter and 1 cm long.

---

*Sponsored in part by the Air Force Office of Scientific Research United States Air Force, under Grant AFOSR 72-2368.

THEORY

When the molecules of a nematic are assumed to rotate freely at the boundaries of their container, Prost and Gasparoux[2] show that for:

$$L = (\gamma_1/\rho)\omega, \quad \omega < \omega_o \tag{1}$$

$$<L> = L_o [\omega/\omega_o - \sqrt{(\omega/\omega_o)^2 - 1}], \quad \omega > \omega_o \tag{2}$$

$$\omega_o = \frac{(1/2)\Delta\chi H^2}{\gamma_1/\rho} \tag{3}$$

where γ_1/ρ is the kinematic twist viscosity in cm^2/sec and ρ is the density in g/cm^3.

For $\omega >> \omega_o$

$$<L> = (1/2) L_o \omega_o/\omega = \frac{(1/8)\Delta\chi^2 H^4)}{\omega(\gamma_1/\rho)}, \tag{4}$$

Here $L_o = \frac{1}{2}\Delta\chi H^2$, (5).

Also when $\omega > \omega_o$ a periodic contribution to the torque is expected, having a frequency

$$\omega_v = 2\omega\sqrt{1 - (\omega_o/\omega)^2}, \tag{6}$$

RESULTS AND DISCUSSION

As predicted by Eq. 1, curves of L vs ω for MBBA at 300K, (Fig. 1), are linear at the origin and have the same initial slope at all fields. Plots of γ_1/ρ vs T are obtained by measuring L vs T at constant H and ω, (Fig. 2), keeping H large and ω small so that $(2\gamma_1\omega/\rho\Delta\chi H^2) < 1$ and thus $L \propto \gamma_1/\rho$. As previously observed[1,3] L does not peak sharply at some value of ω, leaving the method for determining L_o, ω_o from Fig. 1 in question. The peak in the experimental torque L_p, varies linearly with H^2 as predicted for L_o. For frequencies beyond the peak in the torque we find $<L> \sim H^4/\omega$ in accord with the field and frequency dependence of Eq. (4). When ω is increased abruptly to above ω_p, by abruptly decreasing H while ω is held constant, we find that there is a pulsating torque which decreases gradually to zero, (Fig. 3), with a frequency

$$\omega_v = 2\omega \sqrt{1 - (AH^2/\omega)^2} \qquad (7)$$

where A depends upon temperature.

Using data from Fig. 1, curves of $<L>/H^2$ against ω/H^2 superpose, (Fig. 4), as required by the theory given above. The

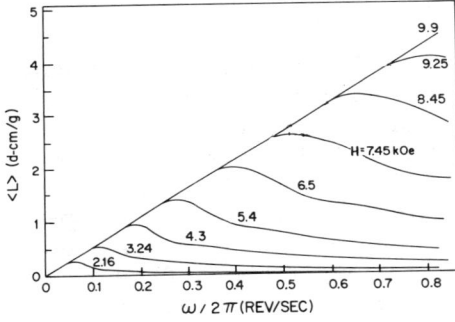

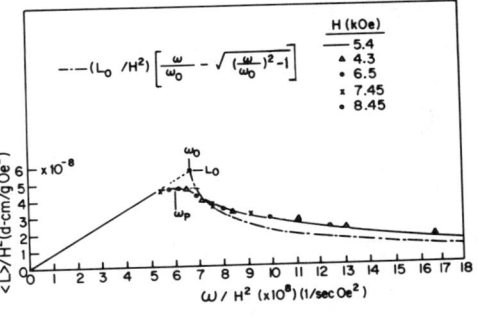

Fig. 1. $<L>$ vs $\omega/2\pi$ for MBBA 300K.

Fig. 3. L vs T for MBBA at 300K after increasing ω/ω_o abruptly from a value below 1 to 2.15 by changing H from 5.4 to 2.95 kOe; $\omega/2\pi = 0.2$ rev/sec and $\omega_o = 1.1\omega_p$.

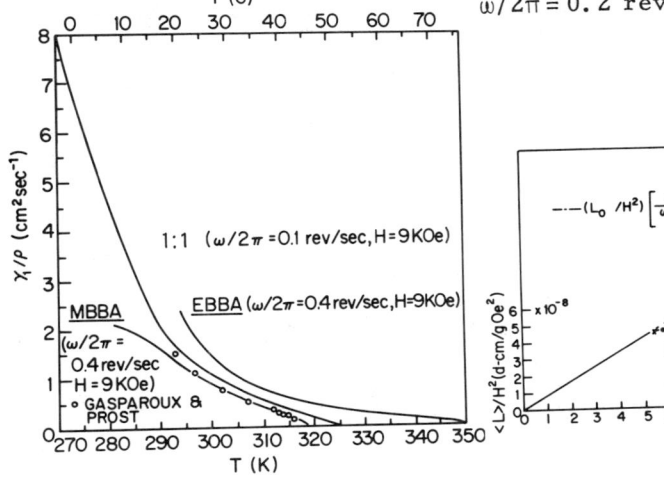

Fig. 2. γ_1/ρ vs T for MBBA, EBBA and a 1:1 mixture by weight.

Fig. 4. $<L>/H^2$ vs ω/H^2 for MBBA at 300K.

agreement with the predicted shape of this curve is however not as good, in line with the previous observation.[2] A further discrepancy with the theory is the decay in the pulsating torque. Also ω_o as derived by comparing (6) and (7) is about 10% higher than ω_p, as shown in Fig. 5. Deviations may be associated with the excitation of translational motion in the liquid which should not occur according to the theory[2] but which seems to be present according to some preliminary observations by us.

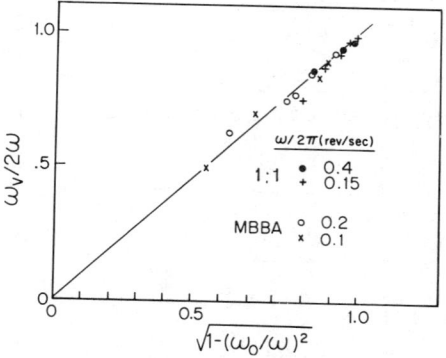

Fig. 5. Theoretical and experimental behaviour of the pulsating torque at 300K in MBBA and a 1:1 mixture of MBBA-EBBA by weight where $\omega_o = 1.1\, \omega_p$.

To find $\Delta\chi(T)$, we assume from Fig. 4 that the initial slope drops off at $0.82\, \omega_o$ where $<L> = 0.82\, L_o$. This choice for ω_o is the same as the value derived by comparing (6) and (7), and as shown below gives a $\Delta\chi(T)$ which is the same as that derived from moment measurements on MBBA[5] and from our own moment measurements on all 3 materials. The torque for a given combination of H and T on an L vs T plot will approach 0.82 ($\frac{1}{2}\Delta\chi H^2$) asymptotically at points on a previously recorded curve made at high H and low ω to find γ_1/ρ, [Fig. 6(a)]; from these points we get $\Delta\chi(T)$, [Fig. 6(b)].

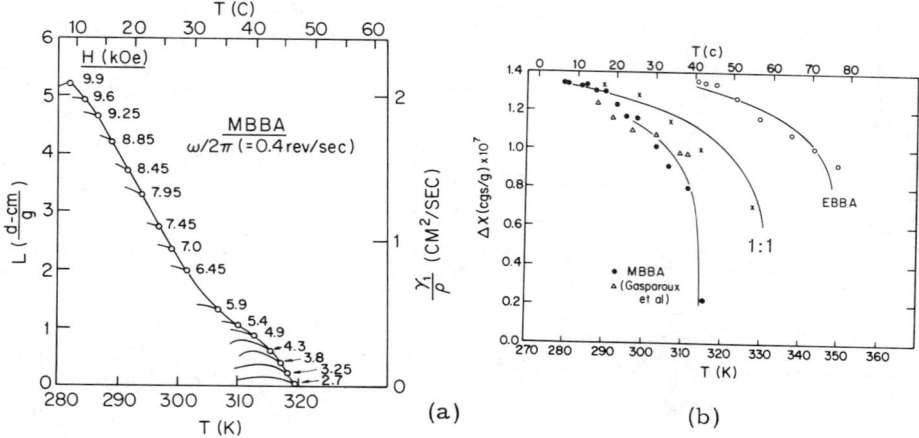

Fig. 6 (a), $<L>$ (H) and γ_1/ρ vs T for MBBA, and (b), $\Delta\chi$ vs T for MBBA, EBBA and a 1:1 mixture.

REFERENCES

1. V. Zvetkov, Acta Physicochimaca, URSS 18, 358 (1943).
2. J. Prost and H. Gasparoux, Phys. Let. 36A, 245 (1971).
3. H. Gasparoux and J. Prost, J. de Phys. 32, 953 (1971).
4. G. T. Croft, F. J. Donahoe and W. F. Love, Rev. Sci. Instr. 26, 360 (1955).
5. H. Gasparoux, B. Regaya and J. Prost, C. R. Acad. Sc. Paris, 272, 1168 (1971).

THE EFFECT OF GASIFICATION ON THE DIAMAGNETISM OF GRAPHITE AND ARTIFICIAL CARBONS

J. J. Santiago[†], L. N. Mulay and P. L. Walker, Jr.
The Pennsylvania State University, University Park, Pa. 16802

ABSTRACT

Magnetic susceptibility and magnetic anisotropy measurements by the Faraday method were carried out on samples of natural graphite, graphitized carbon composites, stress recrystallized pyrolytic graphite and on a graphitized carbon black (Graphon). The total susceptibility at room temperature ranged from -18.75×10^{-6} emu/g for graphon to -21.68×10^{-6} emu/g for the stress recrystallized pyrolytic graphite. The anisotropy ratios ($R = \chi_\perp/\chi_\parallel$) ranged from 1.02 for an extruded lampblack based carbon composite to 40 for the stress recrystallized sample. The temperature dependence of the susceptibility can be accounted for using a two dimensional degenerate electron gas model. From this model the values of T_o, the degeneracy temperature of the electron gas, the number of effective electrons per carbon atom and $\alpha = m/m^*$, the ratio of the electron rest mass to its effective mass were obtained. Within the spirit and limitations of this model the following estimates for T_o were obtained; 344°K for natural graphite, 379°K for stress recrystallized pyrolitic graphite and 500°K for Graphon. A strong correlation exists between T_o and the perfection of the graphitic material. Susceptibility measurements as a function of massive oxidation show that the effect of gasification in dry oxygen upon the susceptibility of the material reflects its perfection and preparation methods.

INTRODUCTION

Graphite possesses one of the largest diamagnetic anisotropies of any material known. When a single crystal of natural graphite is oriented such that the magnetic vector is perpendicular to the basal plane (aromatic plane) the specific susceptibility is $\chi_\perp = -21.75 \times 10^{-6}$ emu/g. If the magnetic field vector is oriented parallel to the basal plane the specific susceptibility is $\chi_\parallel = -0.30 \times 10^{-6}$ emu/g. The ratio $\chi_\perp/\chi_\parallel$ is a good measure of the diamagnetic anisotropy of the material and for selected natural single crystals is as high as 72.[1] The high diamagnetic anisotropy and the underlying physical phenomena have been fairly well known since the pioneering work of Ganguli and Krishnan in 1948.[2] They were able to interpret the high susceptibility for a magnetic field oriented perpendicular to the basal plane as being due to the contribution by quasi-free electrons in the π band.

The current models for the interpretation of the magnetic properties if graphitic materials are due to McClure[3,4] and to Marchand

[†]Current Address, Aerospace Rsch Labs, Wright-Patterson AFB, Ohio

and co-workers.[5] McClure, using the rigorous formalism of band theory, derived explicit expressions for the diamagnetic anisotropy of graphite using the two dimensional band structure approximation. He also obtained computational results for the three dimensional band structure model. Marchand and co-workers, elaborating on a simple model proposed by Ganguli and Krishnan, noticed that the semi-empirical relation y = 1 - exp(-x) proposed to fit the experimental data, could be firmly grounded on existing theories of a two dimensional electron gas. In their theory the above semi-empirical relation takes the form

$$\chi_t/\chi_o = 1 - \exp(-T_o/T) \qquad (1)$$

where $T_o = (\nu N h^2 \alpha)/(4\pi m k S)$ is the degeneracy temperature of the two dimensional electron gas, $\chi_o = [(\nu \beta^2 N)/(kT_o S)](1 - \alpha^3/3)$, N is the total number of carbon atoms, ν the number of holes per carbon atom in the band with effective mass $m^* = m/\alpha$, S the area occupied by the gas, and $\beta = eh/2mC$ the Bohr magneton. Their theory unjustifiably assumes ν to be temperature independent, implying no thermal excitation of carriers from the π band to the conduction band. According to Marchand's simple model, $T_o = \epsilon_o/k$, where ϵ_o is the energy difference from the Fermi level to the top of the valence band and k is the Boltzmann constant. The present work was undertaken to see if magnetic susceptibility measurements could be used to characterize carbon materials and to investigate quantitatively their gasification in oxygen. This approach stemmed from earlier work which showed that the specific diamagnetic susceptibility χ_t is proportional to the position of the Fermi level with respect to the edge of the valence band[3] in carbon materials, and that gasification produces many forms of structural imperfections which can act as traps for electrons thereby lowering the Fermi level.

EXPERIMENTAL

Average and anisotropic magnetic susceptibility measurements by the Faraday method were made on samples of SP-1 (natural graphite), AGKSP, L113SP spectroscopic grade artificial carbons from Union Carbide Co., SRPG (stress recrystallized pyrolytic graphite) and Graphon (a carbon black heat treated to a temperature of 2700°C). From the field dependence of the diamagnetic susceptibility (Honda-Owens plot), a negligible concentration of ferromagnetic impurities was determined. Traces from the diamagnetic susceptibility tensor $(\chi_\perp + 2\chi_\parallel)$ at room temperature ranged from -18.8×10^{-6} emu/g for Graphon to -21.6×10^{-6} emu/g for the SRPG. Other values for the room temperature trace were found to be -20.6, -21.0 and -21.2×10^{-6} emu/g for powdered SP-1, for L113SP and AGKSP extruded rod composites, respectively. The anisotropy ratios $(R = \chi_\perp/\chi_\parallel)$ ranged from 1.02 for the composite L113SP to 38.8 for the mosaic crystal of SRPG. Other values are presented in Table I.

Table I Principal Susceptibilities of Different Graphites and Graphitic Carbons in Units of 10^{-6} emu/g at Room Temperature

Sample	$-\chi_{\parallel}$	$-\chi_{\perp}$	$-\chi$ Total	Anisotropy Ratio
SP-1	Average χ = 6.86		20.58	not Measurable
SRPG	0.54	20.60*	21.68	38.2
AGKSP	8.33	4.60*	21.27	1.84
AGKSP#	Average χ = 7.10		21.30	not Measurable
L113SP	7.06	6.93*	21.06	1.02
L113SP#	Average χ = 7.01		21.03	not Measurable
Graphon	Average χ = 6.25		18.75	not Measurable

*Component parallel to c-axis, across the grain or in extrusion direction; whichever applies. #Powder

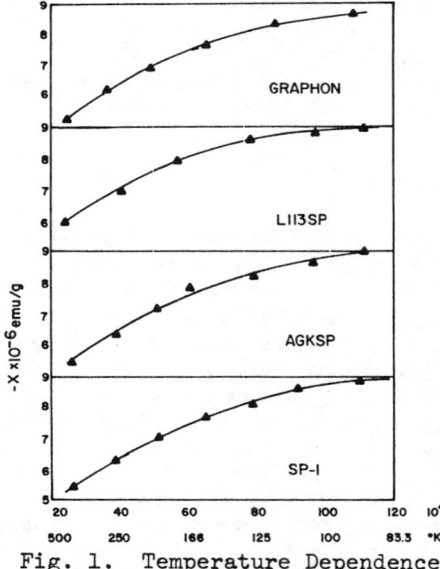

Fig. 1. Temperature Dependence of Diamagnetic Susceptibility.

Diamagnetic susceptibility as a function of temperature was measured from 77°K to 300°K and the results of these measurements are presented in Fig. 1.

The samples were gasified (or massively oxidized) in dry oxygen "in situ" in the Faraday balance. Gasification was carried out at temperatures selected so as to avoid a non-uniform, diffusion controlled reaction. Uniformly gasified samples were obtained, gasified to as much as 80% weight loss. The optimum gasification temperature increases with the perfection of the material to be gasified. These optimum gasification temperatures ranged from 450°C for SP-1 and Graphon to a high of 650°C for SRPG. Susceptibility as a function of gasification was measured for different sample orientation using the Faraday balance. Results of these experiments are presented in Fig. 2.

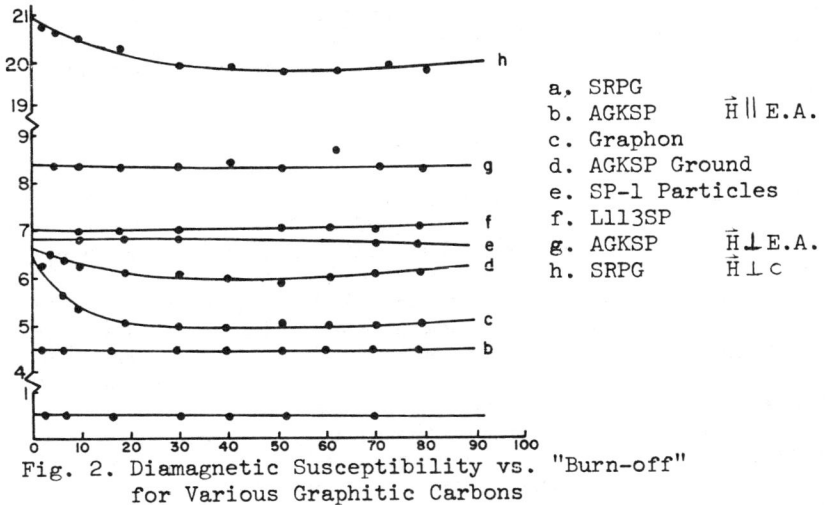

Fig. 2. Diamagnetic Susceptibility vs. "Burn-off" for Various Graphitic Carbons

DISCUSSION

From the susceptibility vs. temperature measurements and using Marchand's model for the susceptibility of a degenerate two dimensional electron gas, values for the ratio $\alpha = m / m^*$ of the rest mass (m) to the effective mass (m^*) of the free carriers, and the degeneracy temperature (T_o) were calculated. The following degenercy temperatures were estimated: For SP-1 powder sample, $T_o = 344.2°K$; Graphon $T_o = 506°K$ and SRPG, $T_o = 379.3°K$. The other carbon materials under study, the composites (filler plus binder) AGKSP and Ll13SP, possessed a degeneracy temperature of $389.4°K$ and $402.0°K$ respectively. The effective masses derived from this model ranged in value from $m^* = 0.003m$ for SP-1 to $m^* = 0.005m$ for Graphon. These values are smaller than the magnitude usually assumed for the effective masses in graphitic materials.

Now that the sign of γ_2 (an overlap integral in the band theory of graphite) has been definitely stablished ($\gamma_2<0$) and the heavy carriers identified as electrons[6], Spain[7] has found from galvanomagnetic data that the light holes close to the H point in the first Brillouin zone (the top of the Fermi surface extending into the next zone) may be identified with the effective minority carriers. It is possible that only the carriers with the smallest effective masses are responsible for the magnetic susceptibility of graphite and graphitic carbons, the contribution of charge carriers with larger m^* values being negligible. This possibility, as previously pointed out by Marchand[5], could account for the small m^* values found in this work and for the temperature independence of ν. One should point out that Schoenbert[8] also found small values for ν of the order of 10^{-5}. Using susceptibility measurements as a characterization tool, one found that the SRPG sample with the lowest T_o of the artificial materials has better over-all metallic properties than the other artifical carbon samples (AGKSP, L113SP and Graphon).

CONCLUSIONS

The results on the effect of gasification upon the susceptibility of the materials were shown to depend on the origin and preparation of the individual samples. For the commercial carbon composites, the susceptibility was found to remain constant with gasification. This behavior was observed for AGKSP and the lampblack based L113SP samples. This was explained on the basis that the electronic effects of carrier trapping would not be significant, since the material was known to contain too many carrier traps, due to their highly disordered structures and created by plastic deformation during fabrication, which mask any gasification effect. For the highly anisotropic SRPG mosaic crystal, when oriented with the field perpendicular to the basal plane, the specific susceptibility ($\chi_\perp$) vs. burn-off (gasification) curve was seen to decrease with burn-off. The susceptibility decreases steadily to 45% burn-off and then remains constant thereafter. However, for the magnetic field parallel to the basal plane of the crystal there was no measurable change in the susceptibility ($\chi_\parallel$) with burn-off. In this material the effect of gasification upon the susceptibility was significant and could be attributed mainly to the formation or enlargement of defects upon gasification which are effective charge carrier traps. Since Graphon has the largest specific surface area for chemisorbing oxygen when activated, it is suggested that the observed decrease in the susceptibility during gasification was a result of chemisorbed oxygen acting as surface trapping sites.

REFERENCES

1. D. E. Soule and C. W. Nezbeda, National Carbon Co. Research Memorandum NRM-99 (1963).
2. W. Ganguli and K. S. Krishnan, Proc. Roy. Soc. (London)-$\underline{117}$, 168 (1941).
3. J. W. McClure, Phys. Rev. $\underline{104}$, 666 (1956).
4. J. W. McClure, Phys. Rev. $\underline{119}$, 606 (1960).
5. A. Pacault, A. Marchand, F. Boy and E. Poquet, Compt. Rend. $\underline{254}$, 1275 (1962).
6. J. A. Woolam, Phys. Rev. B $\underline{4}$, 3393 (1972).
7. I. A. Spain in Chemistry and Physics of Carbon, P. L. Walker, Jr., Ed. Marcel Dekker, Inc. New York (1972).
8. D. Shoenberg, Phill Trans. Roy. Soc. (London) $\underline{245}$, 1 (1952).

Magnetic Properties of Simple Alkali-TCNQ salts

Johan G. Vegter and Jan Kommandeur

Laboratory for Physical Chemistry, University of Groningen,
Zernikelaan, Groningen, the Netherlands.

Abstract

The paramagnetic organic ions of the alkali$^+$-TCNQ$^-$ salts form linear chains. An analysis of the magnetic susceptibility of Li and Cu-TCNQ shows that its variation with temperature can be exactly explained by one-electron theory, leading to symmetry-split ground state bands. The "pairing" in the chains at 0°K is due to the Peierls instability, while Adler Brooks crystallographic phase transitions (which can be first order or continuous) occur at higher temperatures in all alkali salts, except for LiTCNQ. The behavior of the magnetic susceptibility in these salts can be very well understood on the basis of one electron theory, when the pairing of the ions in the chains is allowed to vary with the temperature. Free energy calculations predict the order of the transition and the transition temperature very well.

The transport properties, such as electrical conductivity cannot be explained by this simple model, because they are probably governed by disorder.

Introduction

The magnetic and electrical properties of the salts of tetra-cyano-quinodimethan (TCNQ) have recently been the subject of many investigations (1,2,3). In general, the paramagnetic TCNQ-ions in these crystalline salts stack in linear chains, and therefore, they furnish excellent material for studies of (almost) one-dimensional systems. Also, since the interplanar distances of the TCNQ ions are large by usual "inorganic" standards (3.2-3.3 Å, which, however, is small by "organic" standards), the question has arisen, whether the electronic properties of the crystals should be described by the usual band theory, by a Hubbard model (4), or even by theories appropriate to the Heisenberg linear anti-ferromagnet (5). Finally, since the system can be considered one-dimensional it has been suggested that lattice disorder is the main determining factor (6).

It is the purpose of this paper to show, that the spin susceptibilities of the simple alkali$^+$-TCNQ$^-$ salts can easily be understood on the basis of one-dimensional tight-binding theory together with the crystallographic phase transitions predicted by Adler and Brooks (7).

Theory

The salts all have two TCNQ-ions per unit cell along the chain (8), we therefore find from tight-binding theory for the electronic energy

$$\mathcal{E}(k) = \pm \left[h_1^2 + h_2^2 + 2h_1 h_2 \cos(ak) \right]^{\frac{1}{2}} \quad (1),$$

where h_1 and h_2 are the two inequivalent TCNQ transfer integrals, a the length of the unit cell and $k = 4n\pi/Na$, n from $-\frac{1}{4}N$ to $+\frac{1}{4}N$. After defining $E_c \equiv 2|h_1 h_2|^{\frac{1}{2}}$ and $g \equiv \sinh(\frac{1}{2}\ln|h_1/h_2|)$ we have for half the gap $E_g = gE_c$ and for $\xi(k)$:

$$\xi(k) = \pm E_c \{g^2 + \cos^2(\frac{1}{2}ak)\}^{\frac{1}{2}} \qquad (2)$$

The "dimerization" of the chain comes about through the well-known Peierls instability, which at 0°K allows an electronic energy gain at the cost of some increased lattice repulsion. At higher temperatures the entropy gain of the electrons as well as the increase in the average electronic energy through thermal excitation can lead to an instability of the dimerized phase with respect to the non-alternating phase and thus to a phase transition. This semiconductor to metal transition was first discussed by Adler and Brooks (7) and later by Hallers and Vertogen (9). These authors come to the conclusion, that on the basis of elementary tight binding theory only second-order phase transitions will occur. In the TCNQ-salts the situation is somewhat different than in the theoretical models considered. As will be clear from fig.1., although the interplanar distances alternate, the spatial overlap is such that disappearance of the alternation will not lead to equivalence of the transfer integrals. In the TCNQ salts we therefore expect semiconductor—semiconductor transitions. These can either be first order or continuous (although strictly we cannot speak of a continuous phase transition), since second order is not allowed by symmetry (10).

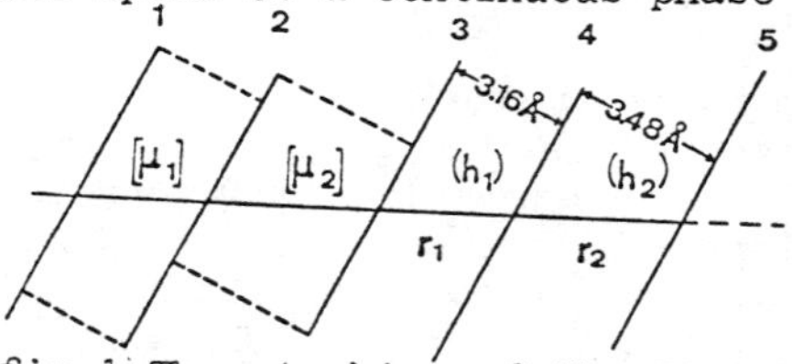

For reasons of clarity we first treat the symmetrical case, where we can allow the transfer integrals to become equal. We define the alternation parameter $\xi \equiv \dfrac{r_2 - r_1}{r_2 + r_1}$,

fig.1. The stacking of the TCNQ-ions.
where r_2, r_1 are the two different interplanar distances. Since the TCNQ ions have p_z orbitals we have for

$$h_{1,2} \approx Z/12(cr_{1,2})^3 \exp(-cr_{1,2}), \text{ where } c = Z/2a_o,$$

a_o is the Bohr radius and Z the effective nuclear charge (Z=3.4) (11). We then find

$$h_{1,2} = h_o(1 \mp \xi)^3 \exp(\pm \tfrac{1}{2} ac\, \xi) \qquad (3)$$

$$E_c = 2|h_o|(1-\xi^2)^3 \approx 2h_o$$

and $g \approx \sinh((\tfrac{1}{2}ac - 3).\xi)$ (4), which indicates that the gap $E_g = gE_c$ is almost linear in the distortion and the total width E_c is insensitive to it. The repulsive lattice energy is given by

$$\Delta E_\ell = N/2(Br_1^{-n} + Br_2^{-n}) = \tfrac{1}{2}NB(\tfrac{1}{2}a)^{-n}\{(1-\xi)^n + (1+\xi)^n\} =$$
$$\equiv \tfrac{1}{2}N\Omega\{(1-\xi)^n + (1+\xi)^n\} \qquad (5)$$

As we will show elsewhere (11), when equations (4) and (5) are approximated by, respectively $g \approx (\tfrac{1}{2}ac - 3).\xi$ (4a) and $\Delta E_\ell \approx \tfrac{1}{2} V K_\xi^{-1}.\xi^2$ (5a) where V is the volume and K_ξ the lattice compressibility, only second order phase transitions are allowed. Inclusion of higher order terms leads to the first-order nature of the transitions.

Neglecting lattice vibrations, the total free energy of the system can now be written as: $F(\xi,T) = F_\zeta(\xi) + F_{el}(\xi,T)$

$$F(\xi,T) = \tfrac{1}{2}N\Omega\left\{(1-\xi)^n + (1+\xi)^n\right\} +$$
$$- \tfrac{2}{\pi}NE_c \int_0^{\frac{1}{2}\pi}\left[(g^2+\cos^2\tfrac{1}{2}ak)^{\frac{1}{2}} + 2\tfrac{kT}{E_c}\ln\left\{1+\exp\left(-E_c/kT(g^2+\cos^2\tfrac{1}{2}ak)^{\frac{1}{2}}\right)\right\}\right]d(\tfrac{1}{2}ak) \quad (6)$$

which expression can be numerically evaluated as a function of g and kT/E_c. To obtain the equilibrium distortion we require
$$(\partial F(\xi,T)/\partial \xi)_{V,T} = 0 \quad (7)$$
which yields the value of ξ at each temperature.

As pointed out above, the situation in the TCNQ-chains is not symmetrical. The projections of the TCNQ ions on each other μ_1 and μ_2 have to be included. The expressions (4) and (5) then change into: $g=\sinh\{(\tfrac{1}{2}ac-3)\cdot\xi -\tfrac{1}{2}\ln\mu_1/\mu_2\}$(4b) and $\Delta E = \tfrac{1}{2}N\Omega\mu_1\{(1-\xi)^{-n} + \tfrac{\mu_2}{\mu_1}(1+\xi)^{-n}\}$(5b). The total free energy $F(\xi,T)$ now also includes linear terms in ξ and this leads to first order phase transitions and to (rapid) continuous changes.

The spin susceptibility can be calculated from $F(\xi,T)$ in the usual fashion(12). The contribution of lattice vibrations was evaluated, but could in this approximation be neglected.

Application to Experiment

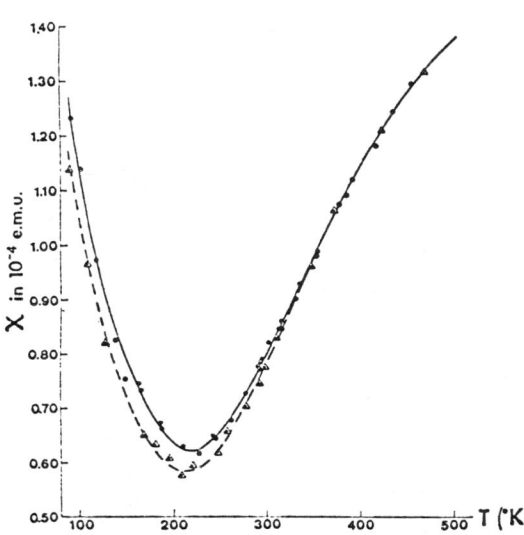

fig.2. Experimental(dots) and Theoretical(lines) spin susceptibilities of Li TCNQ.

Figure 2 gives the measured and calculated spin susceptibility of two samples of crystalline LiTCNQ, obtained from the low temperature susceptibility, being calculated from the Rb-TCNQ crystal structure (8). The sole quantity fitted was E_c/k, the width of the electronic band system, which turns out to be about one half of what one calculates from the known TCNQ wave functions(11). There is no phase transition. A similar result was found for CuTCNQ(12). Figure 3 gives the magnetic susceptibilities calculated for various values of the total bandwidth $2E_c$. The change of this quantity imitates quite well the changes in going from Li to Rb, since the larger size of the kation increases the distance between the TCNQ planes, and thus decreases the quantity E_c. These theoretical results should be compared with the experimental results given in fig.4. It is seen that the behavior of χ can be imitated quite well. Tables I and II collect the relevant parameters in tabular form.

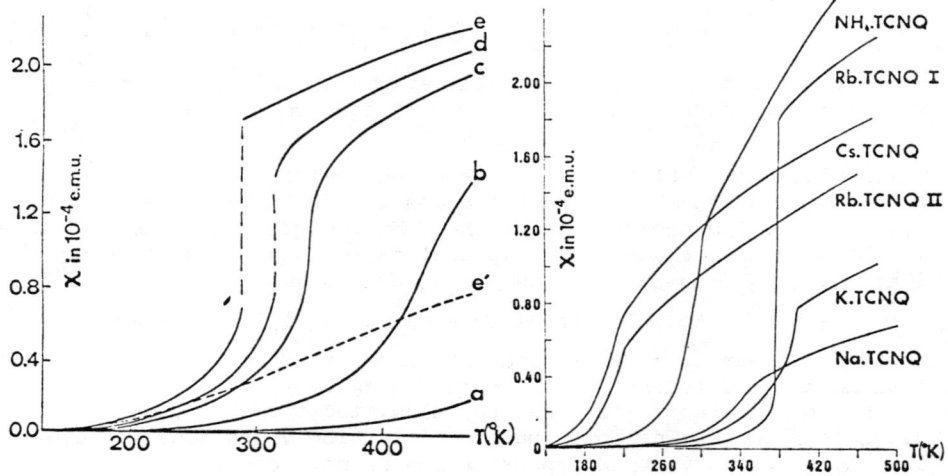

fig.3. Calculated spin susceptibilities with $E_c/k = 3000°K(a)$, $2100°K(b), 1700°K(c)$, $1600°K(d), 1500°K(e)$. e' is case e with constant band parameters.

fig.4. Experimental spin Susceptibilities of a variety of TCNQ salts.

TABLE I Transition data for the M^+TCNQ^- salts. The numbers in parentheses refer to the inflection point of the $\chi(T)$ curve (figure 5.17) for NaTCNQ, exhibiting a continuous change.

	$T_t(°K)$	$\chi T_t/C$	ΔH(cal/mole)	(phase) change
LiTCNQ	no phase transition	or continuous	change observed	
NaTCNQ	(345)	(0.034)	-	continuous
KTCNQ	396	0.082	60	1st- order
NH4TCNQ	299	0.091	≈20	1st- order
RbTCNQ	376	0.105	1010	1st- order
RbTCNQ II	220	0.032	unobservable	1st- order
CsTCNQ	217	0.042	unobservable	1st- order

TABLE II Band parameters in HT phase

	LiTCNQ	CuTCNQ	NaTCNQ	KTCNQ	NH_4TCNQ	RbTCNQ
g	0.66	0.20	0.38	∿0.7	≈0.6-0.3†	≈0.3-0.1†
E_c/k	1185°K	2020°K	2170°K	∿1300°K	∿1000°K	∿1200°K
δn	0.03	0.0005	0.006	0.0004	0.002-0.006	0.0004

†) The g-value still varies continuously above the transition point.

Discussion

The surprise of our measurements and calculations is that one electron theory suffices for these narrow band materials. The reduction of the electron electron interaction is due to the geometry of the TCNQ-molecule, where in the doubly ionized state the charges are located on the "tails"(12.13.) The fact that the experimental bandwidths are only half the theoretical ones may be a remnant of the electron electron interaction.

A real single particle experiment, such as a conductivity would conclusively prove that the one electron theory is correct. Unfortunately, this experiment is hampered by the one dimensionality of the material, since impurities, defects or other disorder reduces the chains to a system of broken strands, where the D.C. conductivity is determined by the process of crossing the breaks. The fact that the magnitude and temperature dependence of the conductivity depend on the frequency of measurement (3) points in this direction.

References

1. J.G.Vegter,T.Hibma and J.Kommandeur,Chem.Phys.Letters 3,427(1969)
2. O.H.LeBlanc,Physics and Chemistry of the Organic Solid State, Vol.III,ch.3 (Interscience,N.Y.1967)
3. I.F.Shchegolev,Phys.stat.sol (a) 12, 9 (1972).
4. A.J.Epstein,S.Etemad,A.F.Garito and A.J.Heeger,Phys.Rev.B5,952 (1972)
5. Z.G.Soos,J.Chem.Phys. 43, 1121 (1965).
6. A.N.Bloch,R.B.Weisman and C.M.Varma,Phys.Rev.Letters 28,753(1972)
7. D.Adler and H.Brooks,Phys.Rev.155, 826 (1967).
8. A.Hoekstra,T.Spoelder and A.Vos,Acta Cryst. B28, 14 (1972).
9. J.J.Hallers and G.Vertogen,Phys.Rev.B4, 2351 (1972).
10. C.Haas,Phys.Rev. 140, A 863 (1965).
11. H.Th.Jonkman and J.Kommandeur,Chem.Phys.Letters 15, 496(1972)
12. J.G.Vegter,P.I.Kuindersma and J.Kommandeur,Proc.2nd Int.Conf. on Low Mobility Materials (Taylors Francis, London, 1971).
13. J.G.Vegter and J.Kommandeur,Proc.5th Mol.Crystal Symposium Philadelphia,Penn.,U.S.A. (1970)

EXTENDED FRENKEL EXCITONS IN THE ONE-DIMENSIONAL BAND SYSTEM OF SIMPLE TCNQ-SALTS

T. Hibma, G. A. Savatsky, and J. Kommandeur

Laboratory for Physical Chemistry, University of Groningen,
Groningen, The Netherlands

ABSTRACT

ESR spectra of many simple TCNQ salts show a dipolar split line due to triplet excitons in addition to a usually much more intense central line. The temperature dependence of these lines is in agreement with a simple semiconductor model in which we identify the dipolar split line with triplet excitons and the central signal with conduction electrons.[1]

We have recently developed the theory of excitons in the region intermediate between the nearly free electron and the localized electron limit for the case of a one-dimensional semiconductor. When only on-site hole interactions are taken into account, an extended exciton state results, which in the limit of zero bandwidth is the normal Frenkel exciton solution.

Using this model we show that the dipolar splitting of the triplet excitons is strongly dependent on the ratio of the electron-hole interaction (I) to the one-electron band-width (w). The model is used to obtain information about w/I for some alkali TCNQ salts. Generally, it appears that w/I is of the order of unity, indicating that although one-electron theory can be satisfactorily used for thermodynamic properties, correlation effects are certainly present.

1. T. Hibma, P. Dupuis, and J. Kommandeur, Chem. Phys. Letters 15, 17 (1972).

ARE DEAD LAYERS REAL?*

L. Liebermann and J. Clinton
University of California, San Diego, La Jolla, Ca. 92037

ABSTRACT

The magnetization of ferromagnetic films formed by electrodeposition has been monitored continuously as the film thickness increases from zero, with surprising results: The magnetization does not increase continuously from zero thickness because of magnetically dead layers; these remain dead although the film thickness increases to bulk. Iron, cobalt, and nickel films exhibit this effect but are quantitatively different.

A number of tests have been devised in an attempt to determine if dead layers represent an intrinsic property of ferromagnets or are associated with the method of film formation. The following tests (on nickel films) are examined in detail: Incorporation of hydrogen or other non-metallic impurities during deposition; diffusion of gold, copper, or silver substrate atoms into the nickel film; discontinuities in the film (islands) leading to superparamagnetism; temperature dependence of dead layers at low ($T/T_c=0.1$) and high ($T/T_c=0.8$) temperatures.

The results of these tests suggest that dead layers are an intrinsic property of magnetic materials. The existence of dead layers is consistent with recent theories of magnetic surface effects.

INTRODUCTION

The study of free surfaces of solids is revealing new and interesting physical properties. Magnetic properties associated with solid surfaces are emphasized in extremely thin films because of their large surface to volume ratio. Recent experiments on extremely thin films suggest one new magnetic property of ferromagnet surfaces: They appear to be magnetically dead[1,2].

The present work is a critical analysis in an attempt to ascertain if the observed dead layers could be an artifact. The experimental observations to be analyzed consist of data on magnetization of films while in the process of formation by electrodeposition. A typical magnetization record[1] during formation of a nickel film is shown in Fig. 1. Two striking facts are immediately apparent: First, it is seen that the film shows no magnetization until it has achieved a thickness of at least five atomic layers. Secondly although the film's magnetization increases linearly with thickness beyond five atomic layers, the linear portion does not intersect the origin as expected; it appears that four layers remain magnetically dead in spite of increasing film thickness. Similar observations are recorded for iron and cobalt films, but in these materials only two atomic layers remain dead.

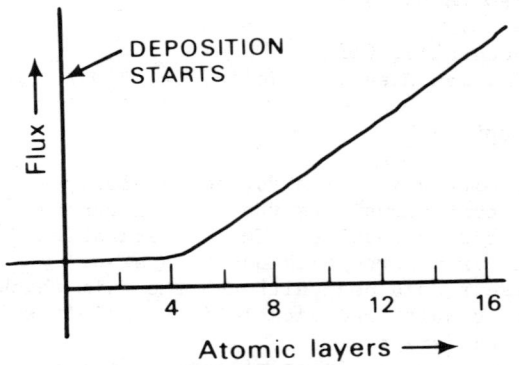

Fig. 1. Recording of the magnetic flux in a nickel film as the thickness is increased by electrodeposition. The thickness is deduced from the electroplating current. The total time of formation of 16 layers is approximately 2 seconds.

It seems reasonable to assume that the dead layers are located at the two surfaces of the film. However, it should be noted that these observations are consistent with, but do not verify this assumption. The data show only that the magnetization equivalent of four layers is lacking; although unlikely, it is possible that the four-layer deficiency is distributed throughout the film rather than a surface effect.

If dead layers exist in thin films, then it appears likely they also occur in thick films and bulk ferromagnets. Hence, the establishment as an intrinsic property of films would have general consequences. The ensuing sections deal with each of a number of tests designed to study possible extraneous effects which could unwittingly produce absent or dead layers in films.

TESTS FOR CONTAMINATION OR MAGNETIC INTERFERENCE

A. SUBSTRATE DIFFUSION

The substrate on which the magnetic films is deposited electrochemically must be an electrical conductor implying, for practical purposes, a metal. This necessary metallic substrate could conceiveably diffuse into the ferromagnetic films and interfere with its magnetic properties to a depth (equivalent) of four atomic layers.

The substrate is prepared by firing a gold film[3] (similar to that used in ceramic dinnerware) at 680°C on a fused quartz microscope slide. This gold film then serves as an initial substrate on which the final substrate is deposited by electrodeposition. The final substrate can be any one of a number of metals: copper, gold, and silver have been used. X-ray analysis has established that the final substrate is either monocrystalline or a mosaic of relatively large monocrystals.

Could significant diffusion occur during the time of formation (½ second) of the dead layers? It might be expected that an answer would be forthcoming by simple calculation of the diffusion time (at room temperature) using known diffusion coefficients (for bulk).

According to calculations of this type diffusion is negligible. However, diffusion in thin films is faster than in bulk because of diffusion along grain boundaries, dislocations and other defects. These are usually found at a higher concentration in films than in bulk. Hence, diffusion cannot be ignored without the further experimental proof presented below.

Copper is soluble in all proportions in nickel, whereas silver is completely insoluble. Hence, if diffusion interference exists in nickel films, copper substrates should exhibit a more prominent effect. However, this is not the case: The number of dead layers in nickel film is consistently four regardless of whether the substrate is copper, silver or gold.

Another direct experiment on diffusion was performed by observing the magnetization of a nickel film (on a copper substrate) with the one free nickel surface in contact with a copper electroplating solution. A decrease in magnetization was looked for when the electroplating current commenced and metallic copper was deposited upon the nickel surface. No diminution of magnetization was observed, implying negligible influence of copper diffusion.

Still another piece of evidence against diffusion is found in the magnetization record of Fig. 1. Note the abrupt transition from dead layers (no magnetization) to a linear increase in magnetization with thickness. If diffusion was responsible, an approximately exponential distribution of the substrate metal would occur in the magnetic film leading to a relatively slow ferromagnetic transition with thickness rather than the abrupt observed transition.

B. INCORPORATION OF HYDROGEN OR OTHER IMPURITIES BY ELECTRODEPOSITION

Conceiveably, the efficiency of the electrodeposition process could be less than 100%. For example, the first four layers could be either lacking completely or contaminated with hydrogen or other cations deposited simultaneously with nickel.

In order to test the overall efficiency of deposition, a series of chemical microanalyses were performed on the first few layers. Two different types of analyses were used: flame absorption spectrometry and a commercially available colorimetric indicator test for nickel[4]. The latter was simpler and easily gave sufficient accuracy to quantitatively analyze one or more atomic layers of nickel deposited on a microscope slide. Fig. 2 shows sample data obtained by Clinton[5] by quantitative analysis of various thin films of nickel. It is seen that the quantity of nickel is known to be deposited to an accuracy within 10% of that predicted from Faraday's Law. Stated in another manner, the maximum amount of contaminant is 10%. Pure electroplating solutions were utilized; hence, any electrodeposited contaminant must necessarily be a nonmetallic ion, possibly hydrogen.

While hydrogen embrittlement of nickel is well known to metallurgists, there is no evidence in the literature to suggest that the magnetization of nickel is markedly diminished by hydrogen. In any case, the effect would have to be extremely intense inasmuch as only 10% or less hydrogen can be present. Hence, the magnetic influence

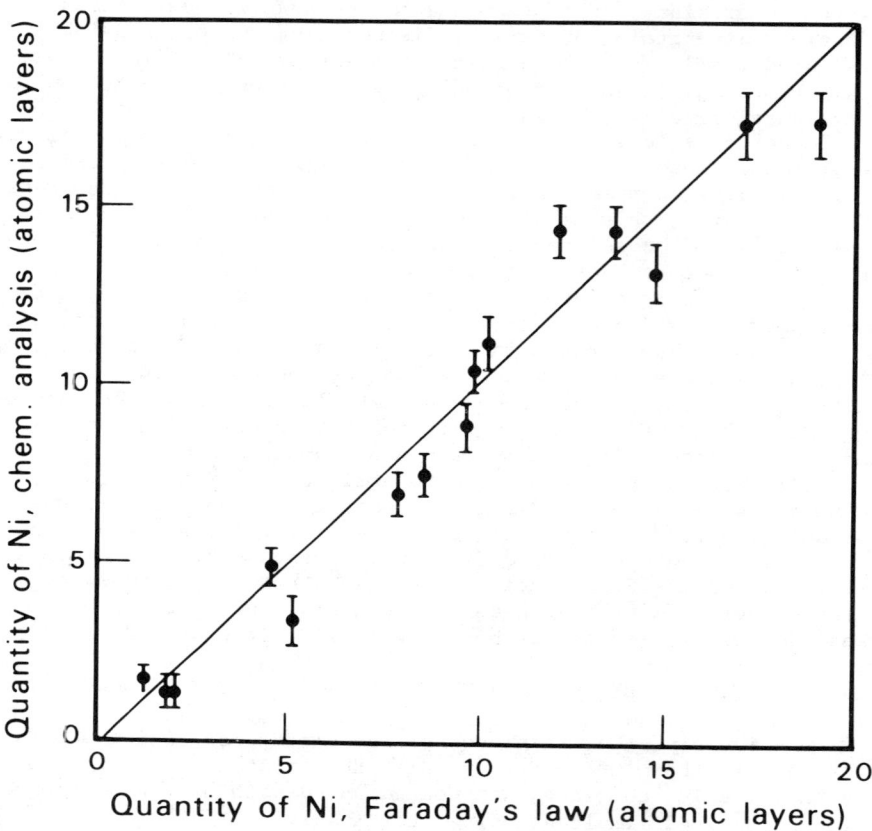

Fig. 2. Analysis of content of thin nickel films by a colorimetric method. Each point represents an individual film whose thickness by quantitative chemical analysis is compared with that obtained from the deposition current.

of these impurities appears negligible.

C. FILM DISCONTINUITIES OR ISLANDS

It is known from experience in vacuum deposition of extremely thin films that the deposited atoms tend to clump in polycrystalline islands rather than deposit uniformly or homogeneously. Islands or clumps, smaller than 100 Å are known to be super-paramagnetic. Hence, dead layers may conceiveably be the result of nonuniform deposition in the initial layers.

Previous work[1] described an electrochemical method for testing

for film uniformity in iron films. This method can be extended to
nickel by utilizing a gold substrate in the following manner: If
gold and nickel are placed in an ionic conducting solution and
electrically connected, nickel will dissolve at a rate which depends
on the area of the gold electrode. Similarly, a thin film of nickel
on a gold substrate will go into solution at a rate which depends
on the area of the gold which is not covered with nickel. Hence,
the time of self-decay of a nickel film electroplated over a gold
substrate is dependent upon the total area of the discontinuities or
breaks in the film exposing the substrate. A quantitative measure
of the area of exposed substrate can be obtained by deliberately
exposing known areas of substrate in the vicinity of the nickel film.
Fig. 3 presents these observations with each datum representing an

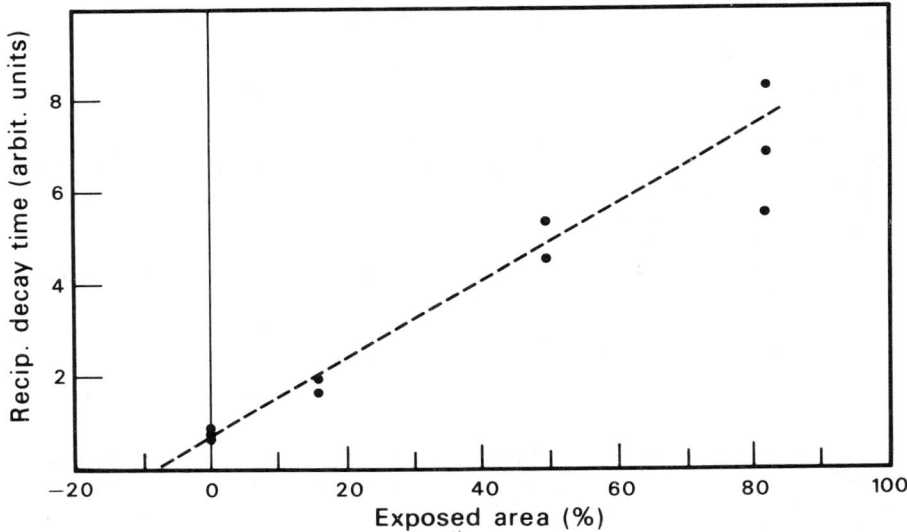

Fig. 3. Data used to analyze the uniformity of deposition of nickel films of two atomic layer thickness. The x-axis intercept indicates a total area uncoated with nickel of 7%, a highly uniform coating.

independent sample of a two atomic layer nickel film deposited on a
gold substrate. Note the linear dependence of the decay rate (reciprocal decay time) on the area of gold exposed. The value for
self-decay lies at the zero of the added area axis. Extrapolating
to zero decay rate gives about 7%. Hence, it can be concluded that,
on the average, a two-layer film has only 7% of the surface uncovered. The electrodeposited film is thus remarkably uniform.

The superior uniformity of these electroplated films compared
to vapor deposition in vacuum arises from the free energy of the
individual atoms incident on the solid surface. Electro-deposited
atoms possess less energy than vapor deposited, upon arrival at the

surface. Paul and co-workers[6] have compared the homogeneity of films deposited electrochemically and by vapor deposition. As might be expected, they found far greater homogeneity and lack of voids in the electrochemically deposited films compared to evaporated films.

D. OXIDATION OF A NICKEL FILM

The possibility of oxidation of a nickel film by dissolved oxygen in the electroplating solution has been considered. Inasmuch as oxidation could occur without electrical current in this reaction, it would be undetectable by the type of analysis of the deposited nickel illustrated in Fig. 2. The oxidation process could occur in either of two ways: a) It could selectively oxidize the first few deposited layers of nickel and then effectively cease, perhaps because of local oxygen exhaustion. b) It could oxidize all deposited layers more or less equally as deposition progresses.

Selective oxidation (of the initial layers) was tested by depositing a film consisting of 10 to 20 layers, stopping deposition for 20 to 30 seconds and again depositing additional layers of nickel. If selective oxidation of the initially deposited layers occurs, additional dead layers should appear after nickel deposition is resumed. No additional dead layers were observed, indicating that selective oxidation is not a factor.

If more or less uniform oxidation occurs, the magnetic properties of the film could be altered, analogous to an alloy of nickel. Whether this "alloy" would exhibit dead layers is an interesting question. Standard external-field spin wave theory of films[7] yields a flux dependence on thickness of the form

$$\phi = M_b D - \alpha$$

where M_b is the bulk magnetization and D is the number of atomic layers in the film. The constant α is equivalent to the number of dead layers. The calculated magnitude of α, according to spin wave theory is much too small to fit observation, but possibly α is anomalously high in the case of oxidized ferromagnets.

In order to test the possibility of uniform oxidation, the electroplating solution was carefully outgassed to remove free oxygen by stirring for twelve hours under vacuum before use. No change in the number of dead layers was observed compared with the use of non-outgassed solution. Thus, it appears doubtful that oxidation of the nickel film is a factor in these experiments.

THERMAL VARIATION OF DEAD LAYERS

The dependence of dead layers on temperature may also shed light on the question of their existence. Because electrodeposition takes place in aqueous solution, it is difficult to produce large temperature changes during the process of formation. However, observations can be made on dead layers after film formation by utilizing a series of films of different but known thicknesses. The magnetization of each individual film then becomes one datum point in the construct of magnetization <u>versus</u> thickness. Hence, each film can be dried and coated for protection in order to study its magnetism at greatly elevated or lowered temperatures.

Observations of magnetization versus thickness from room temperature to 80°K (using the individual films) indicates a decreasing number of dead layers as the temperature is reduced[2]. The reduction reaches a plateau before the lowest temperature is reached, suggesting that two dead layers is the ultimate minimum as $T \to 0$.

As the temperature is raised above room temperature, the number of dead layers increases with increasing temperature. For example, at 240°C the number of dead layers appears to be 10 to 20. Considerable uncertainty arises with elevated temperatures, far more so than at reduced temperatures. The uncertainty at elevated temperatures arises because of severe thermal stresses in the film which affect the magnetization, making reproducible observations difficult.

The more reliable low temperature observations immediately confirm one important fact: Dead layers cannot be explained as the absence of nickel atoms in the film. The revival of two dead layers at low temperatures would preclude this for at least two dead layers. Similarly, the elevated temperature experiments, although less reliable, indicate that the number of dead layers can be increased <u>reversibly</u> with increasing temperatures. In fact, these observations, when extrapolated, imply that a thin film will become non-magnetic because of the proliferation of dead layers before the bulk transition temperature is achieved. It should also be noted that the irreducible minimum of two dead layers at low temperatures is identical with that observed in cobalt and iron at room temperatures. This is consistent with, and to be expected from, the much higher Curie temperatures in the latter ferromagnets.

CONCLUSION

The various tests described above have failed to uncover artifacts associated with the method of observation which might explain dead layers. Admittedly, the phenomenon is exceedingly delicate, involving only a few atomic layers conceivably subject to obscure contamination. However, the observations are precisely reproducible following the given procedures. Additional credence would be given if the phenomenon were observable using widely different procedures, for example, by vacuum evaporation of nickel. Unfortunately, as discussed above, extremely thin evaporated films tend to be non-uniform.

There is no theoretical foundation for skepticism toward the existence of intrinsic dead layers. On the contrary, several recent theories of surface magnetism[2,8,9] predict that the surface layers of nickel are dead.

REFERENCES

* Supported by the U. S. Office of Naval Research.
1. L. N. Liebermann, D. R. Fredkin and H. B. Shore, Phys. Rev. Lett., 22, 539 (1969).
2. L. N. Liebermann, J. Clinton, D. M. Edwards and J. Mathon, Phys. Rev. Lett., 25, 232 (1970).
3. Liquid Bright Gold, type N, obtained from Englehard Industries, Inc., East Newark, N. J.
4. Biquinoline Method obtained from Hach Chemical Co., Ames, Iowa.
5. J. Clinton, unpublished.
6. W. Paul, G. A. N. Connell and N. J. Shevchik, Bulletin of APS, Series II, 16, 347 (1971).
7. I. S. Jacobs and C. P. Bean in Magnetism, edited by G. T. Rado and H. Suhl (Academic Press, N. Y., 1963), Vol. III.
8. K. Levin, A. Liebsch, K. H. Benneman (to be published, Phys. Rev. B).
9. R. E. Watson, P. Fulde, and A. Luther, presented at 18th Annual Conference on Magnetism and Magnetic Materials, Denver, Colo.

A SEARCH FOR MAGNETICALLY DEAD LAYERS IN EVAPORATED IRON FILMS*

J. C. Walker, C. R. Guarnieri, and R. Semper
The Johns Hopkins University, Baltimore, Maryland, 21218

ABSTRACT

Reproducible thin iron films have been made by first producing a single crystal silver substrate by epitaxial growth on cleaved synthetic mica and vacuum depositing Fe^{57} on the silver. The iron deposition takes place in several seconds at pressure less than 10^{-8} torr. The films are immediately overcoated with a layer of silver to prevent oxidation.

Two forms of iron were observed for films of average thickness between 10 Å and 100 Å. One form showed superparamagnetic behavior characteristic of very small crystallites. The other form yielded a Mössbauer spectrum characteristic of ferromagnetic iron with a hyperfine field slightly smaller than that of bulk iron. In most sample preparations either one form or the other was obtained with little tendency for both forms to appear together.

The Mössbauer spectra of both forms were investigated at 293°K, at 77°K, and at 4.2°K. In spite of a sensitivity to non-magnetic iron which was an order of magnitude greater than the anticipated effect, no structure was seen which could be attributed to magnetically "dead" layers.

INTRODUCTION

A primary problem in experimental studies of very thin films (<50 Å thickness) is to assure that the physical nature of the film is well known and reproducible. It is in many cases very important to know whether the film is essentially continuous or made up of discrete platelets or "islands." It is also important to know whether it is made up of large monocrystals or is polycrystalline in nature with very small microcrystallites. To be truly a film it should be approximately atomically flat over a significant area. Not only is it difficult to produce film having definite properties, but it is often even more difficult to determine the properties of the films produced.

THE PRODUCTION OF THIN FILMS

In the work reported here we have used standard techniques for obtaining epitaxial growth of single crystals of silver on cleaved synthetic mica by vacuum deposition.[1,2] Iron enriched to 90% in Fe^{57} was then vacuum deposited on the freshly prepared silver substrate. The sample was overcoated with silver to prevent

---

*Work supported by the National Science Foundation and the U. S. Atomic Energy Commission.

oxidation. The vacuum during silver deposition, iron deposition, and silver overcoating was less than 10^{-8} torr. The techniques used are expected to result in epitaxial growth of the iron on the silver when the silver substrate is single crystalline with the (111) plane parallel to the surface of the mica.[3,4]

We have verified that the silver substrate is, in fact, made of large single crystals by examining a prepared substrate with a highly focussed electron beam. For beam diameters of 5 μ, the scattered electrons produced Laue spots. For defocussed electron beams Debye-Scherrer rings resulted. We therefore can state with certainty that the substrate satisfies the conditions for subsequent epitaxial growth of the iron film. The closest match between the iron and the silver occurs when the (110) plane of the iron is parallel to the plane of the substrate. It is very difficult to make a direct examination of the iron film by electron microscopy because of the silver overcoating. In what follows we are able to infer that in most cases iron films with large single crystals of definite orientation result from these fabrication procedures.

MÖSSBAUER SPECTRA OF THE FILMS

Investigation of the magnetic properties of our iron films has made use of the Mössbauer effect, with the film sample being used as a resonant absorber or scatterer. In most cases the improvement of signal-to-noise ratio possible with a scattering geometry compared to a transmission geometry has dictated the use of resonant scattering for detection of the Mössbauer effect in these very thin samples.[5] A schematic view of the scattering configuration is shown in Figure 1. Because of the large internal conversion coefficient of the 14.4 keV transition in Fe^{57}, the resonantly produced 6.3 keV x-rays of the scatterer were detected instead of the scattered 14.4 keV gamma rays.

The resulting spectra from a large number of film samples were found to fall into two categories. One group showed hyperfine structure resembling that of bulk iron, another group showed a doublet structure. The two groups are represented by examples in Figures 2 and 3. There was only a very slight tendency toward composite spectra, i.e., showing both features (see the top spectrum in Fig. 2, and the bottom spectrum of Fig. 3). Thickness measurements were made using standard x-ray fluorescence techniques and were checked approximately by using the relative strength of the Mössbauer scattering. Although there is a tendency for the thinner films to show doublets and the thicker ones to show normal ferromagnetic spectra, it is seen that either form can result for any thickness between about 10 Å and 40 Å. With

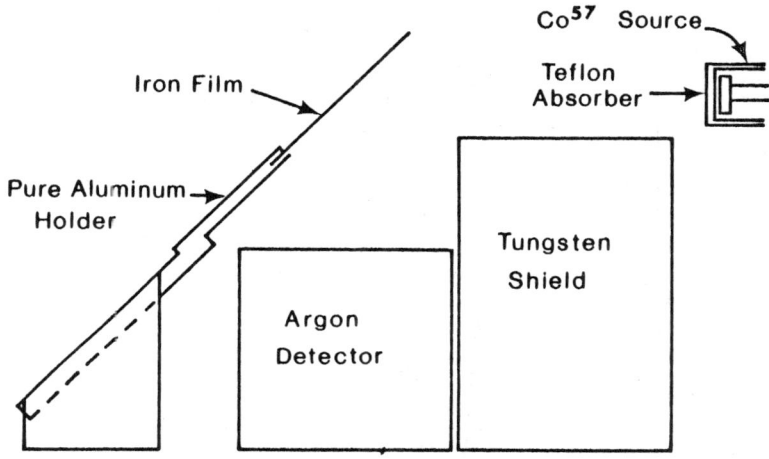

Fig. 1 Mössbauer Scattering Geometry

present techniques only about 40% of the films yield doublets.

A large number of observations of thin film spectra support the following conclusions. The magnetic spectra result from approximately continuous films with large oriented single crystals resulting from epitaxial growth of the iron on the silver substrate. The doublet spectra are characteristic of non-continuous films with small crystallites in the form of islands. These microcrystallites show spectra characteristic of superparamagnetic relaxation and yield room temperature spectra in the form of doublets. The precise reason for narrow doublets rather than singlets is not clear, but an electric quadrupole interaction is suspected. Supporting these conclusions is the following evidence:

(1) The films with magnetic spectra (Fig. 2) show direct evidence of magnetization in the plane of the film. This can be seen by examination of the relative

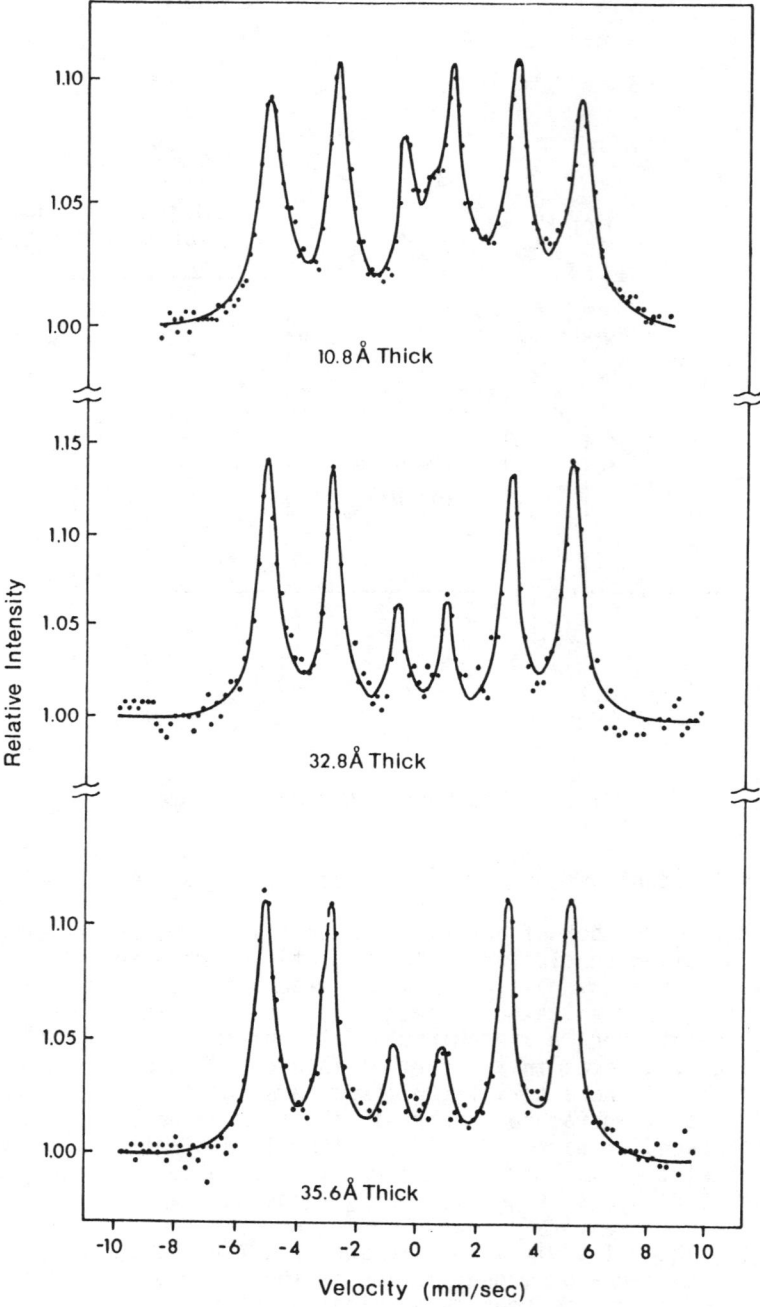

Fig. 2 Spectra for 10.8 Å, 32.8 Å, and 35.6 Å Films

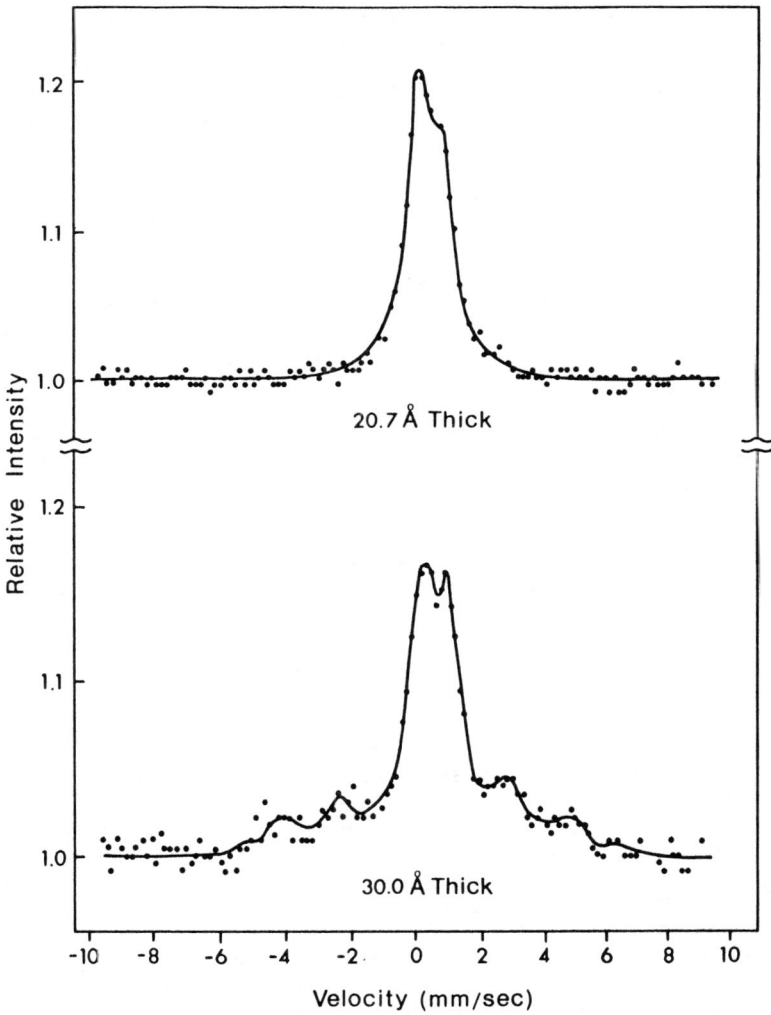

Fig. 3 Spectra for 20.7 Å and 30.0 Å Films

intensities of the various lines in the hyperfine spectrum.[6] These relative intensities change as the sample is rotated in the scattering position about an axis perpendicular to the plane of the film, indicating that the magnetization has a unique direction in the film plane. This would not be expected for a non-continuous film of small crystallites.

(2) The doublet spectra split out as the temperature of the films (Fig. 3) is lowered to 4.2°K, but the resulting hyperfine field is considerably different from bulk iron values. There is no evidence of a preferred direction of magnetization in the plane of the film. This is indicated by the relative intensity of the lines in the spectrum.

(3) Most theories of the magnetization of thin ferromagnetic films indicate a decrease in Curie temperature for films less than 30 Å.[7,8] Our ferromagnetic films show hyperfine fields which are 4 to 6% smaller than bulk values at room temperature, but which equal bulk values at 4.2°K. There is, of course, the fact that it has been previously shown[3,4] that epitaxial growth of iron or silver can be expected under the conditions of our film production.

A SEARCH FOR MAGNETICALLY "DEAD" LAYERS

It can be seen in Figure 2 that we have obtained a ferromagnetic film with average thickness as small as 10.8 Å which is approximately 6 atomic layers. Such a film should show definite evidence of the magnetically "dead" layer phenomenon seen by Liebermann et al.[9,10] For iron which is not magnetically ordered and, indeed, in which the magnetic moment is somehow "quenched" one would expect a large singlet feature in the Mössbauer spectrum. A collapsed hyperfine spectrum should result in a line four times more intense than the most prominent line in a normal hyperfine spectrum for equal amounts of iron. One would surely expect no significant magnetic hyperfine structure from magnetically "dead" iron[6] for two reasons: first, if the atomic moment was not in a magnetically ordered system, relaxation effects would result in the nuclear moment seeing no net field; secondly, if the magnetic moment were somehow quenched the core polarization term which dominates the magnetic hyperfine interaction would be absent. It is possible that a small amount of conduction electron polarization arising from nearby ferromagnetic iron might produce a very small magnetic hyperfine splitting, but this would surely not exceed 10% of the normal hyperfine field[11] and would not affect the conclusion that an effectively unsplit line in the spectrum would be expected for magnetically "dead" layers.

A careful search for such singlet features in those films showing ferromagnetic hyperfine structures has been made at 293°K, at 77°K and at 4.2°K and no evidence for any unusual hyperfine structure has been found. The small amount of anomalous structure seen near the center of the 10.8 Å thick film in Fig. 2 is no doubt the result of some very small amount of superparamagnetic iron as features like this disappear as the films are cooled. In any event, it is much too small to be the result of two magnetic dead layers. Indeed, one would expect the spectrum to be dominated by the singlet feature if magnetic dead layers were present.

Table I shows the expected intensity of a singlet line arising from two magnetically dead layers for a series of ferromagnetic films of various thicknesses. The intensity of the singlet is calculated assuming that it would show no hyperfine splitting greater than 10% of that of normal iron. The numbers in the last column show the ratio of the expected intensity to that of the outer line in the spectrum of normal ferromagnetic iron. When least squares fits to the spectra were performed, no such features were found.

CONCLUSION

From our work we can only conclude that the Mössbauer spectra of our ferromagnetic thin iron films show no evidence of magnetic "dead" layers. The careful preparation techniques and, indeed, the Mössbauer spectra themselves indicate that approximately continuous films of non-oxidized metal consisting of large single crystals have been produced. We, therefore, do not believe that the failure to observe "dead" layer features in our spectra is a result of the nature of the films. It is interesting to note that related experiments have been carried out by Varma and Hoffman[1,2,3] in which a thin film radioactive source was produced under ultra high vacuum conditions. These source films were analyzed using the Mössbauer effect without overcoating of the films, and without removing them from vacuum. The results of these experiments are substantially in agreement with the results reported here.

REFERENCES

1. D. W. Pashley, Phil. Mag. **4**, 316 (1959).
2. H. Jaeger, P. D. Mercer, and R. G. Sherwood, Surface Science **11**, 265 (1968).
3. R. Cinti, J. Devenyi, P. Escudier, R. Montmory, and A. Yelon, C. R. Acad. Sc. Paris **260**, 6849 (1965).
4. L. S. Palatnik and Y. F. Komnik, Soviet Phys. "Doklady" English Transl. **5**, 1072 (1960).

5. B. Cleveland, Dissertation (The Johns Hopkins University, 1970) (unpublished).
6. C. R. Guarnieri, Dissertation (The Johns Hopkins University, 1972) (unpublished).
7. L. Valenta, Czech. J. Phys. $\underline{7}$, 133 (1957).
8. W. Brodkorb and W. Haubenreisser, Phys. Stat. Sol. $\underline{16}$, 577 (1966).
9. L. N. Liebermann, D. R. Fredkin, and H. B. Shore, Phys. Rev. Lett. $\underline{22}$, 539 (1969).
10. L. Liebermann, J. Clinton, D. M. Edwards, and J. Mathon, Phys. Rev. Lett. $\underline{25}$, 232 (1970).
11. K. J. Duff and T. P. Das, Phys. Rev. $\underline{B3}$, 2294 (1971).
12. Varma and R. W. Hoffman, J. App. Phys. $\underline{42}$, 1727 (1971).
13. Varma and R. W. Hoffman, J. Vac Sci Tech $\underline{9}$, 177 (1972).

Table I: Calculated intensities of spectral lines from non-magnetic surface layers.

Film Thickness (Å)	Number of Layers	Relative Height of Non-magnetic Line to Outer Line of Ferromagnetic Spectrum
10.8±1.1	6.3± .6	3.0 ±.4
32.8±3.6	17.2±1.8	.6 ±.1
35.6±3.3	18.6±1.7	.6 ±.1
40.0±3.6	20.7±1.9	.55±.1
42.7±3.7	22.1±1.9	.5 ±.1
45.7±3.7	23.6±1.9	.46±.06
52.6±4.1	27.0±2.1	.39±.04
62.8±4.8	32.1±2.4	.32±.03

TEMPERATURE DEPENDENCE OF MAGNETIC EXCITONS IN SINGLET-TRIPLET SYSTEMS

Richard Silberglitt
National Science Foundation, Washington, D.C.

Earl Callen
The American University, Washington, D.C.

James Cullen
Naval Ordnance Laboratory, Silver Spring, Maryland

ABSTRACT

Recent observations[1] indicate that the magnetic exciton energies in the singlet ground state systems fcc Pr and Pr_3Tl are nearly temperature independent up to several times the ferromagnetic ordering temperature. Since the lowest excited crystal field state is a triplet, assuming that higher levels can be ignored one can describe these systems in terms of two pseudospins 1/2 on each site[2]. We have investigated the temperature dependence of the exciton energies using a decoupling scheme for the pseudospin equations of motion which consistently retains pair correlations, including those between pseudospin operators on adjacent sites. Numerical calculations have been performed using this scheme and a boson formalism for the correlation functions. The results are in agreement with experiment in that the exciton energies do not show any appreciable temperature dependence up to about twice the ordering temperature. The physical reason for this behavior is that $k_B T_c$ is small compared to a typical exciton energy, so that only very long wavelength modes are populated until $T \gg T_c$. In this respect the singlet-triplet system is analogous to lower dimensional magnetic systems, where $k_B T_c$ is small compared to the maximum magnon energy because of the weakness of the coupling in one or more directions.

1. R. J. Birgeneau, J. Als-Nielsen, and E. Bucher, Phys. Rev. B **6**, 2724 (1972).
2. Y. Y. Hsieh and M. Blume, Phys. Rev. B **6**, 2684 (1972).

DIRECT MEASUREMENT OF CRYSTAL-FIELD SPLITTINGS IN PrN*

H. L. Davis and H. A. Mook
Solid State Division, Oak Ridge National Laboratory
Oak Ridge, Tennessee, 37830

ABSTRACT

Inelastic neutron scattering has been employed to measure the crystal-field splittings of the $^3H_4$ ground multiplet corresponding to the Pr^{3+} ion in PrN. An analysis of the resulting data places the Γ_4 crystal-field level at 27.0±1.0 meV, the Γ_3 at 46.3±1.0 meV, and the Γ_5 at 91.±2. meV, above the ground Γ_1. These results can be quantitatively accounted for by a point-charge model with a charge of -3, which sharply contrasts with previous work of others that established the crystal-field levels in other PrX compounds (X=P,As, Sb,Bi,S,Se,Te) could be accounted for by point charges of -2. This difference between PrN and the other PrX compounds strongly implies some major difference in their electronic structures, which may indicate their being, respectively, semiconducting and metallic.

INTRODUCTION

The compound PrN belongs to the large class of NaCl-structured rare-earth monopnictides (N,P,As,Sb,Bi) and monochalcogenides (S,Se, Te). Generally, this class contains a diversity of different magnetic behaviors due to the presence of the rare-earths' 4f electrons, with some compounds being ferromagnetic, others ferrimagnetic, while others are antiferromagnetic. But as is the case for the Pr compounds, not all the compounds are found to undergo magnetic ordering, presumably due to crystal-field effects separating off a nonmagnetic ground level. Be it as it may, irregardless of whether a given compound undergoes magnetic ordering or not, its magnetic properties are profoundly affected by crystal-field effects. Thus, some quantitative information regarding the crystal-field effects in these compounds is important for understanding their magnetic properties.

Since the vast majority of the compounds in the above class have a basically metallic behavior, the standard photon spectroscopy methods, which have been useful for crystal-field studies in insulators, are of little value for the rare-earth compounds. Thus, most previous attempts at extracting crystal-field information for these compounds has been concerned with model analyses of "broad-band-spectral" data, such as specific-heat and/or susceptibility data. A markedly promising exception to the broad-band methods has been the application of neutron spectroscopy. For example, such techniques have been applied to all the PrX compounds except X=N by Turberfield *et al.*[1] (TPBB) and to TmSb by Birgeneau *et al.*[2]

---
*Research sponsored by the U. S. Atomic Energy Commission under contract with Union Carbide Corporation.

For the considered rare-earth compounds, the rare-earth site-symmetry is O_h. Thus, following the notation of Lea et al.[3] (LLW), the crystal-field Hamiltonian may be written

$$H = W\{[x/F(4)]O_4 + [(1-|x|)/F(6)]O_6\} , \qquad (1)$$

with W being an overall splitting factor and x related to the ratio of fourth-and sixth-order terms. From LLW it is seen the ninefold-degenerate $^3H_4$, the ground J-multiplet of Pr^{3+}, will split into two triplets Γ_5 and Γ_4, one doublet Γ_3, and one singlet Γ_1. The energetic ordering of these Γ_i depends on the x value representative of the considered solid; however, a simple point-charge model indicates for PrN they would order, with increasing energy, in the sequence Γ_1 Γ_4, Γ_3, and Γ_5. Indeed, this is the ordering found by TPBB on the Pr compounds they investigated, and it is the order reasonably expected for PrN. At the same time, bulk magnetic data does indicate Γ_1 is the ground level for PrN, but a disagreement exists concerning the W and x values which best reproduce PrN's susceptibility and specific heat data. For example, the work of Junod et al.[4] gives x = -0.8 and W = 0.62 meV, while Stutius[5] estimates x = -1.0 and W = 1.76 meV. Thus, application of neutron spectroscopy to PrN should provide information concerning which values of x and W are more realistic.

The cross-section for scattering neutrons from a single J-multiplet follows the proportionality[6]

$$(\partial^2\sigma/\partial\Omega\partial\omega) \sim (k_f/k_i)F^2(\vec{Q}) \sum_{nm} \rho_n |\langle n|J_\perp|m\rangle|^2 \delta(\varepsilon_{nm} - \hbar\omega) , \qquad (2)$$

which is valid for small momentum transfers. The states $|n\rangle$ and $|m\rangle$ are crystal-field levels having an energy difference ε_{nm}, with the delta-function indicating the scattering neutron will undergo a gain or loss, $\hbar\omega$, in its kinetic energy depending on the sign of ε_{nm}. Also, $J_\perp$ is the component of $\vec{J} \perp$ to the scattering vector $\vec{Q}$, F is the form factor, and k_i and k_f are the momentums of the incident and scattered neutrons. It is important to note, to first order, the relative scattering intensities for the allowed $\hbar\omega$ are proportional to $\rho_n|\langle n|J_\perp|m\rangle|^2$, with ρ_n being a Boltzmann population factor. This proportionality provides a valuable tool for interpreting the neutron data, since any resulting interpretation must provide consistency with any observed changes of intensity with temperature and allowed $\hbar\omega$. In this aspect, the recent documentation of the $|\langle n|J_\perp|m\rangle|^2$ for the LLW Hamiltonian provided by Birgeneau[7] is a valuable tool in the preliminary analysis of neutron crystal-field data.

RESULTS AND DISCUSSION

Our experiments used the Oak Ridge magnetically-pulsed time-of-flight spectrometer,[8] which is capable of producing pulsed monochromatic neutron beams over a wide energy range. For example, in our study of PrN we have utilized beams of energies 13.1, 36.7, 69.8, and 91.6 meV. Then, by using cross-correlation applied between the pulsing signal and the time distribution of neutrons arriving at a given detector, the desired data relating scattered neutron counts vs. time-of-flight channel (or energy) is readily obtained.

The PrN sample consisted of about 3 cc. of powder kindly furnished by D. E. LaValle.[9] Helium temperature data when using 36.7 meV incident neutrons are shown in Fig. 1. The energies plotted in this and later figures correspond to the average times of the time channels, with a positive (negative) energy representing a neutron gaining (losing) that energy. Also, due to the (k_f/k_i) factor of Eq. (2) and the channels' energy-widths being proportional to $(k_f/k_i)^3$, the results of Fig. 1 represent raw data after multiplication by a factor $(k_i/k_f)^4$ and being averaged over the resolution width of the spectrometer. Results obtained at room temperature, using an incident neutron energy of 13.1 meV, are displayed in Fig. 2(a) where raw data is plotted.

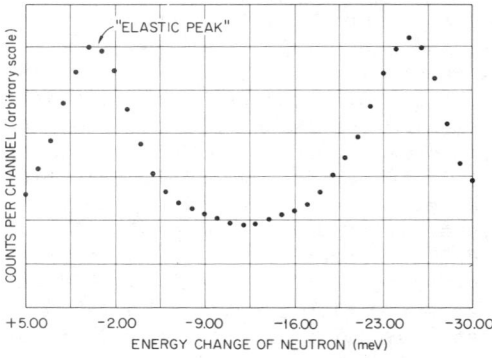

Fig. 1. Helium Temperature Data.

The helium temperature results of Fig. 1 are easily and directly analyzed, since only excitation transitions from the ground Γ_1 will be possible. Also, because the only allowed transition with Γ_1 involves Γ_4, the energy loss peak in the vicinity of 26 meV corresponds to this transition. However, the $\Gamma_1 \to \Gamma_4$ transition energy will be slightly greater than the energy of the peak maximum due to the $F^2(\vec{Q})$ factor in Eq. (2). Two energy gain transitions are contained in the data of Fig. 2(a), with its peak maximum corresponding to the $\Gamma_4 \to \Gamma_1$ transition. At the same time, there is a shoulder on the left of the major peak in Fig. 2(a). Since the LLW results show the ratio $(\Gamma_4 \to \Gamma_1)/(\Gamma_3 \to \Gamma_4)$ is independent of x and equal to 7/5, it is entirely reasonable to assign this shoulder to de-excitation from the Γ_3 to Γ_4.

To test the above assignments, we have written a computer program based on Eqs. (1) and (2). The procedures used were similar to, but not identical to, the ones described by TPBB, and basically involve varying parameters to obtain a "best fit". The parameters used are W and x of Eq. (1), and a level width to introduce an assumed gaussian smearing of the Γ_i levels. For a given set of these parameters the program calculates the cross-section of Eq. (2) as a function of $\hbar\omega$, which is then convoluted with the measured instrumental resolution which varies with $\hbar\omega$. Integrating these results over the channels' energy-widths produces relative values of theoretical counts per channel, and these can be compared with experimental counts after subtraction of a parametric background value which is assumed constant for each considered channel. The results of using the program to obtain the best rms deviation to some of the data of Fig. 2(a) is shown there. The rms deviation between the calculated and experimental points of Fig. 2(a) is less than the statistical counting error. Also, the fit illustrated by Fig. 2(a) leads to the energies quoted for the Γ_4 and Γ_3 levels in Fig. 2(b), which should be considered to have experimental errors of, roughly, ± 1.0 meV.

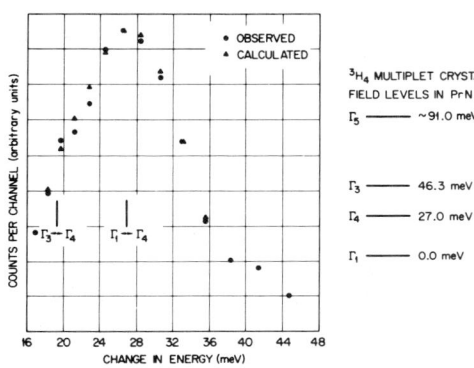

Fig. 2. (a). Room temperature results using 13.1 meV incident neutrons. (b). Energy-level diagram from our total analysis.

The data of Figs. 1 and 2 are insufficient to place the Γ_5 level, which has transitions only with the Γ_3 and Γ_4. To place the Γ_5 we have mainly used room temperature data obtained using 69.8 meV incident neutrons. These results, after multiplication by $(k_i/k_f)^4$ and averaged over the resolution width, are displayed in Fig. 3. The pertinent information is the small peak corresponding to neutrons losing $\simeq$ 45 meV. Since $\rho_n|<n|\vec{J}_\perp|m>|^2$ is about 1.0 for the $\Gamma_3 \to \Gamma_5$ transition and about 12.9 for the sum of the $\Gamma_1 \to \Gamma_4$ and $\Gamma_4 \to \Gamma_3$ transitions, the relative intensity of the small peak is semi-quantitatively consistent with placing the Γ_5 at 91±2 meV. Then in an attempt to observe the expected $\Gamma_4 \to \Gamma_5$ transition expected at $\simeq$ 64 meV, we also collected data using 91.6 meV incident neutrons. Although the decreased resolution present in this last data has prevented isolation of any $\Gamma_4 \to \Gamma_5$ peak, it does contain a strong shoulder-type indication corresponding to the neutrons losing $\simeq$ 65 meV which is qualitatively consistent with the Γ_5 level placement indicated in Fig. 2(b).

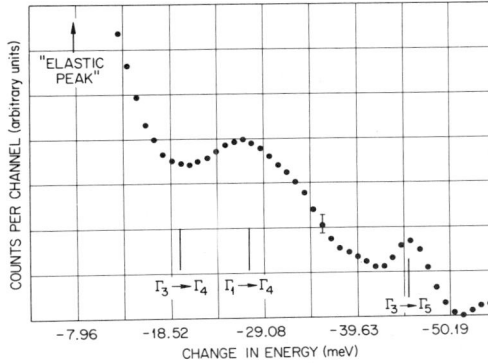

Fig. 3. Room temperature data using 69.8 meV incident neutrons.

The energy values of Fig. 2(b) may be generated, within the quoted error limits, by use of LLW parameters x = -0.97 and W = 1.68 meV. These direct neutron spectroscopy values are quite close to the values estimated by Stutius[5] of x = -1.0 and W = 1.76 meV from specific heat data; thus, we conclude the values x = -0.8 and W = 0.62 estimated by Junod et al[4] do not realistically reflect the actual crystal-field of PrN. At the same time, our value of 27.0±1.0 meV for the Γ_4

is entirely consistent with the value 27.2 meV tabularly quoted by Bucher et al.[10] as having been obtained by neutron spectroscopy. Although no details of their data have been reported, it is our impression Bucher et al. were only able to detect the Γ_4 level.

The W and x values we have obtained for PrN are surprisingly close to the values x = -0.96 and W = 1.69 meV, which would be predicted by a nearest-neighbor point-charge model with charges of -3. To obtain this result we have used the nonrelativistic Hartree-Fock radial integrals $<r^n>$ tabulated by Freeman and Watson[11] for Pr^{3+}. Although relativistic effects can be very significant in the $<r^n>$ integrals, we have used the Freeman and Watson values in order to easily compare our result with the point-charge analysis of TPBB on the other Pr compounds since they used the same $<r^n>$ values. That is, the main point we want to make here is not that our PrN results have an apparent agreement with a point-charge model, which they would not if relativistic $<r^n>$ were used, but that the PrN results do not follow the same trend as found for the other PrX compounds. This is immediately seen from the fact that TPBB found that the same point-charge model we have used to obtain a charge of -3 for PrN requires a charge of -2 for PrX (x=P,As,Sb,Bi,S,Se,Te). Thus, we conclude that this difference in model charge value implies some major difference between the overall electronic structure of PrN and that of the other PrX compounds. We suggest this difference is that PrN is indeed an intrinsic semiconductor as has been implied by the work, e.g., of Sclar,[12] while the other PrX compounds are intrinsically metallic.

REFERENCES

1. K. C. Turberfield, L. Passell, R. J. Birgeneau and E. Bucher, J. Appl. Phys. **42**, 1746 (1971).
2. R. J. Birgeneau, E. Bucher, L. Passell and K. C. Turberfield, Phys. Rev. B **4**, 718 (1971).
3. K. R. Lea, M. J. M. Leask and W. P. Wolf, J. Phys. Chem. Solids **23**, 1381 (1962).
4. P. Junod, A. Menth and O. Vogt, Phys. Kondens. Materie **8**, 323 (1969).
5. W. Stutius, Phys. Kondens. Materie **10**, 152 (1969).
6. P. deGennes, Magnetism, G. T. Rado and H. Suhl, eds. (Academic Press, New York, 1963), Vol. 3, p. 115.
7. R. J. Birgeneau, J. Phys. Chem. Solids **33**, 59 (1972).
8. H. A. Mook and M. K. Wilkinson, p. 173 in Proceedings on Instrumentation for Neutron Inelastic Scattering Research, Vol. III, IAEA, Vienna, 1970.
9. D. E. LaValle, Analytical Chemistry Division, ORNL.
10. E. Bucher, K. Andres, J. P. Maita, A. S. Cooper and L. D. Longinotti, J. Phys. (Paris) **32**, 114 (1971).
11. A. J. Freeman and R. E. Watson, Phys. Rev. **127**, 2058 (1962).
12. N. Sclar, J. Appl. Phys. **33**, 2999 (1962); **35**, 1534 (1964).

EFFECTIVE FIELD APPROXIMATION FOR RARE EARTH IONIC COMPOUNDS DETERMINED FROM NEUTRON SCATTERING EXPERIMENTS

A. Furrer and H. Heer
Delegation AF, Swiss Federal Institute for Reactor Research
CH-5303 Wuerenlingen, Switzerland

ABSTRACT

Because of the well-localized 4f moment, the exchange interaction in rare earth ionic compounds can be described in terms of an effective field theory. It is shown that the effective field parameters are completely determined by combining the results obtained from neutron diffraction in the ordered state and from neutron crystal field spectroscopy in the paramagnetic state. This is exemplified for the cerium monopnictides in the molecular field approximation. To determine the molecular field in the ordered state we used the measured values of the magnetic moment[1,2] at temperatures $T < T_N$. In the paramagnetic state the fluctuation of the total angular momentum $\vec{J}$ gives rise to a fluctuating molecular field whose mean value is zero. Information about this effect (which we call the dynamical Zeeman effect) is contained either in the width or in the intensity of the crystal field transition lines depending on whether the crystal field interaction or the exchange interaction is dominant. The experimental susceptibility data[3-5] are well accounted for by using our earlier crystal field[6-7] and the present molecular field results without any further approximations (in a similar investigation performed by Wang and Cooper[8] details about the magnetic structure and the crystal field splitting in most of the compounds were not yet available). In particular, the anomalous susceptibility behaviour in CeSb can be quantitatively explained.

REFERENCES

1. B. Lebech, P. Fischer, and B. D. Rainford, Conf. Digest No. **3**, Rare Earths and Actinides, Durham (Institute of Physics, London, 1971), p. 204.
2. P. Fischer, private communication.
3. T. Tsuchida and W. E. Wallace, J. Chem. Phys. **43**, 2087, 2885 (1965).
4. G. Busch and O. Vogt, Phys. Letters **20**, 152 (1966).
5. M. Landolt, private communication.
6. B. R. Cooper, A. Furrer, W. Buehrer, and O. Vogt, Solid State Commun. **11**, 21 (1972).
7. A. Furrer, W. Buehrer, H. Heer, W. Haelg, J. Benes, and O. Vogt, to be published in Neutron Inelastic Scattering (IAEA, Vienna, 1972).
8. Y. L. Wang and B. R. Cooper, Phys. Rev. **B2**, 2607 (1970).

SUBLATTICE MAGNETIZATION AND THE EXCHANGE
INTERACTION IN $Tb_zY_{1-z}Sb$

J.W. Cable
Oak Ridge National Lab., Oak Ridge, Tenn. 37830
J.B. Comly, B.R. Cooper, and I.S. Jacobs
General Electric Res. & Dev. Ctr., Schenectady, N.Y. 12301
W.C. Koehler
Oak Ridge National Lab., Oak Ridge, Tenn. 37830
O. Vogt
ETH, Zurich, Switzerland

ABSTRACT

Elastic neutron scattering has been used to measure the sublattice magnetization of $Tb_zY_{1-z}Sb$ for $z=1$, 0.91, 0.66, and 0.64. The moment/Tb at $0°K$ are 7.83, 7.57, 7.43, and 7.07 μ_B; and the T_N are 15.2, 13.4, 8.85, and 8.5°K respectively. These T_N are quite close to those found in the same samples by susceptibility measurements, also used to fix the concentrations. The sublattice magnetizations and T_N are in good agreement with the behavior expected for an exchange field (H_{eff}) proportional to z and varying linearly with the moment/Tb (M). The linear exchange field coefficient (λ) is about 15% larger than that found in earlier bulk magnetic measurements in the paramagnetic regime, $z \leq 0.403$; and in contrast to the earlier work there is little, if any, need for contributions to H_{eff} with higher order dependence on M.

INTRODUCTION

There has been much interest in the magnetic properties of $Tb_zY_{1-z}Sb$ for studying the magnetization and antiferromagnetic ordering process in a crystal-field singlet ground state system.[1,2] The most detailed studies[2,3] have been for $z \lesssim 0.4$ where the alloys are paramagnetic down to $0°K$. Susceptibility studies[2] verified the correctness of the induced magnetism picture for the magnetization process and approach to magnetic ordering in singlet-ground-state systems. High field magnetization experiments[2] in the regime $z \leq 0.403$ suggested the presence of higher degree exchange interaction; and such experiments were used subsequently[3] to measure these higher degree exchange contributions.

The question then arises as to whether the quantitative picture developed for the magnetic behavior of the system $Tb_zY_{1-z}Sb$ in the paramagnetic regime, $z \leq 0.4$, can be carried over to the antiferromagnetic regime, $z \geq 0.4$. (The only comparison of theory and experiment for the regime $z > 0.403$ in Ref.2 was for T_N vs z.) In particular, is there evidence for any change in the values of exchange parameters or for the presence of higher degree exchange in the antiferromagnetic regime? With these questions in mind we have measured the sublattice magnetization of $Tb_zY_{1-z}Sb$, and relate these results to the previous[2,3] study of bulk magnetic properties in the paramagnetic regime.

EXPERIMENTAL PROCEDURE AND RESULTS

The neutron measurements were made at $\lambda=1.07$Å on pillar-shaped single crystal specimens with typical dimensions of 1 x 2 x 3 mm. Tb moment values were obtained from the ratio of the (311) magnetic to the (200) nuclear intensities by assuming the Tb^{3+} free atom form factor, moments parallel to the $\langle 111 \rangle$ direction in each domain (the magnetic structure for the alloys, as for pure[4] TbSb, is MnO-type, i.e. $\pi\pi\pi$-type), and nuclear scattering amplitudes of $b_{Tb}=0.76$, $b_Y=0.79$, and $b_{Sb}=0.54 \times 10^{-12}$ cm. The magnetic intensities were corrected for extinction by use of an effective linear absorption coefficient taken from the pathlength variation of the (200) nuclear intensity. The correction was small in all cases (about 4% in μ_B/Tb) and was applied only to the saturation moment values - not to the relative magnetization-vs-T data. (For the small z=0.64 crystal used to determine the saturation moment such a correction was unnecessary.) The T_N were taken from I-vs-T data on the (311) magnetic reflections.

Susceptibility data for $40°K \leq T \leq 240°K$ fit well to a Curie-Weiss Law, and provide an evaluation of z from the free ion behavior. The peak value of susceptibility provides an additional measurement of the variation of T_N with z, to be compared to the earlier measurements[2] by the same technique and to the present determination by neutron scattering.

Table 1: Tb concentrations, z, investigated and Neel temperatures, T_N.

z	T_N (°K) susceptibility	T_N (°K) neutron
1	15.4	15.2
0.91	13.6	13.4
0.66	9.8	8.85
0.64	9.2	8.5

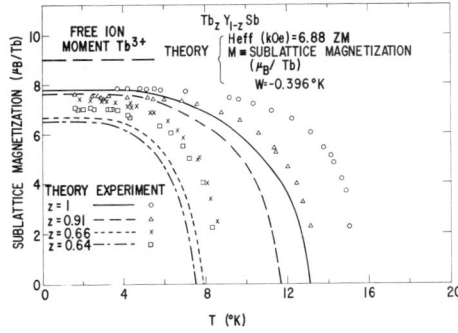

Fig.1 Sublattice moment/Tb vs T compared to linear molecular-field theory for $\lambda/z=6.88$ (kOe/μ_B).

The experimental temperature variation of the sublattice moment/Tb is shown in Fig.1.

ANALYSIS AND DISCUSSION

The experimental decrease of moment/Tb at 0°K and of T_N with decreasing z agree with the behavior expected for a singlet-ground-state system[1].

In Ref.(2) the susceptibility behavior for $z \leq 0.403$ was fit taking

$$H_{eff} = H + \lambda M \qquad (1)$$

with $\lambda/z = -6.97$ (kOe/μ_B) (M is the moment/Tb), and taking the crystal-field parameter W = -0.396°K, corresponding to a splitting of 11.9°K from the Γ_1 singlet ground state to the Γ_4 first excited state. [Analysis of subsequent[3] high field magnetization measurements yielded the slightly different value $\lambda/z = -6.88$ (kOe/μ_B).]

The nearest-neighbor ($\mathcal{J}_1$) and next-nearest-neighbor ($\mathcal{J}_2$) exchange parameters are lumped together differently to give λ for the paramagnetic and antiferromagnetic regimes.[2]

paramagnetic, $\qquad\qquad \lambda/z = 2(12\mathcal{J}_1 + 6\mathcal{J}_2)/g^2\mu_B^2 \qquad\qquad (2)$

MnO-type ordering, $\qquad\qquad \lambda/z = -2(6\mathcal{J}_2)/g^2\mu_B^2 \qquad\qquad (3)$

In Ref.2 the critical value of z (0.403) for the appearance of antiferromagnetism at T = 0 was correctly predicted by taking λ(antiferromagnetic) = $-\lambda$(paramagnetic), implying the $\mathcal{J}_1$ is negligible compared to $\mathcal{J}_2$; and this λ was used to find the variation of T_N for z>0.403. The same model gives values for the sublattice moment/Tb, M, shown in Fig.1. For W = -0.396°K, at a specified T one obtains a nonlinear M vs H_{eff} curve.[5] The intercept of this with the straight line given by (1) (with H=0) gives M.

In Ref.3 one obtained a significantly better fit to the experimental high field anisotropic magnetization data for $z \leq 0.402$ using

$$H_{eff} = H + \lambda M + DM^3 \qquad\qquad (4)$$

with $\lambda/z = -6.88$ (kOe/μ_B) and D = -0.21 (kOe/μ_B^3) than without the term in M^3. As shown in Fig.2, for these values of λ and D (i.e. changing signs), the antiferromagnetic transition would be first order as indicated by the double valued magnetization curves given by the theory. [For $\lambda/z=6.88$ (kOe/μ_B), the antiferromagnetic transition becomes first order for D/z slightly larger than 0.04 (kOe/μ_B^3)].

The lack of agreement of the experimental and theoretical M vs T behavior in Fig.1 corresponds to the fact that the T_N vs z theoretical curve in Fig.6 of Ref.2 falls below the experimental values of T_N at the high z end. In any case, the theoretical behavior shown in Fig.2 is in such qualitative disagreement with the experimental behavior as to eliminate any term $\sim DM^3$ in H_{eff} of the magnitude of that indicated by the paramagnetic

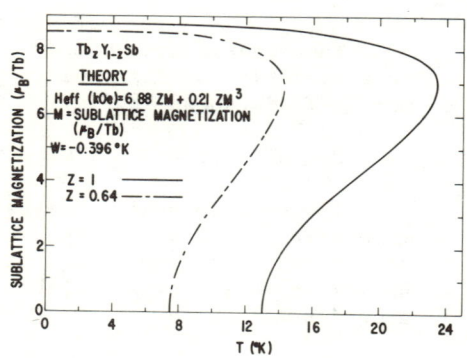

Fig.2 Theoretical sublattic moment/Tb vs T for molecular-field theory with $\lambda/z = 6.88$ (kOe/μ_B) and D/z = 0.21 (kOe/μ_B^3).

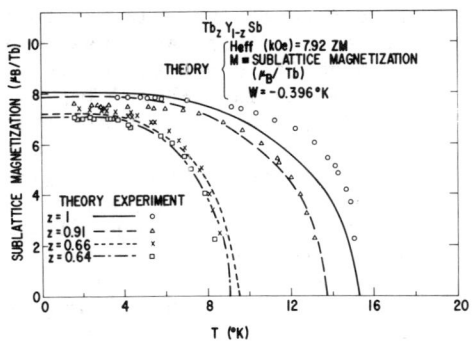

Fig.3 Sublattice moment/Tb vs T compared to linear molecular-field theory for $\lambda/z=7.92$ (kOe/μ_B).

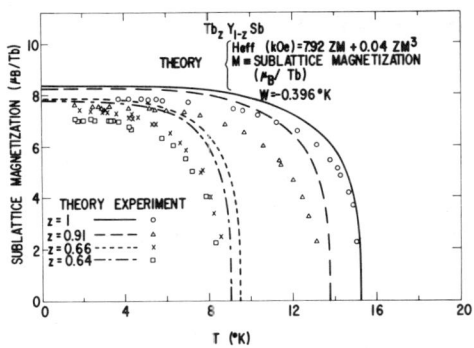

Fig.4 Sublattice moment/Tb vs T compared to molecular-field theory for $\lambda/z=7.92$ (kOe/μ_B) and $D/z=0.04$ (kOe/μ_B^3).

magnetization experiments.[3] On the other hand, increasing λ to match the experimental T_N for pure TbSb, and using an H_{eff} linear in M, as shown in Fig.3 gives reasonably good agreement with the overall experimental behavior. This, in fact, gives better overall agreement than the "squarer" sublattice moment/Tb curves shown in Fig.4, obtained by including a small DM^3 term in H_{eff}.

The presence of a small, positive $\mathcal{J}_1$ ($\mathcal{J}_1=0.007°K$, compared to $\mathcal{J}_2=-0.10°K$) could explain the difference between the paramagnetic λ and the λ required to fit the T_N and sublattice moment/Tb for large z values; however, this does not explain why the paramagnetic λ gives good agreement in Ref.2 for T_N for z values up to $z \approx 0.8$.

As indicated by Fig.5, there may be a question as to the extent to which the experimental variation of T_N with z differs with the theoretical prediction for $\lambda/z=7.92$ (kOe/μ_B). The present experimental results for T_N fall somewhat above those of Ref.2 for the high z values for which the present experiments have been done. It would be interesting to extend the neutron sublattice magnetization experiments to lower z values to see if there is some sharp departure from the sort of agreement shown in Fig.3.

Even if the sublattice moment/Tb and T_N behavior can be explained by λ changing from the paramagnetic value because of nearest-neighbor exchange effects, the question remains as to the apparent presence of higher degree exchange terms in the paramagnetic ($z \leq 0.403$) regime, and the lack of evidence for their existence in the antiferromagnetic regime. One can raise the possibility that the apparent DM^3 contribution to H_{eff} for the high field paramagnetic magnetization experiments is associated with magnetostriction effects, increasing with the applied field and induced magnetization. (If present, such effects would have to be associated with strain effects on the exchange and not on the crystal-

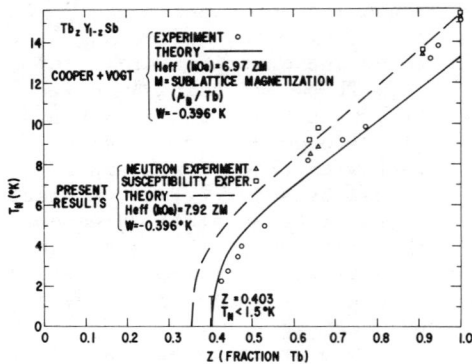

Fig.5 Variation of T_N with z. Cooper and Vogt results are from Ref.2.

field interaction, since magnetostriction effects are absent[5] in the high field magnetization experiments in TmSb and $Tm_{0.53}Y_{0.47}Sb$.) Such magnetostriction effects could be largely absent in the antiferromagnetic regime except for a substantial, essentially discontinuous, distortion at T_N (as in[6] pure TbSb) that might contribute to the change in λ from the paramagnetic regime (i.e. the temperature dependence of the distortion below T_N is small[6]). This does not explain why the paramagnetic λ seems to give the correct T_N for the lower z values in the antiferromagnetic regime, unless the crystal distortion becomes quite small at T_N as z decreases. This suggests that it would be interesting to perform experiments as in Ref.6 to look for correlations between the variation of T_N (i.e. change of λ from the paramagnetic value, necessary to fit T_N in a linear molecular field theory) and crystal-distortion at T_N as z varies.

We wish to thank Miss E. Kreiger for her aid with the numerical calculations.

REFERENCES

1. For a review of singlet-ground-state magnetism see B.R. Cooper and O. Vogt, J. Phys. Radium 32, C1-958 (1971).
2. B.R. Cooper and O. Vogt, Phys. Rev.B 1, 1218 (1970).
3. B.R. Cooper, I.S. Jacobs, C.D. Graham, and O. Vogt, J. Phys. Radium 32, C1-359 (1971).
4. H.R. Child, M.K. Wilkinson, J.W. Cable, W.C. Koehler, and E.O. Wollan, Phys. Rev. 131, 922 (1963).
5. B.R. Cooper and O. Vogt, Phys. Rev. B 1, 1211 (1970).
6. F. Lévy, Physics of Condensed Matter 10, 85 (1969).

CRYSTAL-FIELD EFFECTS IN THE MAGNETIC FORM FACTOR OF TmSb.*

T. O. BRUN and G. H. LANDER, Argonne National Laboratory, Argonne, Ill. 60439

The magnetic form factor of the induced moment on the thulium ion in a single crystal of TmSb has been measured with polarized neutrons. TmSb is a cubic compound (NaCl-type structure) with a singlet as the crystal-field ground state, and previous measurements[1] show that the exchange is negligible. Measurements of the form factor were taken in an applied magnetic field of 12.5 kOe at 5°K. The experimental values with H || <100> fall on a smooth curve as a function of $\sin \theta / \lambda$, while considerable anisotropy is observed with H || <110>. The magnetization density of a single thulium ion is therefore a function of the direction of the external field, whereas the magnetization is isotropic under the experimental conditions. The anisotropies in the magnetization density arise from the interaction of the rare-earth ion with the crystal field. The theoretical magnetic form factor, which has been calculated with the tensor-operator method, predicts this anisotropy. Excellent agreement is obtained between the theoretical and experimental form factors using the $<j_i>$ integrals derived from the experimental form factor of thulium metal.

*Work performed under the auspices of the U. S. Atomic Energy Commission.

[1] B. R. Cooper and O. Vogt, Phys. Rev. B $\underline{1}$, 1211 (1970).

A brief account of this work has been published, Phys. Rev. Letters $\underline{29}$, 1172 (1972). A more complete account is in press in Physical Review.

PRESSURE DEPENDENT EXCHANGE COUPLINGS IN SAMARIUM MONOCHALCOGENIDES

W. M. Walsh, Jr., L. W. Rupp, Jr., and L. D. Longinotti
Bell Laboratories, Murray Hill, N.J. 07974

ABSTRACT

The large exchange induced ESR g-shifts of S-state ions present as trace impurities in the samarium monochalcogenides SmS and SmSe have been investigated as functions of quasihydrostatic pressure at liquid helium temperatures. While roughly comparable with expectations based on intercompound variations the results suggest that local compressibilities may differ significantly from those of the hosts.

INTRODUCTION

At atmospheric pressure the samarium monochalcogenides, SmS, SmSe and SmTe, are Van Vleck paramagnetic semiconductors. Their properties stem primarily from the $^7F_0$ ground state of Sm^{2+} which is only ~ 415 K lower than the magnetic triplet $^7F_1$ state and to the relatively small energy gap from the $4f^6$ configuration to a trivalent state in which one f electron delocalizes into the conduction band.[1] The latter energy gap becomes smaller as the interatomic distance decreases from telluride to sulphide and may even be driven to zero without loss of the NaCl crystal structure by application of pressure.[2] The semiconducting nature of these materials implies rather spread out wave functions for the weakly magnetic f electrons and therefore appreciable exchange coupling is to be expected. The original magnetic susceptibility data of Bucher et al.[1] as well as the pressure dependences of the susceptibilities[3] have been explained by Birgeneau et al.[4] in terms of rapidly varying exchange couplings. This interpretation was considerably strengthened by the observation of large exchange-induced g-shifts in the electron spin resonance spectra of trace S-state impurities in the samarium chalcogenide hosts.[4,5] We have now extended this spectroscopic means of studying exchange couplings by addition of the pressure variable.

EXPERIMENTAL METHOD AND RESULTS

The ESR experiments are performed at ~ 34 GHz using a conical sapphire resonant cavity-pressure window[6] in a piston-cylinder configuration. Crudely hydrostatic pressures up to ~ 6 kbars are generated in silicone grease at the liquid helium temperatures commonly required to avoid spin relaxation by thermally activated conduction electrons. Crystals of SmS and SmSe doped with the desired probe ions are prepared as described earlier.[4] Strain gradients prevent precision measurement of fine and hyperfine structures but allow g-value determinations for Gd^{3+} and Eu^{2+} in SmS and of Gd^{3+} and Mn^{2+} in SmSe. The g-shifts of such spin-only magnetic moments reflect coupling via exchange parameters J_i with Z_i equivalent Sm^{2+} ions:

$$\delta g = -8 \sum_i Z_i J_i / \Delta' \qquad (1)$$

where Δ' is the energy difference between the ground state and the nearby triplet renormalized by the Sm-Sm exchange interactions.[4] Both host-impurity and host-host exchange vary rapidly with lattice parameter a but we present only the raw g-shift data rather than attempt a decomposition. The host compressibilities determined by Chatterjee, et al.[7] are used to express the results in terms of lattice parameter as shown by the short straight lines in the figure and as given in the table:

Fig. 1 g-Value Dependence on Lattice Parameters in the Samarium Monochalcogenides

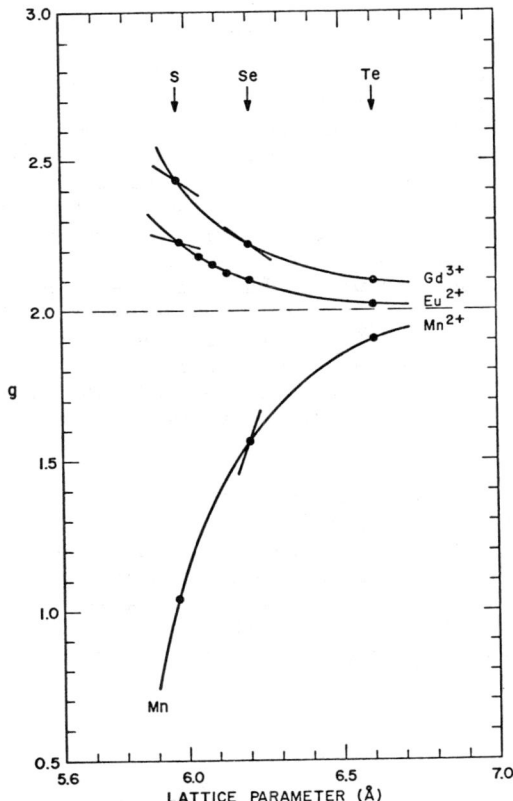

Table I g-Shift Dependence on Interatomic Distance

HOST:ION	g	$\partial \ln \delta g/\partial \ln a$	
		a	b
SmS:Gd	2.43	-21	-9.3
SmS:Eu	2.23	-21	-7.5
SmSe:Gd	2.22	-16	-21
SmSe:Mn	1.567	-20	-41

a   based on intercompound variation

b   pressure measurements reported here

DISCUSSION

The measured pressure dependences of the g-shifts are only very roughly consistent with the behavior expected on the basis of the variation between compounds and alloys. The intercompound curves approximately exhibit a power law behavior: $\partial \ln \delta g/\partial \ln a \sim -20$. Our data for SmSe:Gd are quite consistent with this value but the variation is essentially twice as rapid for SmSe:Mn and roughly half as rapid for SmS:Gd and SmS:Eu. The more rapid variation in the case of SmSe:Mn is reminiscent of the pressure dependence of the Curie temperature of EuO which is much stronger than might be expected from the intercompound behavior.[8] The markedly weak variation of the host-impurity exchange in SmS is more surprising. One might conjecture that the local compressibility is considerably reduced with respect to the pure host value due to stabilization of the divalent state of samarium in the vicinity of a stable impurity.

Finally it should be mentioned that our original intention was to follow the ESR signals in SmS up to the electronic collapse at 6.5 kbar.[2] Unfortunately we are unable to operate successfully much above 1.5 kbar due to loss of cavity Q as the conductivity of the sulphide rises strongly well below the transition pressure. It is not clear whether this is an intrinsic property of the material or is an artifact of nonhydrostatic strain.

ACKNOWLEDGMENTS

We wish to acknowledge contributions to and discussions of the experiments with Drs. R. J. Birgeneau, E. Bucher, A. Jayaraman, D. B McWhan and P. S. Peercy.

REFERENCES

1. E. Bucher, V. Narayanamurti and A. Jayaraman, J. Appl. Phys. $\underline{42}$, 1741 (1971).

2. A. Jayaraman, V. Narayanamurti, E. Bucher and R. G. Maines, Phys. Rev. Letters $\underline{25}$, 368 and 1430 (1971).

3. M. B. Maple and D. Wohlleben, Phys. Rev. Letters $\underline{27}$, 511 (1971).

4. R. J Birgeneau, E. Bucher, L. W. Rupp, Jr. and W. M. Walsh, Jr., Phys. Rev. $\underline{5}$, 3412 (1972).

5. F. Mehran, K. W. H. Stevens, R. S. Title and F. Holtzberg, Phys. Rev. Letters $\underline{27}$, 1368 (1971).

6. A. W. Lawson and G. E. Smith, Rev. Sci. Instr. $\underline{30}$, 989 (1959).

7. A. Chatterjee, A. K. Singh and A. Jayaraman, Phys. Rev. $\underline{B6}$, 2285 (1972).

8. D. B. McWhan, P. C. Souers and G. Jura, Phys. Rev. $\underline{143}$, 385 (1966).

MAGNETIC HYPERFINE INTERACTIONS AND CONDUCTION ELECTRON POLARIZATION AT Eu and Se in $Eu_{1-x}Gd_xSe$

K.Raj and T.J.Burch
Fordham University+, Bronx, N.Y. 10458
and
J.I.Budnick*
National Science Foundation, Washington,D.C. 20550

ABSTRACT

Conduction electrons can be added in a controlled fashion to the magnetic semiconductor EuSe by Gd doping thereby producing large effects in the bulk magnetic properties and in the detailed exchange interactions. We have made a systematic NMR study of both the Eu^{153} and Se^{77} resonances in the system $Eu_{1-x}Gd_xSe$ with $0 \leq x < 0.25$ to examine the role of conduction electrons, and find that the magnitude of the negative hf fields at both Eu and Se nuclei increase steeply for values of $x < 0.03$ and less rapidly for larger Gd concentrations. The observed hf field variations with concentration for both Eu and Se can be largely explained in terms of a RKKY model for both Eu and Se sites.

The substitution of the trivalent impurity Gd^{3+} for Eu^{2+} in the complex magnetically ordered insulator EuSe produces a ferromagnetic conductor in which the number of conduction electrons added approximately equals the number of Gd atoms[1,2]. This system, therefore, provides an opportunity to study the specific contribution of conduction electrons to the hyperfine interaction.

This paper reports a spin-echo NMR study of the concentration dependence of the Eu^{153} and Se^{77} hyperfine field distributions in $Eu_{1-x}Gd_xSe$ alloys, with $0 \leq x \leq 0.25$. The Eu and Se fields shift to higher frequencies with increasing Gd concentration. These shifts are assigned to conduction electrons and are shown to be of the same form as the RKKY functions at the Eu and Se sites. Reports of spectra for some of the lower concentration alloys have been made elsewhere[3].

The samples used in these experiments were made and kindly lent to us by Dr. F. Holtzberg of the IBM Research center. The composition of all samples is nominal. This remark is of particular importance in the case of the 25% Gd sample.

The Se^{77} spin echo spectra at $1.4^\circ K$ and in zero applied magnetic field for pure and doped selenides are shown in Fig.1. The Se resonance is shifted to higher frequencies in the doped samples. We have determined the sign of the Se hyperfine fields in EuSe and in the

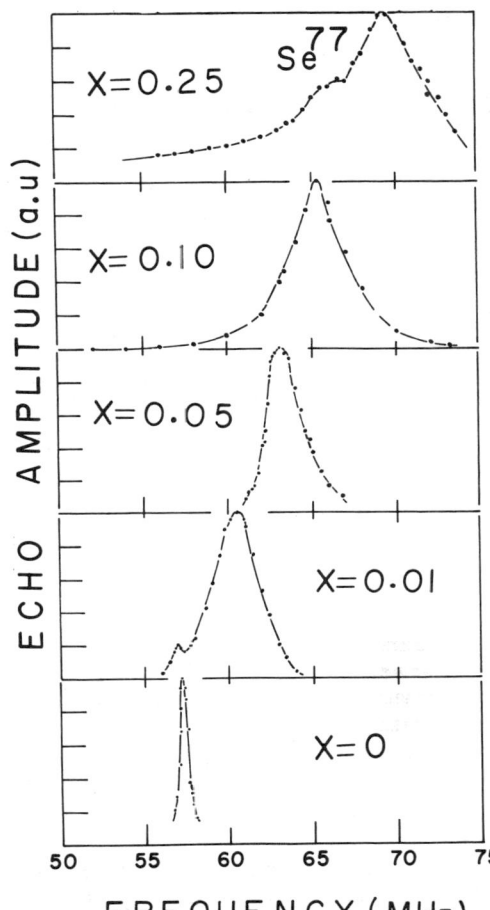

Fig.1. Spectra of Se^{77} in $Eu_{1-x}Gd_xSe$ at 1.4K.

doped materials to be negative[3]. From the observed increase in hyperfine field we conclude that the conduction electrons produce negative spin density at the Se nuclei. Also, the width of the Se line increases with Gd concentration. This line width is approximately the same function of concentration as the Se field shift, implying that this line broadening is a conduction electron effect. The structure and asymmetries in the spectra of doped samples may be due to Gd neighbor effects or to inhomogenities in the samples and are small at low concentrations. The total echo decay time, T_2^*, of Se in all samples is frequency dependent with a minimum of about 100 μs at the center of the line. No correction of the Se spectra was needed for the experimental conditions used.

The Eu^{153} spectra are shown in Fig.2, and show frequency and line width trends similar to those of the Se spectra. In these alloys the total echo decay time T_2^*, of the Eu echoes is frequency dependent and is less than 5 μs. This is rapid enough to appreciably distort the spin echo spectra. The spectra have been corrected to produce a distribution representative of the number of nuclei vs frequency. The variation of T_2^* with frequency and the method of correcting spectra for this effect have been extensively studied and discussed elsewhere[4].

In EuSe we observe in zero applied field two lines[3,5,6] due to Eu^{153} nuclei at 119.0 and 131.0 MHz corresponding to antiferromagnetic (NSNS) and ferrimagnetic (NNS) phases respectively. Only in magnetic fields greater than 7 KG is a third line corresponding to the ferromagnetic (NNN) phase observed at 143.0 MHz. This line at 143.0 MHz is the one which is compared to the ferromagnetic doped samples in which complex changes of phase are not observed. The Eu hyperfine field in all samples is negative and the shifts to higher frequencies with Gd concentration indicates a negative conduction electron contribution to the Eu hyperfine field.

The concentration dependences of the Eu^{153} and Se^{77} hyperfine fields are shown in Fig. 3. The Eu and Se dependences are different. The lines drawn through these fields are the RKKY functions vs electron concentration multiplied by appropriate scale factors. The solid line is the function calculated at Eu sites and the dashed line the function at the Se site. The number of conduction electrons is taken as equal to the Gd concentration. The form of

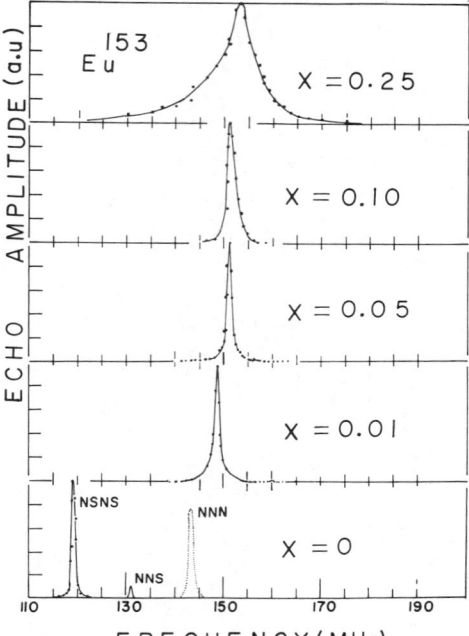

Fig.2. Spectra of Eu^{153} in $Eu_{1-x}Gd_xSe$ at $1.4°K$. Correction has been made for the frequency dependence of T_2^*.

the function used and its relation to the hyperfine field by a constant multiplier is described by Yosida[8]. The effects of all neighbors within five lattice constants are included and the triple degeneracy of the t_{2g} branch of the 5d band of Eu has been taken into account.

Several assumptions are involved in this use of the RKKY function. In the concentrations under consideration small changes in the lattice parameter with increasing Gd concentration are ignored. In calculating the function Eu and Gd are considered equivalent. The number of conduction electrons is equated to the nominal Gd concentration of the alloys. Since Eu vacancies compensate some of the excess Gd^{3+} electrons the actual free electron concentration may be somewhat reduced. Such a reduction was found when the RKKY function was compared to the concentration dependence of the paramagnetic Curie temperature of $EuSe^2$. No independent determination of the actual conduction electron concentration was made for our samples.

We consider the agreement between the shape of

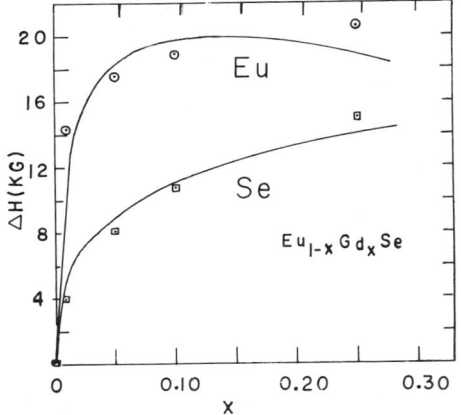

Fig.3 Concentration dependence of Eu^{153} and Se^{77} hyperfine fields in $Eu_{1-x}Gd_xSe$ at 1.4K. The solid and dashed lines are scaled RKKY functions at the Eu and Se sites.

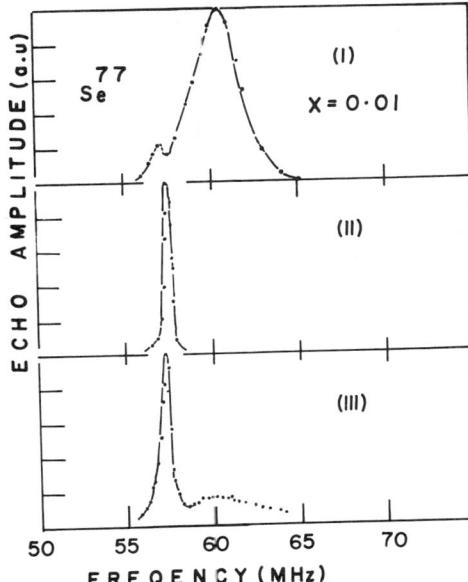

Fig.4. Spectra of Se^{77} in $Eu_{.99}Gd_{.01}Se$ after different chemical treatments.

these calculated curves and the shapes of the field variations of Eu and Se to be strong evidence that the shifts are due to the changing conduction electron concentration. The annealing of a Gd doped sample in a vacuum will remove some of the Eu leaving vacancies and producing a drastic change in electrical conductivity. This process is somewhat reversible and the Eu deficiency can be overcome and the conduction electron replaced by annealing the sample in an atmosphere containing excess Eu metal. These annealing processes are critically dependent on Eu vapor pressure. Dilute samples should respond to the annealing better than more concentrated samples. Fig. 4 I. is that of a conducting 1% Gd sample. Fig 4 II. that of the sample vacuum annealed. The frequency shifts assigned to conduction electrons disappear and the Se and Eu resonance are seen at 57.3 and 119.1 and 131.0 MHz respectively, the frequencies characteristic of the insulator EuSe. Fig.4 III. shows an attempt to return the sample to its original condition by annealing with excess Eu metal that was only partially successful. Lines characteristic of both conducting and insulating regions are observed.

A similar but less pronounced behavior is observed for the 5% Gd sample. Only samples which are conductors show the shifts assigned earlier to conduction electron effects.

Clearly this microscopic probe allows us to study this interesting effect which has not yet been studied systematically.

We wish to thank Dr. F.Holtzberg of the IBM Watson Research Center for the samples used in this study and for many helpful discussions. We wish to thank Dr. S. Skalski of Fordham University for many helpful discussions.The authors wish to thank Mr. W. Rodger and Mr. M. King for technical assistance.

REFERENCES

+Supported by N.S.F.
*On leave from Fordham University

1. S.Methfessel and D.C.Mattis,Magnetic Semiconductors in S.Flugge - H.P.J.Wijn, Handbanch der Physik, Vol.XVIII/I. Springer-Verlag,Berlin,Heidelberg, New York (1968).
2. F.Holtzberg,T.C.McGuire,S.Methfessel and J.C.Suits, Proc.Int.Conf.Magnetism,Nottingham (1964) page 470. Also Phys.Rev.Letters $\underline{13}$, 18 (1964).
3. J.I.Budnick,K.Raj,T.J.Burch and F.Holtzberg, Journal de Physique $\underline{32}$,763 (1971).
4. K.Raj,T.J.Burch and J.I.Budnick,International Journal of Magnetism, in print.
5. W.Zinn,Lectures presented at the conference on Magnetism, 15-28.6.69,Chania,Crete.
6. T.Komaru,T.Hihara andY.Koi, J.Phys.Soc.Japan $\underline{31}$, 1391 (1971).
7. F.Holtzberg and E.A. Giess,Proceedings of the International Conference on Ferites - Japan, July, 1970. Edited by Y.Hoshino, S.Iida and M.Fugimoto University Park Press.
8. K.Yosida,Phys.Rev.$\underline{106}$,893 (1957).

MAGNETIC STRUCTURES OF $Eu_{1-x}Gd_xS$

S. J. Pickart and H. A. Alperin*
Naval Ordnance Laboratory, Silver Spring, Maryland 20910

F. Holtzberg and T. R. McGuire**
IBM Research Center, Yorktown Heights, New York 10598

ABSTRACT

Neutron diffraction measurements were made on the mixed chalcogenide system $Eu_{1-x}Gd_xS$ with x = 0.1, 0.2, 0.3, 0.6 and 0.8. The results indicate that the magnetic structures in this system are inhomogeneous, with short-range ordered clusters of antiferromagnetic Gd spins.

INTRODUCTION

GdS is a metallic antiferromagnet[1] with conduction electron polarization probably providing the exchange mechanism, while EuS is an insulating ferromagnet, ordering under the influence of localized Heisenberg forces. Magnetization measurements[2] made on mixtures of the two showed several interesting properties, among them the fact that the saturation moment decreased linearly with Gd concentration, as if the Gd moment were not contributing, although its spin was $7\mu_B$ as determined from the Curie-Weiss constant. Two models were proposed[2] to explain this behavior, one canted and the other inhomogeneous with local clusters of antiparallel Gd spins. The neutron measurements described here were undertaken to discriminate between these two models.

EXPERIMENTAL RESULTS

Sample preparation was described previously.[2] Neutron diffraction measurements were taken at 1.36A on thin samples prepared by suspending a small amount (∼0.5gm) of the powdered compound, calculated to give 1/e absorption, in a powdered Cu matrix. In spite of the huge absorption cross section of Gd for thermal neutrons, this method was shown previously[1] to give qualitative, but reliable, results on the magnetic ordering.

Typical diffraction data summarizing our results are reproduced in Fig. 1. These are difference patterns, in which data taken at 77K (above the ordering temperatures in all the compounds studied) are subtracted from data taken at 4K, thus displaying the purely magnetic scattering at 4K. Data for samples with x = 0.1 and x = 0.3 were similar to that for x = 0.2.

*Supported by Naval Ordnance Laboratory Independent Research Fund
**Supported in part by Advanced Research Projects Agency

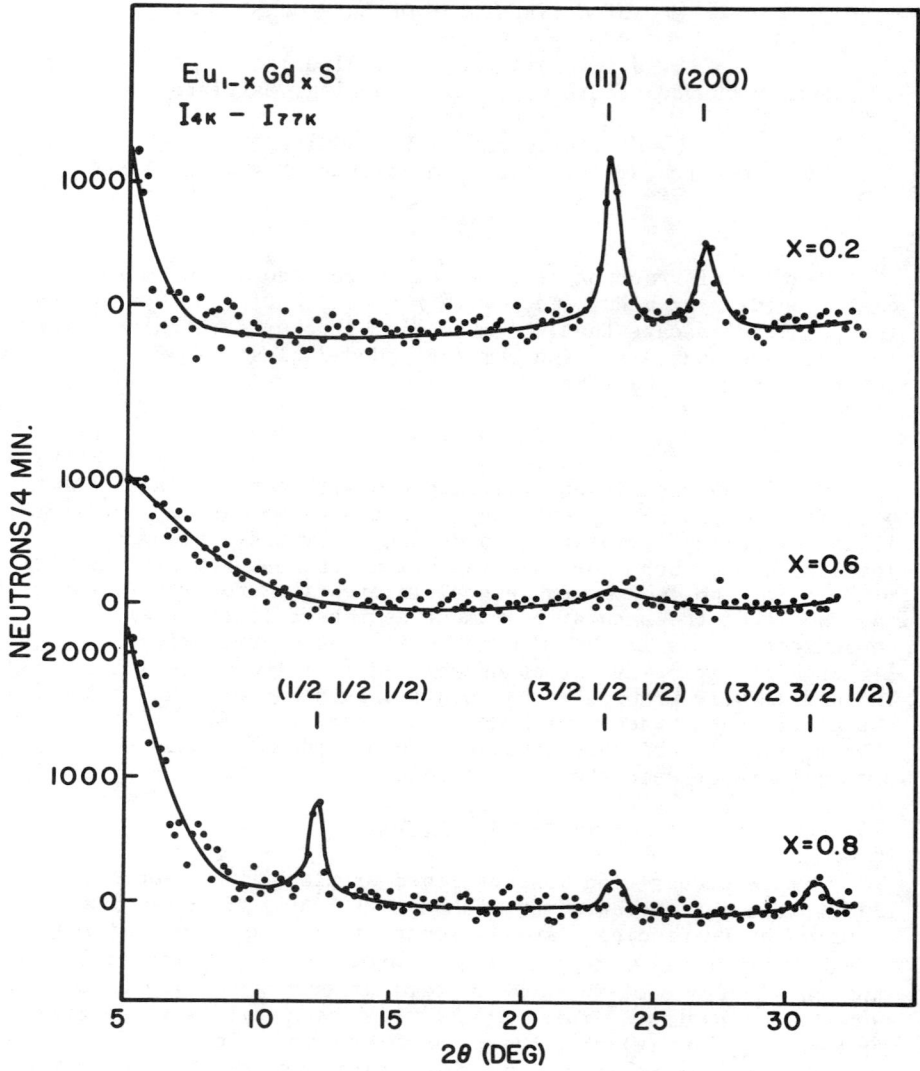

Fig. 1. Difference diffraction patterns (4K data minus 77K data) showing magnetic scattering in $Eu_{1-x}Gd_xS$. The intensities are not normalized to sample volume or corrected for absorption.

The intensities observed at the nuclear peak positions of the fcc cell (integral indices) correspond to ferromagnetic ordering, while those appearing at half-integer positions indicate the second type of antiferromagnetic order observed[1] in pure GdS. The Eu moment in the ferromagnetic region and the Gd in the antiferromagnetic region both correspond to $7\mu_B$ within the accuracy ($\sim 10\%$) obtainable from our low statistical precision. These values are determined by scaling to a nuclear scattering amplitude of $b = 2.2 \pm .2 \times 10^{-12}$ cm determined by the nuclear peaks of all the Gd compounds we have studied thus far.[1]

DISCUSSION

The long-range ordered states observed in this system are thus divided into two regions, separated by an intermediate region with little evidence for any type of long-range order. The region for $x \lesssim 0.4$ shows no antiferromagnetism attributable to the Gd spins, thus excluding a canted spin model, whose transverse component would be expected to give rise to diffraction effects at some superlattice spacing. This would be particularly noticeable for the $x = 0.6$ sample, where the magnetization results[2] show an ordering temperature of 15K and the Gd moment contribution is larger than that of Eu.

On the other hand, these results are consistent with a model of inhomogeneous magnetic structures. This picture is further supported by the extra scattering observed at low angles in Fig. 1. Normally, since the data at 77K include the paramagnetic diffuse scattering, difference patterns like these show a slightly negative background level at angles spanned by the paramagnetic form factor. The anomalous effects in Fig. 1 must therefore arise from scattering from Gd spins which are correlated on a short-range basis only.

The exact nature of these correlations is difficult to determine. We know that their overall moment must be zero from the magnetization results, and a likely situation is the one suggested previously, consisting of local clusters of Gd spins that are antiferromagnetically aligned, but with no correlation between clusters.

We may speculate that this situation of "immiscible" magnetic phases is in fact related to the different origin of the exchange interactions. Evidently, the conduction electron polarization necessary to provide long-range interactions between the Gd spins is not compatible with the exchange interactions between the divalent Eu atoms.

REFERENCES

1. T. McGuire, R. Gambino, S. Pickart and H. Alperin, J. Appl. Phys. 40, 1009 (1969).
2. T. McGuire and F. Holtzberg, AIP Conf. Proc. 5, 855 (1972).

HYPERFINE EFFECTS IN SINGLET GROUND STATE SYSTEMS

R. M. White
Xerox Palo Alto Research Center
Palo Alto, California 94304

B. B. Triplett
Department of Physics, Stanford University
Stanford, California 94305

ABSTRACT

It has recently been shown[1] that in systems characterized by two low-lying electronic singlets that the simultaneous diagonalization of the hyperfine as well as the exchange interaction gives rise to an interference between these two interactions which leads to an enhancement in the magnetic ordering temperature. The nature of the collective modes of such a system will be discussed as well as the phase diagram for such a system with antiferromagnetic exchange interactions.

[1] B. B. Triplett and R. M. White, to be published.

Section 48. Anisotropy and Magnetoelastic Interactions

CONTRIBUTIONS TO THE ANISOTROPY OF Ni^{2+} IONS IN TETRAHEDRAL SITES

A.J. Pointon and G.A. Wetton
The Polytechnic, Portsmouth, England

ABSTRACT

It is shown that the experimentally observed positive anisotropy constant K_1 of tetrahedral site Ni^{2+} ions in nickel ferrite can be explained in terms of either second order spin-orbit coupling or a phenomenological anisotropic exchange. However, while the former requires only a single adjustable parameter, it appears from a simple calculation that four are required to fit the anisotropic exchange Hamiltonian.

INTRODUCTION

Theoretical discussions[1,2] of the magnetocrystalline energy of Ni^{2+} ions on tetrahedral sites in ferrites ($Ni^{2+}(A)$ ions) have predicted that there would be a substantial negative contribution anisotropy constant K_1 due to Jahn-Teller distortion along one of the (100) axes. However, ferrimagnetic resonance (FMR) experiments[3] have actually shown $Ni^{2+}(A)$ ions to give positive values of K_1. In this paper the possible source of this anisotropy, of these ions on what is basically a good cubic site, is discussed.

CALCULATION OF THE MAGNETOCRYSTALLINE ENERGY

The $^3F_4$ ground state of $Ni^{2+}(A)$ is split by a tetrahedral, T_d, crystal field into $^3T_1$, $^3T_2$, $^3A_2$ with $^3T_1$ lowest as shown in Fig. 1. If the splitting of the 9 levels ε_i (i = 1 to 9) of the $^3T_1$ state depend on the direction cosines (n_1, n_2, n_3) of the magnetization vector, there will be a magnetocrystalline free energy for the ion

$$F(n_1, n_2, n_3) = - kT \ln \sum_i e^{-\varepsilon_i(n_1, n_2, n_3)/kT} \qquad (1)$$

The first order anisotropy constant can then be written as

$$K_1 = 4[F(1/\sqrt{2}, 1/\sqrt{2}, 0) - F(0, 0, 1)] \qquad (2)$$

(In the present case $|K_2| \ll |K_1|$ and can be neglected.) Three contributions to the anisotropic part of the free energy are discussed separately.

(a) Anisotropy due to a non-cubic field. Baltzer[1] and others have proposed that the origin of K_1 for Ni^{2+}(A) ions is the splitting of the $^3T_1$ levels by a tetragonal Jahn-Teller distortion. The total free energy then arises from the sum of three terms, one for each of the (100) directions. Not only does this give the sign of K_1 opposite to that found experimentally but we must report that no acceptable fit to the temperature variation of K_1 could be obtained with an axial field term in the Hamiltonian of the form

$$H = g\beta \vec{H}_{ex} \cdot \vec{S} + \lambda \vec{L} \cdot \vec{S} + \Delta_t (1 - L_z^2) \qquad (3)$$

for any choice of exchange field energy $g\beta H_{ex}$, spin-orbit coupling constant λ or non-cubic field splitting Δ_t whether the non-cubic field was taken to be of tetragonal or trigonal symmetry.

(b) Anisotropy due to second order spin-orbit coupling. If we assume that λ is substantially less than the exchange energy and that the spin is quantized along the ζ-direction defined by $\vec{H}_{ex}$, the $^3T_1$ level will split into three orbital triplets with $S_\zeta = 1$, 0 and -1 respectively. With the $^3T_1$ functions taken from Bleaney and Stevens[4] as ϕ_4, ϕ_5 and ϕ_6 and $S_\zeta = 1$ taken as the lowest spin state, the spin-orbit coupling splits these to give three states $\psi_i = \sum_{j=4}^{6} c_j^i |\phi_j, S_\zeta = 1>$; $i = 1,2,3$. The energies (ϵ_i) of these three levels are, in ascending energy, $\epsilon_1 = -3\lambda/2$, $\epsilon_2 = 0$ and $\epsilon_3 = 3\lambda/2$. The coefficients c_j^i are direction dependent as, for example, for the lowest level,

$$\epsilon_1 = -3\lambda/2; \quad c_4^1 = n^+/\sqrt{2}, \quad c_5^1 = (1 + n_3)/2, \quad c_6^1 = n^{+2}/2(1 + n_3), \quad (4)$$

where $n^+ = n_1 + in_2$, ζ is along (n_1, n_2, n_3) and there are similar expressions for $i = 2$ and 3.

The energy level structure, considering only first order effects, is shown in Fig. 1. The second order spin-orbit coupling between the ψ_i and the $|T_2, S_\zeta = 1>$ and $|T_2, S_\zeta = 0>$ states gives orientation dependent contributions to the energies. For the ψ_1

function these are, respectively,

$$-\frac{15\lambda^2}{4\Delta}(1 - 2(n_1^2 n_2^2 + n_2^2 n_3^2 + n_3^2 n_1^2)) \tag{5}$$

and

$$-\frac{15\lambda^2}{8\Delta'}(1 + n_1^2 n_2^2 + n_2^2 n_3^2 + n_3^2 n_1^2) \tag{6}$$

Assuming that the splittings Δ and Δ' of the $|T_2, S_\zeta = 1\rangle$ and $|T_2, S_\zeta = 0\rangle$ states from $|T_1, S_\zeta = 1\rangle$ are substantially the same, the total energy of the lowest state is

$$\varepsilon_1 = -\frac{3}{2}\lambda - \frac{15\lambda^2}{8\Delta} + \frac{45\lambda^2}{8\Delta}(n_1^2 n_2^2 + n_2^2 n_3^2 + n_3^2 n_1^2) \tag{7}$$

This expression predicts $K_1(T = 0)$ as $45\lambda^2/8\Delta$ while, together with equations (1) and (2), and the equivalent expressions for ε_1 and ε_2 it gives the temperature variation of K_1. (Higher order perturbation terms would be required for K_2, which may explain the relative insignificance of this term.)

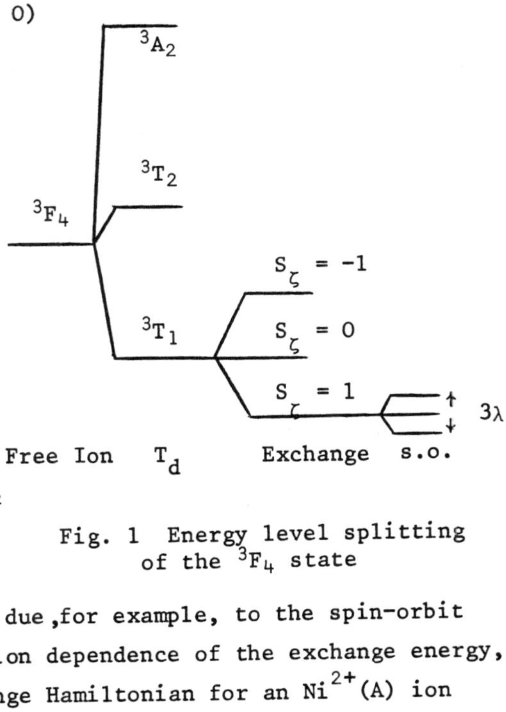

(c) Exchange anisotropy. The $Ni^{2+}(A)$ ion is exchange coupled to its nearest magnetic neighbours through the four oxygen ions situated along alternate cube diagonals. If, due, for example, to the spin-orbit coupling, there is an orientation dependence of the exchange energy, the simplest form of the exchange Hamiltonian for an $Ni^{2+}(A)$ ion would involve four terms $g\beta\vec{H}_{ex}\cdot\vec{S}(1 + \gamma L_{zi}^2)$ where γ is a constant and there are four (111) or D_3 axes for Z_i. We can simplify the calculation by replacing $\vec{H}_{ex}$ in equation (3) by

$$\vec{H}_{ex}[1 + G_2^0(3L_z^2 - L(L+1)) + G_2^2(L_+^2 + L_-^2)] \tag{8}$$

and set $\Delta_t = 0$. G_2^0 and G_2^2 are constants, the second term taking

Fig. 1 Energy level splitting of the $^3F_4$ state

account of any lack of axial symmetry arising from the random distribution of Ni^{2+} and Fe^{3+} ions among the nearest neighbours. The free energy is then summed over the four inequivalent sites.

COMPARISON WITH EXPERIMENT

The temperature dependence of K_1 between 4 and 300K was found for various nickel ferrite, $NiFe_2O_4$, specimens by measurement of the FMR field in the (100), (110) and (111) directions at 9 and 30 GHz. In those specimens quenched from high temperatures, small percentages of Ni^{2+} ions are trapped on tetrahedral sites. (They have an octahedral site preference energy of ~0.8 ev[5].) The effect of the $Ni^{2+}(A)$ ions is then obtained from the difference between the anisotropies of the quenched and slow cooled specimens. The results at X-band are shown in Fig. 2 for different percentages of $Ni^{2+}(A)$.

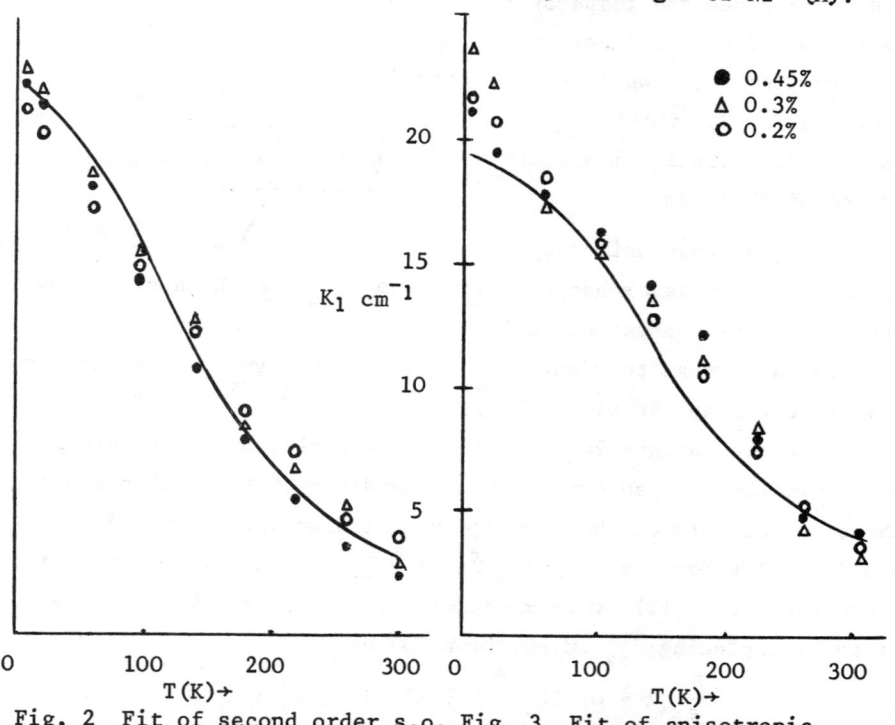

Fig. 2 Fit of second order s.o. K_1 to X-band results (% $Ni^{2+}(A)$ ions as indicated)

Fig. 3 Fit of anisotropic exchange K_1 to Q-band results

Taking the extrapolated value of $K_1(T = 0)$ as ~ 22.5 cm^{-1} gives on assuming the validity of equation (7), $\lambda^2/\Delta = 4$ cm^{-1}. The optically determined value[6] of Δ is 3000 cm^{-1} which gives $\lambda \sim 110$ cm^{-1}. The corresponding fit to K_1 in Fig. 2 must be considered excellent in view of the use of only a single parameter. The value of λ is smaller than generally assumed but this could be due to a reduction in the anisotropy by a Jahn-Teller distortion which, in view of the form of the ground state wave functions, would be orientation dependent.

A reasonable fit to the X-band results is obtained from an anisotropic exchange Hamiltonian based on equation (8). Fig. 3 shows the Q-band results, which are substantially the same as those obtained at X-band, fitted for this case with $g\beta H_{ex} = 310 \pm 5$ cm^{-1}, $\lambda = 250 \pm 10$ cm^{-1}, $G_2^0 = -0.165 \pm 0.05$ and $G_2^2 = 0.23 \pm 0.01$. However, not only do the required values of G_2^0 and G_2^2 appear large but also the value of K_2 predicted by equation (8) is positive (~ 20 cm^{-1}) while the observed K_2 is negative (~ -3 cm^{-1}). At the same time the terms of third order in the spin-orbit coupling predict $K_2 = -\frac{45}{4}\frac{\lambda^3}{\Delta^2}$ which, with the values given for λ and Δ gives $K_2 \simeq -1.6$ cm^{-1}.

REFERENCES

1. P. Baltzer, J. Phys. Soc. Japan, <u>17</u> Suppl. B1, 192 (1962).

2. M.D. Sturge, E.M. Gyorgy, R.C. Le Craw and J.P. Remeika, Phys. Rev., <u>180</u>, 413 (1969).

3. A.J. Pointon, J.M. Robertson and G.A. Wetton, Journal de Phys. <u>32</u>, C1-350 (1971).

4. B. Bleaney and K.W.H. Stevens, Rept. Prog. Phys., <u>16</u>, 109 (1953).

5. J.M. Robertson and A.J. Pointon, Solid State Comm., <u>4</u>, 257 (1966).

6. J.D. Dunitz and L.E. Orgel, J. Phys. Chem. Solids, <u>3</u>, 318 (1957).

A DILEMMA IN EXCHANGE INTERACTION IN EuIG

M. E. Foglio
Southern Illinois University, Carbondale, Illinois 62901

J. H. Van Vleck
Harvard University, Cambridge, Massachusetts 02138

ABSTRACT

The experimental value of the constant

$$\alpha = -(32/5)e^2 (1-R)Q(\beta H^{ex})^2 <r^{-3}>(2E_1 + E_2)/(E_1^2 E_2 h) \qquad (1)$$

was derived by Streever & Caplan (Phys. Rev. B, 3, 2910 (1971)) from the quadrupolar contribution to their nuclear resonance measurements in EuIG. They obtained -54 MHz, whereas the value calculated from (1) was -108MHz. This discrepancy does not seem justified by uncertainty in the parameters appearing in (1); and two attempts to remove it are made in the present paper. In the first, which is unsuccessful, a more precise nuclear Hamiltonian is used, together with large values of the anisotropic exchange parameters. In the second, the discrepancy is removed by adding a biquadratic exchange term to the nuclear Hamiltonian. However the ratio of the required biquadratic to conventional linear exchange is nearly an order of magnitude larger than the usual theoretical estimates.

INTRODUCTION

The interaction between the magnetic dipole and electric quadrupole moments of the nucleus with its own environment, splits the energy levels of the Eu ions in Europium Iron garnet(EuIG). As the interpretation of only low temperature experiments will be discussed here, it suffices to consider the nuclear effective Hamiltonian only for the ground state of the Eu^{3+} ion in EuIG. The magnetic dipole part can be written

$$\mathcal{H}_n = \beta_N \gamma_{Eu} \vec{I} \cdot \vec{H}_n \qquad (1)$$

where $\vec{H}_n$ is an effective field produced by the exchange interaction between the Eu ion and the Fe sublattice of EuIG, (the applied magnetic field can be neglected because it is much smaller than $\vec{H}_n$). The three components of this effective field are

$$H_{nq} = \lambda_q F_q = h_q H_n \quad (q=x,y,z) \qquad (2)$$

where the λ_q and h_q are the direction cosines of the Iron sublattice magnetic moment and of $\vec{H}_n$ respectively, referred to the principal rhombic axes of a given Eu ion; the F_q are the principal values of a tensor that is in diagonal form with respect to the principal axes of Eu. The values of the F_q have been measured using nuclear magnetic resonance[1] and Mössbauer effect [2,3]. Neglecting off-diagonal elements when I is quantized along $\vec{H}_n$, the effective quadrupolar Hamiltonian can be written

$$\mathcal{H}_q = h\nu'(3I_z^2 - I(I+1))/(4I(2I-1)) \qquad (3)$$

where ν' is proportional to the effective field gradient along $\vec{H}_n$ at the Eu nucleus at $0°K$. The value of ν' will be denoted by ν'_q when $\vec{H}_n$ is along a principal axis q of the Eu ion (q= x,y,z), and the quantity $\nu'_{av} = \nu'_x + \nu'_y + \nu'_z$ will be important in the following discussion. Streever and Caplan [1] have derived values for the F_q and ν'_q (q=x,y,z) by fitting (1), (2), and (3) to the n.m.r. spectra they observe when $\vec{H}_n$ has different orientations. There are several contributions to ν' (gradients of electric field) as discussed in the literature [4,5], and it is convenient to write

$$\nu' = \alpha G(\lambda_q, h_q) + \nu''(h_q) \qquad (4)$$

where the first term originates in the Eu-Fe exchange interaction. Both functions $G(\lambda_q, h_q)$ and $\nu''(h_q)$ depend also on the crystal field, energy, and exchange parameters of the Eu ion. Theoretical expressions for these quantities have been derived by Streever and Caplan[1]: they describe the anisotropic exchange as a molecular field with components $\lambda_q H_{ex}(q)$. The relation named 10f in their paper corresponds to (4) in the present work, and the value of α is in our notation

$$\alpha = -(32/5)e^2(1-R)Q(\beta H^{ex})^2 <r^{-3}>(2E_1 + E_2)/(E_1^2 E_2 h) \qquad (5)$$

where R is the Sternheimer shielding factor, (βH^{ex}) is a measure of the isotropic exchange, Q is the quadrupole moment of the nucleus, r is the radius of an f electron, and E_1, E_2 are the energies of the $J=1,2$ levels of Eu^{+++} ion in the absence of exchange. The term ν'' does not contribute to ν''_{av} and thus

$$\nu'_{av} = \alpha G_{av} \qquad (6)$$

where G_{av} is a function of the parameters that enter in the theory used to describe the system.

There are two alternative ways to determine α: by direct calculation from (5) or else from (6), by the use of the experimentally determined value of ν'_{av} together with the value of G_{av} provided by the theory. Streever and Caplan [1] obtained $\alpha = -108$MHz from (5) and $\alpha = -54$MHz from (6). Although one can try to blame the discrepancy on the uncertainties in the parameters in (5), careful consideration indicates [6] that this is rather unlikely. Experimental error also seems improbable [7].

We shall report here two attempts to describe the angular variation of the n.m.r. spectra with the value of α derived from (5).

NUCLEAR HAMILTONIAN

A nuclear Hamiltonian more precise than the one derived by Streever and Caplan was obtained [6] with the use of the following Fe-Eu exchange interaction [6,8,9]:

$$\mathcal{H}_{ex} = 2\beta H^{ex}(\vec{\lambda} \cdot \vec{S})[1 + \sum_{nm} G_n^m O_n^m(L)] \tag{7}$$

Here the G_n^m (n=2,4, m even and m≤n) are the anisotropic exchange parameters, $O_n^m(L)$ are tensor operators as given by Hutchings [11], S and L are the spin and orbital angular momentum of Eu^{3+}, H^{ex} measures the isotropic exchange and $\vec{\lambda} = (\lambda_x, \lambda_y, \lambda_z)$ is the unit vector along the Iron magnetic moment. For simplicity only the J=0,1,2 multiplets of Eu have been considered, and a detailed description of the resulting effective nuclear Hamiltonian will be given elsewhere[6]

It should be more precise than the one given by Streever and Caplan [1] (cf. their formula 10f). The expression of the F_q (Streever and Caplan's H_q) is also given elsewhere [6] as a function of all the parameters.

From the six values of F_q and ν'_q obtained experimentally [1], one can solve the corresponding theoretical relations and derive six parameters; taking arbitrary values of the three G_4^m (e.g. $G_4^m=0$) the natural unknowns in those equations are G_2^0, G_2^2, $\beta H^{ex} <r^{-3}>$, α and the two rhombic crystal field parameters V_2^0 and V_2^2. With these parameters, the angular dependence of the spectra agrees very well with the published one [1], but the value of α is again too small as compared with (5). It seemed then natural to attempt the use of non-zero values of G_{4m}, because α is derived from (6) and our calculation gives

$$G_{av} = \alpha (3/E_2) \left\{ 1 + (220/7)[E_1/(E_2 + 2E_1)](7G_4^0 + G_4^4) + \Delta \right\} \tag{8}$$

where Δ is in lowest order an homogeneous function of V_2^n and G_2^n. The appearance of this particular combination of G_4^m is explained by the symmetry of the problem [6], and it is clear from (9) that the G_4^m will affect α in lowest order than the other parameters will. A trial and error method was first used, and α was obtained for moderate values of the G_4^m. Although values of α in good agreement with (5) were obtained, the corresponding spectra conflicted with the experimental observations; this behaviour is related also to symmetry considerations [6].

Another approach was to consider larger values of the G_4^m; to avoid the cumbersome trial and error procedure, two new equations were added to the six equations discussed above: they forced the theoretical quadrupole spectrum to take required values at two new points. Rather than taking α as unknown, its value α was calculated from (5), and the three G_4^m were considered as three more unknowns in the new system of equations. The computer program worked very well, but it was not possible to satisfy the system of equations and the value of α derived from (5). Solutions were obtained when the imput parameters and α were changed to allow for experimental errors but this approach failed again, because the spectrum was much worse than before at the points that were not fixed by the equations. Also the values of some of the G_4^m were too large and would give unacceptable values for the macroscopic magnetic anisotropy constant K_1.

BIQUADRATIC EXCHANGE

It seems clear that the difficulties in the description of the n.m.r. spectra can be resolved by adding to the Hamiltonian a fairly isotropic term (so that the angular variation of the spectra is not substantially altered) that at the same time gives negative contribution to G_{av} (cf. (6)), so that the α derived from (6) be large enough to satisfy (5). This can be done by adding to (7) a biquadratic exchange term

$$\mathcal{H}_B = J_B (\vec{\lambda} \cdot \vec{S})^2 \qquad (9)$$

This term mixes in first order, states with J=2 into the ground electronic state of Eu^{3+}. When the quadrupolar energy is then calculated with this first order wave function the contribution ν'_B to (4) is easily obtained. In this approximation

$$\nu'_B = \alpha \cdot B \cdot (3(\vec{\lambda} \cdot \vec{h})^2 - 1)/2 \qquad (10)$$

where the dimensionless constant B is given by

$$B = -J_B (E_1/\beta H^{ex})^2 / (2E_2 + 4E_1) \qquad (11)$$

Using the first method of calculation, with $G_{4m} = 0$ and adding (11) to (6), we have examined the angular dependence of the spectra and the value of α for several values of J_B near 3 cm^{-1}. A good agreement with the spectrum was found, and typical values of α where 104MHz for $J_B = 3.3$ cm^{-1} and 109MHz for $J_B = 3.5$cm^{-1} (also $G_2^0 = -0.015$, $G_2^2 = -0.031$, $V_2^0 = 127$cm^{-1}, $V_2^2 = -88$cm^{-1}, $\beta H^{ex} = 15.2$ cm^{-1} and $<r^{-3}> = 7.8$ a.u. for $J_B = 3.5$cm^{-1}).

Is $J_B = 3.5$cm^{-1} a sensible value? For the interaction, between a pair of ions, Anderson writes [10] the exchange Hamiltonian

$$J(\vec{S}_1 \cdot \vec{S}_2) + j(\vec{S}_1 \cdot \vec{S}_2)^2 \qquad (12)$$

and he estimates $j/J \sim 0.01$. For EuIG, most of the Fe-Eu bilinear exchange interaction is produced by the two tetrahedral ions nearest to Eu[14]. Assuming that the same is true for the biquadratic interaction and noting that at low temperatures $<S(Fe)> = 5/2$, one can relate βH^{ex} and J_B to quantities like j and J as follows: $J = (2/5)\beta H^{ex}$ and $j = (2/25)J_B$. With the values given above:

$$j/J = J_B/5\beta H^{ex} = 0.047$$

i.e. some five times larger than the theoretical estimate. This value is rather large, but should not be discarded on the sole basis of the theoretical estimate.

There might still be different mechanisms that would produce an effect similar to that of biquadratic exchange. One example is a crystal field produced by a magnetostrictive deformation of the lattice and with a particular dependence on the direction of the Iron magnetic moment. This model, however, would require completely unreasonable parameters.

REFERENCES

(1) R.L. Streever and P.J. Caplan, Phys. Rev. $\underline{3}$ 2910(1971).
(2) V. Atzmony, E.R. Bauminger, A. Mustachi, I. Nowik, S. Ofer and H. Tassa, Phys. Rev $\underline{179}$ 514 (1969).
(3) E.L. Loh, U. Atzmony and J.C. Walker, AIP Conference Proceedings 5 397 (1971).
(4) R.L. Cohen, Phys. Rev. $\underline{134}$ A94(1964).
(5) J. Blok and D.A. Shirley, Phys. Rev. $\underline{143}$ 278(1966).
(6) M.E. Foglio and J.H. Van Vleck, to be published.
(7) R.L. Streever, private communication.
(8) J.H. Van Vleck, Revista de Matemática y Física Teorica, U.N. Tucuman, $\underline{14}$ 189(1962).
(9) P.M. Levy, Phys. Rev. $\underline{135}$, A155 (1964).
(10) P.W. Anderson, Magnetism, Vol. 1, eds. G.T. Rado and H. Suhl (New York:Academic Press) pp. 25-85.
(11) M.T. Hutchings, Solid State Physics $\underline{16}$ 227(1964).
(12) W.P. Wolf and J.H. Van Vleck, Phys. Rev. $\underline{118}$ 1490(1960).
(13) I. Nowik and S. Oger, Phys. Rev. $\underline{153}$ 409(1967).

MAGNETIC ANISOTROPY OF ORIENTED COBALT FILMS*

B. R. Livesay and S. Spooner
Georgia Institute of Technology
Atlanta, Georgia 30332

ABSTRACT

Cobalt films exhibiting a six-fold magnetic symmetry have been grown by vacuum deposition onto (111) faces of copper single crystals. Anisotropy energies were determined by Fourier analysis of planar torque curves. Cobalt films of 2750 Å, 820 Å and 2100 Å had cubic anisotropy constants, K_2, of -7.3×10^4 erg/cc, -1.6×10^4 erg/cc and -7.7×10^3 erg/cc respectively. Uniaxial anisotropy was present and the constant, K_u, was approximately equal to 10^4 erg/cc in these films. Alloy films of Co -6% Fe were also grown on (111) copper single crystal faces. These films exhibited uniaxial anisotropy only with $K_u \approx 2 \times 10^4$ erg/cc. In both pure and alloyed cobalt, the uniaxial axis was unrelated to substrate orientation.

INTRODUCTION

The small lattice misfit between copper and f.c.c. cobalt results in conditions suitable for the growth of epitaxial cobalt films on copper single crystal substrates. The preparation conditions favorable for the growth of epitaxial cobalt films on other substrates have been investigated.[1,2,3] Investigations reported by Jesser and Matthews[4] and by Fedorenko and Vincent[5] demonstrated that the lattice misfit between cobalt and copper is accommodated by a network of dislocations which slips into the interface in the early stages of the growth of a cobalt film. As the film thickness increases, there is evidence[2,4,5] that the strain energy activates dislocation processes which convert part of the film material to the more stable h.c.p. phase.

Investigations of the magnetocrystalline anisotropy and other magnetic properties of f.c.c. cobalt films grown by either vacuum deposition or by electrodeposition have been reported[6-11]. The cubic anisotropy constants are generally determined from analysis of data taken from either (100) or (110) films. Since the constant K_2 appears alone in the anisotropy energy expression for a cubic (111) plane,[12] attempts were made to grow cobalt films epitaxially on (111) surfaces of copper single crystals by vacuum deposition. The magnetic anisotropy in the plane of the films was then measured using a torque magnetometer to determine the resulting anisotropy constants.

---

*Supported in part by the A.E.C. under Contract No. AT-(40-1)-3674 and by the Air Force Office of Scientific Research under grant No. AFOSR-71-2064.

EXPERIMENTAL

The copper crystal substrates employed in this work were wafers about 0.5 mm thick having a (111) major surface. The wafers were obtained from 99.999% pure, one inch diameter bulk copper crystal rods sectioned parallel to a (111) plane by spark slicing. The oriented bulk section was placed in a reactor core until an integrated neutron flux of 10^{17}-10^{18} neutrons/cm^2 was accumulated. The fast neutrons served to introduce point defects in the crystal for pinning dislocations and thereby minimize damage during subsequent preparation steps. About two months later it was possible to spark slice the crystal again parallel to the (111) surface obtained earlier using techniques which yielded large area thin wafers of uniform thickness.

The wafers were polished mechanically to remove surface irregularities introduced by spark cutting and then electropolished in a solution of 2/3 orthophosphoric acid with 1/3 water. The wafers were vacuum annealed at about 1000°C for 6 to 8 hours to remove the radiation hardening and mechanical damage. Laue X-Ray patterns were then made on each substrate to index the orientation.

The cobalt films were deposited at a rate of about 10 Å/sec by evaporation using alumina coated tantalum boats in a pressure environment of about 2×10^{-6} Torr. The substrate was oriented normal to the deposition beam in a furnace located about 12 cm from the source. The substrate temperature was held at 300°C for more than 6 hours prior to and during the film deposition. Desired film thicknesses were obtained and evaporation rates were determined using a quartz crystal thickness monitor previously calibrated for the specific geometry used during these evaporations. The films were circular with a diameter of 20 mm. No external magnetic field was applied during evaporation.

The measurements of magnetic anisotropy were made using an automatic torque magnetometer described previously.[13] The (111) surface plane was suspended horizontally in the field of an electromagnetic which rotated about a vertical axis. The rotational speed of the magnet was only 5 degrees/sec so that the substrate eddy current torque would be negligible. The Fourier components of the torque curves were analyzed with the aid of a computer program.

RESULTS

Cobalt films grown on copper crystals under the conditions described above were found to have a magnetic torque component with a six fold symmetry. The magnetic easy axes for this six 6θ term were along the ⟨112⟩ crystallographic directions of the substrate crystal. In addition, a relatively large uniaxial torque component corresponding to anisotropy constants, K_u, on the order of 10^4 erg/cc was present in all the cobalt films studied but did not appear to have any particular relationship with the crystallographic orientation of the substrates. Three pure cobalt films deposited on (111) surfaces were found to have a measurable sin 6θ component. Several other films deposited under similar conditions did not exhibit a sin 6θ magnetic anisotropy component. However, two of these were deposited at a lower substrate temperature of 200°C indicating that the higher substrate temperature is more favorable for cobalt films having large oriented f.c.c. grains. Cobalt films were also deposited under similar conditions onto polished polycrystalline copper, glass and mica substrates. These films had only a sin 2θ component with K_u in the range of the values obtained for cobalt films deposited on copper single crystal substrates. The source of the uniaxial anisotropy term must therefore have been associated with some factor other than the crystallography of the substrate.

The magnetocrystalline anisotropy constants given here were evaluated in terms of the anisotropy energy expansion for a cubic crystal structure. In most cases, the cobalt films were a mixture of f.c.c. and h.c.p. phases. However, the electron diffraction patterns indicated that the six fold magnetic symmetry probably resulted from f.c.c. epitaxy even though part of the film material became h.c.p. during growth. The magnetic anisotropy expansion for the torque in the (111) plane of a cubic crystal is[12]

$$-\frac{dE_k}{d\theta} = K_2 \frac{\sin 6\theta}{18} \qquad (1)$$

where the torque, L, is $-dE_k/d\theta$ and θ is measured from the [112] direction in the (111) plane.

Planar torque curves made at three applied fields for a 2750 Å cobalt film are shown in Figure 1. The torque curves, both for positive and negative rotations, made at fields of 4000 oersted and above for this film were coincident so that magnetization vector angle corrections were not needed. Fourier analysis of this curve yielded $K_u = 4.1 \times 10^4$ erg/cc and $K_2 = -7.3 \times 10^4$ erg/cc. A very large rotational hysteresis occurred in the neighborhood of 1000 oersted and the low field torque curves were characterized by a rotatable anisotropy in that the phase of the sin θ torque curve obtained at 20 oersted was completely determined by the direction of the last applied high field of 4000 oersted or greater. The back reflection electron diffraction patterns showed that this particular film was a nearly perfect f.c.c. crystal.

For an 820 Å cobalt film the Fourier analysis of the high field torque curves yielded $K_u = 6.2 \times 10^3$ erg/cc and $K_2 = -1.6 \times 10^4$ erg/cc.

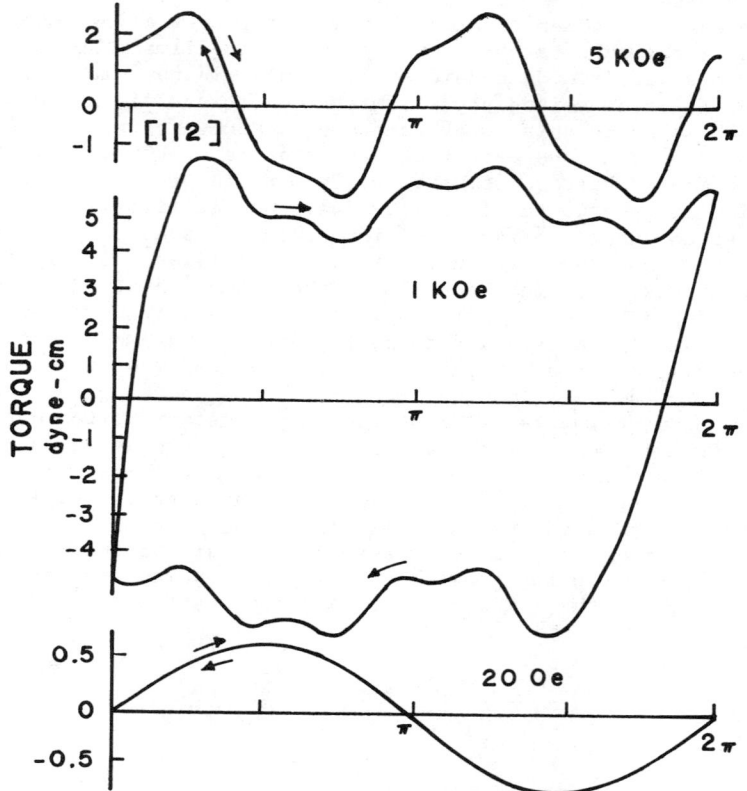

Fig. 1 Torque curves for a 2750 Å f.c.c. cobalt film grown on a (111) copper crystal at the indicated applied fields.

A third film 2100 Å thick had K_u = 5.2 x 10^3 erg/cc and K_2 = -7.7 x 10^3 erg/cc. The back reflection electron diffraction patterns for these films were rings with the f.c.c. reflection spots superimposed. Therefore only part of the film material had the crystallographic orientation of the substrate. It is reasonable to assume that only the oriented f.c.c. material in the cobalt film contributed to the sin 6θ component. No film orientation evidence was detected on any of the electron diffraction patterns for the films which did not have a sin 6θ component.

Additional attempts were made to prepare f.c.c. films which were primarily cobalt by evaporation from the Co - 6% Fe alloy which has a stable f.c.c. structure at room temperature. This series of films was deposited on (111) copper crystal surfaces using the same deposition parameters which resulted in f.c.c. pure cobalt films. None of these films had a sin 6θ component which could be resolved from the resulting uniaxial torque curves. The values of K_u for the Co - 6% Fe films ranged between 1×10^4 and 5×10^4 erg/cc.

DISCUSSION

The anisotropy constants, K_2, measured for the cobalt films deposited here are somewhat smaller than the values reported by Fisher and Goddard.[10] Their films were electrodeposited onto (110) copper single crystal substrates and the values of K_2 were resolved with K_1. Since bulk cobalt does not exist with an f.c.c. structure at room temperature, no comparisons there were possible. It should be noted that the film having the largest value of K_2 was also the one having greatest crystallographic perfection as seen from the electron diffraction patterns.

ACKNOWLEDGEMENTS

We would like to thank Mr. U. L. Brown, Jr. for his assistance in both sample preparation and in the measurements.

REFERENCES

1. C. González and E. Grünbaum, Proc. 5th Internat. Conf. Electron Microscopy, Academic Press, New York, 1962 (DD-1).
2. E. Grünbaum and G. Kremer, Appl. Phys. 39, 347 (1968).
3. K. Otsuka and C. M. Wayman, Phys. Stat. Sol. 22, 559 (1967).
4. W. A. Jesser and J. W. Matthews, Phil. Mag. 17, 461 (1968).
5. A. I. Fedorenco and R. Vincent, Phil. Mag. 24, 55(1971).
6. D. S. Rodbell, J. Phys. Soc. Japan, 17, Suppl. B-I, 313 (1962).
7. J. Goddard and J. G. Wright, Brit. J. Appl. Phys., 15, 807 (1964).
8. J. Goddard and J. G. Wright, Brit. J. Appl. Phys., 16, 1251 (1965).
9. G. P. Pyn'ko, V. G. Pyn'ko and A. S. Komalov, Fiz. Metal. Metalloved., 31, 203 (1971).
10. J. E. Fisher and J. Goddard, J. Phys. Soc. Japan, 25, 413 (1968).
11. U. Gradmann and J. Muller, Z. Angew Phys., 30, 87 (1970).
12. R. M. Bozarth, Ferromagnetism, D. van Nostrand, Inc., New York 1951.
13. B. R. Livesay, L. K. Jordan and E. J. Scheibner, J. Appl. Phys., 37, 1266 (1966).

ANNEALING OF COBALT-FERROUS FERRITE FILMS ON GLASS SUBSTRATES IN THE REMANENT STATE

N. F. Borrelli, S. L. Chen, and J. A. Murphy
Corning Glass Works, Corning, New York 14830

ABSTRACT

It is well known that when a cobalt-ferrous ferrite crystal is annealed in a magnetic field below its Curie temperature, a uniaxial anisotropy will be induced in addition to its normal cubic anisotropy. The magnetic field applied during the annealing has no direct contribution to the induction of the uniaxial anisotropy. Its only function is to align the magnetization in the field direction. It is the direction of the magnetization which determines the direction of the uniaxial anisotropy. Based on this principle, one should be able to induce a magnetic easy axis in cobalt-ferrous ferrite by magnetizing it at room temperature, followed by a heat treatment in its remanent state. This was done on cobalt-iron oxide films on glass substrates. Results on films having different Co/Fe ratios will be discussed.

INTRODUCTION

When cobalt-ferrous ferrites are annealed in a magnetic field, a uniaxial anisotropy is induced. As a result, the rectangularity of the hysteresis loop measured with the applied field parallel to the annealing field increases. Detailed studies on this phenomenon have been made by Bozorth,[1] Wijn,[2] Iida,[3] Penoyer[4,5] and their colleagues.

The magnetic field applied during annealing has no direct relation to the induction of the uniaxial anisotropy. Its only function is to align the magnetization M_s in the field direction. The actual annealing effect is caused by the interaction between M_s and the crystal.[6] Based on this argument, one should be able to induce an easy axis in the material by simply magnetizing it at room temperature prior to a heat treatment. The material, once magnetized, will develop its own easy axis during annealing. (Herein, this process will be referred to as annealing in a remanent state and the term "magnetic annealing" as annealing in a magnetic field.) Some of the results on cobalt-ferrous ferrite films will be discussed in this report.

FILM PREPARATION AND MAGNETIC PROPERTIES

A mixture of Co-Fe-oxide was first deposited on glass substrates by thermal decomposition of iron pentacarbonyl and cobalt nitrosylcarbonyl. The oxide mixture was then reduced to the ferrite structure in H_2O/H_2 atmosphere at 450°C for 1 1/2 hours. The magnetic properties of the "as reduced" films are shown in Fig. 1.

MAGNETIC PROPERTIES OF FILMS ANNEALED IN THEIR REMANENT STATES

Figure 2 shows the hysteresis loops of a sample with Co/Fe ratio of 0.12 after a 147-hour annealing in the remanent state at 150°C in air. Prior to the annealing, they were magnetized in a field of 7500 gauss. Two loops are shown in Fig. 2; one is measured with the field applied parallel to the initial magnetizing field, the other measured at 90° from it. Hereafter, they will be referred to as $\parallel$ and $\perp$ loops, respectively. The $\parallel$ loop is much more rectangular in shape than the $\perp$ loop. This clearly indicates that the same effects observed after a magnetic annealing happen here also.

The coercivity and the squareness ratio are shown in Figs. 3 and 4. The coercivity in the easy direction is about 10 to 20% higher than that in the transverse direction. However, the squareness ratio is about the same. This indicates either that the ratio between the uniaxial anisotropic constant K_u and the cubic anisotropic constant K_1 is low, or the magnetization within each crystallite was not aligned in the proper direction before annealing. If one assumes that the induced uniaxial anisotropy is parallel to one of the cubic axes and plots the total crystalline anisotropic energy in one of the {100} planes, as shown in Fig. 5 for different K_u/K_1 values, it can be seen that unless $K_u/K_1 \geq 1$ the magnetization is energetically stable in both the uniaxial direction and the direction perpendicular to it. The presence of the uniaxial anisotropy merely makes one direction more stable than the other. If one magnetizes the specimen along either the easy or the hard direction, the magnetization will remain in that direction after the applied field is removed, provided that $K_u/K_1 < 1$. This means that the remanences in the two directions will be exactly the same. If one applies a field in an arbitrary direction, the direction which the magnetization M_s takes after the removal of the field will depend on the values of K_u/K_1 and θ. For each K_u/K_1 value there is a critical angle θ_{cr} determined by the maximum of the E vs θ curve. The magnetization M_s will be in the hard direction for $\theta > \theta_{cr}$ and in the easy direction for $\theta < \theta_{cr}$. In a polycrystalline material the axes are randomly distributed with respect to the applied field, and the remanence in the hard direction will depend on K_u/K_1. The higher the latter is, the lower the remanence will be. The remanence in the easy direction is independent of K_u/K_1. The ratio of the remanence in the hard and the easy directions should vary from 1 for $K_u/K_1 = 0$ to 0.39 for $K_u/K_1 \geq 1$, provided that each crystallite possesses no anisotropy other than cubic before annealing.

LOSS OF REMANENCE DUE TO EXTERNAL STRESS

The loss of remanence due to an externally applied stress is very severe before annealing. It increases with the Co/Fe ratio in the film, probably due to the increase of magnetostriction constant λ. After the film is annealed in its remanent state, the loss in remanence becomes direction-dependent. It is much smaller if the film is magnetized in the easy direction before testing than it is when the film is magnetized in the hard direction. If a film is annealed in its demagnetized state, the loss in remanence will be isotropic and will have an intermediate value. These tests were performed by rubbing the surface of the film in its remanent state with a pencil eraser. The remanences, ϕ_1 and ϕ_2, before and after the rubbing, were measured. The remanence loss ΔR is reported as $100(\phi_1-\phi_2)/\phi_1$. Results on some samples are shown in Fig. 6.

CONCLUSION

From the preceding results one can conclude that effects comparable to magnetic annealing on the magnetic properties of a cobalt-ferrous film can be obtained by annealing the film in its remanent state. In order to compare the effects of annealing performed under different conditions, $W_\perp/W_{||}$ is plotted in Fig. 7 against the composition of the film in terms of the Co/Fe ratio, where $W_\perp$ and $W_{||}$ are the areas under the demagnetization curve of the $\perp$ and $//$ loop. The ratio $W_\perp/W_{||} = 1$ for a nonannealed sample and approaches zero in the ideal case for a single crystal. Figure 7 also shows that the highest annealing effects are obtained on films with a Co/Fe ratio around 0.2, corresponding to $x \sim 0.5$ in $Co_xFe_{3-x}O_4$. Also, for films with comparable Co/Fe ratios the effects are higher if the annealing is performed at lower temperature than it is if performed at higher temperature. All of these are consistent with what has been observed in the magnetic annealing. The hysteresis loop of a sample annealed in a magnetic field of 4000 gauss at 150°C for 18 hours is shown in Fig. 8.

ACKNOWLEDGEMENTS

The authors wish to thank Mr. E. Luke and Mr. K. Barnhart for their efforts in film preparation, Dr. J. W. H. Schreurs and Mr. J. D. LaBarre for their effort in measuring the magnetic properties of all samples and for their constant advice.

REFERENCES

1. R. M. Bozorth, E. F. Tilden, and A. J. Williams, Phys. Rev. 99, 1788 (1955).
2. H. P. J. Wijn, H. van der Heide and J. D. Fast, Proc. Inst. Elec. Eng. (London) 104, Part B Supp. 7, 412 (1957).
3. Shuichi Iida, Hisashi Sekizawa, and Yoshimichi Siyama, J. Phys. Soc. of Japan 1358 (1958).
4. R. F. Penoyer and L. R. Bickford Jr., Phys. Rev. 108, 271 (1957).
5. L. R. Bickford Jr., J. M. Brownlow and R. F. Penoyer, J. Appl. Phys. 29, 441 (1958).
6. J. C. Slonczewski, Magnetism, Vol. 1, p. 205, edited by Rado and Suhl, Acad. Press (1963).

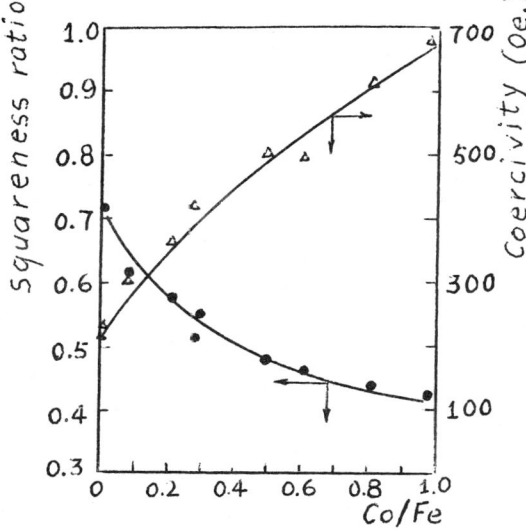

Figure 1

Coercivity and squareness ratio of Co-Fe-oxide films before annealing.

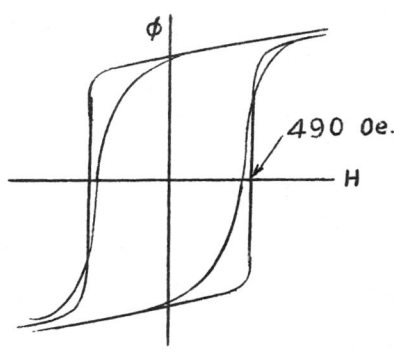

Figure 2

Magnetic hysteresis loop of a Co-Fe-oxide film annealed in the remanent state for 147 hours at 150°C. (Co/Fe = 0.12)

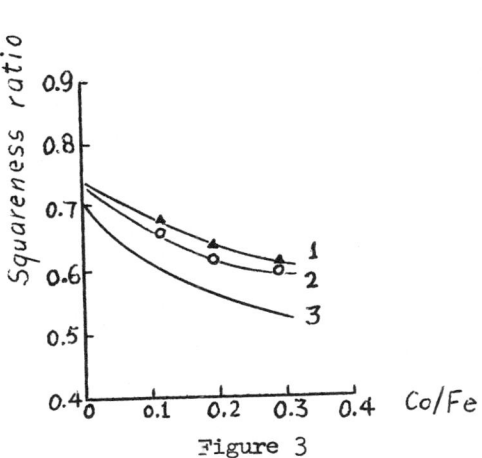

Figure 3

Squareness ratio of Co-Fe-oxide films annealed for 147 hours at 150°C. (1) In the easy direction, (2) in the hard direction, (3) before annealing.

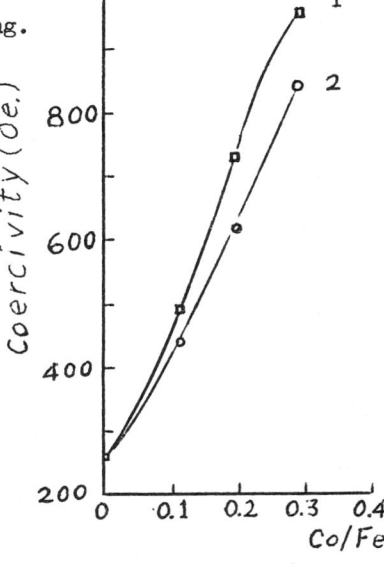

Figure 4

Coercivity of Co-Fe-oxide films annealed for 147 hours at 150°C. (1) In the easy direction, (2) in the hard direction.

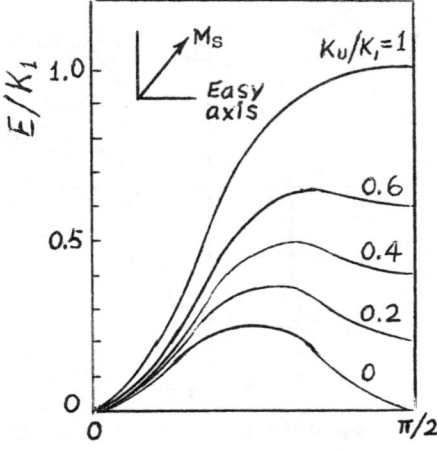

Figure 5

Reduced anisotropic energy E/K_1 as a function of the angle θ between the magnetization M_s and the easy axis in the (100) plane ($E = K_1 \sin^2\theta \cos^2\theta + K_u \sin^2\theta$)

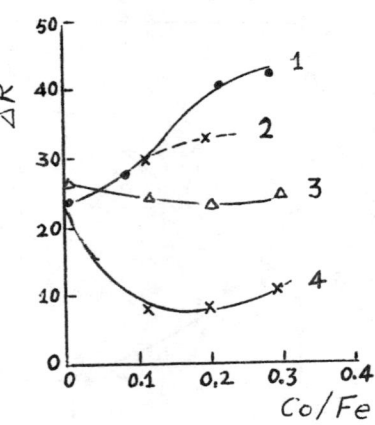

Figure 6

Remanence loss due to an external stress vs Co/Fe ratio. (1) as reduced, (2) annealed and tested in the hard direction (3) annealed in demagnetized state (4) annealed and tested in the easy direction.

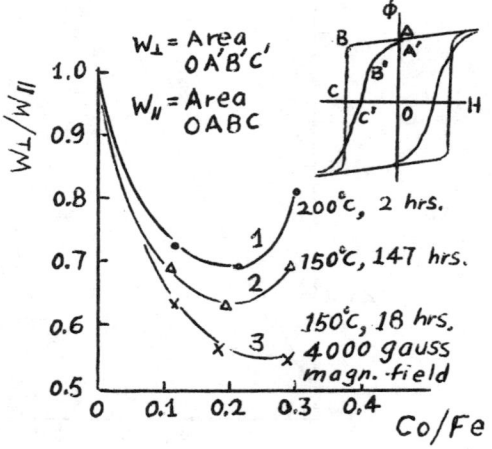

Figure 7

$W_\perp/W_\parallel$ vs Co/Fe for Co-Fe-oxide films (1) and (2) annealed in the remanent state (3) in a magnetic field.

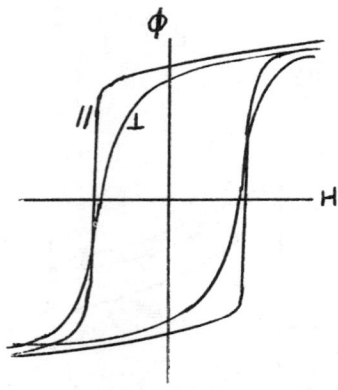

Figure 8

Hysteresis loop of a Co-Fe-oxide film annealed in a magnetic field of 4000 gausses at 150°C for 18 hours.

CHARACTERIZATION OF THE GROWTH INDUCED ANISOTROPY IN A MIXED RARE EARTH IRON GARNET

J. F. Dillon, Jr., E. I. Blount, and E. M. Gyorgy
Bell Laboratories, Incorporated, Murray Hill, New Jersey   07974

ABSTRACT

We have carefully measured the magnetic anisotropy of samples cut from (112) and (110) growth surfaces of flux grown $Gd_{2.34}Tb_{.66}Fe_5O_{12}$. The experimental technique included the use of spherical samples, precision x-ray orientation, correction for the non-collinearity of $\underline{M}$ and $\underline{H}$, and the measurement of torque about all <110> axes. Energy curves were analyzed in terms of the most general second and fourth order terms. We find: 1. The room temperature anisotropy consists only of a cubic K_1 term and a growth induced anisotropy quadratic in the direction cosines; 2. Energy surfaces of two distinctly different orientations are needed to describe the growth anisotropy under different areas of the (110) face; and, 3. Characterization of the growth induced anisotropy in terms of the two parameter model is less straightforward than has appeared to date.

INTRODUCTION

The recent broad interest in the growth induced anisotropy (GIA) of mixed rare earth iron garnets has produced a number of questions about the character of that anisotropy. These include reports of a fourth order contribution to the GIA[1,2], unexpected domain patterns on the (110) faces of flux grown crystals, and anomalies in the orientation of the easy axes under (112) faces. The rough success of the two parameter model in correlating the growth induced anisotropy seen under the (110) and (112) surfaces[3] seemed to call for further examination. In this work we have undertaken careful measurements of the magnetic anisotropy of samples cut from under both growth faces of flux grown $Gd_{2.34}Tb_{.66}Fe_5O_{12}$. The experiments were performed so as to give us confidence in the crystallographic orientation of the observed anisotropy energy surface and so as to be sensitive to the appearance of a fourth order contribution. Space limitations dictate that we truncate our presentation severely. A fuller account will be given elsewhere.

ANISOTROPY EXPRESSION

At the outset it appeared that we would encounter an arbitrarily oriented growth induced anisotropy with second and fourth order terms in the direction cosines. These would be superimposed on the ordinary cubic terms. Thus we have used an expansion of the anisotropy energy which would encompass any such possibility. The particular complete set of orthonormal polynomials chosen is given in Table 1. The coefficient B_k corresponds to the usual first order cubic anisotropy constant K_1. Projection of an arbitrary second and fourth order energy surface on each of these polynomials, gives components which constitute a complete specification.

Table I Anisotropy energy up to fourth order

$$E = A_0(3\alpha_3^2-1)/\sqrt{3} + A_1 \sqrt{2}\alpha_3(\alpha_1+\alpha_2) + A_2 2\alpha_1\alpha_2$$
$$+ M_1 \sqrt{2}\alpha_3(\alpha_1-\alpha_2) + M_2(\alpha_1^2-\alpha_2^2)$$
$$+ B_k[5(\alpha_1^2\alpha_2^2+\alpha_2^2\alpha_3^2+\alpha_3^2\alpha_1^2)-1]/5$$
$$+ B_u[7(2\alpha_3^4-\alpha_1^4-\alpha_2^4)-6(2\alpha_3^2-\alpha_1^2-\alpha_2^2)]/2$$
$$+ B_1 \sqrt{3}\alpha_3(7\alpha_3^2-3)(\alpha_1+\alpha_2)/2$$
$$+ B_2 \sqrt{3}\alpha_1\alpha_2(7\alpha_3^2-1)$$
$$+ B_3 \sqrt{21}\alpha_3(\alpha_1+\alpha_2)(4\alpha_1\alpha_2-1+\alpha_3^2)/2$$
$$+ L_1 \sqrt{3}\alpha_3(7\alpha_3^2-3)(\alpha_1-\alpha_2)/2$$
$$+ L_2 \sqrt{3} (7\alpha_3^2-1)(\alpha_1^2-\alpha_2^2)/2$$
$$+ L_3 \sqrt{21}\alpha_3(\alpha_1-\alpha_2)(4\alpha_1\alpha_2-1+\alpha_3^2)/2$$
$$+ L_4 \sqrt{21}\alpha_1\alpha_2(\alpha_1^2-\alpha_2^2)$$

SAMPLES

Our samples were cut from the same large single crystal of $Gd_{2.34}Tb_{.66}Fe_5O_{12}$ used in several earlier studies.[1,4] Polished spheres with a marked [110] and a [111] direction were prepared by the two pipe method[5] from the desired locations. For each torque curve these were mounted on a ground quartz rod with the desired <110> axis within 0.2° of the rod axis, and the position of a <100> axis in the plane known to a like accuracy. The locations from which the samples were taken were determined largely from an examination of the domain patterns on the face. For the (112) face these indicated that a single sample would be typical of the entire face. The domain pattern on the ($\bar{1}$01) face was divided into four quadrants as indicated in Fig. 1. We took samples which clearly derived from each of four quadrants shown in this figure. The center of this quadrant system is fixed in the crystal, and is associated with a single pitted growth center. Other morphological details must be described elsewhere.

TORQUE MEASUREMENTS

For each of the six spheres, the torque due to magnetocrystalline anisotropy was measured at room temperature about the six <110> axes.

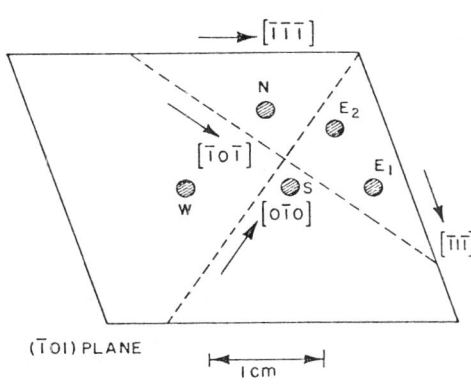

Fig. 1 Sketch of the ($\bar{1}$01) growth surface. The dashed lines represent a division into quadrants which could be clearly seen in the domain patterns. The labeled circles represent the locations from which samples were prepared. Some crystallographic directions in the plane are indicated by arrows.

Since the applied field is not sufficiently large compared to the effective anisotropy field, the magnetization $\underline{M}$ and the applied field $\underline{H}$ are not necessarily collinear. Consequently the experimental torque curve does not immediately give the correct anisotropy constants. The experimental data are corrected for this noncollinearity and the twofold and fourfold terms appropriate for an infinite field are obtained.[6] The coefficients of these terms are linear combinations of the A's, M's, B's, and L's. Thus the energy expansion of Table I and the measured torque curves yield 24 linear equations in the 14 normalized energy coefficients. For each sphere this system was solved to yield a best fit in the least squares sense.

The eigenvectors and **eigen**energies of the GIA for the (112) sample are displayed in Fig. 2(a). In this we have suppressed as nonessential the slight deviation of the hard and easy axes from the ($\bar{1}$10) symmetry plane. The corresponding results for the ($\bar{1}$01) sample designated E_2 are given in Fig. 2(b). For samples N, E_1, and E_2 the hard axis is tilted as shown here and lies near [$\bar{1}\bar{1}$1], whereas for the samples W and S the hard axis is tilted the other way and lies correspondingly close to [$\bar{1}$11].

DISCUSSION

The review paper by Rosencwaig and Tabor[7] gives many references to earlier work on the growth induced anisotropy and its relation to the two parameter phenomenological theory set forth in Ref. 8. In that theory a special axis $\hat{\beta}$ entered the anisotropy through a parameter A insofar as it lay along a 4-fold axis and through a parameter B insofar as it lay along a 3-fold axis. The direction $\hat{\beta}$ was assumed normal to the growth face. The full expression is simply:

$$E = A(\alpha_1^2\beta_1^2 + \alpha_2^2\beta_2^2 + \alpha_3^2\beta_3^2) + B(\alpha_1\alpha_2\beta_1\beta_2 + \alpha_2\alpha_3\beta_2\beta_3 + \alpha_3\alpha_3\beta_3\beta_3) \quad (1)$$

It is interesting to interpret the present data in terms of the two parameter model.

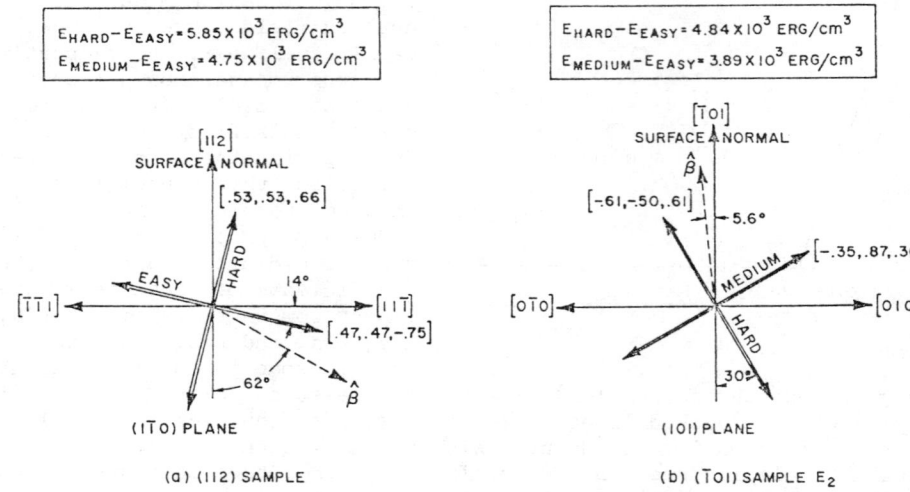

Fig. 2 Eigenvectors and eigenergies of the GIA seen in (a) the (112) sample, and (b) the ($\bar{1}$01) sample E_2. The vectors are shown relative to crystallographic directions. The vectors $\hat{\beta}$ for the two samples are referred to in the "Discussion".

Our observation of the tilted eigenvectors of Fig. 2(b) implies directly that application of the two parameter model requires different directions for the vector $\hat{\beta}$ in the two halves of the face. If $\hat{\beta}$ lies in the (101) plane it can indeed produce tilted eigenvectors as observed. In fact if we regard A, B, and $\hat{\beta}$ = [$\bar{1}$x1] as unknowns, the data of Fig. 2(b) may be used to form a soluble set of three equations in three unknowns. The values obtained from this system are given the first line of Table II.

Table II A and B as derived from data

Source	A (10^3 erg/cm^3)	B (10^3 erg/cm^3)	$\hat{\beta}$
($\bar{1}$01) Data	-2.92	8.70	[-.70,-.16,.70]
(112) Data	-2.10	-14.5	[1, 1,-2.8]

As illustrated on Fig. 2(b) this $\hat{\beta}$ is 5.6° away from the surface normal. A vector $\hat{\beta}$, deviating in the opposite direction from the surface normal would apply to samples S and W. We may treat the (112) data similarly, assume $\hat{\beta}$ has the form [11x] and solve for A, B, and x. The values obtained are given on the second line of

Table II. The special direction β obtained in this solution has no obvious connection with the normal to the face, and the values of A and B are distinctly different from those obtained under $(\bar{1}01)$.

In a survey of seven different mixed rare earth garnets Hagedorn et al.[3] measured the GIA in samples cut from under (112) and (110) growth surface. When their results were cast in the simple A, B formalism, it was found possible to choose values for A and B which roughly yield the medium and hard energies and axes of samples from both faces of each composition. We may parallel their procedure, and use the value of A and B obtained from the $(\bar{1}01)$ face to predict the eigenvalues with $\hat{\beta}$ as [112]. We find that $\hat{\alpha}^{easy} = [.36, .36_3, -.86]$, $E_{hard} - E_{easy} = 4.65 \times 10^3$, and $E_{medium} - E_{easy} = 1.96 \times 10^3$ erg/cm^3. This predicted easy direction is $10°$ from that observed. While the energies are substantially different from those given in Fig. 2(b) they do order correctly.

The lack of discernible correlation between the two lines of Table II should not obscure the broad applicability of the two parameter interpretation implicit in the fact that it does roughly correlate the energy surfaces seen under the two growth faces. A detailed analysis of the relationship between the data now in hand and the very simple two parameter model is beyond the scope of this paper, and must be presented elsewhere. Such an analysis strongly suggests that the two parameter model has some significance, but that a more refined theoretical treatment is clearly called for.

ACKNOWLEDGMENTS

We are pleased to thank E. Heilner for his skillful operation of the torque magnetometer and his creative management of the data. Similarly we thank R. T. Lynch, Jr. for several computer programs and for much of the sample preparation. As before we are grateful to J. W. Nielsen for providing the large sound crystal from which these samples were prepared. We thank F. B. Hagedorn for several discussions.

REFERENCES

1. E. M. Gyorgy, J. F. Dillon, Jr., J. P. Remeika, AIP Conference Proceedings, No. 5, Magnetism and Magnetic Materials-1971 (American Institute of Physics, New York, 1972), p. 680.
2. R. C. LeCraw and R. D. Pierce, Ref. 1, p. 200.
3. F. B. Hagedorn, W. J. Tabor, and L. G. Van Uitert, J. Appl. Phys. (In Press).
4. J. F. Dillon, Jr., E. M. Gyorgy, and J. P. Remeika, Ref. 1, p. 190.
5. W. L. Bond, Rev. Sci. Instr. 25, 401 (1954).
6. This correction was not applied to the experimental data discussed in Ref. 1, and as a result the variation of K_1 reported there is an artifact. The correct results are that K_1 has the value reported here, and is independent of position and thermal history.
7. A. Rosencwaig and W. J. Tabor, Ref. 1, p. 57.
8. E. M. Gyorgy, A. Rosencwaig, E. I. Blount, W. J. Tabor, and M. E. Lines, Appl. Phys. Letters 18, 479 (1971).

STATIC MAGNETIC TORQUE MEASUREMENTS ON A SYSTEM OF FERROMAGNETIC LAYERS, COUPLED BY FEEBLE ANTIFERROMAGNETIC INTERACTIONS ; WEAK FERROMAGNETIC BEHAVIOUR

P.Bloembergen, P.J. Berkhout and J.J.M. Franse
Natuurkundig Laboratorium, University of Amsterdam, The Netherlands

ABSTRACT

Magnetic torque measurements are presented on the compound $(C_2H_5NH_3)_2CuCl_4$, a covert weak ferromagnet, which in the ordered state below 10.2 K consists of an antiferromagnetically ordered assembly of weakly coupled ferromagnetic layers. Detailed information about the magnetic structure can be obtained by measurements in the (0,0,1) and (0,1,0) crystal planes at relatively low fields (H<15kOe)

In this paper we present torque measurements, which were performed in connection with the study of the two-dimensional spin 1/2 Heisenberg ferromagnet by means of the series of layered copper compounds $(C_nH_{2n+1}NH_3)_2CuX_4$, where X stands for Cl or Br and n has been varied from 1 to 10 [1]. The measurements were performed mainly at 1K (~0.1 T_c) in fields up to 15 kOe on thin disk shaped single crystals. The relevant parameters, which constitute the difference between the ideal, 2-dimensional, isotropic system and the real compounds, notably the anisotropy and the interlayer coupling, can be conveniently determined by this method, if the latter has the antiferromagnetic sign, as is the case[2] for $(C_2H_5NH_3)_2CuCl_4$. The orthorhombic unit cell of this compound, containing 4 copper ions, belongs to the space group Pbca; the a and b axes are oriented parallel to the layers containing the copper ions, a and c define the direction of easy and difficult magnetization within and perpendicular to the layers, respectively. Below, we will replace a, b and c by x, y and z, respectively.

Ref.3 contains a preliminary study, in which the results of the measurements in the xy(0,0,1) plane were analyzed in terms of the parameters of a uni-axial antiferromagnet. In this paper we will stress the weak ferromagnetic (WF) properties, which become apparent only by turning the probing field in a plane, containing the z-axis. The experimental results agree with a two-sublattice molecular field model at 0K, with the following expression for the free energy F (per unit of magnetic moment):

$$F/M_0 = - H_E + H_E' \vec{\sigma}_A \cdot \vec{\sigma}_B + \\
+ H_{A1} (\sigma_{Ay}^2 + \sigma_{By}^2)/2 + H_{A1}' \sigma_{Ay} \sigma_{By} + \\
+ H_{A2} (\sigma_{Az}^2 + \sigma_{Bz}^2)/2 + H_{A2}' \sigma_{Az} \sigma_{Bz} - \\
- H_d (\sigma_{Ay} \sigma_{Az} - \sigma_{By} \sigma_{Bz}) - \\
- \vec{H} \cdot (g_A \vec{\sigma}_A + g_B \vec{\sigma}_B)/g_{xx} \qquad (1)$$

Definition of the parameters:

$\vec{\sigma}_A, \vec{\sigma}_B$: unit spin vectors, associated with the spinmomenta $\vec{S}_A$ and $\vec{S}_B$ appropriate to the sublattices A and B.

H_E : <u>intra</u>layer ferromagnetic interaction field.
H_E' : <u>inter</u>layer antiferromagnetic interaction field (primed symbols apply to <u>inter</u>layer coupling terms).
$H_{A1}, H_{A1}', H_{A2}, H_{A2}'$: <u>intra</u>- and <u>inter</u>layer anisotropy fields with respect to spin rotations in the easy (xy) plane (suffix 1) and in the xz plane (suffix 2).
H_d : intralayer anisotropy field, of opposite sign in A and B and representing the main source of the WF properties.
$\vec{H}$: external field.
g_A, g_B: symmetric g-tensors, relating the sublattice momenta $\vec{M}_A$ and $\vec{M}_B$ to $\vec{S}_A$ and $\vec{S}_B$ and containing the diagonal elements g_{xx}, g_{yy} and g_{zz}. The off-diagonal elements $g_{yz}=g_{zy}$- opposite for g_A and g_B-account for about 10% of the WF properties.
M_0 : zero field sublattice magnetization.

The effective fields H_E, H_E', etc. are defined to contain the exchange, as well as the dipole-dipole contributions, including the Lorentz and demagnetizing fields (cf. table I). The Lorentz fields were derived from the Lorentzcoefficients[4], which were calculated for the series of compound by Colpa[5]. The values of the demagnetizing fields in table I apply to a particular sample: weight: 25 mg, diameter: 5.6 mm, thickness 0.6 mm. The numbers in parentheses (1) and (2) in table I refer to susceptibility (De Jongh a.o.[2]) and e.s.r. data (Vega and Maarschall, priv.comm.) respectively.

In the underlying magnetic structure, the antiferromagnetic type of ordering is a consequence of the relative values of the various effective field parameters. In table I, H_c, H_{sy} and H_{sz} mark the transition into the WF phase ($S_{Ax}-S_{Bx}=0$), when the ext. field is directed along the x,y and z axis respectively; H_{sx} represents the transition into the ferromagnetic phase ($\vec{S}_A//\vec{S}_B$) for fields along the x axis. The first order spin-flop transition at $H=H_c$ ($\vec{H}//x$) goes together with a slight tilting out of the xy plane of $\vec{S}_A$ and $\vec{S}_B$ (tilting angle $\phi_x \simeq 2H_d/(H_{A2}+H_{sx}) \simeq 4°$, cf. fig.1a): the covert weak ferromagnetism becomes overt. At $H=H_{sy}$ ($\vec{H}//y$) and $H=H_{sz}$ ($\vec{H}//z$), $\vec{S}_A$ and $\vec{S}_B$ are being turned into the yz plane via a 2nd order transition, with canting angles ϕ_y and ϕ_z with respect to the y and z axis respectively, where ϕ_y (A,B) $\simeq \pm \text{arctg}(H_d/H_{A2}) \simeq \pm 5°$ and $\phi_z(A,B) \simeq \pm \text{arctg}(H_d/H_{A1}) \simeq \pm 60°$ (cf. fig.1b and 1c).

The computations of the torque curves $\vec{L}(\vec{H})$, where $\vec{L}=(\vec{M}_A+\vec{M}_B)\wedge\vec{H}$, were performed numerically by minimizing the free energy F at each field direction by means of an iterative procedure. The full curves of figs 2-6 are calculated with the fitted parameters of table I; the experimental values are indicated by open or closed dots. θ_H represents the angle between $\vec{H}$ and the easy (x) axis, unless stated otherwise; $L_{extr.}$ denotes the extreme value of L(H=const.), attained at a particular value of θ_H.

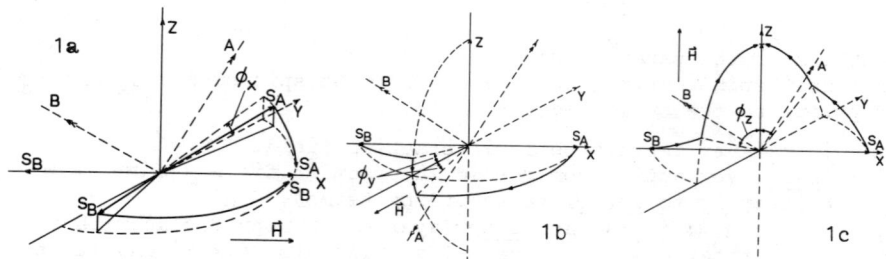

Fig.1. Field dependence of the spin configurations for fields parallel to the x, y and z-axis (Fig.1a: $H > H_c$).

Fig.2 shows the dependence of the torque L_z along the z axis on the external field H, parallel to the xy-plane. The response of the system in this plane is characteristic for a uni-axial antiferromagnet (AF)[3] and behaves quite differently in the field range below H_c, the intermediate range between H_c and H_{sx} and in the high field range above H_{sx}, where it saturates into a sinusoidal relationship between L_z and θ_H.

In the latter region $f_{yy} \equiv g_{yy}/g_{xx}$ and $H_{A1} + H'_{A1}$ can be inferred from the slope and the H=0 intersection of the linear part of $L_{z,extr.}$ vs. H respectively. (Fig.3, insert). In the intermediate field region we have $L_z(\theta_H = \text{const.}) \sim H^2$ (fig.3) and first estimates of H'_E and H_{A1} can be obtained from simple analytical approximations. The influence of the non-zero value of H_d -which can be obtained only from measurements along another direction- has been indicated in fig.3; the resultant slope of curve 2 in the insert of fig.3 is essentially due to the non-zero value of H_d.

Fig.4 shows torque curves along the y axis, $\vec{H}$ being parallel to the xz(0,1,0) plane, calculated from the parameters of table I. Kinks or jumps in the curves mark the transition from the phase, characterized by $S_{Ax} - S_{Bx} \neq 0$, into the WF phase ($S_{Ax} - S_{Bx} = 0$), which occurs in turning the field ($H_c < H < H_{sz}$) from the difficult axis ($\theta_H = \pi/2$) towards the easy axis ($\theta_H = 0$). This transition is of 2nd order between $H = H_{sz}$, and a value of H intermediate between H_{sz} and H_c (H=350 Oe at $\theta_H = 30°$), where it becomes of 1st order[4] (cf.curve 3 in fig.4, in contrast to curves 4-11). Characteristic for the WF phase of a monodomain sample is the non zero value of L_y at fields $H_c < H < H_{sx}$ along the x axis, due to the field induced weak ferromagnetic moment ($\sim H_d$) perpendicular to $\vec{H}$. (See the values attained by L_y at $\theta_H = 0$ for $H > H_c = 320$ Oe in fig.4). Curves 4 and 4' or 10 and 10' show the characteristic differences in behaviour between a covert WF and a simple bi-axial AF ($H_d \equiv f_{yz} \equiv g_{yz}/g_{xx} = 0$, the other parameters being the same; the primed figures apply to the latter case).

Figs.5-7 show some of the experimental and calculated results in the low and high field region, $\vec{H}$ being parallel to the (1,2,0) plane. The zero value of L at $\theta_H = 0$ (fig.5, to the left; θ_H represents the angle between H and the xy plane), in contrast to the calculated

Fig.2. Exp. and calc. torque curves in the xy plane at T=1 K. To the left: $H<H_c$, to the right: $H>H_c$.

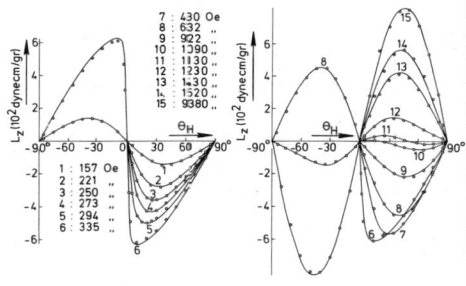

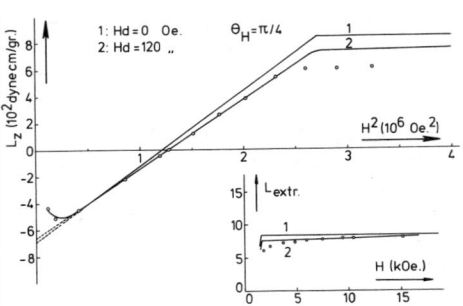

Fig.3. $L_z(\theta_H=\pi/4)$ vs. H^2 and $L_{z,\text{extr}}$ vs. H. Curves 1: $H_d=0$, the other parameters being the same (cf. table I).

Fig.4. Torque curves in the xz plane, calculated with the parameters of table I.

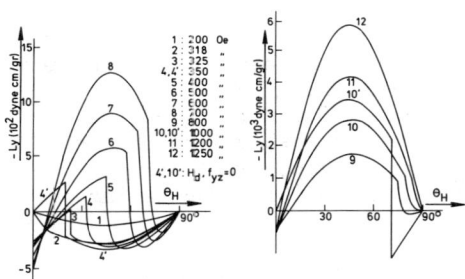

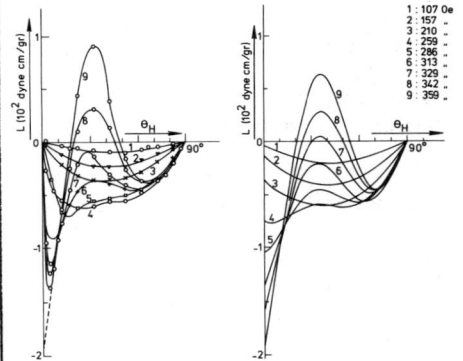

Fig.5. Exp. data (to the left) and calc. torque curves (to the right) in the (1,2,0) plane.

Fig.6. High field range. Exp. data and calc. torque curves in the (1,2,0) plane.

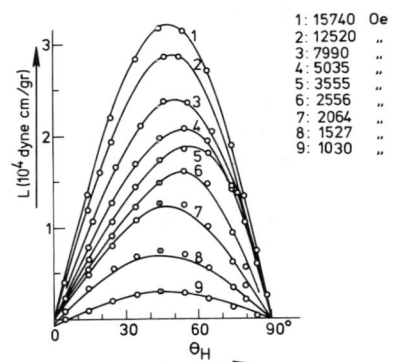

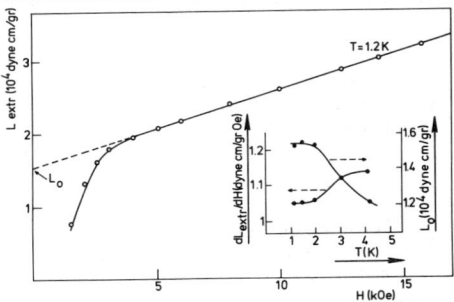

Fig.7. High field range. L_{extr} vs. H in the (1,2,0) plane. Full curve: calc. from the data of table I. Insert: temp. dependence of L_0 and dL_{extr}/dH between 1 and 4 K.

Effective fields	H_E	H_E'	H_{A1}	H_{A2}	H_{A1}'	H_{A2}'	H_d
Exp. values	5.10^5 (1)	830	76	1580	7.5	-45*	119
Demagn. fields	-15	15	-	150	-	150	-1
Lorentz fields	320	-.03	7	690	-	-195	-
Exchange fields	5.10^5	815	69	740	7.5	-	120
g-values				g_{xx}	f_{yy}	f_{zz}	f_{yz}
Exp. values				2.163 (2)	1.000	.9475	.0055*
Calculated transition fields	H_c	H_{sx}	H_{sy}	H_{sz}			
	320	1600	1730	1400			

<u>Table I</u>. Values of the parameter of interest, yielding the best fit between theory and experiment. For the definition of the parameters see the text; field values are in Oe. (*calculated values). The values of the exchange fields are obtained by substracting the dipolar contributions from the exp. values.

results (fig.5, to the right) reflects the presence of a ferromagnetic domain structure in the WF configuration.

Finally, from the high field data (figs 6 and 7) $H_{A2}+H_{A2}'$ and $f_{zz} \equiv g_{zz}/g_{xx}$ can directly be obtained from the slope and the H=0 intersect of the linear part of L_{extr} vs. H, respectively (fig.7). $f_{zz}<1$ corresponds to a non zero slope, H_d in this case having only a negligible influence.

As we have shown, from a simple analysis of torque measurements, detailed information can be obtained about the magnetic properties of antiferromagnetic structures, which are characterized by small values of the anisotropy and of the intersublattice coupling.

This work is part of the research program of the "Stichting voor Fundamenteel Onderzoek der Materie" and was made possible by financial support from the "Nederlandse Organisatie voor Zuiver Wetenschappelijk Onderzoek".

REFERENCES

1. A.R. Miedema, J. de Physique <u>32</u>, suppl. no 2-3, 305 (1971).
2. L.J. de Jongh, W.D. van Amstel and A.R. Miedema, Physica <u>58</u>, 277 (1972).
3. P. Bloembergen and J.J.M. Franse, Solid St. Comm. <u>10</u>, 325 (1972).
4. P. Bloembergen, P.J. Berkhout and J.J.M. Franse, to be published.
5. J.H.P. Colpa, Physica <u>56</u>, 185, 205 (1971).

STUDY OF THE SPIN REORIENTATION IN Co- AND Cr-SUBSTITUTED $YFeO_3$

E. Krén, M. Pardavi, Z. Pokó, E. Sváb and É. Zsoldos
Central Research Institute for Physics,
Budapest, Hungary

ABSTRACT

The effect of substituting Co or Cr for Fe in the systems $YFe_{1-x}Co_xO_3$ and $YFe_{1-x}Cr_xO_3$ has been studied by neutron and x-ray diffraction as well as magnetization measurements. In contrast with the earlier single crystal data, no spin reorientation could be observed at small values of x in polycrystalline samples. In the $YFe_{1-x}Cr_xO_3$ system a rotation of the magnetic moments in the a-c plane was found at around x=0.5. The observation of the spin reorientation in polycrystalline $YFe_{0.98}Co_{0.01}Ti_{0.01}O_3$ is taken as an evidence that the drastic effects of Co-doping can be attributed to the presence of Co^{2+} ions.

INTRODUCTION

Recently there has been renewed interest in the rare earth orthoferrites [1] because of their applicability to bubble domain devices [2,3]. Small bubbles, and hence high packing density, can be obtained in orthoferrites by utilizing the reduction of the uniaxial anisotropy which occurs near the spin reorientation temperature T_R. Above T_R, the Fe spins are aligned antiferromagnetically along the a axis, slightly canted along axis c, while below T_R, the spin alignment is along the c axis, slightly canted along axis a. At T_R the uniaxial anisotropy constant for the a-c plane is zero. Reduced bubble diameters can be obtained slightly above T_R at the expense of some temperature sensitivity.

In $YFeO_3$ such spin reorientation does not occur; the spins are aligned along the a axis and the weak net moment is directed along the c axis at all temperatures below the Néel point. However, small substitutions of Co or Cr cause the reorientation to appear near room temperature in flux-grown single crystals [4]. A strong effect of Co substitution on the reorientation was also observed in single crystals of $ErFeO_3$, $HoFeO_3$ and $DyFeO_3$ [5]. In order to study the effect of substitution, neutron and x-ray diffraction as well as magnetization measurements were performed on polycrystalline samples in the $YFe_{1-x}Co_xO_3$ /x≤0.1/ and $YFe_{1-x}Cr_xO_3$ /0≤x≤1/

systems. The samples were prepared from appropriate mixtures of "specpure" Y_2O_3, Fe_2O_3 and Co_2O_3 or Cr_2O_3. After mixing, the powders were pressed into pellet form and fired for 12 hours at 1300° C in air.

RESULTS

In the $YFe_{1-x}Cr_xO_3$ system samples with the nominal compositions x=0; 0.05; 0.1; 0.3; 0.5; 0.7 and 1 were studied. X-ray diffraction Debye-Scherrer photographs showed the samples to be of single-phase orthorhombic perovskite structure with lattice parameters varying continuously between x=0 and 1. Similarly, the value of T_c determined by both neutron and magnetization measurements changes continuously, as shown in Fig. 1.

Using the neutron diffraction method, the occurrence of spin reorientation can be detected in powdered samples by measuring the intensity ratio of the (011) and (101) magnetic reflections. If the antiferromagnetic moments lie along the a axis /$T > T_R$/, reflection (011) is about three times higher than (101), while if the moments point in the direction of axis c /$T < T_R$/, the two reflections are nearly equal.

In contrast with the earlier reported results [4], no reorientation could be observed at small value of x. However, around x=0.5 a rotation of the magnetic moments in the a-c plane was observed at low temperatures. According to the neutron diffraction measurements at x=0.5, below 200°K the antiferromagnetic moments lie in the a-c plane at an angle of 30±10° to axis a. The temperature dependence of the angle Θ of the weak ferromagnetic moment to axis c, as measured by magnetic methods in a flux-grown single crystal with x=0.38, is plotted in Fig. 2. It can be seen that the rotation of the spins terminates at the intermediate angle of 45°. Similar magnetic structure was observed

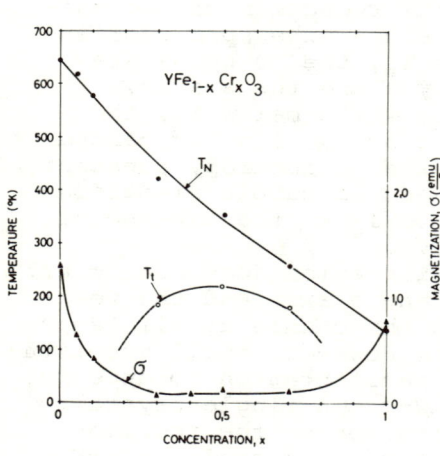

Fig. 1. Magnetic transition temperatures and weak ferromagnetic moment in $YFe_{1-x}Cr_xO_3$.

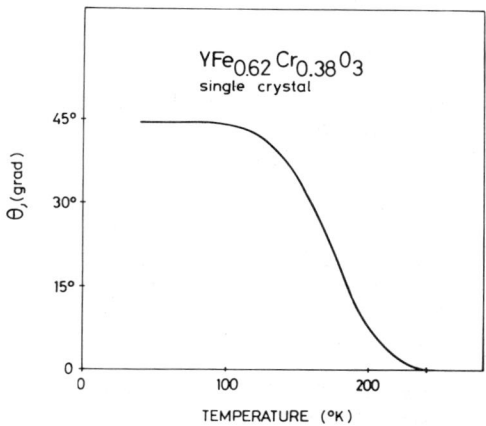

Fig. 2. Temperature dependence of angle Θ of the weak ferromagnetic moment to axis c.

in $TmCrO_3$, $YbCrO_3$ and $LuCrO_3$[6]. The values of T_t in Fig. 1 represent the temperatures below which the magnetic moments turn in the a-c plane. The values of the weak ferromagnetic moment σ measured at 90°K are also shown in Fig. 1; σ is decreased at intermediate compositions, indicating a decrease in the Dzialoshinsky coupling. These observations are in accord with the results reported by Belov et al.[7] from magnetic measurements.

In the $YFe_{1-x}Co_xO_3$ system, polycrystalline samples with the nominal compositions x=0; 0.005; 0.01; 0.05 and 0.1 were investigated. Debye-Scherrer photographs revealed traces of a garnet-type impurity phase. However, the decrease in the Néel temperature, determined from the temperature dependence of the neutron intensities, indicated that the Co ions are at least partly built in the perovskite-type phase. The values of T_N were found to be 650±10° and 600±10°K at x=0 and 0.1, respectively. On raising x, a reduction in the lattice parameters was also observed.

Neutron diffraction experiments demonstrated, in contrast with the single crystal study[4], that no reorientation occurs in the studied concentration range; the antiferromagnetic moments are directed along axis a between 6°K and T_N.

In order to study the effect of substitution of Co^{2+} for Fe^{3+}, a powdered sample with the composition $YFe_{0.98}Co_{0.01}Ti_{0.01}O_3$ was prepared. Neutron diffraction measurements showed that below room temperature the antiferromagnetic moments are parallel to axis c, as in flux-grown single crystals containing the same amount of Co[4], and the reorientation occurs above 230°C. Thus the drastic effects of Co-doping in single crystals[4,5], can be attributed to the presence of Co^{2+} ions which are built into the lattice together with fluorine ions from the flux. This assumption is corroborated by the demonstration of fluorine in flux-

-grown Co-doped single crystals by chemical analysis [8].

ACKNOWLEDGMENTS

The authors are indebted to Prof. L. Pál and Dr. G. Zimmer for valuable discussions and to Dr. A. M. Kadomtseva for providing the single crystal.

REFERENCES

1. R.L. White, J. Appl. Phys. <u>40</u>, 1061 /1969/.
2. A. H. Bobeck, Bell System Tech. J. <u>46</u>, 1901 /1967/.
3. L. G. Van Uitert, R. C. Sherwood, W. A. Bonner, W. H. Grodkiewicz, L. Pictroski and G. Zydzik, Mat. Res. Bull. <u>5</u>, 153 /1970/.
4. L. G. Van Uitert, R. C. Sherwood, E. M. Gyorgy and W. H. Grodkiewicz, Appl. Phys. Letters <u>16</u>, 84 /1970/.
5. L. Holmes, L. G. Van Uitert and R. Hecker, J. Appl. Phys. <u>42</u>, 657 /1971/.
6. E. F. Bertaut, J. Maréschal, G. de Vries, R. Aléonard, R. Pauthenet, J. P. Rebouillat and V. Zarubicka, IEEE Trans. Magnetics <u>MAG-2</u>, 453 /1966/.
7. K. P. Belov, M. A. Belanchikova, A. M. Kadomtseva, T. M. Ledneva, M. M. Lukina, T. L. Ovchinnikova and L. P. Slahina, Fiz. Tverd. Tela <u>14</u>, 944 /1972/.
8. A. M. Kadontseva /private communication/.

MEASURING MAGNETOSTRICTION WITH AN RSM AND RFM

P. J. Flanders
University of Pennsylvania, Philadelphia, Pa. 19104

ABSTRACT

With a rotating sample magnetometer operating at ω, typically between 5 and 50 rev/sec, or with a rotating field magnetometer in which the field spins at an $\omega > 1$ rev/sec, one can measure magnetostriction on single and polycrystal specimens. Although sensitivity is limited by slip-ring noise when the former method is used, high gage factor semiconductor gages ($K \approx 100$) make this limitation of secondary importance. Since the method, which uses a half rather than a full bridge circuit, relies on periodic changes in the gage resistance R_G, the output to first order is insensitive to temperature changes which cause R_G to vary gradually or in a somewhat random fashion. It is for this reason that a large temperature coefficient of resistance for a strain gage does not impose a serious measuring problem.

STANDARD FULL BRIDGE METHOD

For the standard magnetostriction measuring technique in which strain gages are employed as sensors[1], a change in sample length ℓ which occurs as the field rotates in the plane of the gage through an angle ψ relative to one of the gage axes will produce a variation in resistance of the gage that is located in one arm of a balanced full bridge circuit. The subsequent bridge imbalance signal ΔV is proportional to ΔR_G and if the temperature changes during the time required for a measurement there can be a ΔV associated with the ΔT. When $\Delta \ell / \ell \approx 10^{-6}$, temperature variations of more than 0.1 K during a measurement ($\approx 10^2$ sec for a rotation of ψ through 2π) can cause a $\Delta R_G(T)$, due to the temperature coefficient of gage resistance or the linear thermal expansion coefficient of sample and/or gage, which is greater than that produced by magnetostriction, $\Delta R_G(\psi)$.

ROTATING SAMPLE MAGNETOMETER (RSM) AND ROTATING FIELD MAGNETOMETER (RFM)

RSM[2] and RFM[3] measurements are made by detecting an a.c. voltage at $2n\omega$ ($V2n\omega$) appearing across a resistor R_B, usually equal to R_G, that is in series with a battery E_B and a strain gage which is fixed to a sample. The arrangement is commonly referred to as a potentiometer (or half bridge) circuit and uses a single gage (Fig. 1). For the RSM, contact with the gage leads as the

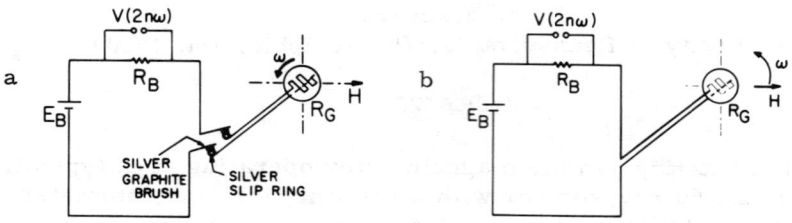

Fig. 1. A half bridge circuit which is used with, (a) an RSM, $\omega \sim 5\text{-}50$ rev/sec, and (b) an RFM, $\omega > 1$ rev/sec, (E is 1-5 volts; R_G and R_B are typically 120 ohms).

sample spins about an axis normal to the field H is made through coin-silver rings and silver-graphite brushes (20% graphite). When measurements are made as a function of temperature a comparable set of rings and brushes is used for connection to a thermocouple. Coherent noise at $2n\omega$ and d.c. thermal voltages, both associated with slip-ring contacts, are respectively $\approx 10^{-6}$ and 10^{-5} volts. An alternative method of rotating the field rather than the sample at constant speed has the advantage of eliminating slip-ring noise. To utilize lock-in techniques most conveniently, the frequency of the rotating field should be at least one rev/sec, this generally limits fields to values which can be generated by a.c. techniques or by spinning a small electromagnet. Since only d.c. is applied to the gage circuit, any a.c. voltage across R_B that is at an even harmonic of ω is due to changes in R_G which are periodic in $1/2n\omega$ ($\approx 10^{-1}, 10^{-2}$ sec); this voltage can be expressed as

$$V(2n\omega) = \frac{E_B R_B R_G}{(R_B + R_G)^2} \left[\Delta R_G(2n\omega t)/R_G \right], \quad (1)$$

$V(2n\omega)$ is measured with an a.c. lock-in amplifier; the lock-in output $S(2n\omega)$ is a full-wave rectified signal having a d.c. value equal to $G\sqrt{2}/4$ times the peak of $V(2n\omega)$ where G is the amplifier gain. The quantity $\Delta R_G(2n\omega t)/R_G$ is equal to $K\Delta\ell(2n\omega t)/\ell$ with an n of 1 for most experiments. Slow variations in temperature produce no output since they are not periodic in $1/2 n\omega$; there are however changes in the magnitude of $V(2n\omega)$ if R_G changes value. When $R_B = R_G$ the peak to peak voltage at 2ω is $E_B \left(R_{G_{||}} - R_{G_\perp} \right)/4R_G$ and the lock-in output is therefore

$$S(2\omega) = \frac{G\sqrt{2}\ E_B \left(R_{G_{\shortparallel}} - R_{G_{\perp}}\right)}{16\ R_G} \qquad (2)$$

To relate magnetostriction constants λ to $S(2\omega)$, the quantity $3K\Delta\lambda/2$ can be substituted for $\left(R_{G_{\shortparallel}} - R_{G_{\perp}}\right)/R_G$ so that

$$S(2\omega) = \left(\frac{G\sqrt{2}\ E_B}{16}\right)\left(\frac{3K\Delta\lambda}{2}\right), \qquad (3).$$

In Table I, values of $\Delta\lambda(\psi)$ are listed for randomly oriented polycrystals and for several of the more useful single crystal orientations.

Table I. $\Delta\lambda(\psi)$ for randomly oriented polycrystals and for several single crystal orientations; $\alpha \cong 2/5$ (ref. 4).

	plane	gage axis	$\Delta\lambda(\psi)$
SINGLE CRYSTAL	(100)	[100]	λ_{100}
		[110]	λ_{111}
	(110)	[100]	λ_{100}
		[111]	λ_{100}
RANDOM POLYCRYSTAL			$\alpha\lambda_{100} + (1-\alpha)\lambda_{111}$

RSM RESULTS

Typical RSM plots of $S(2\omega)$ versus field are given in Fig. 2 using a Kyowa semiconductor gage which is first cemented to an α-Fe_2O_3 crystal and then freed with a solvent. The signal from the free gage is due to its magnetoresistance. ω is 11.4 rev/sec for these experiments. If semiconductor gages are used[5], the limiting factor is not generally signal to noise ratio which is $\approx 10^2$ when the sample has a λ of 10^{-6} but rather it is the accuracy with which a signal due to gage magnetoresistance can be separated from the total output. In our experiments this signal is equivalent to a λ of 2×10^{-7} at 10 kOe when Kyowa gages are used. When using wire gages the limiting factor is sensitivity since K will be almost two orders of magnitude below that for semiconductor gages. Since slip-ring noise is equivalent to a λ of 10^{-6}, rotating wire gages are generally unsuitable for measuring magnetostriction in specimens with small values of λ.

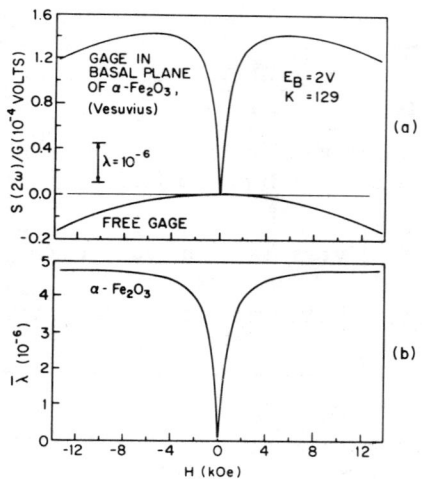

Fig. 2. For α-Fe_2O_3 from Vesuvius, (a) RSM plots (ω=11.4 rev/sec) of $S(2\omega)/G$ when a Kyowa semiconductor gage (K=129) is in the c-plane and then freed, (E_B= 2 volts and $R_G = R_B$ = 120 ohms), (b) the resultant c-plane λ_s.

Fig 3. RFM plots (ω = 1.5 rev/sec) of $S(2\omega)/G$ and $\Delta\lambda$ vs (a) field and (b) temperature, for synthetic α-Fe_2O_3 with a Kyowa wire gage (K = 2.05) in the c-plane (E_B = 4.5 volts and $R_G = R_B$ = 120 ohms.

RFM RESULTS

Since the RFM eliminates slip-ring noise, wire gages with low anisotropic magnetoresistance can be used to measure samples which have small λ. At an ω of 1.5 rev/sec, wire gages have been used to obtain the curves in Fig. 3 where $S(2\omega)/G$ is plotted versus field and temperature for an α-Fe_2O_3 crystal. A non-magnetoresistive Kyowa wire gage (KN-3-Al) was cemented to a c-plane face which was in the plane of rotation. When the gage was removed from this sample the gage signal was equivalent to a

$\lambda \approx 10^{-7}$ which is within the range of gage magnetoresistance reported by Gersdorf[1]. If E_B is reversed one can check for components of V induced by the field at 2ω.

MAGNETORESISTANCE

With gages which exhibit magnetoresistance there is a contribution to the quantity $(R_{G_{\parallel}} - R_{G_{\perp}})/R_G$ in Eq. (2) from anisotropic magnetoresistance; this was found to vary as AH^2 for semiconductor gages designated as BLH (#SPB2-06-12) and Kyowa (#KSP-2-E3). At 295K respective values of A were -5 and -1.9×10^{-12} Oe^{-2} with the signal from the latter equivalent to a λ of -10^{-6} at 10 kOe when $K \approx 120$. For some semiconductor gages an additional complex low field dependence is also observed.

ACKNOWLEDGEMENTS

The author thanks T. Egami who first suggested the RSM for measuring magnetostriction in single crystals of Dy, R. Pearson for proposing the use of a constant speed rotating magnetic field to eliminate slip-ring noise, and C.D. Graham, Jr. for his encouragement and for pointing out the feasibility of spinning a small electromagnet.

REFERENCES

1. R. Gersdorf, Thesis, Amsterdam (1961).
2. P.J. Flanders, Rev.Sci.Instrum. **41**, 697 (1970).
3. F.B. Hagedorn, Rev.Sci.Instrum. **33**, 450 (1967).
4. H.B. Callen and N. Goldberg, J.Appl.Phys. **36**, 976 (1965).
5. W.P. Mason, Crystal Physics of Interaction Processes, Academic Press, New York and London, 1966, Chapter 10.

Section 49. Magnetic Moments and Exchange Enhancement

SPIN FLUCTUATION RESISTIVITY IN A Pd-5%Rh MATRIX CONTAINING DILUTE CONCENTRATIONS OF Ni

D. J. Gillespie
Naval Research Laboratory, Washington, D. C. 20390
and American University, Washington, D. C. 20006
and
A. I. Schindler
Naval Research Laboratory, Washington, D. C. 20390

ABSTRACT

Low temperature (1.3 K to 20 K) electrical resistivity measurements have been made on five Pd-5%Rh alloys containing 0, 1/4, 1/2, 3/4, and 1 at.% Ni; magnetization measurements were also made on the 1% Ni specimen. The host, Pd-5%Rh, is thought to be an exchange-enhanced system like Pd, but, since it exhibits a magnetic susceptibility 70% greater than Pd, it is probably even more strongly enhanced. The results of the measurements indicate that this is the case. We find that the 1% Ni addition increases the susceptibility of the Pd-5%Rh host by more than twice as much as it increases the susceptibility of a pure Pd host. In comparing the resistivities of the two alloys we find, using the local enhancement model of Kaiser and Doniach[1], that the spin-fluctuation temperature of $(Pd-5\%Rh)_{99}Ni_1$ is about half that of $Pd_{99}Ni_1$, which, like the susceptibility data, is an indication of the stronger enhancement of the Pd-5%Rh host.

The spin-fluctuation resistivity contribution from the local enhancement at the Ni sites is obtained experimentally by subtracting the temperature dependent part of the resistivity of the Pd-5%Rh host from that of host plus Ni. This assumes that the small Ni additions have not changed the band structure and phonon effects of the host significantly. The two parameter, Kaiser-Doniach, universal resistivity-temperature function was separately least-squares-fitted to each of the four sets of subtracted data, and an excellent fit was obtained over the entire temperature range in each case. However, the two scaling parameters were functions of Ni concentration, whereas the single-impurity-limit nature of the K-D model implies concentration independent parameters. The resistivity scaling parameter ranged monotonically from 142 nΩ-cm/at.% for the 1/4% sample to 93 nΩ-cm/at.% for the 1% sample, while the temperature scaling parameter (the spin-fluctuation temperature) ranged monotonically from 53 K for the 1/4% to 36 K for the 1% sample. Thus, while the functional dependence of the resistivity upon temperature is that predicted by the model, the resistivity does not scale with concentration. Two possible reasons for this are (1) the host enhancement is significantly increased upon the addition of small amounts of Ni and (2) Ni-Ni interactions are important even at these low concentrations.

1. A. B. Kaiser and S. Doniach, Intern. J. Magnetism **1**, 11 (1970).

ANOMALIES IN THE MAGNETIC PROPERTIES OF Ni_3Ga ALLOYS

C.J. Schinkel, F.R. de Boer and B. de Hon
Natuurkundig Laboratorium der Universiteit van Amsterdam, The Netherlands

ABSTRACT

The magnetic properties at temperatures up to 80K and in fields up to 350 kOe of a number of paramagnetic Ni_3Ga alloys are reported. The results indicate that the magnetization as a function of field can be described very well in the Stoner model with the exception of magnetizations <4 emu./gram at low temperatures. The relatively rapid variation of the susceptibilities with temperature suggests a temperature dependence of the effective magnetic interaction.

INTRODUCTION

An extensive set of experiments has shown[1-4] that the magnetic properties of Ni_3Ga alloys can be described in the Stoner-Edwards-Wohlfarth model[5,3] for itinerant electron paramagnetism. Because of an improvement (sensitivity of the apparatus) and extension (fields up to 350 kOe) of the experimental facilities we repeated the magnetization measurements. The gross features of the results agree with those reported previously[1,2], but some important details will be reported and discussed in this paper.

EXPERIMENTAL

The details of the preparation of the samples are given elsewhere[1,6]. For the magnetization measurements we used three different magnetometers. First a pendulum magnetometer (4.2-80K, 4-11 kGauss), secondly an induction magnetometer (1.2, 4.2 and 77K, 0-60 kGauss), and finally an induction magnetometer combined with the 400 kGauss magnet (4.2 and 77K, 0-340 kGauss). In the overlapping regions the results in the different magnetometers were in good agreement.

RESULTS

The inverse mass susceptibility X_g as a function of the square of the absolute temperature T is given for a number of alloys in figure 1 and table I. The magnetizations σ as a function of applied field H are plotted in figure 2 as σ^2 versus H/σ.

DISCUSSION

The behaviour of the magnetization as a function of field and of temperature is roughly according to the expression for the magnetic isotherms as given by Edwards and Wohlfarth[5]:

$$\frac{H}{\sigma} = \frac{1-IN}{N_g N \mu_B^2} + \frac{\pi^2 k_B^2}{6 N_g N \mu_B^2} \left[\left(\frac{N'}{N}\right)^2 - \frac{N''}{N} \right] T^2 + \frac{1}{2 N_g^3 \mu_B^4 N^3} \left[\left(\frac{N'}{N}\right)^2 - \frac{N''}{3N} \right] \sigma^2 \quad (1)$$

Here I is the effective exchange interaction parameter, N_g is the number of atoms per gram,

N is the density of states per atom at the Fermi energy,
k_B is Boltzmann's constant,
μ_B is the Bohr magneton,
N' and N'' are the first and second derivatives, respectively, of the density of states as a function of energy at the Fermi energy.

The slope of the χ^{-1} versus T^2 curves and that of the σ^2 versus H/σ curves yield two equations in N, $(N'/N)^2$ and N''/N. The density of states can be estimated from the low temperature specific heat[7]. Contributions to the specific heat must be expected from paramagnons, mass enhancement, localized moments etcetera, but a rather safe estimate for the upper limit of N is 3 eV^{-1} per nickel atom. However, solving then for $(N'/N)^2$ and N''/N leads to the unphysical result: $(N'/N)^2 < 0$. In terms of the expression (1) it is a result of the coefficient of T^2 being too large compared with the coefficient of σ^2. It will be clear that the introduction of either a temperature dependence or a magnetization dependence of I (or both) can remove this quantitative problem. The magnetization dependence of I that is needed to make $(N'/N)^2$ at least positive is unreasonable in magnitude, however[6]. Therefore we propose a temperature dependence of I of the form $I = I_0(1+\alpha T^2)$ with $\alpha \simeq 4 \times 10^{-6} K^{-2}$, which means that in the interval 0–70K the variation of I is about 2%.

The assumption that in Ni_3Ga the field dependence of the susceptibility and not its temperature dependence is of true Stoner-Edwards-Wohlfarth origin, i.e. determined by bandparameters only, is supported by the band calculations of Fletcher[4,8,9].

The curves in fig.1 and figure 2 reveal some deviations from the behaviour as to be expected from expression (1). Let us first discuss the σ^2 vs. H/σ curves. The data taken at 4.2K (coinciding with those at 1.2K) show for $\sigma < 4$ emu./gram an increased H/σ with respect to the extrapolations from higher magnetizations. Although we do not know at the moment the origin of these deviations, we can make some remarks.

   I. The anomaly is restricted to low temperatures.
  II. Apparently the magnetization and not the magnetic field is the dominant quantity: all of the alloys behave for $\sigma > 4$ emu./gram in a regular way, while the initial susceptibilities vary an order of magnitude.
 III. The absolute deviations in H/σ at a certain value of σ from the extrapolated straight lines are approximately constant with respect to initial susceptibility, thus it looks like the origin is a variation in the magnetic interaction parameter I of about 0.4%.
  IV. When we take $N=3$ eV^{-1} per nickel atom, then $\sigma = 4$ emu./gram corresponds to a band splitting of about 40 meV, or to 350K. This again is in support of the idea of a temperature dependent interaction as given before: if the alloys do not behave according to simple Stoner theory for band splittings of 350K, why should the susceptibility do so as a function of temperature up to 70K?

The question whether these anomalies, or even most of the magnetic properties of these alloys are determined by inhomogeneous magne-

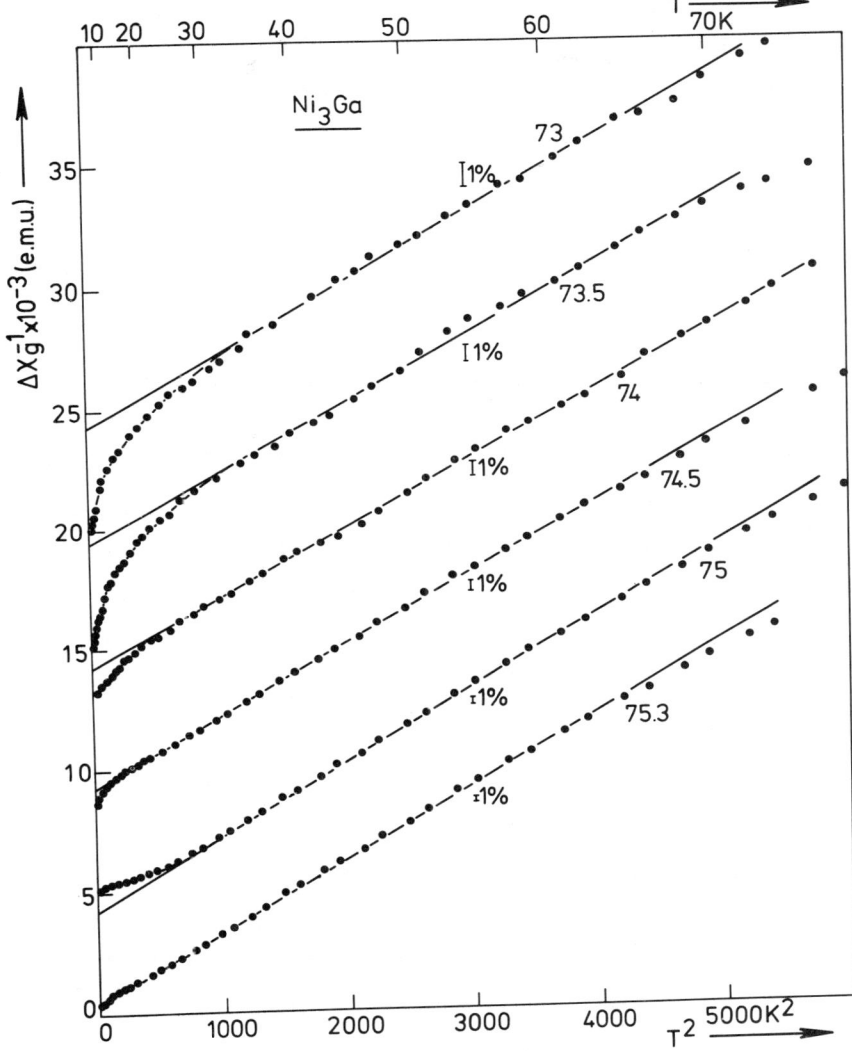

Figure 1. Inverse susceptibilities of a number of Ni$_3$Ga alloys versus the temperature squared. For sake of clarity the curves are shifted arbitrarily in a vertical sense. The absolute values are determined by the values given in the second column of Table I, the curves are labelled with the nickel concentration in at.%.

Table I: Parameters of the magnetic properties of the various Ni_3Ga alloys.

at.% Ni	$\chi_g^{-1}(0)$ extrap. to T=0 10^3 emu.	$d\chi_g^{-1}/dT^2$ emu. K^{-2}	$\frac{dH/\sigma}{d\sigma^2}$, emu. 4.2K	77K
73	89.4	3.08	-	-
73.5	64.6	2.88	-	-
74	48.4	2.80	440	420
74.5	29.2	2.94	360	280
75	14.9	3.00	230	240
75.3	9.4	2.96	240	-
75.5	-	-	-	230

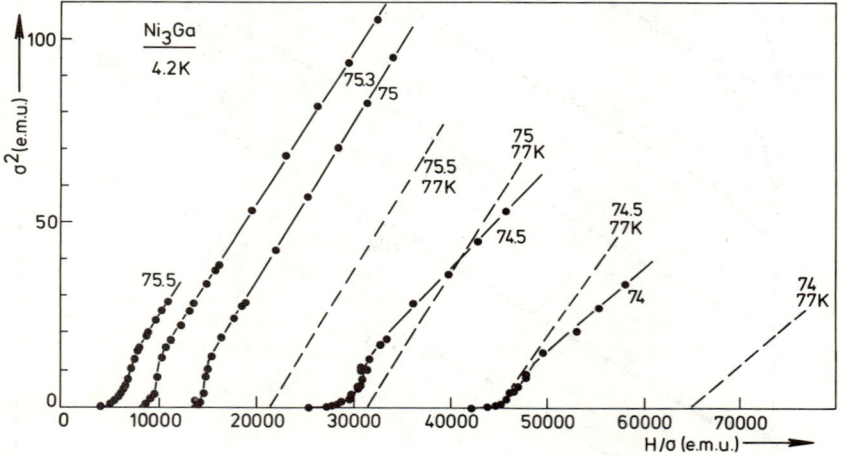

Figure 2: σ^2 versus H/σ for a number of Ni_3Ga alloys. Data points at 4.2K are connected with drawn lines. Dashed lines represent the data taken at 77K. The curves are labelled with the nickel concentration in at.%.

tization (e.g. clusters) will be discussed in more detail elsewhere[6]. On the basis of the magnetic properties, the giant moment inducement [10], low temperature specific heat, effects of cold-working[1] and resistivity measurements[10] the conclusion can be reached that inhomogeneous magnetization plays a minor role: it might cause the rather sharp decrease of H/σ towards vanishing σ.

The curves of χ_g^{-1} vs T^2 deviate from linearity at the highest temperatures, presumably because of terms of higher order in T coming into play. The low temperature deviations can easily be accounted for by slight amount of impurities, such as 10 ppm of dissolved iron. There is, however, a more interesting anomaly which is the most pronounced for the stoichiometric alloy: an increased χ^{-1} relative to the extrapolation from the higher temperatures. Apparently the crystalline order plays an important role, and one is led to the questions would in fully ordered perfectly stoichiometric Ni_3Ga the susceptibility even decrease with decreasing temperature? Or in other words: would the ideal intermetallic compound behave qualitatively like palladium metal? We have approached this problem experimentally from the easiest side and have measured the susceptibility of palladium samples in which we have introduced lattice defects by cold-working at room temperature. The pure palladium (Johnson Matthey, spectrographically standardized) had a residual resistance ratio $\rho(300K)/\rho(4.2K) \simeq$ $\simeq 250$, which was decreased by the cold-working down to 80. While the susceptibilities of the undeformed, the deformed and the annealed specimen at 80K and higher temperatures were practically the same, the susceptibilities of the deformed specimen at liquid helium temperatures were larger by 2.5% than those of the undeformed and the annealed specimens[11].

CONCLUSIONS

The Ni_3Ga alloys can be described as itinerant electron paramagnets. However, the dependence of the susceptibility on temperature and applied magnetic field are not determined by the band shape alone as is assumed in the Stoner model. In particular the temperature dependence of the susceptibility seems to be determined mainly by a small temperature dependence of the effective exchange interaction I. Furthermore, at low temperatures I increases with the magnetization σ for $\sigma < 4$ emu./gram, i.e. for a band splitting smaller than 40 meV. And finally the crystalline ordering tends to decrease the susceptibility at low temperatures as it does in palladium metal. We must say, however, that all these effects of varying effective exchange interactions are small and observable only if the exchange enhancement is very large, as it is in the Ni_3Ga alloys.

We are very grateful for the valuable discussions with Dr.P.F. de Chatel and Dr. W. de Dood.

REFERENCES

1. F.R. de Boer, C.J. Schinkel, J. Biesterbos and S. Proost, J.Appl.Phys. 40, 1049 (1969).
2. F.R. de Boer, thesis Universiteit van Amsterdam (1969).
3. P.F. de Chatel, F.R. de Boer, W. de Dood, J.H.J. Fluitman and C.J. Schinkel, J.Physique 32, C1-999 (1971).
4. K.H. Chang, thesis Universiteit van Amsterdam (1972).
5. D.M. Edwards and E.P. Wohlfarth, Proc.Roy.Soc. A303, 127 (1968).
6. C.J. Schinkel, F.R. de Boer and B. de Hon, to be published.
7. W. de Dood et al., to be published.
8. G.C. Fletcher, Physica 56, 173 (1971).
9. G.C. Fletcher, to be published in Physica.
10. M.S. Schalkwijk, P.E. Brommer, G.J. Cock and C.J. Schinkel, J.Physique 32, C1-997 (1971).
11. C.J. Schinkel, R. Hartog and R. Klijn, not published.

MAGNETIC ORDERING IN THE $Ni_{75}Al_{25-x}Ga_x$ SYSTEM[*]

R. W. Jones
Argonne National Laboratory, Argonne, Illinois 60439 and
University of Notre Dame, Notre Dame, Indiana 46556

G. S. Knapp
Argonne National Laboratory, Argonne, Illinois 60439

C. W. Chu
Cleveland State University, Cleveland, Ohio 44102 and
Argonne National Laboratory, Argonne, Illinois 60439

ABSTRACT

We have made specific heat measurements on ferromagnetic $Ni_{75}Al_{25}$ and paramagnetic $Ni_{75}Al_{15}Ga_{10}$ in the temperature range 2 to 50 K. We find anomalous behavior in the low temperature heat capacities of both alloys which is attributed to magnetic clustering. There is a sharp anomaly in the specific heat due to the magnetic transition in $Ni_{75}Al_{25}$. We have also studied the effect of pressure on the T_c of $Ni_{75}Al_{25}$ and find no change in T_c with pressures up to 15 kbar. The nature of the magnetic order in $Ni_{75}Al_{25}$ is discussed with reference to the extensive data available in the literature.

INTRODUCTION

The intermetallic compound $Ni_{75}Al_{25}$ has been extensively studied[1-5] as an example of a weak itinerant ferromagnet of the Stoner-Wohlfarth type. This interpretation, by deBoer and deChatel,[1] was based on mainly the low ordering temperature (~40 K) and small moment (0.075 μ_B/Ni atom) observed in their investigations. They also found that the field dependence of the magnetization agreed with the prediction of the Wohlfarth theory. However, the recent work of Robbins and Claus[5] indicates that $Ni_{75}Al_{25}$ may contain superparamagnetic clusters at high temperatures, with ordering between clusters at lower temperatures. Their arguments stem mainly from low temperature specific-heat measurements on ferromagnetic $Ni_{75}Al_{25}$ in which, for temperatures below 2 K, they found an upturn in the C/T. The low temperature data was fit to the expression $C(T) = A + \gamma T + \beta T^3$. The constant term A has been ascribed[6-10] to thermally excited magnetic clusters. In an effort to help clarify the nature of magnetic order in $Ni_{75}Al_{25}$, we have made high precision specific-heat measurements from 1.8 to 50 K on $Ni_{75}Al_{25}$ (T_c = 38.5 K) and on $Ni_{75}Al_{15}Ga_{10}$ which shows no long range magnetic order. We too observe anomalous behavior in the low temperature specific heats which is interpreted as a constant term in

---

[*]Work performed under the auspices of the U.S. Atomic Energy Commission and the Center for Educational Affairs at Argonne National Laboratory.

the heat capacity. The magnitude of this term is larger in $Ni_{75}Al_{15}Ga_{10}$ than in $Ni_{75}Al_{25}$. We find a rather sharp ($\Delta T_c \sim 0.5°K$) but small ($\Delta C_p/C_p \sim 0.02$) anomaly in the specific heat of $Ni_{75}Al_{25}$ which can be associated with the magnetic transition.

We have also measured the dependence of the transition temperature of $Ni_{75}Al_{25}$ on pressure by an ac susceptibility technique. We find no change in T_c for pressures up to 15 kbar.

EXPERIMENTAL PROCEDURE AND RESULTS

Nickel (United Materials), aluminum (Alpha Inorganic), and gallium (United Materials), each of 5 N purity, were mixed in proper proportions and arc melted under an argon atmosphere to make $Ni_{75}Al_{25}$ and $Ni_{75}Al_{15}Ga_{10}$. The weight loss on melting was negligible. The samples were annealed at 1000 K for 21 days and water quenched. X-ray patterns of the alloys confirmed their ordered Cu_3Au structure. Susceptibility measurements on $Ni_{75}Al_{25}$ indicated a ferromagnetic transition at $\sim$ 41 K whereas the $Ni_{75}Al_{15}Ga_{10}$ showed no ordering above 1.5 K. Our T_c for $Ni_{75}Al_{25}$ agrees with that of deBoer et al.[2] and Robbins and Claus[5] for alloys of the same composition.

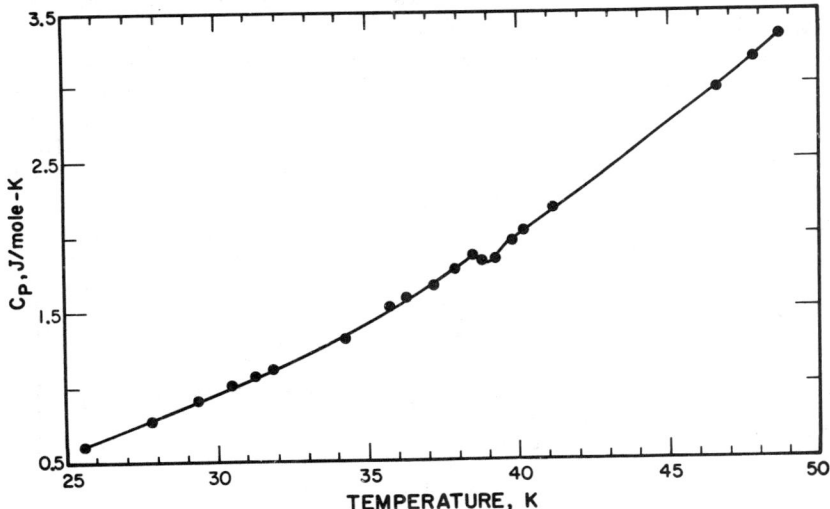

Fig. 1. Heat capacity as a function of temperature for $Ni_{75}Al_{25}$ through the ferromagnetic transition.

Specific heat was measured on $\sim$0.1 mole samples cut from the annealed buttons. The measurements were made with a high-precision differential calorimeter described elsewhere.[11] The precision of the data obtained in this experiment was found to 0.25%. Fig. 1 shows the heat capacity of $Ni_{75}Al_{25}$ plotted vs temperature. The magnitude and sharpness of the anomaly were verified by subsequent experiments. The definition of the transition is well outside experimental uncertainty. Because of the low thermal conductivity of the sample, systematic errors of order 5% may be present in the specific heat for temperatures above 5 K. The heat capacity anomaly occurs at 38.5 K with $\Delta C/C \approx 2\%$ and $\Delta T_c \approx 0.5°K$ as shown in the figure. The sharpness of the transition implies that the ordering is homogeneous on a macroscopic scale. Our results for the low temperature specific heat of $Ni_{75}Al_{25}$ and $Ni_{75}Al_{15}Ga_{10}$ are shown in Fig. 2, where C_p/T is plotted as a function of T^2. For $Ni_{75}Al_{25}$ our results agree with those of Robbins and Claus[5] to within 1%. In this temperature range the specific heat can be written[6,7]

$$C(T) = A + \gamma T + \beta T^3.$$

The constant term is attributed[6-10] to magnetic clusters, thermally excited against the crystal anisotropy field, with $A = Nk$ where N is the number of such clusters and k is Boltzmann's constant. For $Ni_{75}Al_{25}$, Robbins and Claus[5] found $A = 0.4$ mJ/mole-K corresponding to 1 cluster for every 2000 atoms. The upturn in C/T is more pronounced for $Ni_{75}Al_{15}Ga_{10}$ implying that there are more thermally free clusters in that alloy than in $Ni_{75}Al_{25}$. This is understood because the long-range magnetic order in $Ni_{75}Al_{25}$ freezes in many of the clusters and so fewer are available for thermal excitation.

We also measured the dependence of the T_c in $Ni_{75}Al_{25}$ on pressure. Pressure was applied with a self-clamp technique. The pressure medium was a one to one mixture of n-pentane and isoamyl alcohol with a self-sealing teflon seal. The T_c was measured with an ac susceptibility technique described elsewhere.[12] Within the experimental uncertainty of approximately 3°K, we found no change of T_c in $Ni_{75}Al_{25}$ for pressures up to 15 kbar.

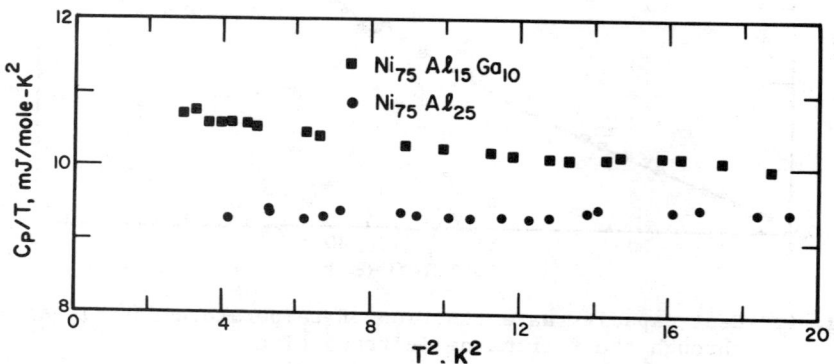

Fig. 2. Low temperature specific heat for $Ni_{75}Al_{25}$ and $Ni_{75}Al_{15}Ga_{10}$.

DISCUSSION

A large number of experimental facts are available from many investigations[1-5,12,13] into the properties of $Ni_{75}Al_{25}$.
(A) deBoer et al.[2] found a low T_c, a small moment per nickel atom and good straight line Arrott plots in $Ni_{75}Al_{25}$. With these results, deChatel and deBoer[1] explained the detailed magnetic behavior of $Ni_{75}Al_{25}$ in terms of the Stoner-Wohlfarth model.
(B) Our specific heat studies near T_c show a very sharp anomaly. The sharpness of the anomaly indicates bulk ordering is taking place. (C) Low temperature specific-heat data of Robbins and Claus[5] and of the present work exhibit anomalous behavior. As discussed above this anomaly can be attributed to thermally excited magnetic clusters. The number of these clusters is small, on the order of one cluster for each 2000 atoms. (D) We find no change ($\pm$ 3°K) in the T_c of $Ni_{75}Al_{25}$ with pressures up to 15 kbar. This result is in contrast with that for $ZrZn_2$ where a pressure of 8 kbar was found[13] to suppress the transition entirely. $ZrZn_2$, a prime example[14] of a weak itinerant ferromagnet, also shows no unusual behavior in the low temperature specific heat. (E) Neutron diffraction work[15] on single crystals of $Ni_{75}Al_{25}$ shows complete atomic order in Cu_3Au structure within the limits of detection of the experiment, $\sim 1\%$. (F) In highly deformed $Ni_{75}Al_{25}$ both deBoer et al.[2] and Robbins and Claus[5] found that superparamagnetic behavior persists down to 4 K.

Whereas much of the above cited data is consistent with the mean field Stoner-Wohlfarth theory, we feel some major discrepancies are apparent. First and foremost, a strong pressure dependence of the T_c in weak itinerant ferromagnets is predicted[16] and has been observed.[13] Also, magnetic inhomogeneities, suggested by the low temperature heat capacity, are not found in the Stoner model where magnetic order results from the mean field splitting of the electronic energy bands.

The superparamagnetic cluster model proposed by Robbins and Claus[5] is able to explain most of the data listed above. In their model, clusters are nucleated by regions of the material in which nickel atoms have more nickel near neighbors than in perfectly ordered $Ni_{75}Al_{25}$. They suggest that these regions are antiphase boundaries. Since such boundaries are spatially extended, many nickel atoms would be involved in nucleating each cluster. The resulting large clusters would be inhomogeneously distributed in the $Ni_{75}Al_{25}$ matrix and magnetic ordering between clusters would be expected to take place at different temperatures in different parts of the sample. Therefore, the specific heat anomaly associated with the transition would be broad in temperature. However, our experiments show a very sharp anomaly. If, on the other hand, a few individual nickel atoms were displaced to aluminum sites, possibly by vacancies in the material,[18] the cluster nuclei would be very small and finely dispersed. Few nickel-aluminum interchanges are needed in this model, in agreement with the neutron work. The neutron-scattering measurements of Ling and Hicks[19] on $Ni_{75}Al_{25}$ doped with iron indicates that the matrix is strongly magnetically polarizable. Therefore, the small nuclei may be strong enough to form the superparamagnetic clusters. The resulting fine dispersion of clusters would be homogeneous on a macroscopic scale and thus inter-cluster ordering would yield a sharp specific-heat anomaly. The small magnitude of the anomaly results from the small number of cluster moments involved in the ordering.

The authors would like to thank Drs. A. Aldred, F. Fradin, G. Felcher, D. Lam, J. Kouvel, B. Coles, A. Miller, and B. Veal for helpful discussions concerning this work.

REFERENCES

1. P. F. deChatel and F. R. deBoer, Physica **48**, 331 (1970).
2. F. R. deBoer, C. J. Schinkel, J. Biesterbos, and S. Proost, J. Appl. Phys. **40**, 1049 (1969).
3. J. H. J. Fluitman, B. R. deVries, R. Boom, and C. J. Schinkel, Phys. Lett. **28A**, 506 (1969).
4. F. R. deBoer, J. Biesterbos, and C. J. Schinkel, Phys. Lett. **24A**, 355 (1967).
5. C. G. Robbins and H. Claus, *Magnetism and Magnetic Materials*, C. D. Graham and J. J. Rhyne, eds. AIP Proc. No. 5, p. 527 (1971).
6. K. Schroeder, J. Appl. Phys. **32**, 880 (1961).
7. J. D. Livingston and C. P. Bean, J. Appl. Phys. **32**, 1964 (1961).
8. E. J. Hayes, A. Hahn, and E. P. Wohlfarth, J. Phys. F **2**, 351 (1972).
9. C. G. Robbins, H. Claus, and P. A. Beck, J. Appl. Phys. **40**, 2269 (1969).
10. K. P. Gupta, C. H. Cheng, and P. A. Beck, Phys. Rev. **133**, A203 (1964).
11. R. W. Jones, G. S. Knapp, and B. W. Veal, to be published.
12. C. W. Chu, R. W. Jones, and G. S. Knapp, to be published.
13. T. F. Smith, J. A. Mydosh, and E. P. Wohlfarth, Phys. Rev.Lett. **27**, 1732 (1971).
14. E. P. Wohlfarth, J. Appl. Phys. **39**, 1061 (1968).
15. Gian Felcher, private communication.
16. E. P. Wohlfarth, J. Phys. C. **2**, 68 (1969).
17. R. Viswanathan, H. L. Lno, and D. O. Massetti, *Magnetism and Magnetic Materials*, C. D. Graham and J. J. Rhyne, eds. AIP Proc. No. 5, p. 1290 (1971).
18. This idea was first suggested to us by B. R. Coles.
19. P. C. Ling and T. J. Hicks, to be published.

THE SPATIAL DISTRIBUTION OF THE MAGNETIZATION AROUND Fe IMPURITIES IN Ni_3Ga*

J. W. Cable and H. R. Child
Solid State Division, Oak Ridge National Laboratory
Oak Ridge, Tennessee 37830

ABSTRACT

We report neutron diffuse scattering measurements of the spatial distribution of the magnetization around Fe impurities in exchange enhanced Ni_3Ga. These show an Fe moment of about 2.9 μ_B and a host matrix magnetization that decreases exponentially with distance from the impurity. The total moment per Fe atom is 9.5 μ_B at 0.9 at.% Fe and 33 μ_B at 0.1 at.% Fe. The host magnetizations extend approximately 10 and 15 Å, respectively, from the impurity sites.

INTRODUCTION

The fcc intermetallic compound Ni_3Ga behaves as an exchange enhanced paramagnet with a Stoner enhancement factor of about 35.[1] Magnetic impurities introduced into this system produce spin polarization around the impurity, giant moments per impurity atom and ferromagnetism at low impurity content.[2] The effects are particularly pronounced with Fe impurities for which ferromagnetism is observed with as little as 0.05 at.% Fe with $T_c \simeq 13°K$ and a spontaneous magnetization of about 35 μ_B/Fe atom. The range and magnitude of the magnetic moment disturbance produced by the Fe are important to the understanding of this remarkable magnetic behavior and these can, in principle, be determined by neutron diffuse scattering measurements. In this paper we report such measurements for Fe doped Ni_3Ga.

EXPERIMENTAL RESULTS

Samples of Ni_3Ga containing 0.9 and 0.1 at.% Fe were prepared in the manner described by De Boer.[3] X-ray and electron probe analyses showed these to be ordered, single phase fcc and macroscopically homogeneous. The neutron measurements were made at the ORR using a neutron wavelength of 4.43 Å. Curie temperatures were determined from the temperature dependence of the small angle scattering which peaks dramatically at T_c. The values obtained, 80°K at 0.9% Fe and 28°K at 0.1% Fe, agree quite well with those previously reported.[2] The magnetic disorder cross section measurements were made by the field off—field on method[4] with a 12 kOe field used to magnetize the sample. The magnetic cross sections obtained for these two compositions at 8°K are shown in Fig. 1.

---

*Research sponsored by the U. S. Atomic Energy Commission under contract with the Union Carbide Corporation.

Both show pronounced K dependencies ($K = 4\pi \sin\theta/\lambda$) that were not present in the nuclear scattering and which are characteristic of spatially extended regions of magnetization.

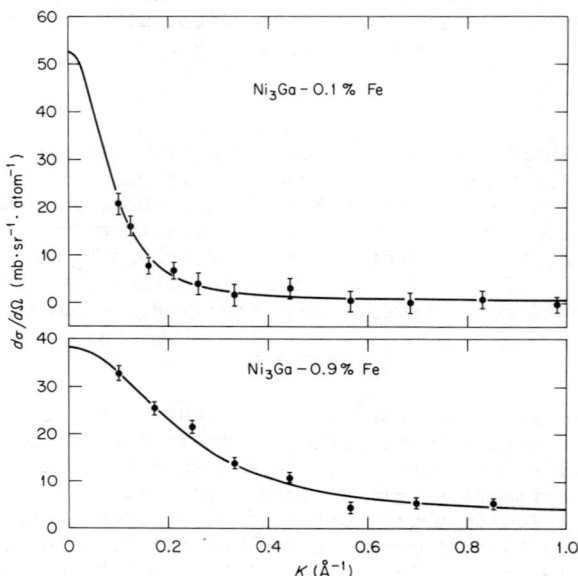

Fig. 1. Magnetic disorder cross sections of Fe doped Ni_3Ga

The magnetic disorder cross section, in millibarns-steradian^{-1}-atom^{-1}, is given by[5]

$$\frac{d\sigma}{d\Omega}(K) = 48.3\, c(1-c)[\mu_{Fe}f_{Fe}(K) - \bar{\mu}_{Ni}f_{Ni}(K) + (1-c)M(K)]^2 \quad (1)$$

in which c is the fractional Fe content, μ_{Fe} is the moment at the Fe site and $\bar{\mu}_{Ni}$ the average moment at the Ni site and the $f(K)$'s are appropriate 3d form factors. $M(K)$ is the Fourier transform of the magnetization induced in the host by the impurity. Following Low,[6] we assume this is proportional to the K dependent susceptibility of the electron gas, which at small K assumes the form

$$M(K) = \frac{M(0)}{1 + K^2/K_1^2} \quad (2)$$

Here, $M(0)$ is the net host magnetization in μ_B and K_1 is an inverse range parameter. Away from the impurity site the moment density, in $\mu_B/\text{Å}^3$, is given by

$$\rho(r) = \frac{M(0)K_1^2}{4\pi r} e^{-K_1 r} \quad (3)$$

The parameters obtained by fitting Eqs. (1) and (2) to the observed cross sections are given in Table I and the fitted curves are compared with the data points in Fig. 1. Clearly the data are well represented by this form for the cross section.

Table I. Moment distribution parameters for Fe Doped Ni_3Ga

at.% Fe	$\bar{\mu}_{Fe} - \bar{\mu}_{Ni}$	M(0)	K_1
0.1	(3)	30 ± 3	0.12 ± 0.01
0.9	2.8 ± 0.3	6.7 ± 0.3	0.28 ± 0.01

DISCUSSION

The cross section parameters in Table I show a moment difference of 2.8 μ_B for the 0.9% compound. With M(0) = 6.7 μ_B and ~80 Ni atoms/Fe atom, $\bar{\mu}_{Ni} \simeq 0.08$ μ_B and therefore, $\mu_{Fe} \simeq 2.9$ μ_B. This compares favorably with previously reported μ_{Fe} values in fcc systems. The total moment of 9.5 μ_B/Fe is, however, significantly lower than the 12.2 μ_B/Fe obtained[2] from the magnetization data at 1% Fe. We have no explanation for this discrepancy. At 0.1% Fe the neutron data do not yield a reliable moment difference. Fortunately, the M(K) parameters are not sensitive to the moment difference at this composition. We assume an Fe moment of 3 μ_B and obtain a total moment of 33 μ_B/Fe in reasonable agreement with the magnetization result[2] of 29 μ_B/Fe.

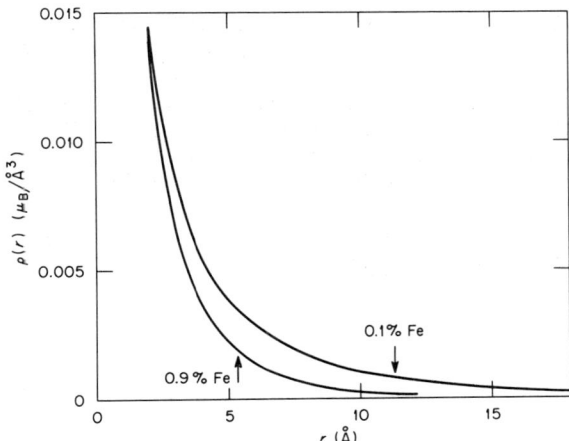

Fig. 2. Moment densities for Fe doped Ni_3Ga

The moment densities corresponding to the M(K) parameters are shown in Fig. 2. At the nearest neighbor distance of 2.5 Å the densities correspond to about 0.15 μ_B/Ni for both compositions. Beyond that, however, the magnetization decreases more rapidly for the 0.9% Fe sample. The arrows in the figure represent half the average distance between impurity atoms at the two compositions and indicate the extent to which the tails of the host magnetizations overlap. Obviously, more dilution is required to avoid the overlapping magnetizations and to approach the isolated impurity case for which the M(K) parameters should represent the response of the host material. Unfortunately, we are near the practical limits of detection at 0.1% Fe. We must therefore rely on theory and Mössbauer results to estimate the M(K) parameters in the dilute limit. According to the exchange enhanced, itinerant electron model both M(K) and $1/K_1^2$ are proportional to $\alpha\chi_o$, where α is the exchange enhancement factor, $(1 - I\chi_o)^{-1}$, and χ_o is the Pauli susceptibility of the non-interacting electron gas. Kim and Schwartz[7] show that the density of states at the Fermi energy, and therefore χ_o, decreases with increasing host magnetization; i.e., Fe content. The resulting variation in α is especially pronounced in systems with large α, such as Ni_3Ga, and this presumably accounts for the large changes in M(0) and $1/K_1^2$ in going from 0.9% Fe to 0.1% Fe. We can use this model to extrapolate from our result for 0.1% Fe to the Mössbauer results[8] for 20 ppm Fe which we accept as the isolated impurity case. At 0.1% Fe, we obtain M(0) = 30 μ_B and $1/K_1^2$ = 70 Å^2 while Maletta and Mössbauer[8] get M(0) $\simeq$ 60 μ_B at 20 ppm. This implies a $1/K_1^2$ value of 140 Å^2 for Ni_3Ga.

REFERENCES

1. F. R. De Boer, C. J. Schinkel and J. Biesterbos, Phys. Letters **25A**, 606 (1967).
2. C. J. Schinkel, F. R. De Boer and J. Biesterbos, Phys. Letters **26A**, 501 (1968); F. R. De Boer, C. J. Schinkel, J. Biesterbos and S. Proost, J. Appl. Phys. **40**, 1049 (1969).
3. F. R. De Boer, Thesis, University of Amsterdam, 1969.
4. C. G. Shull and M. K. Wilkinson, Phys. Rev. **97**, 304 (1955).
5. W. Marshall, J. Phys. C **1**, 88 (1968).
6. G. G. Low, Advances in Physics **18**, 371 (1969).
7. D. J. Kim and B. B. Schwartz, Phys. Rev. Letters **21**, 1744 (1968).
8. H. Maletta and R. L. Mössbauer, Solid State Comm. **8**, 143 (1970); H. Maletta, Z. Physik **250**, 68 (1972).

NMR STUDY OF NiRu and FeRu ALLOYS

J. J. MURPHY
Iona College, New Rochelle, N.Y. 10801

T. J. BURCH
Fordham University,+ Bronx, N.Y. 10458

J. I. BUDNICK*
National Science Foundation, Washington, D.C. 20550

ABSTRACT

Detailed NMR spectra are presented for host and impurity nuclei in dilute alloys of Fe and Ni containing Ru. All shifts in the spectra of the NiRu alloys are toward lower hyperfine fields while FeRu exhibits both positive and negative shifts. Interpretation of the host spectra requires consideration of impurities out to at least the third neighbor shell. The field distribution about the impurity has the same spatial behavior as the moment disturbance deduced from neutron scattering results. Each Ru spectrum is directly compared with its host spectrum and the relationship is discussed.

INTRODUCTION

The magnetic properties of NiRu and FeRu alloys are of more than average interest since the Ru appears to possess a local moment in both hosts [1,2] although the charge screening in each case should be quite different since Ru is isoelectronic with Fe. It has recently been proposed[3] that the Ru moment in Ni (also in Co) is critically dependent on the number of nearest neighbor Ru. This is probably not the case for FeRu. Detailed spin-echo spectra of host and impurity resonances allow some determinations regarding the range and magnitude of neighbor-induced interactions in these alloys.

SPECTRAL CHARACTERISTICS

Typical spectra, taken at helium temperatures, are shown in Figs. 1 and 2. The dashed lines in the Ru spectra indicate the decomposition into ^{99}Ru and ^{101}Ru contributions. The general characteristics of these resonances have been previously reported.[4,5,6] For NiRu shifts to lower resonance frequencies account for all observed spectral structure while for FeRu shifts to both higher and lower frequencies are seen. Ni alloys with concentrations from 0.3 to 2.0 at.% Ru and Fe alloys from 1 to 5 at.% have been studied.

Analysis of the spectra can be begun by considering the concentration dependence of the fraction of the total observed intensity shifted away from the position of the main line. The relative intensities of the shifted contributions in the spectra of NiRu are seen in Fig. 3 where they are compared with the probabilities of at least one impurity atom occurring within a specified set of near neighbor shells. This type of plot is useful when interpreting spectra in terms of hyperfine field shifts arising from a spatially-varying neighbor impurity interaction. By inspection, limits can be set upon the range of the interaction and alternative assignments of the near neighbor shells responsible for the observed shifts are suggested. Implicit in

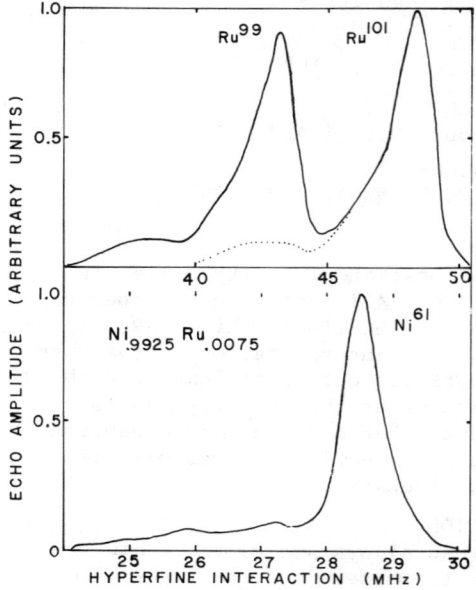

Fig. 1. Spin-Echo Spectra of Ru and Ni in a $Ni_{.9925} Ru_{.0075}$ Alloy at 1.4K. The dotted line is a decomposition of the Ru isotopes.

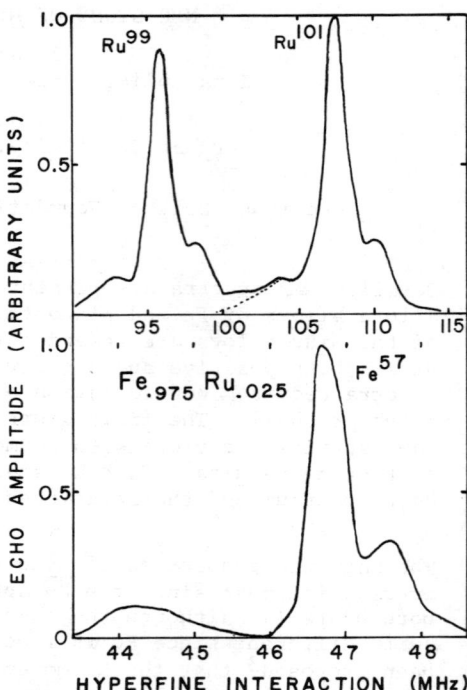

Fig. 2. Spectra of Ru and Fe in a $Fe_{.975}Ru_{.025}$ Alloy at 1.4 K.

the use of such a plot is the assumption that all nuclei, regardless of their near neighbor environment, have actually been observed and contribute to the spectra under investigation. The observed hyperfine field shifts in our spectra necessitate consideration of effects out to at least third near neighbors. This must be the minimal range for hyperfine field shifts $\gtrsim 1\%$ for each of these systems.

HOST SPECTRA

We have fit our Ni and Fe spectra with the following near neighbor shift assignments (expressed as a percentage of the unshifted hyperfine field): for Ni, $\Delta H_1 = -10.5\%$, $\Delta H_2 = -9.2\%$, $\Delta H_3 = -5\%$; for Fe, $\Delta H_1 = -4.7\%$, $\Delta H_2 = -1.1\%$, $\Delta H_3 = +1.4\%$. These assignments compare with the results of neutron scattering experiments[1]. Interpretations of the neutron data give the moment density disturbances in the host matrix as a function of distance from the impurity site. In NiRu these results show a negative moment density which gradually approaches zero from a value of about $-0.2 \mu_{Ni}$ at the nearest neighbor distance. In FeRu the neutron work yields a negative moment density increasing to a positive value with a maximum somewhere in the region between the second and third neighbor shells. The correlation between the NMR and neutron scattering results is striking. Our assignments for Fe are also compatible with interpretations of Mössbauer effect data.[6]

It is not obvious that agreement such as this must exist between hyperfine field shifts and spin density disturbances. The spin density maps are obtained from scattering data assuming isotropy and therefore will not reflect the anisotropy in the near neighbor interactions to which the hyperfine field variations are sensitive. The importance of anisotropy in the field shifts observed for Fe alloys has recently been reemphasized by Cranshaw.[7] The spherical averaging present in the neutron analysis may be the reason that we see no fourth and fifth neighbor field shifts in our Fe data as large as would be anticipated upon inspecting the spin density maps. Indeed, if we were to assume a large anisotropic effect in the (1,1,1) direction, as Cranshaw finds for FeCr, then an assignment of a $\Delta H_5 = -0.9\%$ with $\Delta H_2 = 0$ would not appreciably worsen the quality of our fit.

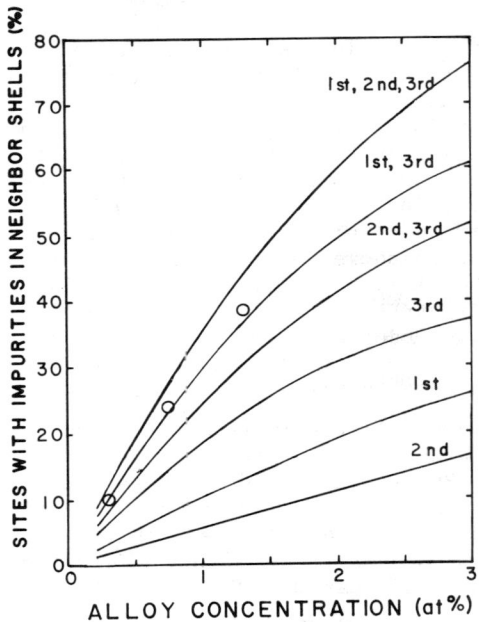

Fig. 3. The percentage of alloy sites in a random alloy with at least one impurity in the indicated neighbor shells. Circles indicate fractional shifted intensity in Ni spectrum.

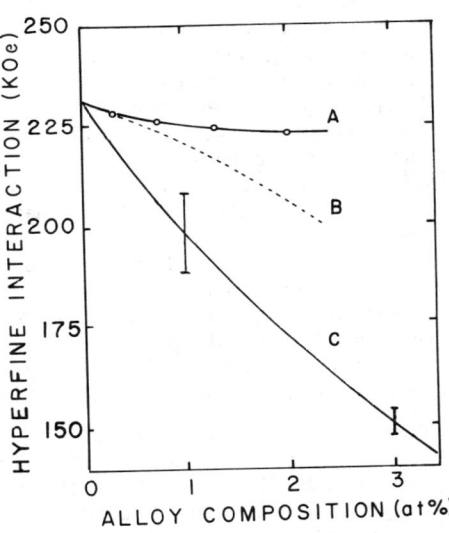

Fig. 4. Concentration dependance of Ru hyperfine field in NiRu. A = field of principal NMR line; B = average NMR field; C = average Mössbauer field.[3]

RUTHENIUM SPECTRA

One can investigate the possibility of a single, common mechanism, such as conduction electron polarization, being responsible for all hyperfine field shifts by being alert for a field shift scaling factor relating the host and impurity spectra in an alloy. We do not find any overall scaling here and conclude no single mechanism is dominant. In general, the presence of overlapping resonances from the two Ru isotopes makes our decompositions less certain than is the case for the host spectra. However, we can present some of the alternative possibilities for understanding the data.

The structure in the spectra of Ru in Ni contains a greater fraction of the observed intensity than is the case for the Ni host. This is indicative of a more widespread near neighbor interaction. Because of the rather severe effect a Ru has on Ni atoms in the first three neighbor shells, it is easily believed that the interaction between Ru-Ru neighbor pairs should have a greater range since the effects occur through the severely disturbed Ni matrix rather than as a direct impurity-impurity interaction. Since there are no partially resolved satellite lines in the shifted portion of these Ru spectra, we prefer to make no specific assignment of near neighbor field shifts.

W. v. Lieres et al. have proposed a dramatic critical moment effect for Ru in Ni.[3] To explain their Mössbauer data on the concentration dependence of the Ru hyperfine field, a model in which the Ru possesses a local moment only if it has a full complement of 12 Ni nearest neighbors is proposed. In our NiRu alloys we have found no low frequency resonances attributable to Ru nuclei with a near-zero local moment. However, if we assume that the critical moment effect does exist, we can then make the following analysis. From our spectra the NMR observed average hyperfine field can be obtained as a function of concentration. This information is shown in Fig. 4 where it is compared with the reported Mössbauer data. The obvious discrepancy between the two measurements can be removed if we assume that we do not observe the resonances of that fraction of the Ru nuclei which have at least one Ru nearest neighbor. Removing the discrepancy by this means places an upper limit on the magnitude of the hyperfine field for Ru nuclei with no local moment. We therefore conclude that, if the proposal of a critical moment effect for Ru in Ni is correct, the magnitude of the hyperfine field for Ru nuclei lacking a moment is about 45 KOe.

We identify the two partially resolved satellites in the Ru spectra for FeRu as first and third neighbor effects. Shifts ΔH_1 = -3.4% and ΔH_3 = +2.6% fit the data. The average Ru hyperfine field varies only slightly with concentration. In view of this, the magnetization data for these alloys, and the generally smaller hyperfine field shifts in Fe as opposed to Ni, we do not anticipate a critical moment effect for Ru in Fe--at least not for small numbers of near neighbor Ru atoms.

The authors wish to thank Dr. W. Potzel of Argonne National Laboratory, and Dr. S. Skalski of Fordham University , for many helpful discussions.

REFERENCES

+ Supported in part by the National Science Foundation
* On leave from Fordham University
1. J.B. Comly, T.M. Holden and G.G. Low, J. Phys. C1, 458 (1968); M.F. Collins and G.G. Low, Proc. Phys. Soc. 86, 535 (1965).
2. D.A. Shirley, S.S. Rosenblum and E. Matthias, Phys. Rev. 170, 363 (1968).
3. W. v. Lieres, H. Maletta, W. Potzel and R.L. Mössbauer; Perspectives for Hyperfine Interactions in Magnetically Ordered Systems. L'Aquila, 1972.
4. J.I. Budnick and S. Skalski, Hyperfine Interactions, ed. A.J. Freeman and R. Frankel, (Academic Press, 1967).

5. H. Kubo, M. Kontani and J. Itoh, J. Phys. Soc. Japan $\underline{22}$, 929 (1967).
6. G. K. Wertheim, V. Jaccarino, J. H. Wernick and D.N.E. Buchanan, Phys. Rev. Letters $\underline{12}$, 24 (1964); I. Vincze, Sol. State Comm. $\underline{10}$, 341 (1972).
7. T.E. Cranshaw, J. Phys. $\underline{F2}$, 615 (1972).

MOMENT FORMATION AND ITINERANT FERROMAGNETISM IN THE FeSi-CoSi MIXED SYSTEM: Co^{59} NMR

S. Kawarazaki, H. Yasuoka and Y. Nakamura
Department of Metal Science and Technology
Kyoto University, Kyoto, Japan

ABSTRACT

The magnetic properties of Co atom in both paramagnetic and ferromagnetic phases of the $(Fe_{1-x}Co_x)Si$ solid solution have been investigated by the Co^{59} NMR. From the spin-echo intensity analysis, it is concluded that the Co atom becomes magnetic when it has one or more Fe atoms in its n.n. metal sites and otherwise remains nonmagnetic.

The hyperfine field at Co atom is about 9 KOe for the alloy with x of 0.3 and increases monotonically with increasing Fe concentration.

INTRODUCTION

Most of the investigations on the ferromagnetism of disordered alloys have been chiefly confined to the alloys between ferromagnetic metals or between ferromagnetic and nonmagnetic metals. It has been restricted within the ordered alloys such as $ZrZn_2$ and Au_4V that the binary alloys between nonmagnetic metals have ferromagnetic phases. There has been no example of a ferromagnetic disordered binary alloy between nonmagnetic metals.

In the pseudo-binary alloy case, however, there are some examples of ferromagnetic disordered alloys between nonmagnetic "metals". The solid solution $(Fe_{1-x}Co_x)Si$ (hereafter denoted as FCSx) is one of them. While FeSi and CoSi are paramagnetic and diamagnetic, respectively, the solid solution FCS has a ferromagnetic phase in an intermediate concentration range (0.2 <x< 0.8) at low temperatures.[1] In this ferromagnetic phase, the saturation magnetization per transition atom is smaller than 0.3 µB and the Curie temperature does not exceed 50 K (see Fig. 1). It is known that before the ferromagnetic phase appears, the diamagnetic Co atom must be converted magnetic with increasing Fe concentrations.[2] So, it is expected that the nonmagnetic-magnetic transition of the Co atom must have some effects on the Co^{59}

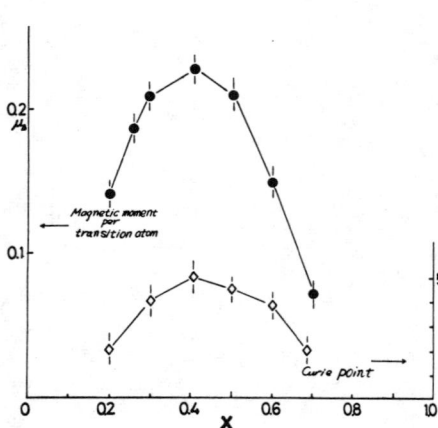

Fig. 1 Saturation magnetization and Curie point of the $(Fe_{1-x}Co_x)Si$ system.

paramagnetic NMR. The hyperfine field at the Fe atom in the FCS0.5 alloy observed in the Mossbauer effect measurement[2] suggested that the magnetic moment at the Fe atom is less than 0.1 μB.

In this paper we describe the magnetic properties of Co atoms in the FCS alloys which appeared in the Co^{59} NMR in both paramagnetic and ferromagnetic states.

EXPERIMENTAL RESULTS AND ANALYSIS

The NMR measurements in the paramagnetic state were made for the FCSx alloys with x between 0.7 and 1.0 with a conventional pulsed NMR apparatus operating at 19 MHz at temperatures of 77 K and 20 K. The integrated amplitude of the spin-echo signals was stored in a box-car integrator and recorded as a function of the external field. The observed spectra have clearly indicated first order quadrupole broadened patterns in powdered specimens (See Fig. 2). With increasing Fe concentration, two remarkable effects were observed; (1) the integrated intensity decreases with decreasing x more rapidly than expected from simple dilution, (2) the central position and the overall line shape of the spectrum do not change drastically with decreasing x. It should be noticed that the intensity of the central transition decreases rapidly with increasing Fe concentration. In order to get further information concerning the alloying effects on Co NMR, we have repeated the same measurements in the NCS system (Fig. 2). In the NCS case, it was found that the spin-echo signals of the central transition are unchanged, while the satellite lines are somewhat wiped out. The fractional decrease of the integrated spin-echo intensity (the area of the spectrum in Fig. 2) is plotted as a function of the concentration in Fig. 3 for both FCS and NCS systems. In Fig. 3, the solid curves (A) and (B) represent the calculated probabilities of finding the Co atoms with no and one or less Fe

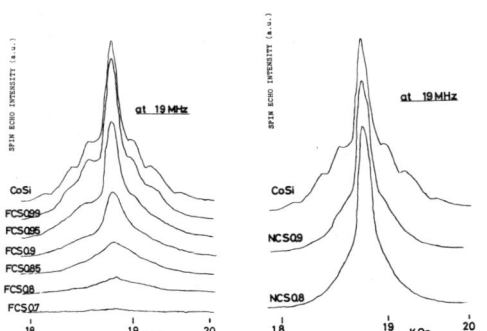

Fig. 2 Spin-echo spectra of Co^{59} NMR in the $(Fe_{1-x}Co_x)Si$ and $(Ni_{1-x}Co_x)Si$ alloys.

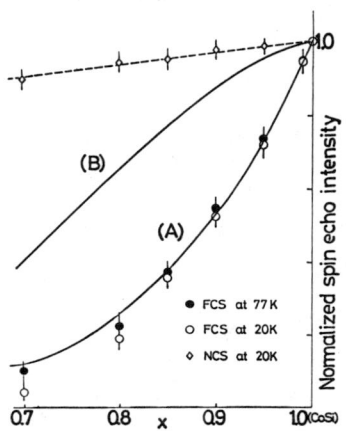

Fig. 3 Fractional decrease of spin-echo intensities of Co^{59} NMR in the FCS and NCS.

impurity atom in its six nearest neighbor metal sites respectively. The experimental data for the FCS system show good agreement with the curve (A). The slight decrease of the intensity for the NCS system may be regarded as due to the quadrupole wipe out in first order.

These results suggest that in the FCS alloys, the Co atoms which have one or more Fe atom in their nearest neighbor metal sites become magnetic and are wiped out magnetically, while the Co atoms having no such Fe atoms (isolated Co) remain nonmagnetic.

The ferromagnetic NMR measurements were made for the FCS alloys with x between 0.3 and 0.6 at 1.5 K. The spin echo signals of Co^{59} in the ferromagnetic FCS's were observed at the frequencies between 10.5 MHz and about 40 MHz. No other separate NMR signals were detected in the higher frequency region up to 200 MHz. The spin-echo spectra, corrected by a factor of $1/\nu$ (ν being the resonance frequency), are shown in Fig. 4, where each spectrum is normalized at its peak value. One sees that each spin echo spectrum has a sharply defined peak and the overall line shape is not of Gaussian type which is usually expected in disordered alloys. A fairly long tail of the spectrum was observed in the higher frequency side for side for each specimen, while the spectrum was sharply cut off in the lower frequency side. The composition dependence of the magnitude of the hyperfine field, $H_n(Co)$, is shown in Fig. 5. Since the peak frequency for FCS0.6 was lower than 10.5 MHz, the lowest limit of our apparatus, it was estimated by extrapolation to be between 8 MHz and 10 MHz. It should be noted that spin-echo spectra are very sensitive to the external magnetic field; an extreme broadening of the spectrum was observed for each specimen when an external field more than 1000 Oe was applied to the specimen under NMR measurement.

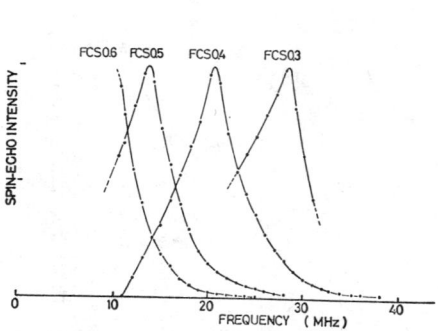

Fig. 4 Spin-echo spectra of Co^{59} NMR in the ferromagnetic FCS alloys.

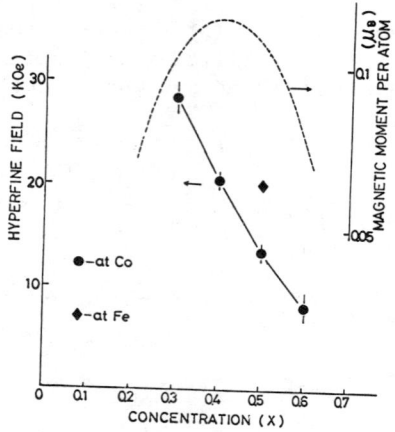

Fig. 5 Concentration dependence of the hyperfine field at Co atom.

DISCUSSION

(i) Energy Band Profile of the FCS Alloys

So far, no band calculation has ever been done for the compounds with B20 crystal structure because of its still complicated manner of the electron bonding. However, we fortunately have a lot of experimental data on the FCS alloys, for example, of the transport properties,[3] the resonance studies[4] and also the low temperature specific heat measurements,[5] which make it possible to depict, even though schematically, an energy band profile available for further discussion on the present experiments: We propose for the FCS system an energy band profile which is composed of two slightly separate d-bands and one s or p band which overlaps with the upper d-band. The lower d-band is completely filled with electron for FeSi, and the upper one is still vacant for FeSi but almost completely filled for CoSi.

(ii) The Magnetic Carriers in the FCS Alloys

Although the magnetic susceptibility measurement of CoSi with dilute Fe impurities have shown that an Fe impurity induces a paramagnetic moment of 3.08 µB at one or more neighborhing Co atoms, the detailed Mossbauer measurement indicated that the Fe atom itself has no localized magnetic moment. In terms of our schematic energy band structure, the d-hole which is carried into CoSi accompanying the Fe atom can move around in the neighborhood of the Fe atom. The result of the Co^{59} NMR in the paramagnetic state indicated that the hopping of the d-hole is confined to the first nearest neighbor metal shell and the Co atoms in this shell become magnetic. One should, however, note that, in the present case, the word "magnetic" is in the NMR sense and means to be very sensitive to the effect of the magnetic field and not necessarily means the existence of localized magnetic moment. From the experimental evidence indicating that the long range ferromagnetic order is just developed at the Fe concentration corresponding to very small probability of finding the isolated Co atoms, it should be also noted that the nearest neighbor Fe-Co pairs play dominant role for the onset of the ferromagnetism in the FCS System.

In Fig. 5, it is seen that Hn(Co) increases monotonically with decreasing x, while the bulk magnetization at 4.2 K has a peak at about x = 0.4. This suggests that the hyperfine field at the Co atom is mainly due to the core polarization induced by the d-electron located at the Co atom and the effect of the bulk magnetization is relatively small. The observed Hn(Co) value of 14 KOe for the FCS0.5 alloy is to be compared with the hyperfine field of about 20 KOe at the Fe atom obtained from the Mossbauer effect measurement of alloy with the same composition.[2] The small magnitude of the hyperfine field at both Fe and Co atoms indicate that the ferromagnetism of the FCS system is, in its nature, due to a weak polarization of itinerant electrons and not due to any other mechanism such as a weak parasitic ferromagnetism in overall antiferromagnetism and the existence of a ferromagnetic second phase.

(iii) Origin of the Ferromagnetism

If one assumes the rigid band approximation (RBA) to the variation of the electronic structure by alloying, one can easily interpret the onset of the ferromagnetism; in terms of the proposed band scheme ferromagnetism can appear when the density of states at the Fermi level in the upper d-band exceeds $1/J$, J being the exchange integral (the Stoner condition). The RBA, however, is too simple to be applied to the 3-d electrons even though it serves well for the primary consideration. Moreover, the RBA cannot interpret the fact that the substitution of Ni for Fe in FeSi never gives rise to ferromagnetism. The fact that the difference of the number of 3-d electrons is one between FeSi and CoSi seems to have a significance. The correlation effects between d-electrons should be taken into consideration.

REFERENCES

1. D. Shinoda, Phys. Stat. Sol. (a) <u>11</u>, 129 (1972).
2. G. K. Wertheim, J. H. Wernick and D. N. B. Buchanan, J. Appl. Phys. <u>37</u>, 3333 (1966).
3. S. Asanabe, D. Shinoda and Y. Sasaki, Phys. Rev. <u>134</u>, A774 (1964).
4. G. K. Wertheim, V. Jaccarino, J. H. Wernick, J. A. Seitchick, H. J. Williams and R. C. Sherwood, Phys. Letters <u>18</u>, 89 (1965).
5. S. Kawarazaki, Y. Nakamura, to be published.

TEMPERATURE DEPENDENCE OF FERROMAGNETIC DISORDER NEUTRON SCATTERING FROM IMPURITIES IN Fe[*]

H. R. Child and J. W. Cable
Solid State Division, Oak Ridge National Laboratory
Oak Ridge, Tennessee 37830

ABSTRACT

We report preliminary measurements of the temperature dependence of the neutron magnetic disorder scattering cross section for 3 at.% impurities in Fe over the T/T_C region 0.3 to 0.8 using 4.4 Å neutrons. The measurements were made by the field off minus field on method and therefore include any field dependent spin correlation scattering present in addition to the disorder scattering. The former effects dominate at $T/T_C = 1$ but data taken on pure Fe could be used to approximately correct the alloy data for $T/T_C \leq 0.8$. The FeV cross section increases at small scattering vector $K = 4\pi\sin\theta/\lambda$ with increasing T/T_C indicating an increase in the range of the magnetic disturbance in the host. The small K cross section of FeSi also increases with increasing temperature but with a more gradual rate of fall with K than FeV indicating a shorter range disturbance. The FeTi cross section has little or no structure for T/T_C as high as 0.8 indicating that the range of the magnetic disturbance, if any exists, remains constant with temperature. Qualitative agreement is found with the molecular field theory of Lovesey and Marshall[1] according to which the impurity induced magnetic moment disturbances vary with temperature in different ways depending on the ratios of impurity to host spins and exchange interactions. However, the theory considers the effect on the impurity site and its first neighbors only whereas our data indicate that the disturbance extends to more than first neighbors in the host. Further studies of these and other alloys are currently in progress. In these new investigations the accuracy of the measurements is increased by about a factor of five as a result of the installation of a position sensitive detector.

---

[*]Research sponsored by the U. S. Atomic Energy Commission under contract with the Union Carbide Corporation.

[1]S. W. Lovesey and W. Marshall, Proc. Phys. Soc. **89**, 613 (1966); S. W. Lovesey, Proc. Phys. Soc. **89**, 625 (1966); Proc. Phys. Soc. **89**, 893 (1966).

THEORY FOR THE SPATIAL CHARACTERISTICS OF MAGNETIC DISTURBANCES IN NICKEL ALLOYS*

J. W. Garland, A. Gonis, and J. S. Kouvel
University of Illinois at Chicago Circle, Chicago, Ill. 60680
and Argonne National Laboratory, Argonne, Ill. 60439

and A. T. Aldred
Argonne National Laboratory, Argonne, Illinois 60439

ABSTRACT

From a local moment point of view, the moment μ_i on any Ni atom depends only on the electronic structure of the other atoms to which it is coupled by d-band hopping terms or interband scattering terms. Although polyvalent impurities may cause charge density perturbations which extend slightly beyond their nearest neighbors, preliminary estimates suggest that any such long range chemical effects are unimportant in determining the range of magnetic disturbances in Ni-base alloys. Here, we consider only binary alloys with nearest neighbor coupling, assume nearest neighbor spins to be aligned parallel, and neglect exchange enhancement and moment formation on impurity sites. For a given type of impurity atom and within these approximations, the moment μ_i on a Ni atom depends almost entirely on n_i, its number of Ni nearest neighbors, and on the moment μ_j on each of them. Then, any long range magnetic disturbance arises from the cooperative dependence of the moments on one another, specified by the function $g_\ell'(n_i,\{\mu_j\}) = (\partial\mu(n_i,\{\mu_j\})/\partial\mu_\ell)_{\ell\epsilon\{j\}}$. The effect of nearest neighbor impurities also depends on the function $g^o(n_i,\{\mu_j\}) = \mu(n_i-1,\{\mu_j\}) - \mu(n_i,\{\mu_j\})$, where the μ_j are held fixed. We have constructed formulas for the parameters $g(R_i)$ of Marshall[1] in terms of average values, g' and g^o, of these functions. Thus, the average magnetic disturbance caused by a substitutional impurity is completely specified by g', g^o, and the average disturbance, $<\mu_{imp} - \mu_{Ni}>$, on an impurity site. An analysis of neutron scattering data in terms of the parameters g' and g^o allows much less ambiguity in fitting the data than does the usual analysis in terms of the $g(R_i)$. The data of Aldred, et al.[2] for Ni-Cu alloys with $0.023 \leq c_{Cu} \leq 0.40$ and that of Comly, et al.[3] for 13 dilute Ni-base alloys can be fit perfectly by choosing g', g^o, and $<\mu_{imp} - \mu_{Ni}>$ as parameters. One finds that g' is proportional to c_{Cu} for dilute Ni-Cu alloys and increases with increasing impurity concentration or valence, reaching a value $g' \approx 0.08$ either for transition-metal impurities or for concentrated Ni-Cu alloys. The observed variation of g' follows qualitatively from first principles theoretical arguments. Those arguments have been used to construct a parameterized model for moment formation; this model also can be accurately fit to the data for the Ni-Cu alloys, and then predicts the experimental results for other Ni alloys without further parameters.

1. W. Marshall, J. Phys. C: Solid State Phys. 1, 966 (1968).
2. A. T. Aldred, B. D. Rainford, T. J. Hicks, and J. S. Kouvel, to be publ.
3. J. B. Comly, T. M. Holden, and G. G. Low, J. Phys. C 1, 458 (1968).

*Based on work performed under the auspices of the USAEC.

SPIN AND ORBITAL SUSCEPTIBILITY OF Co IMPURITIES[*]

A. L. Ritter, J. Bensel, and J. A. Gardner
University of Pennsylvania, Philadelphia, Pa. 19104

ABSTRACT

We have measured the impurity susceptibility and Knight shift of dilute Co impurities in liquid Al-Cu alloys. From these results we can determine the orbital hyperfine field of Co to be 360 ± 30 kG/μ_B. In any case in which core polarization by the 3d spins is relatively small, the Co impurity susceptibility can now be separated into spin and orbital parts. We apply this separation to CuCo and find that 720 to 930×10^{-6}/mole of the total impurity susceptibility of 2400×10^{-6}/mole is due to the orbital susceptibility. Estimating roughly the Co impurity density of states from other data, we find that the orbital enhancement factor for CuCo is of order one.

In the past two or three years it has been demonstrated that the orbital susceptibility of "non-magnetic" 3d impurities in metals can be quite sizeable.[1] These impurities are usually described in terms of the non-magnetic or localized spin fluctuation (LSF) limit of the Friedel-Anderson model. Since the spin and orbital susceptibilities can be simply related to model parameters, it is of considerable interest to be able to separate experimentally the two contributions. If in addition the impurity density of states can be determined, one can in principle find all the model parameters.

It is possible to separate the impurity susceptibility into spin and orbital components from measurement of the impurity susceptibility and Knight shift. The Knight shift is the sum of three components, a contact part which is proportional to the conduction electron susceptibility, an orbital part proportional to the impurity orbital susceptibility, and a core polarization contribution proportional to the impurity 3d spin susceptibility. In most cases of interest the contact Knight shift can be estimated, and one needs to know only the orbital and spin hyperfine fields (i.e. the appropriate constants of proportionality) to separate the impurity susceptibility uniquely. The spin hyperfine field H_{spin} can be measured in the extreme magnetic limit, because the spin susceptibility and core polarization Knight shift are so large that other parts can be neglected. Because of experimental difficulties H_{spin} is presently known only roughly for 3d impurities, however. Measured values are all negative and lie generally in the range 0 to -80 kG/μ_B. The spin hyperfine field of Co apparently is between 0 and -40 kG/μ_B.[1] Theoretical estimates of the 3d impurity orbital hyperfine field are of

---

[*]Work supported in part by the National Science Foundation and the Advanced Research Projects Agency.

order 500 kG/μ_B,[2] but this quantity has not previously been measured experimentally.

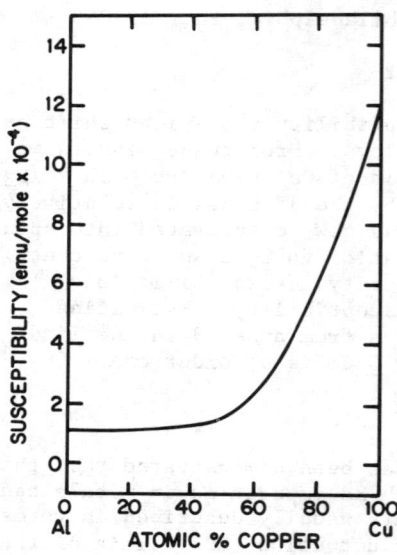

Fig. 1. Molar susceptibility of Co impurities in liquid Al-Cu alloys at 1100° C.

We have been able to find H_{orb} for Co from our measurements of the Co susceptibility and Knight shift in liquid Al and Al-Cu host alloys.[3] The Co susceptibility in liquid Al-Cu is shown in Fig. 1. Below 50 percent Cu, the Co susceptibility is quite small, and we expect to find a sizable orbital contribution to the impurity susceptibility. In this range the Co susceptibility increases rather rapidly with temperature, typically of order 15 percent per 100 degrees. The Co Knight shift also has a positive temperature slope, and in Fig. 2 we show the Co Knight shift K, as a function of Co susceptibility χ. The lack of any explicit temperature or host dependence of the Knight shift has led us to conclude that the change in the susceptibility is caused by a temperature and host dependence of the impurity density of states and that the predominant Knight shift changes are due to the orbital component. The core polarization for this quite non-magnetic state must be small ($|K_{spin}| < 0.05°/°$), and we neglect it.

If the relative proportion of orbital to spin susceptibility can be determined, one can immediately find H_{orb} from the slope of K vs. χ. In the extreme non-magnetic limit the orbital g factor of an impurity should be nearly unity, and it is straightforward to show that the orbital susceptibility of the 3d electrons is then twice

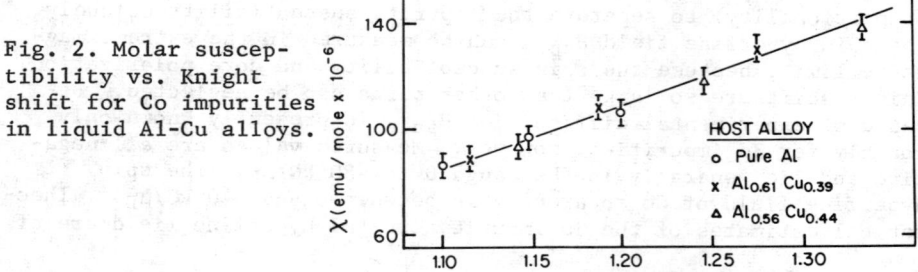

Fig. 2. Molar susceptibility vs. Knight shift for Co impurities in liquid Al-Cu alloys.

their spin susceptibility. Dworin and Narath[4] have extended this result to include localized spin fluctuations, and they find for the susceptibilities of an impurity state having total d density of states ρ_d,

$$\chi_{spin} = \rho_d \mu_B^2 \eta_s \qquad (1)$$

$$\chi_{orb} = 2\rho_d \mu_B^2 \eta_{orb} \qquad (2)$$

where $\eta_s = [1 - (U + 5J) \rho_d/10]^{-1}$ and $\eta_{orb} = [1 - (U-J) \rho_d/10]^{-1}$ are the spin and orbital enhancement factors respectively.

If the two enhancement factors are assumed equal, then $\Delta K/\Delta \chi = (2/3) \Delta K_{orb}/\Delta \chi_{orb}$, and from the slope in Fig. 2 we find $H_{orb} = 360 \pm 30$ kG/μ_B. There are two experimental indications that our assumption of equal spin and orbital enhancement is correct. One is the apparent linearity of K vs. χ which indicates that the ratio of the two enhancement factors is constant (and by implication unity since η_s must diverge more rapidly than η_{orb}). The second is the experimentally observed constant value of the host Knight shift perturbation by Co in this host concentration range.[3] The host perturbation has been shown to be roughly proportional to η_s,[5] and the strong implication is that η_s is unity up to 50 percent Cu.

Since the paramagnetic susceptibility of the impurity s-p band electron will approximately cancel the core diamagnetism, we can determine the contact part of the Co Knight shift by extrapolating Fig. 2 to zero susceptibility. We find $K_{contact} = 0.7 \pm 0.3$ percent and for Co in pure Al near the melting point $K_{orb} = 0.4 \pm 0.1$ percent. The larger error for the contact term arises from uncertainty in the exact value for the "unshifted" gyromagnetic ratio.[6] Narath and Dworin[4] showed that the three Knight shift components obey separate Korringa relations

$$(T_1 T)_i K_i^2 = a_i (\gamma e/\gamma n)^2 (\hbar/4\pi k_B) \qquad (3)$$

where a_i is 1 for the contact term, 5 for the spin and 10 for the orbital Knight shift. These relations provide a useful check on our separation of the Co Knight shift. Clearly only the contact term contributes significantly to the relaxation rate in AlCo. Our experimental value of $(T_1 T)^{-1} = 6 \pm 1.5$/sec. K requires a contact Knight shift of 0.54 percent to satisfy Eq. 3. If the Korringa relations are not greatly enhanced by many body effects, this relation provides a more sensitive determination of $K_{contact}$ than our previous estimate which includes a large uncertainty in the unshifted gyromagnetic ratio. We conclude that the most probable range for $K_{contact}$ is $0.55 \pm .15$ percent. This value is somewhat larger than would be expected in comparison to the copper Knight shift, but it it is clearly in good agreement with all experimental information.

Now that the orbital hyperfine field is known, it is possible to separate the impurity susceptibility for the interesting nearly magnetic impurity state CuCo. The Co susceptibility in Cu at low temperature is 2400 x 10^{-6}/mole,[7] and the Knight shift is 5.2%.[8] If the Co contact Knight shift changes in the same way as do Al and Cu, it will be about 20% smaller in Cu than in Al,[9] and we take $K_{contact}$ = 0.4% for Co in Cu, with the remaining 4.8% due to the orbital and spin Knight shifts. Using 360 kG/μ_B for H_{orb} and the above experimental information, we have computed the orbital and spin components of the Co Knight shift and susceptibility shown below. $(T_1T)^{-1}$ is also computed for each part of the Knight shift using Eq.3. We show the separation for the two extremes (A) H_{spin} = 0 and (B) H_{spin} = -50 kG/μ_B between which the Co spin hyperfine field evidently lies.

Table 1. Susceptibility Separation for CuCo

		Orb	Spin	Contact	Total
(A)	K	4.8	0	0.4	5.2%
	$(T_1T)^{-1}$	48.5		3.4	52/sec. K
	χ	7.2	16.8	---	24 x 10^{-4} mole
(B)	K	6.15	-1.35	0.4	5.2%
	$(T_1T)^{-1}$	80	8	3.4	88/sec. K
	χ	9.3	14.7	---	24 x 10^{-4}/mole

The experimental value of $(T_1T)^{-1}$ = 50 ± 12/sec. K [8] is in best agreement with the assumption of a near zero spin hyperfine field, but agreement with case (B) is also acceptable.

It is clear from the above table that the spin susceptibility is several times larger than the orbital susceptibility. It has often been assumed that the LSF model remains appropriate even for such highly enhanced susceptibilities as these, and it is interesting to estimate the model parameters under this assumption. If ρ_d were known, it would be possible to obtain all parameters using Eqs. 1 and 2. Unfortunately the density of states is not well-known. Low temperature specific heat measurements are dominated by the contribution from Co clusters and/or interactions, and the interpretation is further complicated by uncertainty in the effective mass enhancement by spin fluctuations. For extremely low Co concentration there is some evidence that the one-impurity specific heat dominates. Ignoring any mass enhancement, we obtain from very low concentration specific heat measurement[10] a rough estimate of $\rho_d \cong$ 50 $\rho_{Cu} \cong$ 12 states/ev atom. Another estimate may be obtained by assuming that the width of the Co 3d density of states is about the same as that of Ni. For Ni in Cu the linewidth has been found to be 0.27 ± .02 ev.[11] For an occupancy of 7d electrons appropriate to Co, the density of states at the Fermi level would be 8 states/ev atom. From the susceptibilities given in Table 1, we can calculate $\rho_d \eta_{orb}$ for cases (A) and (B) to be 11.1 and 14.4 states/ev atom respectively. It is striking to note that the orbital enhancement is quite moderate;

in fact within the accuracy of our rough estimates, there may well be no orbital enhancement at all. It remains to be seen whether a spin susceptibility which is enhanced by a factor of 5 or more over its bare band value is properly described by Eq. 1, but the orbital susceptibility does seem to be quite consistent with the band-LSF approximation.

REFERENCES

1. A. Narath, CRC Crit. Reviews in Solid State Physics $\underline{3}$, 1 (1972) and references therein.
2. A. J. Freeman and R. E. Watson, "Magnetism", ed. G. Rado and H. Suhl,(Academic Press, N. Y., 1965).
3. A detailed description of this work is to be published shortly.
4. L. Dworin and A. Narath, Phys. Rev. Letters $\underline{25}$, 1287 (1970).
5. C. P. Flynn, D. A. Rigney, and J. A. Gardner, Phil. Mag. $\underline{15}$, 1255 (1967).
6. R. E. Walstedt, J. H. Wernick, and V. Jaccarino, Phys. Rev. $\underline{162}$, 301 (1967).
7. R. Tournier and A. Blandin, Phys. Rev. Letters $\underline{24}$, 397 (1970).
8. S. Wada and K. Asayama, J. Phys. Soc. Japan $\underline{30}$, 1337 (1971).
9. S. Sotier, R. L. Odle, and J. A. Gardner, Phys. Rev. B $\underline{6}$, 923 (1972).
10. F. J. du Chatenier, thesis, University of Leiden (1964), data quoted by G. J. van den Berg, Proc. of 9th Int. Conf. of Low Temp. Physics, (Plenum Press, N. Y., 1964).
11. H. D. Drew and R. E. Doezema, Phys. Rev. Letters $\underline{28}$, 1581 (1972).

ORIGIN OF THE SINGLE PULSE ECHO IN Co.

Mary Beth Stearns

Ford Motor Co., Scientific Research Staff, Dearborn, Mich. 48121

ABSTRACT

A prominent but completely ignored feature of pulsed NMR in Co is a strong single pulse echo which occurs at a time equal to the pulse length after the end of the pulse. Although it can easily be seen when looking at wall signals in fcc Co, it is strikingly obvious in hcp Co where due to the large anisotropy the signal is mainly from the domains (with a corresponding smaller enhancement factor and broader frequency spectrum due to the demagnetizing fields than the wall signal from fcc Co). In the literature the SPE has been attributed to arising from coherence effects of a broad frequency spectrum. Computer calculation show that this is not the origin of the SPE. In Co two conditions are responsible for the SPE: 1) Since Co is a single isotope with a large nuclear moment it has a strong Suhl-Nakamura interaction ($T_2 \sim 6\mu s$ independent of temperature). Thus under normal operating conditions the usual conditions $\omega_1 T_2 \gg 1$, where ω_1 is the r.f. field strength, is not satisfied. The usual resonance formulae must first be generalized to the case where $T_1 \neq T_2$ and then corrected to keep the steady state magnetization terms. 2) The spread in frequencies of the exciting r.f. pulse due to its pulsed nature must be included in the calculations. This spread is more important than the frequency spread due to a broad spectrum.

The phenomena observable in pulsed NMR experiments on Co are particularily diverse. Because Co is a pure isotope with a large nuclear moment (4.64 nm) it has a rapid transverse relaxation rate due to mutual spin flip of two nucleii by emitting and absorbing virtual magnons; i.e. the Suhl-Nakamura (SN) interaction[1]. We have measured the transverse relaxation time (T_2) to be about the same ($\sim 6\mu s$) in domains and walls and independent of temperature (4.2°K to 298°K). This relaxation time is so short that under normal operating r.f. field strengths the usual condition, $\omega_1 T_2 \gg 1$, is often not satisfied. Where $\omega_1 = \epsilon\gamma B_1 (B_1 = \mu H_1)$ ϵ is the enhancement factor, γ the nuclear gyromagnetic ratio and B_1 the r.f. field strength in the sample. For pulse length, τ, around the order or shorter than T_2 these conditions lead to a single pulse echo (SPE) occurring

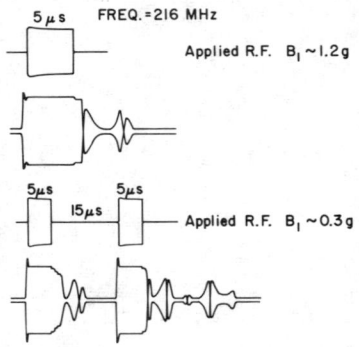

Fig. 1. Typical appearance of single pulse echoes in Co^{59}.

a time τ after the pulse as shown in Fig. 1. In the literature[2] this effect is attributed to coherence effects taking place when exciting a broad line. However, the same SPE also occurs for the narrow line ($\sim \frac{1}{2}$ MH$_z$) wall resonance as well as for the broad line ($\sim$ 18 MH$_z$) domain resonance. Computer calculations show that this is not the origin of the SPE. We shall here derive the correct resonance formulas to use in this case.

We use the same notation and procedure as used by Jaynes[3] to derive the magnetization following an r.f. pulse of frequency ω in the x-direction so $H_x = 2H_1 \cos \omega t$, $H_y = 0$ and H_z is the internal hyperfine field H_o (no external dc fields). We start with the usual Block equation of motion[3] for the nuclear magnetization $\vec{M}(t)$ then generalize Eq. 13 of Ref. 3 to the case of $T_2 \neq T_1$, where T_1 is the longitudinal relaxation time ($\sim$ 120 μs at R.T.). We obtain for the magnetization at time t

$$\vec{M}(t) = e^{-(1/\underline{T}+\underline{\beta})} [\vec{M}(o) - \vec{M}(\infty)] + \vec{M}(\infty) \quad (1)$$

where β is the precession matrix $\vec{B} \times$ ($\times$ is the vector product operation) and $\vec{B}$ has components $[\omega_1, 0, \Delta\omega (= \gamma H_o - \omega)]$. $\underline{T}$ is a matrix whose only non-zero elements are on the diagonal and are (T_2, T_2, T_1). $\vec{M}(o)$ is the magnetization at the beginning of the r.f. pulse given by $[0, 0, M_o]$. $\vec{M}(\infty)$ is the steady state magnetization which during the r.f. pulse is given by

$$\vec{M}(\infty) = M_o/(1+\omega_1^2 T_1 T_2 + \Delta\omega^2 T_2^2) \begin{pmatrix} \omega_1 \Delta\omega T_2^2 \\ \omega_1 T_2 \\ 1 + \Delta\omega^2 T_2^2 \end{pmatrix}. \quad (2)$$

In the calculations of Bloom[2] which are generally used to analyze pulsed NMR experiments the assumption is explicitly made that $\vec{M}(\infty)$ is neglected since in many circumstance it is small compared with the polarization at the start of the interval. However we find that in order to obtain a SPE it is necessary to retain the $\vec{M}(\infty)$ terms.

Let us now consider the case where we apply an r.f. pulse for a time τ_1 and then look at time t_2 after the pulse. During the time t_2 B has components $[0, 0, \Delta\omega]$ and $\vec{M}(\infty) = \vec{M}_o$. We get for the magnetization at time t_2

$$\vec{M}(t_2, \Delta\omega) = e^{-(1/\underline{T}+\underline{\beta}_2)t_2} [\vec{M}(\tau_1, \Delta\omega) - \vec{M}_o] + \vec{M}_o, \quad (3)$$

where $\vec{M}(\tau_1, \Delta\omega)$ is evaluated from Eq. (1). Experimentally we observe the perpendicular component of $\vec{M}$ integrated over the frequency spectrum. $M_x(t_2, \Delta\omega)$ is odd in $\Delta\omega$ and so for a symmetrically shaped spectrum gives no contribution. We therefore consider only $M_y(t_2, \Delta\omega)$ and integrate it over the shape of the spectrum $g(\Delta\omega)$ to get

$$M_y(t_2) = \int_{-\infty}^{\infty} g(\Delta\omega) M_y(t_2, \Delta\omega) d(\Delta\omega). \quad (4)$$

$M_y(t_2, \Delta\omega)$ is given by

$$M_y(t_2,\Delta\omega) = e^{-t_2/T_2} \cos\Delta\omega t_2 M_y(\tau_1,\Delta\omega) - \sin\Delta\omega t_2 M_x(\tau_1,\Delta\omega) \quad (5)$$

where $M_y(\tau_1,\Delta\omega)$ and $M_x(\tau_1,\Delta\omega)$ are complex expressions obtained from Eq. 1. Upon evaluating Eq. 4 on a computer for a broad flat frequency spectra, $g(\Delta\omega)$ = constant, we obtained no SPE which resembled those which are so prominant experimentally. This disagrees with Bloom's[2] conjecture on the origin of the SPE or "edge echo". Another interesting result of these calculations was that in order to get the computer calculations to converge to a definite curve the limit of integration of $\Delta\omega$ had to be at least 100 times ω_1. Thus the often made assertion that only nuclei within a frequency range of a few times ω_1 around the exciting frequency ω contributed to the signal is far from true in this case.

In order to obtain the SPE from the calculations it was necessary to include the fact that we are exciting the nuclei with a frequency spectrum due to the pulsed nature of the r.f. Thus for a square r.f. exciting pulse of length τ_1 and frequency ω, as is well known, the frequency spectrum amplitude is

$$f(\Delta\omega_e) \sim \sin(\Delta\omega_e t_1/2)/(\Delta\omega_e t_1/2) \quad , \quad (6)$$

where $\Delta\omega_e = \omega_e - \omega$. If ω_n is the resonance frequency of some particular nuclei then during the pulse $\Delta\omega$ becomes $(\Delta\omega_n - \Delta\omega_e)$ and after the pulse it is $\Delta\omega_n$ so Eq. 5 becomes

$$M_y(t_2,\Delta\omega_n,\Delta\omega_e) = e^{-t_2/T_2}\left[\cos\Delta\omega_n t_2\, M_y(\tau_1,\Delta\omega_n-\Delta\omega_e) - \sin\Delta\omega_n t_2\, M_x(t_1,\Delta\omega_n-\Delta\omega_e)\right] \quad . \quad (7)$$

To obtain the observed signal we must integrate over $\Delta\omega_e$ so we get

$$M_y(t_2) \sim \int_{-\infty}^{\infty}\int_{-\infty}^{\infty} g(\Delta\omega_n)\,f(\Delta\omega_e)\,M_y(t_2,\Delta\omega_n,\Delta\omega_e)\,d(\Delta\omega_e)d(\Delta\omega_n) \quad . \quad (8)$$

Evaluating this spectrum leads to SPE's as shown in Fig. 2. Here the pulse length is taken as 0.5 T_2 and the ratio of T_1/T_2 was 20. The solid curve is that due to retaining the $\vec{M}(\infty)$ terms and the dashed curve results from leaving out the $\vec{M}(\infty)$ terms as is done in Ref. 2. We see that the $\vec{M}(\infty)$ terms are indeed the main contributors to the SPE under the conditions existing for Co. Under normal operating conditions the domains in the sample are oriented at random to the r.f. field directions. So the domain nuclei experience different r.f. field strengths which vary as $\omega_1 = \epsilon\gamma B_1 \sin\eta$ where η is the angle between the local magnetization direction and the r.f. field direction. Thus we should replace ω_1 by $\epsilon\gamma B_1 \sin\eta$ in Eq. 8 and also average over η. We have also carried out this calculation and the echoes are very similar to those shown in Fig. 2 but now the SPE amplitude maximizes at a somewhat higher value of $\omega_1 T_2$ as expected.

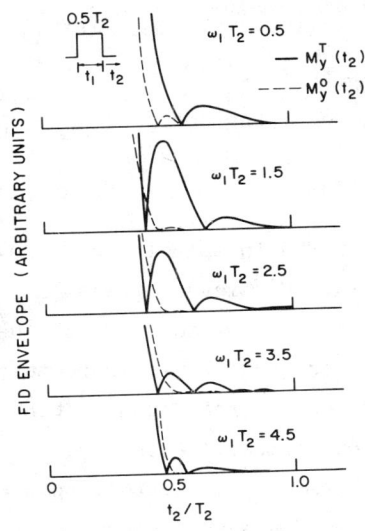

Fig. 2. Calculated SPE's for a pulse of length $0.5T_2$ and $T_1/T_2 = 20$. The solid curve $M_y^T(t_2)$ includes the steady state magnetization terms. The dashed curve $M_y^O(t_2)$ is from the usual formulae used to describe pulsed NMR, as in Ref. 2.

For $\tau_1 = 0.5T_2$ the maximum shifts from about $\omega_1 T_2 = 1$ to $\omega_1 T_2 = 2.5$ upon averaging over η. In most cases an average over ϵ should also be made. This would not be expected to change the echo shape appreciably but again would shift the maximum of the SPE amplitude curve to a higher $\omega_1 T_2$ value.

Care must be taken in T_2 determinations so that the SPE does not interfer with the usual double pulse echo and produce spurious effects at small times.

REFERENCES

1. H. Suhl, Phys. Rev. 109, 606 (1958); J. Phys. Radium 20, 333 (1959); T. Nakamura, Prog. Theoret. Phys. (Kyoto) 20, 542 (1958).
2. A. L. Bloom, Phys. Rev. 98, 1105 (1955).
3. E. T. Jaynes, Phys. Rev. 98, 1099 (1955).

1648 Section 50.   Lattice Effects in
                          Rare-Earth Magnetism

COOPERATIVE JAHN-TELLER EFFECTS IN MAGNETIC MATERIALS

K. A. Gehring

Clarendon Laboratory, Oxford, England.

ABSTRACT

Cooperative Jahn-Teller Effects (CJTE) in spinels and perovskites containing $3d^n$ ions are well-known. For several reasons these systems are not ideally suitable for a detailed study of the effect. The recently observed CJTE in tetragonal crystals containing $4f^n$ ions provide a new and powerful tool in this study. Emphasis is laid on the very wide variety of experimental techniques which have been used. These include calorimetric, spectroscopic and X-ray methods demonstrating the existence of a second order phase transition in which electronic energy levels split and the crystal structure changes. Further experiments demonstrate changes in the elastic and magnetic properties and reveal the nature and detailed behaviour of the coupled modes. The basic simplicity of the CJTE is discussed with reference to $TmVO_4$ as a paradigm. Here the electronic ground state is an isolated orbital doublet and this couples to a single lattice mode. A molecular field theory describing this case is presented and further theoretical aspects are discussed.

1. INTRODUCTION

This review will be concerned with the consequences of the Jahn-Teller (JT) theorem for magnetically concentrated, insulating crystals.

The original theorem[1] applied to polyatomic molecules with an orbitally degenerate state and showed that "stability and degeneracy are not possible simultaneously unless the molecule is a linear one." The proof was group theoretical and consisted of the enumeration of the representations Γ_e of all orbitally degenerate states of all molecular symmetry groups.

It was shown that for all molecular configurations there exists some normal mode of vibration, transforming like Γ_v, for which the symmetric product $[\Gamma_e \times \Gamma_e]$ contains Γ_v. This means that there are matrix elements within the electronic manifold which are linear in the normal vibrational coordinate Q.

$$\mathcal{H}(JT) = -2AQS_z$$

Here A is a constant describing the strength of the coupling and S_z is an operator acting within the electronic manifold. (Henceforth we will refer to S_z as a pseudo-spin operator and take it to act within a pseudo-spin doublet.) If we add to this the potential part $\frac{1}{2}m\omega^2 Q^2$ of the vibrational energy we obtain the two intersecting parabolae shown in Fig. 1. The degenerary at Q = 0 is now not stable and there are energy minima

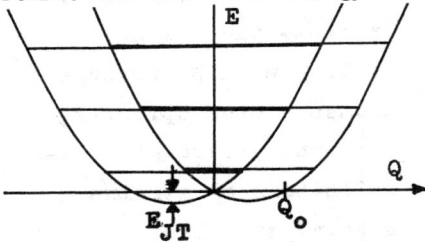

Fig. 1. Potential wells and oscillator energy levels for an isolated JT complex in which an electronic doublet couples to a vibrational singlet.

of depth $E_{JT} = A^2/2m\omega^2$ at $Q_o = \pm A/m\omega^2$.

If the kinetic part, $(-\hbar^2/2m)(\partial^2/\partial Q^2)$, of the vibrational energy is included then the solutions of the vibronic hamiltonian are the uncoupled oscillator energy levels shown in the separated harmonic wells.

A fundamental feature of the JT coupling is that it does not change the degeneracy of the states but it does change the energies of the states and the nature of the wavefunctions. This can be seen by considering the consequence on Fig. 1 of allowing the parameter A to decrease smoothly to zero. If (as is usually the case) there exists some minute symmetry reducing perturbation this will cause tunnelling between the states in the different harmonic wells and there can be no static distortion.

The application of these ideas to a magnetically concentrated solid which distorts cooperatively is conceptually straightforward but there are several complications of detail. The normal modes of a solid are usually described within the framework of phonon theory and the electrons can couple to acoustic and optic phonons. It will also be seen that the coupling to the strain modes (in which the external shape of the crystal changes) is very important. An interaction exists between the JT ions caused by virtual phonon exchange and this leads to a phase transition at a temperature T_D, the cooperative Jahn-Teller effect (CJTE). In some substances orbital degeneracy can be associated with electric quadrupole moments[2] which align at a phase transition[3]. In this model the distortion would be of the magneto-strictive type.

In principle the two JT distortions of the crystal are equivalent and small perturbations can cause tunnelling between them but because we are now dealing with N-particle wave functions the tunnelling time becomes macroscopically enormous and we can see a static distortion. There will also be other effects such as crystal imperfections or strains which lead to a particular distortion being stabilised. Different directions of distortion may be stabilised in different parts of the crystal leading to domains.

2. OLD AND NEW MATERIALS

The existence of cooperative Jahn-Teller effects in spinels and perovskites containing elements from the iron group (with incomplete 3d shells) is well-known and has been reviewed by Sturge[4] and Englman[5]. These materials are both cubic so that orbital states which are doubly ($\Gamma_e = E$) or triply ($\Gamma_e = T_1$ or T_2) degenerate are allowed. If we ignore the breathing mode, A_1, the symmetrised products are $[E \times E] = E$ and $[T_1 \times T_1] = [T_2 \times T_2] = E + T_2$. The orbitally degenerate states therefore couple to the degenerate vibrational modes E or T_2 only and not to singlet modes. The singlet mode model used in Fig. 1 has to be extended to include dynamic effects, simultaneous coupling to more than one mode and higher order coupling before even the simplest properties of these materials can be understood. Much of the complexity in both the experimental data and the theory is a direct consequence of this necessity to elaborate on the singlet mode model.

Over the last two years there has been a great deal of interest in a family of crystals which contain rare earth ions (with incomplete 4f shells) and which exhibit CJTE[6,7]. These have the general formula RXO_4 where R is Tb, Dy or Tm and X is V or As. They crystallise in the tetragonal zircon structure which has the space group symmetry D_{4h}^{19} and magnetic site symmetry D_{2d}. Now the only symmetry allowed orbitally degenerate state in tetragonal symmetry is a doublet ($\Gamma_e = E$) and since $[E \times E] = B_1 + B_2$ this can couple to singlet vibrational modes only. In addition to this basic simplicity there are several further advantages which these tetragonal rare earth crystals possess.

1) The phase transitions occur at low temperatures ($2K < T_D < 34K$) so that the CJTE are not masked by thermal effects.

2) The materials are transparent so that various optical and spectroscopic techniques may be used.

3) It is possible to apply stresses and magnetic fields which cause splittings of the order kT_D.

Some of the properties of these rare earth crystals are summarised in Table 1. Of course not all of them conform exactly to the model discussed above but one of them, $TmVO_4$, will be used as a paradigm because it does have an isolated orbital doublet lowest and can be described by a very simple Ising model. Other members of this family have differences in the patterns of low-lying energy levels which lead to differences in the details of their properties. The most extensively documented material is $DyVO_4$ but since Dy^{3+} has an odd number of electrons and since the symmetry is tetragonal there can be no true orbital degeneracy. Here the effects arise because the lowest states are two Kramers doublets which are accidentally close together. However, provided the strength of the JT coupling is greater than the splitting between the doublets we expect a phase transition and can describe many properties by an Ising model.[7]

	Tb	Dy	Tm
VO_4	$T_D = 33$ $B_{2g}(D_{2h}^{24})$ $A_1(0), E(9), B_1(18)$ Refs: 7, 8, 9, 10.	$T_D = 14.0$ $B_{1g}(D_{2h}^{28})$ $E'(0), E''(9)$. Refs: 7, 11, 12, 13.	$T_D = 2.10$ $B_{2g}(D_{2h}^{24})$ $E(0)$. Refs: 14, 15, 16.
AsO_4	$T_D = 27.7$ $B_{2g}(D_{2h}^{24})$ $A_1(0), E(6), B_1(15)$. Refs: 17, 18.	$T_D = 11.2$ $B_{1g}(D_{2h}^{28})$ $E'(0), E'''(\sim 10)$. Refs: 7, 19.	$T_D = 6.0$ $B_{2g}(D_{2h}^{24})$ $E(0), A?(\sim 10)$. Refs: 16, 20.

Table 1. Summary of some of the properties of tetragonal rare earth crystals exhibiting CJTE. For each substance is shown: the distortion temperature T_D, the distortion symmetry (with the low temperature space group), the symmetries of the low-lying energy levels for $T > T_D$ (with the energy in cm^{-1}), and some selected references. In addition to the materials shown in this table the following rare earth crystals show effects closely related to CJTE: $TbPO_4$, see Lee et al.[21]; $PrAlO_3$, see Cohen et al.[22]; DySb, see Bucher et al.[23].

3. BASIC EXPERIMENTAL RESULTS

It is a useful (though rather arbitrary) procedure to separate certain experimental results and regard them as providing basic information about the CJTE. We include in this group those experiments which demonstrate the existence of a phase transition in which the electronic energy levels split and the crystal structure changes.

It is also useful to select $TmVO_4$ as a material for detailed discussion since its properties are ideally simple. Its ground state is an orbital doublet and there is no other electronic energy level within 50 cm^{-1}.[15] The specific heat of $TmVO_4$ has been measured by Cooke et al.[14] and is shown in Fig. 2 (a). The entropy

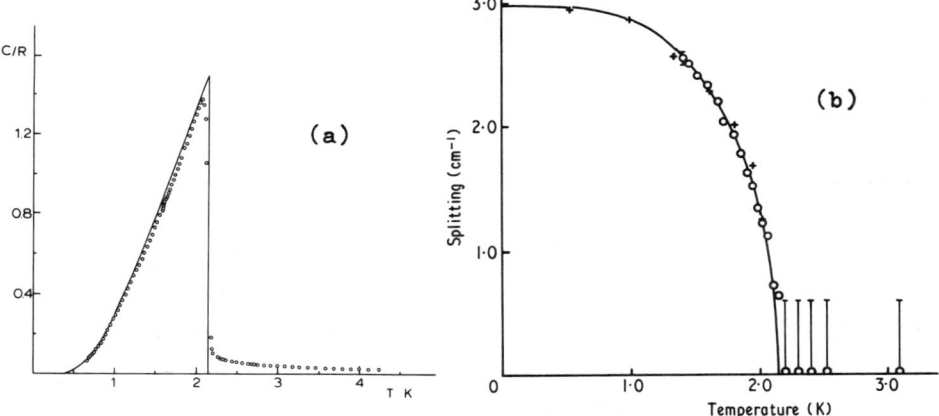

Fig. 2. The specific heat (a) and the electronic energy level splitting (b) of $TmVO_4$ are compared with the results of a molecular field theory (solid lines).

associated with the anomaly is R log 2 showing that there is a second order phase transition in which a doublet splits into two singlets. The solid line is calculated using molecular field theory and is seen to lie **remarkably close to the experimental** points. The splitting of the doublet has been measured directly using optical spectroscopic methods by Becker et al.[15] This splitting is shown in Fig. 2 (b) and is also compared with the results of molecular field theory.

The existence of a change of crystal structure in TmVO$_4$ has been demonstrated by the same workers[15]. They saw changes in an interference fringe pattern generated by the crystal. These changes were associated with the transition from optically uniaxial behaviour (in tetragonal symmetry) to optically biaxial behaviour (in orthorhombic symmetry).

Quantitative evidence for the crystal structure changes in, e.g., TbVO$_4$, has been obtained by Will et al.[9] using X-ray diffraction. Their results are shown in Fig. 3 and it can be seen that

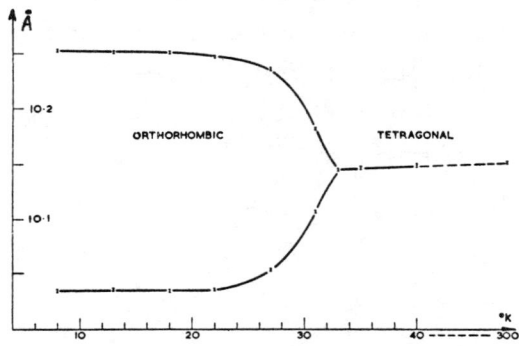

Fig. 3. The **unit cell dimensions of TbVO$_4$ obtained by X-ray diffraction.**[9]

the fractional change in the unit cell parameters are $\pm$ 2.2% with no discontinuous change in the unit cell volume. Further evidence for the change of symmetry is provided by the fact that phonon energy levels, which are degenerate in tetragonal symmetry, split below the transition temperature. This has been observed for optic phonons by Elliott et al.[7] using Raman scattering and for acoustic phonons by Gorodetsky et al.[24] using ultrasonics.

The fact that molecular field theory describes the properties of TmVO$_4$ so well shows that the JT interactions have a very long effective range. This arises because the important interactions are between the electrons and the strain modes of the crystal. As a consequence it is very important to perform experiments on these CJTE materials using unstrained samples. Indeed, Sayetat[25] was able to reduce the size of the crystallographic distortion in TbVO$_4$ and DyVO$_4$ by crushing her samples to a (strained) powder.

4. THEORY

The full hamiltonian for the electron plus lattice system and its solutions have been discussed in detail by Elliott et al.[7] and Gehring et al.[11] and we shall summarize and simplify only a few of the important features. The contributions to the hamiltonian are:

$$\mathcal{H}_1(\text{phonon}) = \sum_{n\underline{k}} (c_n^+(\underline{k})c_n(\underline{k}) + \tfrac{1}{2})\hbar\omega_n(\underline{k})$$

$$\mathcal{H}_2(\text{JT}) = \sum_{n\underline{k}} B_n(\underline{k})(c_n^+(-\underline{k}) + c_n(\underline{k}))\, S_z(\underline{k})$$

$$\mathcal{H}_3(\text{strain coupling}) = CeS_z(\underline{k}=0)$$

$$\mathcal{H}_4(\text{further splittings}) = DS_x(\underline{k}=0)$$

The phonon part of the hamiltonian is written in terms of the usual creation and annihilation operators for optic and acoustic phonons of branch n and wavevector $\underline{k}$. The JT part which we discussed in section 1 has been rewritten in terms of the Fourier transform $S_z(\underline{k})$ of the z-component of the pseudo-spin operators so that the $B_n(\underline{k})$, which measure the strength of the coupling, are related to constants like the A used in section 1. The macroscopic strain, e, couples only the uniform mode $S_z(\underline{k}=0)$ with a strength measured by C. The last term allows us to include the effect of any further splittings of the pseudo-spin doublet. This may be caused by a magnetic field (in the case of a true orbital doublet such as exists in $TmVO_4$) or by a residual crystal field splitting (in the case of the accidental orbital degeneracy in $DyVO_4$).

The first step in obtaining solutions to the hamiltonian is to eliminate terms linear in the phonon operators by means of the substitution

$$d_n^+(\underline{k}) = c_n^+(\underline{k}) + B_n(\underline{k})\, S_z(\underline{k}) / \hbar\omega_n(\underline{k}).$$

This gives

$$\mathcal{H}_1 + \mathcal{H}_2 = \mathcal{H}_1'(\text{displaced phonon}) + \mathcal{H}_2'(\text{CJTE})$$

where

$$\mathcal{H}_1' = \sum_{n\underline{k}} (d_n^+(\underline{k}) d_n(\underline{k}) + \tfrac{1}{2}) \hbar\omega_n(\underline{k})$$

$$\mathcal{H}_2' = -\tfrac{1}{2} \sum_{\underline{k}} J(\underline{k}) \, S_z(\underline{k}) \, S_z(-\underline{k}).$$

A simple approximation consists of solving the electronic part, $\mathcal{H}_2' + \mathcal{H}_3 + \mathcal{H}_4$, of the hamiltonian separately from the displaced phonon part, $\mathcal{H}_1'$. (This ignores the fact that $\mathcal{H}_4$ does not commute with $\mathcal{H}_1'$.) In this approximation the frequency of the displaced phonons is unchanged and there is an effective Ising-like coupling between the pseudo-spins which can lead to a phase transition. The coupling constant $J(\underline{k})$ must be corrected for the self-energy because this is the part which would exist for an assembly of isolated complexes (or a gas of molecules) and cannot contribute to ordering. It is important to realize that either the pseudo-spin interaction via the optic phonons or the strain mode coupling acting alone could cause CJTE. In any particular material it is interesting to discover the relative importance of these two interactions.

By assuming the distortions are uniform (this is necessarily true for strain modes) we can use molecular field theory to obtain a solution to the electronic part of the hamiltonian

$$\mathcal{H}(\text{molecular field}) = -\lambda \langle S_z \rangle S_z + D S_x$$

Here $\lambda = J(\underline{k}=0) + \mu$ where μ represents the amount of coupling which is caused by the strain interaction $\mathcal{H}_3$ ($\mu \propto C^2/K$ where K is the relevant elastic constant). The solutions are

$$W \langle S_z \rangle = \lambda \langle S_z \rangle \tanh(W/kT)$$

where

$$W^2 = \lambda^2 \langle S_z \rangle^2 + D^2.$$

We either have $\langle S_z \rangle = 0$ and $W = D$ (at high temperatures) or $W = \lambda \tanh(W/kT)$ which is the usual molecular field result.

5. NORMAL MODES

A crystal which undergoes CJTE with a macroscopic crystalline distortion is expected to show anomalous elastic properties at the transition. The occurrence of a soft strain mode has been discussed by Pytte[26], and by Elliott et al.[7,27] who also discuss the similarity of this case with the case of the soft mode in hydrogen bonded ferroelectrics.

The normal modes of the coupled electron lattice system have been investigated experimentally by Melcher and Scott[13] using ultrasonics, by Sandercock et al.[10] using Brillouin scattering and ultrasonics and by Elliott et al.[7] using Raman scattering.

The positions of the high energy ($\hbar\omega \gg kT_D$) phonons and electronic states are very little affected by the phase transition even though some have the same symmetry as the distortion. This justifies the separate solution of the parts of the hamiltonian described in the previous section. However, doubly degenerate phonons are split (by $\gtrsim 2$ cm^{-1}) in the distorted phase and this has been interpreted as a consequence of higher order JT interactions[7].

The real interest lies in that acoustic phonon which has the same symmetry as the spontaneous distortion and whose dispersion relation crosses the electronic mode. This phonon may be regarded as a fluctuation in the strain so that the strength of the electron-phonon interaction must tend to the same value as the electron-strain interaction. The frequencies of these coupled modes have been calculated in the random phase approximation[7] taking the non-commutation of the electronic and phonon parts of the hamiltonian into account. This yields the mixed mode dispersion relation shown in Fig. 4. The electronic mode at $\underline{k} = 0$ is shifted upwards from the low temperature molecular field value $\hbar\omega_E = 2W \langle S_z \rangle$ and the velocity of sound is reduced. As the

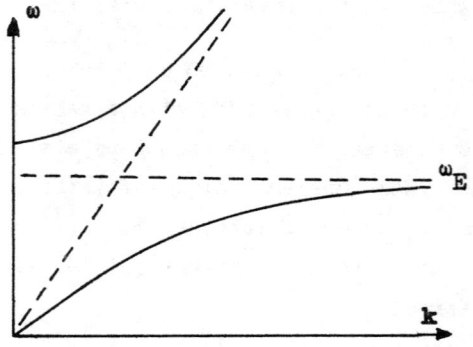

Fig. 4. The mixed mode dispersion relation (solid lines) is compared with the unmixed modes (dotted lines).

temperature is increased $\hbar\omega_E$ falls and forces the lower mode down. The transition temperature is that temperature at which the slope of the mode at the origin, and therefore also the elastic constant, falls to zero. Such an acoustic anomaly has been observed in $DyVO_4$ by Melcher and Scott[13] using an ultrasonic method and their results are shown in Fig. 5. In this material the mixing is caused

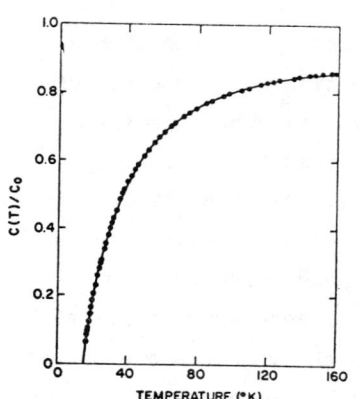

Fig. 5. The reduced elastic constant of $DyVO_4$ (obtained by Melcher and Scott[13]) showing the soft acoustic mode at T_D.

by the residual crystal field term $\mathcal{H}_4 = D\,S_x$.

This model also applied to our paradigm, $TmVO_4$, although the experiments have not yet been performed. However, in this case the mixing can be caused by an external magnetic field $\mathcal{H}_4' = g\beta H S_x$ so that the elastic properties will be strongly dependent on H.

The normal mode analysis described above leads to a value of the elastic constant under conditions of constant population of the states. It is also possible to derive an adiabatic elastic constant by assuming thermal equilibrium and taking derivatives of the free energy. This latter value will be observed in experiments performed at frequencies which are small compared with some electronic relaxation time. In a two-level system the two values are identical. In $TbVO_4$ there are four low-lying electronic energy levels with two of them degenerate for $T > T_D$. Sandercock et al.[10] have used Brillouin scattering and ultrasonics to investigate the consequent frequency dependence of the elastic constant.

6. EXPERIMENTS IN AN APPLIED STRESS OR MAGNETIC FIELD.

It is evident from the form of the contribution to the hamiltonian $\mathcal{H}_3 = C\,eS_z$ that an externally applied stress performs a role in the study of JT systems which is similar to the role played by a magnetic field in the study of magnetic systems. (We choose that stress which has the same symmetry as the spontaneous low temperature distortion.) These effects have been observed by Gehring et al.[11] who used absorption spectroscopy to investigate the behaviour of the low-lying energy levels of $DyVO_4$ as a function of external stress both above and below T_D. They obtained a value for C and showed that the strain coupling was ~ 3 times larger than the optic mode coupling. The existence of a finite optic mode coupling in this material is important because it gives rise to fluctuations and deviations from molecular field theory. They also showed that a sufficiently large stress applied in the distorted phase could convert the sample into one single domain with all distortions parallel. This is, of course, the analogue of the magnetization to saturization of a ferromagnet.

However, JT systems are also sensitive to the application of a magnetic field and the effects are fundamentally different for Kramers and non-Kramers ions. In the case of the non-Kramers ion Tm^{3+} (in $TmVO_4$) the orbital doublet may be split either by a magnetic field or by a crystal strain. The wavefunctions of the states in the two cases are different and are taken to be eigenfunctions of the pseudo-spin operators S_x and S_z respectively. If a strain splitting already exists (e.g. in the distorted phase) then a magnetic field will mix the states. We can think of this as a rotation of the pseudo-spin in the x-z plane. The process is accurately described by the molecular field theory of section 4 with the interpretation of $\mathcal{H}'_4 = g\beta H S_x$. The stable solution is the larger splitting chosen from $W_1 = \pm \frac{1}{2} g\beta H$ and $W_2 = \pm \lambda \tanh(W_2/kT)$. As the field is increased from zero the magnetic moment (measured by $\langle S_x \rangle$) increases[14] and the crystal strain (measured by $\langle S_z \rangle$) decreases[16]. At low fields the splitting of the electronic energy levels is constant at $2W_2$.[15] At a critical field H_c the moment saturates, the distortion is reduced to zero and the splitting begins to increase like $g\beta H$. By fitting the results of Raman scattering experiments in a magnetic field to this theory Harley et al.[16] showed that acoustic mode coupling dominates the behaviour of $TmVO_4$ and that the optic mode coupling is negligibly small.

In the case of Kramers ions such as Dy^{3+} in $DyVO_4$ the Kramers degeneracy cannot be removed by coupling to the lattice so that such materials will still be magnetic in the distorted phase (indeed, $DyVO_4$ becomes antiferromagnetic at $T_N = 3.07$ K [12]). Furthermore $DyVO_4$ has a strong magnetic anisotropy whose direction is determined by the distortion direction. Because of this anisotropy a magnetic field can be used to stabilize a particular distortion direction. In an optical absorption experiment Cooke et al.[6] rotated a magnetic field in the "tetragonal" plane of $DyVO_4$ is the distorted phase. This caused the distortion direction to flip from x, say, to y and then back again, resulting in the Zeeman pattern shown in Fig. 6, with fourfold symmetry. This is a remarkable demonstration of the fact that there is still a

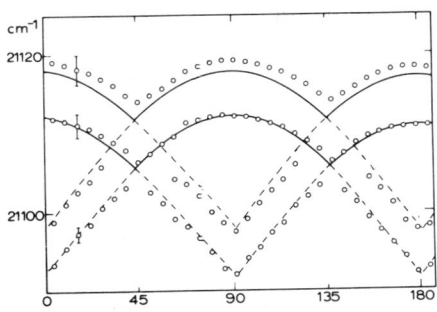

Fig. 6. Part of the Zeeman spectrum of $DyVO_4$ at $T < T_D$ in constant magnetic field as the crystal is rotated about the "tetragonal" axis.

fundamental fourfold symmetry in the properties of these tetragonal crystals even in the distorted phase.

7. CONCLUSIONS

It is probably true to say that none of the phenomena associated with the CJTE which we have been discussing is conceptually new. What is new is the wealth of experimental information which has been gathered in a comparatively short time. This has been possible because of the amenability of the rare earth crystals to a wide variety of experimental techniques and the possibility of ideally simple JT effects in tetragonal symmetry. The results which **were** obtained in these experiments have stimulated a good deal of theoretical activity which has led to a deeper and more detailed understanding of the phenomena.

ACKNOWLEDGMENTS

The rapid progress made in this field has come about through the close collaboration of very many people, both at the Clarendon Laboratory and in other parts of the world. The author is deeply grateful to all of them, especially the authors of refs. 6, 7 and 11.

REFERENCES

1. H. A. Jahn and E. Teller, Proc. Roy. Soc. A $\underline{161}$, 220 (1937).
2. E. Pytte and K. W. H. Stevens, Phys. Rev. Lett., $\underline{27}$, 862 (1971).
3. J. Sivardière and M. Blume, Phys. Rev. B$\underline{5}$, 1126 (1972).
4. M. D. Sturge, Solid State Physics $\underline{20}$, 91 (1967).
5. R. Englman, The Jahn-Teller effect in molecules and crystals, Wiley (1972).
6. A. H. Cooke, C. J. Ellis, K. A. Gehring, M. J. M. Leask, D. M. Martin, B. M. Wanklyn, M. R. Wells and R. L. White, Solid St. Commun., $\underline{8}$, 689 (1970).
7. R. J. Elliott, R. T. Harley, W. Hayes and S. R. P. Smith, Proc. Roy. Soc., A $\underline{328}$, 217 (1972).
8. K. A. Gehring, A. P. Malozemoff, W. Staude and R. N. Tyte, Solid St. Commun., $\underline{9}$, 511 (1971).
9. G. Will, H. Gobel, C. F. Sampson and J. B. Forsyth, Phys. Lett. A $\underline{38}$, 207 (1972).

10. J. R. Sandercock, S. B. Palmer, R. J. Elliott, W. Hayes, S. R. P. Smith and A. P. Young, J. Phys. C, $\underline{5}$, 3126 (1972).

11. G. A. Gehring, A. P. Malozemoff, W. Staude and R. N. Tyte, J. Phys. Chem. Sol., $\underline{33}$, 1487 and 1499 (1972).

12. A. H. Cooke, D. M. Martin and M. R. Wells, Solid St. Commun. 9, 519 (1971).

13. R. L. Melcher and B. A. Scott, Phys. Rev. Lett., $\underline{28}$, 607 (1972).

14. A. H. Cooke, S. J. Swithenby and M. R. Wells, Solid St. Commun, $\underline{10}$, 265 (1972).

15. P. J. Becker, M. J. M. Leask and R. N. Tyte, J. Phys. C, $\underline{5}$, 2027 (1972).

16. R. T. Harley, W. Hayes and S. R. P. Smith J. Phys. C, $\underline{5}$, 1501 (1972).

17. W. Berkhahn, H. G. Kahle, L. Klein and H. C. Schopper, Phys. Stat. Sol. (b), to be published.

18. W. Wuchner, W. Böhm, H. G. Kahle, A. Kasten and J. Laugsch, Phys. Stat. Sol. (b), to be published.

19. H. G. Kahle, L. Klein, G. Müller-Vogt and H. C. Schopper, Phys. Stat. Sol. (b), $\underline{44}$, 619 (1971).

20. B. W. Mangum, J. N. Lee and H. W. Moos, Phys. Rev. Lett., $\underline{27}$, 1517 (1971).

21. J. N. Lee, H. W. Moos and B. W. Mangum, Solid St. Commun., $\underline{9}$, 1139 (1971).

22. E. Cohen, L. A. Riseberg, W. A. Nordland, R. D. Burbank, R. C. Sherwood and L. G. van Uitert, Phys. Rev., $\underline{186}$, 476 (1969).

23. E. Bucher, R. J. Birgeneau, J. P. Maita, G. P. Felcher and T. O. Brun, Phys. Rev. Lett., $\underline{28}$, 746 (1972).

24. G. Gorodetsky, B. Lüthi and B. M. Wanklyn, Solid St. Commun., $\underline{9}$, 2157 (1971).

25. F. Sayetat, Solid St. Commun., $\underline{10}$, 879 (1972).

26. E. Pytte, Phys. Rev. B, $\underline{3}$, 3503 (1971).

27. R. J. Elliott, S. R. P. Smith and A. P. Young, J. Phys. C, $\underline{4}$, L 317 (1971).

SINGLET-GROUND-STATE DYNAMICS

Robert J. Birgeneau
Bell Laboratories
Murray Hill, New Jersey 07974

ABSTRACT

Magnetic systems in which the single ion ground state is a singlet may exhibit a soft mode phase transition corresponding to a polarization instability of the ground state wavefunction. In this article, theory and experiment for such systems is reviewed and discussed in the context of rare earth metals and intermetallic compounds. The theory is reviewed successively for the idealized singlet-singlet, and singlet-triplet models and for real level schemes. The importance of the respective uniaxial, isotropic and typically cubic symmetries is stressed. Inelastic neutron scattering studies of the dynamics in a variety of rare earth singlet ground state systems are discussed. Particular attention is given to three representative systems: TmSb, a model Van Vleck paramagnet, $(Pr_xLa_{1-x})_3Tl$, a family of alloys exhibiting a delicately balanced induced moment ordering for $x > 0.93$, and TbSb, a strongly over-critical antiferromagnet. In these latter two systems the $T = 0$ properties are well understood but existing RPA-type theories are found to be entirely inadequate for the finite temperature properties. Some suggestions for future lines of theoretical and experimental research are made.

I. INTRODUCTION

Over the past 12 years there has been extensive theoretical work on the dynamics of localized magnetic systems in which the single ion crystal field-only ground state is a singlet.[1-14] These calculations have predicted a variety of interesting dynamical effects including in certain cases a soft-mode phase transition corresponding to a polarization instability in the ground state wavefunction. The magnetic excitations in such system are single-ion crystal field transitions which propagate through the lattice via the exchange; theory suggests that they will be well-defined in both the ordered and paramagnetic regimes. Until recently, the development of this area of Magnetism has been severely limited by the lack of complementary experimental studies. In the past two years, however, a number of inelastic neutron scattering experiments on rare earth metals and intermetallic compounds in which the rare earth crystal-field-only ground state is a singlet have appeared.[15-19] The systems studied cover the complete range from very weak to very strong exchange and hence they exhibit the full range of possible behavior from Van Vleck paramagnetism to delicately balanced induced-moment ordering to more conventional ordering with strong anisotropy.

In this paper we shall review these experimental results and shall compare them in detail with existing theory in an attempt to arrive at an understanding of the present status of both theory and experiment. The format of this paper is as follows. In Section II we discuss the theory of the dynamics for each of the singlet-singlet, singlet-triplet (Γ_1-Γ_4) and real-level-scheme systems. In Section III we review experimental results for TmSb, a model Van Vleck paramagnet with very weak exchange, $(Pr_xLa_{1-x})_3Tl$, a family of alloys which exhibit a delicately balanced induced-moment ferromagnetic ordering for $x > 0.93$, and TbSb a singlet-ground state antiferromagnet where the exchange exceeds the critical value necessary for magnetic ordering by a factor of 3.5. We shall also briefly discuss other existing experiments, in particular those on fcc and dhcp praeseodymium. Finally, Section IV contains general discussion and conclusions.

II. THEORY

As a starting point, we take as the Hamiltonian

$$\mathcal{H} = \sum_i V_{ci} - 2 \sum_{i>j} \mathcal{J}_{ij} \vec{J}_i \cdot \vec{J}_j \qquad (1)$$

Here V_{ci} is the single ion crystal field potential appropriate to the ith site. For simplicity we shall confine our discussion to rare earth ions although much of the theory will also apply to transition metal insulators, particularly Ni^{++} and Fe^{++} compounds. The crystal field term, V_{ci}, is most easily represented via the Stevens operator equivalent formalism[20] with the explicit form determined by the local point symmetry. For example, for cubic symmetry

$$V_{ci} = B_4[O_4^0(J_i) + 5\,O_4^4(J_i)]$$
$$+ B_6[O_6^0(J_i) - 21\,O_6^6(J_i)] \qquad (2)$$

B_4, B_6 are numerical coefficients involving a reduced matrix element multiplied by the appropriate coefficient in the tesseral harmonic expansion of the charge potential. B_4, B_6 may be estimated using a simple point charge model (PCM) and, indeed, this method is surprisingly accurate in a number of rare earth intermetallic compounds.[15] Typical level schemes for Tm^{3+} and Tb^{3+} ($J=6$) in cubic symmetry and Pr^{3+} ($J=4$) in cubic and hexagonal symmetry are shown in Fig. 1. In cubic symmetry the ground state may be a Γ_1 singlet with a Γ_4 triplet as the first excited state giving as a first approximation a singlet-triplet model; in hexagonal symmetry one may have both singlet-singlet and singlet-doublet level schemes.

For the exchange in Eq. (1) we have chosen the simplest form possible, that is, isotropic bilinear exchange. For metals the $\mathcal{J}_{ij}$ originate mainly in the RKKY[21] indirect mechanism and hence the

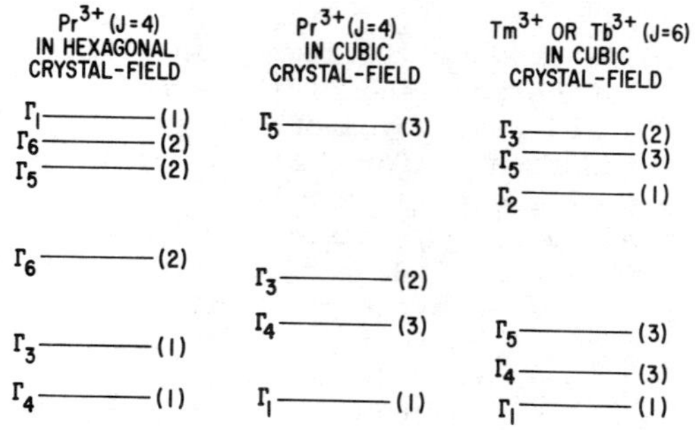

Fig. 1

Crystal-field level schemes for rare-earth ions in sites of hexagonal or cubic symmetry.

```
Pr³⁺ (J=4)              Pr³⁺ (J=4)            Tm³⁺ OR Tb³⁺ (J=6)
IN HEXAGONAL            IN CUBIC              IN CUBIC
CRYSTAL-FIELD           CRYSTAL-FIELD         CRYSTAL-FIELD

Γ₁ ——— (1)              Γ₅ ——— (3)            Γ₃ ——— (2)
Γ₆ ——— (2)                                    Γ₅ ——— (3)
Γ₅ ——— (2)                                    Γ₂ ——— (1)

Γ₆ ——— (2)              Γ₃ ——— (2)
                        Γ₄ ——— (3)            Γ₅ ——— (3)
Γ₃ ——— (1)                                    Γ₄ ——— (3)
Γ₄ ——— (1)              Γ₁ ——— (1)            Γ₁ ——— (1)
```

exchange is likely to be long range and oscillatory in character. We have omitted all anisotropic and higher order exchange and electric multipolar interactions.[22,23] This is a severe approximation which undoubtedly will fail the closest inspection. However, the bilinear isotropic exchange approximation should enable us to understand at least qualitatively the basic features of the induced-moment magnetism. We should also point out that for cubic Γ_1-Γ_4 systems bilinear anisotropic exchange terms make no contributions at long wavelengths while quadrupolar terms do not couple Γ_1 to Γ_4. Thus, only 3rd and higher order multipolar terms will contribute for such systems.

Finally in Eq. (1) we have omitted coupling terms involving both the nuclear spin system and the lattice. The former are essential for understanding the nuclear magnetic properties, but they are only of minor importance for the electronic system.[24,25,26] For rare earths, however, because of the large orbital moment the coupling to the lattice often plays a central role in the dynamics. In particular, any soft mode behavior predicted for the pure electronic system will be modified by the interactions with the acoustic phonon modes of the appropriate symmetry. Indeed these effects have already been shown to be of considerable importance in the closely related KDP and cooperative Jahn-Teller problems.[27] In our case, however, none of the existing experiments either probe the acoustic properties directly or show any explicit effects in the spin system of the coupling to the phonons. At this stage therefore we shall omit all terms in the Hamiltonian involving the lattice.

Calculations of the dynamics of singlet ground state systems have been carried out mainly for the idealized singlet-singlet and, to a lesser extent, singlet-triplet level schemes. Only a very few calculations have been carried out for real systems. We shall therefore review successively the theoretical predictions for each of these three models.

(a) Singlet-Singlet Model

In this model one has two levels, $|0_c\rangle$, $|1_c\rangle$ separated by an energy Δ in the absence of exchange. The axis of quantization is chosen so that only the matrix element

$$\langle 1_c | J^z | 0_c \rangle = \alpha \tag{3}$$

is nonvanishing. Equation (1) then reduces to that of the spin-$\frac{1}{2}$ Ising model in a transverse field

$$\mathcal{H} = -\Delta \sum_i S_i^x - 2 \sum_{i>j} K_{ij} S_i^z S_j^z \tag{4}$$

where

$$K_{ij} = 4\alpha^2 J_{ij} \tag{5}$$

Equation (4) has been proposed as an effective spin Hamiltonian for KDP type order disorder ferroelectrics with Δ the tunnelling frequency and K_{ij} the interactions between the displaced dipoles.[28] It has been used to describe certain rare earth quadrupolar transitions, such as in $DyVO_4$, where in this case Δ represents the initial crystal field splitting and the K_{ij} quadrupole-quadrupole interactions, either direct or phonon-induced.[29] Finally, Kruger et al[4] have used a one-dimensional version of Eq. (4) to describe the collective electronic states in linear chain aromatic molecular crystals. For continuity purposes, we shall use the notation of Eq. (1) rather than (4).

The properties of a system described by Eq. (1) are most easily deduced via molecular field theory for the statics and an effective boson theory for the dynamics. In the paramagnetic region the susceptibility in molecular field theory (M.F.T.) is given simply by

$$\frac{1}{\chi} = \frac{\Delta}{2g^2\mu_B^2\alpha^2} \left[\frac{1+e^{-\Delta/kT}}{1-e^{-\Delta/kT}} - \eta\right] \tag{6}$$

where

$$\eta = 4 \, J(0)\alpha^2/\Delta \tag{7}$$

with

$$J(\vec{k}) = \sum_j J_{ij} \exp[i\vec{k}\cdot(\vec{r}_i - \vec{r}_j)] \tag{8}$$

We assume for simplicity that the coupling is ferromagnetic, that is, $\mathcal{J}(\vec{k})$ has its maximum value for $\vec{k} = 0$. It is immediately apparent from (6) that the susceptibility will diverge for $\eta \geq 1$ at a temperature determined by

$$\frac{1}{\eta} = \frac{1-e^{-\Delta/kT}}{1+e^{-\Delta/kT}} \tag{9}$$

and the phase transition will be second order.

In the ordered region the molecular field eigenstates are

$$|0\rangle = \cos\theta |0_c\rangle + \sin\theta |1_c\rangle$$
$$|1\rangle = -\sin\theta |0_c\rangle + \cos\theta |1_c\rangle \tag{10}$$

with

$$\tan 2\theta = 4\mathcal{J}(0)\alpha \langle J\rangle/\Delta \tag{11}$$

$\langle J \rangle$ must then be obtained self-consistently. At $T = 0$, this reduces to

$$\frac{\langle J\rangle}{\alpha} = \frac{1}{\eta}(\eta^2 - 1)^{\frac{1}{2}} \tag{12}$$

again exhibiting the threshold value $\eta = 1$ for magnetic ordering.

We now consider the dynamics at $T = 0$ as a function of η in the effective boson approximation. This theory has been discussed extensively elsewhere, and we give only a basic outline here. The theory proceeds by first diagonalizing the molecular field Hamiltonian and then assigning fermion operators ($d_{i\alpha}$) to each molecular field state. New operators $a_i^+(\alpha\beta) = d_{i\alpha}^+ d_{i\beta}$ are introduced which couple state β to state α. If one neglects the population in the molecular field excited states and if one takes the occupation number for the molecular field ground state as unity, then the a_i are <u>boson</u> operators. The Hamiltonian can be written in terms of these boson operators and can be diagonalized by standard techniques. For the singlet-singlet system this gives quite generally

$$E_k = \Delta'\left(1 - \eta\left(\frac{\alpha'}{\alpha}\right)^2 \gamma_k\right)^{\frac{1}{2}} \tag{13}$$

where

$$\Delta' = E(1) - E(0) \tag{14}$$
$$\alpha' = \langle 1 | J^z | 0 \rangle \tag{15}$$
$$\gamma_k = \mathcal{J}(k)/\mathcal{J}(0) \tag{16}$$

In the paramagnetic regime, that is, $\eta < 1$, $\Delta' = \Delta$, $\alpha' = \alpha$ so that

$$E_k = \Delta(1-\eta \gamma_k)^{\frac{1}{2}} \qquad (17)$$

whereas in the ordered region at $T = 0$, $\Delta' = \eta\Delta$, $\alpha' = \frac{\alpha}{\eta}$ giving

$$E_k = \Delta(\eta^2 - \gamma_k)^{\frac{1}{2}} \qquad (18)$$

It is evident from Eq. (17) that the system has a <u>soft mode transition as a function of</u> η with $\eta_c = 1$. At $\eta_c = 1$ the dispersion relation is linear in k going to 0 at k = 0 for a ferromagnet.

At $T = 0$ the model is exactly soluble in one dimension with the result[3,4,9]

$$E_k = \Delta\left(1 + \frac{\eta^2}{4} - \eta \cos k\right)^{\frac{1}{2}} \qquad (19)$$

for nearest neighbor interactions. Thus, in one dimension $\eta_c(\text{exact}) = \frac{1}{2}$. Pfeuty and Elliott[9] have studied the dynamics of Eq. (4) using perturbation theory. They develop an expansion for E_k in η for a general lattice with arbitrary dimensionality. They show in addition that this perturbative result reduces to the mean field result, Eq. (17), in the limit that the interaction range goes to infinity.

The temperature dependence of the singlet-singlet magnetic excitons has been studied via the equation of motion technique with RPA or more elaborate decoupling schemes. Using an $S = 1/2$ pseudo-spin representation, Wang and Cooper have deduced results via RPA which are identical to Eqs. (17), (18) but with

$$\eta \to -2\langle S_z\rangle \eta \qquad (20)$$

where $S_z = -1/2$ is associated with $|0_c\rangle$, $S_z = +1/2$ with $|1_c\rangle$. This model then predicts a soft mode phase transition at a temperature determined by

$$2\langle S_z\rangle\eta = -1 \qquad (21)$$

It should be noted that in molecular field theory

$$\langle S_z\rangle = -\frac{1}{2}\frac{1-e^{-\Delta/kT}}{1+e^{-\Delta/kT}} \qquad (22)$$

so that Eq. (21) becomes identical to Eq. (9). Thus in this hybrid RPA-M.F. theory the phase transition is a simple second order soft mode transition. At T_c the dispersion relation should go to·0 at k = 0 linearly in k. Wang and Cooper have carried out calculations for Eq. (20) with $\langle S_z\rangle$ evaluated self-consistently. In this method they find $\eta_c = 1.04$ for a simple cubic lattice; in addition, they find that the transition is first order for $T_c/\Delta > 0.1$. This latter result must be viewed skeptically in the light of series expansion results,[30] which indicate a second order transition with Ising critical exponents for the simple cubic lattice with $\eta > \eta_c = 1.18$.

We shall give explicit numerical examples for these various predictions for the dynamics at the end of this section.

(b) <u>Singlet-Triplet Model</u>

In the singlet-triplet model the crystal field ground state is a singlet, which we denote $|0_c,0\rangle$, with a triplet $|1_c,1\rangle$, $|1_c,0\rangle$, $|1_c,-1\rangle$ at energy Δ. The axes of quantization are chosen such that

$$\langle 1_c,0|J^z|0_c,0\rangle = \alpha \qquad (23)$$

$$\langle 1_c,\pm 1|J^\pm|0_c,0\rangle = \sqrt{2}\,\alpha \qquad (24)$$

and

$$\langle 1_c,\pm 1|J^z|1_c,\pm 1\rangle = \pm\beta \qquad (25)$$

For Pr^{+++}, $J = 4$, for example, for the Γ_1 and Γ_4 states $\alpha = \sqrt{\frac{20}{3}}$, $\beta = 1/2$. In molecular field theory $|0_c,0\rangle$ is coupled only to $|1_c,0\rangle$ so that many of the T = 0 properties for the singlet-singlet model in MF carry over to the singlet-triplet system. For example, the relationship between the induced ground state moment and η, Eq. (12), is identical for both systems. Thus $\eta_c = 1$ also holds for the singlet-triplet model. Similarly the effective boson prediction for the dynamics, Eq. (13), holds for all three exciton modes for $\eta < 1$ and for the longitudinal exciton mode for $\eta \geq 1$. However, both the $T \neq 0$ properties and the transverse modes for $\eta \geq 1$, $T \leq T_c$ will differ for the singlet-triplet model.

The susceptibility at finite temperatures is easily derived and is given by

$$\frac{1}{\chi} = \frac{\Delta}{2g^2\mu_B^2\alpha^2}\left[\frac{1+3\,e^{-\Delta/kT}}{1-e^{-\Delta/kT}+\frac{\beta^2}{\alpha^2}\frac{\Delta}{kT}e^{-\Delta/kT}} - \eta\right] \qquad (26)$$

Thus, in mean field theory the singlet-triplet system will also have a second order transition at a temperature determined by

$$\frac{1}{\eta} = \frac{1-e^{-\Delta/kT_c}+\frac{\beta^2}{\alpha^2}\frac{\Delta}{kT_c}e^{-\Delta/kT_c}}{1+3\,e^{-\Delta/kT_c}} \qquad (27)$$

This is closely analogous to the corresponding formula for the singlet-singlet system, Eq. (10), except for the term in β^2 which originates from the Curie susceptibility in the excited state. We assume throughout this paper that β/α is such that only ordering of the induced moment type will occur.[12]

As noted above at T = 0 the three modes for $\eta < 1$ are given by Eq. (17) while the longitudinal mode is given by Eq. (18) for $\eta \geq 1$. Using an effective boson approach, Grover has found for $\eta > 1$ for the transverse modes

$$\frac{\hbar\omega(T_{1,2})}{\Delta} = \left(\frac{1+\eta}{2}\right)^{\frac{1}{2}} \left(\frac{1+\eta}{2} - \frac{(83\,\eta+77)\gamma_k - (3\,\eta-3)\gamma_k^2}{24\,\alpha^2}\right)^{\frac{1}{2}}$$

$$\pm \left(\frac{\eta^2-1}{16\,\alpha^2}\right)^{\frac{1}{2}} (1-\gamma_k) \quad (28)$$

which in the limit of long wavelengths becomes

$$\frac{\hbar\omega(T_{1,2})}{\Delta} = \left(\frac{1+\eta}{12}\right)^{\frac{1}{2}} \left(\frac{77\,\eta+83}{24\,\alpha^2}\right)^{\frac{1}{2}} a\,k \quad (29)$$

for a simple cubic lattice with nearest neighbor interactions. Thus, the transverse excitons in this system have a linear dispersion relation going to 0 at k = 0.

To discuss the general temperature dependence of the excitations, it is more convenient to go over to the pseudo-spin formalism introduced by Pink;[8] in this scheme two spin 1/2's, $\vec{S}_i, \vec{T}_i$ are assigned to each site; the matrix elements of $\vec{J}$ can be represented by writing

$$\vec{J} \rightarrow a\vec{S} + b\vec{T} \quad (30)$$

where a,b are chosen so as to match the matrix elements of $\vec{J}$ between the singlet and triplet and within the triplet. This gives

$$\alpha = \frac{1}{2}(a-b) \qquad \beta = \frac{1}{2}(a+b) \quad (31)$$

Thus, Eq. (1) becomes

$$\mathcal{H} = \Delta \sum_i{}' \vec{S}_i \cdot \vec{T}_i - 2 \sum_{i>j} \mathcal{J}_{ij}(a\vec{S}_i + b\vec{T}_i) \cdot (a\vec{S}_j + b\vec{T}_j) \quad (32)$$

closely analogous to that for a two sublattice ferrimagnet.[12] It is important to note that this system has <u>isotropic symmetry</u>. In the induced moment phase transition one specific direction, the z direction, is selected, thus giving a case of <u>broken symmetry</u>. Therefore, the nonrelativistic analog of Goldstone's theorem will apply.[31] This means that there will always be a branch of excitations which goes to 0 as $k \rightarrow 0$ which in the limit k = 0 corresponds to that mode which rotates the induced moment from one direction to another. It is clear from the physical nature of this system that the Goldstone bosons are just the transverse excitons in the ordered state. Thus, the transverse exciton branches are required to go to 0 at k = 0 at all temperatures below T_c.

The general temperature dependence of the excitations in the singlet-triplet system has been discussed by several authors.[10-12] In the RPA in the paramagnetic regime

$$E_k = \Delta(1+4\eta \langle S^z T^z \rangle \gamma_k)^{\frac{1}{2}} \qquad (33)$$

In mean field theory

$$\langle S^z T^z \rangle = -\frac{1}{4} \frac{1-e^{-\Delta/kT}}{1+3e^{-\Delta/kT}} \qquad (34)$$

Thus in a simplified hybrid RPA-MF theory the singlet-triplet model will exhibit a soft mode phase transition at a temperature determined by

$$\frac{1}{\eta} = \frac{1-e^{-\Delta/kT_c}}{1+3e^{-\Delta/kT_c}} \qquad (35)$$

This differs from the corresponding susceptibility expression for T_c, Eq. (30), by the factor $\frac{\beta^2}{\alpha^2} \frac{\Delta}{kT_c} e^{-\Delta/kT_c}$. Smith[11] has interpreted this difference in T_c to mean that the singlet-triplet model will not show a soft-mode transition. However, the theory is so oversimplified that probably no conclusions at all about the phase transition can be drawn from it. In particular, even within the RPA it is at least necessary to carry out the calculations self-consistently. From symmetry grounds alone, however, we can say that on approaching T_c from below, the longitudinal mode must go to 0 as $T \to T_c^-$ provided that the transition is second order. Briefly, by Goldstone's theorem, the transverse excitons must go to 0 at k = 0 for $T \leq T_c$, but at T_c all three modes will coincide, at least in a simple elementary excitation picture; hence the necessity for a "soft longitudinal mode" below T_c.

So far we have not discussed either excitations within the triplet or lifetime effects. For induced moment systems in which $\eta \gtrsim 1$, so that $k_B T_c \ll \Delta$, the triplet excitations are likely to be only of minor importance in the dynamics of the phase transition. However, there will be anticrossing-effects with the singlet-triplet modes which will alter the form of the dispersion relations at long wavelengths. As of yet, there is no convincing evidence either from theory or experiment as to whether the triplet modes will be propagating or overdamped in form. Lifetime considerations for the singlet-triplet excitons are probably always important. Indeed, it seems quite possible that for $\eta > 1$, the $k = 0$ excitons will be overdamped for all $T \geq T_c$. Correspondingly, the actual transition may be characterized by the divergence of a central peak of the sort discussed, for example, by Shirane and Axe[32] for the structural phase transition in Nb_3Sn; in this case the central peak could originate in elastic or quasielastic scattering processes within the triplet. However, these aspects of the problem have not been explored properly either by theory or by experiment. Finally, Fulde and Peschel[10] have suggested that in metallic singlet ground state systems paramagnon effects may be important at long wavelengths.

(c) Real Level Schemes

Theoretical work on real level schemes has by its very nature been largely limited to those systems which have actually been explored experimentally. We shall therefore postpone most such discussion until the experimental section. However, for illustrative purposes, we shall discuss briefly work by Cooper[13] and the present author on Pr^{+++} in a cubic field.

$Pr^{+++}(4f^2, 3H_4)$ has the energy level scheme shown in Fig. 1. The energies are $E(\Gamma_1) = 0$, $E(\Gamma_4) = \Delta$, $E(\Gamma_3) = \frac{12}{7}\Delta$, $E(\Gamma_5) = \Delta'$ where the factor $\frac{12}{7}$ for $E(\Gamma_3)$ is symmetry-determined. Inspection of the wavefunctions as tabulated by Lea, Leask and Wolf[33] shows that $|\Gamma_1\rangle$ is connected only to one of the $|\Gamma_4\rangle$ and $|\Gamma_3\rangle$ states by J^z. Since their relative energies are fixed, the molecular field properties may be solved for universally.

The molecular field criterion for ordering is still $\eta \geq 1$. However, for the real level scheme the relationship between η and the ground state induced moment is found to be

$$\frac{1}{\eta^2} = 1 - 0.1837 \left(\frac{\langle J \rangle}{\alpha}\right)^2 - 0.115 \left(\frac{\langle J \rangle}{\alpha}\right)^4 + \ldots \quad (36)$$

where $\alpha = \langle \Gamma_4 | J^z | \Gamma_1 \rangle$ compared with $\frac{1}{\eta^2} = 1 - \left(\frac{\langle J \rangle}{\alpha}\right)^2$, Eq. (12), in the absence of the $|\Gamma_3\rangle$ state. Thus, the $|\Gamma_3\rangle$ admixtures have a drastic effect on the relationship between $\langle J \rangle/\alpha$ and η. Indeed any property which depends sensitively on the higher order terms in the molecular field will be strongly affected by the additional levels beyond the singlet and the triplet.

The dynamics at $T = 0$ have been discussed extensively by Cooper[13] using an effective boson approximation. The dispersion relations for $\eta < 1$ reduce to those for the singlet-singlet mode, Eq. (17). For $\eta \geq 1$ the problem must be solved numerically for the particular values of η and Δ. For $\eta \gtrsim 1$ the coupling to the Γ_3 state is weak so that the Γ_1-Γ_4 longitudinal excitation energy reduces simply to Eq. (13) with $|0\rangle = |\Gamma_1\rangle'$, $|1\rangle = |\Gamma_4\rangle'$, the molecular field states. The gap at k = 0 is found to be ~1.08 times that in the absence of the Γ_3 state. The Γ_1-Γ_3 longitudinal exciton is predicted to be essentially dispersionless with an energy equal to the molecular field splitting.

The transverse dynamics at T = 0 are rather more complicated. In particular, they depend sensitively on the relative values of Δ' and Δ. We therefore refer the reader to Cooper's exposition for a complete discussion of this problem. The salient qualitative result of Cooper's analysis, however, is that the admixtures of the higher states into the $|\Gamma_1\rangle$, $|\Gamma_4\rangle$ levels introduce a <u>gap</u> into the transverse Γ_1-Γ_4 exciton spectra. This is due simply to the fact that these admixtures reduce the effective symmetry of the Γ_1-Γ_4 system from isotropic to cubic so that Goldstone's theorem no longer applies. It has not yet been explored as to whether this introduction of a gap in real systems removes the apparent necessity of a longitudinal mode going to 0 as $T \rightarrow T_c^-$. Certainly in a mean field description the Γ_1-Γ_4 system becomes more isotropic as $T \rightarrow T_c^-$ and hence we might expect to return to the pure singlet-triplet problem. However, this argument may break down if one goes beyond mean field theory or any simple RPA description.

Calculations on the dynamics of real Van Vleck paramagnets at finite temperature using a mean field description of the dynamic susceptibility, have been reported by Peschel et al.[10] They also consider Pr^{+++} in a cubic field; they find that for $k_B T \ll \Delta$ only the Γ_1-Γ_4 modes have appreciable dispersion. Modes involving transitions between the Γ_4, Γ_3, Γ_5 levels seem to be important only to the extent that they "anticross" the Γ_1-Γ_4 excitons.

(d) <u>Model Calculations</u>

In order to illustrate the various predictions of the theories discussed in the previous subsections, we consider explicitly the case of $Pr^{+++}(^3H_4)$ situated on a face-centered-cubic lattice. The energy level diagram is as in Fig. 1 with $E(\Gamma_5) = \Delta' > \Delta$. In particular, for reasons which will be clear in the next section, we take $\Delta' = 2.42 \Delta$; this corresponds to an LLW parameter[33] of x = -0.877. We assume, in addition, that the exchange is ferromagnetic and confined to nearest neighbors alone. For $\vec{k}$ in the $\langle 110 \rangle$ direction, this gives

$$\vec{k} \parallel \langle 110 \rangle \qquad \mathcal{J}(k) = 2 J_{nn}\left[\cos\left(\frac{ka}{\sqrt{2}}\right) + 1 + 4 \cos\left(\frac{ka}{2\sqrt{2}}\right)\right] \qquad (37)$$

Typical theoretical results at T = 0 for η below and just above the critical value of 1 are shown in Fig. 2. The real level scheme results are taken directly from Cooper.

For η = 0.892 the system is a simple Van Vleck paramagnet; in the effective boson treatment for both the singlet-triplet and real level schemes, the dispersion relation for the Γ_1-Γ_4 excitons reduces to the singlet-singlet result Eq. (17). For $\eta = 1.014$ the system is a ferromagnet with an induced-moment, $\langle J \rangle / \alpha = 0.371$. For the singlet-triplet model the dispersion relation for the longitudinal mode again reduces to the singlet singlet result Eq. (18). The transverse modes are calculated from Grover's result, Eq. (28). As discussed previously, for the idealized singlet-triplet model the transverse mode goes to 0 at $k = 0$. However, away from $k = 0$, for η near 1, the longitudinal and transverse modes have very similar dispersion relations. For the Pr^{+++} real level scheme, the longitudinal mode differs only slightly from the singlet-triplet result. However, we note that a very large gap has opened up in the transverse Γ_1-Γ_4 exciton spectra. This large gap should have a marked effect on the low temperature thermodynamic properties.

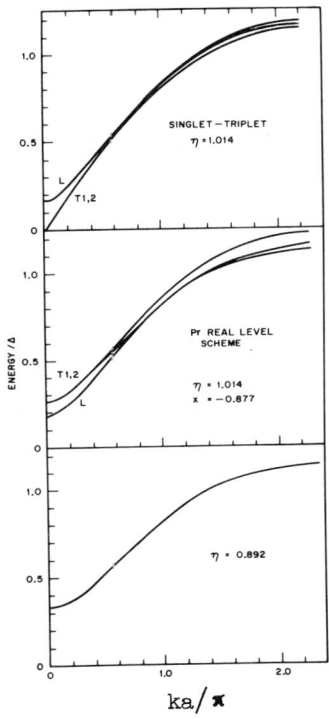

Fig. 2

Model calculations for singlet-triplet and Pr real level schemes. The latter results are taken from Ref. 13

The temperature dependence of the excitons for $\eta = 0.892$ using the R.P.A.-M.F. result, Eqs. (34), (35) for the singlet-triplet model are shown in Fig. 3. Similar results will be obtained from a self-consistent RPA calculation. We note that for $kT/\Delta \sim 1$ the dispersion has nearly vanished. For the real level scheme the results are very similar to those shown in Fig. 3. The only important qualitative difference is that at higher temperatures modes involving Γ_3, Γ_5 levels will appear. As mentioned previously, according to Peschel et al,[10] these could exhibit anticrossing effects with the singlet-triplet (Γ_1-Γ_4) modes which would make them readily observable. For $\eta > 1$ for the singlet-triplet model the longitudinal mode will drop to zero at T_c giving three degenerate modes which vary linearly with k near $k = 0$. With increasing temperature in the RPA a gap will open up at $k = 0$; the dispersion will thus decrease with increasing temperature in a manner quite similar to that illustrated in Fig. 3. For the real level scheme in an RPA-

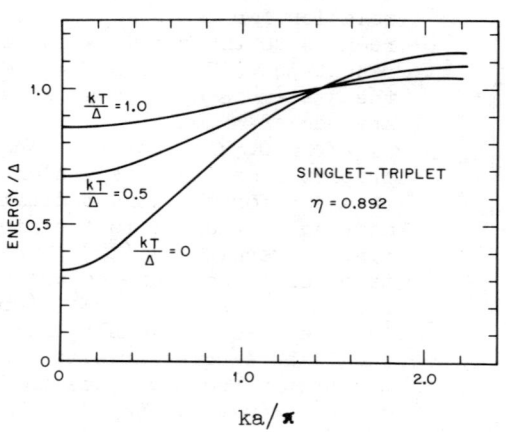

Fig. 3

Theoretical temperature dependence of singlet-triplet excitons with $\eta = 0.892$ in the RPA-MF approximation (Eqs. (33), (34)).

type description similar effects are predicted although in this case both the longitudinal and transverse modes will go to zero at k = 0 in unison at $T = T_c$.

III. EXPERIMENT

(a) $\eta \ll 1$

For systems in which $\eta \ll 1$ the dispersion is negligible and hence all experiments reduce to a study of crystal field effects alone. However, since we are interested mainly in rare earth metals and intermetallic compounds, this is by no means a trivial part of the whole problem. Indeed, it is essential to establish that a crystal field description is indeed valid for isolated rare-earth ions in metals. In addition, it is most valuable to have a model "zero-exchange" compound on which one can test the various experimental techniques to be used for the study of compounds with $\eta \sim 1$.

The most thoroughly investigated singlet ground state system with weak exchange is the rock salt intermetallic compound TmSb.[16,34-37] In this compound the Tm ion is in the 3^+ charge state with the configuration $4f^{12}$. The ground state multiplet is $^3H_6$; the next multiplet $^3H_4$ is at ~7900 K and hence it may be ignored in the discussion of the much smaller splitting of the ground state multiplet. Since the ground state manifold of the Tm^{3+} ion has L = 5, S = 1, the total angular momentum J = 6 is largely orbital in character; hence, one might anticipate that crystal field effects could easily dominate over exchange effects.

The point symmetry at the Tm^{3+} site in TmSb is O_h so that the crystal field Hamiltonian, Eq. (2), is appropriate. Instead of B_4, B_6 it is more convenient to use the Lea, Leask, Wolf[33] (LLW) parameters x,w defined by

$$B_4 F(4) = Wx \tag{38}$$

$$\frac{B_4}{B_6} = \frac{x}{1-|x|} \frac{F(6)}{F(4)} \tag{39}$$

where $F(4) = 60$, $F(6) = 7560$ for $J = 6$ are numerical constants. A simple PCM calculation for TmSb with $Z(Sb) = -3$ gives $x = -0.955$, $W \sim -0.71$ K; these parameters, in turn, give an energy level diagram similar to that shown for $J = 6$ in Fig. 1. From single crystal anisotropic magnetization measurements Cooper and Vogt[34] have shown that x must lie in the range $-1 \leq x \leq -.6$ with the ground state the Γ_1 singlet and the Γ_4 triplet 26 K; this is consistent with Schottky anomaly specific heat measurements[35] which indicate that TmSb is a Van Vleck paramagnet with the first excited state a triplet at 27 K. Cooper and Vogt have also carried out susceptibility measurements on $Tm_{.53}Y_{0.47}Sb$ where to a first approximation the Yttrium acts as a simple dilutent. They find χ/Tm ion is identical to that found in TmSb and hence they postulate that TmSb has negligible exchange.

The excitations in rare earth intermetallic compounds can be most directly studied via inelastic neutron scattering techniques. This method has been discussed extensively elsewhere,[15] and we give here only the salient features. The neutron scattering cross section at small momentum transfers for an assemblage of <u>noninteracting</u> ions is given simply by

$$\frac{\partial^2 \sigma}{\partial \Omega_f \partial \omega_f} = N \left(\frac{1.91 e^2}{2mc^2} g_J \right)^2 \frac{k_f}{k_i} F^2(\vec{k})$$

$$\times \sum_{n,m} \rho_n |\langle n | \vec{J}_\perp | m \rangle|^2 \delta \left(\frac{E_n - E_m}{\hbar} - \omega \right) \tag{40}$$

where $\vec{J}_\perp$ is the component of the total angular momentum perpendicular to the scattering vector $\vec{k}$. The other symbols have their usual meaning. Thus, at small momentum transfers, the scattering should obey magnetic dipole selection rules. Birgeneau[37] has tabulated the $|\langle \Gamma_i | \vec{J}_\perp | \Gamma_j \rangle|^2$ for all values of x and for all J. For a singlet-singlet system with dispersion, Eq. (40) becomes

$$\frac{\partial^2 \sigma}{\partial \Omega_f \partial \omega_f} = \left(\frac{1.91 \, e^2}{2 \, mc^2} g_J \right)^2 \frac{k_f}{k_i} F(\vec{k}) \sin^2 \theta_f \frac{\alpha'^2 \Delta'}{E_{\vec{k}}} \delta \left(\frac{E_{\vec{k}}}{\hbar} - \omega \right)$$

$$\times \left(\begin{array}{c} n_{\vec{k}} + 1 \\ n_{\vec{k}} \end{array} \right) \tag{41}$$

where $\theta_{\vec{k}_f}$ is the angle $\vec{k}_f$ makes with the z axis, Δ', α' are defined by Eq. (14), (15), and we assume that the boson population factor $n_{\vec{k}} \ll 1$. The analogous result holds for the singlet-triplet paramagnet and for the longitudinal mode in a singlet-triplet ferromagnet. For real level schemes the corresponding scattering cross section must be worked out in each case.

Inelastic neutron scattering studies of TmSb have been carried out by Birgeneau et al[16] on polycrystalline samples. For $|k|$ within the first Brillouin zone, this gives a spherically-averaged dispersion relation whereas for larger k it gives a peculiarly averaged density of states. Typical time-of-flight spectra as a function of temperature are shown in Fig. 4. We shall not discuss time-of-flight neutron scattering here, but instead we shall confine ourselves to consideration of the final results. At 4.9 K a single sharp peak in energy loss (excitation creation) is observed at -2.22 meV = 26 K; the width is essentially that of the instrument at all k vectors; furthermore, the peak position is found to be independent of k. Since $\vec{J}$ transforms like Γ_4, at low temperatures the scattering must originate in the Γ_1-Γ_4 singlet-triplet excitons. At 12 K the Γ_1-Γ_4 excitations can be seen in both energy gain and energy loss with a relative intensity precisely that predicted by Boltzman statistics. Thus the neutron scattering shows that in TmSb the low temperature dynamics are characterized by sharp dispersionless excitations which correspond to simple magnetic dipole crystal field transitions. As the temperature is increased, transitions between the higher states appear and indeed well-defined transitions have been observed up to room temperature. At 80 K the intrinsic width of the peaks is about 4 K.

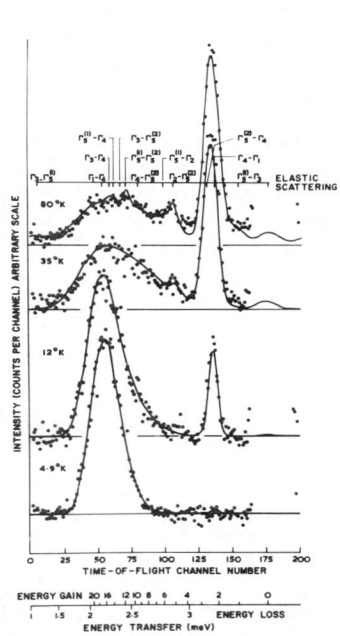

Fig. 4

Time-of-flight neutron spectra for TmSb as a function of temperature. Data taken from Ref. 16.

The solid lines in Fig. 4 have been calculated using Eq. (40). The relative intensities, line spacing and temperature dependence of the spectra are well accounted for; it is clear therefore that the crystal field eigenfunctions and eigenvalues for the energy levels are good approximations for this compound. The actual crystal field parameters deduced from Fig. 4 are x = -0.785, W = -0.993 K in reasonable agreement with the nearest neighbor point charge values of -0.995,-0.71 K. Furthermore, the measured x,W give an excellent fit to the anisotropic magnetization.[36]

Similar results have been obtained by Turberfield et al[15] in the praseodymium monopnictides and monochalcogenides. In these compounds Pr^{+++} has an energy level diagram similar to that shown in Fig. 1. At room temperature, all of the allowed magnetic dipole transitions (Γ_1-Γ_4, Γ_4-Γ_3, Γ_4-Γ_5, Γ_3-Γ_5) are observed and their relative intensities and energies can be accurately accounted for using Eq. (40). Single crystal neutron scattering measurements on dhcp praseodymium have been performed by Rainford and Houmann.[17] In dhcp Pr the packing sequence is ABAC; the A layers have a local environment of approximately cubic symmetry, while in the B and C layers the atoms have a hcp arrangement of nearest neighbors. To a first approximation the cubic site Γ_1-Γ_4 excitons can be decoupled from the hexagonal site excitations; Rainford and Houmann have found that $\eta \ll 1$ for the cubic site, and from their dispersion they extract an explicit $\mathcal{J}(\vec{k})$.

In summary, in all singlet ground state rare earth intermetallic systems studied so far, it is found that for $\eta \ll 1$ simple crystal field theory gives a good account of the dynamics over a wide range of temperature. Various relaxation processes play a role, but they do not seem to cause excessive broadening of the lines, at least up to temperatures comparable with Δ.

(b) $\eta \sim 1$

From the discussion given in Section II, it is evident that the most interesting region in the singlet ground state phase diagram is for $\eta \gtrsim 1$. In this case one anticipates a delicately balanced induced-moment ordering with radical changes in the bulk properties occurring for very small changes in η. Bucher and coworkers[18,38] have explored in detail the alloy system $(Pr_x La_{1-x})_3 Tl$ and have shown that it exhibits nearly ideal $\eta \sim 1$ behavior. We therefore shall discuss this system in some detail.

Pr_3Tl has the Cu_3Au structure; this is equivalent to fcc Pr with the corner Pr atoms replaced by Tl. The Pr^{3+} point symmetry is strictly tetragonal; however, if the Tl charge is taken equal to that of Pr^{3+}, then in the PCM the effective symmetry is cubic. Similarly, the true primitive lattice is simple cubic with 3 Pr^{3+} atoms per unit cell. We shall instead simply regard Pr_3Tl as a dilute fcc structure; this avoids the algebraic complications associated with the true lattice while still reflecting the overall cubic symmetry.

As discussed by Bucher et al,[38] in the PCM the Pr^{3+} ground state will be a Γ_1 singlet provided only that the Pr^{3+}, Tl^{3+} charges are taken as positive. This has been confirmed by measurements of the susceptibility and specific heat across the entire system. These results are summarized in Fig. 5. La_3Tl is a strong coupling superconductor with $T_s = 8.95$ K. With the addition of Pr^{3+}, T_s decreases in the manner characteristic of singlet ground state impurities;[39] in this case the destruction of Cooper pairs occurs via virtual inelastic processes between the Γ_1 and Γ_4 states. Susceptibility measurements show that for $x \to 0$ in $(Pr_xLa_{1-x})_3Tl$, $\chi_{vv}(x=0) = 0.048$ cm^3/mole Pr giving $\Delta = E(\Gamma_4) - E(\Gamma_1) = 68$ K. Specific heat Schottky anomaly measurements for $0.03 < x < 0.25$ show that the first excited state is a triplet at $\Delta = 78 \pm 10$ K; thus, the first excited state must be the Γ_4 level with $68 \leq \Delta \leq \sim 78$ K. The Γ_3 level then lies at about 125 K; the position of the Γ_5 level is undetermined except that it must lie above Γ_4. We note, however, that in all of the Pr monopnictides and monochalcogenides[15] $\Delta' = E(\Gamma_5) \gg \Delta$; the PCM predicts $\Delta' = 2.42 \Delta$.

With increasing Pr concentration the $T = 0$ Van Vleck susceptibility/Pr ion increases markedly indicating that the exchange is ferromagnetic and that $\eta \to 1$. Finally, at $x \sim 0.93$ spontaneous ordering appears. For $x = 1$, that is, for pure Pr_3Tl, a ferromagnetic ordering occurs with a Curie temperature $T_c = 11.3 \pm 0.3$ K and a moment of ~ 0.75 μ_B. Thus $(Pr_xLa_{1-x})_3Tl$ appears to be an ideal singlet ground state system in which η can be tuned through 1 by varying x through 0.93. It therefore should be a model candidate for testing the predictions for the dynamics discussed in Section II.

Elastic and inelastic neutron scattering studies of Pr_3Tl and $(Pr_{.88}La_{.12})_3Tl$ have been carried out by Birgeneau, Als-Nielsen and Bucher.[18] The elastic scattering studies show that Pr_3Tl exhibits a ferromagnetic transition which is at least nearly second order at $T_c = 11.6 \pm 0.3$ K in good agreement with the bulk property value. No spontaneous ordering could be detected in $(Pr_{.88}La_{.12})_3Tl$ again in agreement with Bucher et al.[38] The inelastic studies were performed on polycrystalline samples around the (000) reciprocal lattice position; for an fcc structure this gives a spherically-averaged dispersion relation which for dominant

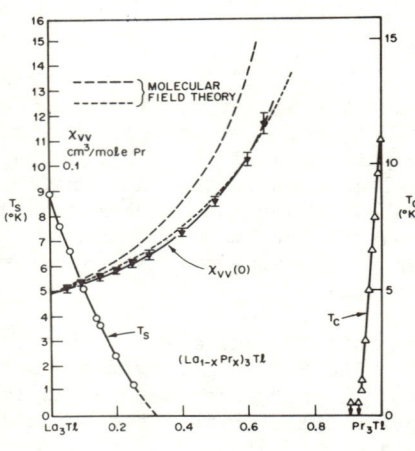

Fig. 5

Magnetic and superconducting phase diagram of $(Pr_xLa_{1-x})_3Tl$. Data taken from Ref. 38.

near neighbor interaction turns to be very close to that in the $\langle 110 \rangle$ direction. The dispersion relations measured at 4.5 K are shown in Fig. 6. Qualitatively, they correspond very closely to those anticipated from the model calculations in II(d). For Pr_3Tl it did not prove possible to resolve separate longitudinal and transverse excitons and instead the mean energy is obtained.

We now compare these results quantitatively with the T = 0 effective boson theory of Cooper. From the measured moment of $\sim 0.75\mu_B$, one obtains $\langle J \rangle/\alpha = 0.371$ which from the molecular field relation, Eq. (36), gives $\eta = 1.014$. For $(Pr_{.88}La_{.12})_3Tl$ we make the simplistic assumption that the effect of adding 12% La is simply to reduce η by 12%, that is, $\eta((Pr_{.88}La_{.12})_3Tl) = 0.892$. From the macroscopic measurements we have $68 < \Delta < 78$ K for $(Pr_{1-x}La_x)_3Tl$; to match the zone boundary energy in Pr_3Tl exactly, we take $\Delta = 77$ K which lies within this range. Having specified Δ, η and assuming nearest neighbor interactions alone, one may then calculate the dispersion relations with no further adjustable parameters. The solid lines in Fig. 6 are the theoretical $\langle 110 \rangle$ dispersion relations so-obtained. For Pr_3Tl we have given the mean of the transverse and longitudinal energies since these are not separately resolved. The agreement is clearly excellent for Pr_3Tl and relatively good for $(Pr_{.88}La_{.12})_3Tl$; in the latter case a slightly smaller value of Δ would make the fit nearly exact.

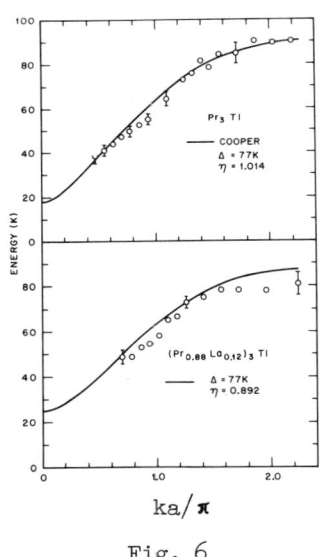

Fig. 6

Γ_1-Γ_4 spherically-averaged dispersion relations in Pr_3Tl and $(Pr_{.88}La_{.12})_3Tl$. The solid lines are the effective boson theory dispersion curves as discussed in the text. Data taken from Ref. 18.

We now consider the temperature dependence of the excitations. Typical results for both Pr_3Tl and $(Pr_{.88}La_{.12})_3Tl$ at $k = 0.6$ Å^{-1} ($\frac{ka}{\pi} = 0.94$) are shown in Figs. 7,8. Quite surprisingly, in both systems the excitation energies are essentially temperature independent up to quite high temperatures. The excitations seem simply to decrease continuously in intensity. Simultaneously, a broad distribution of scattering extending out to about 10 meV grows up around the exciton peaks; at 78 K in Pr_3Tl, the exciton merges into this background scattering and is no longer observable. These results are clearly in marked

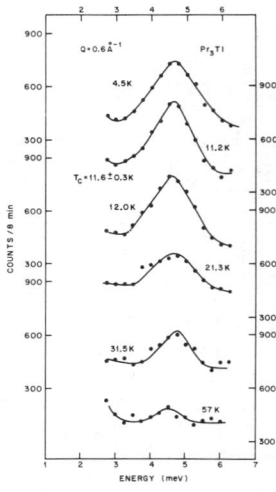

Fig. 7

Typical exciton scans as a function of temperature at k = 0.6 Å^{-1} ($\frac{ka}{\pi}$ = 0.94) in Pr_3Tl. Similar behavior was found at all other wave vectors; instrumental width = 1.25 meV. Data taken from Ref. 18.

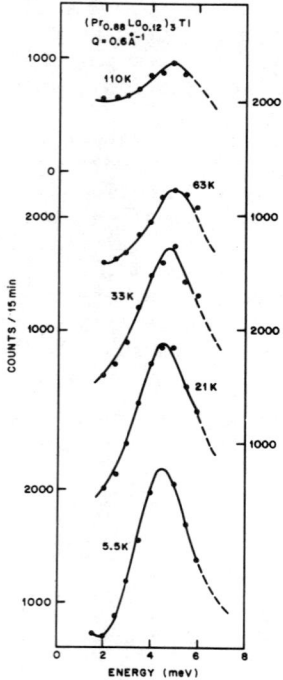

Fig. 8

Exciton scans as a function of temperature at k = 0.6 Å^{-1} ($\frac{ka}{\pi}$ = 0.94) in $(Pr_{.88}La_{.12})_3Tl$. Data taken from Ref. 18.

disagreement with the RPA-MF predictions discussed in IIC and illustrated in Fig. 3. Indeed all existing theories predict that for $k_BT \sim \Delta$, the dispersion should have largely vanished, whereas there is no sign of any renormalization whatsoever.

Birgeneau et al also carried out a complete survey of the dynamics at all accessible k-vectors around T_c in Pr_3Tl. For k > 0.3 Å^{-1} no effect of the phase transition on the dynamics is observed at all and in particular, there is no evidence of the type of mode softening anticipated from the RPA theory. However, if the temperature dependence of the mode energies was confined to k vectors less than about 0.2 Å^{-1}, it would not have been observed in the powder experiment. Thus, these experiments rule out the simple type of soft mode behavior predicted by RPA, but they do not preclude more complex soft-mode dynamics. We shall comment on these results more extensively in the final section.

Similar experiments have been performed on polycrystalline fcc Pr and on single crystal dhcp Pr both of which have $\eta \sim 1$. In fcc Pr, dynamics quite similar to those discussed above are observed; in particular, there is no observable temperature dependence of the exciton energies. However, the effective boson theory using η, Δ derived from bulk properties does not fit nearly so well as in Pr_3Tl. This may be connected with the fact that susceptibility measurements indicate $T_c \sim 8$ K whereas elastic neutron scattering gives $T_c \sim 20$ K with a rather odd-shaped magnetization curve. Polycrystalline dhcp Pr is thought to exhibit induced moment antiferromagnetic ordering at 25 K. However, single crystal dhcp Pr does not appear to order down to 1.5 K. Rainford and Houmann[17] have studied the hexagonal-site excitons in dhcp Pr as a function of temperature. They observe a 15% decrease in the Γ' point exciton energy in going from 18 K to 4.2 K, and they interpret this as the beginning of a soft mode phase transition. However, given the general anomalous behavior of dhcp Pr, it is not clear how significant this 15% effect actually is.

(c) $\eta \gg 1$

Perhaps the most thoroughly investigated singlet ground state system is the intermetallic alloy $Tb_{1-x}Y_xSb$.[19,40] Here Tb^{3+} is the singlet ground state ion while Y^{3+} is thought to act as a simple dilutent. The ground multiplet for $Tb^{3+}(4f^8)$ is $^7F_6$ so that L=3, S=3, J=6. The Tb^{3+} energy level diagram is as in Fig. 1 with $\Delta(\Gamma_1-\Gamma_4) \sim 11.9$ K. The Neel temperature varies from 15.1 K for x=0 to 0 K at x = 0.60. Thus, for pure TbSb, $T_N/\Delta \sim 1.3$ so that $\eta \gg 1$. Cooper and Vogt[40] have analyzed the bulk susceptibilities across the $Tb_{1-x}Y_xSb$ series, and they find that the results can be fitted surprisingly well with simple molecular field theory with nearest neighbor exchange $J_1 \sim 0$, the next-nearest neighbor exchange $J_2 \sim -0.09°K$.

A detailed inelastic neutron scattering study of the magnetic excitations in TbSb has been reported by Holden, Svensson, Buyers and Vogt.[19] TbSb orders with the MnO spin structure which has [111] ferromagnetic planes of spins with successive sheets along the (111) axis antiparallel; this structure requires $J_2 > 1/2\ J_1$. In TbSb the spins point along the [111] directions. There are in all four such directions so that in a multidomain sample the four different sets of modes will be superimposed. Two well-resolved excitation branches are observed at 4.5 K with the dispersion relations shown in Fig. 9. Note that in this diagram, energies are in units of THz where 1 THZ = 48 K. Holden et al have analyzed these results using an effective boson approach similar to that discussed in Section II. They find that the scattering is dominated by the transverse excitons; the parentage of these modes is primarily $\Gamma_1 \leftrightarrow \Gamma_4$ although there is considerable admixture of the higher lying level wavefunctions,

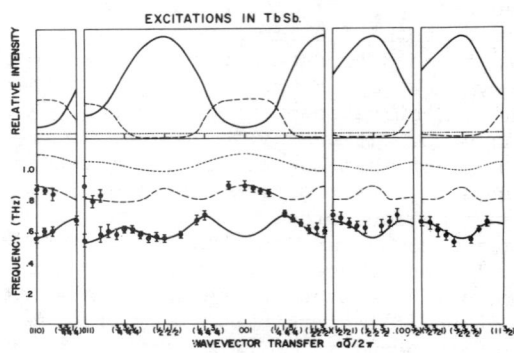

Fig. 9

Dispersion relations and corresponding intensities for three lowest branches of magnetic excitons in TbSb at 4.4 K. The excitations are generated successively by J^-, J^+, J^z. The lines are the results of least squares analysis in terms of effective boson theory. Data taken from **Ref**. 19. Figure kindly supplied by T. M. Holden. Energies are in THz where 1 THz = 48 K; wave vectors in nuclear reciprocal lattice units.

particularly $|\Gamma_5^{(2)}\rangle$, due to the large exchange. The solid lines are the calculated dispersion relations averaged over the four domains. The parameters so-obtained are Δ = 13.4 K, J_1 = +0.05 K, J_2 = -0.14 K. These are similar to those deduced by Cooper and Vogt although the agreement is by no means exact. This difference may arise either from the inadequacy of molecular field theory and/or higher order exchange and multipolar effects neglected in Eq. (1).

The most surprising feature of the TbSb results, however, is the temperature dependence of the scattering. Typical results at the position (1.45, 1.45, 1.45) corresponding to ka/π = 0.043 are shown in Fig. 10. As the temperature is raised towards T_N = 15.1 K, the exciton peak at 0.54 THZ = 26 K moves to lower energies and intense critical scattering appears around E = 0. Above T_N the exciton can no longer be resolved and instead the scattering is dominated by a strong quasielastic peak centered at E = 0. Similar results are obtained at other wavevectors although at certain positions such as (0.75, 0.75, 0.75) for $T > T_N$ there is a distinct shoulder at the expected crystal field transition energy. However, nowhere do Holden et al observe well-defined excitations for $T > T_N$ of the sort found in Pr_3Tl. In this system η = 56(6J_2)/Δ = 3.4. Thus Holden et al have shown that for an η of this magnitude, the concept of well-defined crystal field excitons in the paramagnetic regime is simply invalid.

IV. DISCUSSION AND CONCLUSIONS

From the experimental results discussed in the previous section, it is apparent that the dynamics of rare earth metallic singlet ground state systems vary dramatically as η goes from $\ll$ 1 to $\gg$ 1.

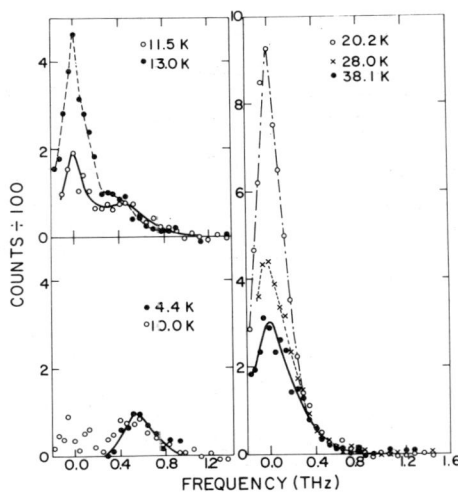

Fig. 10

Scattered-neutron distributions observed at (1.45, 1.45, 1.45) at several temperatures as indicated; the nonmagnetic background scattering has been subtracted. Data taken from Ref. 19 and figure kindly supplied by T. M. Holden.

In most cases the $T = 0$ dynamics can be adequately described using a simple effective boson theory. However, the finite temperature behavior, particularly for $\eta \geq 1$ in the paramagnetic regime, is rather unexpected. For $\eta \ll 1$ a simple crystal field description with Boltzman statistics appears to be entirely adequate. There is some broadening of the levels at higher temperatures due to phonon and conduction electron relaxation processes, but normally these are not large enough to smear out the excitations. For $\eta \gg 1$ the situation is more closely analogous to a normal antiferromagnet or ferromagnet with large anisotropy. In particular, a crystal field exciton description, especially one involving soft modes, does not seem to be applicable.

For $\eta \sim 1$ the situation is more complex. Well-defined excitons are observed in both the paramagnetic and ordered regions. However, at higher temperatures the excitons vanish into the background of quasielastic scattering before their energies are appreciably renormalized. This is in marked contradiction with RPA-type theories which predict appreciable renormalization of the dispersion for $kT \sim \Delta$. There are a variety of explanations which have been suggested for this apparently anomalous behavior. First, one could imagine that there would be contributions arising from the terms omitted from Eq. (32); in particular one might anticipate repulsion from higher lying phonon or exciton modes which would inhibit the renormalization. Second, a characteristic feature of most decoupling schemes is that they do not incorporate short range order effects properly. Thus, the apparent temperature independence of the dispersion relation could simply be a manifestation of the "renormalization by the energy" behavior already well-known in antiferromagnets.[41] Indeed, there are close qualitative analogies between Pr_3Tl and such low-dimensional antiferromagnets as $(CD_3)_4N\ MnCl_3$ (TMMC).[42] In both cases well-defined excitations exist in the paramagnetic regime and in each case the phase transition temperature is much less than the characteristic coupling energy in the system. In Pr_3Tl the transition is inhibited by the singlet-triplet splitting Δ whereas in TMMC it is inhibited by the long wavelength fluctuations. Concomitantly, in both systems most of the entropy is removed before three-dimensional long range order sets in. There are, of course, some caveats in this analogy but it is at least suggestive that the true answer to the Pr_3Tl anomaly lies in correlation effects.[43]

We have not cast any light on the soft-mode aspects of the phase transition problem except to show that for singlet-triplet systems with $\eta \sim 1$ any softening which does occur must be confined to long wavelengths. It should be noted, however, that soft acoustic phonon modes have been observed in the cooperative Jahn-Teller system $DyVO_4$[44] and in the hydrogen-bonded ferroelectric KDP.[45] Both of these problems can be cast in the form of an $S = 1/2$ Ising model in a transverse field, that is, Eq. (4). The acoustic phonon mode then is forced to have zero gradient at T_c by a soft exciton mode to which it is coupled. In each of these two systems $\eta \gg 1$, and indeed in KDP the tunnelling mode is always of an overdamped form[46] for $T > T_c$.

It is clear that inelastic neutron scattering studies on <u>single crystals</u> of $\eta \sim 1$ singlet ground state systems would be invaluable. Such studies should elucidate the behavior of the critical spin dynamics around T_c; indeed it seems likely that the nature of the phase transitions in these materials will only be clarified by experiment which can then guide any attempts at a microscopic theory. In turn, it is clear that singlet-ground state magnetic systems offer a cornucopia of interesting and challenging theoretical problems. Series expansion calculations on the singlet-triplet spin Hamiltonian Eq. (32) would be invaluable. For the paramagnetic regime spin dynamics calculation using techniques such as those developed by Blume and Hubbard[47] for the Heisenberg paramagnet could prove most interesting. It also seems possible that exact results might be obtainable for the singlet-triplet problem in one dimension. Further investigation of the consequences of cubic rather than isotropic symmetry for real singlet ground state systems is also required.

Finally, in this article so far we have not mentioned what may be the most interesting aspect of the singlet-singlet problem. Based on exact results in one dimension together with series expansion results in two and three dimensions, Pfeuty and Elliott[9] have conjectured that the critical behavior of the three-dimensional singlet-singlet system at $T = 0$ with respect to Δ may be identical to that of the four-dimensional Ising model w.r.t. temperature. Thus, singlet ground state magnetism may open up the four-dimensional world to experimental study - a most enticing possibility, to say the least!

ACKNOWLEDGMENTS

I would like to thank Jens Als-Nielsen, Klaus Andres, Martin Blume, Ernst Bucher, William Brinkman, Bernard Cooper, Thomas Holden and my coworkers at Bell Laboratories, Brookhaven and Risö for many enjoyable and enlightening discussions on the subject matter of this paper.

REFERENCES

1. R. M. Bozorth and J. H. Van Vleck, Phys. Rev. 118, 1493 (1960).
2. G. T. Trammell, J. Appl. Phys. 31, 3625 (1960).
3. S. Katsura, Phys. Rev. 127, 1508 (1962). A. S. Pikin and V. M. Tsukernik, Sov. Phys. JETP 23, 914 (1966).
4. J. I. Krugler, C. G. Montgomery and H. M. McConnell, J. Chem. Phys. 41, 1 (1964).
5. Y. Kitano, F. Specht and G. T. Trammell in "Proceedings of the International Conference on Magnetism, Nottingham 1964" (Institute of Physics and the Physical Society 1965), p. 480.
6. B. Grover, Phys. Rev. 140, A1944 (1965).
7. Y. L. Wang and B. R. Cooper, Phys. Rev. 172, 539 (1968), ibid. 185, 696 (1969).
8. D. A. Pink, J. Phys. C 1, 1246 (1968).
9. P. Pfeuty, Ann. Phys. (N.Y.) 57, 79 (1970). P. Pfeuty and R. J. Elliott, J. Phys. C 4, 2370 (1971).
10. P. Fulde and I. Peschel, Adv. in Phys. 21, 1 (1972); I. Peschel, M. Klenin and P. Fulde, J. Phys. C 5, L194 (1972).
11. S. R. P. Smith, J. Phys. C 5, L157 (1972).
12. Y. Y. Hsieh and M. Blume, Phys. Rev. B6, 2708 (1972); Y. Y. Hsieh (to be published).
13. B. R. Cooper, Phys. Rev. B6, 2730 (1972).
14. D. P. Chock, P. Resibois, G. Dewel and R. Dagonnier, Physica 53, 364 (1971).
15. K. C. Turberfield, L. Passell, R. J. Birgeneau and E. Bucher, Phys. Rev. Letters 25, 752 (1970); J. Appl. Phys. 42, 1746 (1971).
16. R. J. Birgeneau, E. Bucher, L. Passell and K. C. Turberfield, Phys. Rev. B4, 718 (1971).
17. B. D. Rainford and J. C. G. Houmann, Phys. Rev. Lett. 26, 1254 (1971); B. D. Rainford, "Magnetism and Magnetic Materials 1971" (AIP Conference Proceedings, No. 5, Ed. by C. D. Graham, Jr. and J. J. Rhyne) p. 591.
18. R. J. Birgeneau, J. Als-Nielsen and E. Bucher, Phys. Rev. Letters 27, 1530 (1971), Phys. Rev. B6, 2724 (1972) and (to be published).
19. T. M. Holden, W. J. L. Buyers, E. Swenson and O. Vogt, Bull. Amer. Phys. Soc. 16, 325 (1971) and (to be published).
20. K. W. H. Stevens, Proc. Phys. Soc. (London) A65, 209 (1952); M. T. Hutchings, Solid-State Phys. 16, 227 (1964).
21. M. A. Ruderman and C. Kittel, Phys. Rev. 96, 99 (1954).
22. T. A. Kaplan and D. H. Lyons, Phys. Rev. 129, 2072 (1962).
23. R. J. Birgeneau, M. T. Hutchings, J. M. Baker and J. D. Riley,

J. Appl. Phys. 40, 1070 (1969).
24. T. Murao, J. Phys. Soc. Japan 31, 683 (1971), "ibid" 33, 33 (1972).
25. B. B. Triplett and R. M. White (to be published).
26. K. Andres, Phys. Rev. B1 (to be published).
27. R. J. Elliott, A. P. Young, S. R. P. Smith, J. Phys. C 4, L317 (1971).
28. P. G. de Gennes, Solid State Comm. 1, 132 (1963).
29. R. J. Elliott, G. A. Gehring, A. P. Malozemoff, S. R. P. Smith, W. S. Staude and R. N. Tyte, J. Phys. C 4, L179 (1971).
30. R. J. Elliott and C. Wodd, J. Phys. C 4, 2359 (1971).
31. R. V. Lange, Phys. Rev. 146, 301 (1966). The applicability of Goldstone's theorem to the singlet-triplet problem was first pointed out by M. Blume (private communication).
32. G. Shirane and J. D. Axe, Phys. Rev. Letters 27, 1803 (1971).
33. K. R. Lea, M. J. M. Leask and W. P. Wolf, J. Phys. Chem. Solids 23, 1381 (1962).
34. B. R. Cooper and O. Vogt, Phys. Rev. B 1, 1211 (1970).
35. E. Bucher, K. Andres, J. P. Maita, A. S. Cooper and L. D. Longinotti, J. Phys. (Paris) 32, 114 (1971).
36. S. Foner, B. R. Cooper and O. Vogt, Phys. Rev. B 6, 2040 (1972).
37. R. J. Birgeneau, J. Phys. Chem. Solids 33, 59 (1972).
38. E. Bucher, J. P. Maita and A. S. Cooper, Phys. Rev. B 6, 2709 (1972). K. Andres, E. Bucher, S. Darack and J. P. Maita, Phys. Rev. B 6, 2716 (1972).
39. P. Fulde, L. L. Hirst and A. Luther, Z. Physik 230, 155 (1970).
40. B. R. Cooper and O. Vogt, Phys. Rev. B 1, 1218 (1970).
41. T. Oguchi, Phys. Rev. 117, 117 (1960). F. Keffer and R. Loudon, J. Appl. Phys. 32, 25 (1961).
42. M. T. Hutchings, G. Shirane, R. J. Birgeneau and S. L. Holt, Phys. Rev. B 5, 1999 (1972).
43. See for example, R. Silberglitt, E. Cullen and J. R. Cullen (this conference).
44. R. L. Melcher and B. A. Scott, Phys. Rev. Lett. 28, 607 (1972).
45. E. M. Brody and H. Z. Cummins, Phys. Rev. Lett. 21, 1263 (1968).
46. I. P. Kaminow and T. C. Damen, Phys. Rev. Lett. 20, 1105 (1968).
47. M. Blume and J. Hubbard, Phys. Rev. B 1, 3815 (1970); J. Hubbard, J. Phys. C 4, 53 (1971); F. B. McLean and M. Blume, Phys. Rev. (to be published).

EXCITON DYNAMICS AND COHERENCE IN MnF_2*

R. M. Macfarlane and A. C. Luntz
IBM Research Laboratory, San Jose, Ca. 95114

ABSTRACT

The coherence of Frenkel excitons is limited by the rate at which they scatter from one wavevector state $\underline{k}$, to another within a given branch. We have used a technique of resonant optical pumping and monochromatic fluorescence detection to study the dynamics and coherence of Frenkel excitons in MnF_2 having fairly well defined wavevectors.[1] The technique also appears to be very promising for a study of exciton coherence in molecular crystals.

The lowest $^4T_1$ exciton in MnF_2 at 18418.5 cm^{-1} was excited by a nitrogen laser pumped, repetitively pulsed tunable dye laser, having a bandwidth <1 cm^{-1}. Experiments were performed around 1.9°K. Three modes of excitation were used: (i) direct creation of $\underline{k} \approx 0$ excitons, (ii) exciton-magnon (e-m) pumping at 18475 cm^{-1} which, from the decay of e-m states, creates excitons with $\underline{k}$ near the zone boundary, and (iii) exciton-phonon pumping at 18524 cm^{-1}. The number of excitons in a given region of $\underline{k}$-space was monitored by observing either intrinsic, pure exciton emission ($\underline{k}_{ex} \approx 0$), or the peak of the exciton-magnon emission ($\underline{k}_{ex} \approx$ zone boundary).

When creating excitons with large k and observing $k \approx 0$, or creating them with $k \approx 0$ and observing at large k we find a fluorescence rise time of ~ 1µsec which we interpret as the time taken for an exciton to scatter across the Brillouin zone. This is much less than the exciton trapping time (250µsec), and is consistent with our observation of equal lifetimes for the decay of zone center and zone boundary excitons.[2] In the other configurations, the fluorescence rise time is <100 nsec, i.e., the decay of (e-m) and (e-p) states is <100 nsec.

*This research was supported in part by the Air Force Office of Scientific Research under contract F44620-71-C-0081.

REFERENCES

1. For a study of sublattice exciton dynamics without wavevector resolution, see J. F. Holzrichter, R. M. Macfarlane and A. L. Schawlow, Phys. Rev. Letters **26**, 652 (1971).
2. Under different conditions this may not always be the case, see R. E. Dietz, A. E. Meixner, H. J. Guggenheim and A. Misetich, Phys. Rev. Letters **21**, 1067 (1968).

MAGNETIC CIRCULAR DICHROISM OF A DAVYDOV COUPLED TRANSITION IN ANTI-FERROMAGNETIC CoF_2*

Y. H. Wong, C. D. Pfeifer and W. M. Yen
Dept. of Physics, Univ. of Wisconsin, Madison, Wisc. 53706

ABSTRACT

Results of a study of the magnetic circular dichroism (MCD) properties of the $^4T_1 \rightarrow {}^2A$, $^2T_1(^2P)$ absorption of CoF_2 located at 22,769 cm^{-1} and corresponding to two center magnon sideband are reported. Magnetic field dependences and polarity of the MCD signals allow us to identify this transition as involving a Davydov coupled Γ_2^+ exciton as its origin. This is determined by a dependence of the magnetic field splitting and the sense of the splitting on $|\vec{H}|$. The extrapolated magnitude of the inter-sublattice coupling is of the order of 10^{-2} cm^{-1} in agreement with expectations. The effective splitting factor of the compound excitation is found to be .015. Results of related measurements are reported including the observation of a finite CD signal as well as observation of temperature renormalization effects on the splitting factors.

Recent studies of the rutile structured antiferromagnetic insulators indicate that magnetic circular dichroism (MCD) techniques can be used to probe the optical spectra of these materials with high sensitivity.[1,2] The MCD studies serve as a useful complement to existing absorption and Raman Scattering data in these materials. In this paper, we wish to report measurements of the CD and MCD of the one-magnon sideband[3] at 22,769 cm^{-1} of the $^4T_1 \rightarrow {}^2A$, $^2T_1$ (^2P) excitonic transition in antiferromagnetic CoF_2. We identify the origin of this sideband as a Davydov coupled Γ_2^+ exciton. Also we conclude that the observed zero magnetic field dichroism is due to an inequivalence of the magnetic sublattices similar to the case of MnF_2.[4]

Experimentally, absorption and dichroism were measured along the c-axis of the CoF_2 sample using MCD modulation techniques.[5] The sample was located in a magnetic field parallel to the c-axis. MCD spectra were measured from 0 to ± 41 kOe at temperatures ranging from 4 to 14°K. Data were taken for the c-axis of the sample parallel and antiparallel to the propagating light vector.

Typical absorption spectra of the one-magnon sideband together with its corresponding MCD signal are shown in Fig. 1. The phase of the MCD identifies the polarity of the splitting $\Delta(=\varepsilon_R-\varepsilon_L)$ of the exciton-magnon excitation. The

*Supported by the National Science Foundation Grant G.P. 15426.

peak to peak amplitude is a measure of the amplitude of the splitting. The field dependence of Δ for the two crystal orientations are plotted in Fig. 2.

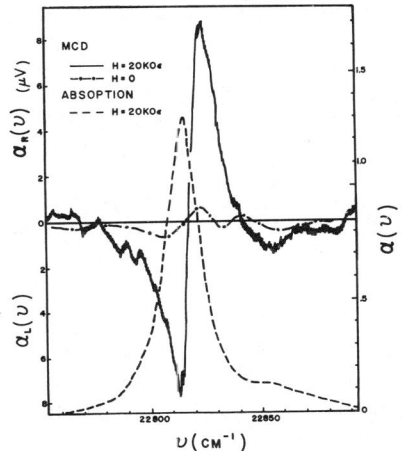

Fig. 1. Absorption and MCD signals of the one-magnon sideband of $^4T_1 \to {}^2A, {}^2T_1\,(^2P)$ transition at 22,769 cm^{-1}.

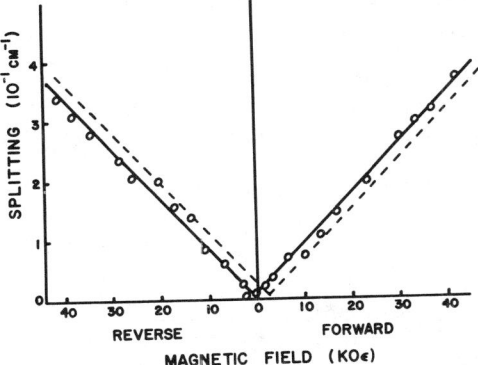

Fig. 2. Magnetic field dependence of splitting for both crystal orientations. The sign of the splitting (dotted line) is the reverse of that shown.

The data represented in Fig. 1 and 2 exhibit the following salient features: for fields up to 41 kOe, the splitting (Δ) displays a linear dependence on $|\vec{H}|$. The minimum splitting occurs at H_0 = 1.5 kOe. The magnitude of this minimum is $1.5\pm.3\times 10^{-2}$ cm^{-1}. The polarity of the splitting Δ and the shift H_0 is reversed in sign when the crystal is rotated 180° about an axis ⊥ to the c-axis.

The linear dependence of the splitting on $|\vec{H}|$ indicates that a Davydov type of coupling between the two sublattices exists for this exciton-magnon excitation in CoF_2. This can be seen by examining the secular equation for the simple rutile antiferromagnet with applied magnetic field parallel to the c-axis. The splitting Δ is given by

$$\Delta = \varepsilon_R - \varepsilon_L = [\,(H_{AA} - H_{BB} + 2g\beta H)^2 + 4H_{AB}^2\,]^{1/2} \quad (1)$$

If the inter-sublattice interaction is finite i.e.

$$H_{AB} \simeq H_{AA} - H_{BB} + 2g\beta H, \quad (2)$$

then terms in Eq. (1) would yield a splitting proportional to $[H^2 + H_{AB}^2]^{1/2}$.

This splitting would be independent of the directionality of $\vec{H}$ and would depend only on the magnitude $|\vec{H}|$. Taking into account the CD, i.e. the difference $H_{AA}-H_{BB}$, which is to be discussed shortly, the behavior denoted in Fig. 2 shows a dependence on $|\vec{H}|$ and thus substantiates our argument. It is also consistent with the predictions of Loudon[7] that the Γ_2^+ exciton origin of the sideband transition possesses a finite Davydov interaction which in our case is given by the minimum value of the V shape curve (Fig.2), i.e. .015 ± .003 cm^{-1}. This is to be contrasted with the linear dependence of the splittings observed in MnF$_2$ on $\vec{H}$, where no Davydov splitting is expected.[7]

The observed splittings occur at H_0 = ± 1.5 kOe rather than at $\vec{H}$ = 0 as expected from a pure Davydov coupling scheme. The fact that the splitting at H = 0 is larger than that at H_0 implies that the two sublattices are intrinsically inequivalent, i.e. $H_{AA} \neq H_{BB}$ in Eq. (1). A similar effect has been observed in MnF$_2$.[4] This inequivalence can be described by an effective magnetic field $\vec{H}_i$ which is intrinsic to the crystal. In other words, Eq. (1) can be written as,

$$\Delta = \{[2g\beta(\vec{H} - \vec{H}_i)]^2 + 4H_{AB}^2\}^{1/2} \tag{3}$$

In addition to its effect on the magnitude of the zero field splitting such a field will also account for the polarity change observed in H_0 and Δ when the sample is rotated 180°. In CoF$_2$ this field has magnitude H_0 along the c-axis.

We have examined the MCD of the materials MnF$_2$,[4] FeF$_2$, and CoF$_2$ in search of the zero field circular dichroism (CD). The effect is present in MnF$_2$ and CoF$_2$ but can not be observed in FeF$_2$. This would seem to imply that the intrinsic inequivalence of the two sublattices mentioned above is related to the Kramers or non-Kramers character of the impurity ion. Sugano[8] has suggested that this inequivalence may be due to a Jahn-Teller related effect in the Kramers ion, leading in turn to uniaxial lattice distortions.

A study of the splitting as a function of magnetic field, yields an effective g factor for the combined exciton-magnon excitation to be $g_{eff}^{exc-mag}$ = 0.015 ± 1% at 4°K. Following the scheme of Sell et.al. with a g_{eff}^{exc} = 4.900[3] for the Γ_2^+ exciton, we conclude that

$$g_{eff}^{mag} = 4.9 - .015 \simeq 4.88$$

The splitting observed at a constant applied field H, shows that the effective splitting factor, g_{eff}, is temperature dependent. Results are shown in Fig. (3) and may be attributed to the temperature renormalization of the g-factor.

$$\Delta g(T) = \frac{g(T) - g(0)}{g(0)} \tag{4}$$

The changes in $\Delta g(T)$ exhibit a T^n dependence where n = 1.3. Existing theories[10] as well as experimental results indicate that the renormalization is proportional to the susceptibilities $\chi_\parallel$ and $\chi_\perp$, neither of which show the observed power dependence.[10] The implication is that the discrepancy arises from the temperature dependence of the exciton and magnon interaction and that it must be included in the renormalization of the magnon sideband or compound excitation splitting factors.

We would like to acknowledge the able assistance of F. L. Scarpace and discussions with S. Sugano.

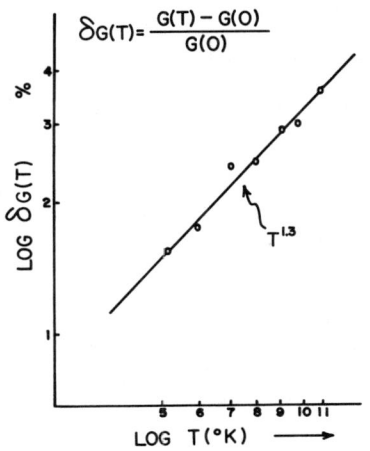

Fig. 3. Temperature dependence of the g_{eff} renormalization for the combined excitation.

REFERENCES

1. Ming Y. Chen, F. L. Scarpace, M. W. Passow and W. M. Yen, Phys. Rev. 48, 132 (1971).
2. F. L. Scarpace, Ming Y. Chen and W. M. Yen, Jour. of Appl. Phys. 42, 1655 (1971).
3. J. P. van der Ziel, and H. J. Guggenheim, Phys. Rev. 166, 479 (1968).
4. Y. H. Wong, F. L. Scarpace, Ming Y. Chen and W. M. Yen in Magnetism and Magnetic Materials - 1971, AIP Conference Proceeding No. 5, edited by C. D. Graham, Jr., and J. J. Rhyne (American Institute of Physics, New York, 1972), p. 655, and to be published.
5. S. N. Jasperson and S. E. Schnatterly, Rev. of Sci. Inst. 40, 761 (1969).
   F. L. Scarpace Thesis (U. of Wisconsin, 1971)unpublished.
6. A. S. Davydov, "Theory of Molecular Exciton", Plenum Publishing Co., New York.
7. R. Loudon, Adv. Phys. 17, 243 (1968).
8. S. Sugano (private communication).
9. D. D. Sell, R. L. Green and R. M. White, Phys. Rev. 158, 489 (1967).
10. M. W. Passow, D. L. Huber and W. M. Yen, Phys. Rev. Lett. 23, 477 (1969) and references therein.

FAR INFRARED ABSORPTION SPECTRUM OF $CoBr_2 \cdot 2H_2O$: MAGNONS, MAGNON BOUND STATES, PHONONS, AND MAGNON+PHONON EXCITATIONS*

J. B. Torrance
IBM Research, Yorktown Heights, N. Y. 10598

K. A. Hay
Materials Division, CERL, Leatherhead, Surrey, England

ABSTRACT

The excitation of single, k=0 magnons in the antiferromagnetic, ferrimagnetic, and ferromagnetic phases of $CoBr_2 \cdot 2H_2O$ at $\sim 2°K$ are described and fit to a five parameter spin wave theory. In addition, 2-magnon and 3-magnon bound states are observed in the ferromagnetic phase. At higher energies, another series of single spin excitations is observed and identified as due to the simultaneous excitation of a magnon plus a phonon. It is shown that these magnons are excited near the Brillouin zone boundary in the b-direction.

INTRODUCTION

In most of its magnetic properties,[1] $CoBr_2 \cdot 2H_2O$ is very similar to the more familiar compound, $CoCl_2 \cdot 2H_2O$. They are crystallographically isomorphic[2] with monoclinic symmetry; the Co^{++} ions form linear $-CoBr_2-$ chains along the c-axis. In both compounds the exchange interaction J_o between Co^{++} spins within the same chain is ferromagnetic and much stronger than the interactions between chains. Furthermore, the exchange interactions have extreme longitudinal anisotropy, with an "easy" b-axis.[3,4] This strong anisotropy and the weak exchange interactions between chains give rise to very small magnon dispersion in the a-b plane, i.e. perpendicular to the chains.

Both $CoBr_2 \cdot 2H_2O$ and $CoCl_2 \cdot 2H_2O$ are metamagnetic[1] and presumably have the same spin structures, which are shown in Fig. 1.[3,5] This figure represents a typical cross section through the chains of spins that are coming out of the plane of the figure. The exchange interactions between chains are also shown: J_1 and J_2 are antiferromagnetic and J_3 is so small[4] that we will neglect it. Below T_N, these materials are two sublattice antiferromagnets [Fig. 1(a)]. In the presence of a large magnetic field along the axis of magnetization (b-axis), there is a transition at $H_o = H_{C1}$ to a ferrimagnetic spin structure [Fig. 1(b)]. At $H_o = H_{C2}$ there is a second metamagnetic transition to the ferromagnetic phase [Fig. 1(c)]. The values of T_N,

Fig. 1
Spin structures for the three metamagnetic states of $CoBr_2 \cdot 2H_2O$

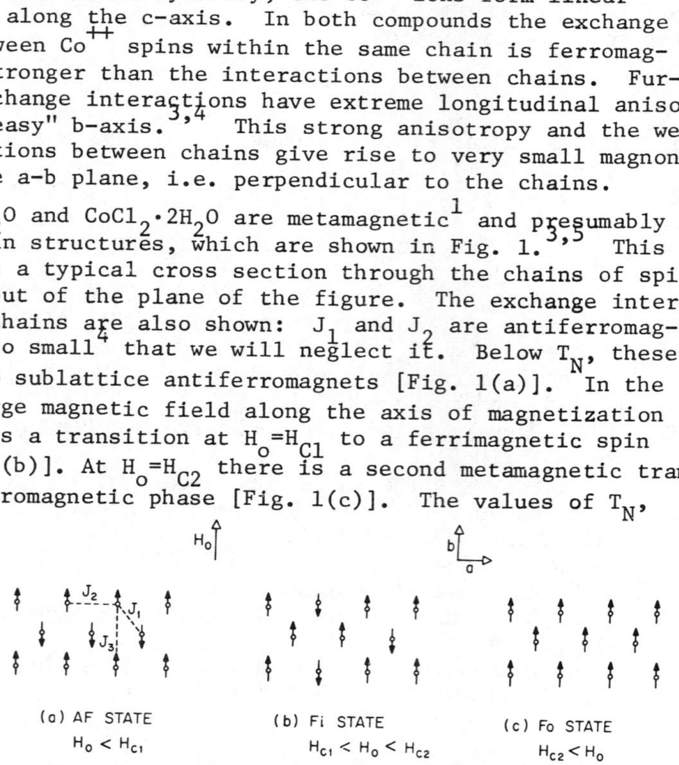

(a) AF STATE
$H_o < H_{C1}$

(b) Fi STATE
$H_{C1} < H_o < H_{C2}$

(c) Fo STATE
$H_{C2} < H_o$

*The experimental work reported here was performed while the authors were at the Div. of Engineering & Applied Physics, Harvard University.

H_{C1} and H_{C2} for $CoBr_2 \cdot 2H_2O$ and $CoCl_2 \cdot 2H_2O$ are compared in Table I.

There is one dramatic difference between these two compounds, however. If a crystal of $CoBr_2 \cdot 2H_2O$ is immersed into liquid nitrogen and then returned to room temperature, it shatters into millions of tiny fibers or bundles of linear chains. This transition can be avoided by tightly encasing the sample in a holder to prevent it from shattering. Our sample was fitted into a hole milled through a small copper plate and both were then sandwiched between two wedged quartz substrates and glued together with G.E. 7031 varnish. Far infrared transmission measurements at $\sim 2°K$ were then performed on this sample using techniques similar to those described in Ref. 4.

FAR INFRARED ABSORPTION SPECTRUM

The transmission data were smoothed and normalized to a background obtained largely from measurements at several different magnetic fields, where no absorption lines in that frequency region were observed. These transmission spectra exhibited several absorption lines at each field. The energies of the field dependent lines are plotted as the open circles in Fig. 2. In addition, several field independent lines were also observed (at 41.8, 53, 62, 71, 78 and 93 cm^{-1}), but are not shown in Fig. 2. These lines are presumably phonon and/or multi-phonon excitations.[7] Above $\sim 100 cm^{-1}$ the sample became opaque, presumeably due to strong phonon absorption. All the data were taken at $\sim 2°K$, without a polarizer.

Note that the observed absorption lines a_1, b_1, c_1, d_1, and e_1 have g-values of approximately 6, the expected value for Co^{++} in this symmetry. These excitations, therefore, undoubtedly correspond to single magnons or spin waves. Comparing this spin wave spectrum to that observed[4] in $CoCl_2 \cdot 2H_2O$, the spectra are qualitatively very similar, with that of the bromide lying ~ 10 cm^{-1} lower. This reduced energy is consistent with the lower values of T_N, H_{C1} and H_{C2} (Table I) and is simply a manifestation of the generally weaker exchange interactions in the bromide.

In addition to the spin waves, two extra lines are observed in the ferromagnetic phase (e_2 and e_3 in Fig. 2) which have g-values 2 and 3 times the g-value of the spin waves. These must, therefore, correspond to the excitation of 2 (and 3) magnons, but the energies are much less than 2 (or 3) times the single magnon energy. These are then bound states of magnons, and crudely correspond to the excitation of neighboring spin deviations. Multiple magnon bound states were first ob-

TABLE I: Comparison of T_N and the five spin wave parameters. The values of H_{C1} and H_{C2} in brackets are from magnetization data.

	$T_N(°K)$	H_{C1}(kOe)	H_{C2}(kOe)	$g_\parallel$	$j^\perp$	$E_o(cm^{-1})$
$CoBr_2 \cdot 2H_2O$	9.5[1]	13.7[1] [13.7][1]	29.25[1] [29.8][1]	6.19	.33	15.5
$CoCl_2 \cdot 2H_2O$	17.5[3]	31.0[4] [31.3][6]	44.9[4] [44.9][6]	6.81[4]	.28[4]	21.4[4]

served in $CoCl_2 \cdot 2H_2O$,[4] where the relative energies of e_1, e_2 and e_3 are very similar to those in Fig. 2 (supporting our identification). These magnon bound states will not be discussed in detail here, since we have only observed two of them. (Presumably this is due to the lack of strong transverse exchange anisotropy, which gives rise to their non-zero intensity.)[4] The excitations near 40 cm^{-1} are discussed near the end of this paper.[8]

THEORETICAL ANALYSIS OF SPECTRUM

In order to attempt to fit the spectra to a theoretical calculation, we first assume that all the lines have the same g-value (with that of the 2- and 3-magnon bound state being 2 and 3 times as large).

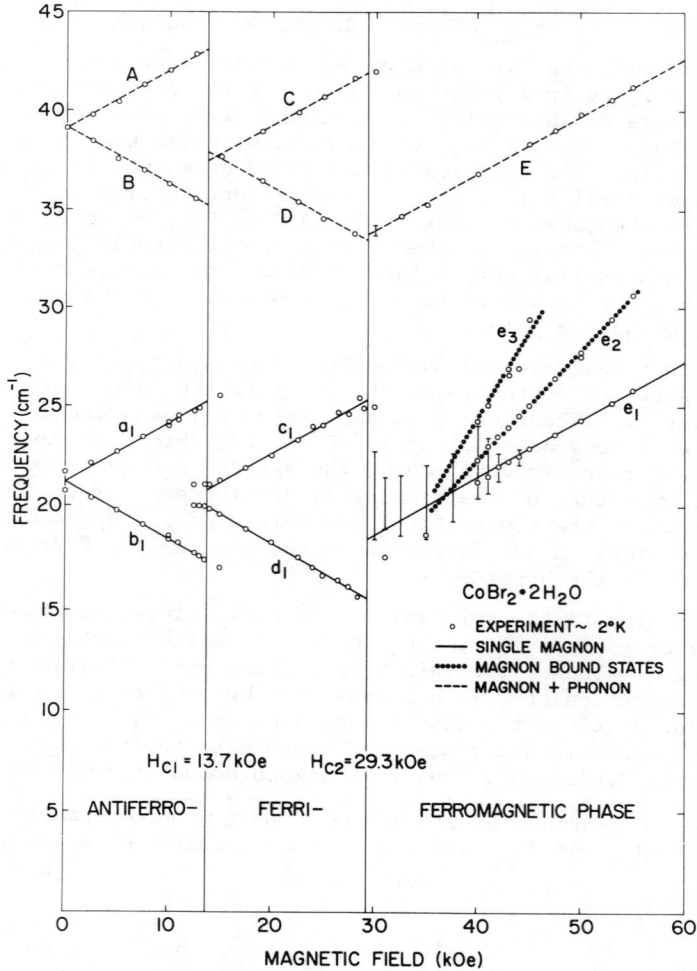

Fig. 2 Experimental and theoretical far infrared absorption spectra. The widths of the broad lines near 20 cm^{-1} are indicated by error bars.

Using $g_\parallel=6.19$, we obtain the fit shown by the solid, dashed and dotted lines in Fig. 2. Since all of the lines appear to have the same g-value, we need only determine their relative energies.

We now concentrate on the spin waves: i.e. a_1-e_1. The spin waves in these spin structures (Fig. 1) correspond to antiferro-, ferri- and ferro-magnetic resonance, in the respective phase. For the cases of $CoCl_2 \cdot 2H_2O^4$ and $FeCl_2 \cdot 2H_2O^9$, the spin wave frequencies have already been calculated in terms of the exchange interactions. These results may be written in the following form:

$$h\nu_{a,b}=E_o+g_\parallel\mu_B(\pm H_o+\tfrac{1}{3}H_{C1}[2-j^\perp]+\tfrac{1}{3}H_{C2}[1+j^\perp])-\delta_{AF} \quad (1)$$

$$h\nu_c=E_o+g_\parallel\mu_B(H_o+j^\perp H_{C2}/2)-\delta_{Fi}$$
$$h\nu_d=E_o+g_\parallel\mu_B(-H_o+H_{C2})-\delta_{Fi} \quad (2)$$

$$h\nu_e=E_o+g_\parallel\mu_B(H_o-H_{C2}[1-j^\perp]) \quad (3)$$

where we have defined $J^\perp \equiv \tfrac{1}{2}(J^{xx}+J^{yy})$ and assumed $J_1^\perp/J_1^{zz}=J_2^\perp/J_2^{zz}\equiv j^\perp$ (the longitudinal exchange anisotropy). Also, $\delta_{Fi}\equiv(j^\perp g_\parallel \mu_B H_{C2})^2/4E_o$ and $\delta_{AF}\equiv(j^\perp g_\parallel \mu_B [H_{C1}+2H_{C2}])^2/18E_o$, and E_o is related to J_o by $E_o \equiv 2(J_o^{zz}-J_o^\perp)$. (We have neglected any transverse exchange anisotropy).^{4}

The spin wave frequencies are thus determined by five parameters: $g_\parallel$, H_{C1}, H_{C2}, $j^\perp$, and E_o. Since a_1 and b_1 are assumed to intersect at $H_o=0$, these five parameters are uniquely determined by the five absorption lines (Fig. 2). The values thus obtained are shown on Table I. The values of H_{C1} and H_{C2} are in excellent agreement with those (in brackets) obtained from magnetization measurements. The values of $g_\parallel$ and $j^\perp$ compare well with the corresponding values observed in $CoCl_2 \cdot 2H_2O$,^{4} while the lower value of E_o is indicative of the weaker exchange interactions in the bromide. The agreement with the data points (Fig. 2) is excellent, with two exceptions: near $H_o=0$ and $H_o=H_{C1}$. At zero field the two resonance lines are not degenerate as predicted by Eq. (1). This splitting is caused by the transverse anisotropy of the exchange interaction between chains,^{4} which could have been included by using another parameter. Near H_{C1}, a_1 and b_1 persist into the ferrimagnetic region, while c_1 and d_1 persist below H_{C1}, a phenomenon also observed in $CoCl_2 \cdot 2H_2O^4$ and $FeCl_2 \cdot 2H_2O$.^{9}

We now turn to the absorption lines observed in the 33-43 cm^{-1} region of Fig. 2. These lines also have g-values of 6.19 (see fit Fig. 2) corresponding to the excitation of a <u>single</u> Co^{++} spin deviation. However, their energies are much higher than any of the magnons, even those at the zone boundary, and much lower than the lowest crystal field level.^{7} Our interpretation is that these lines correspond to the simultaneous excitation of a magnon and non-magnetic excitation, for example a phonon.

Note that the relative positions of these lines are not the same as for the spin waves. For example, E and D intersect at H_{C2} at almost the same energy, while e_1 is almost 3 cm^{-1} above d_1 at H_{C2}. The difference between E-D and e_1-d_1 at H_{C2} is due to the fact that they are excited from different regions of k-space. Using simple spin wave theory, we can calculate <u>for arbitrary k</u> the energy difference between the relevant ferromagnetic and ferrimagnetic resonance modes. This difference is approximately given by:

$$h\nu_E - h\nu_D \simeq \frac{j^\perp g_\parallel \mu_B}{3} \left\{ (H_{C2} - H_{C1}) \cos \vec{k} \cdot \vec{a} \right. \\ \left. + (2H_{C2} + H_{C1}) \cos \frac{\vec{k} \cdot \vec{a}}{2} \cos \frac{\vec{k} \cdot \vec{b}}{2} \right\} \quad (4)$$

For the spin waves, k=0 and the difference is $+j^\perp g_\parallel \mu_B H_{C2} = 2.8 \text{cm}^{-1}$, as we observe for e_1-d_1 (Fig. 2). Note that for $\vec{k}$ perpendicular to the a-b plane, the difference remains 2.8cm^{-1}, while the observed difference is only 0.5cm^{-1}.

From Eq. (4) the E-D splitting is predicted to be much smaller for $\vec{k}$ in the a-b plane. For example, if $\vec{k}$ is at the zone boundary in the $\vec{a}$ or $\vec{b}$ direction, the difference is -0.5 or +0.5cm^{-1}. Since E does lie ∼0.5cm^{-1} <u>above</u> D, this result would suggest that k∼π/b in the b-direction. In order to check this assertion, we have calculated the spin wave energies at k=π/b using the <u>same</u> five parameters used in the k=0 fit (Table I), resulting in the dashed lines in Fig. 2 – an essentially perfect fit. This fit required <u>only one</u> additional parameter – the energy of the non-magnetic excitation, which was found to be 18.0cm^{-1}. We conclude, therefore, that the A-E spectra correspond to the excitation of spin waves very near k=π/b plus a non-magnetic excitation (a phonon[8]) near k=-π/b with an energy of 18.0cm^{-1}.

REFERENCES

1. A. Narath, J. Phys. Soc. Japan **19**, 2244 (1964).
2. B. Morosin, J. Chem. Phys. **47**, 417 (1967).
3. A. Narath, Phys. Rev. **136**, A766 (1964).
4. J. B. Torrance and M. Tinkham, Phys. Rev. **187**, 595 (1969); and D. F. Nicoli and M. Tinkham (to be published).
5. D. E. Cox, G. Shirane, B. C. Frazer, and A. Narath, J. Appl. Phys. **37**, 1126 (1966).
6. A. Narath and J. E. Schirber, J. Appl. Phys. **37**, 1124 (1966).
7. Low lying crystal field levels are expected at higher frequencies (∼150 cm^{-1}) with a g_1-value of ∼1. (A. Narath, Phys. Rev. **140**, A552 (1965)).
8. A more detailed discussion of the magnon plus phonon excitation will be given in J. B. Torrance and K. A. Hay (to be published).
9. K. A. Hay and J. B. Torrance, Phys. Rev. B **2**, 746 (1970).

TEMPERATURE BROADENING OF FAR INFRARED SPIN-CLUSTER EXCITED STATES IN LINEAR ISING $CoCl_2 \cdot 2H_2O$ *

D.F. Nicoli and M. Tinkham
Harvard University, Cambridge, Mass. 02138

ABSTRACT

The nature of the spin-cluster ("multi-magnon") excitations which have been observed in $CoCl_2 \cdot 2H_2O$ at liquid helium temperatures is reviewed. Using the far infrared laser lines of HCN and DCN, the authors measured the spin-cluster resonance linewidths as a function of sample temperature and found excellent agreement to a simple exponential broadening law with a single characteristic energy independent of cluster size. It seems possible that the optic phonon which intersects the Ising "fan" of excitations is playing the dominant role in the relaxation mechanism giving rise to the T-dependent energy widths.

INTRODUCTION

The far infrared E vs H spectrum of $CoCl_2 \cdot 2H_2O$ ("CC2") at $1.5°K$ was obtained by Torrance and Tinkham[1] using a tunable monochromator operating in the range of 20 to 60 cm^{-1}, with applied fields up to 50 kOe. They found in each of the three metamagnetic phases an excitation spectrum which closely resembled the ideal Ising "fan" of lines, seen in Fig. 1. This fan represents the creation of "clusters" (in the localized picture) of adjacent flipped spins on a linear ferromagnetic S=1/2 chain. The spectrum which results from this purely Ising part of the general nearest-neighbor exchange Hamiltonian is

$$E_n = 2J_o^{zz} + n g^{zz} \mu_B (H-H_{crit}) \qquad (1)$$

The size of the spin-cluster is n. J_o^{zz} is the large ferromagnetic intrachain exchange ($\simeq 12.7$ cm^{-1}), g^{zz} the unusually large g-factor along the easy z-axis ($\simeq 6.8$), and H_{crit} the effective field origin determined by the interchain exchanges.

The large rhombic anisotropy exchange within a given chain, $J_o^a \equiv |J^{xx} - J^{yy}|/2$, was recognized as playing the crucial role in allowing any of the clusters larger than the ordinary n=1 "spin wave" to be excited from the ground state by radiation. The general single-chain Hamiltonian, the diagonalization of which produced the very successful experimental agreement for Torrance and Tinkham, is indicated schematically by

$$\mathcal{H}(\text{single chain}) = -\sum_i \vec{S}_i \cdot \underline{J} \cdot \vec{S}_{i+1} + \text{Zeeman energy}$$
$$\equiv \mathcal{H}^{\text{Ising}} + \mathcal{H}^\perp + \mathcal{H}^a \qquad (2)$$

where $\mathcal{H}^a = -J_o^a \sum_i S_i^+ S_{i+1}^+ + \text{h.c.} \qquad (3)$

* Supported in part by the NSF, ARPA and ONR.

$\mathcal{H}^a$ couples the spin-clusters $|n\rangle$ and $|n+2\rangle$; one can therefore speak of the separate odd and even "manifolds" of excitations. The transverse polarization then allows excitation of all <u>odd</u> manifold lines from the ground state. The parallel polarization permits excitation of all <u>even</u> manifold lines, since $\mathcal{H}^a$ perturbs the fully-aligned ground state by admixing some 2-cluster component. This "selection rule" is indicated in Fig. 1 also.

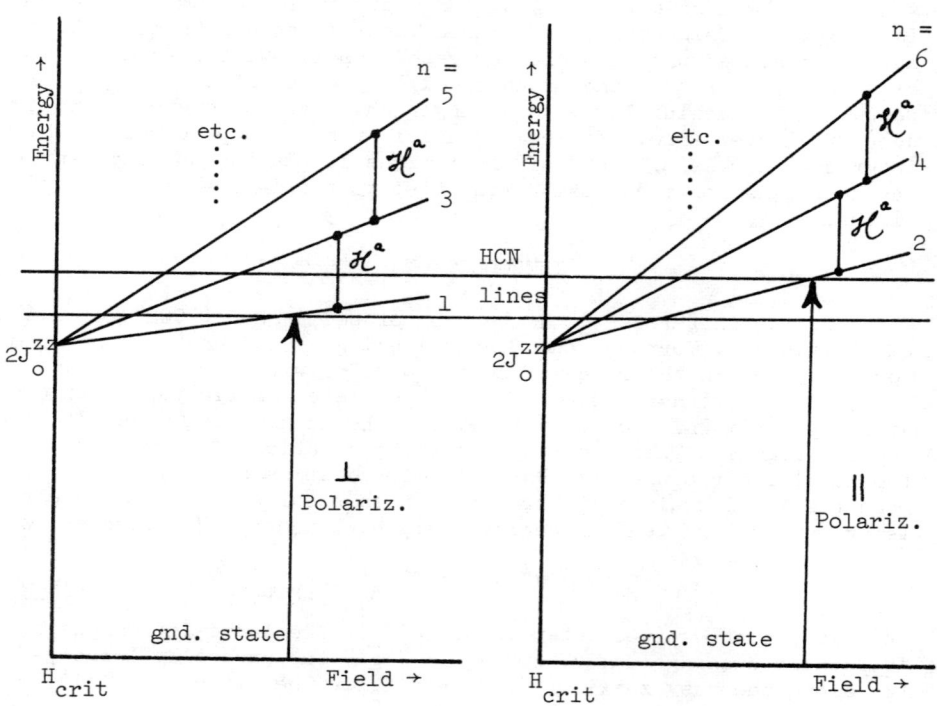

Fig. 1. The linear-Ising E <u>vs</u> H spectrum; manifold coupling by $\mathcal{H}^a$ and selection by radiation.

The present authors[2,3] subsequently have examined in considerable detail the excitation spectrum of CC2 using the four major laser lines of HCN and DCN, which intersect the Ising-like fans at convenient energies. The HCN lines at 29.71 and 32.16 cm^{-1} are indicated schematically in Fig. 1; the two analogous lines of DCN near 50 cm^{-1} were also used. Whereas Torrance and Tinkham had with difficulty observed spin-clusters of 4 or 5 spins at the largest, the laser spectroscopy revealed clusters of size up to 14 (effective-g almost 100!). The lineshapes were exceptionally clean and easily resolved from the baseline, with nearly complete manifold separation according to polarization, allowing one to study in detail for the first time the dynamical properties -- linewidths and absolute intensities -- of these unusual excitations. The object of this paper is to report on the nature of the observed linewidth broadening with temperature in the range 1.3 to $14°K$. (CC2 Néel point: $\simeq 17°K$.)

THERMAL LINEWIDTH BROADENING

The observed lineshapes [in field or energy] were found to be highly Lorentzian in many cases; in others, the extent and shape of the line asymmetries could be well accounted for by a calculation of the transmission through the finite sample slab near resonance with interference included -- a method used by G.T. Rado[4] to explain similar line skewing in $DyPO_4$.

The data can be divided into two categories. With 1-mm thick crystals (difficult to make very thin because of fragility) the n=1 and 2 resonances are so strong that they "bottom" at zero transmission near the peaks; only at about $8-9°K$ have the peak strengths fallen sufficiently to permit accurate width measurement, from there to about $14°K$. In the other category, $n \geq 3$, the low-T widths can be accurately determined, but these weaker lines become too spread out for effective measurement beyond about $9°K$. Also, some adjustment of the width values is usually necessary at the elevated temperatures due to the merging together of the resonance skirts for a given Ising fan. One finds that the full width Γ is essentially unchanged in going from $1.3°K$ to between 3.5 and $4.5°K$, depending on the resonance. This low-T width we designate by Γ_o. Beyond this point the widths all increase rather dramatically with T. Energy widths are calculated from the measured field quantities, using the local slopes dE/dH at the laser crossings from the full CC2 spectrum.

Fig. 2 contains representative data plots of $\log_e(\Gamma-\Gamma_o)$ vs $1/T$ for the n=1 cluster, AF phase (0<H<31 kOe), and the n=2 cluster, Fo phase (H>45 kOe). Similar plots result for the same n's in different phases as well as for $n \geq 3$. Our finding is that those lines which can be observed over some substantial temperature range (viz., for $n \leq 3$) obey very well the simple exponential thermal broadening law,

$$\Gamma(T) = \Gamma_o + A \exp(-\Delta/kT) \qquad (4)$$

where a <u>universal</u> Δ of about 30 cm^{-1} ($\pm$ 2 cm^{-1} or so) is obtained. This value holds not only for the variety of spin-clusters excited at 29.7 and 32.2 cm^{-1} (HCN lines), but also for the 2-cluster excited at 51.4 and 52.7 cm^{-1} (DCN lines). (Unfortunately, only the n=2 cluster has adequate intensity at 50 cm^{-1}; n=1 cannot quite be reached in the Fo phase with a field of 125 kOe.) Virtually the identical Δ results, for instance, from the n=3 line in the AF phase, for T from 5 to $9°K$, and the n=1 excitation (same phase), T from 9 to $14°K$. The same choice for Δ seems consistent with the data for n=4 as well, but over the much shorter temperature range, 4 to $6°K$.

Γ_o is found to vary between 0.2 and 0.6 cm^{-1}, with no obvious correlation to cluster size or magnetic phase. Field inhomogeneity broadening is not important. The size prefactor A <u>does</u> depend on the cluster size: it ranges between 35 cm^{-1} and several hundred cm^{-1} (and

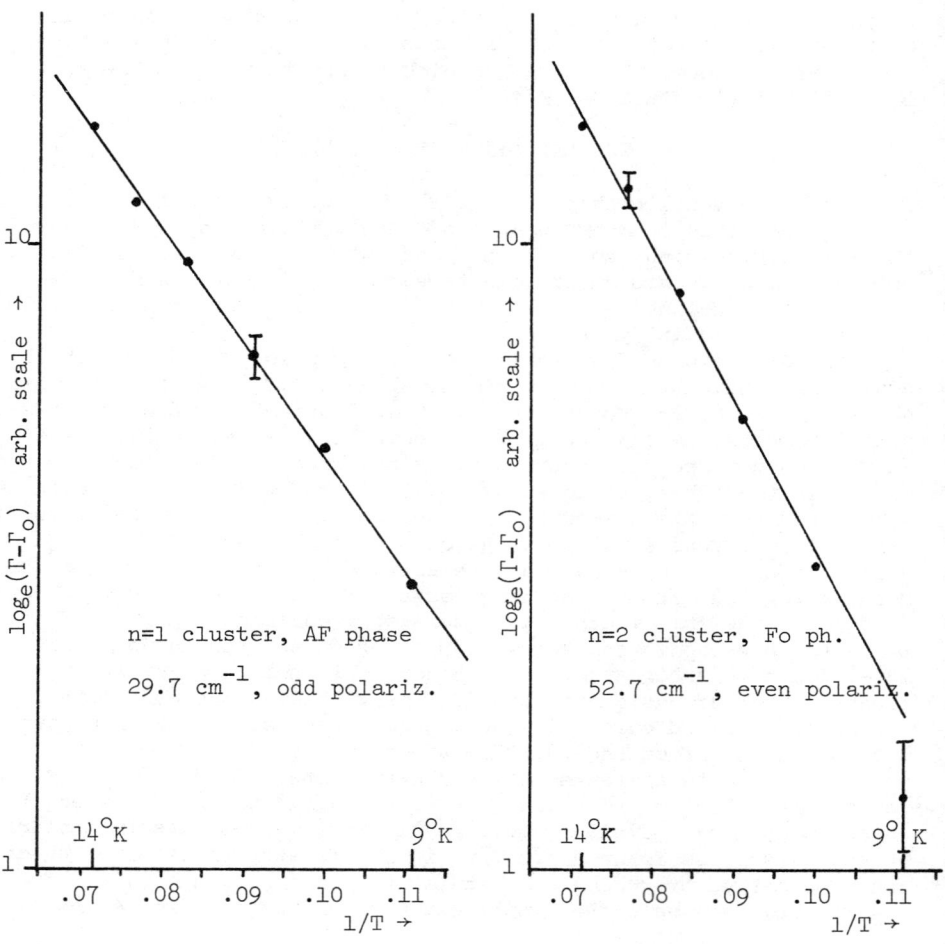

Fig. 2. Plots of $\log_e(\Gamma-\Gamma_o)$ vs $1/T$ for representative n=1,2 clusters.

larger), increasing rapidly with increasing n. Also, it seems to strictly decrease with increasing excitation (laser) frequency, although the selection of frequencies is very limited. The qualitative observation is certainly that the larger is n, the "faster" does the resonance broaden with temperature, and effectively disappear into the baseline.

There is another experimental quantity which would appear to provide verification that the linewidths are being correctly measured (with merging corrections): namely, the integrated line intensity. This is proportional to $\Gamma \cdot \log_e(\tau_o/\tau_p)$, where τ_o is the background transmission (far from resonance) and τ_p is the value at the peak. It turns out that for the n=3 resonances (both AF and Fo phases), which can be accurately measured over the whole range 1.3 to about 8°K, the above quantity is essentially <u>constant</u> over the entire temperature range (no systematic deviations). With the stronger n=1 and 2 lines the integrated intensity does begin to drop off after about 9°K. We expect it to be constant below about 9°K, however, from rough calculations of the ground state "population" -- because the basic exchange cost, $2J_o^{zz}$, for the thermal creation of spin-clusters of any size in the ground state chains is so large, $\simeq 25$ cm^{-1}.

An important observation is that the characteristic energy Δ for all practical purposes coincides with the energy of the optic phonon, 29.3 cm^{-1}. The latter was identified by Torrance and Tinkham by the bending and splitting of the n=1 excitation ("spin wave") as well as n=3,5,... due to $\mathcal{H}^a$-coupling. (It was positively verified by D.N. using deuterated CC2, producing a 5% decrease in phonon energy.) However, the even-clusters do <u>not</u> show admixture with this phonon. The characteristic energy Δ certainly appears to be about the <u>same</u> for <u>both</u> even- and odd-size clusters. Although the spin Hamiltonian (2) clearly separates these excitations into two non-communicating manifolds, there may well exist a universal relaxation mechanism operating on both even- and odd-clusters which depends on thermal occupation of this optic phonon. If one considers quite generally the direct processes of [acoustical] phonon emission and absorption by the CC2 fan, one arrives at temperature dependences approximating (for our T range) $\exp(-h\nu/kT)$, where $h\nu$ is the laser energy; this clearly does not yield a universal Δ value, contradicting the 50 cm^{-1} data for n=2. An approximate calculation of these processes yielded a parameter A which was too small.[5]

Another characteristic energy for the CC2 system near 30 cm^{-1} is the basic cluster exchange cost, $2J_o^{zz} \simeq 25$ cm^{-1}, which might be consistent with the Δ found. This parameter cannot be related to a universal decay mechanism for the multi-magnons in any obvious way. Another possibility instead of relaxation <u>per se</u> is the idea that there may be some <u>dispersion</u> occurring for the spin-cluster energies, (1), at elevated temperatures due to interactions <u>between</u> chains. A search is under way for the explanation of this broadening behavior.

REFERENCES

1. J.B. Torrance and M. Tinkham, Phys. Rev. <u>187</u>, 587, 595 (1969).

2. D.F. Nicoli, Harvard Univ. Dissertation, unpublished, 1972.

3. D.F. Nicoli and M. Tinkham, to be published.

4. G.T. Rado, Phys. Rev. <u>B5</u>, 1021 (1972).

5. H. Fogedby, private communication.

SINGLE-ION-INDUCED MAGNON SIDEBAND IN THE OPTICAL ABSORPTION SPECTRUM OF $GdCl_3$

R. S. Meltzer
University of Georgia, Athens, Georgia 30601

ABSTRACT

We report the observation of a magnon sideband in the $^6P_{7/2} \leftarrow {}^8S_{7/2}$ optical absorption spectrum of Gd^{+3} in the ferromagnetic insulator $GdCl_3$. The transition energy and its dependence on magnetic field demonstrate this to arise from the simultaneous excitation of a magnon and optical exciton. The decrease in its intensity with increasing magnetic field suggest the source of its intensity is directly connected with the zero-point spin deviations which occur in a ferromagnet possessing significant magnetic dipolar interactions. The magnitude of the intensity is shown to be consistent with a single-ion transition mechanism which because of the zero-point spin deviations allows a small probability for a double-excitation process. This transition mechanism is distinct from that responsible for the previously observed magnon sidebands in the transition-metal magnetic insulators.

INTRODUCTION

We recently reported the determination of the exciton dispersion of several ^6P excited states of Gd^{+3} in $GdCl_3$ by fitting the lineshapes of the single-ion-induced magnon→exciton lineshapes to a calculated lineshape function.[1] This paper will be referred to as I. The optical absorption also contains several weak transitions whose energies relative to the single-excitation transitions and energy dependence as a function of magnetic field indicate the simultaneous excitation of an exciton and magnon: i.e. magnon sideband. We will show that these transitions occur because of the zero-point spin deviations present in $GdCl_3$ and that their intensity is consistent with a single-ion rather than an exchange-coupled transition mechanism.

EXPERIMENT AND INTERPRETATION

High resolution ultraviolet spectra of $GdCl_3$ were obtained on a 9 meter Ebert spectrometer built at the University of Georgia. A 15,000 line/inch grating used in fifth order gave a resolution of 400,000. The spectra were recorded on an EMI 9558 phototube. The phototube was scanned repetitively in the focal plane of the spectrometer and the data stored in a multichannel analyzer. The procedure is described in more detail in I. All spectra were obtained with the helium bath at 1.17±0.03°K and with a magnetic field directed along the c axis. The linearly polarized light beam was directed perpendicular to the field.

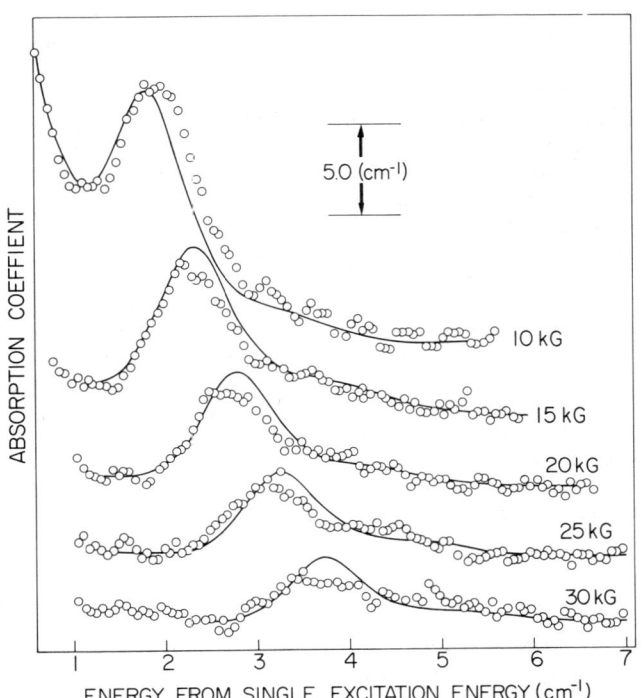

Fig. 1. Magnon sideband absorption for the state $^6P_{7/2}, M_J = -5/2$ as a function of magnetic field at $1.17 \pm 0.03°K$. The circles are the experimental data in the σ polarization: the solid curves are calculated lineshapes of Eq. (3) to arbitrary scale using the exchange parameters described in I. Energy is measured relative to the exciton-only transition.

This paper will deal exclusively with the $^8S_{7/2} \rightarrow {}^6P_{7/2}$, $M_J = -5/2$ transition. This state is nearly pure $M_J = -5/2$. The observed σ-polarized absorption spectra at several magnetic fields are shown in Fig. 1. Energy is measured relative to the single-excitation ground→exciton ($\vec{k}=0$) transition hereafter denoted as exciton-only transition. There are three points to note in the figure. (1) The transition energy is higher than that of the exciton-only transition. (2) The energy relative to the exciton-only transition increases linearly as a function of field according to $\Delta E_{obs} = (1.8 \pm 0.2)\beta H$. (3) The transition intensity falls off rapidly with increasing magnetic field. It is also found that the transition intensity is independent of temperature.

All the $^6P_{7/2}$ excited states in $GdCl_3$ have been observed, and the nearest excited states to $^6P_{7/2} M_J = -5/2$ in the absence of a magnetic field are at least 20 cm^{-1} away.[1] Therefore the observed transition cannot arise from any single-excitation process. However its energy relative to the exciton-only transition and the dependence of the energy separation on magnetic field are consistent with the assumption that it is a magnon sideband. The zero-field-extrapolated energy of the lineshape maximum relative to that of the exciton-only transition is 1.05 ± 0.1 cm^{-1}, which compares favorably with the energy of a magnon at the top face of the Brillouin zone where the density of states is highest. The relative energy as a function of field is expected to be $\Delta E_{calc} = 2\beta H$ which is close to $\Delta E_{obs} = (1.8 \pm 0.2)\beta H$.

The Gd^{+3} ground state is an octet whereas all excited states are at most sextets. Since the Gd^{+3} ions are aligned ferromagnetically, it appears that any magnon sideband must involve a $\Delta M_S \geq 2$ transition. Electric dipole processes allow only $\Delta M_S = 0$ transitions. Electric dipole magnon sidebands in $GdCl_3$ resulting from the exchange coupling of a pair of Gd^{+3} ions appears forbidden, except perhaps if spin-orbit coupling is invoked in second order. In the transition-metal antiferromagnets the exciton and magnon were created on opposite sublattices, satisfying the $\Delta M_S = 0$ electric dipole selection rule.

However in a ferromagnet like $GdCl_3$ for which the magnitude of the exchange and magnetic dipole interactions are comparable, zero-point spin deviations are present. This means that the crystal states cannot be characterized by a pure value of the single-ion M_S. We shall show that the zero-point spin deviations provide an alternate transition mechanism for the magnon sidebands. The observed decrease in absorption strength with increasing magnetic field is consistent with this explanation since an increase in magnetic field drives up the magnon branches and reduces the zero-point spin deviations.

LINESHAPE CALCULATIONS

Magnon sideband oscillator strengths in the 3d transition-metal antiferromagnets resulting from the exchange coupling of a pair of ions is typically 10^{-8} to 10^{-6}.[3,4] These will be greatly reduced in $GdCl_3$ due to (1) much weaker exchange in the 4f salts and (2) their dependence on the fraction of zero-point spin deviations, roughly 0.3%. Since the observed magnon sideband oscillator strengths in $GdCl_3$ are 10^{-10} we prefer instead to consider an alternate mechanism.

We start with the single-ion transition moment operator for the transition from the first spin-excited component of the ground state manifold to the electronically excited state $^6P_{7/2}$ $M_J = -5/2$. We write it as

$$\vec{M}_f = \vec{m}_f \sum_{i(p)} A^\dagger_{i(p)} a_{i(p)} \qquad (1)$$

where $\vec{m}_f$ is the single-ion transition moment, $A^\dagger_{i(p)}$ is the creation operator for the electronic excitation on site i of sublattice p and $a_{i(p)}$ is the destruction operator for a spin excitation on that site.

The single-ion operators are transformed to a basis in the crystal state space by diagonalizing the magnon and exciton Hamiltonians.[1,5] We define crystal magnon creation and destruction operators $b^\dagger_{\vec{k}\mu}, b_{\vec{k}\mu}$ and exciton creation and destruction operators

$B_{\vec{k}\nu}^{\dagger}, B_{\vec{k}\nu}$. Here μ and ν take on the values 1,2 corresponding to the two magnon and exciton branches. In terms of the crystal state operators $\vec{M}_f$ can be written

$$\vec{M}_f = \vec{m}_f \sum_{\vec{k}} \sum_{\mu\nu} [(S_{1\mu}R_{3t}+S_{2\mu}R_{4t})B_{\vec{k}\mu}^{\dagger} b_{\vec{k}\nu} + (S_{1\mu}R_{3\nu}+S_{2\mu}R_{4\nu})B_{\vec{k}\mu}^{\dagger} b_{-\vec{k}\nu}^{\dagger}] \quad (2)$$

with $t = \nu+2$. R and S are transformation matrices defined in I which depend on the interionic interactions, external magnetic field, and $\vec{k}$. The terms $B_{\vec{k}\mu}^{\dagger} b_{\vec{k}\nu}$ destroy a magnon and create an exciton of wavevector $\vec{k}$. They give rise to the temperature-dependent magnon→exciton transition discussed in I. The terms $B_{\vec{k}\mu}^{\dagger} b_{-\vec{k}\nu}^{\dagger}$ simultaneously create both an exciton and magnon of opposite wavevector. Its contribution to the lineshape function defined below is temperature independent since the exciton and magnon creation operators act on the ground state. Defining the exciton and magnon energies $\varepsilon_{\vec{k}\mu}^{ex}$ and $\varepsilon_{\vec{k}\nu}^{mag}$, the lineshape function is

$$\alpha(\varepsilon) = C|\vec{m}_f|^2 \sum_{\vec{k}} \sum_{\mu,\nu} |S_{1\mu}R_{3\nu}+S_{2\mu}R_{4\nu}|^2 \delta(\varepsilon - \varepsilon_{\vec{k}\mu}^{ex} - \varepsilon_{\vec{k}\nu}^{mag}) \quad (3)$$

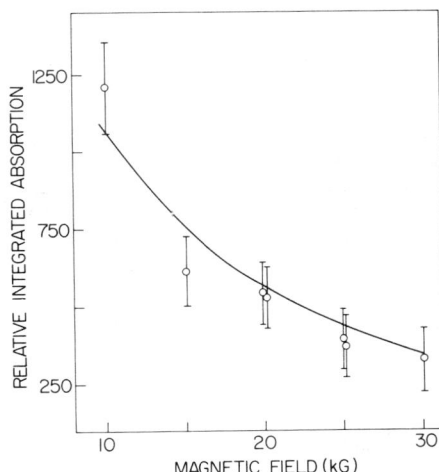

Fig. 2. Relative integrated absorption coefficient of the $^6P_{7/2}$, $M_J = -5/2$ magnon sideband as a function of magnetic field at $1.17\pm0.03°K$. The circles are the experimental integrated absorption coefficients of Fig. 1: the solid curve is the calculated integrated lineshapes of Fig. 1.

where C is a constant appropriate to a magnetic dipole process.

The lineshape function was then summed over 4000 randomly selected points in the Brillouin zone. The contribution from each term in the sum was given a Gaussian distribution about its transition energy ε (Gaussian half width 0.5 cm).[6] The calculated lineshapes are shown by the curves in Fig. 1. Their absolute magnitude is chosen to best fit the observed lineshapes. The lineshapes, their energy relative to the exciton-only transitions, and the dependence of the energy on magnetic field are all in quite good agreement with the observed lineshapes. The experimentally determined integrated absorption coefficient at several magnetic fields is shown in Fig. 2 where it is compared to the calculated integrated lineshape function. Both are to arbitrary scales so only the relative magnetic field dependence is significant. Note the satisfactory agreement.

Finally the absolute magnitude of the integrated absorption coefficient can be compared with the magnon→exciton single-excitation integrated lineshape since both lineshape functions are proportional to the square of the same single-ion transition moment. The result of this comparison is that the magnon sideband has half the predicted integrated absorption coefficient, a very encouraging result in view of the difficulties involved in the experiments.

CONCLUSIONS

We have identified a magnon sideband in the ferromagnet $GdCl_3$. The mechanism for this weak transition is a single-ion process which appears as a double excitation because of the zero-point spin deviations. This mechanism is distinct from the mechanism responsible for magnon sidebands observed in transition-metal salts.

An unresolved problem is the appearance of a π-polarized absorption with half the intensity of and similar magnetic field behavior to that of the σ-polarized absorption. A single-ion transition mechanism allows only a polarization identical to that of the magnon→exciton transition. Several other components of the ^6P manifold show similar sidebands.

Single-ion-induced magnon sidebands can occur in both ferromagnets and antiferromagnets whenever the presence of zero-point spin deviations allows the spin selection rules which forbid an electric dipole exchange-coupled pair transition to be broken down in lower order by a single-ion transition. Each case must be examined separately on the basis of the states involved and extent of zero-point spin deviations. The intensity of the magnon sideband relative to the single-excitation transitions provides an absolute measure of the zero-point spin deviations present in the ground state. This may be compared with predictions based on spin-wave calculations.

A more complete analysis of these single-ion induced magnon sidebands will be reported elsewhere

REFERENCES

1. R. S. Meltzer and H. W. Moos, Phys. Rev. $\underline{B6}$, 264 (1972).
2. Above 30kG this is not a good approximation for this component of $^6P_{7/2}$. See Ref. 1.
3. R. S. Meltzer, Marian Lowe, and Donald S. McClure, Phys. Rev. $\underline{180}$, 561 (1969).
4. R. M. MacFarlane and J. W. Allen, Phys. Rev. $\underline{B4}$, 3054 (1971).
5. C. D. Marquard, Proc. Phys. Soc. (London) $\underline{92}$, 650 (1967).
6. The k=0 ground→exciton transition has a width not insignificant compared to the sideband width even though it represents essentially one point in the Brillouin zone. The contribution of each point in the lineshape function was given a similar distribution.

MUFFIN-TIN RADII AND HYPERFINE INTERACTIONS IN FERROUS COMPOUNDS

Y. Hazony
Princeton University Computer Center
Princeton, N.J. 08540

ABSTRACT

The hyperfine interactions in octahedral iron compounds are discussed in terms of the effective volumes available for the Fe^{2+} ion in the lattice, as defined by a touching non-overlapping muffin-tin model. The data indicate

$$\langle r^{-3}(t_{2g})\rangle \propto \langle r\rangle^{-3}$$

varying by a factor of 3:1 over the covalency range between FeF_2 and FeS. The $3d(e_g)$ orbitals are considerably less expanded than the $3d(t_{2g})$ ones, displaying a ratio of

$$\langle r^{-3}(e_g)\rangle / \langle r^{-3}(t_{2g})\rangle \sim 2.5$$

over the same range of covalency. This effect signifies the importance of the asphericity of the boundary conditions in a touching non-overlapping muffin-tin model.

BREAKING OF DILATATIONAL SYMMETRY OF DIPOLE INTERACTIONS BY OCTUPOLE FORCES

M. Duda, T. Erber, R. Olenick
Physics Department, I.I.T., Chicago, Ill. 60616

H. G. Latal
University of Graz, Graz, Austria

ABSTRACT

We have constructed a 7 × 7 square Ewing lattice which can be symmetrically dilated. The breaking of the dilatational invariance of dipole interactions by octupole admixtures can be explicitly demonstrated. The discontinuous structural transitions associated with this symmetry breaking permit us to assign a sharply defined range to the octupole forces. We have shown by a combination of experimental, analytical, and digital computer methods that the range of the octupole forces increases with extreme rapidity as the complexity of the arrays is augmented. We conjecture that large Ewing arrays cannot be structurally stabilized solely by dipole interactions, and that the "compulsory" intervention of dilatationary non-invariant forces is the origin of the intrinsic scales of length which are associated with domain structure.

It is generally taken for granted that the physical configurations of composite systems are essentially determined by the nature of the strongest and/or longest range forces which mediate the interactions between the constituents. In the present note we show that two dimensional magnetic Ewing arrays represent a practical counter-example. In this instance the strong, long-range forces correspond to dipole interactions, and the weak forces which ultimately rise to dominance in complex arrays are the octupole (resp. n-pole) interactions. The essential diagnostic which permits us to prove that the range of the octupole forces actually increases with increasing complexity of the arrays is the breaking of the dilatational invariance of the dipole interactions. Experimentally we can check this feature through the mechanical variation of the lattice constant of a square Ewing array: Any disturbance of a locally stable magnet configuration during a dilation automatically signals the presence of non-dipole forces.

The magnets used in this work are precisely machined cylinders of Cunife I: length 0.889 ± 0.002 cm, diameter 0.318 ± 0.004 cm, mass 0.587 ± 0.015 g, magnetic moment 7.26 ± 0.03 G-cm^3. They are mounted on bearings, free to rotate in a horizontal plane, and will respond to torques as small as 0.15 dyne-cm. The magnetic array consists of a square 7 × 7 lattice which can be mechanically dilated from a spacing of 11 mm to 63 mm. The overall positional accuracy is ± 0.05 mm. Extreme smoothness in deformation is obtained with matched sets of precision rack and pinion drives. The entire array is leveled and mounted inside a double molybdenum permalloy shield. Background field intensities are < 0.04 G. Systematic bias due to background fields can be checked by rotating the entire assembly.

The potential at a distance r from the center of a uniformly magnetized right circular cylinder, to second order, is given by

$$V(r,\theta) = \mu \frac{\cos\theta}{r^2} \left[1 + \frac{d^2}{2r^2} (5\cos^2\theta - 3) \right] \quad (1)$$

where θ is the angle between $\vec{\mu}$ and $\vec{r}$. The dipole moment ($\vec{\mu}$) and the equivalent magnetic length (2d) are related by $\mu = 2md$, where m is the pole strength. It is clear that either the dipole component or the octupole component---but not their sum---is homogeneous in r. Since the total energy of a Ewing array is a linear superposition of terms derived from $V(r,\theta)$, we immediately arrive at the

Theorem: The angular configurations of all locally stable states of a square lattice of dipoles are invariant under symmetric dilations of the lattice.

Figs. 1 and 2 show the situation for 2 × 2 and 3 × 3 systems. The ordinates represent lattice spacing, the abscissas correspond to a telescoping of all configuration coordinates. The diagrams are projections of the phase-space "road maps" which display the pattern variations as the lattice is dilated. Fig. 1 shows that beyond the AF → N fork at 16.8 mm, and the extinction of the F pattern at 21.6 mm, only the dilatationally invariant neutral pattern remains. These features have been checked by analytical and digital computer methods,[1,2]; a typical comparison is given in Table I.

The square boxes at the bottom of Fig. 2 are keyed to the "dipole + octupole" patterns shown on Fig. 3. The circles indicate another family of patterns which correspond to "dipole + octupole + 32nd pole" forces.

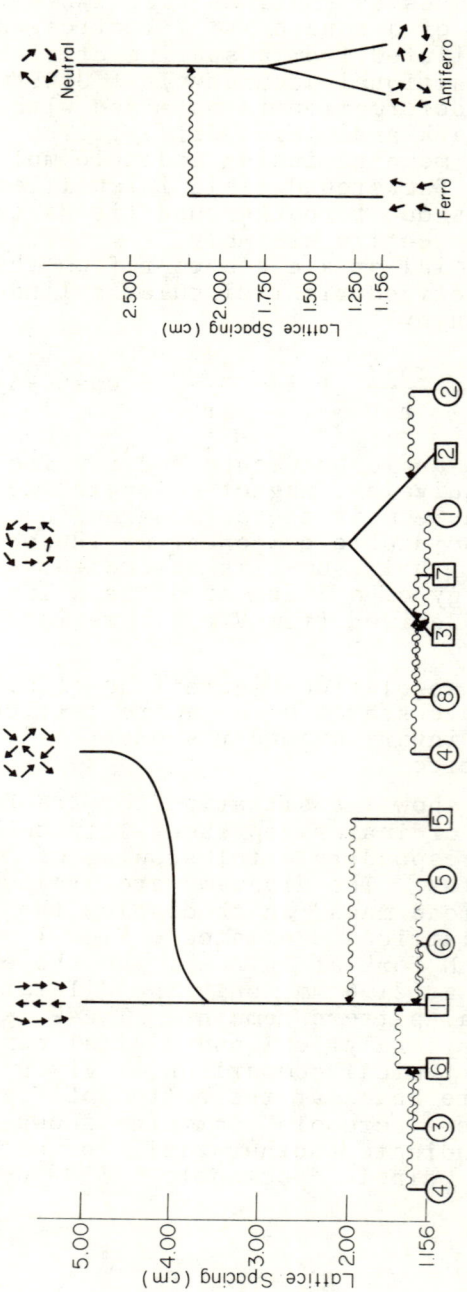

Fig. 1 2x2 Dilation

Fig. 2 3x3 Dilation

Table I. Results for 2 × 2 AF Pattern (a = 1.156 cm)

	θ_1	θ_2	θ_3	θ_4
Exp.	-162.0	-15.0	164.0	18.0
Comp.	-164.9	-14.9	164.9	14.8
Anal. Slt.	-165.2	-14.8	165.2	14.8

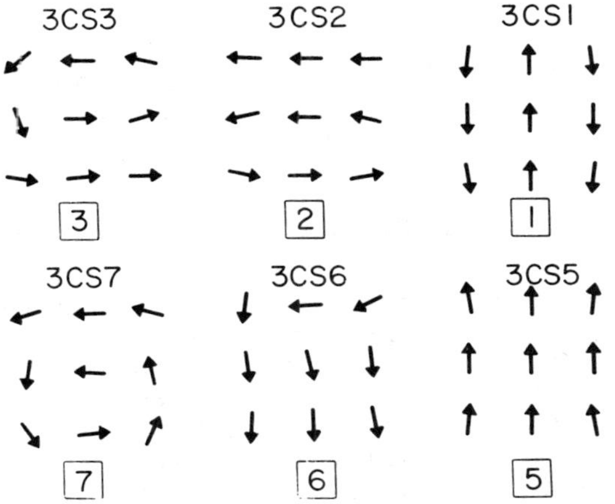

Fig. 3   3 × 3 Patterns

The pattern variations of an n × n system are obtained by photographing the array, and for a given pattern, $\mathcal{P}_i$, measuring the orientation of each magnet: $\theta_1^{(i)}, \ldots, \theta_{n^2}^{(i)}$. We can then define a <u>pattern distance</u>

$$d(\mathcal{P}_i, \mathcal{P}_j) = (N\pi)^{-1} \sum_{\lambda=1}^{N} |\theta_\lambda^{(i)} - \theta_\lambda^{(j)}| \qquad (2)$$

and invoke the
<u>Theorem</u> (A. Sklar): The set $\mathcal{P}$ is a metric space under d. Obviously $d(\mathcal{P}_i, \mathcal{P}_j) = 0$, for all $\mathcal{P}_i$ and $\mathcal{P}_j$ which are connected by dilatationally invariant paths. A practical application of this diagnostic is shown on Fig. 4:

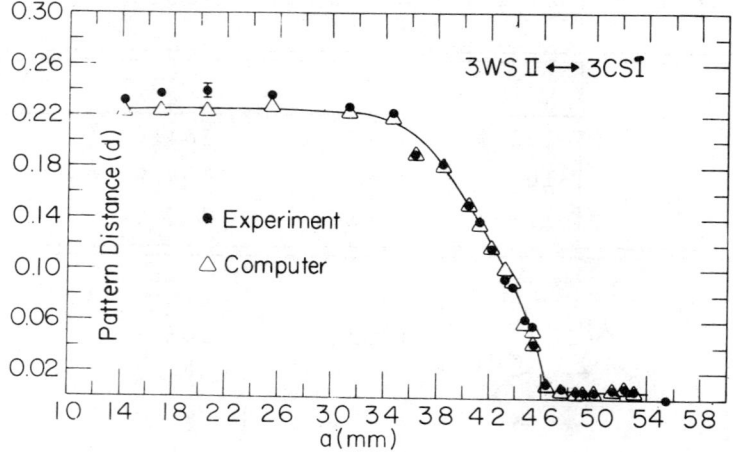

Fig. 4   Pattern Distance vs Lattice Spacing

The pattern variation corresponds to the curved path shown in Fig. 2---it is the confluence of the dipole patterns 3WSI and 3WSII. All three dipole states shown on Fig. 2 lie on an essentially flat energy (hyper) surface; their energies agree to within 1%. The previous observation that the magnetic circulation is a topological invariant is borne out by all observations on dilated systems.[2] Our results are summarized on Table II.

Table II.   Effective Range of Octupole Forces

Number of Magnets	1	4	9	16
Range of Octupole Forces (mm)	∼5	22	46	>60

A complete account of this work will be published elsewhere.

The support of the Research Corporation and the National Science Foundation is gratefully acknowledged.

REFERENCES

1. T. Erber and H. G. Latal, Acta Phys. Austriaca, $\underline{34}$, 313-30 (1971).
2. T. Erber, H. G. Latal, and B. N. Harmon, Adv. in Chem. Phys. $\underline{20}$, ed. by I. Prigogine and S. A. Rice, Wiley & Sons, pp. 71-134 (1971).

AUTHOR INDEX

Abeledo, C.R.	1168
Adam, J.D.	150
Ageev, A.N.	155
Ahn, K.Y.	1423
Akselrad, A.	408
Alderson, J.E.A.	1330
Aldred, A.T.	83,88,1638
Allen, R.P.	582
Almasi, G.S.	207
Alperin, H.A.	1569
Altman, R.F.	1163
Anderson, J.R.	46
Anderson, R.L.	1445
Archer, J.L.	197,222
Argyle, B.E.	344,403
Arko, A.J.	192
Arnold, G.	658
Arora, B.L.	870
Arrott, A.S.	822,899,941
Austin, A.E.	1049
Babel, D.	664
Bachmann, K.	578
Bagus, P.S.	547
Bailey, G.C.	130
Bajorek, C.H.	212,1445
Banks, E.	1158
Barrett, P.H.	75
Bartel, J.J.	1393
Bartel, L.C.	530
Baumann, L.S.	779
Bekebrede, W.R.	309
Belson, H.S.	749
Beni, G.	1493
Benningfield, L.U.	769
Bensel, J.	1639
Benz, M.G.	583
Berkhout, P.J.	1598
Berkner, D.D.	894
Berkowitz, A.E.	966
Bertram, H.N.	1440
Bhagat, S.M.	125
Birgeneau, R.J.	1664
Blackstead, H.A.	769,774
Blank, S.L.	256,498
Blazek, Z.	991
Blazey, K.W.	735
Bloembergen, P.	1598
Bloomberg, D.S.	941
Bloomfield, P.E.	779,796
Blount, E.I.	1593
Blume, M.	503
Bobeck, A.H.	202,442,498
Bongers, P.F.	1418
Borrelli, N.F.	398,1588
Boschetti, P.L.	1435
Bowman, A.	658

Braginski, A.I.	354
Brewer, L.	1
Brodsky, M.B.	192,1076
Brotzen, F.R.	764
Brown, B.R.	1423
Brown, H.A.	558
Brun, T.O.	1559
Bruni, F.J.	1384
Brya, W.J.	729
Budnick, J.I.	905,1564,1627
Buehrer, W.	1059
Bullock, D.C.	483
Burch, T.J.	1564,1627
Burmeister, R.A.	304
Burn, R.A.	1434
Busch, G.	689
Butera, R.A.	1065, ,1081
Buyers, W.J.L.	841
Byrom, E.	1294
Cabib, D.	1504
Cable, J.	1360,1554,1623,1637
Callen, E.	1547
Candela, G.A.	1494
Cannella, V.	541,785
Cannon, J.A.	905
Carnall, Jr., E.	740
Carr, Jr., W.J.	369
Carnes, J.G.	1178
Carriker, R.C.	608,633
Casey, M.	227
Catalano, A.	1355
Catalano, E.	179
Caton, R.	791
Chang, T.S.	880,889
Charap, S.H.	354
Chaudhari, P.	403
Chen, D.	1435
Chen, H.H.	553,842
Chen, J.H.	1143
Chen, S.L.	398,1588
Child, H.R.	1319,1623,1637
Chin, G.Y.	593
Chock, E.P.	138
Chu, C.W.	1618
Ciak, F.J.	202
Clark, A.E.	749
Clinton, J.	1531
Clover, R.B.	388,488
Coburn, T.J.	740
Collins, J.H.	150
Comly, J.B.	1554
Cone, R.L.	1039
Constantin, C.	1389
Cooke, J.F.	1218
Cooper, B.R.	1554
Copeland, J.A.	383,393

Cowan, D.O.	1494
Cox, D.E.	659,674,684
Craig, R.S.	905
Cullen, J.R.	1547
Culvahouse, J.W.	1044
Cunningham, R.M.	41
Cutler, L.S.	388,488
Dachs, H.	664,854
Daido, K.	1416
Darby, Jr., J.B.	1325
Das, D.K.	628,638
Das, S.G.	1304
Davidov, D.	138
Davis, H.L.	1218,1548
DeBoer, F.R.	1613
DeBonte, W.J.	349
Degani, J.	143
DeGraaf, A.M.	510
DeHon, B.	1613
DeJongh, L.J.	561
Dellinger, W.G.	790
Della Torre, E.	232
Dillinger, J.R.	1029
Dillon, Jr., J.F.	1593
Dionne, G.F.	169
Dodd, R.E.	149
Donoho, P.L.	764,769
Doyle, W.D.	227
Duda, M.	1710
Duff, K.J.	541
Dugautier, C.	1034
Dunlap, B.D.	83,88
Dunmyre, G.R.	1361
Dwight, K.	1355
Edelstein, A.S.	784
Egami, T.	759
Ehrenfreund, E.	1499,1509
Eibschutz, M.	684
Elliott, M.T.	774
Elliott, R.J.	841
Ellis, D.E.	1294
Enders, B.	174,179
Endoh, Y.	98
Erber, T.	1710
Erdos, P.	566,1070
Eremenko, V.V.	734
Erskine, J.L.	747
Escorne, M.	1374
Etemad, S.	1499,1509
Evans, B.J.	1369,1398
Everett, G.E.	102
Falk, H.	503
Fedro, A.J.	1060
Fehlner, F.P.	398
Ferer, M.	836
Ferrari, J.M.	1417
Ferraris, J.P.	1494
Field, W.G.	603
Fischer, R.F.	339
Fisher, M.E.	817
Flanders, P.J.	759,1515,1607
Fleury, P.A.	729
Foglio, N.E.	1578
Folen, V.J.	1027
Fomin, V.I.	734
Foner, S.	1115,1514
Foster, K.	971
Fowlis, D.C.	393
Fradin, F.Y.	192
Francis, C.L.	1044
Frankel, R.B.	1168
Franse, J.J.M.	1598
Frederick, W.G.D.	573,582
Freeman, A.J.	547,1294,1300
	1304,1309
Fukuyama, G.	1127
Fulde, P.	32,35
Furrer, A.	1059,1553
Gaglione, S.	174
Gale, A.A.	638
Gardner, J.A.	801,1639
Garito, A.F.	1476
Garland, J.W.	1638
Garrett, H.J.	573,582
Gehring, K.A.	1648
Geldart, D.J.	893
George, P.K.	197,222
Gergis, I.S.	51
Ghazali, A.	1374
Gibart, P.	1148,1153
Gillespie, D.J.	1612
Gittleman, J.I.	961
Glinka, C.J.	659
Goldstein, R.M.	383
Gonis, A.	1638
Goodenough, J.B.	1355
Goodings, D.A.	754
Gose, W.A.	1178
Graham, Jr., C.D.	759
Green, M.L.	593
Grimberg, A.J.T.	1324
Grundy, P.J.	364
Guarnieri, C.R.	1539
Guentert, O.J.	419
Guertin, R.P.	1115
Guntherodt, G.	1284
Guntherodt, H.J.	689
Gurevich, A.G.	155
Gurmen, E.	41
Gutzwiller, M.C.	1197
Gyorgy, E.M.	1148,1593
Haelg, W.	1059
Hanak, J.J.	961
Hankey, A.	880,889

Hansen, P.	423	Kaplan, T.A.	1504
Harbus, F.	884	Karnezos, N.	801
Harmer, R.S.	613	Kasai, M.	373,377
Harmon, B.N.	1309	Kawarazaki, S.	1632
Harris, D.H.	164	Keefe, G.E.	207
Harrold, W.J.	598	Keig, G.A.	237
Hartings, M.	613,618	Keller, D.A.	905
Hartog, R.	1365	Keller, J.	514
Harvey, A.	88	Kenan, R.P.	875
Harvey, A.R.	1076	Kestigian, M.	309
Hay, K.A.	1694	Ketcham, R.A.	102
Hazony, Y.	1709	Khanna, S.K.	1509
Heeger, A.J.	1476	Khattak, C.P.	659,674
Heer, H.	1059,1553	Kikuchi, R.	505
Heiman, N.	118	Kim, D.J.	535
Heinrich, B.	822,941	Kinsner, W.	232
Heller, P.	93	Kita, Y.	373,377
Henderson, R.G.	510	Klokholm, E.	271,319
Hester, R.K.	118	Knapp, G.S.	1618
Hewitt, B.S.	256	Kobayashi, T.	493
Hibma, T.	1530	Koehler, W.C.	1314,1319,1554
Hiskes, R.	304	Koelling, D.D.	1300,1304
Hoffmann, F.	113	Kommandeur, J.	1525,1530
Holmes, L.M.	1153	Konishi, S.	986
Holtzberg, F.	1259,1279,1569	Kooyer, R.L.	1454
Honda, S.	986	Kotthaus, J.P.	57
Hone, D.	1493	Kouvel, J.S.	904,1638
Hooper, H.O.	702	Kren, E.	1379,1603
Horn, P.M.	910,915	Krishnan, R.	112
Hoshikawa, K.	1416	Kronmuller, H.	1006
Hsu, F.S.L.	684	Krueger, D.A.	816
Hu, H.L.	319	Krumme, J.P.	423
Huber, D.L.	504,1029	Kunesh, C.J.	1065
Huber, J.G.	1075	Kusuda, T.	986
Hudak, J.J.	46		
Huffman, G.P.	1361,1368	Lacey, R.F.	388,488
Hufner, S.	723	Lahut, J.A.	966
Hull, G.W.	593,1153	Lam, D.J.	83,88
Hurault, J.P.	1459	Lambeth, D.	831
Hurd, C.M.	1330	Landau, D.P.	870,1163
Hutchings, M.Y.	1403	Lander, G.H.	88,1138,1559
		Latal, H.G.	1710
Iida, S.	567	Lawrence, J.M.	910
Inatsu, M.	1450	Lee, K.	1429
Inose, F.	373,377	Lee, T.H.	740
Ito, K.	1450	LeGall, H.	746
		Leoni, F.	724
Jaccarino, V.	57	Leroux-Hugon, P.	1374
Jacobs, I.S.	107,1554	Lesensky, L.	638
Jamet, J.P.	746	Lessoff, H.	149
Janssen, M.M.	314	Levinson, L.M.	1138
Jarrett, H.S.	521	Levinstein, H.J.	498
Jasnow, D.	817	Levy, P.M.	548,553,842
Johnson, W.A.	339	Lewis, J.L.	1299
Jones, G.A.	364	Lichtenwalner, C.P.	119
Jones, R.W.	1618	Liebermann, L.	1531
Josephs, R.M.	286,329	Linz, A.	93
		Litster, J.D.	894
Kachi, S.	714	Liu, L.L.	811
Kafalas, J.A.	1355	Liu, N.L.H.	1238
Kaplan, N.	143	Liu, S.H.	748

Livesay, B.R.	1583
Livingston, J.D.	643
Lomer, W.M.	17
Longinotti, L.D.	1560
Lowde, R.D.	1403
Lubitz, P.	125
Luntz, A.C.	1689
Lurie, N.A.	93
Lutes, O.S.	1454
Luther, A.	32,35
Lyu, S.L.	740
MacFarlane, R.M.	1689
Mackintosh, A.R.	1256
Maglic, R.C.	56
Mahoney, J.P.	159
Malozemoff, A.P.	344,458
Maple, M.B.	138,1075
Martin, D.L.	583
Martin, T.W.	740
Matthews, J.W.	271
Mayadas, A.F.	212
McCollum, B.C.	324
McCurrie, R.A.	588
McGuire, T.R.	1279,1289,1569
McLane, L.B.	769
McNiff, Jr., E.J.	1115
Meier, H.A.	689
Meiklejohn, W.H.	1102
Melcher, R.L.	184,520
Meltzer, R.S.	1704
Menth, A.	578
Menyuk, N.	1355
Meyer, J.S.	526
Micklitz, H.	75
Mildrum, H.G.	618,623
Mills, R.E.	875
Milne, A.D.	414
Minkiewicz, V.J.	659
Minnaja, N.	1001,1435
Misetich, A.	1168
Moch, P.	1034
Mokhir, A.P.	734
Mook, H. A.	1548
Moon, R.M.	1314,1319
Moore, Jr., G.E.	217
Moser, F.	740
Motoya, K.	1411
Mueller, F.M.	1304
Mueller, M.H.	88
Muellner, W.C.	904
Muhlestein, L.D.	41
Mulay, L.N.	1520
Muller, M.W.	981
Murphy, J.A.	398,1588
Murphy, J.J.	1627
Mydosh, J.A.	785
Nagy, I.	916
Nakamura, Y.	1411,1632
Narasimhan, K.S.V.L.	1065,1081
Nellis, W.J.	1076
Nereson, N.	658,669
Nesbitt, E.A.	593
Nicoli, D.F.	1649
Nielsen, J.W.	256
Nishida, H.	493
Noakes, J.E.	822,899
Nobile, M.	1001
North, J.C.	334,339
Nowik, I.	83,88
O'Kane, D.F.	319
Oeffinger, T.R.	354
Olenick, R.	1710
Olivei, A.	745
Oosterhuis, W.T.	98
Orbach, R.	138,1238
Owens, J.M.	150,414
Pal, L.	916
Paladino, A.E.	638
Pardavi, M.	1603
Parks, R.D.	910,915
Patterson, R.W.	354,981
Patton, C.E.	135
Paulikas, A.P.	40
Pearce, G.W.	1178
Pearlman, D.	740
Pepper, D.E.	841
Perlstein, J.H.	1494
Pfeiffer, C.D.	1690
Pfeiffer, L.	119
Pfeuty, P.	817
Pickart, S.J.	1569
Pincus, P.	1493
Pisarev, R.V.	1034
Pizzarello, F.A.	413
Plaskett, T.S.	271,319
Plovnick, R.H.	184
Pointon, J.A.	1573
Poko, Z.	1603
Popkov, Y.A.	734
Popma, T.J.A.	1418
Porteseil, J.L.	991
Portis, A.M.	120
Praddaude, H.C.	1115
Prinz, G.A.	1299
Pulliam, G.R.	413
Purwins, H.G.	1059
Rado, G.T.	1417
Raj, K.	1564
Rao, V.U.S.	1081
Rashidi, A.S.	608
Ratnam, D.V.	568
Ray, A.E.	613
Reed, W.A.	1148
Reid, W.R.	638
Reiff, W.M.	1514
Remeika, J.P.	1411
Reno, R.C.	1350

Rettori, C.	138
Reynolds, W.T.	976
Rice, T.M.	566
Richard, T.G.	893
Richards, P.L.	174,179
Richards, P.M.	187,729
Riedel, E.K.	865
Ritter, A.L.	1639
Robbins, M.	1148,1153
Roberts, S.	107
Robertson, J.	314,1418
Robinson, J.M.	1070
Rockwood, A.C.	583
Rodot, H.	1374
Rosenberg, M.	1389
Rosencwaig, A.	442
Rossol, F.C.	359
Rubinstein, M.	1384
Rupp, Jr., L.W.	1560
Rybaczewski, E.F.	1509
Sablik, M.J.	548
Salama, K.	764
Salamon, M.B.	187
Sampson, J.L.	603
Sankar, S.G.	905
Santiago, J.J.	1520
Sarachik, M.P.	791
Sasaki, F.	547
Sato, H.	505,679
Sattler, K.	1274
Saunderson, D.H.	1403
Savage, R.O.	159
Savage, W.R.	790
Savatsky, G.A.	1530
Sawatzky, E.	1423,1428
Schelleng, J.H.	1054
Schindler, A.I.	1612
Schinkel, C.J.	1324,1365,1613
Schlapbach, L.	689
Schlomann, E.	478
Schnettler, F.J.	1148
Schryer, N.L.	1026
Schwartz, B.B.	535
Schwee, L.J.	378,996
Schweitzer, J.W.	526,790
Searle, C.W.	573
Seidel, J.	971
Sellmyer, D.J.	806
Semper, R.	1539
Shafer, M.W.	1173,1289
Shanfield, Z.	75
Shanley, C.W.	623
Shaughnessy, T.P.	1143
Shen, L.	546
Sherwood, R.C.	593,684,1153
Shick, L.K.	256
Shilling, J.W.	971,976
Shirane, G.	93
Shtrikman, S.	1515
Siegmann, H.C.	1274
Silber, L.M.	174
Silberglitt, R.	1547
Sill, L.R.	1060
Skalyo, Jr., J.	98
Slonczewski, J.C.	458
Smith, A.B.	309
Smith, D.H.	334
Snow, S.R.	1060
Solomons, B.	414
Southern, B.W.	754
Spiwak, R.R.	339
Spooner, S.	1163,1360,1583
Stacy, W.T.	314
Stampfel, J.P.	98
Stankoff, A.	113
Stanley, H.	811,831,880,884,889
Stauffer, D.	827,836
Stauss, G.H.	1384
Stearns, M.B.	27,1644
Stein, B.F.	329
Steiner, M.	664
Steinitz, M.O.	1138
Stern, E.A.	747
Stone, D.R.	46
Strangway, D.W.	1178
Strauss, W.	202
Street, G.B.	1428
Stringfellow, M.W.	1403
Strnat, K.J.	613,618,623
Stutius, W.	1029
Subba Rao, G.V.	1173
Sugaya, H.	1086
Sugimoto, M	1335
Sugita, Y.	493
Suits, J.C.	1429
Suran, G.	113
Svab, E.	1603
Swartzendruber, L.J.	1350,1369
Swift, W.M.	976
Syllaios, A.	515
Tabor, W.J.	442
Tai, K.L.	217
Tamagawa, N.	749
Tang, C.H.	598
Tao, L.	1173
Tauber, A.	159,1158
Tchernev, D.I.	515
Teitelbaum, H.H.	548
Tennant, W.E.	174,179
Teranishi, T.	1450
Terlep, K.D.	207
Thomas, G.A.	910

Thomas, L.K.	806
Thompson, D.A.	1445
Thompson, E.D.	51
Thornley, S.J.	414
Thorpe, M.F.	503
Timofeev, Yu A.	155
Tinkham, M.	1699
Tocci, L.R.	197
Tommet, T.	504
Torrance, J.	1279,1289,1694
Triplett, B.B.	1572
Tront, J.	618
Tsay, Y.C.	546
Tsui, J.B.Y.	623
Tucciarone, A.	724
Turk, R.A.	144
Turner, P.A.	217
Uemura, C.	1416
Uffer, L.F.	553
VanHook, H.J.	419
VanHout, M.J.G.	314
VanVleck, J.H.	1578
Varnerin, L.J.	339
Veal, B.W.	40
Vegter, J.G.	1525
Vella-Coleiro, G.P.	424,442
Vergne, R.	991
Vittoria, C.	130,149
Voegeli, O.	458
Vogt, O.	1554
VonMolnar, S.	1259
Wachter, P.	1284
Wagner, R.J.	1299
Waites, R.F.	488
Wakiyama, T.	921
Walatka, V.V.	1494
Walker, E.	1059
Walker, J.C.	1539
Walker, L.R.	1026
Walker, Jr., P.L.	1520
Walsh, Jr., W.M.	1560
Wang, F.F.Y.	674
Wang, J.T.	774
Wang, Y.L.	540
Watson, R.E.	32,35
Weber, M.	1168
Weeks, S.P.	118
Weihrauch, P.F.	638
Wells, R.G.	568
Welsh, L.B.	1325
Werner, S.A.	679
Wernick, J.H.	593
Wertheim, G.K.	723
West, F.G.	483
West, R.G.	169
Westrum, Jr., E.F.	1393
Wetton, G.A.	1573

Wettstein, E.C.	638
White, R.M.	1572
Wieder, H.	1434
Wigen, P.E.	144
Wilson, K.G.	843
Wilson, L.K.	149
Windsor, C.G.	1403
Wittekoek, S.	1418
Wohlleben, D.	1075
Wolf, W.P.	1039
Wolfe, R.	339
Wong, Y.H.	1690
Wood, V.H.	1049
Wortis, M.	836
Wunsch, P.K.	769
Yakoviev, E.N.	155
Yasuoka, H.	1411,1632
Yen, W.M.	1690
Yu, J.T.	144
Yessik, M.	679
Zimmer, G.J.	1379
Zsoldos, E.	1603

AIP Conference Proceedings

	L.C. Number	ISBN
No. 1 Feedback and Dynamic Control of Plasmas (Princeton 1970)	70-141596	0-88318-100-2
No. 2 Particles and Fields - 1971 (Rochester 1971)	71-184662	0-88318-101-0
No. 3 Thermal Expansion - 1971 (Corning 1971)	72-76970	0-88318-102-9
No. 4 Superconductivity in d- and f- Band Metals (Rochester 1971)	74-188879	0-88318-103-7
No. 5 Magnetism and Magnetic Materials - 1971 (2 parts) (Chicago 1971)	59-2468	0-88318-104-5
No. 6 Particle Physics (Irvine 1971)	72-81239	0-88318-105-3
No. 7 Exploring the History of Nuclear Physics (Brookline 1967, 1969)	72-81883	0-88318-106-1
No. 8 Experimental Meson Spectroscopy - 1972 (Philadelphia 1972)	72-88226	0-88318-107-X
No. 9 Cyclotrons - 1972 (Vancouver 1972)	72-92798	0-88318-108-8
No. 10 Magnetism and Magnetic Materials - 1972 (Denver 1972)	72-623469	0-88318-109-6
No. 11 Transport Phenomena - 1973 (Brown University Conference)		0-88318-110-X

QC
761
C6
1972
pt. 2